Signals and Systems

Principles and Applications

Shaila Dinkar Apte

CAMBRIDGE
UNIVERSITY PRESS

4843/24, 2nd Floor, Ansari Road, Daryaganj, Delhi - 110002, India

Cambridge University Press is part of the University of Cambridge.

It furthers the University's mission by disseminating knowledge in the pursuit of education, learning and research at the highest international levels of excellence.

www.cambridge.org
Information on this title: www.cambridge.org/9781107146242

© Shaila Dinkar Apte 2016

This publication is in copyright. Subject to statutory exception and to the provisions of relevant collective licensing agreements, no reproduction of any part may take place without the written permission of Cambridge University Press.

Hardback first published 2016

Printed in India by Shree Maitrey Printech Pvt. Ltd., Noida

A catalogue record for this publication is available from the British Library

Library of Congress Cataloguing in Publication data
Apte, Shaila Dinkar.
Signals and systems : principles and applications / Shaila Dinkar Apte.
Delhi, India ; New York : Cambridge University Press, 2016.
   Includes bibliographical references and index.
Identifiers: LCCN 2015046420 | ISBN 9781107146242 (hardback : alk. paper)
Subjects: LCSH: Signal processing.
Classification: LCC TK5102.9 .A68 2016 | DDC 621.382/2--dc23 LC record available at http://lccn.loc.gov/2015046420

ISBN 978-1-107-14624-2 Hardback

Cambridge University Press has no responsibility for the persistence or accuracy of URLs for external or third-party internet websites referred to in this publication, and does not guarantee that any content on such websites is, or will remain, accurate or appropriate.

To

my beloved husband Late Mr Dinkar
my grandchildren Aarohi, Shriya, Shreyas and Shruti
my students from W.C.E., Sangli and R.S.C.O.E., Pune

Contents

Preface		*xiii*
1.	**Introduction to Signals**	
	1.1 Introduction to Signals	1
	1.2 Sampling Theorem	3
	1.3 Sampling of Analog Signals (Case I)	3
	1.4 Recovery of Analog Signals (Case I)	6
	1.5 Sampling of Analog Signals (Case II)	8
	1.6 Recovery of Analog Signals (Case II)	9
	1.7 Analytical Treatment	11
	1.8 Analytical Examples	13
	1.9 Anti-Aliasing Filter	19
Summary		20
Multiple Choice Questions		21
Review Questions		22
Problems		22
2.	**Signals and Operations on Signals**	
	2.1 Signals	24
	2.2 Graph Terminology and Domains	27
	2.3 Applications of Signals and Systems	33
	2.3.1 Basic communication system	33
	2.3.2 Basic control system	35
	2.4 Classification of Signals	36
	2.4.1 Analog signals	36
	2.4.2 Discrete time signals (DT signals)	41
	2.4.3 Digital signals	45
	2.5 Elementary Signals used for Testing of Systems	47
	2.5.1 Reasons for using test or standard inputs	47

		2.5.2	Standard analog signals	48
		2.5.3	Standard DT signals	59
	2.6	Classification of Signals Based on Signal Properties		72
		2.6.1	Even and odd signals	72
		2.6.2	Periodic and aperiodic signals	81
		2.6.3	Causal and non-causal signals	96
		2.6.4	Deterministic and random signals	99
		2.6.5	Energy and power signals	100
	2.7	Operations on Signals		118
		2.7.1	Time shifting	118
		2.7.2	Time reversal	120
		2.7.3	Time and amplitude scaling	124
		2.7.4	Addition, subtraction and multiplication	135

Summary   147
Multiple Choice Questions   149
Review Questions   153
Problems   154

3. CT and DT Systems

	3.1	Properties of CT and DT Systems – Linearity and Shift Invariance		181
		3.1.1	Linearity property	182
		3.1.2	Time invariance / shift invariance property	194
	3.2	Properties of CT and DT Systems – Causality and Memory		201
		3.2.1	Causality property	201
		3.2.2	Memory	206
	3.3	Properties of CT and DT Systems – Invertibility and Stability		209
		3.3.1	Invertibility	209
		3.3.2	Stability	215
	3.4	System Representation as Interconnection of Operations		219
	3.5	Series and Parallel Interconnection of Systems		224
		3.5.1	Series interconnection of systems	225
		3.5.2	Parallel interconnection of systems	227

Summary   235
Multiple Choice Questions   236
Review Questions   238
Problems   239

4. Time Domain Response of CT and DT LTI Systems

    4.1   Response of CT Systems   246
           4.1.1   Zero input response   247
    4.2   System Representation as Impulse Response   249
           4.2.1   Representation of signals in terms of impulses   250
           4.2.2   Calculation of impulse response of the system   252
    4.3   Convolution Integral for CT Systems   256
           4.3.1   Zero state response   269
    4.4   Response of DT Systems   271
           4.4.1   Zero input response   274
           4.4.2   Impulse response of DT system   275
           4.4.3   Zero state response of DT system   277
    4.5   Representation of DT Signals in Terms of Delta Functions   278
           4.5.1   Convolution sum for DT systems   279
           4.5.2   Convolution using MATLAB   291
    4.6   Unit Step Response of CT and DT LTI Systems   295
    4.7   Properties of LTI DT Systems   303
           4.7.1   Memory property of CT and DT LTI systems   303
           4.7.2   Condition of causality for CT and DT LTI systems   305
           4.7.3   Stability for CT and DT LTI systems   309
    4.8   Series and Parallel Interconnection of Systems   315

Summary   319
Multiple Choice Questions   321
Review Questions   323
Problems   323

5. Fourier Series Representation of Periodic Signals

    5.1   Signal Representation in Terms as Sinusoids   332
           5.1.1   Orthogonality property   332
           5.1.2   Basis functions   337
    5.2   FS Representation of Periodic CT Signals   339
           5.2.1   Evaluation of fourier coefficients of trigonometric FS   340
           5.2.2   Exponential FS representation of periodic CT signals   342
    5.3   Application of Fourier Series Representation   347
    5.4   Properties of Fourier Series for CT Signals   378
    5.5   Recovery of CT Signal from FS   390

		5.5.1	Gibbs phenomenon	391
	5.6	FS Representation of DT Periodic Signals		394

Summary   398

Multiple Choice Questions   400

Review Questions   402

Problems   403

6. Fourier Transform Representation of Aperiodic Signals

	6.1	Fourier Transform Representation of Aperiodic CT Signals		410
		6.1.1	Evaluation of magnitude and phase response using hand calculations	412
	6.2	Fourier Transform of Some Standard CT Signals		423
		6.2.1	Use of dirac delta function	435
		6.2.2	Applications of dirac delta function	436
	6.3	Fourier Transforms of Periodic CT Signals		441
	6.4	Inverse Fourier Transform		444
	6.5	FT of Aperiodic DT Signals (DTFT)		449
		6.5.1	FT of standard aperiodic DT signals	456
	6.6	Properties of FT and DTFT		470
	6.7	FT and DTFT of Signals using FT/DTFT Properties		483
	6.8	Analysis of LTI System using FT and DTFT		495

Summary   499

Multiple Choice Questions   501

Review Questions   506

Problems   506

7. Laplace Transform

	7.1	Definition of Laplace Transform		513
	7.2	Laplace Transform of Some Standard Functions		519
	7.3	Properties of LT		527
	7.4	Solved Examples on LT		539
		7.4.1	LT of standard aperiodic signals	551
		7.4.2	LT of standard periodic signals	558
		7.4.3	LT of signals using properties of LT	564
		7.4.4	Properties of ROC	571
	7.5	Inverse LT		572
		7.5.1	Transform analysis of LTI systems	586
		7.5.2	Total response of the system using LT	592
		7.5.3	Stability considerations in *S* domain	598

		Summary	603
		Multiple Choice Questions	604
		Review Questions	609
		Problems	610
8.	**Z Transform**		
	8.1	Physical Significance of a Transform	619
	8.2	Relation between LT and ZT	619
	8.3	Relation between Fourier Transform (FT) and Z Transform	621
	8.4	Solved Problems on Z Transform	622
	8.5	Properties of ROC	630
	8.6	Properties of Z Transform	642
	8.7	Relation between Pole Locations and Time Domain Behavior	655
	8.8	Inverse Z Transform	660
		8.8.1 Power series method/long division method	660
		8.8.2 Partial fraction expansion method	664
		8.8.3 Residue method	674
	8.9	Solution of Difference Equation using Z Transform	678
		8.9.1 Applications of ZT and IZT	685
		Summary	685
		Multiple Choice Questions	687
		Review Questions	688
		Problems	689
9.	**Random Signals and Processes**		
	9.1	Probability	693
		9.1.1 Conditional probability	696
		9.1.2 Bayes theorem	701
	9.2	Random Variable	706
		9.2.1 Cumulative distribution function (CDF)	707
		9.2.2 Probability density function (pdf)	710
	9.3	Statistical Properties of Random Variables	716
	9.4	Standard Distribution Functions	722
		9.4.1 Probability distribution functions for continuous variables	722
		9.4.2 Probability distribution functions for discrete variables	731
		9.4.3 Functions for finding moments	738
	9.5	Central Limit Theorem and Chi Square Test, K-S Test	740
	9.6	Random Processes	744

9.7	Estimation of ESD and PSD		747
	9.7.1	Computation of energy density spectrum of deterministic signal	747
	9.7.2	Estimation of power density spectrum of random signal	756

Summary — 759

Multiple Choice Questions — 760

Review Questions — 762

Problems — 763

Index — 767

Preface

It gives me immense satisfaction in presenting this book to my students who have been eagerly waiting to see it. The difficulties encountered by students in understanding the physical significance of different concepts inspired me to write a student-friendly book, rich in technical content. The subject Signals and Systems is essential for undergraduate students of Electronics Engineering, Electrical Engineering, Computer Engineering, and Instrumentation Engineering disciplines. The subject has a diverse range of applications. A thorough knowledge of different transforms studied in Mathematics is essential for an understanding of the subject.

The subject involves a number of complex algorithms which require in-depth domain knowledge. If a concept is explained with concrete examples and programs, the reader will definitely take interest in the field. The reader will experience great joy when she observes tangible outcomes of an experiment on her computer screen. A number of signals occurring in nature like speech, ECG, EEG etc. have some random components. Research in this field is somewhat difficult. Despite significant progress in an understanding of the subject, there remain many things which are not well grasped. There is room for explaining the basic concepts of signals and systems, in a better manner. Different concepts are illustrated here using MATLAB programs. The outputs of the MATLAB programs are given in the form of graphs. With extensive experience in signals and systems research, I observed, that people require significant amount of time before they can begin grasping the subject. However with proper guidance, a person can acquire considerable knowledge in this field. The motivation behind this book has been, that a new comer be provided information about where and how to start and how to proceed.

There was a request from my students as well as from well wishers that I write a book on signals and systems. It was also suggested that the book include the theory of random signals. Several concrete examples have been given to illustrate the concepts pertaining to random signals. For better understanding, we have included output from many MATLAB programs. Interpretations of the results are also explained, which will not only help the reader but also enable instructors to incorporate the examples into their classroom teaching.

The book seeks to rigorously examine the subject of Deterministic and Random Signals and Systems. It will provide a solid foundation for specialized courses in signal processing at both the undergraduate and postgraduate levels and will serve as a basic practical guide for PhD students.

I have enjoyed whatever little work I have done in signals and systems. Although I have stated that I undertook this project for new comers to the field, it was also done with a desire to further my own knowledge.

Acknowledgments

I am thankful for the encouragement that my late husband, Dinkar gave me. I have few words to describe my gratitude for his inspiration. I am also grateful for the inspiration provided by my students from Walchand College of Engineering, Sangli and from Rajarshi Shahu College of Engineering, Pune. I sincerely thank all the students who worked under my guidance for their undergraduate, postgraduate or PhD projects. Their painstaking efforts and sincerity encouraged me to seek newer challenges. We carried out lots of experiments in Signal Processing, which besides enriching the knowledge of my students also improved my understanding of the subject. The curiosity of the students to understand the physical significance of each concept prompted me to undertake this project after completion of projects titled "Digital Signal Processing", "Speech and Audio Processing" and "Advance Digital Signal Processing". The constant demand by my students that I write a book on "Signal and Systems" accelerated my work.

I got a lot of help and inspiration from my family members especially from my late husband Dinkar, my son Anand, daughter-in-law Aditi, daughter Amita and son-in-law Prasad. I thank my colleagues Bhalke and Tirmare for giving me useful suggestions; my student Prashul for helping me type some of the examples; the reviewers from different countries for reviewing and giving me concrete suggestions.

The idea of publishing the book materialized when Manish Choudhary from Cambridge University Press accepted the proposal for this book. I am grateful to him. I also express my gratitude to the team at Cambridge University Press for the cooperation and strong support during the review and proofreading stages.

Finally, I wish to thank all who helped me directly and indirectly for their support during the long process of producing this book.

I look forward to feedback on this book at: *sdapte@rediffmail.com*

1

Introduction to Signals

> **Learning Objectives**
> - Definition of a signal.
> - CT and DT signals.
> - Sampling theorem in time domain.
> - Aliasing in frequency domain.
> - Interpolation formula – a sync function.
> - Recovery of signal from signal samples.
> - Phase reversal of the recovered signal.
> - Anti-aliasing filter.

We discuss the basic definitions of analog/continuous time (CT) and discrete time (DT) signals in this chapter. We need to understand the basics of discrete time signals, i.e., the sampled signals. The theory of sampled signals is introduced in this chapter. The analog signal is first interfaced to a digital computer via analog to digital converter (ADC). ADC consists of a sampler and a quantizer. We will mainly discuss the sampler in this chapter. The analog signal, when sampled, gets converted to discrete time (DT) signal. Here, the time axis is digitized with a constant sampling interval T. The inverse of T is the sampling frequency. The sampling frequency must be properly selected for faithful reconstruction of the analog signal.

1.1 Introduction to Signals

Any physical quantity that carries some information can be called a signal. The physical quantities like temperature, pressure, humidity, etc. are continuously monitored in a process. Usually, the information carried by a signal is a function of some independent variable, for example, time. The actual value of the signal at any instant of time is called its amplitude. These signals are normally plotted as amplitude vs. time graph. This graph is termed as the waveform of the signal. The signal can be a function of one or more independent variables. Let us now define a signal.

Definition of a signal A signal can be defined as any physical quantity that varies with one or more independent variables.

Let us consider temperature measurement in a plant. The measured value of temperature will be its amplitude. This temperature changes from one instant of time to another. Hence, it is a function of time, which is an independent variable, as it does not depend on anything else. Temperature can be measured at two different locations in a plant. The values of temperature at two different locations may be different. Hence, the temperature measured depends on the time instant and also on the location. We can say that temperature is a function of two independent variables, namely time and location in a plant. This temperature can take on any continuous value, like 30.1, 30.001, 31.3212, etc. The time axis is also continuous i.e. the temperature is noted at each time value. The signal is then called continuous time continuous amplitude (CTCA) signal.

Definition of a CTCA/Analog signal When the signal values are continuous and are noted at each continuous time instant (the independent variable) the signal is said to be a CTCA signal. Analog signal is shown in Fig. 1.1.

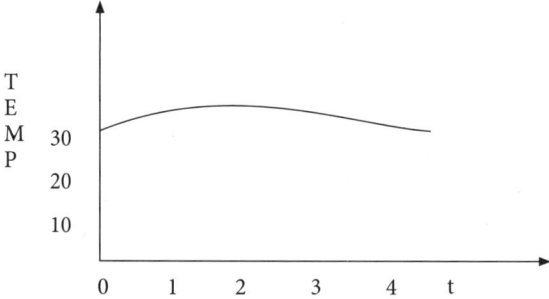

Fig. 1.1 Plot of the analog signal

If we now decide to note the temperature after one second, for instance, we will get continuous values of temperature at discrete instants. We will now define Discrete Time Continuous Amplitude (DTCA) signal.

Definition of a DTCA/DT signal When the signal values are continuous and are noted at each discrete time instant, namely the independent variable, the signal is said to be a DTCA signal. This DT signal is shown in Fig. 1.2.

Definition of a DTDA or Digital signal A digital signal, also called as DTDA signal, is obtained when amplitude axis is also discretized.

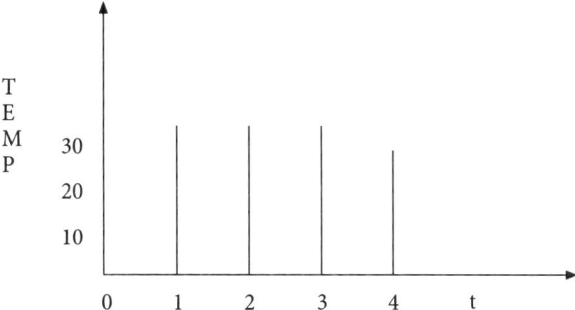

Fig. 1.2 Plot of DT signal

1.2 Sampling Theorem

When any signal is to be sampled, one must have some prior knowledge about the signal, namely, its frequency contents. This information is generally known to us. For example, a telephone grade speech signal contains useful information only up to 3.4 kHz. The signal may then be filtered to remove the frequency contents, if any, above 3.4 kHz to ensure that the signal bandwidth is within 3.4 kHz. We will now define *Sampling Theorem*. Consider any band-limited signal of bandwidth W.

Theorem When any energy signal that is band-limited with a bandwidth W is sampled using a sampling frequency ≥ 2W, it is possible to reconstruct signal from signal samples.

The sampling rate of 2W is called Nyquist rate. The sampling frequency selected is denoted as F_S. We will denote the Nyquist frequency $F_S/2$ as F_N.

When an analog signal is sampled, the most important factor is the selection of the sampling frequency. In simple words, *Sampling Theorem* may be stated as, "Sampling frequency is appropriate when one can recover the analog signal back from the signal samples." If the signal cannot be faithfully recovered, then the sampling frequency needs correction. Let F_S denote the sampling frequency. We will consider two different cases to understand the meaning of proper sampling.

Concept Check

- What is sampling theorem in time domain?
- How do we find the Nyquist rate?

1.3 Sampling of Analog Signals (Case I)

Consider a signal with a bandwidth of 10 Hz. We sample it using F_S = 100 Hz. That is

$$F_s > 2W(2 \times 10\,\text{Hz})\,\text{Hz}$$

We will concentrate on the highest frequency component of the signal, namely, $F = 10$ Hz. There will be 10 samples obtained per cycle of the waveform.

To get a deeper insight into the sampling process, we introduce the theoretical concept of an impulse train. In the time domain, the sampled signal is obtained by multiplying the sinusoidal waveform with a train of impulse with separation of $T_S = 1/F_S = 1/100$, as shown in Fig. 1.3.

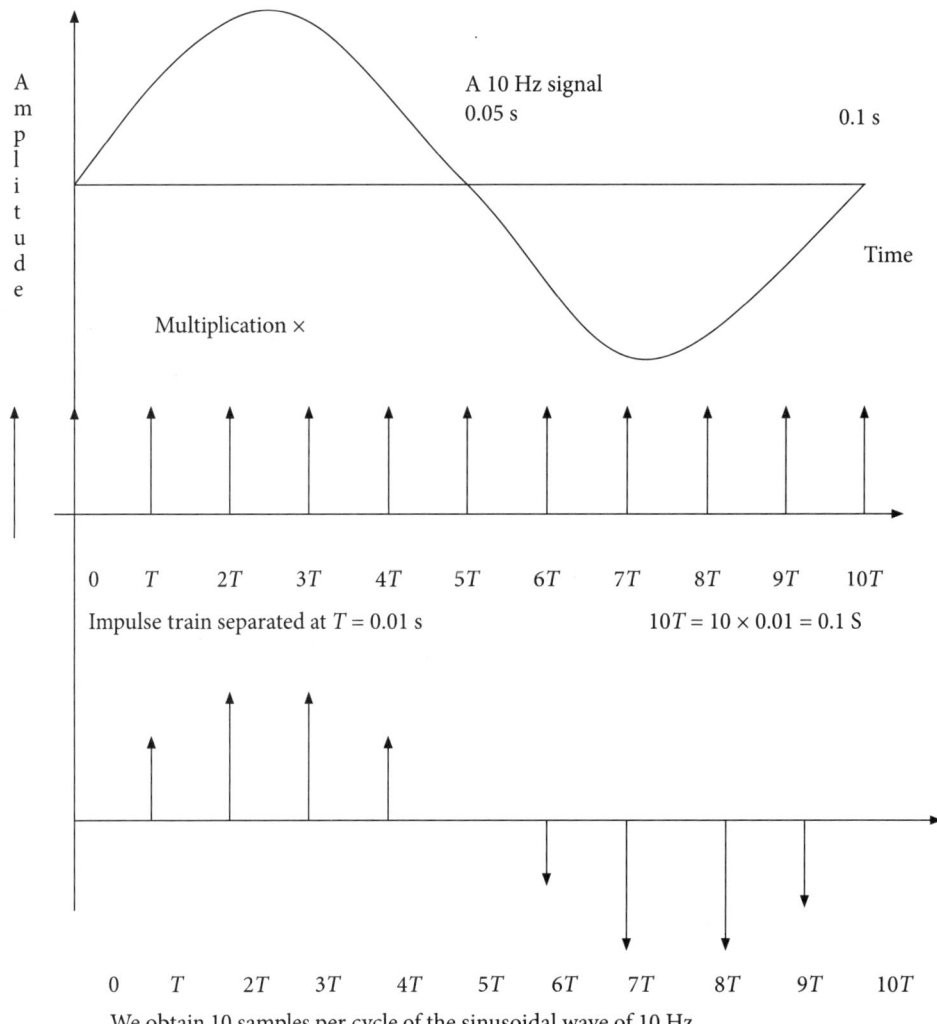

Fig. 1.3 Samples of the wave obtained by multiplication of the wave with an impulse train

Multiplication of two signals in time domain is equivalent to convolution of their transforms in the frequency domain. (This is discussed in detail in Chapter 5.) When the signal is sampled using a sampling frequency of

$10F = 10*10 = 100$ Hz, the signal of frequency F gets multiplied by a train of impulses with a separation interval of $T = 1/F_S$, that is, 1/100. In the frequency domain, the Fourier Transform (FT) of the signal and FT of the impulse train will convolve with each other. The FT of a sinusoid with a 10 Hz frequency is a spectral line at 10 Hz, that is, at $0.1F_S$. The FT of an impulse train is again the impulse train with a separation interval of $1/T_S$, that is, F_S (100 Hz). Hence, the convolution of two transforms results in a signal spectrum replicated around zero frequency and frequencies that are multiples of 100 Hz, as shown in Fig. 1.4.

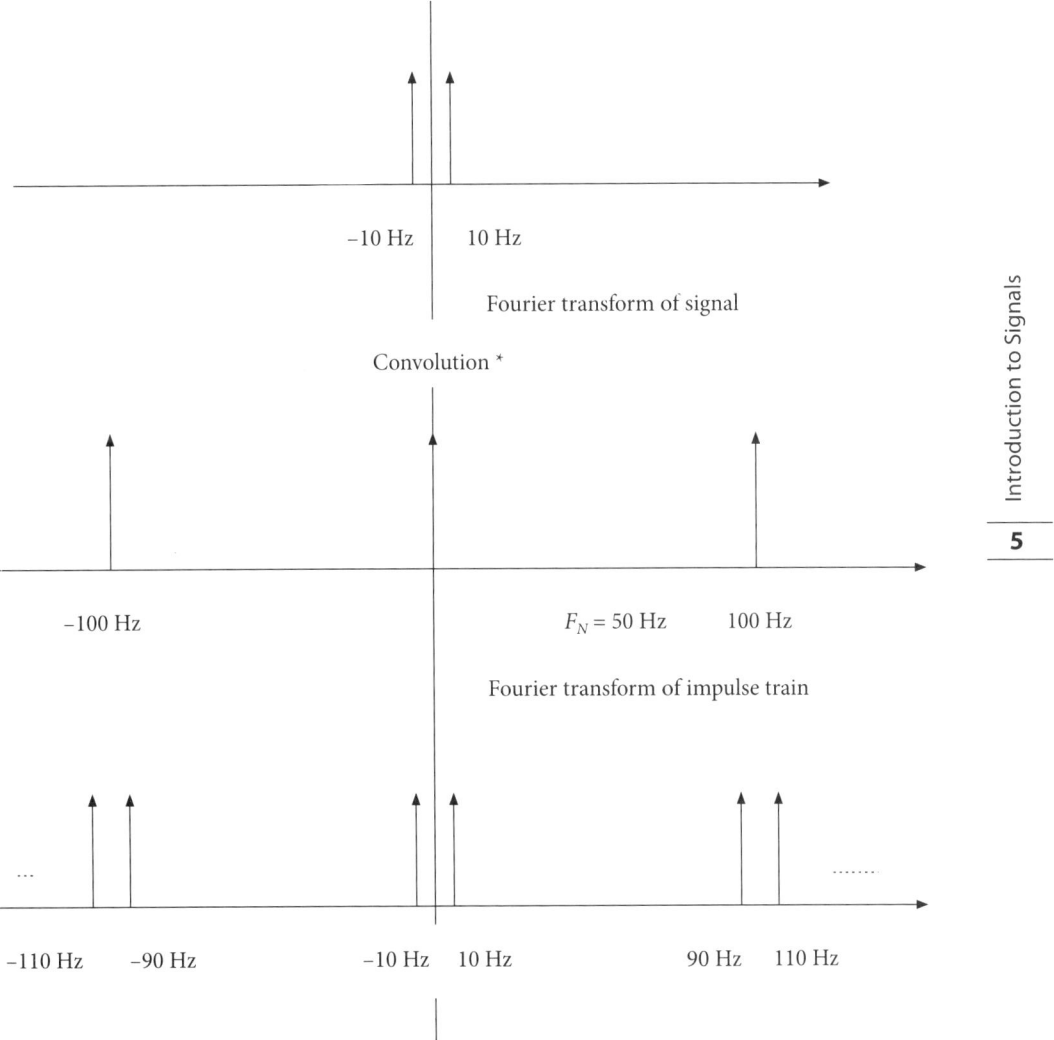

Fig. 1.4 Convolution of two transforms (FT of signal and FT of impulse train)

Concept Check

- What must be the sampling frequency if the signal bandwidth is 50 Hz?
- What is the Fourier transform of the impulse train?
- What is the Fourier transform of a sinusoid with frequency of 40 Hz?

1.4 Recovery of Analog Signals (Case I)

The convolved spectrum in Fig. 1.5 shows replicated copies of the signal spectrum around zero frequency and multiples of the sampling frequency. The replicas are well separated, that is, there is no mixing of adjacent replicas. The signal spectrum around zero frequency can be easily isolated if the convolved spectrum is passed via a rectangular window between −50 Hz and 50 Hz ($-F_N$ to $+F_N$), as shown in Fig. 1.5. The resulting isolated signal spectrum is then inverse Fourier transformed to recover the original signal of 10 Hz.

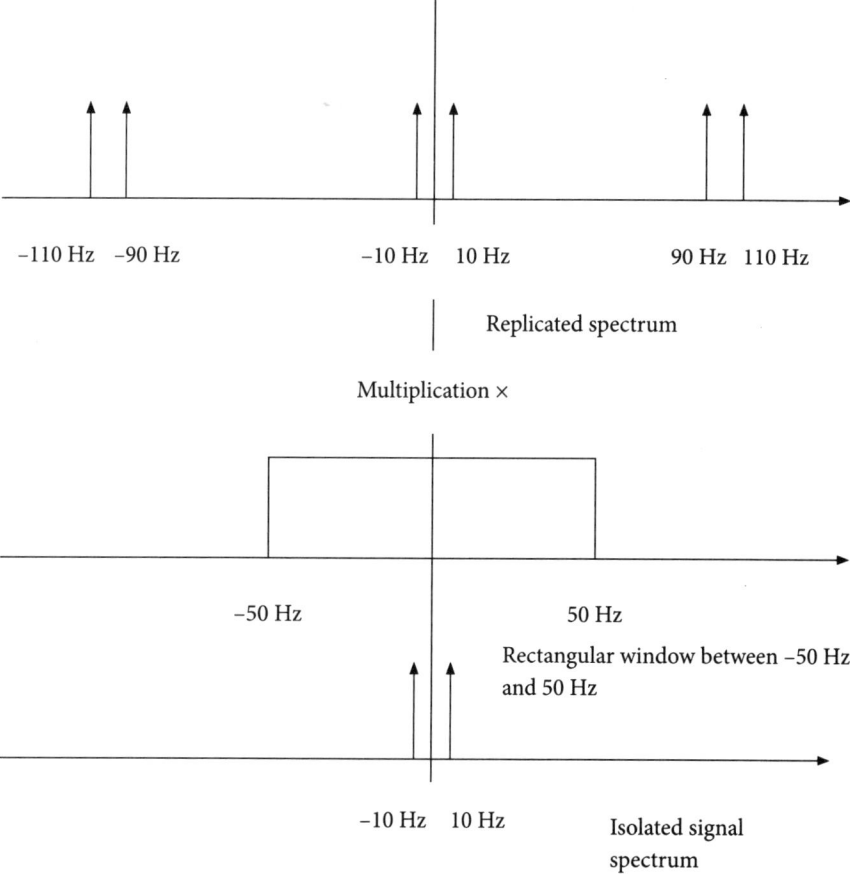

Fig. 1.5 Isolation of signal spectrum by passing the convolved spectrum via a rectangular window

There is a multiplication of convolved spectrum and a rectangular window in Fourier domain, as shown in Fig. 1.3. This is equivalently a convolution of the signal samples with the impulse response of the ideal (brick wall) low-pass filter, that is, a sync function, as shown in Fig. 1.6.

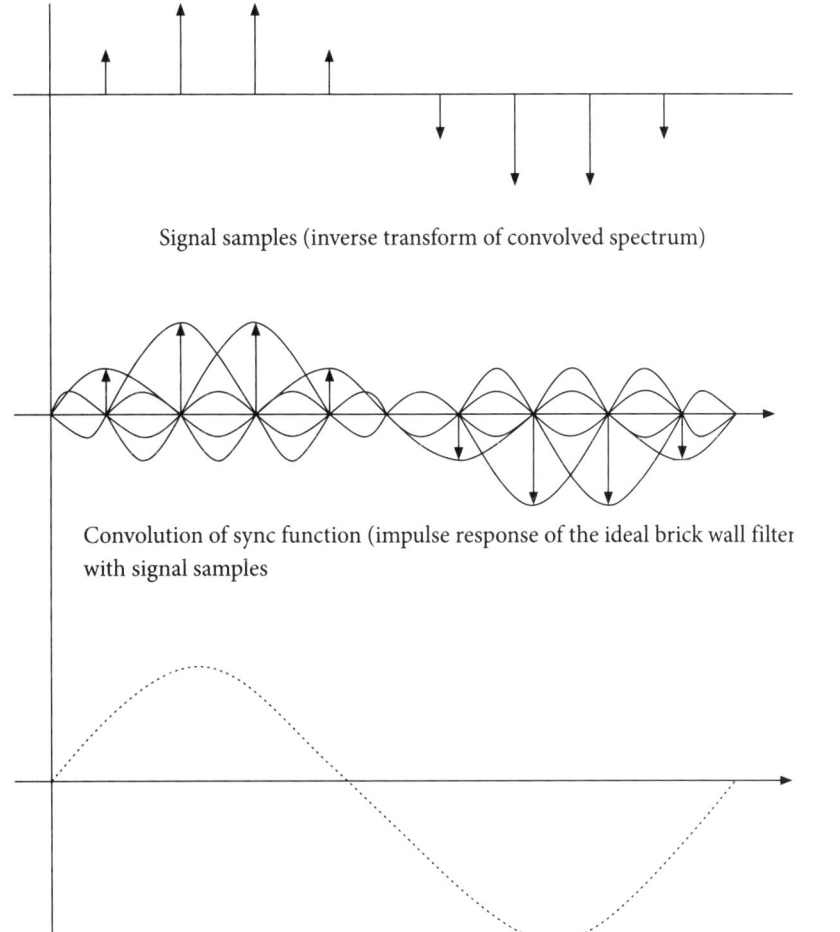

Signal samples (inverse transform of convolved spectrum)

Convolution of sync function (impulse response of the ideal brick wall filter with signal samples

Interpolated signal is shown with dotted line. It is the result of addition of infinite weighted sum of scaled and shifted sync functions at every instant.

Fig. 1.6 Reconstruction of the 10 Hz signal component from signal samples using sync function as the interpolation function

We say that the sync function is used as an interpolation function to reconstruct the signal from the signal samples. The interpolation function is a sync function given by

$$g(t) = \text{sync}(2\pi Wt) = \frac{\sin(2\pi Wt)}{2\pi Wt} \tag{1.1}$$

Here W is the bandwidth of the signal. In our example, it is equal to 10 Hz. The signal can be recovered as

$$X(t) = \sum_{n=-\infty}^{\infty} X(n) g\left(t - \frac{n}{F_S}\right) \qquad (1.2)$$

where F_S is the sampling frequency, equal to 100 Hz in our example.

The reconstruction process is complicated involving the infinite weighted sum of the function $g(t)$ and its shifted versions. The reconstruction formula is only of theoretical interest. Practically, the signal is just low-pass filtered with the cut-off frequency of the signal bandwidth to recover the signal.

Concept Check

- When the two signals are multiplied in Fourier domain, what happens in the time domain?
- What is a sync function?
- What is the meaning of interpolation function?

1.5 Sampling of Analog Signals (Case II)

Now, let us consider the 90 Hz signal component present in a signal with a bandwidth of 90 Hz. Let it be sampled using a sampling frequency of 100 Hz. The sampling frequency F_S is below the Nyquist rate (2 × bandwidth = 180 Hz). We will get 10 samples per 9 cycles of the waveform, as shown in Fig. 1.7. The sampled signal is obtained as a multiplication of the 90 Hz signal with a train of impulse with a period of 0.01 s.

In the transform domain, there will be convolution of the transform of the 90 Hz signal, that is, a spectral line at 90 Hz and the transform of the impulse train, that is, an impulse again with the period of F_S (100 Hz), as shown in Fig. 1.8. We can notice from the figure that the replicated spectral lines overlap or cross each other in the convolved spectrum.

Concept Check

- What will happen if the sampling frequency is below the Nyquist rate?
- What happens to the spectrum of signal sampled below the Nyquist rate?
- What is aliasing?

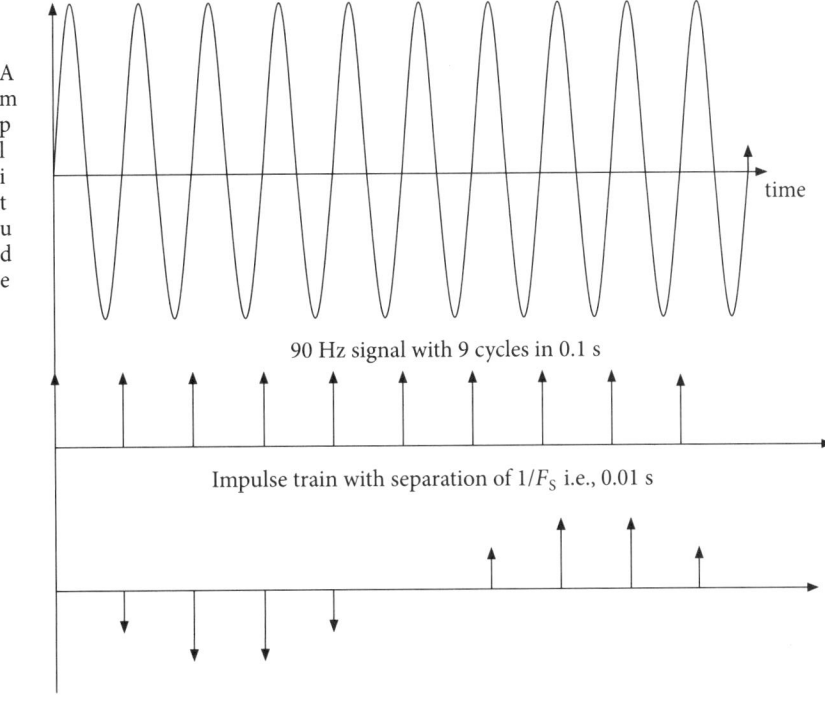

Fig. 1.7 Ten samples per 9 cycles of waveform when the 90 Hz signal is sampled with 100 Hz F_S

Note The signal that will be reconstructed from these samples is a 10 Hz signal with a 180° phase shift.

1.6 Recovery of Analog Signals (Case II)

The convolved spectrum in Fig. 1.6 shows replicated copies of the signal spectrum around zero frequency and multiples of the sampling frequency, that is, 100 Hz. The replicas are not well separated and there is mixing of adjacent replicas. The signal spectrum around zero frequency has 90 Hz and –90 Hz components, whereas the replica around 100 Hz has 10 Hz and 190 Hz components and the replica around –100 Hz has –190 Hz and –10 Hz components. The replica of 90 Hz around 100 Hz appears as a 10 Hz signal. We say that 90 Hz is aliased as 10 Hz. The overlapping of the spectrum does not allow the isolation of the 90 Hz signal. We can use a low-pass filter of 50 Hz, that is, F_N to filter the spectrum and it will then pass 10 Hz, which is an aliased version of 90 Hz.

The phenomenon of high-frequency components in the spectrum of the original signal taking the identity of low-frequency in the spectrum of its sampled version is called aliasing effect. The high frequencies are aliased as low frequencies.

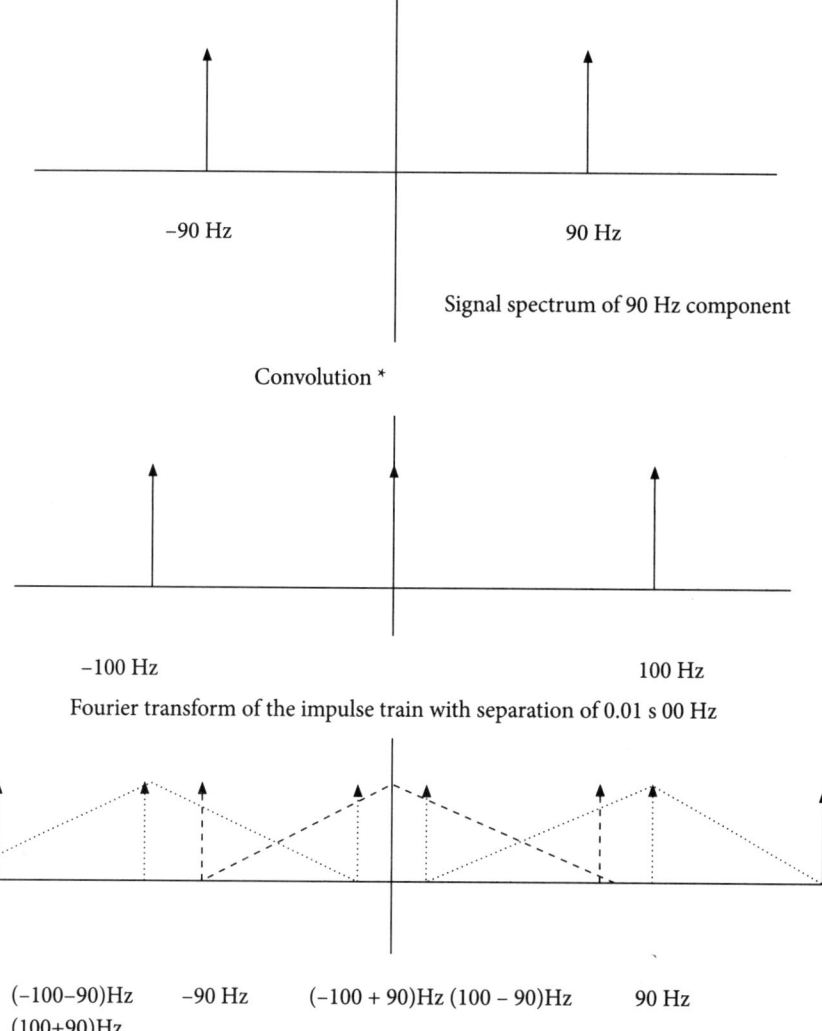

Fig. 1.8 Overlapped replicated spectrum (a replica of 90 Hz around 100 Hz) is seen as 10 Hz (dotted lines). The 90 Hz signal line is the replica of 90 Hz around 0 Hz (dashed lines). The 90 Hz signal is aliased as a 10 Hz signal when the spectrum is passed via a 50 Hz window (50 Hz being Nyquist frequency)

Concept Check
- What is aliasing effect?
- What will be the aliased frequency if a signal of 70 Hz is sampled with a 100 Hz sampling frequency?

1.7 Analytical Treatment

We can explain both the above cases analytically. Let us consider a signal with a frequency of 10 Hz. Let it be represented as

$$X(t) = \sin(2\pi \times 10 \times t) \tag{1.3}$$

When the signal is sampled, time will be digitized with a sampling interval of T. Every next sample is obtained after time T. So, the nth sample will be obtained at $t = n \times T$. Substituting $t = n \times T$ in Eq. (1.3) we get

$$x[nT] = \sin(2\pi \times 10 \times nT) \tag{1.4}$$

Now, put $T = 1/F_s$, that is, 1/100 Hz. T is normally dropped in the representation of X as T is a constant.

$$x[n] = \sin\left(2\pi \times \frac{10}{100} \times n\right) \tag{1.5}$$

$$x[n] = \sin\left(2\pi \frac{1}{10} n\right) \tag{1.6}$$

The generalized equation representing Discrete Time signal is

$$x[n] = \sin(2\pi f n) \tag{1.7}$$

Here f represents the digital frequency given by $f = F/F_s$ where F is the analog frequency and F_s is the sampling frequency (discussed in detail in Chapter 3). Now comparing Eqs (1.6) and (1.7), we find that $f = 1/10$ will recover the 10 Hz signal from its sampled version.

Now, we will consider a signal with a frequency of 90 Hz. Let it be represented as

$$X(t) = \sin(2\pi \times 90 \times t) \tag{1.8}$$

When the signal is sampled, t will be digitized with a sampling interval of T. Substituting $t = nT$ in Eq. (1.8), we get

$$x[nT] = \sin(2\pi \times 90 \times nT) \tag{1.9}$$

Now, put $T = 1/F_s$, that is, 1/100 Hz. Dropping T in the representation, from Eq. (1.9) we obtain,

$$x[n] = \sin\left(2\pi \times \frac{90}{100} \times n\right) \qquad (1.10)$$

$$x[n] = \sin\left(2\pi \times \frac{9}{10} \times n\right) \qquad (1.11)$$

$$x[n] = \sin\left(2\pi n - 2\pi \times \frac{1}{10} \times n\right) \qquad (1.12)$$

As $\sin(x)$ is the same as $\sin(x + 2\pi)$, we can simplify the above expression to

$$x[n] = \sin\left(-2\pi \frac{1}{10} n\right) = -\sin\left(2\pi \frac{1}{10} n\right) \qquad (1.13)$$

Comparing Eqs (1.13) and (1.7), we get $f = -1/10$. This will recover the 10 Hz signal from its sampled version with a phase shift of 180° (refer to the note in Fig. 1.6). We find that sinusoids up to 50 Hz frequency (Nyquist frequency) are properly recovered. Sinusoids between 50 Hz and 100 Hz, that is, between Nyquist frequencies and the sampling frequency are shifted in phase and are aliased as (100 – the original frequency). Here, 90 Hz gets aliased as 10 Hz with 180° phase shift. The analog and the corresponding digital frequencies with phase can be represented in a graph form, as shown in Fig. 1.9.

We can now conclude that the analog frequencies above $0.5F_S$ are always aliased and take the value between 0 and $0.5F_S$. The frequency $0.5F_S$ is called the folding frequency. Undersampling, that is sampling below the Nyquist rate, not only changes the frequency of the original signal but also changes the phase. Only two phase values are possible, 0° and 180°. If the frequency of the analog signal is between 0 and $F_S/2$, the phase is zero. If the analog frequency is between $F_S/2$ and F_S, the phase value is 180°, as shown in Fig. 1.9.

Concept Check

- When the analog signal of 60 is sampled using a sampling frequency of 100 Hz, what is the aliased frequency and what is the phase value of the aliased output?
- When is the phase value zero and when is it 180°?

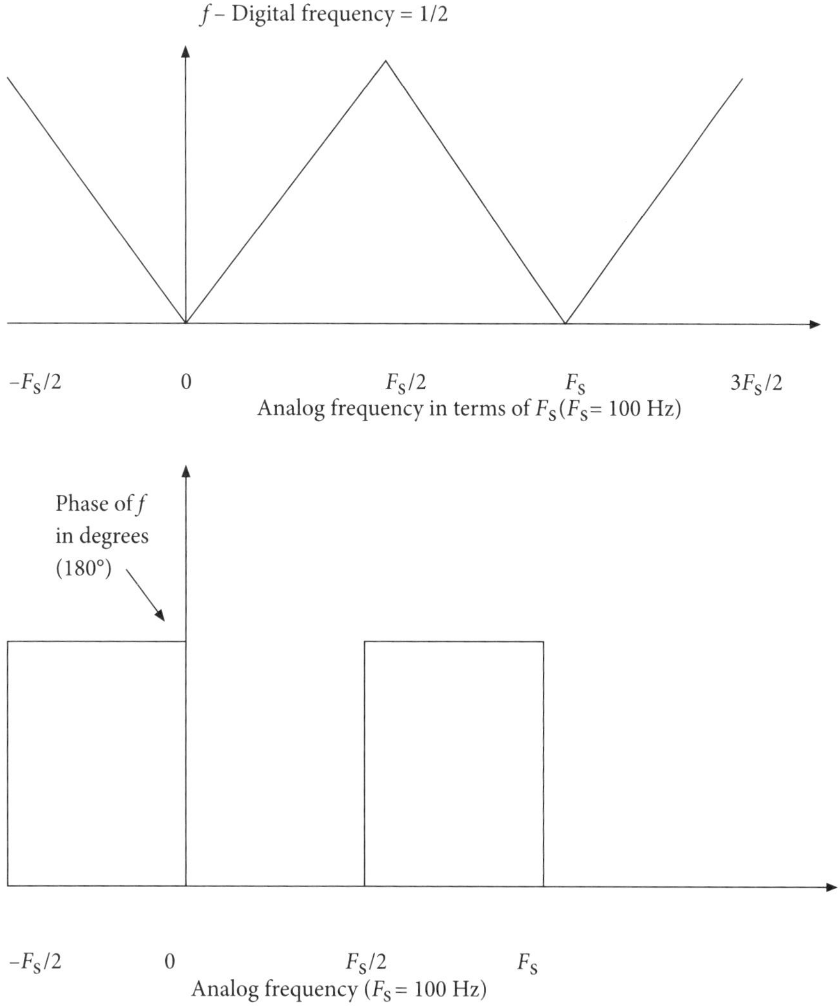

Fig. 1.9 Variation of digital frequency and phase with analog frequency with a sampling frequency of F_s

1.8 Analytical Examples

Example 1.1

Let the analog signal be represented as

$$X(t) = 3\cos(50\pi t) + 2\sin(300\pi t) - 4\cos(100\pi t) \tag{1.14}$$

What is the Nyquist rate for this signal? If the signal is sampled with a sampling frequency of 200 Hz, what will be the DT signal obtained after sampling? What will be the recovered signal?

Solution

The signal contains three frequencies, namely, 25 Hz, 150 Hz and 50 Hz. The highest frequency is 150 Hz. The recommended rate of sampling is 300 Hz (2×150 Hz).

If the signal is a sampling frequency of 200 Hz, substituting $t = n \times T_s = n/F_s$ in Eq. (1.14), we get,

$$X[nT_s] = 3\cos\left(50\pi \frac{n}{200}\right) + 2\sin\left(300\pi \frac{n}{200}\right) - 4\cos\left(100\pi \frac{n}{200}\right) \quad (1.15)$$

$$x[n] = 3\cos\left(\pi \frac{n}{4}\right) + 2\sin\left(3\pi \frac{n}{2}\right) - 4\cos\left(\pi \frac{n}{2}\right) \quad (1.16)$$

The standard equation for the discrete signal is

$$x[n] = A\sin(2\pi f n) \quad (1.17)$$

The expression for the angle is suitably modified so that the angle is less than π. So Eq. (1.16) is modified as

$$x[n] = 3\cos\left(\pi \frac{n}{4}\right) + 2\sin\left(2\pi - \pi \frac{n}{2}\right) - 4\cos\left(\pi \frac{n}{2}\right) \quad (1.18)$$

$$x[n] = 3\cos\left(\pi \frac{n}{4}\right) - 2\sin\left(\pi \frac{n}{2}\right) - 4\cos\left(\pi \frac{n}{2}\right) \quad (1.19)$$

Now, the signal has three digital frequencies, namely, $f = 1/8, f = -1/4, f = 1/4$. Equation (1.19) represents the DT signal obtained after sampling.

The recovered signal frequencies are $f = 1/8$, that is, $F = 25$ Hz; $f = -1/4$, that is, $F = 50$ Hz with a phase shift of 180°; $f = ¼$, that is $F = 50$ Hz. We see that the 150 Hz signal is aliased as $200 - 150$ Hz, that is, 50 Hz with 180° phase shift. The recovered signal contains only two frequencies, namely, 25 Hz and 50 Hz, whereas the original signal had three frequencies 25 Hz, 150 Hz and 50 Hz.

Example 1.2

Find the recovered signal if the signal of frequency 60 Hz is sampled using a sampling frequency of 80 Hz. What is the phase value?

Solution

Let $x(t) = \sin(2\pi F t)$

Putting $F = 50$ Hz and $t = nT = n/F_s$, we get

$$x[n] = \sin\left(\frac{2\pi \times 60n}{80}\right) = \sin\left(\frac{6\pi n}{4}\right)$$

$$= \sin\left((2\pi - \frac{\pi}{2})n\right) = \sin\left(-\frac{2\pi \times n}{4}\right) \quad (1.20)$$

Compare the equation with $x[n] = \sin(2\pi fn)$ where f is the recovered digital frequency given by $f = F/F_s$. Here $f = -1/4 = -20/80$, so the analog frequency recovered is 20 Hz. There is a negative sign so the phase shift of 180° is introduced.

Example 1.3

Let the analog signal be represented as

$$x(t) = \cos(150\pi t) + 2\sin(300\pi t) - 400\cos(600\pi t)$$

What is the Nyquist rate for this signal? If the signal is sampled with a sampling frequency of 400 Hz, what is the DT signal obtained after sampling? What is the recovered signal?

Solution

The signal is given by

$$x(t) = \cos(150\pi t) + 2\sin(300\pi t) - 400\cos(600\pi t) \quad (1.21)$$

The maximum frequency in the signal is 300 Hz. So the Nyquist frequency is 600 Hz. If the signal is sampled using a sampling frequency of 400 Hz, by putting $t = n/400$ in the equation we get

$$\begin{aligned}
x[n] &= \cos(150\pi n/400) + 2\sin(300\pi n/400) - 400\cos(600\pi n/400) \\
&= \cos(3\pi n/8) + 2\sin(3\pi n/4) - 400\cos(3\pi n/2) \\
&= \cos(3\pi n/8) + 2\sin(3\pi n/4) - 400\cos((2\pi - \pi/4)n) \\
&= \cos(3\pi n/8) + 2\sin(3\pi n/4) - 400\cos(\pi n/4)
\end{aligned} \quad (1.22)$$

Compare the equation with $x[n] = \sin(2\pi fn)$ where f is the recovered digital frequency given by $f = F/F_s$. $f = 3/8, 3/4, 1/4 = 75/400, 150/400$ and $100/400$, so the analog frequencies recovered are 75 Hz, 150 Hz and 100 Hz, respectively. Here the 300 Hz component will be aliased as a 100 Hz component.

Example 1.4

Let the analog signal be represented as

$$x(t) = \sin(10\pi t) + 2\sin(20\pi t) - 2\cos(40\pi t)$$

What is the Nyquist rate for this signal? If the signal is sampled with a sampling frequency of 20 Hz, what is the DT signal obtained after sampling? What is the recovered signal?

Solution

The signal is given by

$$x(t) = \sin(10\pi t) + 2\sin(20\pi t) - 2\cos(40\pi t) \tag{1.23}$$

The signal contains 5 Hz, 10 Hz and 20 Hz. The maximum frequency in the signal is 20 Hz, so the Nyquist frequency is 40 Hz.

If the signal is sampled using a sampling frequency of 20 Hz, by putting $t = n/20$ in the equation we get

$$\begin{aligned} x[n] &= \sin(10\pi n/20) + 2\sin(20\pi n/20) - 2\cos(40\pi n/20) \\ &= \sin(\pi n/2) + 2\sin(\pi n) - 2\cos(2\pi n) \end{aligned} \tag{1.24}$$

Compare the equation with $x[n] = \sin(2\pi f n)$ where f is the recovered digital frequency given by $f = F/F_S$. $f = 1/4, 1/2, 1 = 5/20, 10/20, 20/20$ so the analog frequencies recovered are 5 Hz, 10 Hz and 0 Hz, respectively. The 20 Hz signal is aliased as a dc component.

Example 1.5

Find the recovered signal if the signal of frequency 150 Hz is sampled using sampling frequencies of 400 Hz and 200 Hz, respectively. What is the phase value in each case?

Solution

Let $x(t) = \sin(2\pi F t)$. Putting $F = 150$ Hz and $t = nT = n/F_S$, $F_S = 400$ Hz we get

$$x[n] = \sin\left(\frac{2\pi \times 150 n}{400}\right) = \sin\left(\frac{2\pi \times 3}{8 \times n}\right) = \sin\left(\frac{3\pi}{4} \times n\right) = \sin\left(2\times\frac{3\pi n}{8}\right) \tag{1.25}$$

Compare the equation with $x[n] = \sin(2\pi f n)$ where f is the recovered digital frequency given by $f = F/F_S$. Here $f = 3/8 = 150/400$, so the analog frequency recovered is 150 Hz. There is a positive sign so the phase shift of 0° is introduced.

Let $x(t) = \sin(2\pi F t)$. Putting $F = 150$ Hz and $t = nT = n/F_S$, $F_S = 200$ Hz we get

$$\begin{aligned} x[n] &= \sin\left(\frac{2\pi \times 150 n}{200}\right) = \sin\left(\frac{2\pi \times 3}{4} \times n\right) \\ &= \sin\left(2\pi - \frac{2\pi}{4} \times n\right) = \sin\left(-\frac{2\pi n}{4}\right) \end{aligned} \tag{1.26}$$

Compare the equation with $x[n] = \sin(2\pi fn)$ where f is the recovered digital frequency given by $f = F/F_S$. Here $f = -1/4 = -50/200$, so the analog frequency recovered is 50 Hz. There is a negative sign so the phase shift of 180° is introduced.

Example 1.6

Find the recovered signal if the signal of frequency 150 Hz and 250 Hz are sampled using a sampling frequency of 200 Hz. What is the phase value?

Solution

Let $x(t) = \sin(2\pi Ft)$. Putting $F = 150$ Hz and $t = nT = n/F_S$, $F_S = 200$ Hz we get

$$x[n] = \sin\left(2\pi \times \frac{150n}{200}\right) = \sin\left(2\pi \times \frac{3}{4}n\right)$$

$$= \sin\left((2\pi - \frac{\pi}{2})n\right) = \sin\left(-\frac{2\pi n}{4}\right) \quad (1.27)$$

Compare the equation with $x[n] = \sin(2\pi fn)$ where f is the recovered digital frequency given by $f = F/F_S$. Here $f = -1/4 = -50/200$, so the analog frequency recovered is 50 Hz. There is a negative sign so the phase shift of 180° is introduced.

Let $x(t) = \sin(2\pi Ft)$. Putting $F = 250$ Hz and $t = nT = n/F_S$, $F_S = 200$ Hz we get

$$x[n] = \sin\left(2\pi \times \frac{250n}{200}\right) = \sin\left(2\pi \times \frac{5}{4}n\right)$$

$$= \sin\left((2\pi + \frac{\pi}{2})n\right) = \sin\left(\frac{2\pi n}{4}\right) \quad (1.28)$$

Compare the equation with $x[n] = \sin(2\pi fn)$ where f is the recovered digital frequency given by $f = F/F_S$. Here $f = 1/4 = 50/200$, so the analog frequency recovered is 50 Hz. There is a positive sign so the phase shift of 0° is introduced.

Example 1.7

Consider the analog sinusoidal signal

$$x(t) = 5\sin(500\pi t)$$

a. The signal is sampled with $F_S = 1500$ Hz. Find the frequency of the DT signal.
b. Find the frequency of the DT signal if $F_S = 300$ Hz.

Solution

a. Let

$x(t) = 5\sin(2\pi F t)$. Putting $F = 250$ Hz and $t = nT = n/F_S$, $F_S = 1500$ Hz we get

$$x[n] = \sin\left(2\pi \times \frac{250n}{1500}\right) = \sin\left(2\pi \times \frac{1}{6}n\right) = \sin\left(\left(\frac{2\pi}{3}\right)n\right) = \sin\left(\frac{2\pi n}{3}\right) \quad (1.29)$$

Compare the equation with $x[n] = \sin(2\pi f n)$ where f is the recovered digital frequency given by $f = F/F_S$. Here $f = 1/6 = 250/1500$, so the analog frequency recovered is 250 Hz. There is a positive sign so the phase shift of 0° is introduced.

b. Let $x(t) = \sin(2\pi F t)$. Putting $F = 250$ Hz and $t = nT = n/F_S$, $F_S = 300$ Hz we get

$$x[n] = \sin\left(2\pi \times \frac{250n}{300}\right) = \sin\left(2\pi \times \frac{5}{6}n\right)$$

$$= \sin\left(\left(2\pi - \frac{2\pi}{6}\right)n\right) = \sin\left(-\frac{2\pi n}{6}\right) \quad (1.30)$$

Compare the equation with $x[n] = \sin(2\pi f n)$ where f is the recovered digital frequency given by $f = F/F_S$. Here $f = -1/6 = -50/300$, so the analog frequency recovered is 50 Hz. There is a negative sign so the phase shift of 180° is introduced.

Example 1.8

An analog signal given by $x(t) = \sin(200\pi t) + 3\cos(250\pi t)$ is sampled at a rate of 300 sample/s. Find the frequency of the DT signal.

Solution

Given that $x(t) = \sin(200\pi t) + 3\cos(250\pi t)$ and the frequencies in the signal are 100 Hz and 125 Hz. The sampling frequency $F_S = 300$ Hz.

Let us put $F = 100$ Hz and $t = nT = n/F_S$, $F_S = 300$ Hz, we get

$$x[n] = \sin\left(2\pi \times \frac{100n}{300}\right) = \sin\left(2\pi \times \frac{1}{3}n\right) = \sin\left(\frac{2\pi n}{3}\right) \quad (1.31)$$

Compare the equation with $x[n] = \sin(2\pi f n)$ where f is the recovered digital frequency given by $f = F/F_S$. Here $f = 1/3 = 100/300$, so the analog frequency recovered is 100 Hz. There is a positive sign so the phase shift of 0° is introduced.

Let us put $F = 125$ Hz and $t = nT = n/F_S$, $F_S = 300$ Hz we get

$$x[n] = \sin\left(2\pi \times \frac{125n}{300}\right) = \sin\left(2\pi \times \frac{5}{12}n\right) = \sin\left(\frac{2 \times 5\pi n}{12}\right) \quad (1.32)$$

Compare the equation with $x[n] = \sin(2\pi fn)$ where f is the recovered digital frequency given by $f = F/F_S$. Here $f = 5/12 = 125/300$, so the analog frequency recovered is 125 Hz. There is a positive sign, so the phase shift of 0° is introduced.

Concept Check

- When the analog signal consisting of 60 Hz and 40 Hz is sampled using a sampling frequency of 100 Hz, what are the frequencies in the recovered signal?

1.9 Anti-Aliasing Filter

It is clear from the above examples that when the signal is sampled with a sampling frequency of F_S, the frequencies above $F_S/2$ are aliased. The signal bandwidth is known; however, the signal may contain some spurious noise above $F_S/2$. The noise frequencies above $F_S/2$ will then be aliased and will affect the original signal frequencies. To ensure that the spurious noise does not affect the signal components, it is recommended that the signal is passed via a low-pass filter of a cut-off frequency $F_S/2$ prior to sampling. The low-pass filter is termed as anti-aliasing filter. This confirms that the signal bandwidth is really $F_S/2$.

Example 1.9

Design an anti-aliasing filter for a signal represented as

$$x(t) = \sin(80\pi t) + \sin(100\pi t) - 6\cos(150\pi t)$$

Solution

We have to design an anti-aliasing filter for a signal represented as

$$x(t) = \sin(80\pi t) + \sin(100\pi t) - 6\cos(150\pi t) \quad (1.33)$$

Compare the equation with $x(t) = \sin(2\pi Ft)$ where F is the analog frequency. We find that the signal contains frequencies as 40 Hz, 50 Hz and 75 Hz. So the required sampling frequency is 150 Hz and the Nyquist frequency is 75 Hz. Anti-aliasing filter must have a cut-off frequency of 75 Hz.

Example 1.10

Design an anti-aliasing filter for a signal represented as

$$x(t) = \cos(170\pi t) + \cos(190\pi t) - 3\cos(250\pi t)$$

Solution

We have to design an anti-aliasing filter for a signal represented as

$$x(t) = \cos(170\pi t) + \cos(190\pi t) - 3\cos(250\pi t) \qquad (1.34)$$

Compare the equation with $x(t) = \sin(2\pi Ft)$ where F is the analog frequency. We find that the signal contains frequencies as 85 Hz, 95 Hz and 125 Hz. So the required sampling frequency is 250 Hz. The Nyquist frequency is 125 Hz. The anti-aliasing filter must have a cut-off frequency of 125 Hz.

Concept Check

- Why is an anti-aliasing filter needed?
- What will happen if the anti-aliasing filter is not used?

Summary

In this chapter, we have described and explained the following important concepts.

- We have started with the statement of sampling theorem in time domain. When a time domain signal is sampled, the sampling frequency must be such that it must be possible to recover the signal from the signal samples. The theorem says that the sampling frequency must be selected as greater than or equal to twice the maximum frequency present in the signal, that is, the bandwidth of the signal. This indicates that only band-limited signals can be sampled.
- Sampling is explained using the concept of impulse train. When a signal is sampled, it is multiplied by the impulse train. In Fourier domain, there will be convolution of the spectrum of the signal with Fourier Transform (FT) of impulse train, which is again the impulse train. To recover the signal from signal samples, we multiply the convolved spectrum with a rectangular window between $-F_N$ and F_N. In time domain, there will be convolution of signal samples with a sync function (inverse Fourier transform of rectangular window). The sync function acts as an interpolation function and the signal can be reconstructed provided the adjacent replicated spectra do not overlap, that is, the sampling frequency $F_S \geq 2 \times$ Bandwidth. When the sampling frequency is smaller than $2 \times$ bandwidth, the spectrum passed via a rectangular window between $-F_N$ and F_N is for the aliased frequency. The high frequencies are aliased as low frequencies. The phase reversal occurring when the analog frequency is greater than F_N and less than F_S was explained using the analytical treatment. The aliased frequency is then F_S-analog frequency.
- The need for anti-aliasing filter is emphasized. When the signal is passed via a low-pass filter with a cut-off frequency equal to the bandwidth,

the high-frequency components are eliminated to ensure the signal bandwidth. This eliminates the possibility of aliasing.

Multiple Choice Questions

1. Analog signal has
   (a) time and amplitude axis continuous (b) time axis continuous
   (c) amplitude axis continuous (d) none of the axis continuous
2. DT signal has
   (a) only time axis continuous (b) only time axis discrete
   (c) amplitude axis continuous (d) amplitude axis discrete
3. Digital signal has
   (a) time axis discrete (b) amplitude axis discrete
   (c) time and amplitude axis discrete (d) both axis continuous
4. Sampling a signal is equivalent to multiplying it with
   (a) a sync function (b) a train of impulse
   (c) a train of sync functions (d) a rectangular window
5. Aliasing occurs when a signal of bandwidth W is sampled with a sampling frequency F
   (a) greater than W (b) less than W
   (c) greater than $2W$ (d) less than $2W$
6. Fourier transform of a rectangular window is
   (a) a sync function (b) an impulse train
   (c) a modified sync function (d) a rectangular window
7. Fourier transform of a train of impulse is
   (a) a sync function (b) a train of impulse
   (c) a modified sync function (d) a rectangular window
8. When aliasing occurs
   (a) low-frequency becomes a high-frequency
   (b) high-frequency becomes a mid-frequency
   (c) high frequencies above signal bandwidth become low frequencies below signal bandwidth
   (d) none
9. Anti-aliasing filter is a low-pass filter with a cut-off frequency of
   (a) signal bandwidth (b) 2 × signal bandwidth
   (c) 1/2 of signal bandwidth (d) 3 × signal bandwidth

Review Questions

1.1 Define analog signal and give appropriate example.
1.2 Define DT signal and illustrate it with a plot.
1.3 What is uniform sampling? What is a DT signal?
1.4 State the sampling theorem in time domain. Explain the aliasing effect.
1.5 What is Nyquist rate of sampling?
1.6 How does one decide the sampling frequency for any signal to be sampled? Can a signal with infinite bandwidth be sampled?
1.7 Explain the process of sampling using the concept of a train of pulses.
1.8 Explain the occurrence of replicated spectrum when the signal is sampled.
1.9 How can one recover the signal from signal samples using a sync function
1.10 Explain the aliasing effect for a signal which is sampled below the Nyquist rate.
1.11 What is a practical way to recover the signal from signal samples?
1.12 When does a phase reversal occur for a recovered aliased signal?
1.13 What is the need for an anti-aliasing filter?

Problems

1.1 Find the recovered signal if the signal of frequency 50 Hz is sampled using a sampling frequency of 80 Hz. What is the phase value?

1.2 Let the analog signal be represented as

$$x(t) = \cos(150\pi t) + 2\sin(300\pi tt) - 400\cos(600\pi t)$$

What is the Nyquist rate for this signal? If the signal is sampled with a sampling frequency of 300 Hz, what is the DT signal obtained after sampling? What is the recovered signal?

1.3 Let the analog signal be represented as

$$x(t) = \sin(10\pi t) + 2\sin(20\pi t) - 2\cos(30\pi t)$$

What is the Nyquist rate for this signal? If the signal is sampled with a sampling frequency of 20 Hz, what is the DT signal obtained after sampling? What is the recovered signal?

1.4 Find the recovered signal if the signal of frequency 150 Hz is sampled using sampling frequencies of 450 Hz and 200 Hz, respectively. What is the phase value in each case?

1.5 Design an anti-aliasing filter for a signal represented as

$$x(t) = \sin(80\pi t) + \sin(100\pi t) - 6\cos(150\pi t) + 8\cos(200\pi t)$$

1.6 Design an anti-aliasing filter for a signal represented as

$$x(t) = \cos(170\pi t) + \cos(190\pi t) - 3\cos(250\pi t) - \sin(400\pi t)$$

1.7 Find the recovered signal, if the signal of frequencies 150 Hz and 250 Hz are sampled using a sampling frequency of 400 Hz. What is the phase value?

1.8 Consider the analog sinusoidal signal

$$x(t) = 5\sin(500\pi t)$$

(c) The signal is sampled with F_S = 2000 Hz. Find the frequency of the DT signal.

(d) Find the frequency of the DT signal if F_S = 300 Hz.

1.9 An analog signal given by $x(t) = \sin(200\pi t) + 3\cos(250\pi t)$ is sampled at a rate 200 sample/s. Find the frequency of DT signal.

Answers

Multiple Choice Questions

| 1 (a) | 2 (b) | 3 (c) | 4 (b) | 5 (d) |
| 6 (a) | 7 (b) | 8 (c) | 9 (a) | |

Problems

1.1 30 Hz, 190°.

1.2 600 Hz, DT signal will contain 75 Hz, 150Hz. 300 Hz component will be aliased as a DC component.

1.3 30 Hz, DT signal will contain 5 Hz, 10 Hz. 15 Hz component will be aliased as a DC component.

1.4 150 Hz, 0°, 50 Hz, 180°.

1.5 An LPF with a cut-off frequency of 100 Hz.

1.6 An LPF with a cut-off frequency of 200 Hz.

1.7 50 Hz, 0°, 150 Hz, 180°.

1.8 250 Hz, 0°, 50 Hz, 180°.

1.9 100 Hz, 0°, 75 Hz, 180°.

2

Signals and Operations on Signals

Learning Objectives

- Definition of a signal.
- Graph terminology and domains – time, DT, spatial.
- Applications of signals.
- Elementary signals.
- Classification of signals.
- What is a sequence?
- Properties of signals.
- Even and odd signals.
- Periodic and aperiodic signals.
- Energy and power signals.
- Operations on signals.

This chapter aims to introduce basic concepts related to CT and DT signals and operations on the signals such as translation, scaling, addition, multiplication, etc.

Natural signals like speech, seismic signal, Electro Cardio Gram (ECG) and Electro Encephalogram (EEG) signals are processed using sophisticated digital systems. All these information-carrying signals are functions of the independent variables, namely, time and space. We first deal with basic definition and classification of signals.

2.1 Signals

We have gone through the basics of signals in chapter 1. We will go through it in detail in this chapter. Let us first define a signal.

Definition of a signal A signal is defined as a single valued function of an independent variable, namely, time that conveys some information. At every instant of time there is a unique value of the function. The unique value may be real or complex.

A signal X is represented as $X = F(t)$, where F is some function and t represents time, the independent variable. Here, time is a continuous variable and X also takes on continuous values. X is then termed as continuous time, continuous amplitude (CTCA) signal that is the analog signal. When a signal is written as some mathematical equation, it is deterministic in the sense that the signal value at any time t can be determined using the mathematical equation. A signal is called scalar-valued if its value is unique for every instant of time. Most of the naturally occurring signals are scalar-valued, real, analog and random in nature. Speech signal is an example of such naturally occurring random signals. Speech signal is generated when a person utters a word. Researchers are still trying to generate a model for speech signal and exact mathematical equation cannot be written for a speech signal. It has a random signal component of some strength. It is called as random signal. If a signal has multiple values at any instant of time, it is called vector-valued. For example, color images, where at every pixel there is a three-dimensional vector representing RGB color components. In this chapter we will refer to scalar-valued and deterministic signals only. Hence, the signals will always be represented in the form of an equation.

Example 2.1

Consider the equation given by $x(t) = \cos(2\pi ft)$ where f is a frequency of a cosine wave. Is $x(t)$ a signal? Is it a scalar signal?

Solution

Here, time t is an independent variable. x is some function of an independent variable namely time. So $x(t)$ represents a signal. The signal x is real valued and is a scalar signal, as x has a unique value given by $\cos(2\pi ft)$ at every time instant t. The plot of the signal is represented in Fig. 2.1.

Example 2.2

Consider the equation given by $x(t) = t + jt$. Is $x(t)$ a signal? Is it a scalar signal?

Solution

Here, time t is an independent variable. x is some function an independent variable namely time. So $x(t)$ represents a signal. The signal x is complex valued and is a scalar signal, as x has a unique value given by $t + jt$ at every time instant t. The plot of the signal is represented in Fig. 2.2. A separate plot for magnitude and angle is plotted because the signal values are complex.

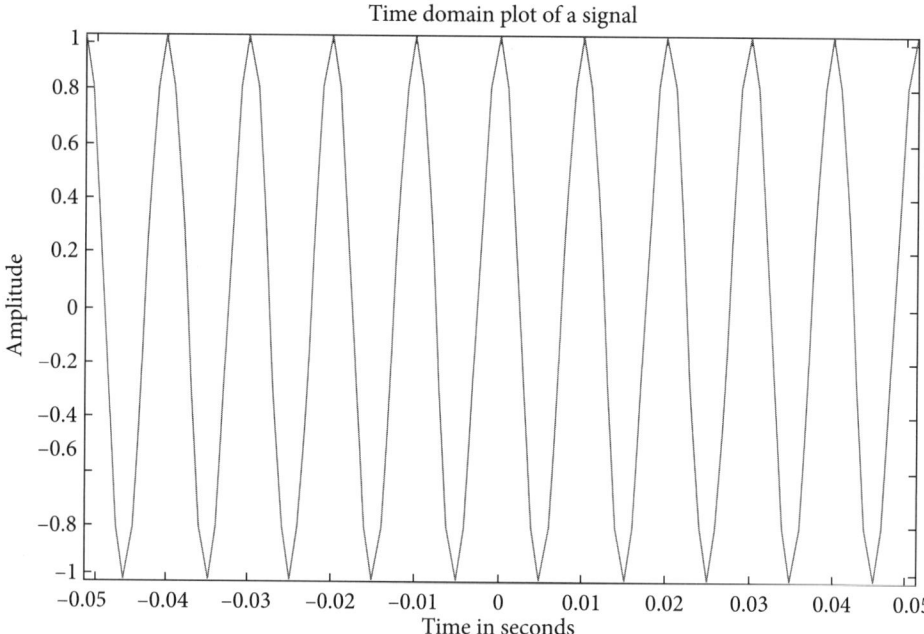

Fig. 2.1 Time domain plot of a cosine signal

Example 2.3

Consider the equation given by $x(t) = (t\bar{i}, 2t\bar{j}, t^2\bar{k})$. Is $x(t)$ a signal? Is it a scalar signal?

Solution

Here, time t is an independent variable. x is some function of an independent variable namely time. So $x(t)$ represents a signal. The signal x is a triplet and is a vector signal.

Concept Check

- What is a signal?
- If $y(t) = \sin^2(2\pi f t)$, can $y(t)$ represent a signal?
- If $y(t) = t$, is it a scalar-valued signal?
- What is a random signal?
- If $x(t) = (t\bar{i}, 2t\bar{j})$ Is $x(t)$ a scalar signal? Can we call it a vector signal?

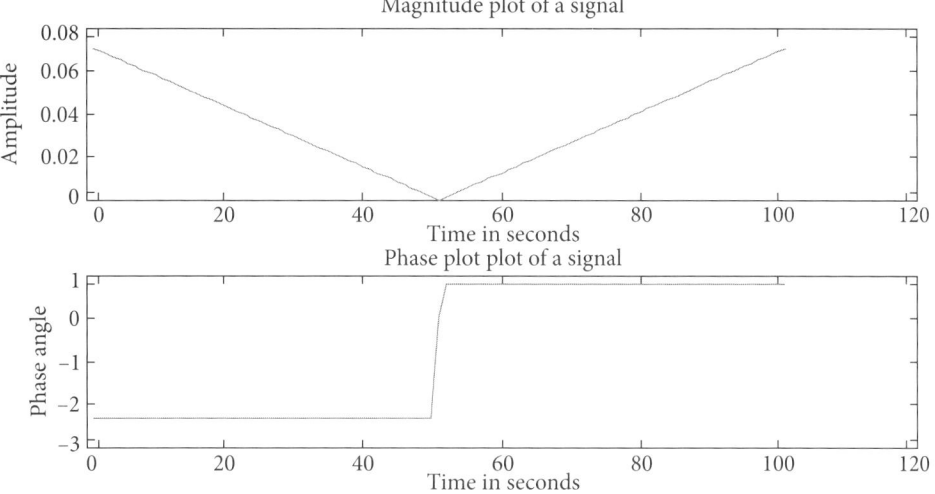

Fig. 2.2 Magnitude and phase plot of the signal

2.2 Graph Terminology and Domains

Consider a simple example of a signal, for example, a speech signal. A speech signal is recorded using a microphone. Output voltage of a microphone changes with time, in accordance with the speech signal. We may represent this signal in the form of a graph as voltage versus time. Voltage is a dependent variable and time is an independent variable. The independent variable is plotted on X-axis and is called as the domain.

Example 2.4

Consider a speech signal represented as a function of time, it is in time domain. Plot the signal as a CT signal and a DT signal.

Solution

The plot of the signal is shown in Fig. 2.3. A MATLAB program given below opens a file containing speech and plots it. When we use a command plot, we get a continuous time graph, as shown in Fig. 2.3. When we use a command stem, we get a discrete time graph, as shown in Fig. 2.4. When the speech signal is passed via A/D converter and interfaced to a digital computer, it is first sampled at a regular fixed time interval of T. We say that we have digitized the time axis and it now becomes discrete time (DT) signal. It will be represented as a variation of voltage with respect to discrete time. X-axis is now discrete and it is said to be represented in the DT domain, as shown in Fig. 2.4. A MATLAB program to read and plot a speech file is given below.

```
clear all;
fp=fopen('watermark.wav','r');
fseek(fp,42244,-1);
a=fread(fp,80);
a=a-128;
plot(a);
xlabel('time');
ylabel('amplitude');
title('analog plot of the utterance');
figure;
stem(a);xlabel('sample number');
ylabel('amplitude');
title('DT plot of the utterance');
```

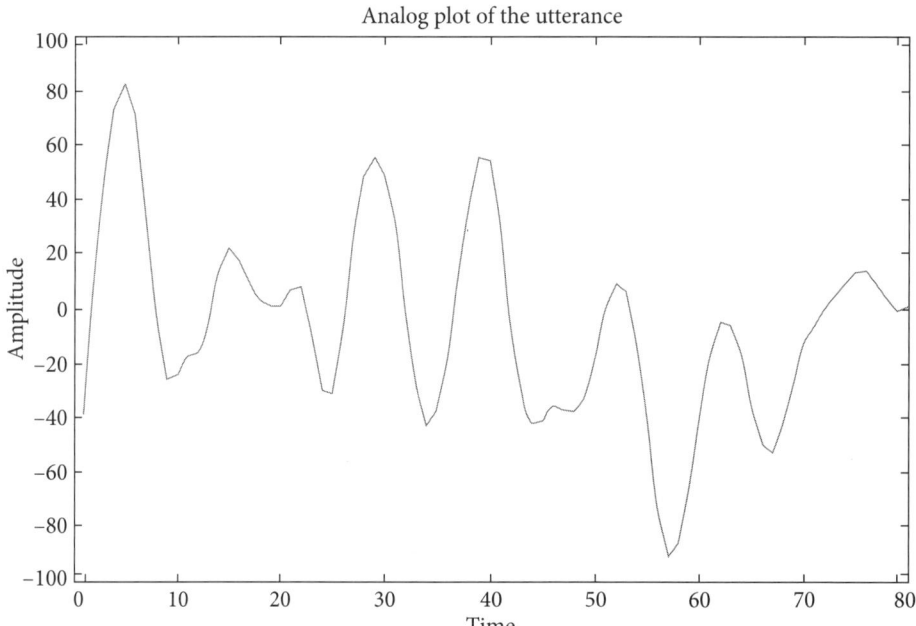

Fig. 2.3 Analog plot of a speech signal

Another commonly used domain for representing signals is a frequency domain. Here, the signal will be represented in the form of magnitude against frequency, that is, the resulting graphical representation will have a frequency on X-axis. This represents the amplitudes of different frequencies present in the signal. The frequency is now the independent variable and the domain is the frequency domain. The frequency domain is used when we take FT of a signal.

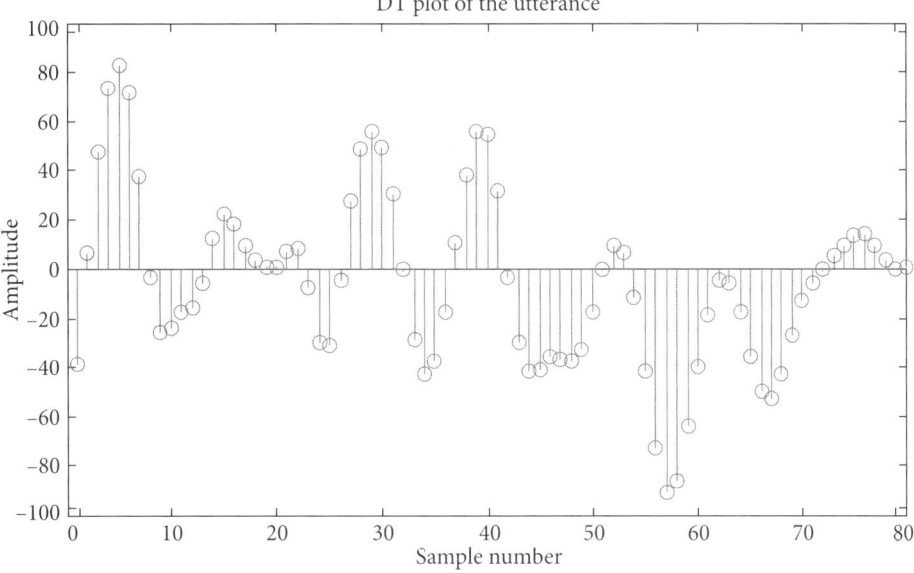

Fig. 2.4 DT plot of the same speech signal

Example 2.5

Consider the exponential signal. Plot the FT of the signal.

Solution

The magnitude plot and the phase plot of an FT of an exponential signal is represented as absolute amplitude verses frequency and angle verses frequency, respectively, as shown in Fig. 2.5. We will discuss the use of this domain in chapter 5 on Fourier series representation and chapter 6 on Fourier transform representation of the signals in detail. A MATLAB program is given below for finding magnitude and phase of FT of exponential signal.

```
clear all;
t=0:0.1:40;
x=exp(-3*t);
for i=1:20,
 y(i)=abs(1/sqrt((9+(i)*(i))));
end;
z(21)=1/3;
for i=1:20,
 z(i+21)=y(i);
end
for i=1:20,
 z(i)=y(21-i);
end
```

```
figure;
subplot(2,1,1);
s=-20:1:20;
plot(s,z);title('Magnitude plot of Fourier Transform
of exponential signal'); xlabel('frequency');ylabel
('amplitude');
subplot(2,1,2);
for i=1:20,
 y1(i)=angle(1/(3+1j*i));
end;
z1(21)=0.0;
for i=1:20,
 z1(i+21)=y1(i);
end
for i=1:20,
 z1(i)=-y1(21-i);
end;
plot(s,z1);title('Phase plot of FT of exponential
signal');xlabel('frequency');ylabel('angle');
```

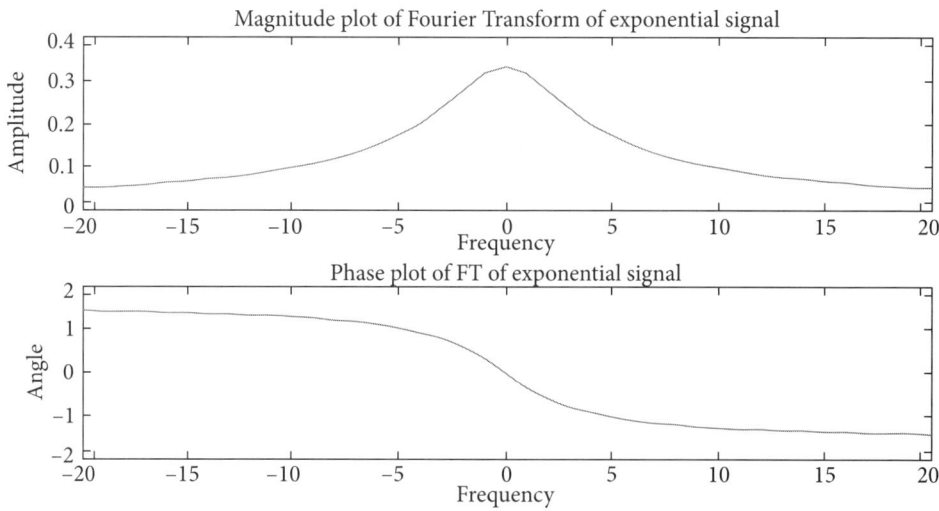

Fig. 2.5 Plot of magnitude and phase of FT of the exponential signal

Consider image as a signal. See how it is represented in the graph form. It is a three-dimensional (3D) representation with two independent variables, namely, distance along X direction and distance along Y direction. When the independent variable is spatial distance, the domain is named "spatial domain". The image signal is represented as a variation of magnitude versus spatial distance along two directions. It is said to be represented in spatial domain and is a two-dimensional signal.

Example 2.6

Consider a cameraman image. Plot the gray scale image.

Solution

The image is a gray scale image, which means that the intensity value of each pixel is represented as a gray scale value from 0 to 255. The two dimensions are the *x* coordinate value and *y* coordinate value. The intensity value of each pixel is represented as a third-dimensional value namely gray scale value ranging from 0 to 255.

Fig. 2.6 Plot of a spatial domain signal

Example 2.7

If the signal is given by $x[n] = \sin(2\pi f_D nT)$. Here, T is the sampling interval and f_D is a digital frequency namely $f_D = \dfrac{\text{analog frequency } f}{\text{sampling frequency } f_S}$ where $f_S = \dfrac{1}{T}$, what is its domain?

Solution

Here, the signal is a function of the independent variable namely discrete time. The domain is a DT domain. The plot of the signal is a stem plot, as shown in Fig. 2.7. The samples are taken at regular interval of T. The signal sample value at $t = nT$ is represented as $x[nT]$.

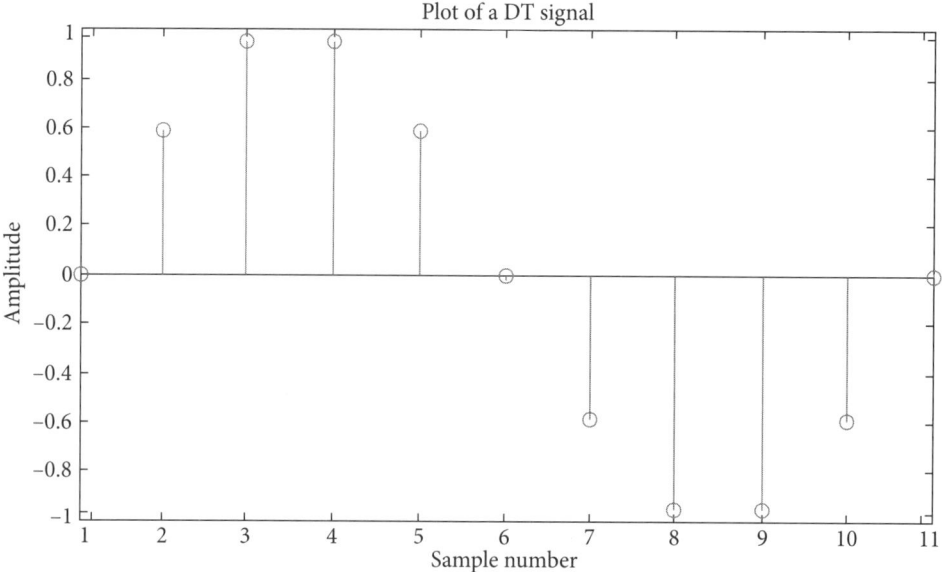

Fig. 2.7 Plot of a DT signal

Example 2.8

If the signal is given by $x(t) = \sin(2\pi Ft)$ what is its domain? If the signal is written as $x(F) = \sin(2\pi Ft)$, here F is the analog frequency. What is its domain?

Solution

$x(t) = \sin(2\pi Ft)$. Here, the signal is a function of the independent variable namely time. Time takes on all continuous values. The domain is a time domain. $x(F) = \sin(2\pi Ft)$ Here, the signal is a function of the independent variable namely frequency. Frequency takes on all continuous values. The domain is a frequency domain.

Example 2.9

If $f(x, y)$ represents image signal, what is its domain?

Solution

Here, the signal is a function of the independent variables namely x and y i.e., the spatial co-ordinates. The domain is a spatial domain.

Concept Check

- What is a domain?
- How does one recognize a domain?
- What is the domain for any image signal?
- If we are measuring temperature during a day, what will be the domain?

2.3 Applications of Signals and Systems

Let us understand the significance of signals and the significance of the study of signals. We will discuss different applications of signals in fields of communication, control systems, etc., which are the subjects for further study in engineering. We would like to understand the need for the course on signals and systems. You will come across the signals in the medical field: for example Electro Cardio Gram (ECG), Electro Encephalon Gram (EEG), Magnetic Resonance Imaging (MRI), X-rays, etc. To extract the information from these signals, they are required to be further processed. The doctors can draw certain conclusions only after getting results of certain processing on these signals. You are acquainted with multimedia communication, mobile communication, satellite communication, RADAR communication, Radio communication, TV, etc. These communication systems involve signals like speech, music, images, videos, etc. The signals are processed to extract some meaningful information. We will define a system.

Definition of a system The blocks that operate on signals as input and produce some meaningful output again in the form of signal are termed as signal-processing blocks or systems.

Let us introduce the communication systems in general form.

2.3.1 Basic communication system

The communication system will consist of a transmitter, a channel and a receiver, as shown in Fig. 2.8. Each of this unit can be called as a system. Let us understand the functioning of each block.

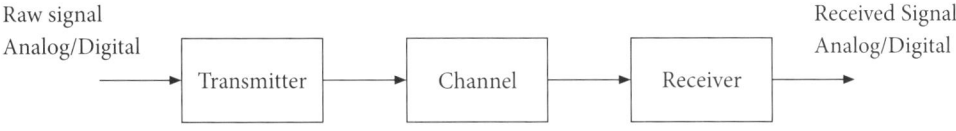

Fig. 2.8 Basic communication system

We will first consider basic analog communication system, for example, radio broadcasting. Here, the signal is the music or the speech recorded in the studio.

1. **Transmitter** These signals are processed in the transmitter. The analog signal is filtered to remove noise. The signal is then modulated using modulation like amplitude modulation (AM) or frequency modulation

(FM). Modulation will allow the signal to reach a longer distance using transmitting antenna. In case of amplitude modulation or frequency modulation, the message signal changes the amplitude or the frequency, respectively, of the carrier signal which is a high-frequency sinusoidal signal.

2. **Channel** The signal transmitted from the antenna is then passed over the channel. The channel can be free space or wire. When it is transmitted over free space in case of radio transmission, TV, satellite and mobile communication, etc it is often broadcasting. Wired communication is used for telephones. This is usually point-to-point communication.

3. **Receiver** The receiver will consist of the demodulator for AM or FM signal.

We will now consider the digital communication system.

1. **Transmitter** In case of digital communication, the transmitter will consist of a sampler, quantizer and coder. The raw signal like speech, video is always analog signal. The analog signal is first sampled using the sampling frequency decided by the sampling theorem. The reader may refer to chapter 1 on sampling for further details. The sampled signal is called as a DT signal. This is then quantized and converted to digital signal. It is further coded using lossless or lossy coding techniques. The reader will understand these coding techniques in the course on Digital communication. The coded signal is in the form of voltage levels like in case of Pulse code modulation used in telecommunication.

2. **Channel** The transmitted signal then passes over the channel. The band width of the channel is always limited. Hence, the channel functions as a low-pass filter. The signal gets distorted when it passes over the channel. The external noise also gets added in the signal. The channel equalizers are used to overcome the problem of the band limitation of the channel. The equalizers are used at the receiver end. The channel coding will include the techniques like error-correcting codes such as the inclusion of cyclic redundancy check characters (CRC).

3. **Receiver** The receiver will have a channel equalizer at the start to correct for channel errors. The receiver for analog communication will consist of demodulators and decoders. The receiver for digital system will include a use CRC to correct the received bits. Then it will decode the bits using PCM demodulator and use a low-pass filter to recover the analog signal. The quantization cannot be undone. Hence, the received signal will always include quantization noise. We can keep quantization noise to its minimum by adaptively selecting the step size. The reader will understand these concepts in detail in the course on digital communication.

Another broad class of systems is control systems, which find applications in the small, medium and large-scale industries. We will now discuss the basics of control systems.

2.3.2 Basic control system

The basic analog signal control system may be depicted, as shown in Fig. 2.9.

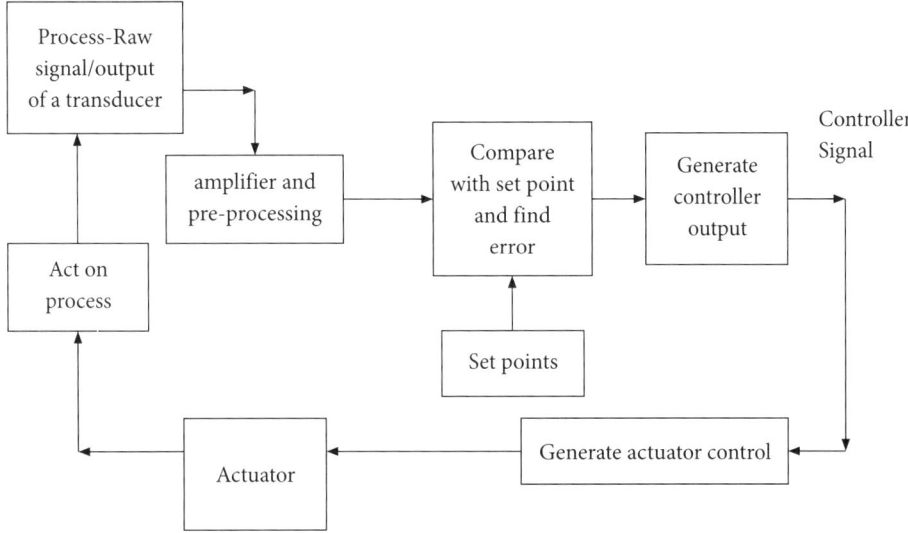

Fig. 2.9 Basic process control system

Functions of different blocks can be described as follows. Consider a process control system that is used for controlling the parameters like temperature, pressure, humidity, etc, for example, in some plant or some green house or the agricultural area. Here, the raw signal is the output of the transducers used for the measurement of temperature, pressure and humidity. There will be some set points decided for each parameter and if the value of the parameter goes above or below the threshold or the set point, then the control action is taken.

The control system will find the error between the actual parameter value and the set point and will calculate the controller output using, for example, the on-off type of controller or the proportional controller that generates the output proportional to the error. The control action may consist of switching of the heater, for example, for temperature and so on. It may be a closed loop system that gives the output of the controller to some actuator that acts on the actual controlling part, for example, the heater in this case. In case of digital control system, only the controller either on-off or proportional controller will be implemented digitally. The error signal is digitized and given to the controller. The output of the controller is converted back to analog signal and given to actuator control. When actuator acts on the process the control loop is closed and it is termed as a closed loop control system.

Concept Check
- What are the main blocks of any communication system?
- What is the function of a transmitter?

- How does the channel affect the signal?
- How does the receiver take care of errors?
- What is a control system?
- How is the signal preprocessed in a process control system?
- How will you implement a controller in case of CT and DT systems?
- What is a closed loop control system?

2.4 Classification of Signals

Basically, the signals are of three types. Signals can be classified as analog, Discrete Time (DT), or Digital signals.

2.4.1 Analog signals

As we have discussed, all naturally occurring signals are analog in nature. Analog signals are continuous amplitude and continuous time (CTCA) signals. Analog signals arise when any physical signal is converted to the electrical signal using a transducer such as temperature sensor, pressure cells, photovoltaic cells, etc. An analog signal can take any real value from zero to infinity. Let us consider some examples of analog signals.

Example 2.10

(Analog signal) Consider a signal given by $x(t) = \sin(2\pi Ft)$ where F is the frequency of the signal equal to 5 Hz. Plot a signal.

Solution

The signal exists for all t. This is an analog signal. The signal is plotted in Fig. 2.10, which is the result of the output of a MATLAB program. The signal exists for all continuous values of time and takes on any continuous value of amplitude. The plot is shown only for some finite duration.

Let us do it practically. We will use MATLAB to generate and plot this analog signal. A MATLAB program is given below. We will generate a vector of time values with sampling interval of 0.001 seconds. Using a plot command, it joins all successive time values to get the appearance of a continuous signal.

```
clear all;
f=5;
w=2*pi*f;
t=-1:0.001:1;
s=sin(w*t);
plot(t,s);
title('plot of sine wave-approximation to analog sine wave is plotted');
xlabel('time'); ylabel('amplitude');
```

The output of MATLAB program is plotted in Fig. 2.10.

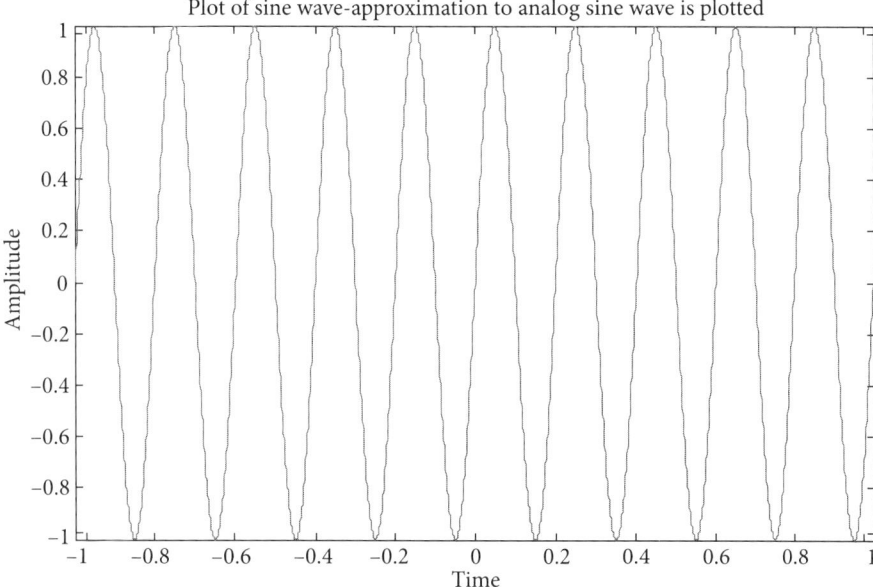

Fig. 2.10 Plot of a sine wave using MATLAB

Example 2.11

Plot an analog signal given by

$$x(t) = 0.01 \times t \text{ for } 0 \leq t \leq 5 \text{ seconds}$$
$$= 0 \text{ otherwise}$$

Solution

The signal is defined only for values of *t* between 0 to 5 seconds. The signal plot can be represented, as shown in Fig. 2.11.

Let us write the program to generate this signal. We will again generate a vector of time values with sampling interval of 0.01 seconds. Using a plot command, it joins all successive *x* values to get the appearance of a continuous signal. The output of the program is plotted in Fig. 2.11.

A MATLAB program can be written as

```
clear all;
t=0:0.01:5;
x=0.01*t;
plot(t,x);
title('plot of signal x');
xlabel('time'); ylabel('amplitude');
```

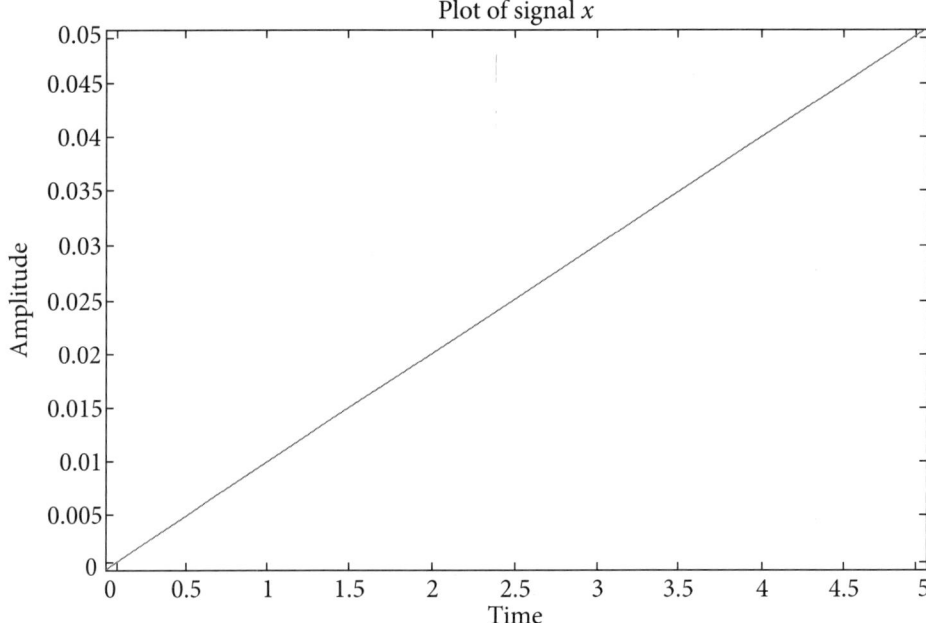

Fig. 2.11 Plot of signal x with slope of 0.01

Example 2.12

Plot analog signal given by

$$x(t) = 1 \text{ for } 0 \leq t \leq 1 \text{ seconds}$$
$$= -1 \text{ for } 1 < t \leq 2 \text{ seconds}$$

Solution

The signal is defined only for values of t between 0 to 2 seconds. The signal plot can be represented, as shown in Fig. 2.12.

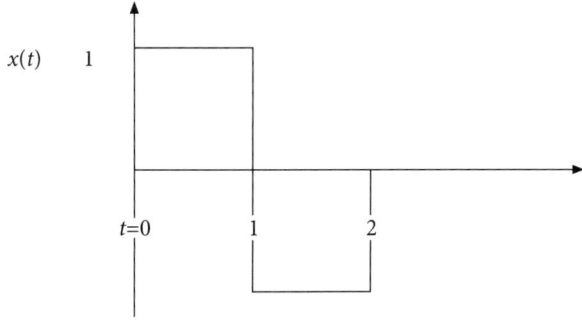

Fig. 2.12 Plot of signal $x(t)$ defined between $t = 0$ to $t = 2$ seconds

Let us write the program to generate this signal. We will again generate a vector of time values with a sampling interval of 0.01 seconds. Using a plot command, it joins all successive *x* values to get the appearance of a continuous signal. The output of the program is plotted in Fig. 2.13. Here, the signal values are discontinuous. This makes the program difficult to write in discrete time domain. To generate an index as an integer in MATLAB, time values are multiplied by 10.

A MATLAB program can be written as

```
clear all;
t=0:0.01:1;
i=1+t*100;
for i=1:10,
 x(i)= 1;
end
t=1:0.01:2;
i=11+(t-1)*100;
for i=11:20,
 x(i)=-1;
end
plot(x);
title('plot of signal x');
xlabel('time'); ylabel('amplitude');
```

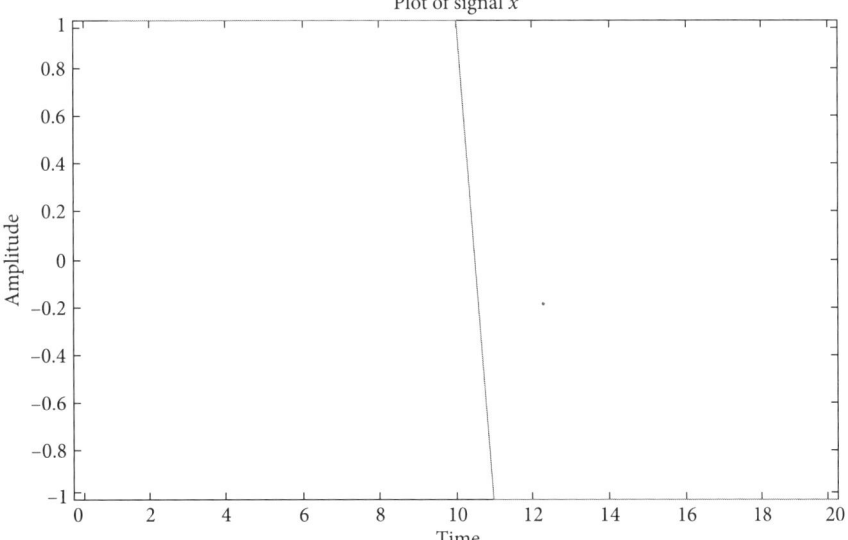

Fig. 2.13 Plot of signal *x*

Consider a simple sinusoidal signal in Example 2.8. The signal has a frequency of 5 Hz. Two analog sinusoids of different frequencies are always distinct. The frequency of a sinusoid is purely determined by the number of oscillations per second.

Concept of negative frequency

The frequency as determined by number of oscillations per second is definitely a positive quantity. We need to introduce negative frequencies for mathematical convenience. The sinusoid of angular frequency Ω with peak amplitude A may be represented as

$$x(t) = A\cos(\Omega t) = \frac{A}{2} e^{j\Omega t} + \frac{A}{2} e^{-j\Omega t} \qquad (2.1)$$

The co-sinusoid is formed using two constant amplitude complex conjugate exponential signals or unit phasors rotating in opposite directions, as is indicated by Eq. (2.1) shown in Figs 1.14 and 1.15, respectively. When two phasors are added, the real parts will add up and the imaginary parts will cancel each other to get only the cosine term in Eq. (2.1). The amplitude of phasors is unity as

$$\begin{aligned} |e^{j\Omega t}| &= |\cos(\omega t) + j\sin(\omega t)| \\ &= \sqrt{\cos^2(\omega t) + \sin^2(\omega t)} = 1 \end{aligned} \qquad (2.2)$$

A phasor rotating in counter-clockwise direction is said to have a positive frequency and the one rotating in clockwise direction is said to have a negative frequency. The mathematical calculations can be simplified if we use exponential signal representation as against the sinusoidal representation. With the introduction of negative frequency, we will now consider the range of analog frequencies between $-\infty$ and $+\infty$. The analog sinusoidal signal can have any continuous value of frequency from zero to infinity. We will denote the analog frequency and the analog angular frequency by capital letters such as F and Ω, respectively.

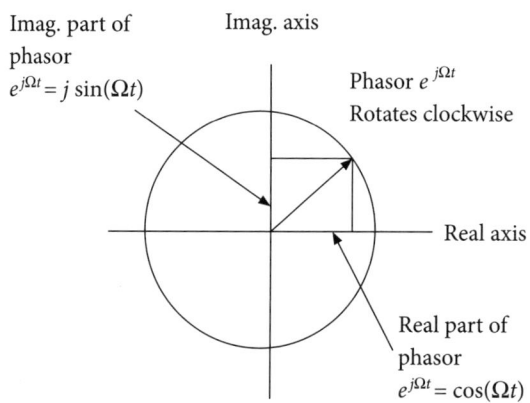

Fig. 2.14 Phasor $e^{j\Omega t}$ rotating clockwise

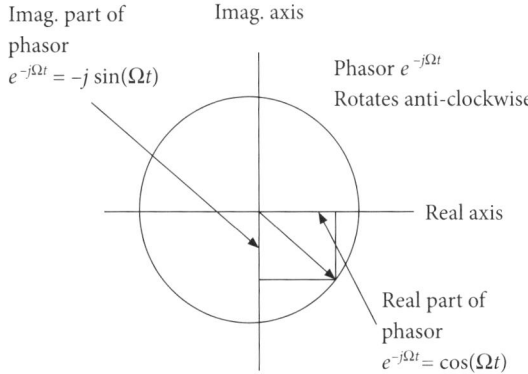

Fig. 2.15 Phasor $e^{-j\Omega t}$ rotating anti-clockwise

The concept of negative frequency will be again useful in understanding Fourier series and Fourier Transform.

2.4.2 Discrete time signals (DT signals)

Let us introduce the concept of DT signals. The natural signal is interfaced to a digital computer via the analog-to-digital (A/D) converter. The A/D converter is a sampler followed by a quantizer. A set of samples collected from a sequence of signal values at discrete time values is called a discrete time signal or a DT signal. The sample values are still continuous. A DT signal is a discrete time, continuous amplitude (DTCA) signal. When the time axis is digitized, the signals are called discrete time (DT) signals. The X-axis in the graph-form representation is named as sample number. Generally, the sampling interval is constant, that is it is uniform sampling. When the X-axis is the sample number, the signal representation is normally in time domain. Let the signal $X(t)$ be sampled. The samples are collected at $t = 0, t = T, t = 2T$ and so on. The samples are represented as $X[0], X[1T], X[2T], ...$ and so on. As T is constant, it is dropped in the representation and the signal samples are written as $X[0], X[1], X[2], ...$. The nth sample will be represented as $X(n)$. The signal then consists of signal samples, namely, $\{X[0], X[1], X[2], ...\}$. This set of signal samples is also called as a sequence $x[n]$.

Let us go through some examples of DT signals.

Example 2.13

Let us represent the sampling frequency as f_s then $T = 1/f_s$, where T = sampling interval. The signal can be written as $x[n] = \sin(2\pi f n T)$. Plot the signal. The digital domain frequency is represented by a small letter f and angular frequency by ω.

Solution

The sampled signal represented as $x[n]$ exists for discrete time values i.e., at $t = 0, 1*T, 2*t$, etc. The nth sample is represented as $x[n]$. Hence, it is termed as a discrete time signal or DT signal. It does not mean that the signal has a zero value at all other values of time. The signal is simply not defined for values of time other than $t = 0, 1 * T, 2 * T$, etc. Fig. 2.16 shows a plot of a signal.

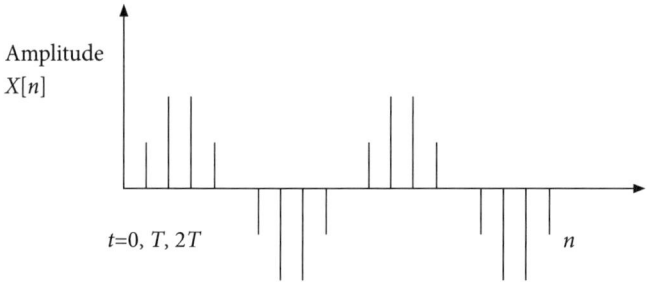

Fig. 2.16 A Discrete time signal $x(n)$ plotted for different values of n

Let us write the program to generate this signal. A MATLAB program to generate a signal is given below. The plot of the signal is shown in Fig. 2.17.

```
clear all;
f=10;
T=0.01;
for n=1:21,
x(n)=sin(2*pi*f*(n-1)*T);
end
stem(x);title('plot of DT signal x');
xlabel('sample number');ylabel('amplitude');
```

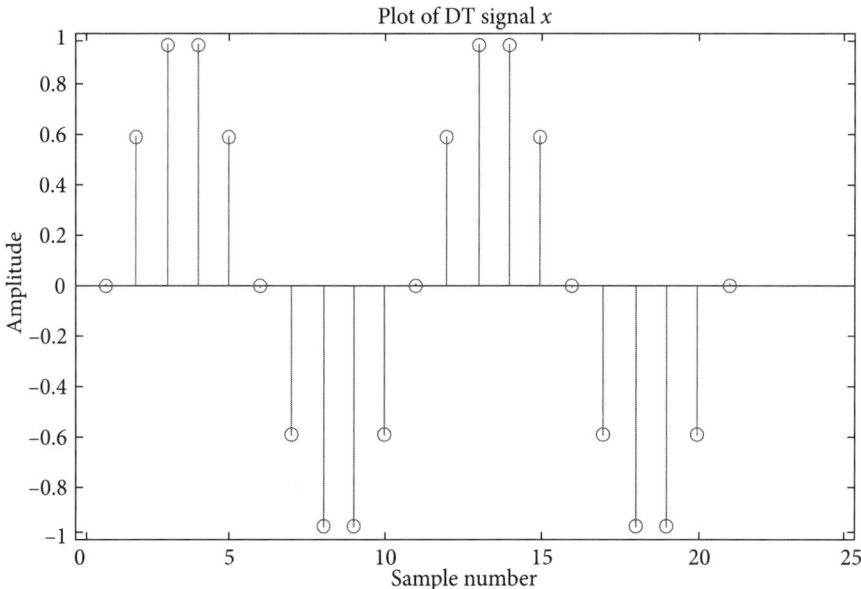

Fig. 2.17 Plot of signal x

Example 2.14

(Discrete time signal) Plot a sampled signal

$x(n) = 1$ for $0 \leq n \leq 6$

$\quad\quad\,\, = 0$ otherwise

Solution

The signal is defined as 1 for values of n from 0 to 6 only. It is equal to for all other values of n. The signal plot is shown in Fig. 2.18.

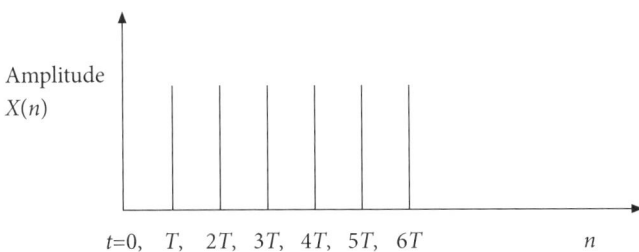

Fig. 2.18 Plot of signal $x(n)$ in Example 2.14

Let us write the program to generate this signal. A MATLAB program to generate a signal is given below. The plot of the signal is shown in Fig. 2.19. MATLAB does not use the index as zero. So, we have to do a trick. Generate the s variable from 0 to 6 and plot the values of x against s.

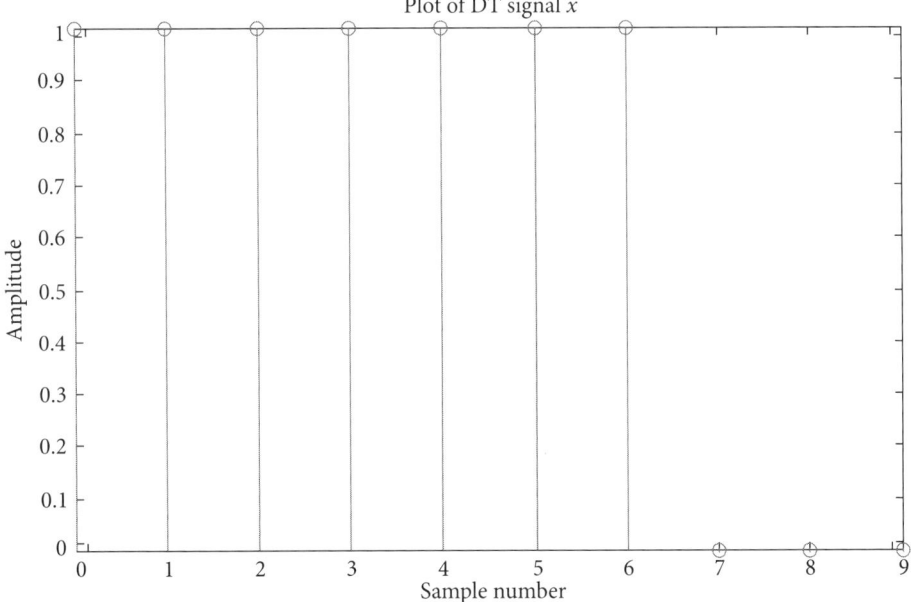

Fig. 2.19 Plot of signal x

```
clear all;
for n=1:7,
 x(n)=1;
end
for n=8:10,
 x(n)=0;
end
s=0:1:9;
stem(s,x);title('plot of DT signal x');
xlabel('sample number');ylabel('amplitude');
```

Example 2.15

(Discrete time signal) Plot a sampled signal

$x(n) = 1$ for $0 \leq n \leq 3$

$ = 1$ for $-2 \leq n \leq -1$

$ = 0$ otherwise

Solution

The signal is defined as 1 for values of *n* from −2 to 3 only. It is equal to for all other values of *n*. The signal plot is shown in Fig. 2.20.

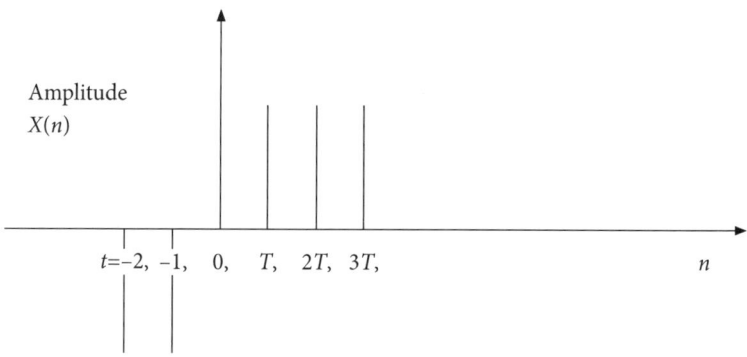

Fig. 2.20 Plot of signal *x*(*n*) for Example 2.15

Let us write the program to generate this signal. A MATLAB program to generate a signal is given below. The plot of the signal is shown in Fig. 2.21. MATLAB does not use the index as zero. So, we have to do a trick. Generate the s variable between −2 and 3 and plot the values of *x* against *s*.

```
clear all;
for n=1:2,
 x(n)=0;
```

```
end
for n=3:4,
  x(n)=-1;
end
for n=5:8,
  x(n)=1;
end
for n=9:10,
  x(n)=0;
end
s=-4:1:5;
stem(s,x);title('plot of DT signal x');
xlabel('sample number');ylabel('amplitude')
```

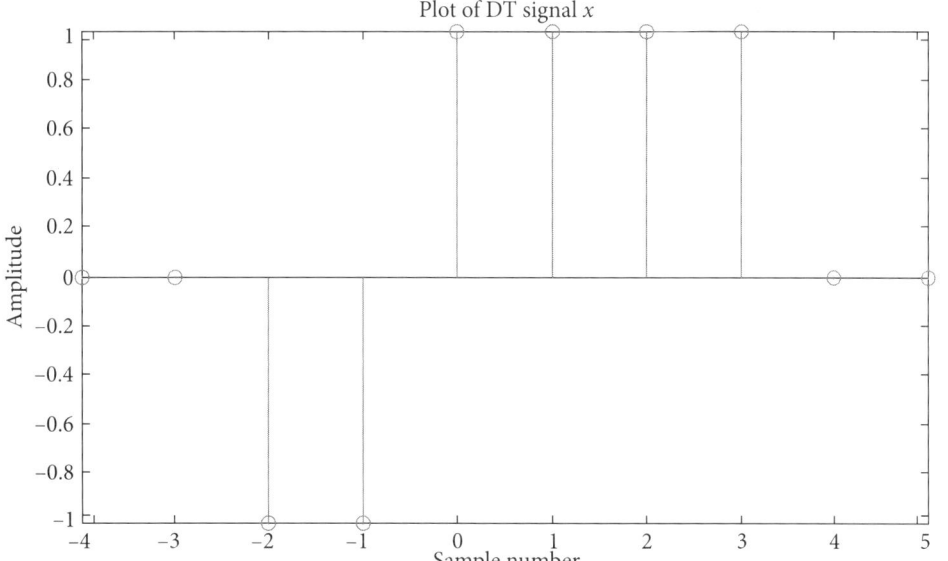

Fig. 2.21 Plot of signal *x(n)* for Example 2.15 using MATLAB

2.4.3 Digital signals

When the amplitudes of the DT signal samples are further quantized, the signal so formed is a digital or discrete time, discrete amplitude signal.

Example 2.16

Digital signal: In case of a DT signal the time axis is digitized; however, the amplitude axis is continuous. For a digital signal, the amplitude is quantized and is represented using, for example, a 4-bit with most significant bit as a sign bit. Let the signal be represented as $X = \{0, 0.25, 0.5, 0.25, 0.75\}$

Solution

The digital signal X will be represented as

$$X = \{0.000, 0.010, 0.100, 0.010, 0.110\} \tag{2.3}$$

Consider 4-bit signed coding. The range of values between −1 and +1 is divided in 16 levels. The quantization step size is 0.125. The centre level 0.000 is representing a zero. Each level is indicating a value of 0.125. So, the second level i.e., if third bit is 1, i.e., 0.010 will represent a value 0.25. Level 3 i.e., if the second bit is 1, i.e., 0.100 will represent a value of 0.75, as shown in Table 2.1 for quantization levels.

Table 2.1 Correspondence between level value and digital output for 4-bit coding

Level value	Digital output
0.875	0.111
0.75	0.110
0.625	0.101
0.5	0.100
0.375	0.011
0.25	0.010
0.125	0.001
0.000	0.000
−0.125	1.111
−0.25	1.110
−0.375	1.101
−0.5	1.101
−0.625	1.011
−0.75	1.010
−0.875	1.001
−1.0	1.000

The reader can easily evaluate the output for any analog input.

Concept Check

- What is a random signal?
- Define analog signal.
- What is a negative frequency?
- State the range of analog frequencies.
- How will you obtain a DT signal from analog signal?
- How will you convert DT signal to digital signal?

- When you use a digital computer, how will you generate the appearance of continuous time using MATLAB?
- For a DT signal, what is the value of the signal at non-sampling points of time?

2.5 Elementary Signals used for Testing of Systems

The communication system or any control system is designed using pen and paper. This theoretically designed system is tested using simulated standard signals. The response of the system to standard inputs is tested. The standard signals are selected so that the system is tested for all worst-case conditions and adverse situations. The sudden spike change in the input is tested using impulse. Sudden rise in input is tested using unit step and rectangular pulse. The smooth change in input is tested using a ramp signal and so on. If the system performance is satisfactory, then the system is actually implemented. This saves the cost, as the system response for standard input prompts the designer to change the parameters of the system for giving required output. The system is also tested for stability. Let us study standard analog and DT signals used for testing of a system. Many times, the complex signals are represented in terms of the standard signals to simplify the representation and analysis. Hence, standard signals are useful for the analysis of complex input signals. Let us first understand the reasons for using particular inputs as standard input.

2.5.1 Reasons for using test or standard inputs

In time-domain analysis, the dynamic response of the system is tested in response to various inputs. The system response is analyzed at different intervals of time after the input signal is applied. In practice, the input signals appearing at the input of a measurement system are not known a priori. Many times, the actual input signals vary randomly with respect to time. These cannot be mathematically defined due to random nature. However, for the purpose of analysis and design, designers assume some basic types of input signals that can be easily defined mathematically, so that the performance of a system can be analyzed with these standard signals. The actual input signal can be represented as a combination of these standard inputs. Hence, once the system is tested for the standard inputs, the system is ready for implementation.

In the time-domain analysis, the following standard test signals are used. We will discuss the reason for using some of the standard inputs.

1. Consider the step input signal. It represents the sudden application of the input at specified time. In some mechanical system, there may be a shaft and the step input will represent the sudden rotation of the shaft. If the electrical supply is suddenly switched on, or a valve is suddenly opened or closed, it will resemble the step input.

2. If the input signal is changing at a constant rate, it resembles ramp input.

3. Sudden shocks for example sudden high voltage due to lightning or short circuit will represent the impulse as input.
4. For testing the system for non-linear inputs, a parabolic signal is used.

Steady state errors will decide the accuracy of the system. Measurement of steady state error is an important performance measure of the system. These errors arise due to the nature of inputs and due to non-linearities of the system components. Steady state errors can be estimated by judging the steady state errors in response to step, ramp and parabolic inputs.

2.5.2 Standard analog signals

The standard analog signals used for the testing of analog systems are listed below.

1. **CT Impulse signal**

   The unit impulse signal is denoted by $\delta(t)$ and is also called as Dirac delta function or simply delta function. It represents a very sudden change in the input signal for testing of a system. It is defined as follows.

   $$\delta(t) = 0 \quad \text{for} \quad t \neq 0$$

   such that $\int_{t=-\infty}^{t=\infty} \delta(t)dt = 1$ (2.4)

   It is very difficult to visualize the delta function. Imagine the rectangular pulse with a very small width of ε. The height of the rectangle is now $1/\varepsilon$ as the area enclosed by the rectangle is 1. Now we allow the width of the rectangle to shrink and increase the height such that the area enclosed remains 1. When ε tends to zero, $1/\varepsilon$ will tend to infinity and still area enclosed will be 1. This visualization is depicted in Fig. 2.22. We can visualize it as a very *tall-and-thin* spike having unit area at the origin, as shown in Fig. 2.22. The impulse function may also be visualized as an exponential pulse, Gaussian pulse or even the triangular pulse. A delta function is defined by small width tending to zero and keeping the area at unity. The shape of the pulse is not important.

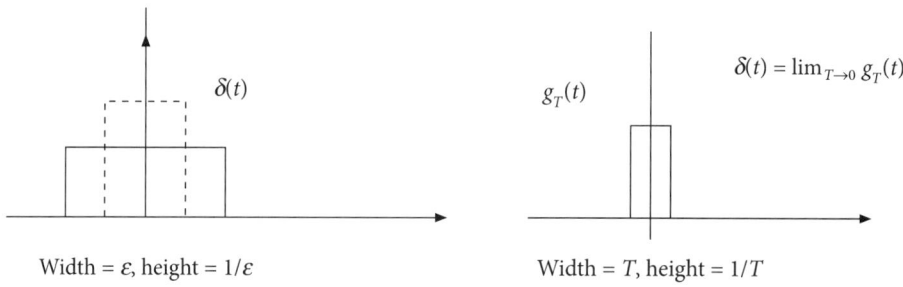

Fig. 2.22 Plot of delta function (with visualization as a rectangular pulse of almost zero width and infinite height)

Usually, people try to find the impulse response of the system and it is used for characterization of the system. The input signal is decomposed in terms of scaled and translated unit impulses. The response of the system to any input is then found by convolving the input with the impulse response of the system. This is possible due to superposition principle. The reader will study the details of the system response in further chapters 3 and 4.

Let us study properties of CT delta function.

1. It is an even function of time i.e.,

$$\delta(t) = \delta(-t) \qquad (2.5)$$

   This means that the delta function is symmetrical with respect to the y axis i.e., amplitude axis.

2. The second property is related to the evaluation of the signal at different time values using a delta function. The delta function exists at $t = 0$. If we want to evaluate the value of the signal at $t = 0$, we will use the first equation out of 2.6 stated below. To find the value of the signal at any time $t = t_0$, we will use the shifted delta function at $t = t_0$ second equation in 2.6. The equation says that the delta function **samples or shifts out** the value of the function at $t = t_0$. This is called as the **sampling or shifting property of delta function**. This defines the delta function in the generalized format. We do not say anything about what the impulse function looks like, instead we specify its effect on the test function. The delta function has the mathematical meaning only when it appears as the integrand inside the integral.

$$\int_{t=-\infty}^{t=\infty} x(t)\delta(t)dt = x(0)$$

$$\int_{-\infty}^{\infty} x(t)\delta(t-t_0)dt = x(t_0) \qquad (2.6)$$

   as $\quad \int_{-\infty}^{\infty} \delta(t-t_0)dt = 1 \qquad$ for $\quad t = t_0$

3. The third property is related to the time scaling of the delta function. If we represent the delta function as the rectangular pulse with its width tending to zero in the limiting case, as shown in Fig. 2.15, $\delta(at)$ represents a rectangular pulse scaled in time by a factor of 'a' with amplitude unchanged. To keep the integral value of unity, the amplitude must be scaled by the same factor 'a'. Refer to Fig. 2.23.

$$\delta(at) = \lim_{T \to 0} g_T(at)$$

$$g'_T(at) = ag_T(at)$$

$$\delta(at) = \frac{1}{|a|}\lim_{T\to 0} g'_T(at) = \frac{1}{|a|}\delta(t) \tag{2.7}$$

This is represented by the equation as $\delta(at) = \dfrac{1}{|a|}\delta(t)$ (2.8)

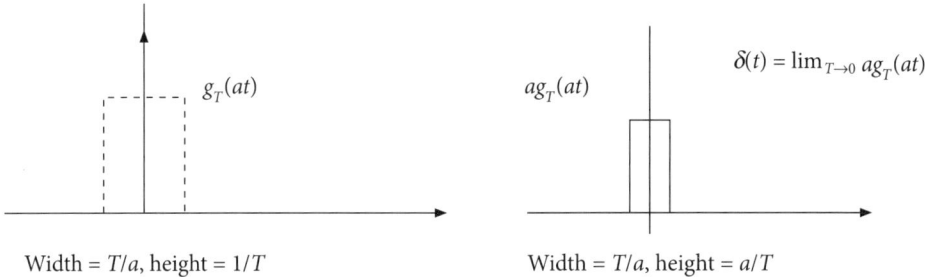

Width = T/a, height = $1/T$ Width = T/a, height = a/T

Fig. 2.23 Scaling property of delta function

4. The integral of the impulse function is the unit step function and the derivative of unit step function is a delta function. $du/dt = \delta(t)$. Strictly speaking, the unit step function has a discontinuity at zero, and its derivative will not exist. Let us define the unit step function as the integral of the delta function.

$$\int_{-\infty}^{t} \delta(t)dt = \begin{cases} 0 \text{ for } t < 0 \\ 1 \text{ for } t \geq 0 \end{cases} = u(t) \tag{2.9}$$

Physical significance of impulse We cannot generate the physical impulse. However, it is useful for providing an approximation to a physical signal in terms of short duration pulses. The reader will understand this fact further when convolution is discussed in further chapters. This fact is utilized in finding the characterization of the system in terms of its impulse response. As a test signal it is useful for finding the response of the system for sudden changes like voltage spikes.

2. **Unit step function**

   The continuous time (CT) or analog unit step is defined as follows.

$$u(t) = \begin{cases} 1 & \text{for } t \geq 0 \\ 0 & \text{otherwise} \end{cases} \tag{2.10}$$

A time-shifted unit step function, shifted towards right by a time units, can be written as

$$u(t-a) = \begin{cases} 1 & \text{for } t \geq a \\ 0 & \text{otherwise} \end{cases} \qquad (2.11)$$

The CT unit step function is continuous except for a discontinuity at $t = 0$. Fig. 2.24 shows the unit step function.

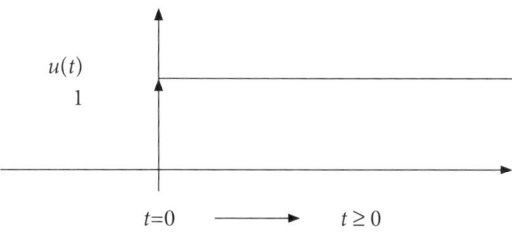

Fig. 2.24 Unit step function

Physical significance of unit step function   The unit step is particularly easy to realize as it represents the closing of a switch for applying a battery voltage. As a test signal, it has significance related to the response time of the system to a quick input. We can analyze as to how quickly the system responds to sudden changes. This is measured in terms of the rise time for the output waveform.

3. **Rectangular Pulse**

   The rectangular pulse is denoted as a rect function and is defined as

   $$\text{rect}\left(\frac{t}{\tau}\right) = \begin{cases} 1 & \text{for } |t| \leq \tau/2 \\ 0 & \text{for } |t| > \tau/2 \end{cases} \qquad (2.12)$$

   The plot of the function is shown in Fig. 2.25. This occurs naturally in logic devices and is encountered in digital switching circuits. Fourier Transform of the rectangular function is a sinc function. The rectangular pulse contains a wide range of harmonics. Owing to infinite sharp edges in time domain, the frequency domain representation has infinite bandwidth.

   Note   The reader will understand the concepts well in chapter 6 on Fourier transform (FT). The FT of the rectangular pulse is shown to be a sinc function.

   Physical significance of the rectangular pulse   The rectangular pulse is used for square wave testing of an amplifier. The sudden rising edge represents all frequencies from zero to infinity. The time response of the system is tested using rise time measurements. The flat edge is representative

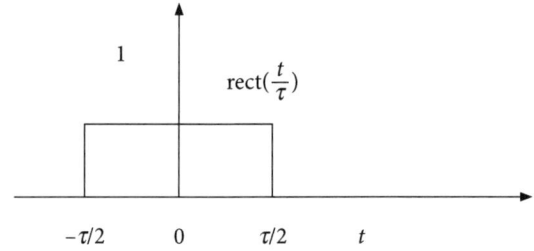

Fig. 2.25 Plot of rectangular pulse

of small frequencies. It is used to find the low-frequency response of the amplifier. It is measured as the tilt in the flat top of the square wave at the output of the amplifier.

4. **Ramp function**

The CT ramp signal is denoted as $r(t)$ and is defined as follows.

$$r(t) = tu(t) = \begin{cases} t & \text{for } t \geq 0 \\ 0 & \text{for } t < 0 \end{cases} \quad (2.13)$$

A plot of the signal is shown in Fig. 2.26. It represents a lowly varying input signal. Unit ramp function is the integration of the unit step function.

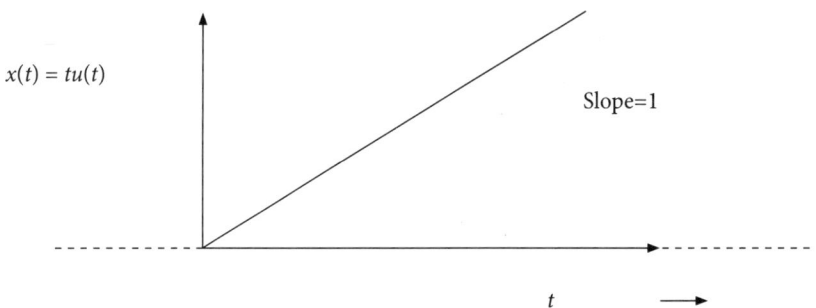

Fig. 2.26 Plot of analog ramp function

Physical significance of the ramp function The ramp function represents a slow rise. As a test signal it is useful to find the response of the system to slow changes in the input. The response will be detected as the slope of the ramp at the output of the system.

5. **Signum function**

The signum function denoted by sgn(t) is defined as follows.

$$\text{sgn}(t) = \begin{cases} 1 & \text{for } t > 0 \\ 0 & \text{for } t = 0 \\ -1 & \text{for } t < 0 \end{cases} \quad (2.14)$$

The plot of the signal is shown in Fig. 2.27. It extracts the sign of the real number or the sign of the function, hence, also called as sign function.

$$\text{sgn}(f(t)) = \begin{cases} 1 & \text{for } f(t) > 0 \\ 0 & \text{for } f(t) = 0 \\ -1 & \text{for } f(t)t < 0 \end{cases} \quad (2.15)$$

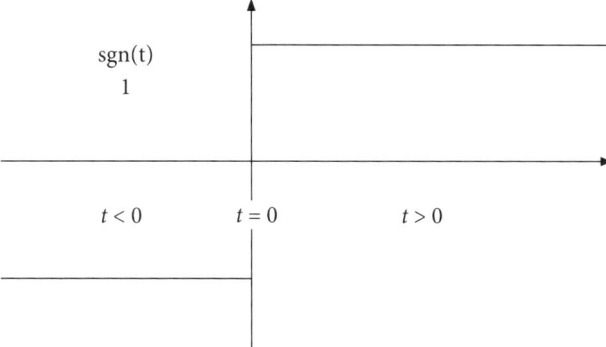

Fig. 2.27 Plot of signum function

Physical significance of the signum function It extracts the sign of the signal or a function.

6. **Sinc Function**

   The CT-normalized sinc function is defined as

   $$\sin c(\omega_0 t) = \frac{\sin(\pi \omega_0 t)}{\pi \omega_0 t} \quad (2.16)$$

   The three reasons for using normalized sinc function can be listed as follows.

   1. Firstly, the definite integral of the function is one.
   2. Secondly, all zeros of the normalized sinc function is integer value of the argument and thirdly, the normalized sinc function is the Fourier Transform of the rectangular function with no scaling.
   3. The third reason is the main concept behind using a sinc function in reconstructing the continuous bandlimited signal from uniformly spaced samples of that signal.

   Definition of normalized sinc function The sinc function is said to be normalized when it includes π in the sine argument and also in the denominator. The plot of a normalized sinc function will have a maximum value of 1 i.e., the amplitude is normalized to 1.

Definition of un-normalized sinc function The un-normalized sinc function is defined as

$$\sin c(\omega_0 t) = \frac{\sin(\omega_0 t)}{\omega_0 t} \qquad (2.17)$$

This function does not contain π in the sine argument and also in the denominator.

The plot of normalized sinc signal is shown in Fig. 2.28. We can see that the maxmum amplitude is normalized to one.

The plot is generated using the MATLAB program given below.

```
Clear all;
t=-10:.01:10;
y=sinc(t);
axis([-10 10 -2 2]);
plot(t,y)
grid;
title('plot of a sinc function'); xlabel('time');yla
bel('amplitude');
```

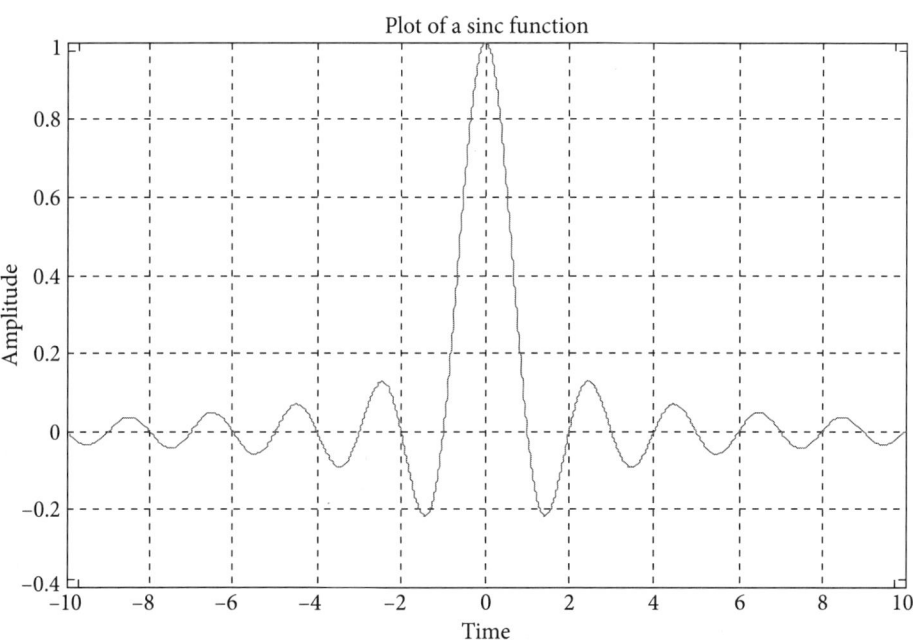

Fig. 2.28 Plot of a sinc function

Physical significance of the sinc function This function is used to reconstruct the analog signal from the signal samples. Sinc function is used as an interpolation function. Fourier Transform (FT) of a rectangular pulse is sinc function. The reader may refer to chapter 5 on FT for details of the sinc function properties.

7. **Sinusoidal function**

    The continuous time sinusoid of frequency f or angular frequency ω is defined as follows.

    $$x(t) = \sin(2\pi ft + \theta) = \sin(\omega t + \theta) \tag{2.18}$$

    The plot of a sinusoidal function is shown in Fig. 2.29. Let frequency be 5 Hz and the angle θ be 45 degrees. A MATLAB program to plot the sinusoid is given below.

```
clear all;
f=5;theta=45;
w=2*pi*f;
t=0:0.001:1;
s=sin(w*t+((theta*pi)/180));
plot(t,s);
title('plot of sine wave with phase shift theta of 45 degrees');
xlabel('time'); ylabel('amplitude');
```

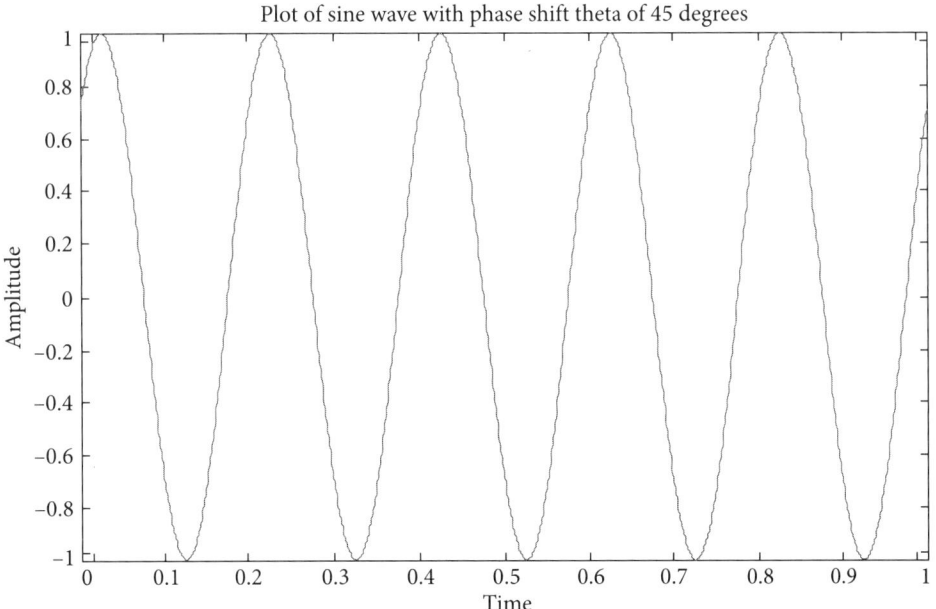

Fig. 2.29 Plot of a sinusoidal function

Physical significance of a sine wave function Any signal can be decomposed in terms of linear combination of scaled and shifted sine waves (Fourier series or Fourier transform representation). This enables one to find the response of the system to any input. This is called as the frequency response of the system. The sine wave function is useful as a test signal. We use sine wave testing for finding the frequency response of the amplifiers. If the system produces a sine wave in response to a sine wave input, it indicates that the system has no phase distortion.

8. **Exponential function**

A CT exponential function is written as

$$x(t) = \exp(st) = e^{st} = e^{(\sigma + j\omega)t} \tag{2.19}$$

if $\sigma = 0$, $x(t) = e^{j\omega t} = \cos(\omega t) + j\sin(\omega t)$

Here, s represents the complex frequency. The real part and imaginary part are represented as σ and $j\omega$, respectively. If the real part is zero, it reduces to the complex sinusoid. Refer to Figs 2.14 and 2.15 for imagination of complex sinusoids. If the imaginary part is zero, it is represented in Fig. 2.30 with real part greater than zero, less than zero and equal to zero. A MATLAB program to plot the signal is given below.

```
clear all;
t=0.05;
for n=1:101,
y(n)=exp(2*t*n);
end
subplot(3,1,1);plot(y);title('plot of exponential function with real part greater than zero');xlabel('time');ylabel('Amplitude');
 t=0.05;
for n=1:101,
y(n)=exp(-2*t*n);
end
subplot(3,1,2);plot(y);title('plot of exponential function with real part less than zero');xlabel('time');ylabel('Amplitude');
 t=0.05;
for n=1:101,
y(n)=exp(0*t*n);
end
subplot(3,1,3);plot(y);title('plot of exponential function with real part equal to zero');xlabel('time');ylabel('Amplitude');
```

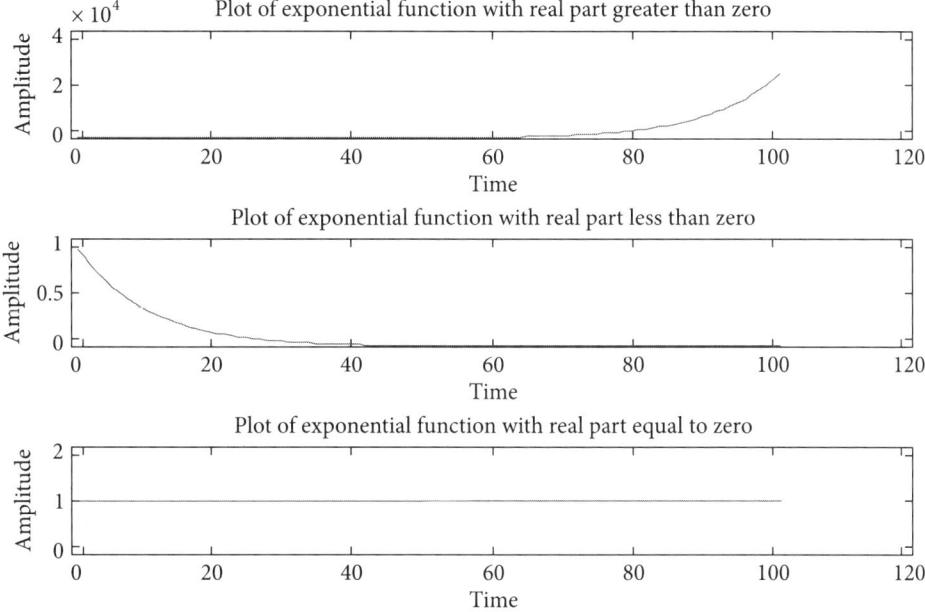

Fig. 2.30 Plot of exponential signal with imaginary part = 0 and real part positive, negative and zero respectively

Physical significance of exponential function The exponential function can be used to model many physical signals occurring in nature, for example, charging and discharging of a capacitor. Exponential distribution is one of the standard distributions defined for continuous random variables. Complex exponential function can be used as a building block for many complex signals.

9. **Parabolic Signal**

   The unit parabolic signal is defined as

   $$x(t) = \frac{t^2}{2} \quad \text{for } t \geq 0 \tag{2.20}$$

   and = 0 otherwise

   As the signal exists only for $t >= 0$, it can be represented as

   $$x(t) = \frac{t^2}{2} u(t) \text{ where } u(t) \text{ is a unit step function.} \tag{2.21}$$

   The parabolic function is used for testing the nonlinearity of a system. The function can be plotted using a MATLAB program. The plot of the parabolic function is shown in Fig. 2.31.

```
clear all;
for m=1:200,
 x(m)=m^2/2;
end
plot(x);title('plot of parabolic function');xlabel('
time');ylabel('amplitude');
```

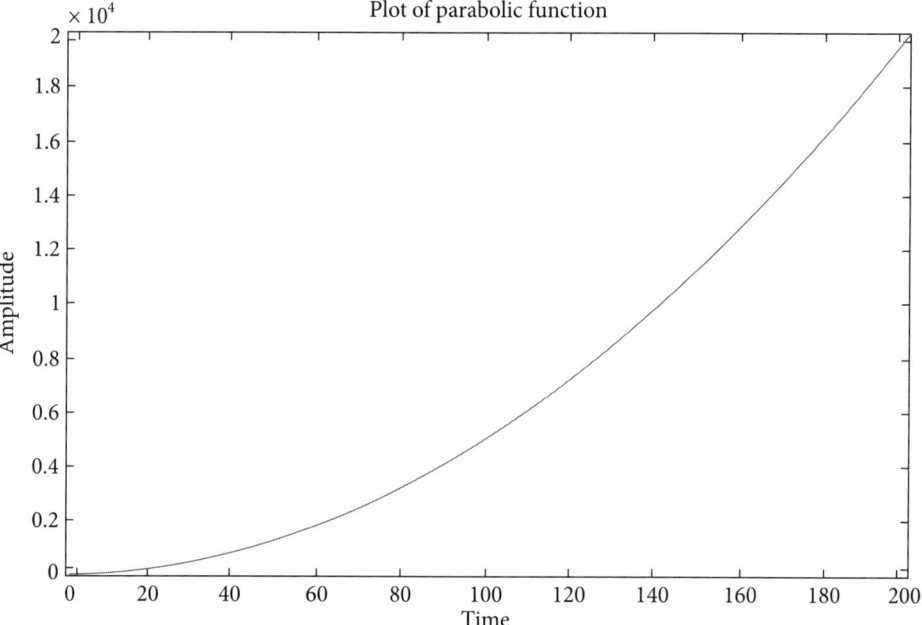

Fig. 2.31 Plot of parabolic function

Physical significance of parabolic function The parabolic function is basically a non-linear function. It is used for testing the system for non-linearities.

10. **Gaussian Function**

The Gaussian signal is defined as

$$g(t) = \frac{1}{\sqrt{2\pi}\sigma} e^{-(t-a)^2/2\sigma^2} \quad \text{for} -\infty \leq t \leq \infty \tag{2.22}$$

σ – standard deviation, a – mean

The function can be plotted using a MATLAB program. The plot of the parabolic function is shown in Fig. 2.32.

```
clear all;
sigma=2;
mean=0;
x=(-10:0.1:10);
amp=1/(sigma*sqrt(2*pi));
a=(x-mean).*(x-mean);
y=amp*exp(-a/2*4);
plot(x,y);xlabel('x');ylabel('pdf');title('Gaussi
an pdf');
```

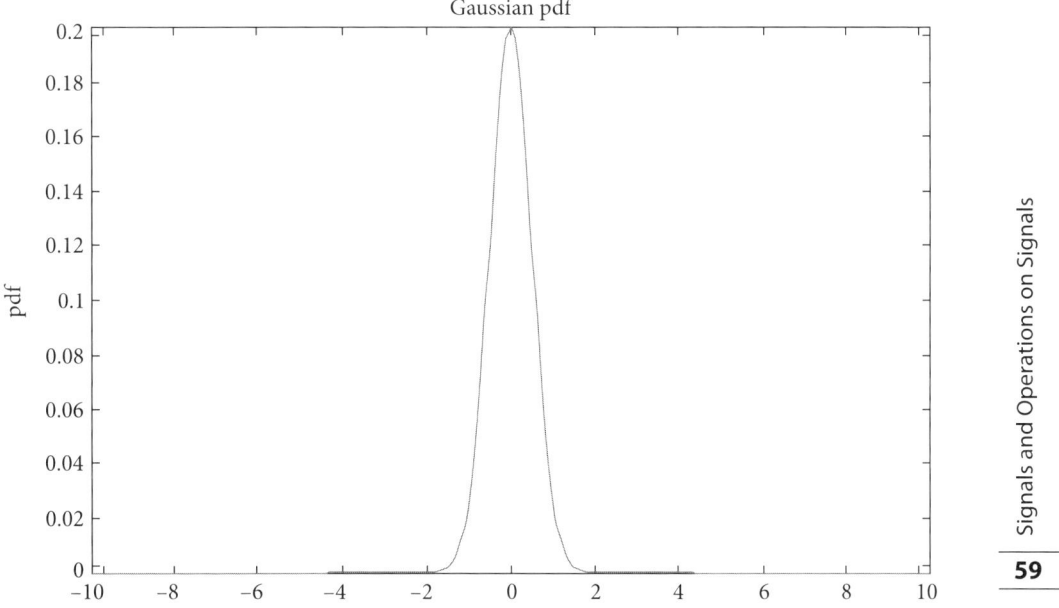

Fig. 2.32 Plot of Gaussian function

Physical significance of Gaussian functions The function is useful in probability theory. It represents the standard probability distribution used for analysis of continuous random signals. The reader may refer to chapter 9.

2.5.3 Standard DT signals

There are some elementary DT signals normally used for analyzing the systems. The system behavior can be tested against these standard inputs. DT signal is also called as a sequence. **The physical significance of digital test signals is same as that of the analog counterparts.**

1. **Unit Impulse/DT delta function**

$$\delta[n] = \begin{cases} 1 \text{ for } n=0 \\ 0 \text{ otherwise} \end{cases} \quad (2.23)$$

$$\delta[n-a] = \begin{cases} 1 \text{ for } n-a=0 \\ 0 \text{ otherwise} \end{cases} \quad (2.24)$$

Figure 2.33 shows unit impulse sequence. It is also called as a normalized delta function.

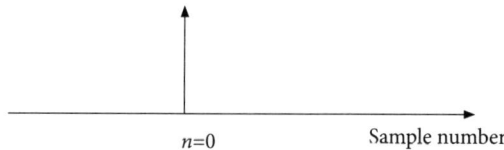

Fig. 2.33 Unit impulse sequence

Let us list properties of DT delta function.

1. $\delta[n] = u[n] - u[n-1]$ \quad (2.25)

Proof

Let us plot $u[n]$ and $u[n-1]$ and subtract $u[n-1]$ form $u[n]$. We can verify that the result is a DT delta function. Refer to Fig. 2.34.

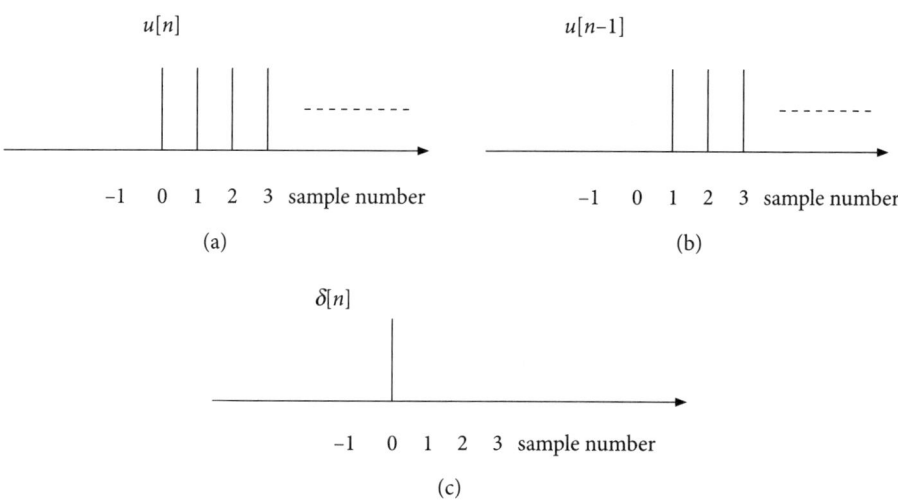

Fig. 2.34 a) Plot of $u[n]$, b) Plot of $u[n-1]$, c) Plot of $u[n]-u[n-1] = \delta[n]$

2. $\sum_{k=-\infty}^{\infty} x[k]\delta[n-k] = x[n]$ (2.26)

Proof

$\delta[n-k] = \begin{cases} 1 \text{ for } n=k \\ 0 \text{ otherwise} \end{cases}$

When k varies from $-\infty$ to ∞, the delta function is zero except when $n = k$. Put $n = k$ in the expression for $x[k]$, we get $x[n]$.

2. **Unit Step**

$u[n] = \begin{cases} 1 \text{ for all } n \geq 0 \\ 0 \text{ otherwise} \end{cases}$ (2.27)

Figure 2.35 shows unit step sequence.

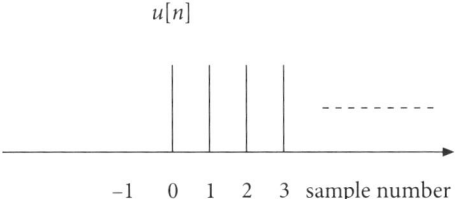

Fig. 2.35 Unit step sequence

Any function that exists only for $n \geq 0$ is usually appended by $u[n]$.

3. **Rectangular Pulse**

The rectangular pulse is denoted as a rect function and is defined as

$\text{rect}\left[\dfrac{k}{2N+1}\right] = \begin{cases} 1 & \text{for } |k| \leq N \\ 0 & \text{for } |k| > N \end{cases}$ (2.28)

The plot of the function is shown in Fig. 2.36 for $k = 3$.

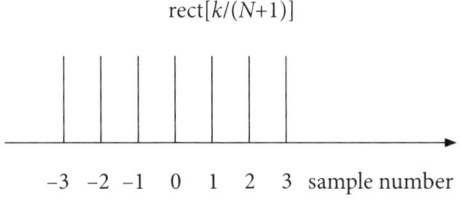

Fig. 2.36 The plot of function rect

4. **Unit Ramp**

$$u_R[n] = \begin{cases} n \text{ for all } n \geq 0 \\ 0 \text{ otherwise} \end{cases} \quad (2.29)$$

Figure 2.37 shows unit ramp sequence.

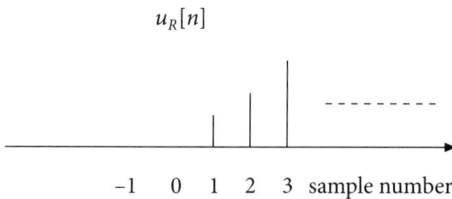

Fig. 2.37 Plot of unit ramp function

5. **Signum Function**

The DT signum function denoted by sgn[k] is defined as follows.

$$\text{sgn}[k] = \begin{cases} 1 & \text{for } k > 0 \\ 0 & \text{for } k = 0 \\ -1 & \text{for } k < 0 \end{cases} \quad (2.30)$$

The plot of the signal is shown in Fig. 2.38.

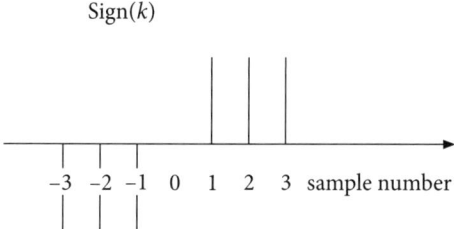

Fig. 2.38 Plot of DT signum function

6. **Sinc Function**

The CT sinc function is defined as

$$\sin c[\omega_0 k] = \frac{\sin(\pi \omega_0 k)}{\pi \omega_0 k} \quad (2.31)$$

The un-normalized sinc function is defined as

$$\sin c[\omega_0 k] = \frac{\sin(\omega_0 k)}{\omega_0 k} \quad (2.32)$$

The plot of normalized sinc signal is shown in Fig. 2.39. The plot is generated using the MATLAB program given below.

```
clear all;
k=-40:1:40;
omega=0.2;
y=sinc(omega*k);
axis([-40 40 -2 2]);
stem(k,y)
grid;
title('plot of a sinc function'); xlabel('sample number');ylabel('amplitude');
```

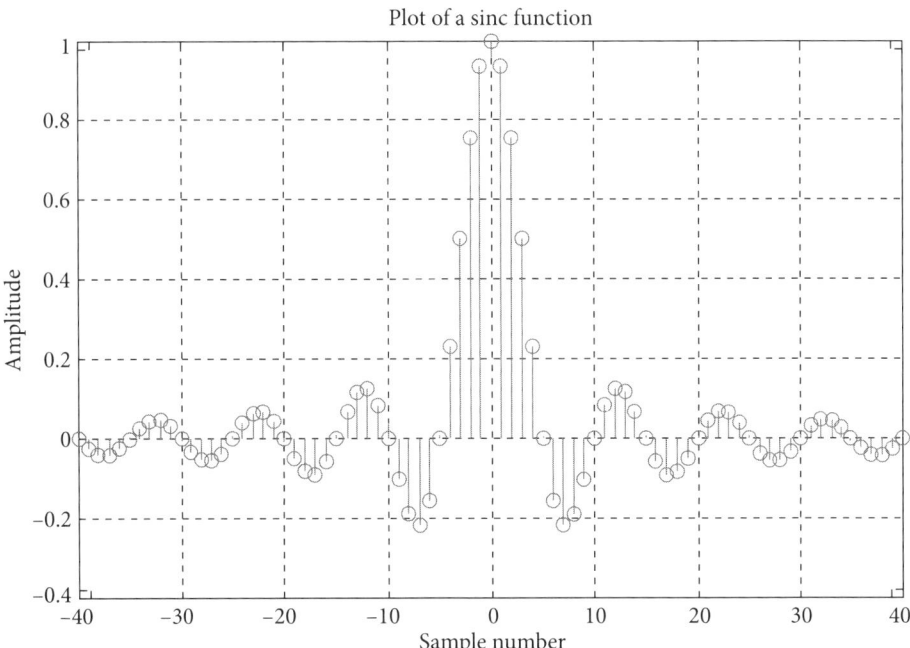

Fig. 2.39 Plot of a DT sinc function

7. **Sinusoidal Function**

The DT sinusoid of frequency f or angular frequency ω is defined as follows.

$$x[n] = \sin(2\pi f nT) \text{ general format} \qquad (2.33)$$

let $x[n] = \sin(0.05\,\pi n)$

T is the sampling interval. The plot of a sinusoidal function is shown in Fig. 2.40. Let frequency be 5 Hz and the angle ϑ be 0 degrees. Let the sampling frequency be 200 Hz. $x(n) = \sin(2^*\,pi^*\,5^*\,1/200^*n) = \sin(0.05\,\pi n)$ A MATLAB program to plot the sinusoid is given below.

```
clear all;
f=5;
T=0.005;
for n=1:41,
x(n)=sin((2*pi*f*(n-1)*T));
end
s=-20:1:20;
stem(s,x);title('plot of DT sine function');xlabel
('sample number');ylabel('Amplitude');
```

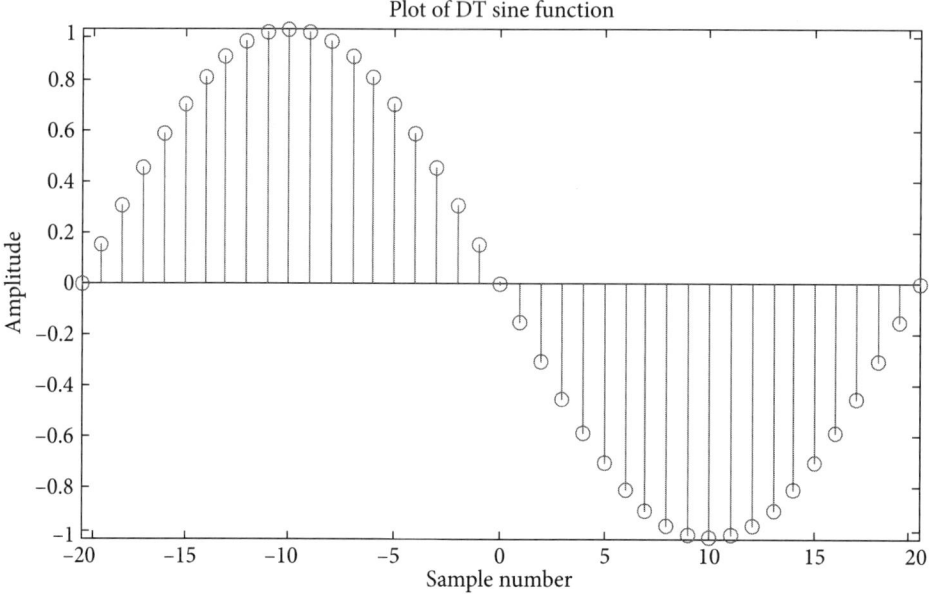

Fig. 2.40 Plot of a DT sinusoidal function

8. **Exponential Function**

A DT exponential function is written as

$$x[n] = \exp(snT) = e^{snT} = e^{(\sigma + j\omega)nT}$$

if $\sigma = 0$, $x(n) = e^{j\omega nT} = \cos(\omega nT) + j\sin(\omega nT)$ (2.34)

Here, s represents the complex frequency. The real part and imaginary part are represented as σ and $j\omega$, respectively. If the real part is zero, it reduces to the complex sinusoid. If the imaginary part is zero, it is represented in Fig. 2.41 with real part greater than zero, less than zero and equal to zero. A MATAB program to plot the signal is given below.

```
clear all;
T=0.05;
for n=1:21,
y(n)=exp(2*T*n);
end
subplot(3,1,1);stem(y);title('plot   of   exponential
function with real part greater than zero');xlabel
('sample number');ylabel('Amplitude');
 T=0.05;
for n=1:21,
y(n)=exp(-2*T*n);
end
subplot(3,1,2);stem(y);title('plot   of   exponential
function with real part less than zero');xlabel('sample
number');ylabel('Amplitude');
 T=0.05;
for n=1:21,
y(n)=exp(0*T*n);
end
subplot(3,1,3);stem(y);title('plot   of   exponential
function with real part equal to zero');xlabel('sample
number');ylabel('Amplitude');
```

9. **Exponential Signal**

$x[n] = a^n$ for all n (2.35)

Figure 2.42a shows exponential signal sequence. ($a > 1$).

The dotted lines in the Figure indicate that the sequence extends up to infinity.

If the value of 'a' is less than 1, then the sequence will be, as shown in Fig. 2.42b.

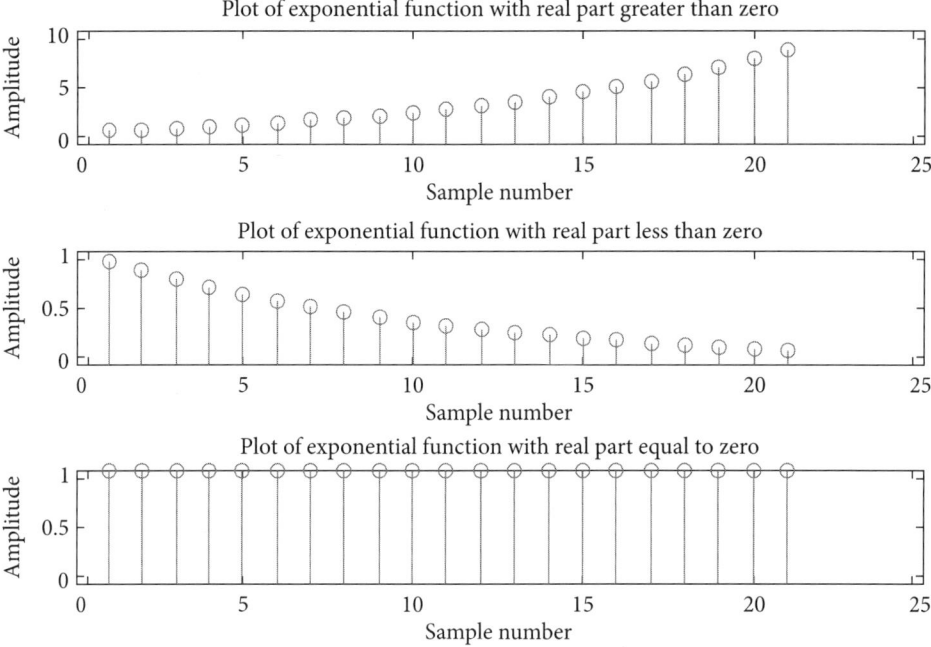

Fig. 2.41 Plot of exponential DT signal for imaginary part of $s = 0$ and real part of $s > 0$, $s < 0$ and $s = 0$

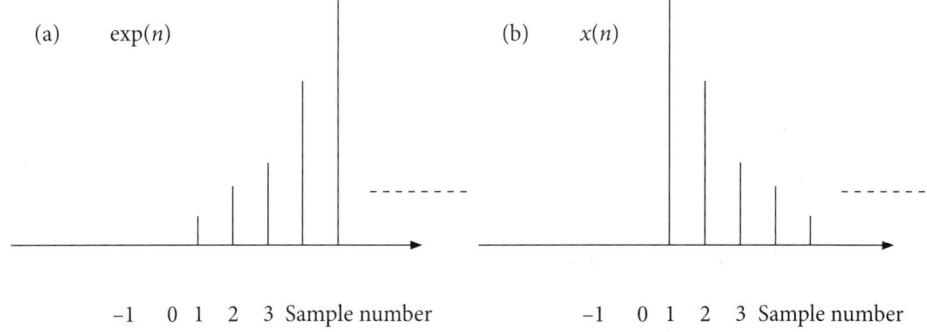

Fig. 2.42 a) DT exponential sequence $a > 1$, b) Plot of $x(n)$ for $a < 1$

10. **Parabolic Function**

The DT parabolic function can be written as

$$x[n] = \frac{n^2}{2} \quad \text{for } n \geq 0 \tag{2.36}$$

As the signal exists only for $n >= 0$, it can be represented as

$$x[n] = \frac{n^2}{2} u[n] \text{ where } u[n] \text{ is a unit step function.} \tag{2.37}$$

The function can be plotted using a MATLAB program. The plot of the parabolic function is shown in Fig. 2.43.

```
clear all;
for i=1:50,
 x(i)=(i^2)/2;
end
stem(x);title('plot of parabolic function');xlabel
('sample number'); ylabel('amplitude'
```

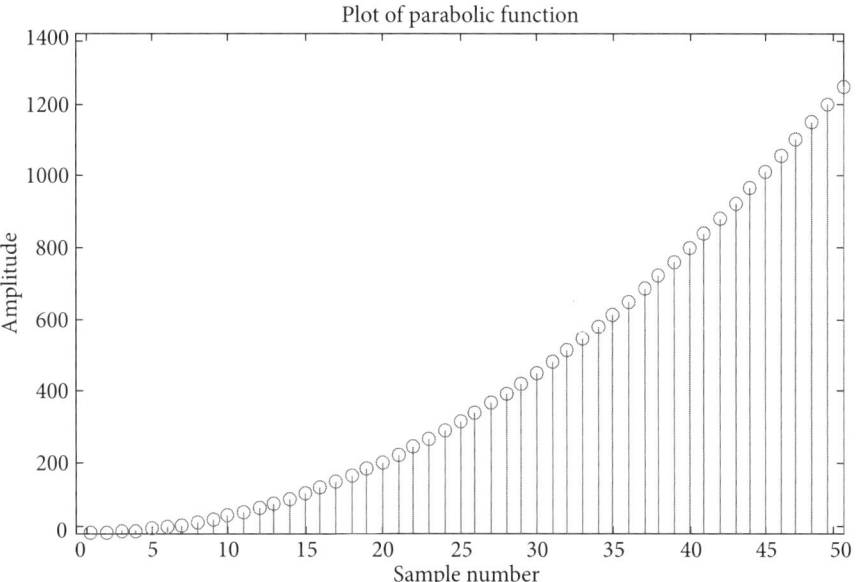

Fig. 2.43 Plot of DT parabolic function

Let us go through some simple examples to illustrate the properties of standard signals.

Example 2.17

Evaluate the integral $\int_{-\infty}^{\infty} e^{-5t^2} \delta(t-4) dt$.

Solution

Let us first define the delta function.

$$\delta(t-4) = \begin{cases} 1 & \text{for } t = 4 \\ 0 & \text{otherwise} \end{cases} \tag{2.38}$$

$$\int_{-\infty}^{\infty} e^{-5t^2} \delta(t-4) dt = [e^{-5t^2}]\Big\downarrow_{t=4} = e^{-5 \times 16} = e^{-80} \tag{2.39}$$

The integral exists only when we put the value of t as 4. The delta function pulls the value of exponential function at $t = 4$.

Example 2.18

Evaluate the integral $\int_{-\infty}^{\infty} t^2 \delta(t-4) dt$.

Solution

Let us first define the delta function.

$$\delta(t-4) = \begin{cases} 1 & \text{for } t = 4 \\ 0 & \text{otherwise} \end{cases} \tag{2.40}$$

$$\int_{-\infty}^{\infty} t^2 \delta(t-4) dt = [t^2]\Big\downarrow_{t=4} = 16 \tag{2.41}$$

The value of the integral exists for $t = 4$. The delta function pulls the value of t^2 at $t = 4$.

Example 2.19

Evaluate the integral $\int_{-\infty}^{\infty} \sin(3\pi t) \delta(t) dt$.

Solution

Let us first define the delta function.

$$\delta(t) = \begin{cases} 1 & \text{for } t = 0 \\ 0 & \text{otherwise} \end{cases} \tag{2.42}$$

$$\int_{-\infty}^{\infty} \sin(3\pi t) \delta(t) dt = \sin(3\pi t)\Big\downarrow_{t=0} = 0 \tag{2.43}$$

The integral value is found out by putting $t = 0$ as the delta function has zero value for all other values of time. The delta function pulls the value of sine function at $t = 0$.

Example 2.20

Evaluate the integral $\int_{-\infty}^{\infty} (t-2)^2 \delta(t-2) dt$.

Solution

Let us first define the delta function.

$$\delta(t-2) = \begin{cases} 1 & \text{for } t = 2 \\ 0 & \text{otherwise} \end{cases} \tag{2.44}$$

$$\int_{-\infty}^{\infty} (t-2)^2 \delta(t-2) dt = 0 \tag{2.45}$$

Here, delta function exists only for $t = 2$. The delta function pulls the value of function at $t = 2$.

Example 2.21

Evaluate the integral $\int_{-\infty}^{\infty} [\cos(4t)\delta(t) + \sin(4t)\delta(t-4)] dt$.

Solution

Let us first define the delta function.

$$\delta(t-4) = \begin{cases} 1 & \text{for } t = 4 \\ 0 & \text{otherwise} \end{cases} \tag{2.46}$$

$$\delta(t) = \begin{cases} 1 & \text{for } t = 0 \\ 0 & \text{otherwise} \end{cases} \tag{2.47}$$

$$\int_{-\infty}^{\infty} [\cos(4t)\delta(t) + \sin(4t)\delta(t-4)] dt = \cos(0) + \sin(16) = 1 + \sin(16) \tag{2.48}$$

The first term of the integral is existing for $t = 0$ and the second term is non-zero only for $t = 4$. The delta function pulls the value of cosine and sine function at $t = 0$ and $t = 4$, respectively.

Example 2.22

Evaluate the integral $\int_{-\infty}^{\infty} e^{2j\omega t} \delta(t) dt$.

Solution

Let us first define the delta function. (Refer to Eq. (2.44) repeated here.)

$$\delta(t) = \begin{cases} 1 & \text{for } t = 0 \\ 0 & \text{otherwise} \end{cases}$$

$$\int_{-\infty}^{\infty} e^{2j\omega t} \delta(t) dt = \int_{-\infty}^{\infty} \delta(t) dt = 1 \tag{2.49}$$

$e^{2j\omega \times 0} = 1$

The integral has value of 1 for $t = 0$. The delta function pulls the value of exponential function at $t = 0$.

Example 2.23

Evaluate the summation $\sum_{n=-\infty}^{\infty} e^{5n}\delta[n]$.

Solution

Let us define DT delta function.

$$\delta[n] = \begin{cases} 1 & \text{for } n = 0 \\ 0 & \text{otherwise} \end{cases} \quad (2.50)$$

$$\sum_{n=-\infty}^{\infty} e^{5n}\delta[n] = 1 \quad (2.51)$$

The delta function exists for $n = 0$. The delta function pulls the value of exponential function at $n = 0$.

Example 2.24

Evaluate the summation $\sum_{n=-\infty}^{\infty} \sin(2n)\delta[n-3]$.

Solution

Let us first define the delta function.

$$\delta[n-3] = \begin{cases} 1 & \text{for } n = 3 \\ 0 & \text{otherwise} \end{cases} \quad (2.52)$$

$$\sum_{n=-\infty}^{\infty} \sin(2n)\delta[n-3] = \sin(6) \quad (2.53)$$

The summation has a non-zero value for $n = 3$. The delta function pulls the value of sine function at $n = 3$.

Example 2.25

Evaluate the summation $\sum_{n=-\infty}^{\infty} e^n \delta[n+3]$.

Solution

Let us first define the delta function.

$$\delta[n+3] = \begin{cases} 1 & \text{for } n = -3 \\ 0 & \text{otherwise} \end{cases} \quad (2.54)$$

$$\sum_{n=-\infty}^{\infty} e^n \delta[n+3] = e^{-3} \qquad (2.55)$$

Here, the delta function exists only for $n = -3$. The delta function pulls the value of exponential function at $n = -3$.

Example 2.26

Prove that $\int_{-\infty}^{\infty} x(t)\delta(t-\tau)dt = x(\tau)$

Solution

Let us first define the delta function.

$$\delta(t-\tau) = \begin{cases} 1 & \text{for } t = \tau \\ 0 & \text{otherwise} \end{cases} \qquad (2.56)$$

$\int_{-\infty}^{\infty} x(t)\delta(t-\tau)dt = x(\tau)$ (The intergral exists only for $t = \tau$) $\qquad (2.57)$
we put $t = \tau$

The delta function pulls the value of x at $t = \tau$.

Concept Check
- Why the standard signals are used for testing of a system?
- What is the use of impulse as a standard input?
- How will you test response of a system to a sudden change in input signal?
- How will you test response of a system to a continuous change in input signal?
- How will you estimate the steady state errors?
- List different analog input signals used as standard signals.
- Define a Dirac delta function.
- How will you characterize a system using impulse as input?
- State the properties of Delta function.
- What is the use of a rectangular pulse for system testing?
- Define the relation between unit step and unit ramp function.
- Why a signum function is also called as a sign function?
- State three reasons for using a sinc function.
- Why we use a sine wave for testing an amplifier?
- What is the use of a parabolic function?
- What is the significance of a Gaussian function?
- Define a unit impulse sequence.

- Define a unit step sequence.
- Evaluate the integral $\int_{-\infty}^{\infty} t^2 \delta(t-3) dt$.

2.6 Classification of Signals Based on Signal Properties

The analog, DT and Digital signals exhibit properties namely even and odd property, periodicity property, randomness, etc. We give below the classification of the signals based on these properties.

2.6.1 Even and odd signals

We will first treat the even and odd property for analog signals. A continuous time, continuous amplitude signal or analog signal is said to be even if $x(t) = x(-t)$ for all t. The even signals are symmetrical with respect to the origin or a vertical i.e., amplitude axis.

Further, the signal is said to be an odd signal if it satisfies $x(t) = -x(-t)$ for all t. The odd signals are anti-symmetrical with respect to the origin or a vertical i.e., amplitude axis. Any signal $x(t)$ can be written as a summation of even part $x_e(t)$ and odd part $x_o(t)$. Let us substitute $t = -t$ in the expression for $x(t)$, we will get

$$x(t) = x_e(t) + x_o(t)$$

$$x(-t) = x_e(-t) + x_o(-t) = x_e(t) - x_o(t)$$

$$x_e(t) = \frac{1}{2}[x(t) + x(-t)] \text{ and}$$

(2.58)

$$x_o(t) = \frac{1}{2}[x(t) - x(-t)]$$

Here, we have assumed that the signal $x(t)$ is real. If the signal is complex valued, it can be represented as

$$x(t) = a(t) + jb(t), \text{ if } x(-t) = x^*(t)$$

(2.59)

* represents a complex conjugate

It can be easily verified that the signal $x(t)$ is conjugate symmetric, if $a(t)$ is even and $b(t)$ is odd.

Example 2.27

Consider an analog signal given by
$x(t) = A$ for $-T/2 \leq t \leq T/2$. Find if the signal is even.

Solution

We will plot the signal and will find out if $x(-t) = x(t)$. The plot of a signal is shown in Fig. 2.44. We observe that $x(t) = x(-t)$ for all t. It is symmetrical with respect to the origin or a vertical i.e., amplitude axis. Hence, the signal is even. The same signal can also be written as

$$x(t) = A \, \text{rect}(t/T) = 1 \text{ for } |t| \leq T/2 \tag{2.60}$$
$$= 0 \text{ for } |t| > T/2$$

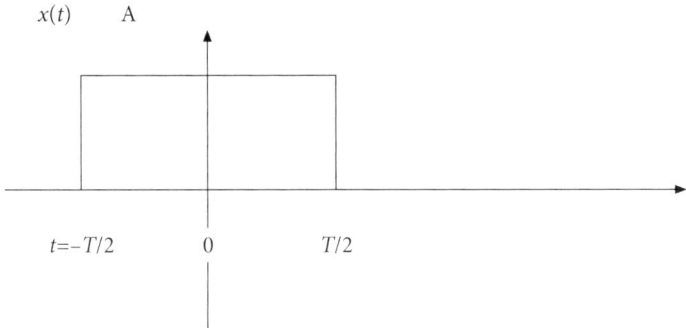

Fig. 2.44 Plot of signal $x(t)$ for Example 2.27

Example 2.28

Consider an analog signal given by

$x(t) = -A$ for $-T/2 \leq t \leq 0$ and
$x(t) = A$ for $0 < t \leq T/2$

Find if it is even or odd.

Solution

Let us plot the signal. The plot is shown in Fig. 2.45. We observe that $x(t) = -x(-t)$ for all t. The signal is anti-symmetrical with respect to the origin or a vertical i.e., amplitude axis. Hence, the signal is an odd signal.

Example 2.29

Consider a signal given by

$x(t) = e^{j\omega t}$ Find the even and odd part of the signal.

Solution

We will write the signal as

$$x(t) = e^{j\omega t} = \cos(\omega t) + j\sin(\omega t) \tag{2.61}$$

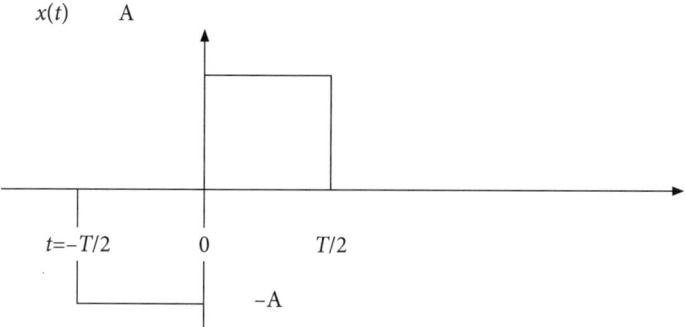

Fig. 2.45 Plot of signal $x(t)$ for Example 2.28

So, $\quad x(-t) = e^{-j\omega t} = \cos(-\omega t) + j\sin(-\omega t) = \cos(\omega t) - j\sin(\omega t)$ (2.62)

$$x_e(t) = [x(t) + x(-t)]/2 = \cos(\omega t)$$ (2.63)

$$x_o(t) = [x(t) - x(-t)]/2 = j\sin(\omega t)$$

We can conclude that analog cosine function is an even function and analog sine function is an odd function. Let us plot the sine function and cosine function using MATLAB and confirm that cosine is an even function and sine is an odd function. A MATLAB program to plot sine and cosine function is given below. Figure 2.46 shows the output of MATLAB program for plot of sine and cosine functions. The plot clearly indicates that cosine function is an even function and sine function is an odd function.

```
clear all;
f=10;
T=0.005;
for n=1:41,
x(n)=sin((2*pi*f*(n-1)*T));
y(n)=cos(2*pi*f*(n-1)*T);
end
s=-20:1:20;
subplot(2,1,1);
plot(s,x);title('plot of sine function for positive
and negative angles');xlabel('angle pi divided in 20
points');ylabel('Amplitude');
subplot(2,1,2);
plot(s,y);title('plot of cosine function for positive
and negative angles');xlabel('angle pi divided in 20
points');ylabel('Amplitude');
```

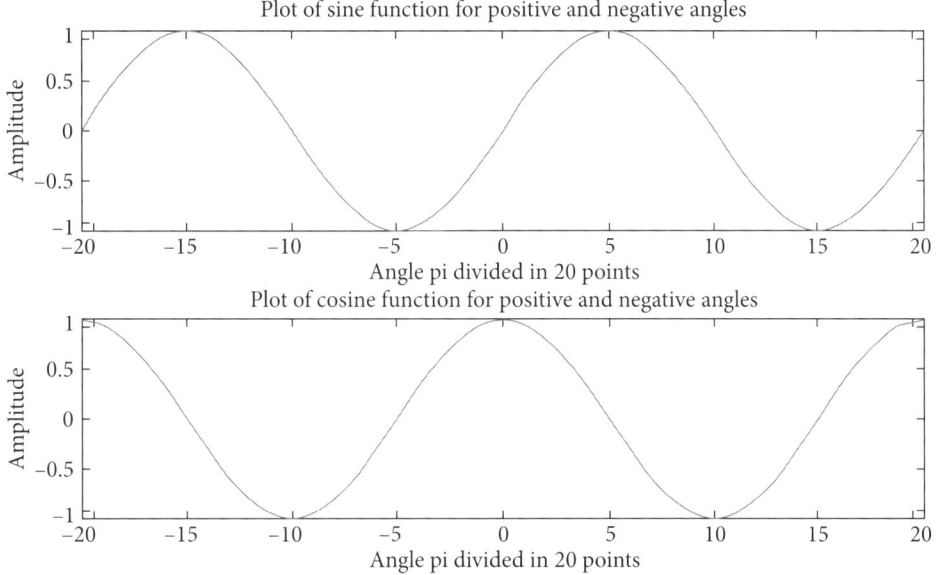

Fig. 2.46 Plot of sine and cosine function for positive and negative angles

If the signal is defined as $x(t) = e^{j\omega t}$ for all $t \geq 0$, then the signal is neither even nor odd. The signal is simply undefined for all negative values of t.

Example 2.30

Consider a signal given by $x(t) = \cos(t) + \sin(t) + \cos(t)\sin(t)$. Find the even part and odd part of the signal.

Solution

$$x(t) = \cos(t) + \sin(t) + \cos(t)\sin(t) \tag{2.64}$$

$$x(-t) = \cos(-t) + \sin(-t) + \cos(-t)\sin(-t) \tag{2.65}$$

$$= \cos(t) - \sin(t) - \cos(t)\sin(t)$$

$$x_e(t) = [x(t) + x(-t)]/2 = \cos(t) \tag{2.66}$$

$$x_o(t) = [x(t) - x(-t)]/2 = \sin(t) + \cos(t)\sin(t)$$

We can conclude that the product of even (cosine) and odd (sin) functions is an odd function.

Example 2.31

Consider a signal given by $x(t) = 1 + t + 4t^2$. Find the even and odd parts of the signal.

Solution

$$x(t) = 1 + t + 4t^2 \tag{2.67}$$

$$x(-t) = 1 + (-t) + 4(-t)^2 = 1 - t + 4t^2 \tag{2.68}$$

$$x_e(t) = [x(t) + x(-t)]/2 = 1 + 4t^2 \tag{2.69}$$

$$x_o(t) = [x(t) - x(-t)]/2 = t$$

We can conclude that the terms containing even power of t like t^2, t^4, t^6, etc. are even functions and terms containing odd orders of t like t, t^3, t^5, etc. are odd functions.

Example 2.32

Consider a signal given by $x(a) = \cos(a) + \sin^2(2a)$. Find the even part and odd part of the signal.

Solution

$$x(a) = \cos(a) + \sin^2(2a) \tag{2.70}$$

$$x(-a) = \cos(-a) + \sin^2(-2a) = \cos(a) + \sin^2(2a) \tag{2.71}$$

$$x_e(t) = [x(t) + x(-t)]/2 = \cos(a) + \sin^2(2a) \tag{2.72}$$

$$x_o(t) = [x(t) - x(-t)]/2 = 0$$

We can conclude that the square term of the odd function is always an even function.

Example 2.33

Consider a signal given by $x(a) = a^2 \cos(a) + a \sin(2a)$. Find the even part and odd part of the signal.

Solution

$$x(a) = a^2 \cos(a) + a \sin(2a) \tag{2.73}$$

$$x(-a) = (-a)^2 \cos(-a) + (-a)\sin(-2a) = a^2 \cos(a) + a\sin(2a) \qquad (2.74)$$

$$x_e(t) = [x(t) + x(-t)]/2 = a^2 \cos(a) + a\sin(2a) \qquad (2.75)$$

$$x_o(t) = [x(t) - x(-t)]/2 = 0$$

We can conclude that the product term of two even functions is always an even function and the product of two odd functions is always an even function. *Note that we may use any other variable in place of t.*

Example 2.34

Consider a signal given by $x(t) = 1 + t\cos(t) + t^2 \sin(t) + t^3 \sin(t)\cos(t)$. Find the even part and odd part of the signal.

Solution

$$x(t) = 1 + t\cos(t) + t^2 \sin(t) + t^3 \sin(t)\cos(t) \qquad (2.76)$$

$$x(-t) = 1 + (-t)\cos(-t) + (-t)^2 \sin(-t) + (-t)^3 \sin(-t)\cos(-t) \qquad (2.77)$$

$$= 1 - t\cos(t) - t^2 \sin(t) + t^3 \sin(t)\cos(t)$$

$$x_e(t) = [x(t) + x(-t)]/2 = 1 + t^3 \sin(t)\cos(t) \qquad (2.78)$$

$$x_o(t) = [x(t) - x(-t)]/2 = t\cos(t) + t^2 \sin(t)$$

We can conclude that the second and the third term are products of one even term and one odd term, so the product term is an odd function. The fourth term is a product of 2 odd functions and one even function that results in an even function.

Example 2.35

Consider a signal given by $x(t) = (1 + t^3)\cos(9t)$. Find the even part and odd part of the signal.

Solution

$$x(t) = (1 + t^3)\cos(9t) = \cos(9t) + t^3 \cos(9t) \qquad (2.79)$$

$$x(-t) = (1 + (-t)^3)\cos(-9t) = \cos(-9t) + (-t)^3 \qquad (2.80)$$

$$\cos(-9t) = \cos(9t) - t^3 \cos(9t)$$

$$x_e(t) = [x(t) + x(-t)]/2 = \cos(9t) \qquad (2.81)$$

$$x_o(t) = [x(t) - x(-t)]/2 = t^3 \cos(9t)$$

We can conclude that to find the even and odd parts of the signal, one needs to expand the brackets if any.

Example 2.36
Consider a signal given by $x(t) = (t + t^3) \cos(3t)$. Find the even part and odd part of the signal.

Solution

$$x(t) = (t + t^3)\cos(3t) = t\cos(3t) + t^3 \cos(3t) \tag{2.82}$$

$$x(-t) = (-t) + (-t)^3)\cos(-3t) = -t\cos(3t) - t^3 \cos(3t) \tag{2.83}$$

$$x_e(t) = [x(t) + x(-t)]/2 = 0$$

$$x_o(t) = [x(t) - x(-t)]/2 = t\cos(3t) + t^3 \cos(3t) \tag{2.84}$$

We can conclude that the function consists of sum of two terms that are the products of one even term and one odd term. So, both the terms are odd.

Example 2.37
Consider a signal given by $x(t) = (t + t^3)\tan(3t)$. Find the even part and odd part of the signal.

Solution

$$x(t) = (t + t^3)\tan(3t) = t\tan(3t) + t^3 \tan(3t) \tag{2.85}$$

$$x(-t) = (-t) + (-t)^3)\tan(-3t) = t\tan(3t) + t^3 \tan(3t) \tag{2.86}$$

$$x_e(t) = [x(t) + x(-t)]/2 = t\tan(3t) + t^3 \tan(t) \tag{2.87}$$

$$x_0(t) = [x(t) - x(-t)]/2 = 0$$

The tan function is the ratio of sine (odd function) and cosine (even function). The ratio will be an odd function. We can conclude that the function consists of the sum of two terms that are the products of two odd terms. So, both the terms are even.

Example 2.38
Consider a signal given by $x(t) = t \times (1 + t + t^2)$. Find the even part and odd part of the signal.

Solution

$$x(t) = t \times (1 + t + t^2) = t + t^2 + t^3 \tag{2.88}$$

$$x(-t) = (-t) \times (1 + (-t) + (-t)^2) = -t + t^2 - t^3 \tag{2.89}$$

$$x_e(t) = [x(t) + x(-t)]/2 = t^2$$

$$x_o(t) = [x(t) - x(-t)]/2 = t + t^3 \tag{2.90}$$

We can conclude that the function consists of sum of three terms out of which two terms are odd and one term is even. This fact is evident when we expand the bracket.

We will now consider even and odd classification for DT signals. We will illustrate this with the help of following examples.

Example 2.39

Consider a discrete time signal given by

$x[n] = 1$ for $-2 \le n \le 2$

$ = 0$ otherwise

Find if the signal is even or odd.

Solution

Let us plot the signal. A plot is shown in Fig. 2.47. The signal is symmetrical with respect to the origin and with respect to the amplitude axis. $x[-n] = x[n]$ for all n. Hence, the signal is even.

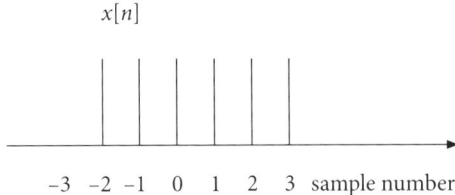

Fig. 2.47 Plot of signal for Example 2.39

Example 2.40

Consider a discrete time signal given by

$x[n] = 1$ for $1 < n \le 2$

$ = -1$ for $-1 \le n \le -2$

Find if the signal is even or odd.

Solution

Let us plot the signal. A plot is shown in Fig. 2.48. The signal is anti-symmetrical with respect to the origin and with respect to the amplitude axis. $x[-n] = -x[n]$ for all n. Hence, the signal is odd.

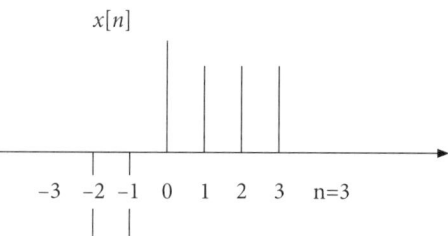

Fig. 2.48 Plot of signal for Example 2.40

Example 2.41

Consider a discrete time signal given by

$x[n] = 1$ for $1 \leq n \leq 5$

$ = 0$ otherwise

Find if the signal is even or odd.

Solution

Let us plot the signal. A plot is shown in Fig. 2.49. The signal is having a value zero for all negative values of n. It is neither symmetric nor anti-symmetrical with respect to the origin and with respect to the amplitude axis. Hence, the signal is neither even nor odd.

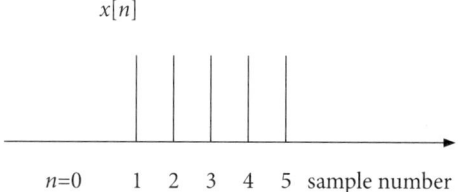

Fig. 2.49 Plot of signal for Example 2.41

We conclude that if the signal is neither symmetric nor anti-symmetric with respect to the amplitude axis, then the signal is neither even nor odd.

Physical significance of even and odd property If the signal is even, its mirror image is same as itself. If the signal is odd, its mirror image is $x(-t) = -x(t)$. This is useful when we take the convolution of one signal with the other. Here, the first step for convolution, we have to take the mirror image of the signal. (The reader may refer to chapter 4 for convolution)

2.6.2 Periodic and aperiodic signals

Let us go through the classification of signals based on periodicity property.

Periodicity of analog Signals

Let us consider the periodic and aperiodic classification for analog signals. The periodic signal $x(t)$ is a function for which there exists some value T that satisfies the condition $x(t + T) = x(t)$ for all t. T denotes the period. For example, consider a sinusoidal signal given by $x(t) = A \sin(t)$. Here, the signal is periodic with the period of 2π. For some signals no such value T exists for which $x(t + T) = x(t)$ holds good. Such a signal is termed as an aperiodic signal.

Example 2.42

Consider the equation given by $x(t) = \cos(2\pi f t)$. Is $x(t)$ a periodic signal? What is the period of this signal?

Solution

Here, we will check if there is some T for which $x(t) = x(x + T)$

Put $t = t + T$ in $x(t) = \cos(2\pi f t)$ (2.91)

We obtain

$$x(t+T) = \cos(2\pi f t + 2\pi f T) = \cos(2\pi f t) \text{ if } 2\pi f T = 2\pi \text{ i.e., } T = \frac{1}{f} \quad (2.92)$$

Hence, there exists some T for which $x(t) = x(x + T)$. The signal is periodic. $1/f$ is the period.

We can conclude that any continuous (Analog) cosine or sine function is periodic with period of $1/f$ where f is the frequency of the cosine or a sine wave. The value of the frequency must be finite. It may be rational or irrational. Periodicity has nothing to do with rationality.

So the signal $x(t) = \cos(2t)$ is also periodic with period equal to $x(t + T) = \cos(2t + 2T) = \cos(2t)$ $2T = 2\pi$ i.e., $T = \pi$ and frequency equal to $1/\pi$. Here, the frequency is irrational.

Example 2.43

Consider a signal $x(t) = \cos(4\pi t)$, find if the signal is periodic and find the period.

Solution

$$x(t+T) = \cos(4\pi t + 4\pi T) = \cos(4\pi t) \text{ if } 4\pi T = 2\pi \text{ i.e. } T = 1/2. \quad (2.93)$$

So, period of signal is ½ seconds and frequency f is 2. Now if we define the signal only over a finite duration as $x(t) = \cos(4\pi t)$ for $|t| \leq 2$, then the signal will become truly aperiodic; however, for a defined duration, it will be treated as periodic.

A periodic signal by definition must strictly start from $-\infty$ and end at ∞. However, in practice, the signals are available only for a finite duration. So, we relax the periodicity condition. The signal will be treated as periodic if it is periodic over the finite duration for which it is defined.

Example 2.44

Consider a signal $x(t) = \cos(3t)$, find if the signal is periodic and find the period.

Solution

$$x(t+T) = \cos(3t+3T) = \cos(3t) \text{ if } 3T = 2\pi \text{ i.e., } T = 2\pi/3. \tag{2.94}$$

So, period of signal is $2\pi/3$ seconds and frequency f is $3/2\pi$. The value of the period is not rational, still the signal is periodic. Every analog sinusoidal signal is periodic.

Example 2.45

Consider the equation given by $x(t) = t + \cos(2\pi ft)$. Is it a periodic signal?

Solution

Here, we will check if there is some T for which $x(t) = x(t + T)$. Put $t = t + T$ in $x(t) = t + \cos(2\pi ft)$. We obtain

$$x(t+T) = t + T + \cos(2\pi ft + 2\pi fT). \tag{2.95}$$

The component t increases as t increases so no such T will exist for which the signal will be periodic. Hence, the signal is aperiodic.

Example 2.46

Consider the equation given by $x(t) = (\cos(2\pi ft))^2$. Is it a periodic signal?

Solution

Here, we will check if there is some T for which $x(t) = x(t + T)$. Put $t = t + T$ in $x(t) = (\cos(2\pi ft))^2$. We obtain

$$x(t+T) = (\cos(2\pi f(t+T)))^2 = (\cos(2\pi ft))^2 \text{ when } T = \frac{1}{2f}. \tag{2.96}$$

The function is periodic with period equal to $1/2f$.

Example 2.47

Consider the equation given by $x(t) = |\cos(2\pi ft)|$. Is $x(t)$ a periodic signal?

Solution

Here, we will check if there is some T for which $x(t) = x(t + T)$. Put $t = t + T$ in $x(t) = |\cos(2\pi ft)|$. We obtain

$$x(t+T) = |\cos(2\pi f(t+T))| = |\cos(2\pi ft)| \text{ when } T = \frac{1}{2f}. \qquad (2.97)$$

The function is periodic with period equal to 1/2f. Let us verify this by writing a MATLAB program. A plot of the function is shown in Fig. 2.50.

```
clear all;
f=10;
T=0.005;
for n=1:41,
y(n)=abs(cos(2*pi*f*(n-1)*T));
end
s=-20:1:20;
plot(s,y);title('plot of absolute value of cosine
function for positive and negative angles');xlabel
('angle pi divided in 20 points');ylabel('Amplitude');
```

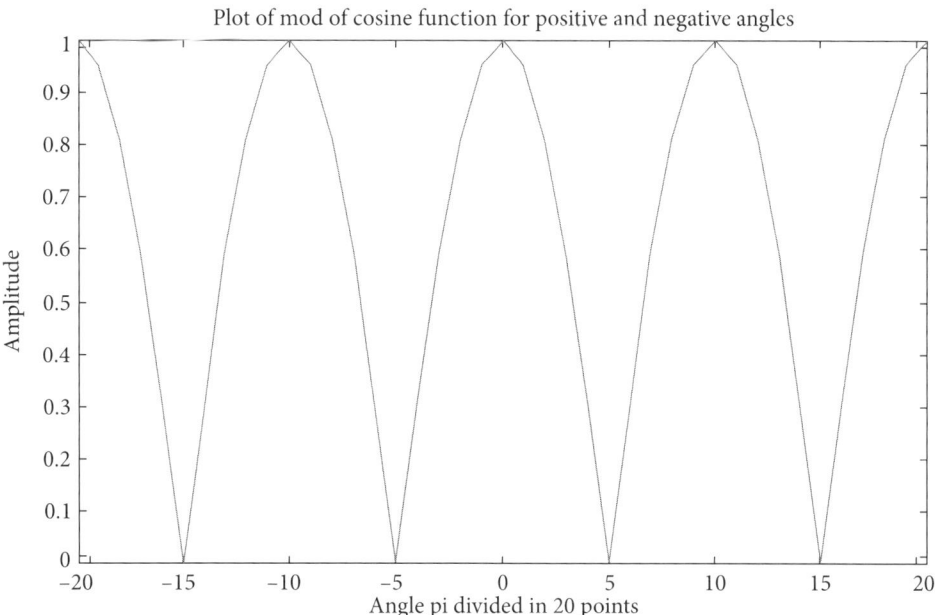

Fig. 2.50 A plot of a signal for Example 2.47

Example 2.48

Consider the signal shown in Fig. 2.51. Is $x(t)$ a periodic signal?

Solution

The signal exists only over a small duration. It is not repeating. Hence, the signal is not periodic.

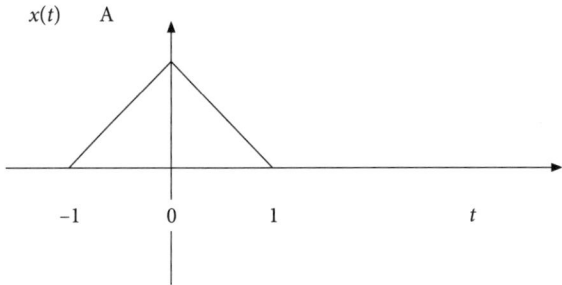

Fig. 2.51 A signal for Example 2.48

Example 2.49

Consider the signal $x(t)$ shown in Fig. 2.52. If $y(t) = \sum_{k=-5}^{5} x(t-2k)$ is $y(t)$ a periodic signal ?

Solution

Let us plot the signal to find if the signal is periodic. The plot of the signal is shown in Fig. 2.52.

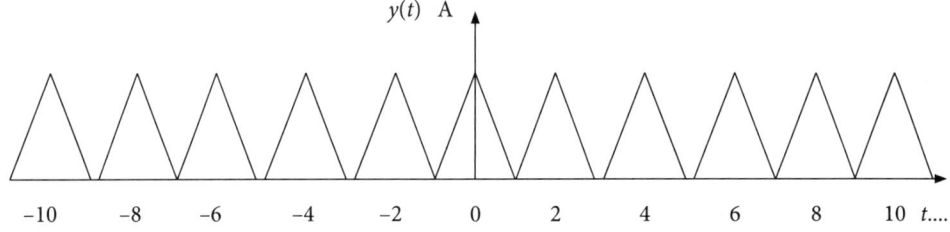

Fig. 2.52 A signal for Example 2.49

Referring to the Fig. 2.52, we can see that the signal repeats after $t = 2$. However, the signal does not exist before $t = -11$ and after $t = 11$. So, the signal can be considered as periodic over the period for which it is defined.

Example 2.50

Consider the signal $x(t)$ shown in Fig. 2.53. If $y(t) = \sum_{k=-\infty}^{\infty} x(t-2k)$ is $y(t)$ a periodic signal ?

Solution

Let us plot the signal to find if the signal is periodic. The plot of the signal is shown in Fig. 2.53.

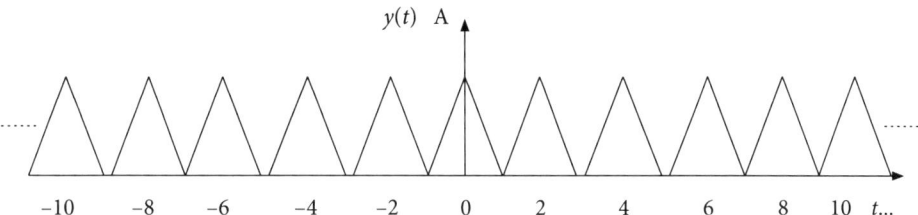

Fig. 2.53 A signal for Example 2.50

Referring to Fig. 2.53, we can see that the signal repeats after $t = 2$. The signal is defined and it exists from $-\infty$ to $+\infty$. So, the signal can be considered as truly periodic.

Example 2.51

Consider the equation given by $x(t) = e^{-2t}$. Is $x(t)$ a periodic signal?

Solution

Let us plot the signal to find if it is periodic. A MATLAB program is given below. Figure 2.54 shows a plot that indicates that the function is aperiodic.

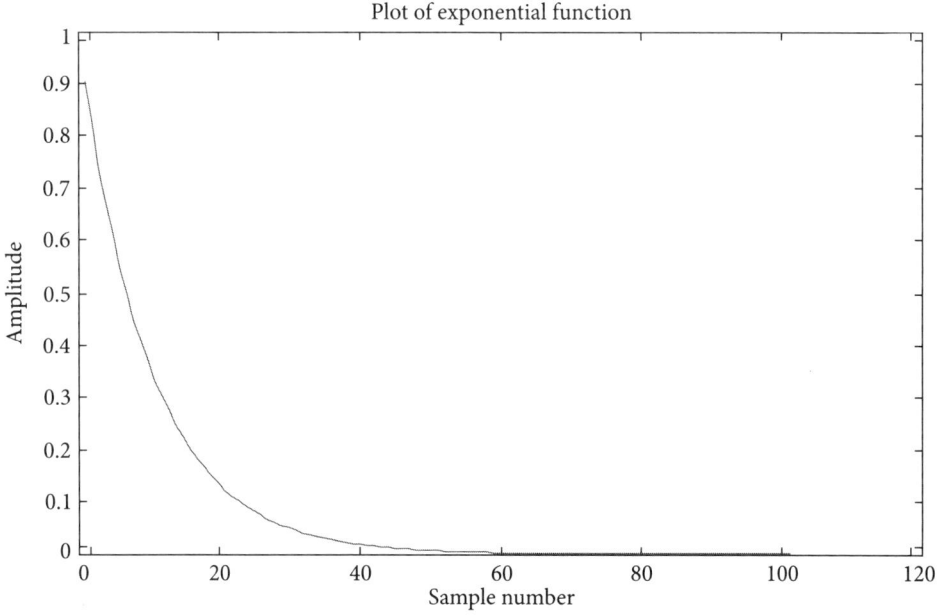

Fig. 2.54 Plot of signal for Example 2.51

```
clear all;
t=0.05;
for n=1:101,
y(n)=exp(-2*t*n);
end
plot(y);title('plot of exponential function');xlabel
('sample number');ylabel('Amplitude');
```

Periodicity of DT signals

Let us now illustrate the periodicity for DT signals.

Example 2.52

Consider the signal $x[n]$ given by $x[n] = (-1)^n$ is $x[n]$ a periodic signal?

Solution

Let us evaluate the values of the signal for different values of n and plot the signal. The value of the signal is +1 for all even values of n and is equal to −1 for all odd values of n for positive as well as negative values of the exponent i.e., n.

$$x(0) = (-1)^0 = 1, x(1) = (-1)^1 = -1, x(2) = (-1)^2 = 1 \text{ and so on}$$

$$x(-1) = (-1)^{-1} = \frac{1}{(-1)} = -1, x(-2) = (-1)^{-2} = \frac{1}{(-1)^2} = 1 \text{ and so on} \quad (2.98)$$

The signal plot is shown in Fig. 2.55. Hence, the signal is periodic with the period equal to 2 as the signal repeats after every 2 samples.

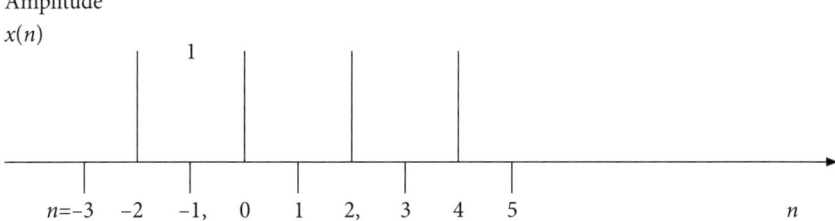

Fig. 2.55 Plot of signal for Example 2.52

Example 2.53

Consider the signal $x[n]$ given by $x[n] = (-1)^{n^2}$. Is $x[n]$ a periodic signal?

Solution

Let us evaluate the values of the signal for different values of n and plot the signal. The value of the signal is +1 for all even values of n and is equal to −1 for all odd values of n for positive as well as negative values of the exponent i.e., n.

$$x(0) = (-1)^{0^2} = 1, x(1) = (-1)^{1^2} = -1, x(2) = (-1)^{2^2} = 1 \text{ and so on}$$

$$x(-1) = (-1)^{(-1)^2} = \frac{1}{(-1)^1} = -1, x(-2) = (-1)^{(-2)^2} = \frac{1}{(-1)^4} = 1 \text{ and so on} \quad (2.99)$$

The signal plot is shown in Fig. 2.56. Hence, the signal is periodic with a period equal to 2.

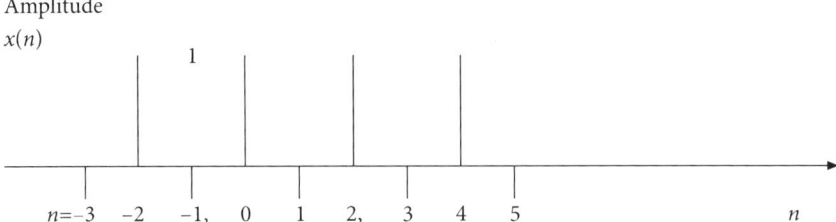

Fig. 2.56 Plot of the signal for Example 2.53

Example 2.54

Consider the signal shown in Fig. 2.57. Is $x[n]$ a periodic signal ?

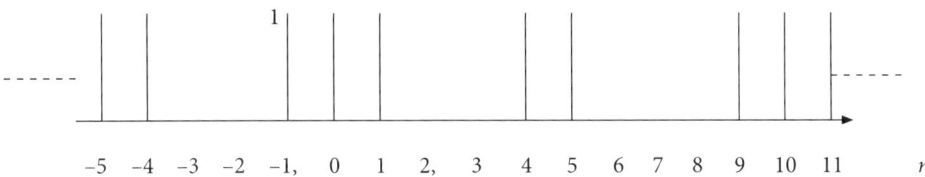

Fig. 2.57 Plot of signal for Example 2.54

Solution

We can refer to the plot of the signal to see that the signal repeats itself after every 10 samples (i.e., from $n = -1$ to $n = 9$). So, the period is 10 samples.

Example 2.55

Consider the signal shown in Fig. 2.58. Is $x(t)$ a periodic signal ?

Solution

We can observe from the signal plot that even though the positive square waveform between −4 and −3 repeats between −2 to −1 and that from 1 to 2 repeats between 3 to 4, it is negative for some duration i.e., from −0.5 to 0.5. Hence, the signal is not periodic.

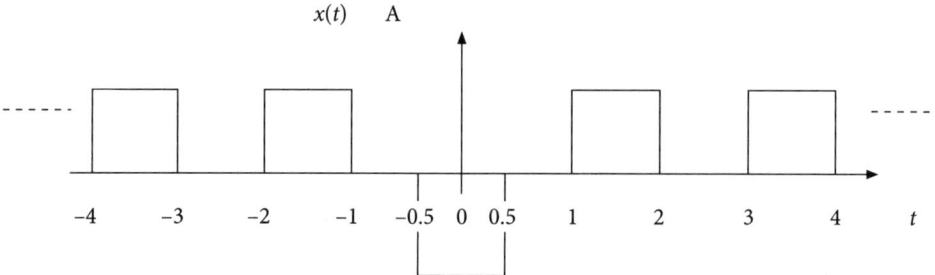

Fig. 2.58 Signal $x(t)$ for Example 2.55

Periodicity of DT Sinusoidal Signals

The CT sinusoidal signals are all periodic. We need to discuss periodicity of DT sinusoids only. The periodic analog signal when sampled may generate a DT signal, which is one of the following:

Case 1 Periodic.
Case 2 Periodic with different reconstructed frequency (aliased frequency).
Case 3 It may become aperiodic.

We will first see the criterion for a DT signal to be periodic.

The DT signal $x[n]$ is periodic if and only if its digital frequency f is a rational number (ratio of two integers). To test if it is periodic we have to check if

$$x[n] = x[n + N] \text{ for some integer } N \tag{2.100}$$

That is

$$x[n] = A\cos(2\pi f n) = A\cos[2\pi f(n+N)] \tag{2.101}$$

Equation (2.101) is true if there exists some integer value k for which
$2\pi f N = 2\pi k$, because $\cos(2\pi k + 2\pi f N) = \cos(2\pi f N)$ if k is integer

$$f = \frac{k}{N} \tag{2.102}$$

The fundamental period of the sinusoid is N when k and N are relatively prime. N represents the number of samples in one period of the sinusoid i.e., frequency in terms of number of samples. We will illustrate three cases with the help of examples solved below.

Example 2.56

(Case 1) This Example illustrates case 1.

Consider a sinusoid of frequency 10 Hz sampled with a sampling frequency F_s of 30 Hz. The resulting DT signal is

$$x[n] = A\cos\left(2\pi \frac{10}{30} n\right)$$

$$= A\cos\left(2\pi \frac{1}{3} n\right) \qquad (2.103)$$

The generalized representation for a digital signal is

$$x[n] = A\cos(2\pi f n) \qquad (2.104)$$

Comparing Eq. (2.103) with Eq. (2.104), we find that $f = 1/3$. It is rational and is lying within the range of $-1/2$ to $1/2$. It is perfectly recoverable in analog domain as per the sampling theorem. The DT signal retains the properties of the original sinusoid. It is periodic with digital period of three samples.

Example 2.57

(Case 2) This Example illustrates case 2.

Discrete time sinusoids whose frequencies are separated by an integral multiple of 2π are identical. Any sequence resulting from a sinusoid with a frequency of $|w| > \pi$ or $|f| > 1/2$ is identical to the corresponding sequence obtained from a sinusoidal signal with frequency $|f| < 1/2$. Hence, the frequencies in the range $-1/2 < f < 1/2$ are unique and all frequencies $f > 1/2$ fold back to $|f| < 1/2$. This indicates that the highest rate of oscillations for a DT sinusoid is attained when $|f| = 1/2$. The range of DT signal frequencies is between $-1/2$ and $1/2$. (Refer Fig. 1.7 in Chapter 1 for details of sampling.)

Let us consider the same signal with a bandwidth of 10 Hz. Here, the sampling frequency as per the sampling theorem is 20 Hz. We will sample the signal with the sampling frequency below the Nyquist rate, that is, $F_S = 7.5$ Hz. Let us concentrate on the 10 Hz sinusoid. We will see that the DT sinusoid will be periodic; however, when reverted back in time domain, the signal is aliased. The DT signal is represented as

$$x[n] = A\cos\left(2\pi \frac{10}{7.5} n\right) = A\cos\left(2\pi \frac{4}{3} n\right) \qquad (2.105)$$

$$x[n] = A\cos\left(\frac{8}{3}\pi n\right) = A\cos\left(2\pi + \frac{2}{3}\pi n\right) \qquad (2.106)$$

As $\cos(x) = \cos(x + 2\pi)$, the signal becomes

$$x[n] = A\cos\left(\frac{2}{3}\pi n\right) \qquad (2.107)$$

Here F/F_S is equal to 1/3. It is a rational number, so the DT sinusoid is periodic. If we put F_S as 7.5 Hz, F turns out to be 2.5. Here, the 10 Hz signal is aliased as a 2.5 Hz signal. We conclude that unless the value of F_S is known, the analog signal cannot be reconstructed from the signal samples. (Note that the DT signals in Examples 2.56 and 2.57 have the period N of 3 samples. However, both have different sampling frequencies namely 30 Hz and 7.5 Hz respectively. In Example 2.57, the recovered signal is aliased.)

Example 2.58

(Case 3) This example illustrates case 3.

Let us consider the same signal again with a bandwidth of 10 Hz. Here the sampling frequency as per the sampling theorem is 20 Hz. We will sample the signal with a sampling frequency above the Nyquist rate, that is, $F_S = 30$ Hz. Let us concentrate on $10/\pi$ Hz sinusoid. The analog sinusoidal signal of $10/\pi$ Hz is periodic. When sampled with a sampling frequency of 30 Hz, the resulting DT signal will turn out to be aperiodic. The DT signal is

$$x[n] = A\cos\left(2\pi \frac{10}{\pi} \frac{n}{30}\right) = A\cos\left(\frac{2}{3}n\right) \quad (2.108)$$

To test if it is periodic we have to check if f is rational.
Comparing Eq. (2.107) with Eq. (2.108) we get

$$\frac{2}{3}n = 2\pi f n$$

$$f = \frac{1}{3\pi} \quad (2.109)$$

Since f is not rational this implies that the analog signal of frequency $10/\pi$ that was periodic is now converted in DT domain to an aperiodic signal.

Ideally, one may consider the sampling frequency as $90/\pi$, then the DT signal will become

$$x[n] = A\cos\left(2\pi \frac{10}{\pi} \frac{n\pi}{90}\right) = A\cos\left(\frac{2}{9}\pi n\right) \quad (2.110)$$

To test if it is periodic we have to check if f is rational.
Comparing Eq. (2.110) with Eq. (2.107) we get

$$\frac{2}{9}\pi n = 2\pi f n \Rightarrow f = \frac{1}{9} \quad (2.111)$$

Since f is rational, the DT signal will be periodic. Only problem is practically, a sampling frequency of $90/\pi$ will not be possible.

The reader can note the additional points stated below.

1. *A very small change in frequency of the analog signal may drastically change the fundamental period of discrete time signal.*

Let us represent the frequency of the digital signal as f and frequency of the analog signal as F. Let F_s represent the sampling frequency as decided by the sampling theorem. Then $f = \dfrac{F}{F_s}$. (For details of sampling and sampling theorem, the reader may refer to chapter 1 on sampling.)

To illustrate this let us consider one numerical example.

Example 2.59

Consider a sinusoid of frequency 10 Hz sampled using a sampling frequency of 30 Hz. The signal is represented by $x[n] = A\cos\left(2\pi \dfrac{10}{30} n\right) = A\cos\left(2\pi \dfrac{1}{3} n\right)$ by putting value of $F = 10$ Hz and $F_s = 30$ Hz with $f = 1/3$ (note that we have made F and F_s relatively prime) and has a period of three samples. Now change the frequency of a sinusoid by 1 Hz. Let it be 11 Hz (for example). Sample it using a sampling frequency F_s of 30 Hz. The DT signal frequency f is 11/30 and the period of DT signal is 30.

2. *A very small change in the sampling frequency can result in a large change in fundamental period of DT signal. This fact is illustrated in the numerical example below.*

Example 2.60

Consider a sinusoid of frequency 10 Hz sampled using a sampling frequency of 30 Hz. The signal is represented with $f = 1/3$ as $f = \dfrac{F}{F_s}$ (note that we have made F and F_s relatively prime) and has a period of three samples. Now, change the sampling frequency by 1 Hz. Let it be 31 Hz (for example). The discrete time signal frequency $f = 10/31$ and the period of DT signal is 31.

Let us go through additional examples for better understanding of the concept of periodicity.

Example 2.61

Consider the signal $x[n] = \cos(0.01 n\pi)$. Is $x[n]$ a periodic signal?

We have to check if

$x[n] = x[n + N]$ for some integer N

That is

$$x[n] = \cos(0.01 n\pi) = \cos(0.01\pi(n + N)) \tag{2.112}$$

i.e., $2\pi \dfrac{1}{200} N = 2\pi k \Rightarrow \dfrac{1}{200} = \dfrac{k}{N} \Rightarrow$ period is $N = 200$ \qquad (2.113)

As k and N are relatively prime, the fundamental period of the sinusoid is N = 200 samples.

Example 2.62

Consider the signal $x[n] = \cos\left(\dfrac{30n}{105}\pi\right)$. Is $x[n]$ a periodic signal?
We have to check if

$x[n] = x[n + N]$ for some integer N

That is

$$x[n] = \cos\left(\frac{30}{105}n\pi\right) = \cos\left(\frac{30}{105}\pi(n+N)\right) \qquad (2.114)$$

i.e., $2\pi\dfrac{30}{210}N = 2\pi k \Rightarrow \dfrac{1}{7} = \dfrac{k}{N} \Rightarrow$ period is $N = 7$ \qquad (2.115)

As k and N are relatively prime, the fundamental period of the sinusoid is N = 7.

Example 2.63

Consider the signal $x[n] = \cos(3\pi n)$. Is $x[n]$ a periodic signal?
We have to check if

$x[n] = x[n + N]$ for some integer N

That is

$$x[n] = \cos(3n\pi) = \cos(2\pi n + \pi n) = \cos(\pi n) = \cos(\pi(n+N)) \qquad (2.116)$$

i.e., $2\pi\dfrac{1}{2}N = 2\pi k \Rightarrow \dfrac{1}{2} = \dfrac{k}{N} \Rightarrow$ period is $N = 2$ \qquad (2.117)

As k and N are relatively prime, the fundamental period of the sinusoid is N = 2.

Example 2.64

Consider the signal $x[n] = \sin(3n)$. Is $x[n]$ a periodic signal?
We have to check if

$x[n] = x[n + N]$ for some integer N

That is

$$x[n] = \sin(3n) = \sin(3(n+N)) \qquad (2.118)$$

i.e., $2\pi\dfrac{3}{2\pi}N = 2\pi k \Rightarrow \dfrac{3}{2\pi} = \dfrac{k}{N}$ \qquad (2.119)

As k and N are not relatively prime, hence the signal is aperiodic.

Example 2.65

Consider the signal $x[n] = \sin\left(\dfrac{62n}{10}\pi\right)$. Is $x[n]$ a periodic signal?

We have to check if

$x[n] = x[n + N]$ for some integer N

That is

$$x[n] = \sin\left(\frac{62n}{10}\pi\right) = \sin\left(6n\pi + \frac{2}{10}n\pi\right)$$

(2.120)

$$= \sin\left(\frac{2}{10}n\pi\right) = \sin\left(\frac{2}{10}\pi(n+N)\right)$$

i.e., $2\pi\dfrac{2}{20}N = 2\pi k \Rightarrow \dfrac{2}{20} = \dfrac{k}{N} = \dfrac{1}{10}$ (2.121)

As k and N are relatively prime, hence, the period of the signal is $N = 10$.

Example 2.66

Consider the signal $x[n] = 3\cos\left(5n + \dfrac{\pi}{6}\right)$. Is $x[n]$ a periodic signal?

We have to check if

$x[n] = x[n + N]$ for some integer N

That is

$$x[n] = 3\cos\left(5n + \frac{\pi}{6}\right) = 3\cos\left(5(n+N) + \frac{\pi}{6}\right)$$

(2.122)

i.e., $2\pi\dfrac{5}{2\pi}N = 2\pi k \Rightarrow \dfrac{5}{2\pi} = \dfrac{k}{N}$ (2.123)

As k and N are not relatively prime, hence, the signal is aperiodic.

Example 2.67

Consider the signal $x[n] = 2\exp\left(j\left(\dfrac{n}{6} - \pi\right)\right)$. Is $x[n]$ a periodic signal?

We have to check if

$x[n] = x[n + N]$ for some integer N

That is

$$x[n] = 2\exp\left(j\left(\frac{n}{6} - \pi\right)\right) = 2\exp\left(j\left(\frac{(n+N)}{6} - \pi\right)\right) \qquad (2.124)$$

i.e., $2\pi \dfrac{1}{12\pi} N = 2\pi k \Rightarrow \dfrac{1}{12\pi} = \dfrac{k}{N}$ \qquad (2.125)

As k and N are not relatively prime, hence, the signal is aperiodic.

Example 2.68

Consider the signal $x[n] = \cos\left(\dfrac{\pi}{8}\right)\sin\left(\dfrac{n\pi}{8}\right)$. Is $x[n]$ a periodic signal?

We have to check if

$x[n] = x[n + N]$ for some integer N

That is

$$x[n] = \cos\left(\frac{\pi}{8}\right)\sin\left(\frac{n\pi}{8}\right) = \cos\left(\frac{\pi}{8}\right)\sin\left(\frac{(n+N)\pi}{8}\right) \qquad (2.126)$$

i.e., $2\pi \dfrac{1}{16} N = 2\pi k \Rightarrow \dfrac{1}{16} = \dfrac{k}{N}$ \qquad (2.127)

As k and N are relatively prime, hence, the signal is periodic with period $N = 16$.

Linear combination of signals

Example 2.69

Consider the signal $x[n] = \cos\left(\dfrac{n\pi}{2}\right) - \sin\left(\dfrac{n\pi}{8}\right) + 3\cos\left(\dfrac{n\pi}{4} + \dfrac{\pi}{3}\right)$. Is $x[n]$ a periodic signal?

We have to check if

$x[n] = x[n + N]$ for some integer N for all 3 terms.

That is

$$\begin{aligned} x[n] &= \cos\left(\frac{n\pi}{2}\right) - \sin\left(\frac{n\pi}{8}\right) + 3\cos\left(\frac{n\pi}{4} + \frac{\pi}{3}\right) \\ &= \cos\left(\frac{(n+N)\pi}{2}\right) - \sin\left(\frac{(n+N)\pi}{8}\right) + 3\cos\left(\frac{(n+N)\pi}{4} + \frac{\pi}{3}\right) \end{aligned} \qquad (2.128)$$

i.e., for the first term, $2\pi \dfrac{1}{4} N = 2\pi k \Rightarrow \dfrac{1}{4} = \dfrac{k}{N}$ \qquad (2.129)

As *k* and *N* are relatively prime, hence, the signal is periodic with period $N = 4$ for the first term.

i.e., for the second term, $2\pi \dfrac{1}{16} N = 2\pi k \Rightarrow \dfrac{1}{16} = \dfrac{k}{N}$ \hfill (2.130)

As *k* and *N* are relatively prime, hence, the signal is periodic with period $N = 16$ for the second term.

i.e., for the third term, $2\pi \dfrac{1}{8} N = 2\pi k \Rightarrow \dfrac{1}{8} = \dfrac{k}{N}$ \hfill (2.131)

As *k* and *N* are relatively prime, hence, the signal is periodic with period $N = 8$ for the second term. Considering the periods for all three terms, the period for the signal is the highest common divisor i.e., LCM value of *N* for all terms which is equal to 16.

Example 2.70
Consider a linear combination of two analog sinusoidal functions. $x(t) = 3\sin(4\pi t) + \sin(3\pi t)$. Find if the signal is periodic.

Solution
We will check if $x(t + T) = 3\sin(4\pi(t + T)) + \sin(3\pi(t + T)) = x(t)$ for some *T*.

$$x(t + T) = 3\sin(4\pi (t + T)) + \sin (3\pi (t + T))$$

$$= 3\sin(4\pi t + 4\pi T) + \sin(3\pi t + 3\pi T)$$

$3\sin(4\pi t + 4\pi T) = 3\sin(4\pi t)$ if $4\pi T = 2\pi \Rightarrow T = 1/2$ \hfill (2.132)

$\sin(3\pi t + 3\pi T) = \sin(3\pi t)$ if $3\pi T = 2\pi \Rightarrow T = 2/3$

common period *T* can be found using $\dfrac{M}{N} = \dfrac{1/2}{2/3} = \dfrac{3}{4}$

Where the symbols '*M*' and '*N*' stand for the period of the first term and the period of the second term, respectively.

Hence, the period for the combination signal can be found as follows.
The period for the linear combination of two terms is

$$4 \times M = 3 \times N = 4 \times \dfrac{1}{2} = 3 \times \dfrac{2}{3} = 2 \text{ seconds}$$

Example 2.71
Consider a linear combination of two analog sinusoidal functions. $x(t) = 3\cos(4\pi t) + \sin(5t)$. Find if the signal is periodic.

Solution

We will check if $x(t+T) = 3\cos(4\pi(t+T)) + \sin(5(t+T))$ for some T.

$$x(t+T) = 3\cos(4\pi(t+T)) + \sin(5(t+T))$$

$$= 3\cos(4\pi t + 4\pi T) + \sin(5t + 5T)$$

$$3\cos(4\pi t + 4\pi T) = 3\cos(4\pi t) \text{ if } 4\pi T = 2\pi \Rightarrow T = 1/2 \quad (2.133)$$

$$\sin(5t + 5T) = \sin(5T) \text{ if } 5T = 2\pi \Rightarrow T = 2\pi/5$$

common period T can be found using $\dfrac{M}{N} = \dfrac{1/2}{2\pi/5} = \dfrac{5}{4\pi}$

Where M and N stand for the period of the first term and the second term, respectively.

Hence, the period for the combination signal can be found as follows.
The period for the linear combination of two terms is

$$4\pi \times M = 5 \times N = 4\pi \times \frac{1}{2} = 5 \times \frac{2\pi}{5} = 2\pi \text{ seconds}. \quad (2.134)$$

The period is not a rational number. However, because it is an analog sinusoid, the combination signal is periodic.

Physical significance of property of periodicity When the signal is periodic, it repeats itself from minus infinity to plus infinity i.e., it extends over infinite duration. Hence, its total energy is infinite. If the average power is finite, it becomes a power signal. To find the frequency contents of the signal, we need to find power spectral density (PSD). If the signal is aperiodic and if it exists over a finite duration, it is an energy signal. We need to find the energy spectrum density (ESD) to analyze the frequency contents of the signal. (The reader may refer to chapter 9 on random signals for ESD and PSD.)

2.6.3 Causal and non-causal signals

The CT and DT signals are characterized as causal and non-causal signals. They are also called right-handed and left-handed signals and sequences, respectively. The CT signal is said to be causal if $x(t) = 0$ for all $t < 0$. It is also called as a right-handed signal. The CT signal is said to be anti-causal if it exists only for negative values of time. It is also called as a left-handed signal. The sequence $x[n]$ is called as a right-handed sequence if it is zero for all $n < 0$. The right-handed sequence is said to be a causal sequence. The sequence $x[n]$ is called as a left-handed sequence if it exists only for $n < 0$. The left-handed sequence is said to be a non-causal sequence. Non-causal sequences are basically the signal samples obtained off-line and are useful for off-line systems when the signal ahead of time is required to be known.

Example 2.72
Consider a CT signal given by $x(t) = e^{-2t}$ for $t \geq 0$, is it a causal signal?

Solution
The signal exists only for $t \geq 0$, and is equal to zero for all $t < 0$. According to the definition of causality, the signal is causal and is also called as right-handed sequence.

Example 2.73
Consider a CT signal given by $x(t) = e^{3t}$ for $t < 0$, is it a causal signal?

Solution
The signal exists only for $t < 0$, and is equal to zero for all $t \geq 0$. According to the definition of causality, the signal is non-causal and is also called as left-handed sequence.

Example 2.74
Consider a CT signal given by $x(t) = e^{2t}u(t-1)$. Find if the signal is causal.

Solution
Let us first write the definition of u(t–1).

$$u(t-1) = \begin{cases} 1 & \text{for } t \geq 1 \\ 0 & \text{otherwise} \end{cases} \quad (2.135)$$

As the signal is appended by u(t–1), it exists for positive values of $t \geq 1$, and is zero for all $t < 0$. Hence, it is causal.

Example 2.75
Consider a signal given by $x(t) = 3\sin c(5t)$. Find if the signal is causal.

Solution
The sinc function exists from minus infinity to infinity. Hence, the signal exists for negative values of t and is anti-causal.

Example 2.76
Consider a CT signal $x(t) = e^{2t}[u(t+2) - u(t-2)]$. Find if the signal is causal.

Solution
Let us write the definitions of the u functions.

$$u(t-2) = \begin{cases} 1 & \text{for } t \geq 2 \\ 0 & \text{otherwise} \end{cases} \quad (2.136)$$

$$u(t+2) = \begin{cases} 1 & \text{for } t \geq -2 \\ 0 & \text{otherwise} \end{cases} \quad (2.137)$$

The signal is a both-sided signal. The signal exists for negative values of t as well and hence, it is anti-causal.

We will now consider the examples of DT signals.

Example 2.77

Consider a unit step sequence $u[n]$. Find if it is causal.

Solution

The sequence is defined as

$$u[n] = \begin{cases} 1 & \text{for all } n \geq 0 \\ 0 & \text{otherwise} \end{cases} \quad (2.138)$$

This is a right-handed sequence and is a causal sequence.

Example 2.78

Consider the following sequence

$$u[-n-1] = \begin{cases} 1 & \text{for all } (-n-1) \geq 0 \text{ i.e., } n < -1 \\ 0 & \text{otherwise} \end{cases} \quad (2.139)$$

Solution

The signal exists for negative values of n. This is a left-handed and a non-causal sequence. The sequence can also be a both-sided sequence; however, then it is a non-causal sequence.

Example 2.79

Consider a DT signal $x[n] = \left(\frac{1}{4}\right)^n u[n+2] - \left(\frac{1}{2}\right)^n u[n-2]$. Find if the signal is causal.

Solution

Let us write the definitions of the unit step functions.

$$u[n-2] = \begin{cases} 1 & \text{for } n \geq 2 \\ 0 & \text{otherwise} \end{cases} \quad (2.140)$$

$$u[n+2] = \begin{cases} 1 & \text{for } n \geq -2 \\ 0 & \text{otherwise} \end{cases} \quad (2.141)$$

The sequence exists from $n = -2$ to $n = 1$. It exists for negative values of n as well and is anti-causal.

Example 2.80

Consider a signal $u[-n]$. Find if the signal is causal.

Solution

Let us write the definition of the function.

$$u[-n] = \begin{cases} 1 & \text{for } -n \geq 0 \text{ i.e., } n < 0 \\ 0 & \text{otherwise} \end{cases} \qquad (2.142)$$

The signal exists for all negative values of n. Hence, the sequence is anti-causal.

Physical significance of causality When a signal is causal, it extends from $t = 0$ /$n = 0$ onwards. When the signal is non-causal, it exists for all negative values of t. the signal can also be both-sided. When the signal values are acquired from $t = 0$ / $n = 0$ onwards, it is said to be real time signal and when we have signal values also available for negative values of time, it is said to be off line signal.

2.6.4 Deterministic and random signals

For a deterministic signal, there is no uncertainty with respect to its value at any instant t. The signal can be specified by writing a mathematical expression. For example, a signal represented as $x(t) = \sin(2\pi ft)$ is a deterministic signal. Using the equation, it is possible to determine the signal value at any time instant. On the other hand, unlike deterministic signals, random signals take different values with different probabilities at any point in time. *Almost all naturally occurring signals* **contain random signal of some strength**.

In case of random signals, the next sample cannot be predicted with precision. Researchers are putting forth different models for the random signals like speech that allow us to write the mathematical expression for a short duration signal. So, even though we write mathematical expression for a random variable, it is after modeling it for a short duration signal. The mathematical model only fits closely to the signal and is not the exact representation. The reader will understand the details in chapter 9 on random signals.

Example 2.81

Consider the equation given by $x(t) = t + \sin(4\pi ft) + \cos(2\pi ft)$. Is $x(t)$ a deterministic signal ?

Solution

The equation for the signal is provided. We can use this equation to find the value of the signal at any time t. Hence, the signal is deterministic.

Example 2.82

Consider the signal $x[n] = \cos\left(\dfrac{n\pi}{2}\right) + 3\cos\left(\dfrac{n\pi}{4} + \dfrac{\pi}{3}\right)$. Is x[n] a deterministic signal?

Solution

The equation for the signal is provided. We can use this equation to find the value of the signal at any time instant n. Hence, the signal is deterministic.

Note Any signal for which we can write a mathematical equation to find the value of a signal at any time t or any time instant n is a deterministic signal. Researchers are putting forth different models for different random signals like speech that allow them to write the mathematical expression for a short duration signal. Almost all naturally occurring signals have some random component of signal. *The natural signals are like elephants. The blind persons (researchers) around the elephant try to describe the elephant by touching some body parts. If the researcher analyzes the tail part of the elephant, he/she will put forth a model describing the tail of the elephant and so on. No model so far exists for the entire elephant!!*

Physical significance of deterministic and random property When the signal is deterministic, the next samples are easily calculable. This allows us to find all statistical properties like mean and variance, etc. When the signal is a random signal, further signal samples cannot be exactly predicted. We need to only estimate the statistical properties. (The reader may refer to the chapter on random signals -chapter 9)

2.6.5 Energy and power signals

Let us now define the measure to find the size of the signal. The size of the signal can be measured in terms of its energy because it is a square quantity and hence, is positive. It in a way defines the strength of the signal. For a signal $x(t)$, x represents the amplitude or magnitude or value of the signal. The instantaneous power is given by $|x(t)|^2$. The instantaneous power integrated over its duration gives the total energy. The generalized format for representing the total energy of a signal $x(t)$ is defined as

$$E = \int_{-\infty}^{\infty} |x(t)|^2\, dt \qquad (2.143)$$

or

$$E = \lim_{T \to \infty} \int_{-T}^{T} |x(t)|^2\, dt \qquad (2.144)$$

In order that the total energy is measurable, the signal value must tend to zero as the T approaches positive or negative infinity, otherwise the integral will not converge.

Definition of energy signal The signals that have finite or measurable energy are termed as energy signals. This means that the energy signal must exist over a finite duration. It is indicating the amount of energy that can be extracted from the signal when the load is applied.

Power is the energy per unit time. Now, to calculate power over a period (2T), we divide the energy over a period $2T$ by $2T$. The average power is then calculated as the limit of power over the period as the period tends to infinity. Mathematically, this is given by

$$P = \lim_{T \to \infty} \frac{1}{2T} \int_{-T}^{T} |x(t)|^2 \, dt \tag{2.145}$$

Definition of instantaneous power of the signal Instantaneous power is defined as the power of the signal at a particular instant. It is defined as $P_x(t) = |x(t)|^2$.

The definition of the power divides the energy by T and T tends to infinity. Hence, energy signals have finite energy and zero average power.

For some signals the amplitude of the signal does not go to zero as time t tends to infinity. Here, the total energy becomes infinite. The meaningful measure of the size of such signals is the time average of energy i.e., the power of the signal. Such signals extending up to infinity have infinite energy. If the average power is finite, then it is termed as the power signal.

Definition of Power signal The signals that have infinite or non-measurable energy and finite average power are termed as power signals. This means that the power signal must exist over infinite duration. In order to have finite average power, the signal must be periodic or it must have statistical regularity.

If the average power does not exist, then the signal is neither energy nor a power signal. Signals for which both energy and average power are infinite are neither energy signals nor power signals. Example: White noise (idealized form of noise, the power spectral density of which is independent of the operating frequency). The adjective white is used in the sense that white light contains equal amount of all frequencies within the visible band of electromagnetic radiation. Periodic signals are power signals. They exist over infinite time and hence, total energy of the signal is infinite. The average power for a period is finite. Usually, random signals are also power signals. For example, a sinusoidal signal is a power signal and a short-lived aperiodic signal is an energy signal.

Let us define energy and power for a DT signal. In case of DT signals, the integral in the analog signal equation will be replaced by the summation. The total energy of the signal is defined as

$$E = \sum_{n=-\infty}^{\infty} x^2[n] \tag{2.146}$$

The average power and instantaneous power $P_x(t)$ are defined as

$$P = \lim_{N \to \infty} \frac{1}{2N} \sum_{n=-N}^{N} x^2[n] \qquad (2.147)$$

$$P_X(t) = |x[n]|^2$$

Example 2.83

Consider the sinusoid of frequency 1 KHz. Is it a power signal?

Solution

The signal is a sinusoid that exists from $-\infty$ to ∞. Hence, the energy of a signal is infinite. Let us calculate the average power. The period of a signal $2T$ is $1/1\text{KHz} = 0.001$. So, $T = 0.0005$

Using Eq. (2.145) we get

$$P = 1000 \int_{-0.0005}^{0.0005} |\sin(2\pi 1000 t)|^2 dt = 500 \int_{-0.0005}^{0.0005} |1 - \cos(4\pi 1000 t)| dt \qquad (2.148)$$

$$P = \frac{500}{1}\left[t - \sin(4\pi\, 1000 t)/4\pi \times 1000\right]_{-0.0005}^{0.0005} = 500 \times 2 \times 0.0005 = 0.5 \qquad (2.149)$$

Here, the sin term integrated over one complete cycle gives zero.

The average power is finite. The signal is a power signal.

Example 2.84

If $x(t) = \sin(2\pi 100 t)$ for $0 \le t \le 1/2$, is $x(t)$ an energy signal?

Solution

The signal is a sinusoid existing only over a finite period between $t = 0$ to $\tfrac{1}{2}$. Hence, the energy of a signal is finite and the signal is an energy signal. The period of the signal is $T = \dfrac{1}{f} = \dfrac{1}{100} = 0.01$ seconds. Let us find the total energy of the signal.

$$E = \int_{t=0}^{0.5} \sin^2(2\pi 100 t)\, dt$$

$$= \frac{1}{2}\int_0^{0.5}(1 - \cos(2\pi 200 t))\, dt$$

$$= \frac{1}{2}\left[t\Big|_0^{0.5} - \frac{\sin(2\pi 200 t)}{400\pi}\Big|_0^{0.5}\right] \qquad (2.150)$$

$$= \frac{1}{2}[0.5 - 0] = \frac{1}{4}$$

Let us find the average power of the signal.

$$P = \lim_{T\to\infty} \frac{1}{T} \int_0^{0.01} |\sin^2(2\pi 100t)| \, dt$$

$$= \lim_{T\to\infty} \frac{1}{2T} \int_0^{0.01} |1 - \cos(2\pi 200t)| \, dt \qquad (2.151)$$

$$= \lim_{T\to\infty} \frac{1}{2T} \left[t \Big\downarrow_0^{0.01} - \frac{\sin(2\pi 100t)}{400\pi} \Big\downarrow_0^{0.01} \right] = \lim_{T\to\infty} \frac{1}{0.02}[0.01] = 0.5$$

The average power is finite. The signal is a power signal over the interval it is defined.

Example 2.85

Consider a signal defined as

$$x(t) = \begin{cases} t & \text{for } 0 \leq t \leq 1 \\ 2-t & \text{for } 1 \leq t \leq 2 \\ 0 & \text{otherwise} \end{cases}$$

Find the energy and power of the signal and classify it as an energy or a power signal.

Solution

Let us find the total energy of the signal.

$$E = \int_{t=0}^1 t^2 \, dt + \int_{t=1}^2 (2-t)^2 \, dt \qquad (2.152)$$

$$E = \frac{t^3}{3} \Big\downarrow_{t=0}^{t=1} + 4t \Big\downarrow_{t=1}^{t=2} + \frac{t^3}{3} \Big\downarrow_{t=1}^{t=2} - 2t^2 \Big\downarrow_1^2$$

$$= \frac{1}{3} + 4 + \frac{7}{3} - 6 \qquad (2.153)$$

$$= \frac{8}{3} - 2 = \frac{2}{3}$$

The total energy is finite. So, the signal is an energy signal. The signal exists only for a finite duration.

Let us find the average power of the signal.

$$P = \lim_{T\to\infty} \frac{1}{2T} \int_{-T}^{T} |x(t)|^2 \, dt = \lim_{T\to\infty} \frac{1}{2T}\left[\int_0^1 t^2 dt + \int_1^2 (2-t)^2 dt\right]$$

$$= \lim_{T\to\infty} \frac{1}{2T}\left[t^3/3 \Big\downarrow_0^1 + 4t \Big\downarrow_1^2 - 2t^2 \Big\downarrow_1^2 + t^3/3 \Big\downarrow_1^2\right] \tag{2.154}$$

$$= \lim_{T\to\infty} \frac{1}{2T}\left[\frac{1}{3} + 4 - 6 + 7/3\right] = \lim_{T\to\infty} \frac{1}{2T} \times \frac{2}{3} = 0$$

The average power of the signal is zero. Hence, the signal is an energy signal.

Example 2.86
Consider a signal defined as

$$x(t) = \begin{cases} 3\cos(\pi t) & \text{for } -1 \le t \le 1 \\ 0 & \text{otherwise} \end{cases}$$

Find the energy and the power of the signal and classify it as an energy or a power signal.

Solution
Let us find the total energy of the signal. The frequency of the signal is ½ and period is 2.

$$E = \int_{t=-1}^{1} 9\cos^2(\pi t)\,dt = \frac{9}{2}\int_{-1}^{1}[1+\cos(2\pi t)]\,dt \tag{2.155}$$

$$E = \frac{9}{2}[t \Big\downarrow_{-1}^{1} + \sin(2\pi t)/2\pi \Big\downarrow_{t=-1}^{t=1}$$

$$= \frac{9}{2}[2 + (\sin(2\pi)/2\pi) - (\sin(-2\pi)/2\pi)]$$

$$= 9$$

and $P_X = \lim_{T\to\infty} \frac{9}{T} = 0$ \hfill (2.156)

The total energy is finite. So, the signal is an energy signal. Signal exists only for a finite duration. Hence, the period tends to infinity and average power is zero.

Example 2.87

Consider a signal defined as

$$x(t) = \begin{cases} 5 \text{ for } -2 \leq t \leq 2 \\ 0 \text{ otherwise} \end{cases}$$

Find the energy and the power of the signal and classify it as an energy or a power signal.

Let us first find the instantaneous power of the signal.

Instantaneous power is given by

$$P_X(t) = |x(t)|^2 = 25 \text{ for } -2 \leq t \leq 2 \quad (2.157)$$

$$= 0 \text{ otherwise}$$

The total energy of the signal is given by

$$E = \int_{-2}^{2} 25 \, dt = 25 \times t \Big|_{-2}^{2} = 25 \times 4 = 100 \quad (2.158)$$

The average power is given by

$$P_X = \lim_{T \to \infty} \frac{100}{T} = 0 \quad (2.159)$$

The total energy is finite and average power is zero. Hence, the signal is an energy signal.

Example 2.88

Consider a signal given by

$$x(t) = \{5\cos(\pi t) + \sin(\pi t) \text{ for } -\infty \leq t \leq \infty\}$$

Find the energy and the power of the signal and classify it as an energy or a power signal.

The total energy of the signal is given by

$$E = \int_{-\infty}^{\infty} \left[25 \cos^2(\pi t) + \sin^2(\pi t) \right] dt$$

$$= \int_{-\infty}^{\infty} [12.5(1 + \cos(2\pi t)) + [0.5(1 - \cos(2\pi t))] dt \quad (2.160)$$

$$= \infty$$

The period of the signal is

$$T = \frac{1}{f}, \quad 2\pi ft = \pi t, \quad f = \frac{1}{2}, \quad T = 2. \qquad (2.161)$$

The average power is given by

$$P_X = \frac{E_X}{2} = (25+1)/2 = 13 \qquad (2.162)$$

$$P_X = \frac{1}{2}\int_0^2 \left[25\cos^2(\pi t) + \sin^2(\pi t)\right] dt$$

$$= \frac{1}{2}\int_0^2 [12.5(1+\cos(2\pi t)) + [0.5(1-\cos(2\pi t))] dt$$

$$= \frac{1}{2}\left[12.5\left(t \downarrow_0^2 + \frac{\sin(2\pi t)}{2\pi} \downarrow_0^2\right)\right] + \left[0.5\left(t \downarrow_0^2 - \frac{\cos(2\pi t)}{2\pi} \downarrow_0^2\right)\right] \qquad (2.163)$$

$$= \frac{1}{2}[25 + 0 + 1 + 0] = 13$$

The signal has infinite energy and finite average power. Hence, the signal is a power signal.

Example 2.89

Consider a periodic DT sinusoid given by

$$g[k] = 3\cos\left(\frac{\pi k}{10}\right)$$

Find if the signal is an energy signal or a power signal.

Solution

Let us first find the period of the signal.
 We have to check if

$g[n] = g[n + N]$ for some integer N, i.e.,

$$g[k+N] = 3\cos\left(\frac{\pi k}{10} + \frac{\pi N}{10}\right) \qquad (2.164)$$

i.e., $2\pi \dfrac{1}{20} N = 2\pi k' \Rightarrow \dfrac{1}{20} = \dfrac{k'}{N}$ \qquad (2.165)

As k' and N are relatively prime, the signal is periodic with a period $N = 20$.

The signal exists for all k. Hence, the total energy of the signal is infinite. Let us find the average power for one cycle.

$$P = \frac{1}{20}\sum_{k=0}^{19} 9\cos^2\left(\frac{\pi k}{10}\right) = \frac{9}{20}\sum_{k=0}^{19}\frac{1}{2}\left[1+\cos\left(\frac{2\pi k}{10}\right)\right] = \frac{9}{20}\frac{20}{2} = 4.5 = (3)^2/2 \quad (2.166)$$

3 – peak amplitude of the signal

Note The summation of the cos term over one period is zero. We find that the average power is finite. Hence, the signal is a power signal.

Example 2.90

Consider an analog periodic signal sketched in Fig. 2.59. Find if the signal is an energy signal or a power signal?

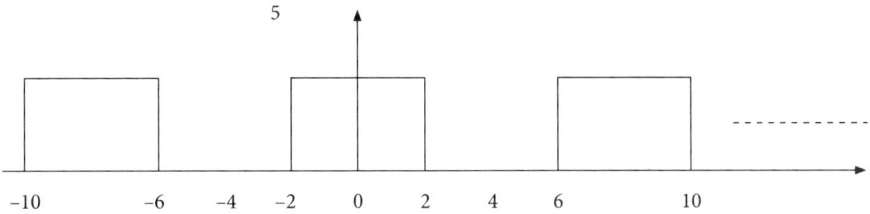

Fig. 2.59 Plot of signal for Example 2.90

Solution

The figure shows that the signal is a periodic signal with period from -4 to $+4$ i.e., period is 8. The signal varies from minus infinity to plus infinity. So, the total energy of the signal will be infinity. Let us find the average power of the signal. The average power is given by

$$P_X = \frac{1}{8}\int_{-2}^{+2} 25\,dx = \frac{1}{8}\times 25\times x\Big|_{-2}^{2} = \frac{25}{8}\times 4 = \frac{25}{2} = 12.5. \quad (2.167)$$

The average power is finite. So, the signal is a power signal.

Example 2.91

Consider an analog periodic signal sketched in Fig. 2.60. Find if the signal is an energy signal or a power signal?

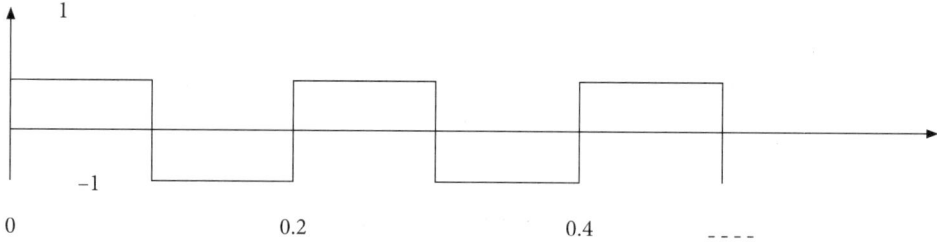

Fig. 2.60 Plot of signal for Example 2.91

Solution

The figure shows that the signal is a periodic signal with period from 0 to 0.2 i.e., period is 0.2 seconds. The signal varies from minus infinity to plus infinity. So, the total energy of the signal will be infinity. Let us find the average power of the signal. The average power is given by

$$P_X = \frac{1}{0.2}[\int_0^{0.1} 1 \times dx + \int_{0.1}^{0.2} (-1)^2 \times dx = \frac{1}{0.2} \times [0.1 + 0.1] = 1. \qquad (2.168)$$

The average power is finite. So, the signal is a power signal.

Example 2.92

Consider an analog periodic signal, a triangular wave sketched in Fig. 2.61 existing between minus infinity and plus infinity. Find if the signal is an energy signal or a power signal?

Solution

The figure shows that the signal is a periodic signal with period from 0 to 0.2 i.e., period is 0.2 seconds. The signal varies from minus infinity to plus infinity. So, the total energy of the signal will be infinity. Let us find the average power of the signal. The average power is given by (the slope of the straight line between 0 to 0.1 is 20 and that between 0.1 to 0. is −20)

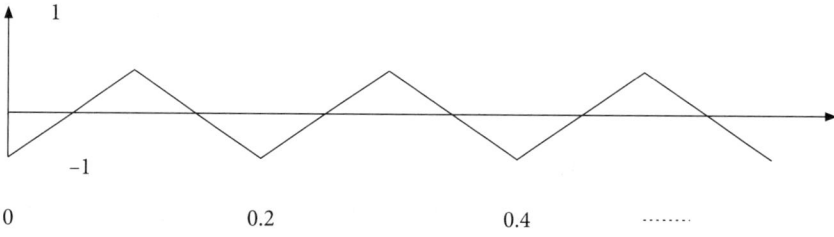

Fig. 2.61 Plot of signal for Example 2.92

$$P_X = \frac{1}{0.2}\left[\int_0^{0.1}[20x-1]^2\,dx + \int_{0.1}^{0.2}[3-20x]^2\right]$$

$$= \frac{1}{0.2}\left[\int_0^{0.1}(400x^2-40x+1)dx + \int_{0.1}^{0.2}(9-120x+400x^2)dx\right]$$

$$= \frac{1}{0.2}\left[400\frac{x^3}{3}\Big\downarrow_0^{0.1} -40\frac{x^2}{2}\Big\downarrow_0^{0.1} +x\Big\downarrow_0^{0.1} +9x\Big\downarrow_{0.1}^{0.2} -120\frac{x^2}{2}\Big\downarrow_{0.1}^{0.2}\right.$$

$$\left.+400\frac{x^3}{3}\Big\downarrow_{0.1}^{0.2}\right] \quad (2.169)$$

$$= \frac{1}{0.2}\left[\frac{0.4}{3} - 0.2 + 0.1 + 0.9 - \frac{3.6}{2} + \frac{2.8}{3}\right]$$

$$= \frac{1}{0.2} \times \frac{0.2}{3} = \frac{1}{3}$$

The average power is finite. So, the signal is a power signal.

Example 2.93

Consider a DT periodic signal, as shown in Fig. 2.62. Find if the signal is an energy signal or a power signal? Find the average power.

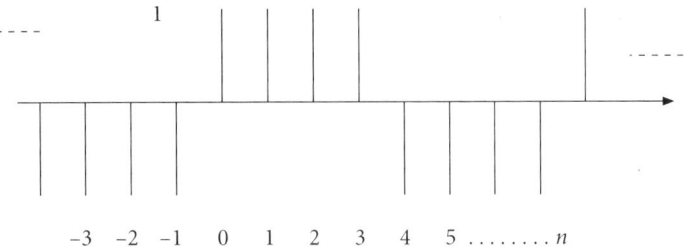

Fig. 2.62 Plot of signal for Example 2.93

Solution

The signal is periodic with period equal to 8 samples. The signal extends over infinite duration. Hence, it has infinite energy. The average power is the power for one period.

$$P = \frac{1}{8}\sum_{n=0}^{n=7} 1 = 1. \quad (2.170)$$

Example 2.94

Consider a DT periodic signal, as shown in Fig. 2.63. Find if the signal is an energy signal or a power signal?

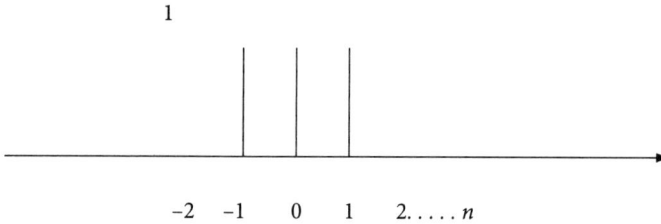

Fig. 2.63 Plot of signal for Example 2.94

Solution

The signal exists only for a finite duration. Hence, the signal is the energy signal. Let us find the total energy of the signal.

$$E = \sum_{n=-1}^{n=1} 1 = 3. \tag{2.171}$$

The period of the signal is infinity. The average power is given by

$$P = \frac{1}{T}\sum_{n=-1}^{n=1}(1)^2 = \frac{3}{\infty} = 0. \tag{2.172}$$

Example 2.95

Consider a signal defined as

$$x[n] = \begin{cases} n \text{ for } 0 \leq n \leq 5 \\ 10 - n \text{ for } 6 \leq n \leq 10 \\ 0 \text{ otherwise} \end{cases}$$

Find the energy and power of the signal and classify it as an energy or a power signal.

Solution

Let us plot the signal first. The plot of the signal is shown in Fig. 2.64. The total energy of the signal is given by

$$E = \sum_{0}^{5} n^2 + \sum_{5}^{10}(10-n)^2 = 1+4+9+16+25+16+9+4+1 = 85 \tag{2.173}$$

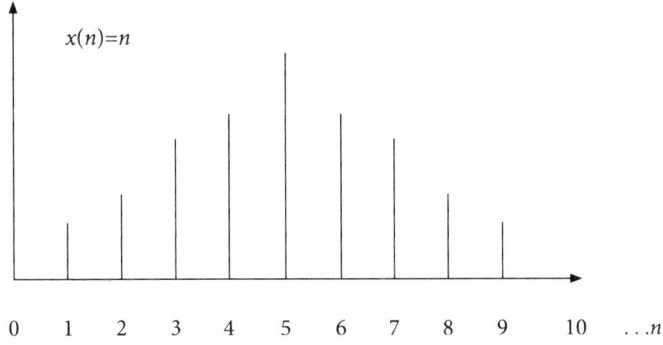

Fig. 2.64 Plot of signal for Example 2.95

Energy is finite. Hence, signal is an energy signal.
The power of the signal is zero as the period is infinite.

Example 2.96

Consider a signal defined as

$$x[n] = \begin{cases} \cos(\pi n) \text{ for } -4 \leq n \leq 4 \\ 0 \text{ otherwise} \end{cases}$$

Find the energy and power of the signal and classify it as an energy or a power signal.

Solution

Let us plot the signal first. The period of the signal is

$$2\pi \frac{1}{2} N = 2\pi k \Rightarrow \frac{k}{N} = \frac{1}{2}, \text{period is 2 samples.} \tag{2.174}$$

The total energy of the signal is given by

$$E = \sum_{n=-4}^{4} \cos^2(\pi n) = 1+1+1+1+1+1+1+1+1 = 9. \tag{2.175}$$

The energy is finite and the signal is an energy signal. The average power is zero as the signal extends only over a small duration and hence, the period tends to infinity.

$$P_X = \lim_{N\to\infty} \frac{1}{N} \sum_{n=0}^{1} \cos^2(\pi n) = 0 \qquad (2.176)$$

The signal is the energy signal.

Example 2.97

Consider a signal defined as

$$x(n) = \begin{cases} \sin(\pi n) & \text{for } -4 \le n \le 4 \\ 0 & \text{otherwise} \end{cases}$$

Find the energy and power of the signal and classify it as an energy or a power signal.

Solution

The period of the signal is

$$2\pi \frac{1}{2} N = 2\pi k \Rightarrow \frac{k}{N} = \frac{1}{2}, \text{period is 2 samples.} \qquad (2.177)$$

The total energy of the signal is given by

$$E = \sum_{n=-4}^{4} \sin^2(\pi n) = 0.$$

The energy is finite and the signal is an energy signal.

The average power is the power over one period

$$P_X = \lim_{N\to\infty} \frac{1}{N} \sum_{n=0}^{1} \sin^2(\pi n) = 0 \qquad (2.178)$$

Example 2.98

Consider a trapezoidal signal given by

$$X(t) = \begin{cases} t+5 & -5 \le t \le -4 \\ 1 & -4 \le t \le 4 \\ 5-t & 4 \le t \le 5 \\ 0 & \text{otherwise} \end{cases}$$

Find the total energy of the signal.

Solution

The total energy is given by

$$E = \int_{-5}^{-4}(25+10t+t^2)dt + \int_{-4}^{4} 1 \times dt + \int_{4}^{5}(25-10t+t^2)dt$$

$$= 25 + 10\frac{t^2}{2}\Big|_{-5}^{-4} + \frac{t^3}{3}\Big|_{-5}^{-4} + 8 + 25 - 10\frac{t^2}{2}\Big|_{4}^{5} + \frac{t^3}{3}\Big|_{4}^{5} \tag{2.179}$$

$$= 58 - 45 + 61/3 - 45 + 61/3$$

$$= 26/3$$

Example 2.99

Find if the signal $x[n] = \left(\frac{1}{2}\right)^n u[n]$ is an energy signal or a power signal?

Solution

Energy of the signal $E = \lim_{N \to \infty} \sum_{m=-N}^{N} |x(m)|^2$ \hfill (2.180)

$$E = \lim_{N \to \infty} \sum_{m=-N}^{N} \left|\left(\frac{1}{2}\right)^m u(m)\right|^2 \tag{2.181}$$

$$E = \lim_{N \to \infty} \sum_{m=0}^{N} \left|\left(\frac{1}{4}\right)^m\right| \tag{2.182}$$

$$E = \sum_{m=0}^{\infty} \left|\left(\frac{1}{4}\right)^m\right| = \frac{1}{1-\frac{1}{4}} = \frac{4}{3} \text{ joules} \tag{2.183}$$

The power of the signal $P = \lim_{N \to \infty} \frac{1}{2N+1} \sum_{m=-N}^{N} |x(m)|^2$ \hfill (2.184)

$$P = \lim_{N \to \infty} \frac{1}{2N+1} \sum_{m=-N}^{N} \left(\frac{1}{4}\right)^m \tag{2.185}$$

$$P = \lim_{N \to \infty} \frac{1}{2N+1} \sum_{m=0}^{N} \left(\frac{1}{4}\right)^m \qquad (2.186)$$

$$P = \lim_{N \to \infty} \frac{1}{2N+1} \left[\frac{1-\left(\frac{1}{4}\right)^{N+1}}{1-\left(\frac{1}{4}\right)} \right] \to 0 \qquad (2.187)$$

The signal has finite energy and zero power. So, the signal is an energy signal.

Example 2.100
Find if the signal $x[n] = u[n]$ is an energy signal or a power signal?

Solution

$u[n] = 1$ for all $n \geq 0$

The signal has infinite samples. The energy of the signal is

$$E = \sum_{n=0}^{\infty} 1 = \infty \qquad (2.188)$$

The average power of the signal is

$$P = \lim_{N \to \infty} \frac{1}{2N+1} \sum_{n=0}^{N} 1 = \frac{N+1}{2N+1} \to \frac{1}{2} \text{ is finite} \qquad (2.189)$$

So, the signal is a power signal.

Example 2.101
Find if the signal $x[n] = u[n] - u[n-4]$ is an energy signal or a power signal.

Solution

$u[n] = 1$ for all $n \geq 0$ and $u[n-4] = 1$ for all $n \geq 4$.

$x[n] = u[n] - u[n-4] = 1 \text{ for } 0 \leq n \leq 3$.

The total energy of the signal is

$$E = \sum_{n=0}^{3} 1 = 4 \text{ is finite.} \qquad (2.190)$$

The total energy is finite. The period is infinite. So, the average power is zero. So, the signal is an energy signal.

Example 2.102

Find the power of the signal given by $x(t) = e^{j2t} \cos(5t)$

Solution

$x(t) = (\cos(2t) + j\sin(2t))\cos(5t)$

$$= \frac{1}{2}[\cos(7t) + \cos(3t)] + j\frac{1}{2}[\sin(7t) - \sin(5t)] \qquad (2.191)$$

The period of the signal is 2*pi/5. The peak value of the exponential component of the signal is 1. Average power of the signal is given by

$$P = 4\left(\frac{1}{2}\right)^2 / 2 = \frac{1}{2} \qquad (2.192)$$

Example 2.103

Determine if the signal is an energy signal or a power signal. Find the energy or power of the signal given by $x(t) = \sin(4t)[u(t-1) - u(t-9)]$.

Solution

The signal exists only over a finite duration from $t = 1$ to $t = 9$. Hence, the signal is an energy signal. Let us find the total energy of the signal.

$$E = \int_1^9 |\sin(4t)|^2 dt = \int_1^9 (1 - \cos^2(4t))dt = \int_1^9 \left[1 - \frac{1}{2}(1 - \cos(8t))\right] dt \qquad (2.193)$$

$$= \frac{1}{2}\int_1^9 dt = 4\, joules$$

The period of the signal is infinite. So, the average power is zero and the signal is an energy signal.

Example 2.104

Find the energy and average power of the signal given by $x[n] = e^{j[(\pi/3)n + \pi/2]}$

Solution

The energy of the signal is

$$P = \lim_{N \to \infty} \frac{1}{2N+1} \sum_{-N}^{N} |e^{j[(\pi/3)n + \pi/2]}|^2 = \lim_{N \to \infty} \frac{1}{2N+1} \sum_{-N}^{N} 1 = \frac{2N+1}{2N+1} = 1$$

$$E = \lim_{N \to \infty} \sum_{-N}^{N} |e^{j[(\pi/3)n+\pi/2]}|^2 = \lim_{N \to \infty}(2N+1) = \infty \qquad (2.194)$$

Note that $|e^{j[(\pi/3)n+\pi/2]}| = 1$

The average power is finite. Total energy is infinite. Hence, the signal is a power signal.

Example 2.105

Find if the signal is an energy or a power signal.

$$x[n] = \begin{cases} n^2 & \text{for } 0 \le n \le 3 \\ 10-n & \text{for } 4 \le n \le 6 \\ n & \text{for } 7 \le n \le 9 \\ 0 & \text{otherwise} \end{cases}$$

Solution

The signal exists only for a small duration. Hence, it is an energy signal. Let us find the energy.

$$E = \sum_{n=0}^{3} |n^2|^2 + \sum_{n=4}^{6}(10-n)^2 + \sum_{n=7}^{9} n^2 = (1+16+81)+(36+25+16) \qquad (2.195)$$

$$+ (49+64+81)$$

$$= 98 + 77 + 194 = 369$$

The energy is finite.

$$P = \lim_{N \to \infty} \frac{1}{2N+1}\left[\sum_{n=0}^{3}|n^2|^2 + \sum_{n=4}^{6}(10-n)^2 + \sum_{n=7}^{9} n^2\right]$$

$$= \lim_{N \to \infty} \frac{1}{2N+1}[(1+16+81)+(36+25+16)+(49+64+81)] \qquad (2.196)$$

$$= \lim_{N \to \infty} \frac{1}{2N+1}[98+77+194] = \lim_{N \to \infty} \frac{1}{2N+1}[369] = 0$$

The average power is zero. Hence, the signal is an energy signal.

Example 2.106

Find the energy of the signal shown below in Fig. 2.65.

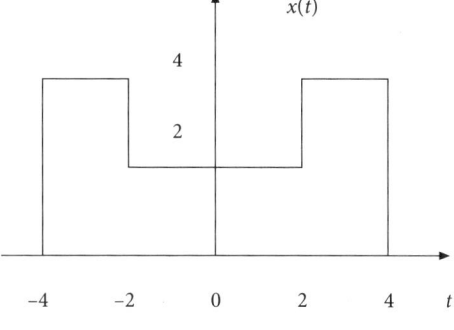

Fig. 2.65 Plot of signal for Example 2.106

Solution

The energy of the signal is given by

$$E = \int_{-4}^{-2} 4^2 \, dt + \int_{-2}^{2} 2^2 \, dt + \int_{2}^{4} 4^2 \, dt \quad (2.197)$$

$$= 32 + 16 + 32 = 80$$

Physical significance of energy and power signals When the signal is an energy signal, it exists for a finite duration. We need to find ESD to analyze the frequency contents of the signal. When the signal is a power signal, it has infinite energy and finite average power. We need to find PSD to analyze the frequency contents of the signal. (The reader may refer to chapter 9 for ESD and PSD)

Concept Check

- Define even and odd properties of a signal.
- Define one even and one odd trigonometric function.
- Will the product of even and odd functions be odd?
- Will the function be odd if it contains even powers of t?
- Will the square function of odd function be an odd function?
- Will the product of two odd functions be an even function?
- Define periodicity for a CT sinusoid.
- Will you treat the signal as periodic if it exists for a finite duration?
- Define periodicity for a DT sinusoid.
- How will you find a period for a combination of two CT sinusoids?
- Define causality for a CT and DT signal.
- Define deterministic and random signals.
- Define energy signals and power signals.

- If $y(t) = \sin^2(2\pi f t) + \cos(6\pi f t)$, is $y(t)$ a deterministic signal?
- If $y(t) = \cos(6\pi t)$ for $0 \leq t \leq 1/3$, is $y(t)$ an energy signal?

2.7 Operations on Signals

Typical signal-processing operations are time and amplitude scaling, multiplication, addition and shifting. Advanced signal-processing operations include convolution, filtering, interpolation and decimation. We will illustrate and discuss amplitude scaling, addition, multiplication, time scaling, time shifting and folding operations. We will then describe a precedence rule that will help the reader to implement the operations on the signals successively. Let us go through simple operations on signals one by one.

2.7.1 Time shifting

We will illustrate the time shifting operation for CT signal by considering a numerical example.

Example 2.107

Consider time-shifting operation for CT signal. Shift the signal shown in Fig. 2.66 towards right and left by one time unit.

Solution

Let the time shift be of T time units. The shifted signal can be represented as

$$y(t) = x(t - T) \tag{2.198}$$

Let T be positive and equal to 1. Consider the signal $x(t)$ as plotted in Fig. 2.66. The shifted signal is represented in Figs 2.67 and 2.68 for positive and negative value of T.

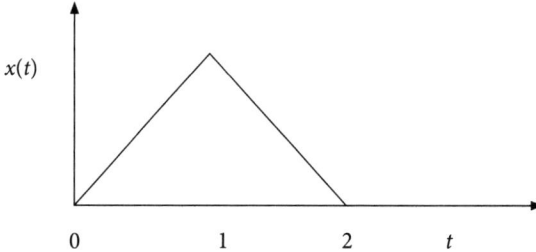

Fig. 2.66 Plot of signal $x(t)$

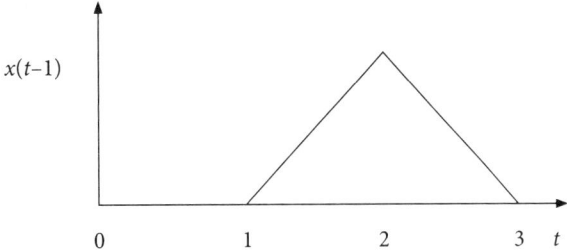

Fig. 2.67 Plot of x(t − 1)

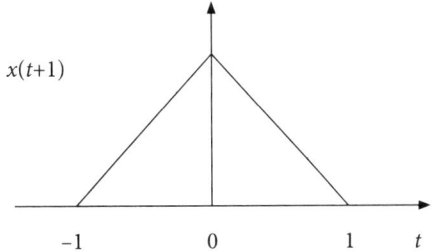

Fig. 2.68 Plot of x(t + 1)

The reader may note that $x(t-T)$ for positive T shifts the signal towards right by T units and for T negative, it shifts the signal towards left by T units. When T is negative, the signal becomes $x(t + T)$.

Physical significance of time shifting Many times when we compare the two signals for recognizing the pattern, we find that the signals are identical except that one signal is shifted with respect to the other. We need to slide and shift one signal over the other to find if they are matching. Time-shifting operation is useful for pattern matching and recognition.

We will illustrate time shifting for DT signals.

Example 2.108

Consider time-shifting operation in case of a DT signal $x[n]$ shown in Fig. 2.69. We will draw $x[n - 2]$ and $x[n + 2]$. The shifted signals are shown in Fig. 2.70 and Fig. 2.71. The signal $x[n]$ is given by

$$\begin{aligned} x[n] &= 2 \quad \text{for } n = 1, 2 \\ &= 1 \quad \text{for } n = 3 \\ &= 0 \quad \text{elsewhere} \end{aligned} \qquad (2.199)$$

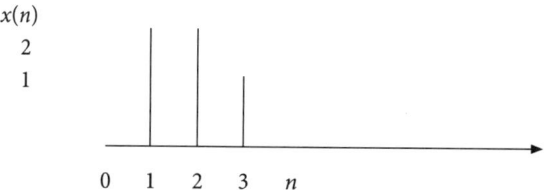

Fig. 2.69 Plot of signal x[n]

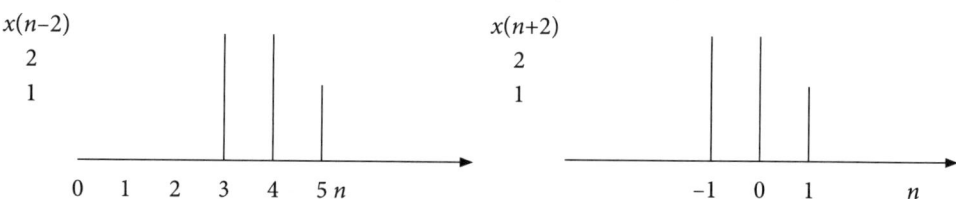

Fig. 2.70 x(n) shifted right by 2 positions **Fig. 2.71** x(n) shifted left by 2 positions

The reader may note that x[n − k] shifts the signal towards right by k samples for positive values of k and shifts the signal towards left by k positions for negative values of k. Here, we have considered k = 2 for the above example. When k is negative, the signal is represented as x[n + k]. For k = 2, we have to shift the signal x[n] towards right by 2 samples. This means, the sample at n = 1 will appear at n = 1 + 2 i.e., n = 3 (Refer to Fig. 2.70). For k = −2, we have to shift the signal x[n] towards left by 2 samples. This means, the sample at n = 1 will appear at n = 1 − 2 i.e., at n = −1 (refer to Fig. 2.71).

2.7.2 Time reversal

We will illustrate the time reversal operation for CT signal by considering numerical example.

Example 2.109

Consider time reversal operation for CT signal. Perform time reversal of the signal shown in Fig. 2.66 and shift it towards right and left by one time unit.

Solution

Consider time reversal of signal represented in Fig. 2.66. Let us plot x(−t). Let the time shift be of T time units. The shifted signal can represented as

$$y(t) = x(-t + T) \tag{2.200}$$

Let T be positive. Consider the signal x(−t) as plotted in Fig. 2.72. Note that it is the mirror image of the signal x(t). The shifted signal is represented in Figs 2.73 and 2.74 for positive and negative value of T = 1.

The reader may note that $x(-t + T)$ for positive T shifts the signal towards right by T units and for T negative, it shifts the signal towards left by T units. Compare this with $x(t - T)$.

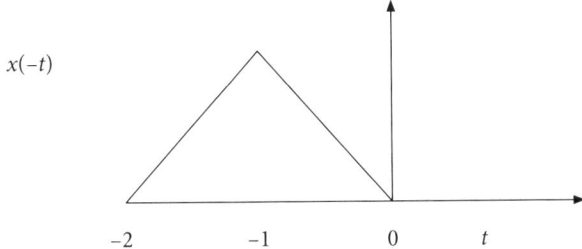

Fig. 2.72 Plot of signal $x(-t)$

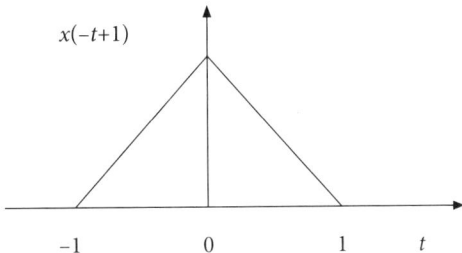

Fig. 2.73 Plot of $x(-t + 1)$

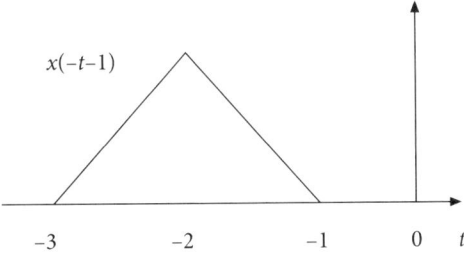

Fig. 2.74 Plot of $x(-t - 1)$

Example 2.110

Let us now consider the time reversal and shifting for DT signal $x[n]$ shown in Fig. 2.69.

Solution

Let us plot $x[-n]$. This is a time-reversed signal. Time-reversed signal is plotted in Fig. 2.75. Note that it is the mirror image of signal $x[n]$. Let it be shifted by

k units to get $x[-n + k]$ with $k = 2$. This shifts the signal $x[-n]$ by 2 samples towards right for positive values of $k = 2$ and shifts it towards left by 2 samples for negative values of $k = -2$. The shifted signals are shown in Figs 2.76 and 2.77, respectively.

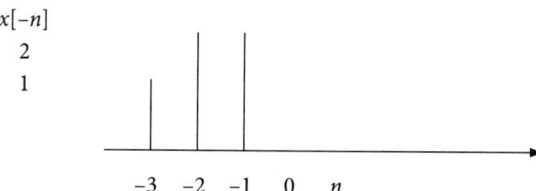

Fig. 2.75 Plot of time-reversed signal $x[-n]$

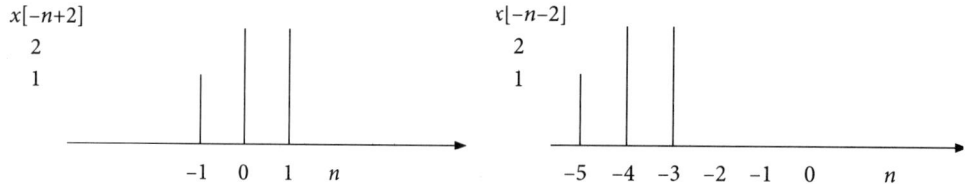

Fig. 2.76 $x[-n]$ shifted right by 2 samples   **Fig. 2.77** $x[-n]$ shifted left by 2 samples

Physical significance of time reversal Sometimes, the two signals are identical, except that they are mirror images of each other. The time reversal operation will allow one to check if one signal is a mirror image of the other. Time-shifting operation is required to be performed as a first step in convolution operation. (The reader will come across convolution in chapter 4.)

Let us solve some numerical problems based on these properties.

Example 2.111

Consider the following signal given by

$$x(t) = \begin{cases} t+2 & -2 \leq t \leq -1 \\ 1 & -1 \leq t \leq 1 \\ -t+2 & 1 \leq t \leq 2 \\ 0 & \text{elsewhere} \end{cases}$$

Plot the signal $x(t-3)$, $x(-t + 2)$.

Solution

Let us first plot the signal $x(t)$. Figure 2.78 shows a plot of $x(t)$. Let us now shift $x(t)$ towards right by 3 units. The shifted signal is plotted in Fig. 2.79.

Let us execute time reversal and plot x(–t). Figure 2.80 shows plot of x(–t). Owing to symmetry about y axis, x(t) = x(–t). Let us shift it to the right by 2 units to get x(–t + 2). Figure 2.81 shows the plot of x(–t + 2).

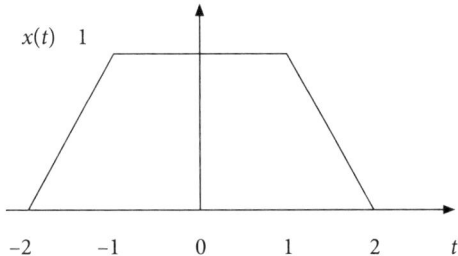

Fig. 2.78 Plot of signal x(t)

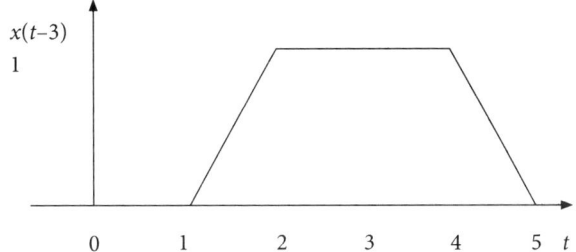

Fig. 2.79 Plot of signal x(t – 3)

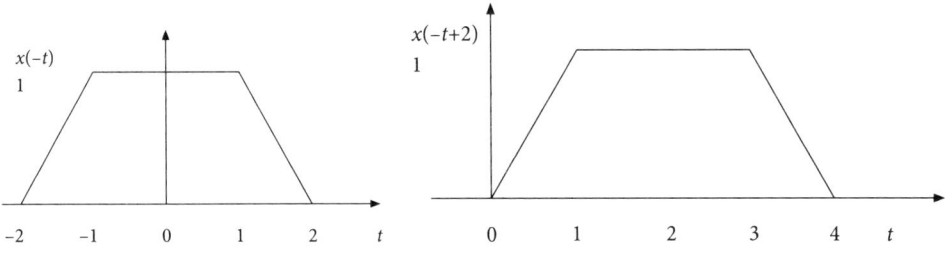

Fig. 2.80 Plot of x(–t)   **Fig. 2.81** Plot of x(–t + 2)

Example 2.112

Consider the following DT signal given by $x[n] = u[n] - u[n-4]$. Plot $x[n]$, $x[-n]$, $x[n+3]$, $x[-n-2]$.

Solution

Let us plot $x[n]$.

$$x[n] = 1 \quad \text{for} \quad 0 \leq n \leq 3 \tag{2.201}$$
$$= 0 \quad \text{otherwise}$$

Plot of $x[n]$ is shown in Fig. 2.82. Figure 2.83 plots $x[-n]$.

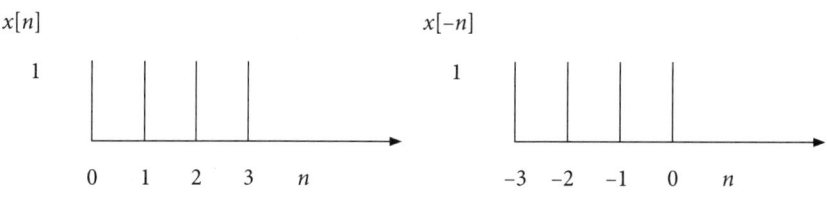

Fig. 2.82 Plot of $x[n]$ **Fig. 2.83** Plot of $x[-n]$

Let us now plot $x[n + 3]$. This is a signal $x(n)$ shifted left by 3 samples, as shown in Fig. 2.84. The signal $x[-n-2]$ is a signal $x(-n)$ shifted left by 2 samples. It is shown in Fig. 2.85.

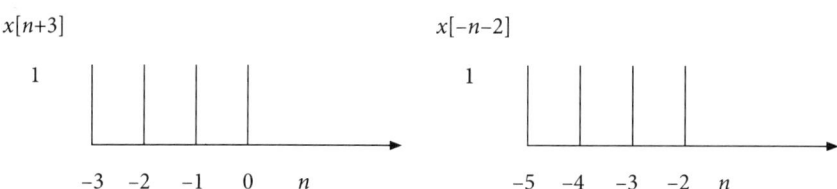

Fig. 2.84 Plot of $x[n + 3]$ **Fig. 2.85** Plot of signal $x[-n-2]$

2.7.3 Time and amplitude scaling

Let us consider amplitude scaling of a signal $x(t)$.

Example 2.113

Consider signal $x(t)$ given below.

$$x(t) = \begin{cases} t+2 & -2 \leq t \leq -1 \\ 1 & -1 \leq t \leq 1 \\ -t+2 & 1 \leq t \leq 2 \\ 0 & \text{elsewhere} \end{cases}$$

The signal is plotted in Fig. 2.78. Let us plot scaled signals $2x(t)$ and $½x(t)$.

Solution

Let us plot both these signals as shown in Figs 2.86 and 2.87, respectively. Note that the signal $2x(t)$ has double the amplitude at each time instant as compared to $x(t)$ and the signal $½ x(t)$ has half the amplitude at each time instant as compared to $x(t)$ (refer to Fig. 2.78).

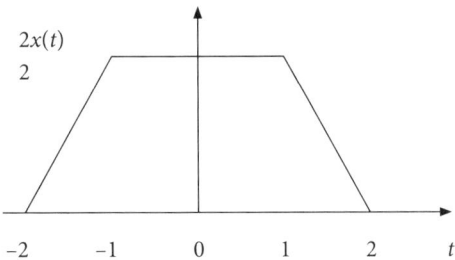

Fig. 2.86 Plot of signal $2x(t)$

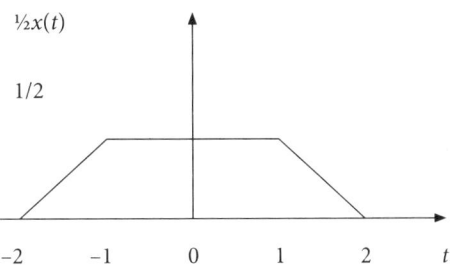

Fig. 2.87 Plot of signal $\frac{1}{2} x(t)$

Let us introduce time-scaling operation.

Example 2.114

Consider signal $x(t)$ of Example 2.113 above, as shown in Fig. 2.78. Plot $x(2t)$ and $x(t/2)$.

Solution

The signal $x(2t)$ is a signal compressed by a factor of 2 and $x(1/2t)$ is a signal dilated (elongated) by a factor of 2. The plots for compressed and elongated signals are shown in Figs 2.88 and 2.89, respectively. Note the time scale along x axis in elongated and compressed signals.

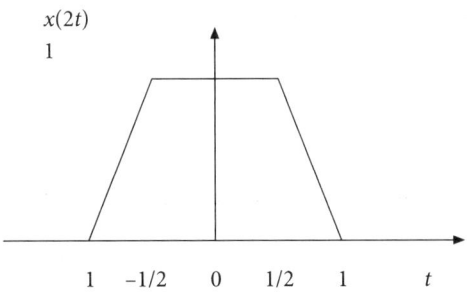

Fig. 2.88 Plot of $x(2t)$

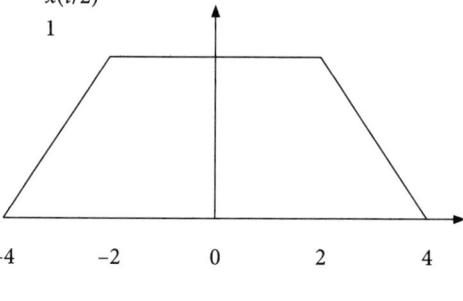

Fig. 2.89 Plot of $x(t/2)$

The reader may remember that multiplication of t (the argument of x) by a factor 'a' compresses the signal along x axis by the same factor. Division of t by a factor elongates the signal along x axis by the same factor.

Example 2.115

Consider signal $x[n]$ shown below in Fig. 2.90. Plot $x[2n]$ and $x[n/2]$.

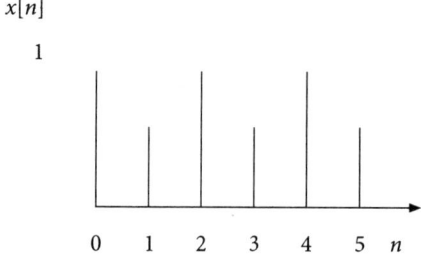

Fig. 2.90 Plot of signal $x[n]$

Solution

$x[2n]$ will be a compressed signal. Here, in DT domain, compression by a factor of 2 will actually decimate the signal by 2 i.e., we have to collect alternate samples only. The signal $x[2n]$ is shown plotted in Fig. 2.91. We can observe that the samples with value equal to 1 are lost when we collect alternate samples. The signal $x[n/2]$ is plotted in Fig. 2.92. This is a dilated signal. Here, the sample at $n = 1$ is shifted to $n = 2$ and a sample value of zero appears at $n = 1$. Similarly, samples at all odd values of n will be zero and the samples of $x[n]$ will appear at all even values of n successively.

Physical significance of amplitude and time scaling When we are comparing the two signals, one may just be the amplified version of the other. In this case, we normalize the signals and then compare them with each other. Here, we need to find the scaling factor for amplitude scaling. Sometimes, the time scale is different for two signals. Consider the two speech signals having utterance of the same word two times. When a person utters the same word two times, the signal duration differs for both the signals. When we want to compare these two signals, we need to find the time scaling factor for time scaling. In practice, for speech signals, we use time warping (other name given for time scaling.)

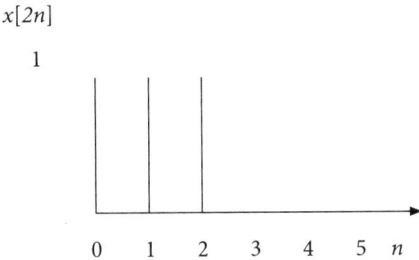

Fig. 2.91 Plot of $x[2n]$

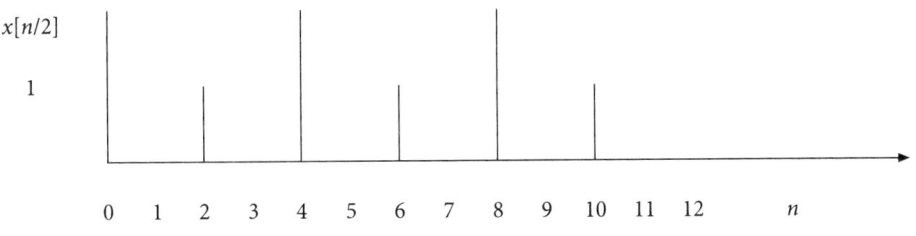

Fig. 2.92 Plot of signal $x[n/2]$

Let us now combine the time scaling and time-shifting operations. We will illustrate this with a numerical problem.

Example 2.116

Consider signal $x(t)$ given below.

$$x(t) = \begin{cases} t+2 & -2 \le t \le -1 \\ 1 & -1 \le t \le 1 \\ -t+2 & 1 \le t \le 2 \\ 0 & \text{elsewhere} \end{cases}$$

Plot $x(2t - 3)$.

Solution

We have to execute time shifting first and then time scaling. This is called as precedence rule.

Let us first plot $x(t-3)$, which is a signal shifted towards right by 3 samples. The signal plot is shown in Fig. 2.93. Now execute time scaling i.e., compression by a factor of 2. The signal $x(2t-3)$ is shown in Fig. 2.94.

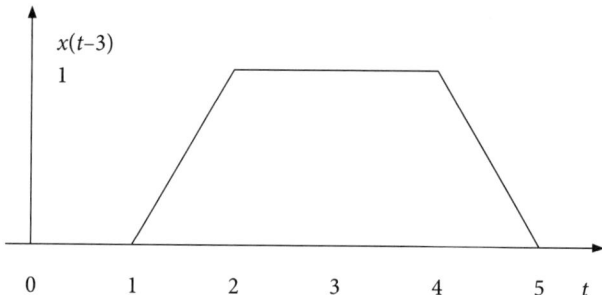

Fig. 2.93 Plot of $x(t-3)$

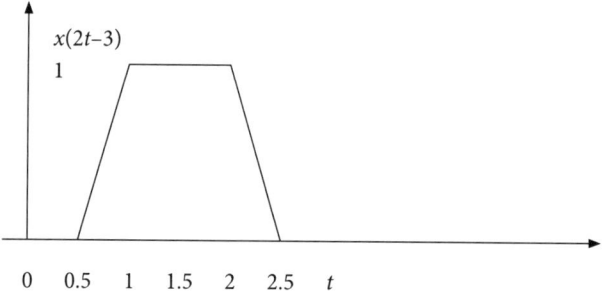

Fig. 2.94 Plot of $x(2t-3)$

If we implement scaling and then time shifting later, for example, find $x(2t)$ first and then translate by 3 units, we will get a wrong result. Let us verify this. We will plot $x(2t)$ and then find $x(2t-3)$ by shifting operation. The plots are shown in Figs 2.94 and 2.95, respectively.

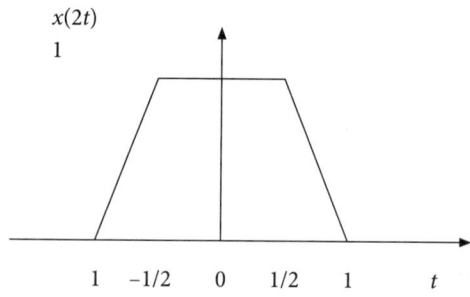

Fig. 2.95 Plot of $x(2t)$

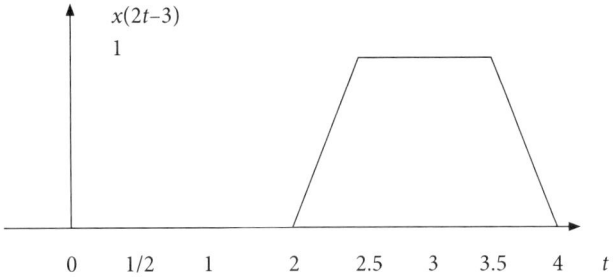

Fig. 2.96 Plot of x(2t − 3)

What went wrong?

We will now find what went wrong when we implemented scaling first. Let us examine the problem again. We have to find x(2t−3). We may write it as x(2(t-3/2)). This means if we scale first, we have to translate by x shift of 3/2 rather than 3. Let us do it. We will shift x(2t) towards right by 3/2 units. The signal can be plotted as shown in Fig. 2.97.

The reader can observe that the result is the same as we obtained when we used a precedence rule. So, if we take proper care, we can implement scaling first and then the translation.

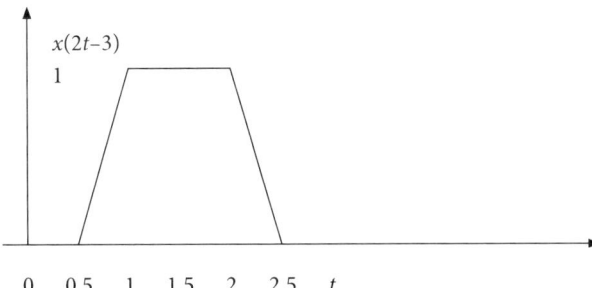

Fig. 2.97 Plot of x(2(t−3/2))

Example 2.117

Consider a rectangular pulse given by

$$x(t) = \text{rect}\left(\frac{t}{2}\right) = \begin{cases} 1 & \text{for } |t| \leq 1 \\ 0 & \text{for } |t| > 1 \end{cases}$$

Draw the following functions derived from the rectangular pulse.
x(3t), x(3t + 2), x(−2t − 1), x(2(t + 2)), x(2(t − 2)), x(3t) + x(3t + 2).

Solution

Let us first draw $x(t)$. It is shown in Fig 2.98. We will now time scale it by a factor of 3 to get $x(3t)$. It is a compressed signal, as shown in Fig. 2.99.

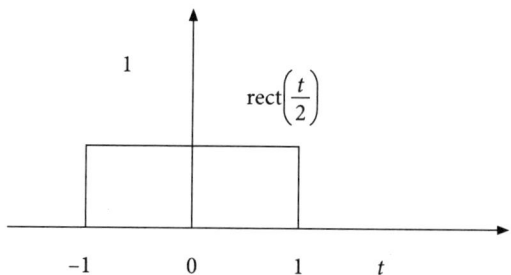

Fig. 2.98 Plot of $x(t)$

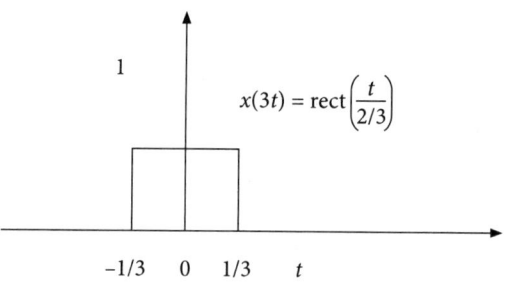

Fig. 2.99 Plot of $x(3t)$

$x(3t + 2) = x(3(t + 2/3))$ is $x(3t)$ signal shifted left by 2/3 time units, as shown in Fig. 2.100.

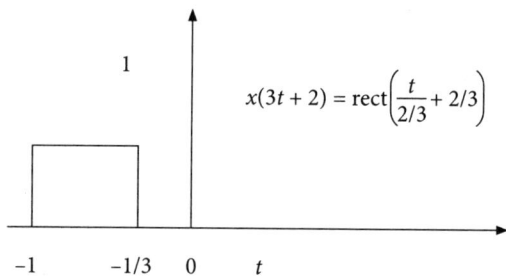

Fig. 2.100 Plot of $x(3t + 2)$

Let us plot $x(2t)$ and invert it to get $x(-2t)$. We will find that $x(-2t)$ is the same as $x(2t)$, as the signal is symmetrical about y axis. It is shown in Fig. 2.101. $x(-2t-1) = x(-2(t + 1/2))$ is signal $x(-2t)$ shifted towards left by ½ time units. Let us draw $x(-2t -1)$. It is shown in Fig. 2.102.

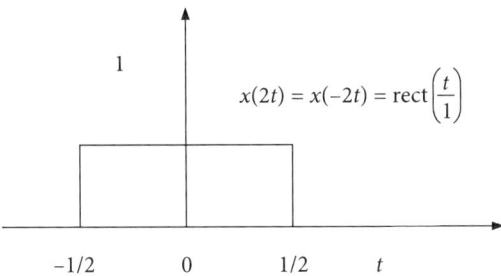

Fig. 2.101 Plot of $x(-2t)$

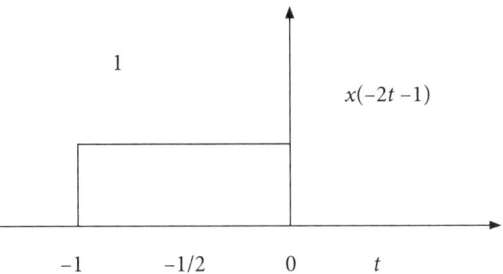

Fig. 2.102 Plot of $x(-2t-1)$

We have already plotted $x(2t)$. So, let us shift it to left by 2 time units to get $x(2(t+2))$. It is shown in Fig. 2.103. $x(2(t-2))$ is $x(2t)$ shifted towards right by 2 time units, as shown in Fig. 2.104.

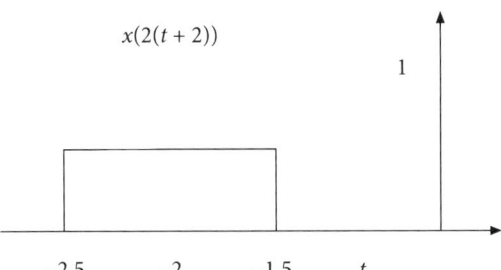

Fig. 2.103 Plot of $x(2(t+2))$

We have already plotted $x(3t)$. We will now plot $x(3(t+2/3))$ to get $x(3t+2)$, which is a shifted signal $x(3t)$ towards left by 2/3 time units. It is plotted in Fig. 2.105. Let us add $x(3t)$ to $x(3t+2)$ to get $x(3t) + x(3t+2)$. It is shown in Fig. 2.106.

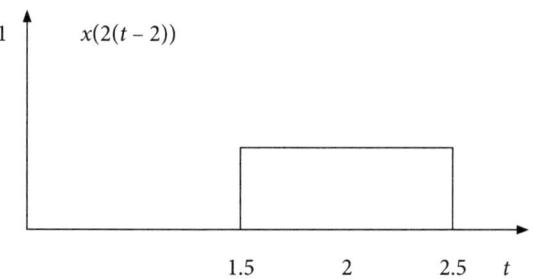

Fig. 2.104 Plot of signal $x(2(t - 2))$

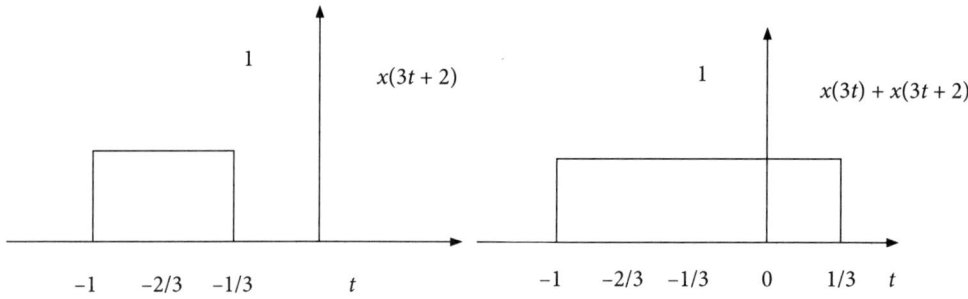

Fig. 2.105 Plot of $x(3t + 2)$ **Fig. 2.106** Plot of $x(3t) + x(3t + 2)$

Let us solve the same problem using precedence rule.

Example 2.118

Consider a rectangular pulse given by

$$x(t) = \text{rect}\left(\frac{t}{2}\right) = \begin{cases} 1 & \text{for } |t| \leq 1 \\ 0 & \text{for } |t| > 1 \end{cases}$$

Draw the following functions derived from the rectangular pulse. Use precedence rule.

$x(3t + 2), x(-2t - 1), x(2(t + 2))$

Solution

Refer to Fig. 2.101 for the plot of $x(t)$. To plot $x(3t + 2)$ using precedence rule, we will first shift the signal $x(t)$ towards left by 2 time units. It is shown in Fig. 2.107. Now let us compress it by a factor of 3 to get $x(3t + 2)$. It is shown in Fig. 2.108. The reader may verify that the Fig. 2.108 is same as Fig. 2.105.

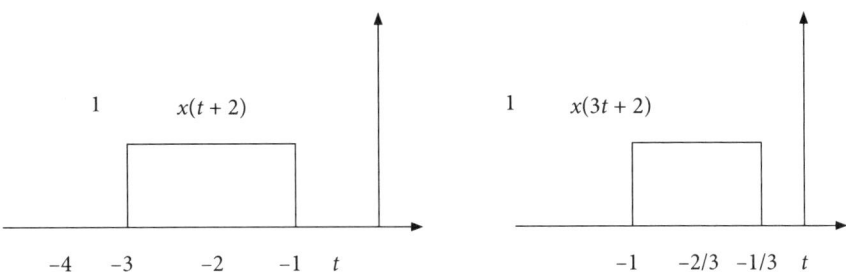

Fig. 2.107 Plot of $x(t+2)$ **Fig. 2.108** Plot of $x(3t+2)$

Owing to the symmetric nature of $x(t)$, $x(-t)$ is same as $x(t)$. To plot $x(-2t-1)$ using precedence rule, we will first shift $x(-t)$ towards left by 1 time unit to get $x(-t-1)$. It is depicted in Fig. 2.109. Then we will compress it by 2 to get $x(-2t-1)$, as shown in Fig. 2.110. The reader may compare Fig. 2.110 with Fig. 2.102 to confirm that they are the same.

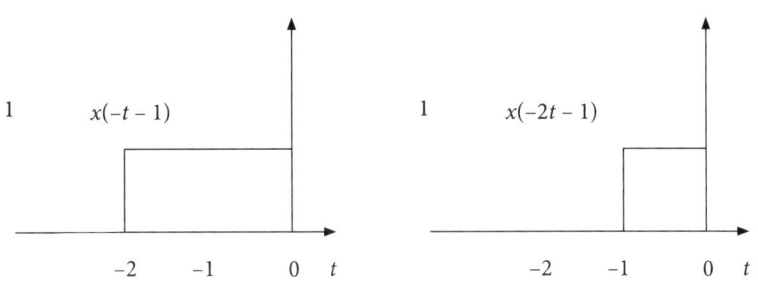

Fig. 2.109 Plot of $x(-t-1)$ **Fig. 2.110** Plot of $x(-2t-1)$

To find $x(2(t+2))$ using precedence rule, we have to first expand the bracket to write it as $x(2t+4)$. We will translate it towards left by 4 positions to get $x(t+4)$, as shown in Fig. 2.111. Then we will compress it by a factor of 2 to get $x(2t+4)$, as depicted in Fig. 2.112. The reader can compare Fig. 2.112 with the plot of Fig. 2.103.

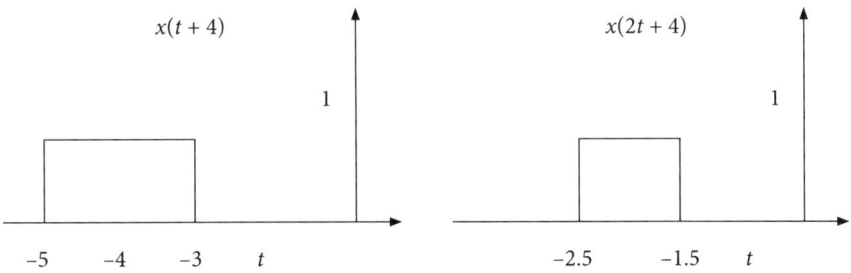

Fig. 2.111 Plot of $x(t+4)$ **Fig. 2.112** Plot of $x(2t+4)$

The reader is encouraged to plot $x(2(t-2))$, $x(3t) + x(3t+2)$ using a precedence rule.

Let us consider similar operations on DT signals. We will illustrate the concepts with the help of a numerical example.

Example 2.119

Consider a DT signal $x[n]$, as shown in Fig. 2.113. Draw $x[2n]$, $x[2n+2]$, $x[n/2]$, $x[n+2]$, $x[-n+2]$.

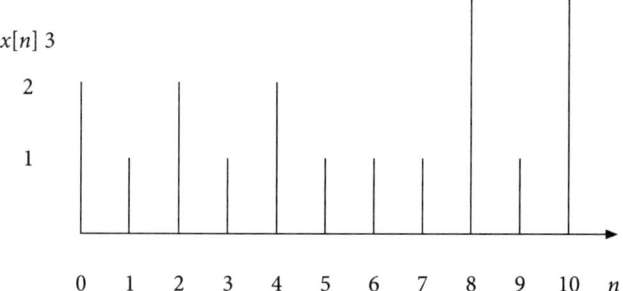

Fig. 2.113 Plot of $x[n]$ for Example 2.119

Solution

Let us plot $x[2n]$, which is obtained by taking alternate samples of $x[n]$. It is plotted in Fig. 2.114. This operation is also termed as down sampling by a factor of two.

Down sampling by a factor of M requires deleting $M-1$ samples after every sample that is retained. Consider $x[2n+2] = x[2(n+1)]$. This is obtained by shifting $x[2n]$ towards left by one sample. It is shown in Fig. 2.115. Let us plot $x[n/2]$ now. It is obtained by interpolating $x[n]$ by inserting zeros between every two samples, as shown in Fig. 2.116. This process is also termed as up sampling by a factor of two. Up sampling by a factor of M is done by inserting $M-1$ zeros between every two samples.

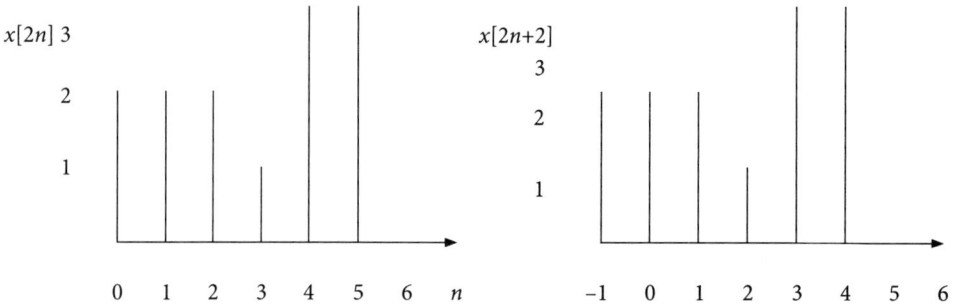

Fig. 2.114 Plot of $x[2n]$ **Fig. 2.115** Plot of $x[2n+2]$

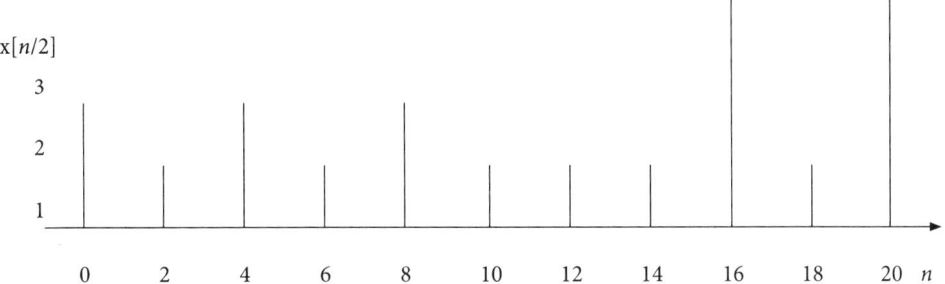

Fig. 2.116 Plot of $x[n/2]$

Let us plot $x[n + 2]$, $x[-n + 2]$. $x[n + 2]$ is obtained by shifting $x[n]$ towards the left by 2 samples, as shown in Fig. 2.117, and $x[-n + 2]$ is obtained by shifting $x[-n]$ towards the right by 2 samples, as shown in Fig. 2.118.

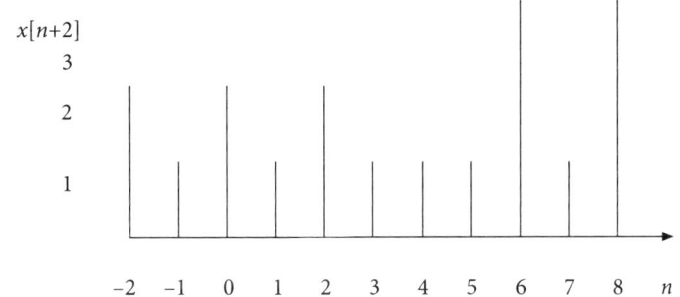

Fig. 2.117 Plot of $x[n + 2]$

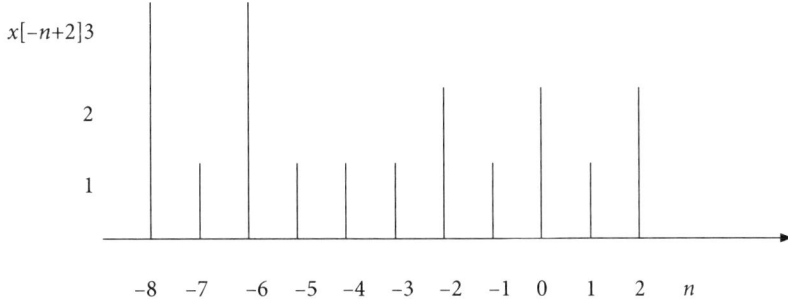

Fig. 2.118 Plot of $x[-n + 2]$

2.7.4 Addition, subtraction and multiplication

This section illustrates different operations such as addition, subtraction and multiplication on CT and DT signals. We will explain these operations with the help of suitable numerical examples. We will consider 2 CT signals. We will implement addition, subtraction and multiplication of $x(t)$ and $y(t)$.

Example 2.120

Consider two CT signals $x(t)$ and $y(t)$, as shown in Fig. 2.119. Find $x(t) + y(t)$, $x(t) - y(t)$ and $x(t)y(t)$, $x(t)y(t-1)$, $x(t+1)y(t-2)$, $x(t-1)y(-t)$, $x(t)y(-t-1)$, $x(2t)y(-t+2)$, $x(2t) + y(2t)$.

Solution

Plot of $x(t)$, $y(t)$ and $x(t) + y(t)$ is shown one below the other in Fig. 2.119. Plot of $x(t)$, $y(t)$ and $x(t) - y(t)$ is shown one below the other in Fig. 2.120. Let us now plot $x(t)y(t)$ in Fig. 2.121. $y(t-1)$ is a signal $y(t)$ shifted towards the right by 1 time unit. We have plotted $x(t)$, $y(t-1)$ and $x(t) y(t-1)$ in Fig. 2.122. $x(t+1)$ is a signal $x(t)$ shifted left by 1 time unit and $y(t-2)$ is a signal $y(t)$ shifted towards the right by 2 time units. We have plotted $x(t+1)$, $y(t-2)$ and the product $x(t+1) y(t-2)$ in Fig. 2.123. $x(t-1)$ is a signal $x(t)$ shifted right by 1 time unit and $y(-t)$ is the time-reversed signal $y(t)$. The product $x(t-1) y(-t)$ is shown in Fig. 2.124. $y(-t-1)$ is a signal $y(-t)$ shifted left by 1 time unit. $x(t) y(-t-1)$ is plotted in Fig. 2.125. $x(2t)$ and $y(2t)$ both are compressed signals by a factor of 2. The plot of $x(2t) + y(2t)$ is shown in Fig. 2.126.

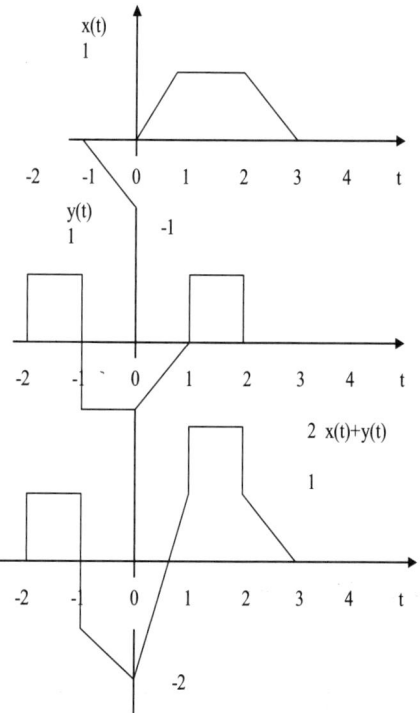

Fig. 2.119 Plot of $x(t) + y(t)$

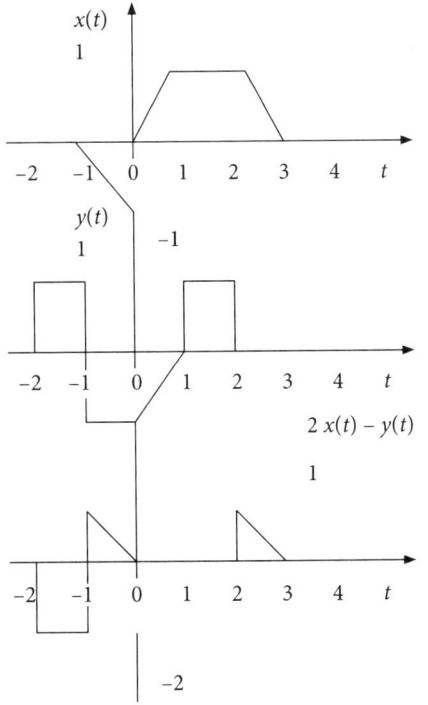

Fig. 2.120 Plot of $x(t) - y(t)$

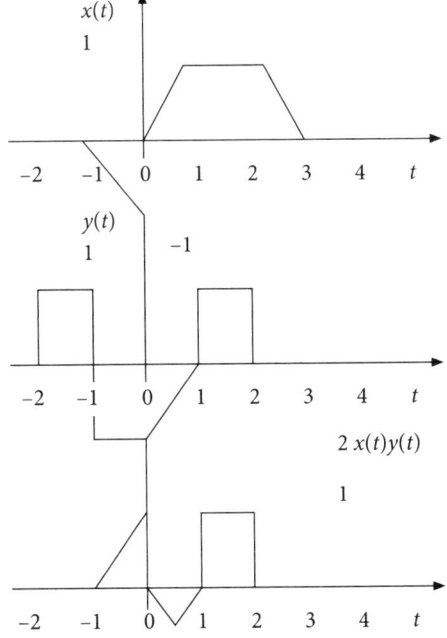

Fig. 2.121 Plot of $x(t)$, $y(t)$ and $x(t)\,y(t)$

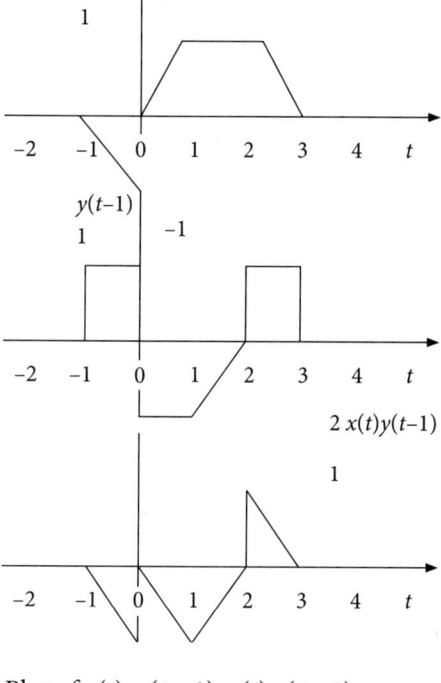

Fig. 2.122 Plot of $x(t)$, $y(t-1)$, $x(t)\,y(t-1)$

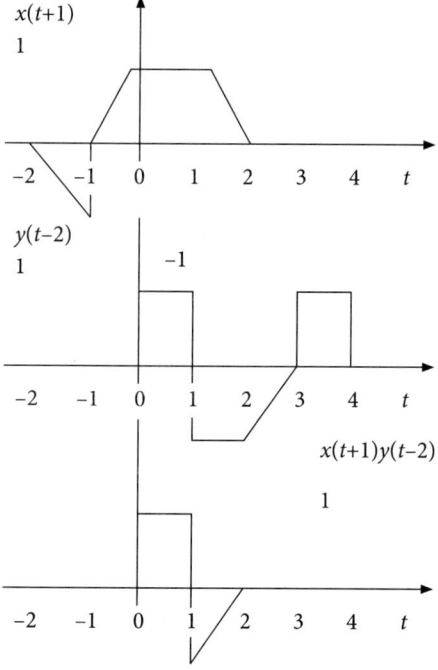

Fig. 2.123 Plot of $x(t+1)\,y(t-2)$

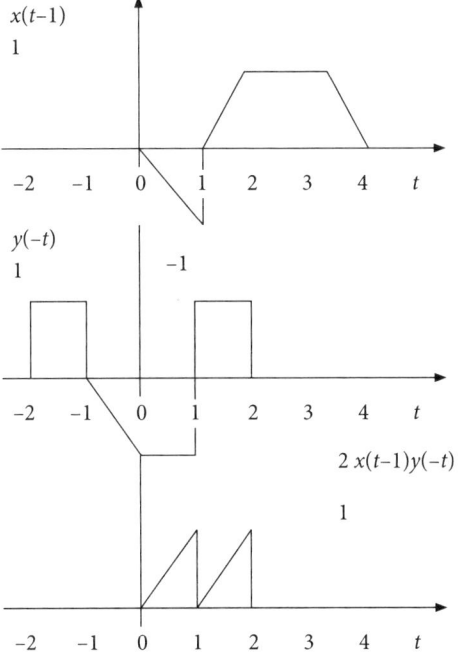

Fig. 2.124 Plot of $x(t-1)\,y(-t)$

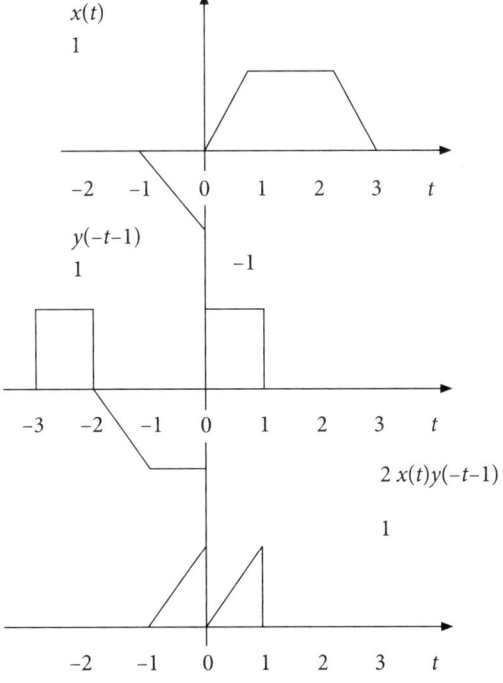

Fig. 2.125 Plot of $x(2t)\,y(-t+2)$

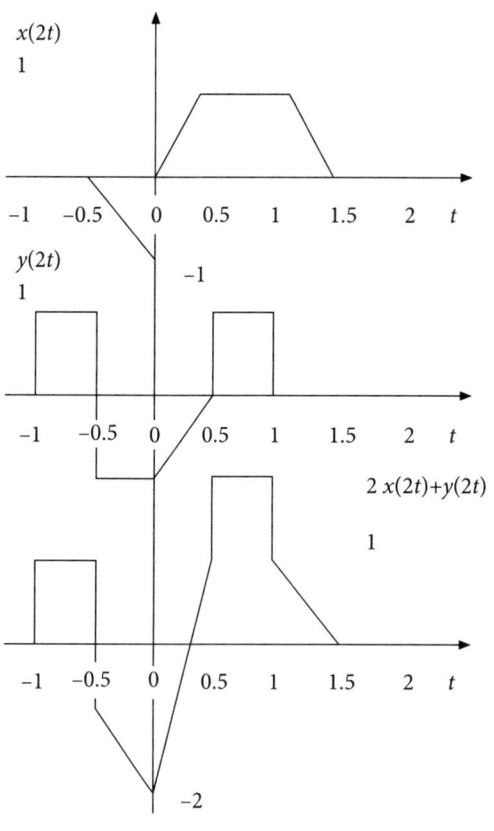

Fig. 2.126 Plot of $x(2t)\, y(3t)$. (Note scale on x-axis)

Example 2.121

Consider DT signals $x[n]$ and $y[n]$, as shown in Figs 2.127 and 2.128. Plot $x[n] + y[n]$, $x[n]\,y[n]$, $x[2n]\,y[n]$ and $x[n-1]\,y[n+2]$.

$$x[n] = \{2,1,2,1,2,1,2,1\},\ y[n] = \{2,1,1,1,1,1,1,0\}$$

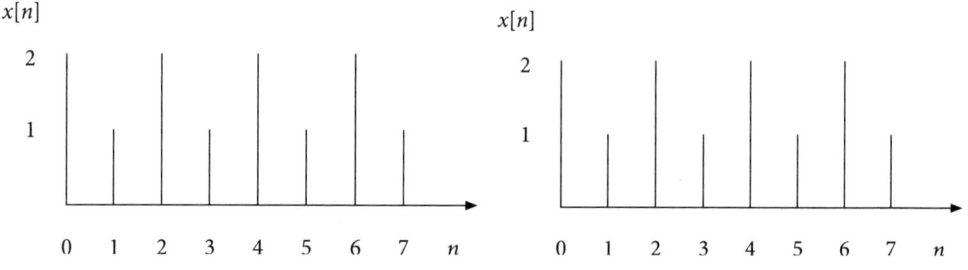

Fig. 2.127 Plot of $x[n]$ **Fig. 2.128** Plot of $y[n]$

Solution

Let us plot $x[n] + y[n]$. We have to add sample by sample. i.e., first sample of $x[n]$ will be added to the first sample of $y[n]$ and so on. The signal $x[n] + y[n]$ is shown in Fig. 2.129. $x[n] + y[n] = \{4,2,3,2,3,2,3,1\}$. The multiplication of the two DT signals is a multiplication taken sample by sample. The reader may verify that

$x[n]\, y[n] = \{4,1,2,1,2,1,2,0\}$, it is plotted in Fig. 2.130. We can obtain $x[2n]$ by taking alternate samples of $x[n]$ i.e., $x[2n] = \{2,2,2,2\}$. The reader may verify the result.

$x[2n]\, y[n] = \{4,2,2,2,0,0,0,0\}$. The plot of $x[2n]y[n]$ is shown in Fig. 2.131. The signal $x[n-1]$ is a shifted signal $x[n]$, shifted towards the right by 1 sample and $y[n+2]$ is signal $y[n]$ shifted left by 2 samples. Plot of $x[n-1]\, y[n+2]$ is shown in Fig. 2.132.

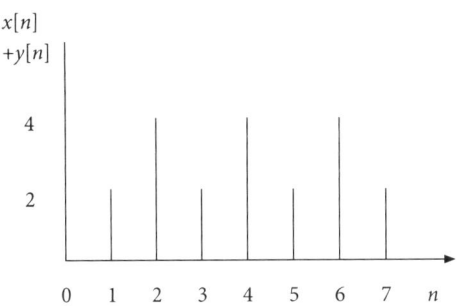

Fig. 2.129 Plot of $x[n] + y[n]$

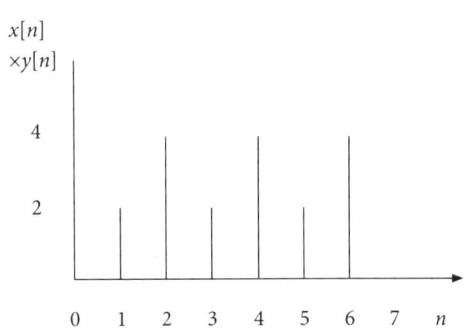

Fig. 2.130 Plot of $x[n]\, y[n]$

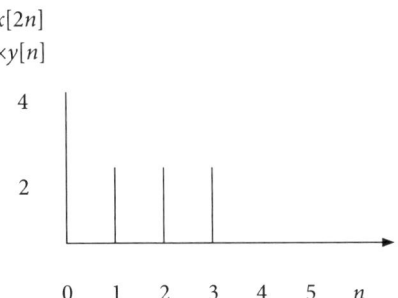

Fig. 2.131 Plot of $x[2n]\, y[n]$

Let us solve some more examples for practice.

Example 2.122

Sketch the waveforms given by following equations where $u(t)$ is a unit step function and $r(t)$ is a unit ramp function.

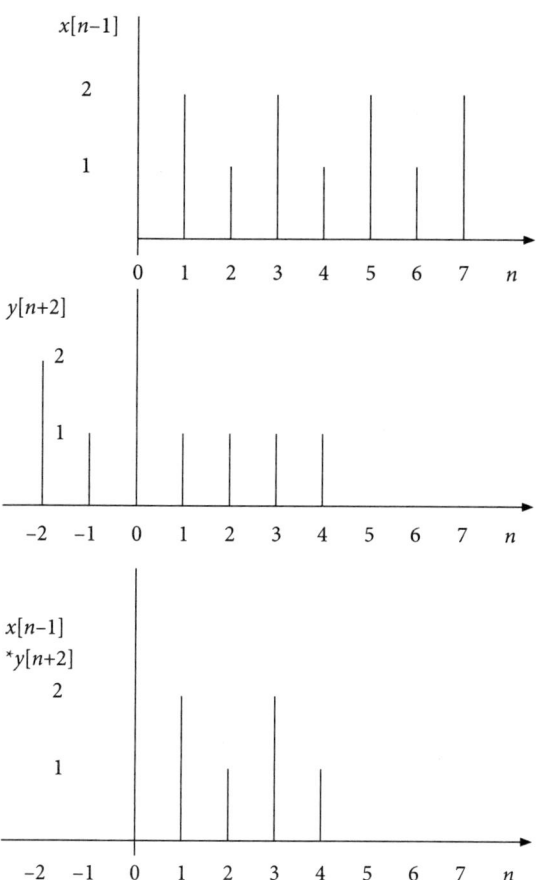

Fig. 2.132 Plot of $x[n-1]$, $y[n+2]$ and $x[n-1]\,y[n+2]$

i. $x(t) = u(t) - u(t-2)$
ii. $x(t) = u(t+1) - 2u(t) + u(t-1)$
iii. $x(t) = -u(t+3) + 2u(t+1) + u(t-3)$
iv. $x(t) = -r(t+3) - r(t) + r(t-2)$
v. $x(t) = r(t) - r(t-1) - r(t-3) + r(t-4)$

Solution

Let us draw each component of $x(t)$ one below the other and then draw $x(t)$.

Consider the signal given by $x(t) = u(t) - u(t-2)$. It is drawn in Fig. 2.133. $x(t) = u(t+1) - 2u(t) + u(t-1)$ is plotted in Fig. 2.134. $x(t) = -u(t+3) + 2u(t+1) + u(t-3)$ is depicted in Fig. 2.135. $x(t) = -r(t+3) - r(t) + r(t-2)$ is plotted in Fig. 2.136. Note that slope is -1 at $t = -3$ and slope is equal to -2 at $t = 1$, slope is -1 at $t = 2$.

$x(t) = r(t) - r(t-1) - r(t-3) + r(t-4)$ is plotted in Fig. 2.137. Note that slope is zero at $t = 1$ and slope is equal to -1 at $t = 3$, slope is zero at $t = 4$.

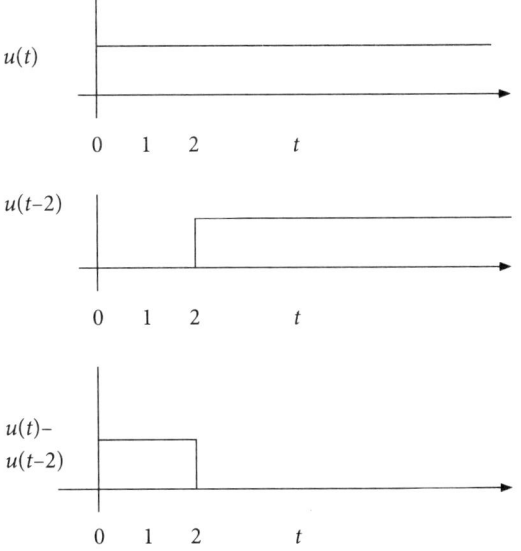

Fig. 2.133 Plot of i) $x(t) = u(t) - u(t-2)$

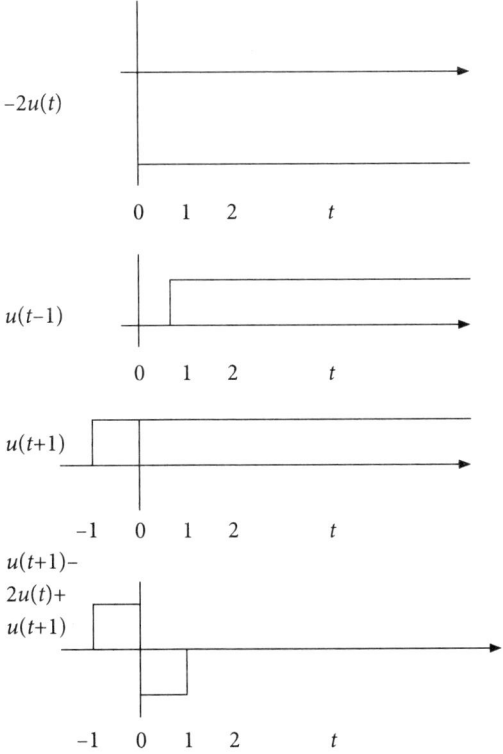

Fig. 2.134 Plot of ii) $x(t) = u(t-1) - 2u(t) + u(t+1)$

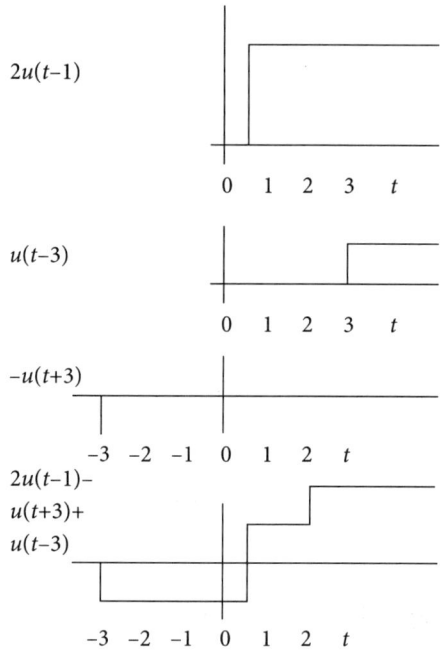

Fig. 2.135 Plot of $2u(t-1) - u(t+3) + u(t-3)$

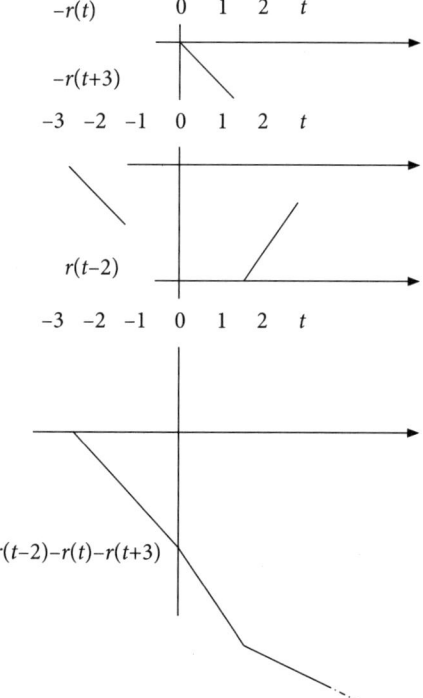

Fig. 2.136 Plot of $r(t-2) - r(t) - r(t+3)$

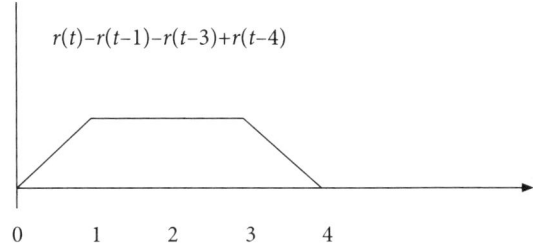

Fig. 2.137 Plot of $r(t) - r(t-1) - r(t-3) + r(t-4)$

Example 2.123

Express the signals $y(t)$, $z(t)$ and $w(t)$ shown in Figs 2.138, 2.139 and 2.140, respectively, in terms of $x(t) = \text{rect}\left(\dfrac{t}{2}\right)$.

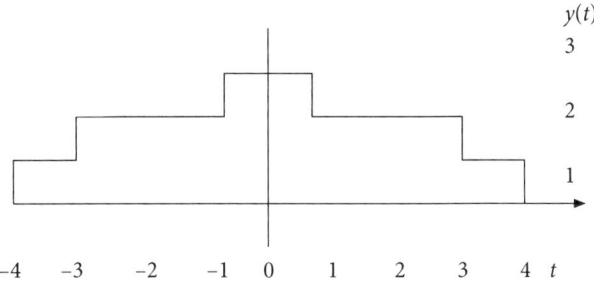

Fig. 2.138 $y(t)$ for Example 2.101

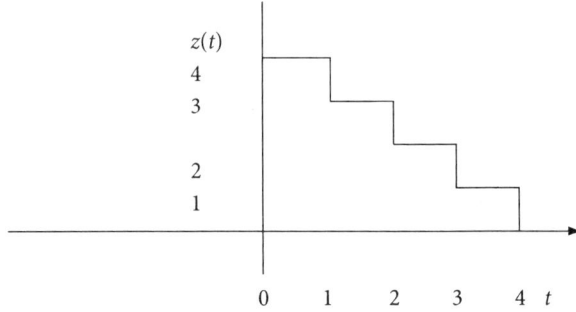

Fig. 2.139 $z(t)$ for Example 2.101

Solution

Let us express signal in Fig. 2.138 in terms of rect function.

$y(t) = \text{rect}(t/4) + \text{rect}(t/3) + \text{rect}(t)$. Refer to Figs 2.141 and 2.142 for $\text{rect}(t/3)$ and $\text{rect}(t/4)$.

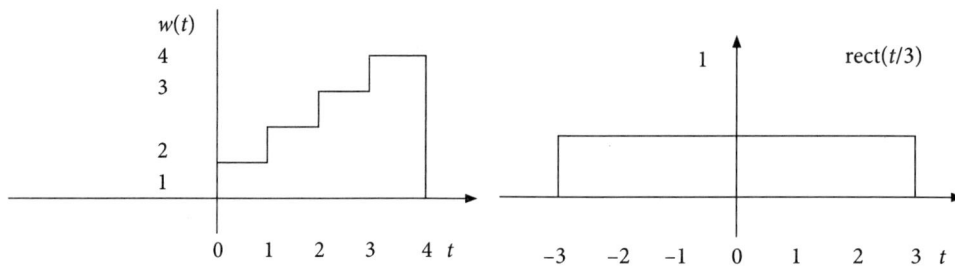

Fig. 2.140 w(t) for Example 2.101   **Fig. 2.141** Plot of rect(t/3)

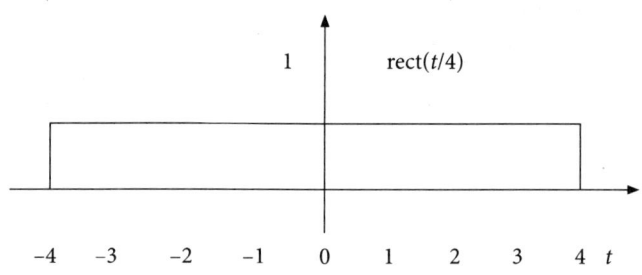

Fig. 2.142 Plot of rect(t/4)

Let us express the signal in Fig. 2.139 in terms of the rect function. Figures 2.143, 2.144, 2.145, 2.146 and 2.147 show the signals rect(2t), rect(2t − 1), rect(2t − 3), rect(2t − 5), rect(2t − 7), respectively.

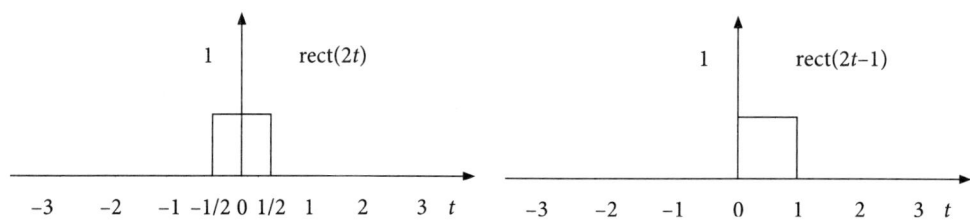

Fig. 2.143 Plot of rect(2t)   **Fig. 2.144** Plot of rect(2t − 1)

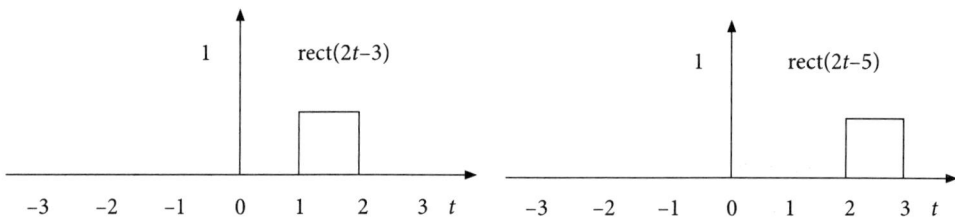

Fig. 2.145 Plot of rect(2t − 3)   **Fig. 2.146** Plot of rect(2t − 5)

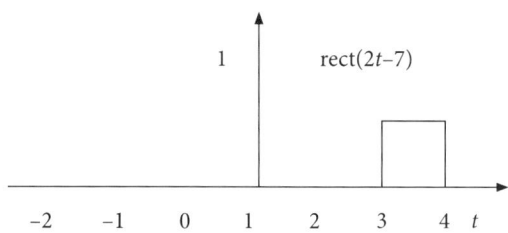

Fig. 2.147 Plot of rect(2t − 7)

The reader is encouraged to verify that

$$z(t) = 4\text{rect}(2t - 1) + 3\text{rect}(2t - 3) + 2\text{rect}(2t - 5) + \text{rect}(2t - 7). \quad (2.202)$$

$$w(t) = \text{rect}(2t - 1) + 2\text{rect}(2t - 3) + 3\text{rect}(2t - 5) + 4\text{rect}(2t - 7)$$

Physical significance of addition and multiplication of the two signals Both these operations are performed in a convolution process. Especially for DT signals, the process of adding and multiplying sample by sample is useful for understanding the process of convolution. (The reader may refer to chapter 4 for convolution.)

Concept Check

- Define time shifting for a CT signal.
- If signal is represented as $x(t - T)$ with T positive, is it a signal $x(t)$ shifted towards left?
- If signal is represented as $x(-t - T)$ with T positive, is it a signal $x(t)$ shifted towards left?
- Is $x(2t)$ a dilated version of $x(t)$?
- Is $x(t/2)$ a dilated version of $x(t)$?
- How will you implement $x[2n]$ for a DT signal?
- How will you implement $x[n/2]$ for a DT signal?
- Define a precedence rule.
- How will you solve your problem without using a precedence rule?
- How will you multiply the two sequences?

Summary

In this chapter, we have described signals and their classification.
- A signal may have one or multiple values for a particular time. If it has a single value for every time instant over which it is defined, it is a single-valued signal called 'scalar signal'. If it has multiple values, it is a multiple-valued signal called as a vector signal'. If a signal can be represented using a mathematical expression, it is said to be deterministic, otherwise it is random. Naturally occurring signals are random signals.

- The signal is plotted with dependent variable on y axis and independent variable on x axis. The domain is decided by the independent variable. If we plot temperature for different time instants the signal is said to be in time domain. When different frequency components of the signal are plotted with frequency on x axis, it is said to be in frequency domain. The image signal is in spatial domain. Basically the signal can be analog or continuous time continuous amplitude (CTCA) signal, discrete time signal or digital i.e., discrete time discrete amplitude (DTDA) signal.

- When the system is designed on paper, it is tested by applying simulated standard inputs. If the performance of the system is satisfactory, then it is implemented. The system can be tested for worst case situations by applying am impulse, unit step or ramp as standard inputs. The standard CT and DT signals are listed and their significance is emphasized for each signal listed. Some examples based on delta functions are solved to illustrate the use of delta function in case of deterministic signals.

- Signals can be either even or odd. If the signal is symmetrical with respect to the amplitude axis, it is even and if it is anti-symmetric with respect to the amplitude axis, then it is odd. The product of an even and odd signal is odd. The signal is even if it contains all even orders of t. The product of two odd signals is an even signal. The section discusses the classification of signals based on signal properties. Signals can be periodic or aperiodic. It is emphasized that the analog sinusoids are always periodic irrespective of if the period is rational or irrational. Several examples are solved to explain the basic concepts. The practical signals are of finite duration. If the signal is periodic over a finite duration, we relax the periodicity condition and treat it as periodic even if it is not extending up to infinity. The DT sinusoid is periodic if the period is rational and can be represented as the ratio of two prime numbers, for example, k/N. The period of the DT signal is then N samples. A periodic CT sinusoid can become aperiodic DT signal if the CT signal has a frequency f/π. Periodic signals are repetitive and aperiodic signals non-repetitive. Signals are also classified as either energy or power signals. An energy signal has finite energy and zero average power, whereas a power signal has finite average power but infinite energy. Some signals have infinite energy and infinite average power and are said to be neither energy nor a power signal.

- Different signal operations are explained in the next section. Time-shifting operation for shift left and shift right is explained for signal $x(t)$ and the time-reversed signal $x(-t)$. The amplitude scaling is straightforward; however, time scaling is a bit complicated. When the signal is time-scaled by a factor of ½, it gets dilated and when time-scaled by a factor of 2, it gets compressed. When the time scaling as well as time shifting is involved, it must be done with due precaution. To solve it correctly, the precedence rule helps the students better. We emphasize that we can do time scaling first and then shifting by properly putting the operations in bracket. For example, $x(2t - 3)$ will be written as $x(2(t - 3/2))$ for proper implementation. The sample by sample addition and multiplication is so illustrated for both CT and DT signals. Time scaling by ½ and 2 means interpolation and decimation, respectively, in case of DT signals. All operations are explained and illustrated for both CT and DT signals. Some miscellaneous problems are solved for the benefit of students. The physical significance of each standard signal, each property and each operation on the signals is explained.

Multiple Choice Questions

1. A signal has
   (a) single value for single time instant
   (b) single value for multiple time instants
   (c) can have any number of values for any number of time instants
   (d) none

2. A DT signal has
   (a) continuous time continuous amplitude
   (b) continuous time discrete amplitude
   (c) discrete time continuous amplitude
   (d) discrete time discrete amplitude

3. The signal is represented in frequency domain if
   (a) X-axis represents frequency
   (b) X-axis represents time
   (c) Y-axis represents time
   (d) Y-axis represents frequency

4. Sine wave signal is a
   (a) energy signal
   (b) power signal
   (c) short-lived signal
   (d) neither energy nor power signal

5. Consider an image signal. It is a
   (a) spatial domain signal
   (b) time domain signal
   (c) frequency domain signal
   (d) vector signal

6. A plot of FT of speech signal will represent
   (a) a time domain signal
   (b) frequency domain signal
   (c) spatial domain signal
   (d) a magnitude plot

7. A signal is said to be even signal if
   (a) it is symmetric with respect to y-axis
   (b) it is asymmetric with respect to y-axis

(c) it is symmetric with respect to x-axis

(d) it is anti symmetric with respect to x-axis

8. A signal is said to be periodic if

(a) $x(t) = x(t + 2T)$ for all t

(b) $x(t) = x(2t + T)$ for all t

(c) $x(t) = x(t + NT)$ for all t

(d) $x(t) = x(t + T)$ for all t

9. A signal given by $x(t) = 5 \sin(4\pi t)$ for all t is

(a) deterministic

(b) non-deterministic

(c) random

(d) stationary random

10. A signal given by $x(t) = u(t - 2)$ is

(a) left-handed sequence

(b) causal sequence

(c) anti-causal sequence

(d) both-sided sequence

11. A signal given by $x(t) = \sin(2\pi f t)$ for all t is

(a) energy signal

(b) neither energy signal nor a power signal

(c) power signal

(d) power signal over a small duration

12. A signal given by $x(t) = u(t + 2) - u(t - 2)$ is

(a) an energy signal

(b) a power signal

(c) neither energy nor power signal

(d) none of the above

13. A signal represented as x = [1010 0101 0001 1001 1100 0011 0001 0101] is

(a) a DT signal

(b) a CT signal

(c) a vector signal

(d) a digital signal

14. When impulse is given as input to the system, the output of the system is called

(a) input response

(b) impulse response

(c) steady response

(d) stable response

15. Analog sinusoid is periodic

(a) irrespective of the period being rational or irrational

(b) only if period is rational

(c) only if period is irrational

(d) period is a ratio of two prime numbers

16. A DT sinusoid is periodic if

(a) the period is a ratio of two integers

(b) the period is a ratio of two prime numbers

(c) the period is rational and is ratio of two prime numbers

(d) the period is irrational

17. A DT signal given by $x[n] = \sin(\pi n / 8)$ for all n has period equal to
    (a) 16
    (b) 8
    (c) 4
    (d) 1/16

18. A signal given by $x(t) = (t\bar{i}, t\bar{j}, 5t\bar{k})$ is
    (a) a scalar signal
    (b) a vector signal
    (c) a time domain signal
    (d) a frequency domain signal

19. A product of two odd signals is an
    (a) odd signal
    (b) even signal
    (c) is neither odd nor even
    (d) none of the above

20. When the signal is time scaled by a factor of ½, time axis is
    (a) dilated
    (b) compresses
    (c) elongated
    (d) enhanced

21. Differentiation of the ramp function is
    (a) is a unit step function
    (b) is an impulse
    (c) is exponential function
    (d) is again a ramp function

22. A precedence rule uses
    (a) time shifting first and then time scaling
    (b) time scaling first and then time shifting
    (c) scaling and shifting in any order
    (d) time operations first

23. The shifting property of the delta function states that
    (a) $\int_{-\infty}^{\infty} x(t)\delta(t-t_0)dt = x(0)$
    (b) $\int_{-\infty}^{\infty} x(t)\delta(t-t_0)dt = x(t_0)$
    (c) $\int_{-\infty}^{\infty} x(t)\delta(t-t_0)dt = x(t)$
    (d) $\int_{-\infty}^{\infty} x(t)\delta(t-t_0)dt = x(t-t_0)$

24. The transmitter, channel and receiver are the basic blocks of the
    (a) Control system
    (b) Any processing system
    (c) Communication system
    (d) Any system

25. If the power supply is suddenly switched on, it is equivalent to
    (a) step input
    (b) ramp input
    (c) impulse input
    (d) sine wave input

26. A sudden shock is equivalent to
    (a) step input
    (b) ramp input
    (c) impulse input
    (d) sine wave input

27. The signal $x(t) = a(t) + j\,b(t)$ has conjugate symmetry if
    (a) $a(t)$ is even and $b(t)$ is even
    (b) $a(t)$ is even and $b(t)$ is odd
    (c) $a(t)$ is odd and $b(t)$ is odd
    (d) $a(t)$ is odd and $b(t)$ is even

28. Use shifting property of delta function to find $\sum_{n=-\infty}^{\infty} e^{2n}\delta(n+2)$

   (a) $\sum_{n=-\infty}^{\infty} e^{-4}$ (c) $\sum_{n=-\infty}^{\infty} e^{2}$

   (b) $\sum_{n=-\infty}^{\infty} e^{4}$ (d) $\sum_{n=-\infty}^{\infty} e^{-2}$

29. A CT signal given by $x(t) = e^t u(t-1)$
   (a) is anti causal
   (b) is causal
   (c) is left handed
   (d) is both sided

30. A DT signal given by $x(n) = u(-n-2) - u(-n-5)$ is
   (a) is anti causal
   (b) is causal
   (c) is left handed
   (d) is both sided

31. If the signal $x(t)$ is shifted left by 2 time units, it will be represented as
   (a) $x(t-2)$
   (b) $x(-t+2)$
   (c) $x(-t-2)$
   (d) $x(t+2)$

32. If the signal $x(-t)$ is shifted right by 2 time units, it will be represented as
   (a) $x(t-2)$
   (b) $x(-t+2)$
   (c) $x(-t-2)$
   (d) $x(t+2)$

33. If the signal $x(t)$ is dilated by a factor of 3, it will be represented as
   (a) $x(t/3)$
   (b) $x(3t)$
   (c) $x(t-3)$
   (d) $x(3t-3)$

34. If a DT signal $x[n]$ is dilated by a factor of 2, we have to
   (a) interpolate the signal by inserting two zeros between every two samples
   (b) interpolate the signal by inserting a zero between every two samples
   (c) delete alternate samples
   (d) delete two samples alternately

35. To generate DT signal $x[2n]$, we have to
   (a) interpolate the signal $x[n]$ by inserting two zeros between every two samples
   (b) interpolate the signal $x[n]$ by inserting a zero between every two samples
   (c) delete alternate samples of $x[n]$
   (d) delete two samples of $x[n]$ alternately

36. A DT signal $x[3n-3]$ is a signal $x[n]$
   (a) compressed 3 times and shifted towards right by one sample
   (b) compressed 3 times and shifted towards right by three samples

(c) dilated 3 times and shifted towards right by one sample
   (d) dilated 3 times and shifted towards right by one sample
37. A CT signal $x(2t - 4)$ is a signal $x(t)$
   (a) compressed 2 times and shifted towards right by one sample
   (b) compressed 2 times and shifted towards right by two samples
   (c) dilated 2 times and shifted towards right by one sample
   (d) dilated 2 times and shifted towards right by two samples
38. Up sampling by a factor of M is done by
   (a) Inserting $M - 1$ zeros between every two samples
   (b) Inserting M zeros between every two samples
   (c) Deleting $M - 1$ zeros between every two samples
   (d) Deleting $M - 1$ zeros between every two samples
39. The normalized sinc function is defined as
   (a) $\sin c(\omega_0 t) = \dfrac{\sin(\pi \omega_0 t)}{\pi \omega_0 t}$
   (c) $\sin c(\omega_0 t) = \dfrac{\sin(\pi \omega_0 t)}{\omega_0 t}$
   (b) $\sin c(\omega_0 t) = \dfrac{\sin(\omega_0 t)}{\omega_0 t}$
   (d) $\sin c(\omega_0 t) = \dfrac{\sin(\omega_0 t)}{\pi \omega_0 t}$
40. FT of the rectangular pulse is
   (a) a signum function
   (b) a sign function
   (c) a parabolic function
   (d) a sinc function

Review Questions

2.1 What is a signal?
2.2 What are scalar-valued signals and multiple-valued signals?
2.3 Why is analog signal called continuous time, continuous amplitude signal?
2.4 What is a DT signal?
2.5 What is a domain? What is a spatial domain?
2.6 Why are naturally occurring signals random signals?
2.7 Can any short-lived signal be a power signal?
2.8 Can a periodic signal be a power signal?
2.9 State different applications where you come across signals.
2.10 Define a system. Explain the use of communication system.
2.11 Explain how the control system processes the signals.
2.12 What is the need for standard signals?
2.13 Explain use standard signals for worst-case testing of a system.

2.14 How will you make use of standard signals for system testing before it is actually implemented?

2.15 List various standard CT signals and their significance.

2.16 What is a unit impulse? What practical situation it will represent?

2.17 Draw unit step and unit ramp and state the relation between the two.

2.18 What is a signum function? Why is it called as a sign function?

2.19 What is the significance of a sinc function? How is it related to a rectangular pulse?

2.20 List various standard DT signals and plot them.

2.21 List the physical significance of all standard signals.

2.22 Define a periodic signal. Explain the procedure for finding the period of a signal.

2.23 Explain the situation when a periodic CT sinusoid after sampling becomes aperiodic DT sinusoid.

2.24 Explain the periodicity condition for a DT sinusoid.

2.25 How will you find the period for a combination signal?

2.26 Define even and odd property of a signal. Define different combinations of even and odd signals and state whether they will be even or odd.

2.27 Define energy signal. Define a power signal. Can a signal be neither energy nor a power signal?

2.28 State the physical significance of different properties of the signals.

2.29 Define different operations on signals such as time scaling and amplitude scaling.

2.30 Explain time-shifting operation. Explain the steps for implementing time scaling followed by time sifting or vice versa.

2.31 How will you implement time scaling for a DT signal? Explain multiplication and addition of two DT signals.

2.32 What is precedence rule? Why you need to use it? How to solve the problem correctly without using a precedence rule?

2.33 Explain the physical significance of different operations on the signals.

Problems

2.1 Find the signal value at $t = 0.2$ s for a signal given by $y(t) = \sin(0.25\pi t)$.

2.2 Find the average power for a signal $y(t) = \sin(0.25\pi t)$.

2.3 Find the domain for the two-dimensional image signal.

2.4 Generate DT signal samples at $t = 0.1, 0.2$ and 0.3 for a signal $y(t) = 4t$.

2.5 If $x(t) = (t\bar{i}, 2t\bar{j}, 5j\bar{k})$. Can we call it a vector signal?

2.6 Consider a signal given by $x(t) = \sin(2\pi ft)$ where f is the frequency of the signal equal to 100 Hz. Plot a signal using MATLAB program.

2.7 Plot an analog signal given by

$x(t) = 0.1 \times t$ for $0 \le t \le 10$ seconds

$= 0$ otherwise

Write a MATLAB program to generate a signal.

2.8 Plot analog signal given by

$x(t) = -1$ for $0 \le t \le 2$ seconds

$= 1$ for $2 < t \le 4$ seconds

and write MATLAB program to plot the signal.

2.9 Consider a sampled signal $x(t) = \sin(2\pi t)$ where a sample is taken at $t = 0, T, 2T, 3T$, etc. T represents a sampling time given by $T = 1/f_s$ = sampling frequency. The signal can be written as $x(n) = \sin(20\pi nT)$. Plot the signal using a MATLAB program.

2.10 (Discrete time signal) Plot a sampled signal

$x(n) = 1$ for $0 \le n \le 3$

$= -1$ for $4 \le n \le 6$

$= 0$ otherwise

2.11 (Discrete time signal) Plot a sampled signal

$x(n) = -2$ for $1 \le n \le 3$

$= 1$ for $-2 \le n \le 0$

$= 0$ otherwise

2.12 $x = \{0, 0.125, 0.5, 0.25, 0.125\}$ Write a digital signal values using 4-bit representation with first bit as the sign bit.

2.13 Evaluate the following integrals i) $\int_{-\infty}^{\infty} e^{-2t^2} \delta(t-1) dt$ ii) $\int_{-\infty}^{\infty} t^2 \delta(t-6) dt$

iii) $\int_{-\infty}^{\infty} \sin(\pi t) \delta(t-1) dt$ iv) $\int_{-\infty}^{\infty} (t-1)^2 \delta(t-1) dt$

v) $\int_{-\infty}^{\infty} [\sin(2t)\delta(t) + \sin(2t)\delta(t-2)] dt$ vi) $\int_{-\infty}^{\infty} e^{4j\omega t} \delta(t) dt$

2.14 Evaluate the summation i) $\sum_{n=-\infty}^{\infty} e^n \delta(n)$ ii) $\sum_{n=-\infty}^{\infty} \cos(3n)\delta(n-2)$.

iii) $\sum_{n=-\infty}^{\infty} e^{2n} \delta(n+1)$

2.15 Consider an analog signal given by

$x(t) = A$ for $-3/2 \leq t \leq 3/2$. Find if the signal is even.

2.16 Consider an analog signal given by

$x(t) = -4$ for $-3/2 \leq t \leq 0$ and

$x(t) = 4$ for $0 < t \leq 3/2$

Find if it is even or odd.

2.17 Consider a signal given by

$x(t) = e^{j3\omega t}$. Find the even and odd part of the signal.

2.18 If the signal is defined as $x(t) = e^{j3\omega t}$ for all $t \geq 0$, find if the signal is even or odd.

2.19 Find the even and odd parts of the following signals.

i) $x(t) = \cos(2t) + \cos(3t) + \cos(t)\sin(2t)$ ii) $x(t) = 1 + 3t + t^2 + t^3$

iii) $x(a) = \cos(2a) + \sin^3(a)$ iv) $x(a) = a^2 \cos(2a) + a^3 \sin(2a)$

v) $x(t) = 1 + 2t\cos(t) + t^2 \sin(3t) + t^3 \sin(2t) \cos(5t)$

vi) $x(t) = (1 + t^2) \cos(4t)$ vii) $x(t) = (t + t^3) \sin(5t)$

viii) $x(t) = (t + t^3) \tan(t)$ ix) $x(t) = t \times (1 + t^2 + t^3)$

2.20 Consider a discrete time signal given by

$x(n) = 1$ for $-3 \leq n \leq 3$

$= 0$ otherwise

Find if the signal is even or odd.

2.21 Consider a discrete time signal given by

$x(n) = 1$ for $1 \leq n \leq 4$

$= -1$ for $-1 \leq n \leq -4$

Find if the signal is even or odd.

2.22 Consider a discrete time signal given by

$x(n) = 1$ for $1 \leq n \leq 3$

$= 0$ otherwise

Find if the signal is even or odd.

2.23 Find if the signal is periodic and find the period.

i) $x(t) = \sin(2t)$ ii) $x(t) = 2t + \cos(4\pi t)$

iii) $x(t) = (\sin(4\pi t))^2$ iv) $x(t) = |\cos(4\pi t)|$

2.24 Consider the signal shown in figure below. Is $x(t)$ a periodic signal?

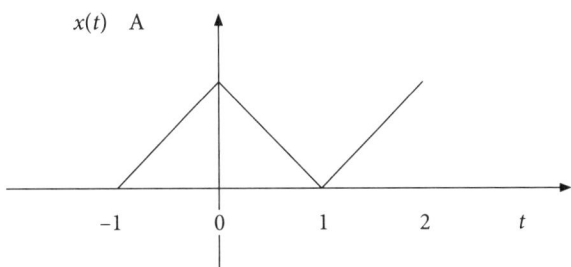

2.25 Consider the signal $x(t)$ shown in figure below. If $y(t) = \sum_{k=-6}^{6} x(t-2k)$, is $y(t)$ a periodic signal?

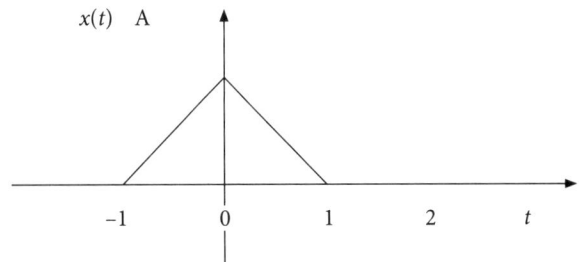

2.26 Consider the signal $x(t)$ shown in figure below. If $y(t) = \sum_{k=-\infty}^{\infty} x(t-4k)$, is $y(t)$ a periodic signal?

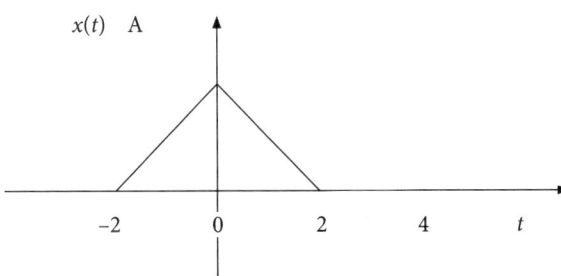

2.27 Consider the equation given by $x(t) = e^{-5t}$. Is $x(t)$ a periodic signal?
2.28 Consider the signal $x(n)$ given by $x(n) = (-1/2)^n$. Is $x(n)$ a periodic signal?

2.29 Consider the signal $x(n)$ given by $x(n) = (-1)^{n^3}$. Is $x(n)$ a periodic signal?

2.30 Consider the signal shown in figure below. Is $x(n)$ a periodic signal?

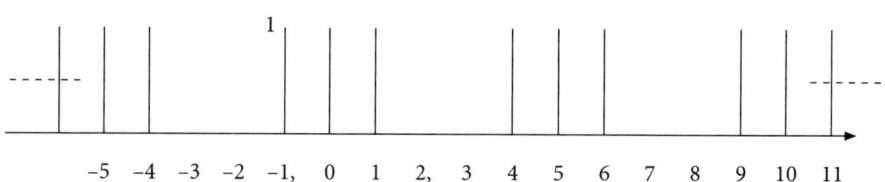

2.31 Consider the signal shown in figure below. Is $x(t)$ a periodic signal?

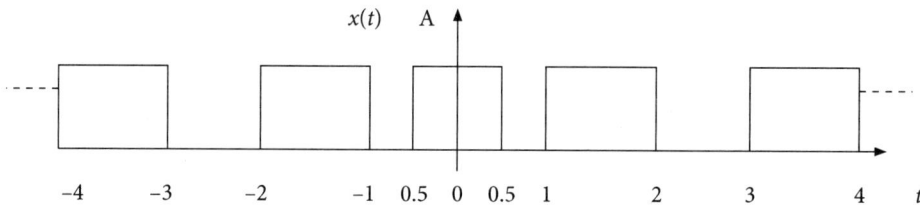

2.32 Find if the following DT signals are periodic

i) $x(n) = \cos(0.03n\pi)$ ii) $x(n) = \cos\left(\dfrac{10n}{105}\pi\right)$ iii) $x(n) = \cos(5\pi n)$

iv) $x(n) = \sin(2n)$ v) $x(n) = \sin\left(\dfrac{82n}{10}\pi\right)$ vi) $x(n) = 5\cos\left(3n + \dfrac{\pi}{4}\right)$

vii) $x(n) = 2\exp\left(j\left(\dfrac{n}{4} - \dfrac{\pi}{3}\right)\right)$ viii) $x(n) = \cos\left(\dfrac{\pi}{2}\right)\sin\left(\dfrac{n\pi}{8}\right)$

2.33 Consider a linear combination of two analog sinusoidal functions. $x(t) = 3\sin(6\pi t) + \cos(4\pi t)$. Find if the signal is periodic.

2.34 Consider a linear combination of two analog sinusoidal functions. $x(t) = 2\cos(3\pi t) + \sin(3t)$. Find if the signal is periodic.

2.35 Consider a $x(n) = u(n) - u(n-8)$. Find if it is causal.

2.36 Consider the following sequence $x(n) = u(-n - 1) - u(-n - 5)$ find if the signal is causal.

2.37 Consider a CT signal given by $x(t) = e^{5t} u(t - 1)$. Find if the signal is causal.

2.38 Consider a signal given by $x(t) = 2\sin c(7t)$. Find if the signal is causal.

2.39 Consider a CT signal $x(t) = e^t[u(t+4) - u(t-3)]$. Find if the signal is causal.

2.40 Consider a DT signal $x(t) = \left(\frac{1}{2}\right)^n u(n+5) - \left(\frac{1}{3}\right)^n u(n-4)$. Find if the signal is causal.

2.41 Find if the following signals are deterministic signals.

i) $x(t) = 2t + \sin(3\pi ft) + \cos(4\pi ft)$ ii) $x(n) = \cos\left(\frac{n\pi}{4}\right) + 3\sin\left(\frac{n\pi}{3} + \frac{\pi}{5}\right)$

2.42. Consider the sinusoid of frequency 2 KHz. Is it a power signal?

2.43 If $x(t) = \sin(2\pi 50 t)$ for $0 \leq t \leq 1/2$, is $x(t)$ an energy signal?

2.44 Consider a signal defined as

$$x(t) = \begin{cases} 2t \text{ for } 0 \leq t \leq 1 \\ 4 - 2t \text{ for } 1 \leq t \leq 2 \\ 0 \text{ otherwise} \end{cases}$$

Find the energy and power of the signal and classify it as an energy or a power signal.

2.45 Consider a signal defined as

$$x(t) = \begin{cases} 2\cos(\pi t / 2) \text{ for } -1 \leq t \leq 1 \\ 0 \text{ otherwise} \end{cases}$$

Find the energy and power of the signal and classify it as an energy or a power signal.

2.46 Consider a signal defined as

$$x(t) = \begin{cases} 2 \text{ for } -3 \leq t \leq 3 \\ 0 \text{ otherwise} \end{cases}$$

Find the energy and power of the signal and classify it as an energy or a power signal.

2.47 Consider a signal given by

$$x(t) = \{3\cos(\pi t) + 2\cos(\pi t) \text{ for } -\infty \leq t \leq \infty\}$$

Find the energy and power of the signal and classify it as an energy or a power signal.

2.48 Consider a periodic DT sinusoid given by

$$g(k) = 5\cos\left(\frac{\pi k}{20}\right).$$ Find if the signal is an energy signal or a power signal.

2.49 Consider an analog periodic signal sketched in figure below. Find if the signal is an energy signal or a power signal?

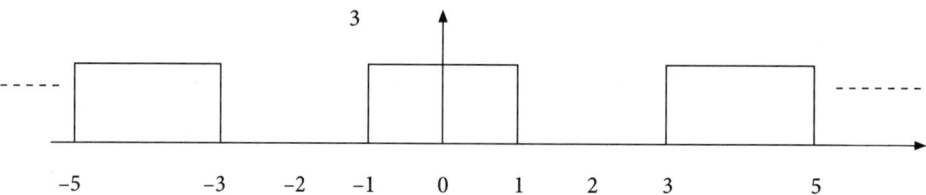

2.50. Consider an analog periodic signal sketched in figure below. Find if the signal is an energy signal or a power signal?

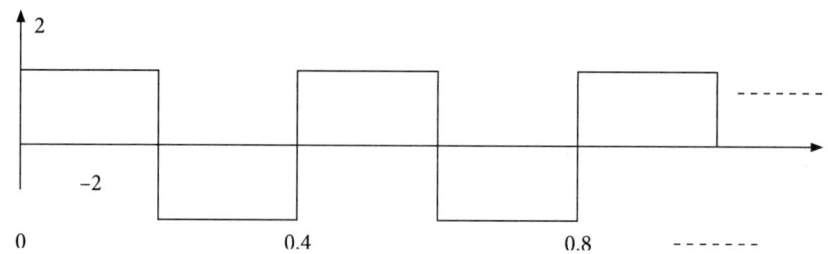

2.51. Consider an analog periodic signal, a triangular wave sketched in figure below. Find if the signal is an energy signal or a power signal?

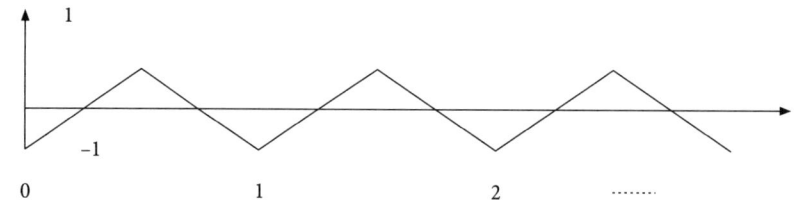

2.52 Consider a DT periodic signal, as shown in the figure below. Find if the signal is an energy signal or a power signal? Find the average power.

2.53 Consider a DT periodic signal, as shown in the figure below. Find if the signal is an energy signal or a power signal?

[Figure: impulses of height 1 at n = -1, 0, 1 on axis labeled -2 -1 0 1 2 ..... n]

2.54 Consider a signal defined as

$$x(n) = \begin{cases} n & \text{for } 0 \leq n \leq 4 \\ 9-n & \text{for } 5 \leq n \leq 9 \\ 0 & \text{otherwise} \end{cases}$$

Find the energy and power of the signal and classify it as an energy or a power signal.

2.55 Consider a signal defined as

$$x(n) = \begin{cases} \sin(\pi n/2) & \text{for } -4 \leq n \leq 4 \\ 0 & \text{otherwise} \end{cases}$$

Find the energy and power of the signal and classify it as an energy or a power signal.

2.56 Consider a trapezoidal signal given by

$$X(t) = \begin{cases} t+5 & -5 \leq t \leq -3 \\ 2 & -3 \leq t \leq 3 \\ 5-t & 3 \leq t \leq 5 \\ 0 & \text{otherwise} \end{cases}$$

Find the total energy of the signal.

2.57 Find if the signal $x(n) = \left(\dfrac{1}{3}\right)^n u(n)$ is an energy signal or a power signal?

2.58 Find if the signal $x(n) = 3u(n)$ is an energy signal or a power signal?

2.59 Find if the signal $x(n) = u(n) - u(n-7)$ is an energy signal or a power signal.

2.60 Find the Power of the signal given by $x(t) = e^{j5t} \cos(3t)$

2.61 Determine if the signal is an energy signal or a power signal. Find the energy or power of the signal given by $x(t) = \sin^2(3t)$

2.62 Determine if the signal is an energy signal or a power signal. Find the energy or power of the signal given by $x(t) = \sin(2t)[u(t-1) - u(t-5)]$.

2.63 Find the Energy and average Power of the signal given by $x(n) = e^{j[(\pi/2)n + \pi/6]}$

2.64 Find if the signal is an energy or a power signal.

$$x(n) = \begin{cases} n & \text{for } 0 \leq n \leq 4 \\ 10-n & \text{for } 6 \leq n \leq 9 \\ 0 & \text{otherwise} \end{cases}$$

2.65 Find energy of the signal shown below.

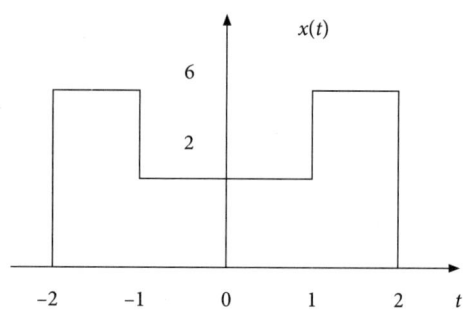

2.66 Consider a rectangular pulse given by

$$x(t) = \text{rect}\left(\frac{t}{4}\right) = \begin{cases} 1 & \text{for } |t| \leq 2 \\ 0 & \text{for } |t| > 2 \end{cases}$$

Draw the following functions derived from the rectangular pulse.

$x(3t)$, $x(3t + 4)$, $x(-2t - 2)$, $x(2(t + 2))$, $x(2(t - 2))$, $x(3t) + x(3t + 4)$.

2.67 Solve problem 66 using a precedence rule.

2.68 Consider signal $x(n)$ shown below. Plot $x(2n)$ and $x(1/2n)$.

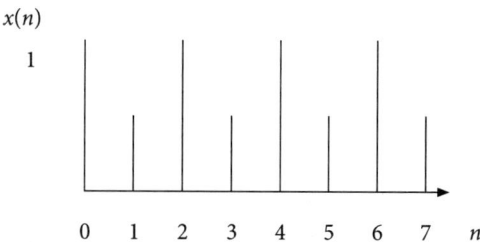

2.69 Consider two CT signals $x(t)$ and $y(t)$, as shown in the figure below. Find $x(t) + y(t)$, $x(t) - y(t)$ and $x(t)y(t)$, $x(t)y(t - 1)$, $x(t + 1)y(t - 2)$, $x(t - 1)y(-t)$, $x(t)y(-t - 1)$, $x(2t)y(-t + 2)$, $x(2t) + y(2t)$.

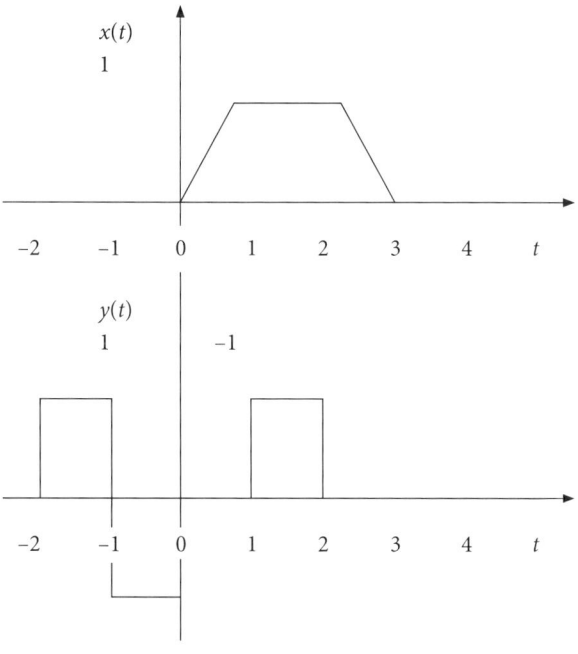

2.70 Consider DT signals $x(n)$ and $y(n)$, as shown in Figures 1 and 2. Plot $x(n) + y(n)$, $x(n)y(n)$, $x(2n)y(n)$, $x(n-1)y(n+2)$.

$x(n) = \{1,1,2,1,1,1,2,1\}$, $y(n) = \{2,1,2,1,1,1,2,0\}$

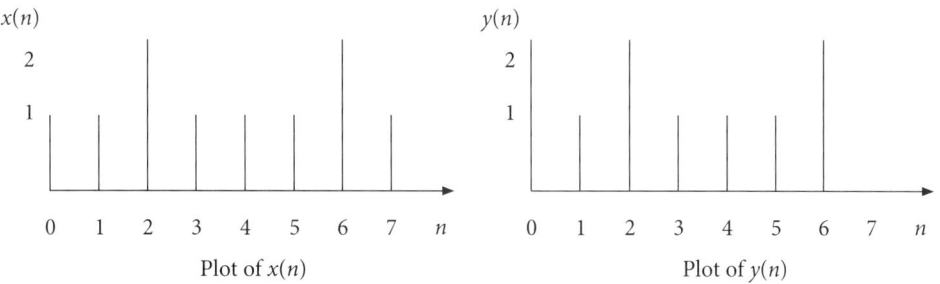

2.71 Sketch the waveforms given by the following equations where $u(t)$ is a unit step function and $r(t)$ is a unit ramp function.

i) $x(t) = u(t) - u(t-4)$

ii) $x(t) = u(t+2) - 2u(t) + u(t-2)$

iii) $x(t) = -u(t+2) + 2u(t+1) + u(t-2)$

iv) $x(t) = -r(t+2) - r(t) + r(t-2)$

iv) $x(t) = r(t) - r(t-2) - r(t-3) + r(t-4)$

Answers

Multiple Choice Questions

1 (a)	2 (c)	3 (a)	4 (b)	5 (a)
6 (b)	7 (a)	8 (d)	9 (a)	10 (b)
11 (c)	12 (a)	13 (d)	14 (b)	15 (a)
16 (c)	17 (a)	18 (b)	19 (b)	20 (a)
21 (a)	22 (a)	23 (b)	24 (a)	25 (a)
26 (c)	27 (b)	28 (a)	29 (b)	30 (a)
31 (d)	32 (b)	33 (a)	34 (b)	35 (c)
36 (a)	37 (b)	38 (a)	39 (a)	40 (d)

Problems

2.1   1

2.2   0.25

2.3   Spatial domain

2.4   0.4, 0.8 1.2

2.5   a vector signal

2.6
```
clear all;
f=100;
w=2*pi*f;
t=0:0.0001:0.1;
s=sin(w*t);
plot(t,s);
title('plot of sine wave-approximation to analog sine wave is plotted');
xlabel('time'); ylabel('amplitude');
```

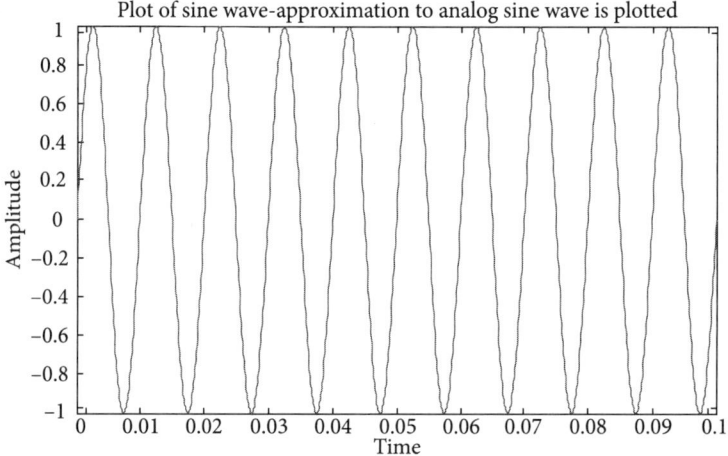

2.7 ```
clear all;
t=0:0.1:10;
x=0.01*t;
plot(t,x);
title('plot of signal x');
xlabel('time'); ylabel('amplitude');
```

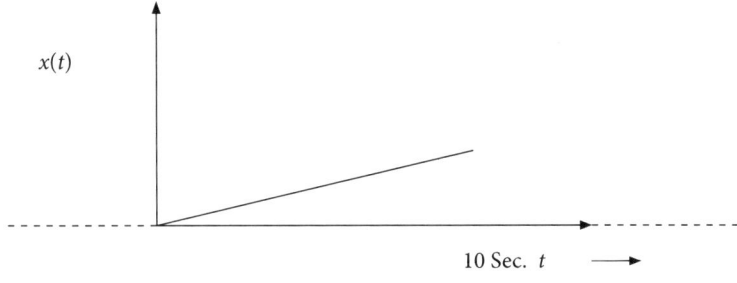

Plot of signal $x(t)$ with slope = 0.1

2.8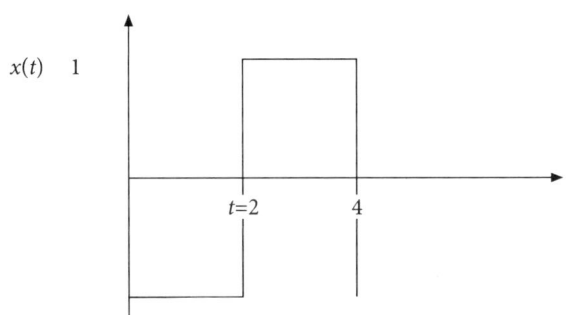

Plot of signal $x(t)$ defined between $t = 0$ to $t = 4$ seconds

A MATLAB program can be written as

```
clear all;
t=0:0.01:2;
i=1+t*10;
for i=1:20,
 x(i)= -1;
end
t=2:0.01:4;
i=11+(t-1)*10;
for i=21:40,
 x(i)= 1;
end
plot(x);
title('plot of signal x');
xlabel('time'); ylabel('amplitude');
```
```

2.9
```
clear all;
f=10;
T=0.01;
for n=1:21,
x(n)=sin(2*pi*f*(n-1)*T);
end
stem(x);title('plot of DT signal x');
xlabel('sample number');ylabel('amplitude');
```

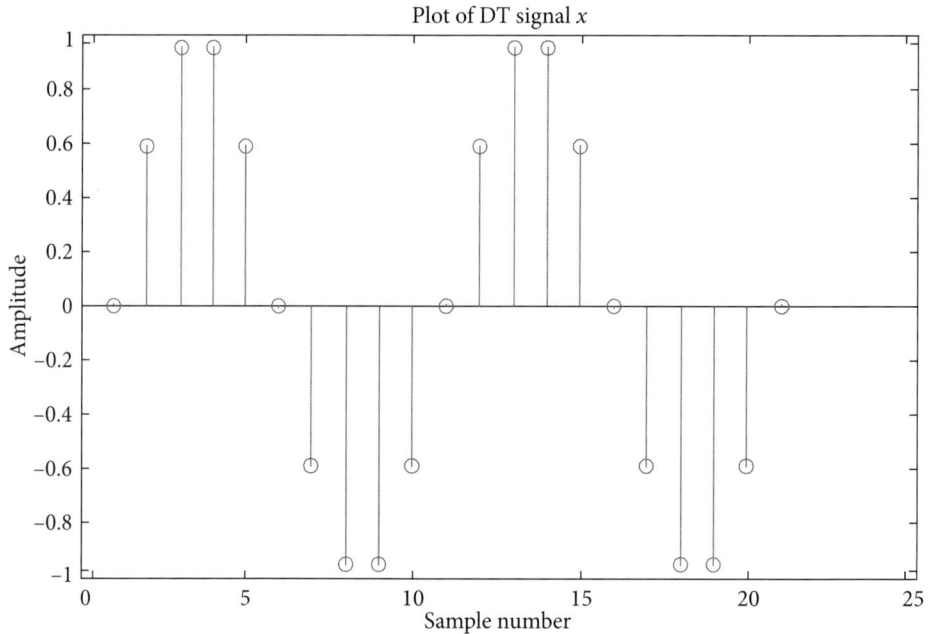

2.10

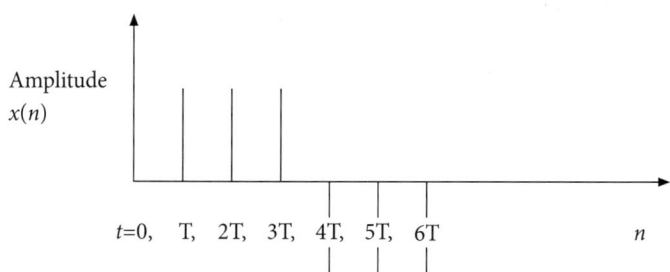

Plot of signal $x(n)$ in Example 2.11

```
clear all;
for n=1:4,
x(n)=1;
```

```
end
for n=5:7,
 x(n)=-1;
end
s=0:1:6;
stem(s,x);title('plot of DT signal x');
xlabel('sample number');ylabel('amplitude');
```

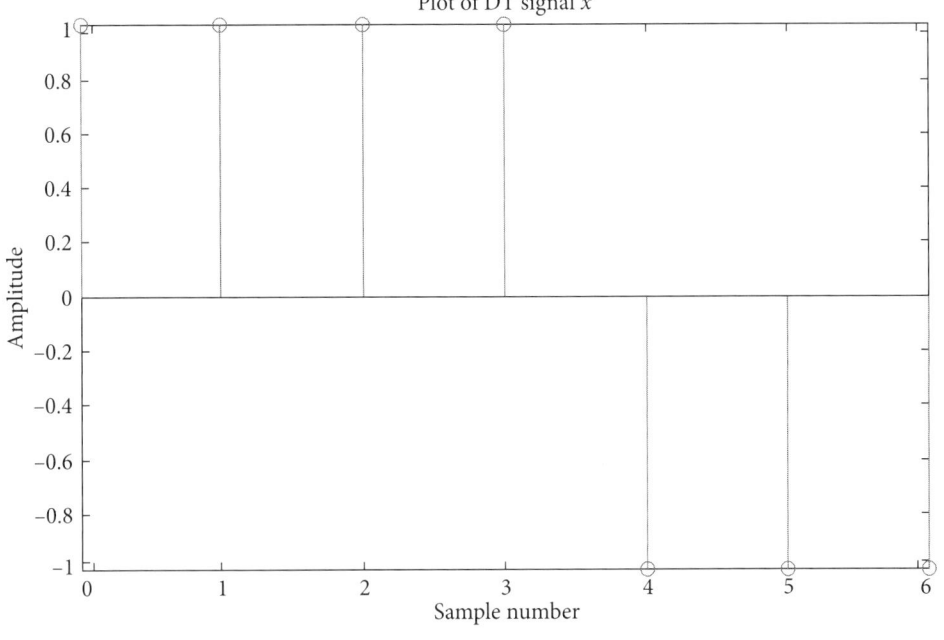

Plot of signal x

2.11 Solution: The signal plot is shown in Fig. 13.

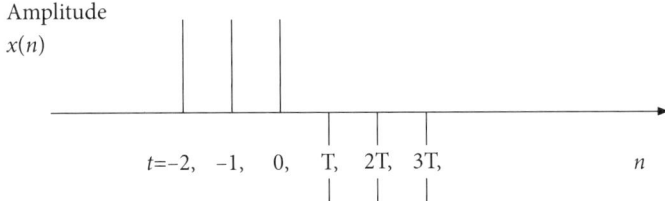

Plot of signal $x(n)$ for Example 2.11

2.12 $X = \{1000, 1001, 1100, 1010, 1001\}$

2.13 i) $\int_{-\infty}^{\infty} e^{-2t^2} \delta(t-1) dt = [e^{-2t^2}] \downarrow t = 1 = e^{-2 \times 1} = e^{-2}$

ii) $\int_{-\infty}^{\infty} t^2 \delta(t-6) dt = [t^2] \downarrow t = 6 = 36$

iii) $\int_{-\infty}^{\infty} \sin(\pi t) \delta(t-1) dt = \sin(\pi) \downarrow t = 1 = 0$

iv) $\int_{-\infty}^{\infty} (t-1)^2 \delta(t-1) dt = 0$

v) $\int_{-\infty}^{\infty} [\sin(2t)\delta(t) + \sin(2t)\delta(t-2)] dt = \sin(4)$

vi) $\int_{-\infty}^{\infty} e^{4j\omega t} \delta(t) dt = \int_{-\infty}^{\infty} \delta(t) dt = 1$

note: $e^{4j\omega \times 0} = 1$

2.14 i) $\sum_{n=-\infty}^{\infty} e^n \delta(n) = 1$ ii) $\sum_{n=-\infty}^{\infty} \cos(3n)\delta(n-2) = \cos(6)$ iii) $\sum_{n=-\infty}^{\infty} e^{2n} \delta(n+1) = e^{-2}$

2.15 Yes, signal is even.

2.16 The signal is odd.

2.17 $x_e(t) = [x(t) + x(-t)]/2 = \cos(3\omega t)$

$x_o(t) = [x(t) - x(-t)]/2 = j \sin(3\omega t)$

2.18 The signal neither even nor odd. The signal is simply undefined for all negative values of t.

2.19 i) $x_e(t) = [x(t) + x(-t)]/2 = \cos(2t) + \cos(3t)$

$x_o(t) = [x(t) - x(-t)]/2 = \cos(t)\sin(2t)$

ii) $x_e(t) = [x(t) + x(-t)]/2 = 1 + t^2$

$x_o(t) = [x(t) - x(-t)]/2 = 3t + t^3$

iii) $x_e(t) = [x(t) + x(-t)]/2 = \cos(2a)$

$x_o(t) = [x(t) - x(-t)]/2 = \sin^3(a)$

iv) $x_e(t) = [x(t) + x(-t)]/2 = a^2 \cos(2a) + a^3 \sin(2a)$

$x_o(t) = [x(t) - x(-t)]/2 = 0$

v) $x_e(t) = [x(t) + x(-t)]/2 = 1 + t^3 \sin(2t)\cos(5t)$

$x_o(t) = [x(t) - x(-t)]/2 = 2t \cos(t) + t^2 \sin(3t)$

vi) $x_e(t) = [x(t) + x(-t)]/2 = \cos(4t) + t^2 \cos(4t)$

$x_o(t) = [x(t) - x(-t)]/2 = 0$

vii) $x_e(t) = [x(t) + x(-t)]/2 = t\sin(5t) + t^3 \sin(5t)$

$x_o(t) = [x(t) - x(-t)]/2 = 0$

viii) $x_e(t) = [x(t) + x(-t)]/2 = t\tan(t) + t^3 \tan(t)$

$x_o(t) = [x(t) - x(-t)]/2 = 0$

ix) $x_e(t) = [x(t) + x(-t)]/2 = t^4$

$x_o(t) = [x(t) - x(-t)]/2 = t + t^3$

2.20 The signal is symmetric with respect to the y axis and hence, is even.

2.21 The signal is anti-symmetric with respect to the y axis and hence, is odd.

2.22 The signal is having a value zero for all negative values of n. It is neither symmetric nor anti-symmetrical with respect to the origin and with respect to the amplitude axis. Hence, the signal is neither even nor odd.

2.23 i) Signal is periodic with period equal to π. ii) Signal is aperiodic.

iii) Signal is periodic with period = ¼. iv) Periodic. Period = ¼.

2.24 Signal is not periodic.

2.25 The signal $y(t)$ exists only between $t = -13$ to $t = 13$. Over this limited period of time, the signal is periodic.

2.26 The signal is periodic.

2.27 Signal is aperiodic.

2.28 Signal is aperiodic.

2.29 Signal is periodic with period equal to 2 samples.

2.30 Signal is periodic with period equal to 5 samples.

2.31 Signal is aperiodic.

2.32 i) periodic-32 samples, ii) period-21 samples, iii) period-2 samples, iv) signal is aperiodic, v) period is 10 samples, vi) aperiodic, vii) aperiodic, viii) period-16 samples.

2.33 Period for combination is 1 second.

2.34 Period or combination is $2\pi/3$ seconds.

2.35 Signal is a causal signal as it is zero for all negative values of n.

2.36 Signal is left-handed and non-causal sequence.

2.37 Signal is causal.

2.38 Signal is non-causal.

2.39 Signal is both-sided sequence and hence, is non-causal.
2.40 Signal is both-sided sequence and hence is non-causal.
2.41 Both the signals are deterministic signals.
2.42 Signal is a power signal with average power of 0.5.
2.43 Signal is a power signal with average power of 0.25.
2.44 Signal exists over a finite duration. It is an energy signal with total energy of 2 units.
2.45 Signal is an energy signal as it exists for finite duration with total energy of 4 units.
2.46 Energy signal with energy of 24 units.
2.47 Power signal. Average power is 6.5 units.
2.48 Power signal with average power of 12.5 units
2.49 Power signal with average power of 4.5 units.
2.50 Power signal with average power of 4 units.
2.51 Power signal with average power of 25/3 units.
2.52 Power signal with average power of 3/4 units.
2.53 Energy signal. Total energy is 4 units. Average power is zero.
2.54 Energy signal. Total energy is 60 units.
2.55 Energy signal. Total energy is 4 units.
2.56 Energy is 52/3 units.
2.57 Energy signal with total energy of 9/8 units.
2.58 Power signal with average power equal to 3/2.
2.59 Energy signal. Energy equal to 7 units.
2.60 Average power is ½ units.
2.61 Power is ½ units.
2.62 Energy is 2 units.
2.63 Average power is 1 unit. Energy is infinite.
2.64 Energy signal. Energy = 60 units.
2.65 Energy is 80 units.
2.66

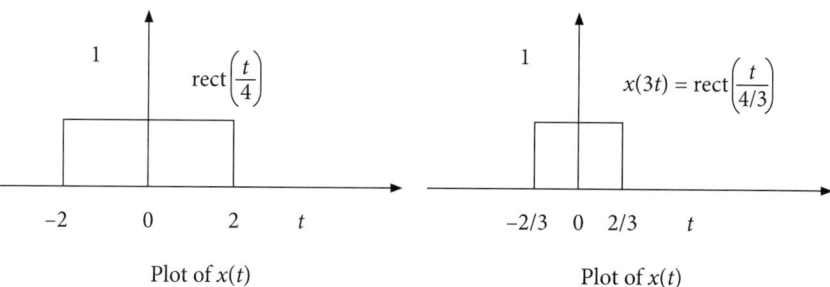

Plot of $x(t)$ Plot of $x(t)$

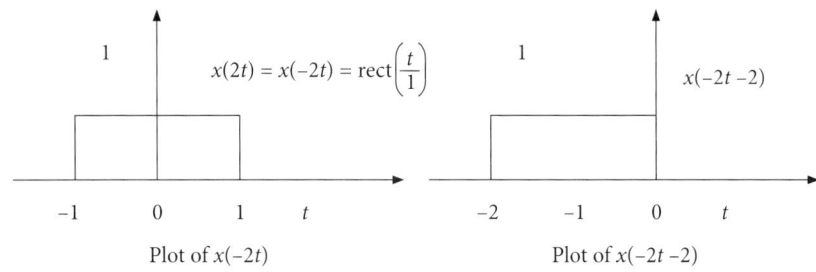

Plot of $x(3t + 4)$

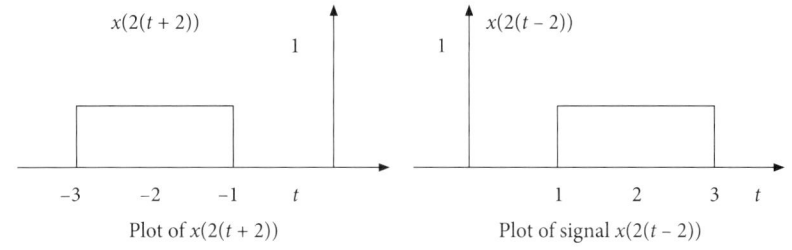

Plot of $x(-2t)$; Plot of $x(-2t - 2)$

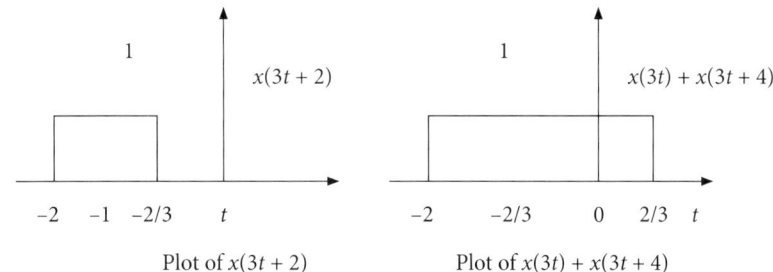

Plot of $x(2(t + 2))$; Plot of signal $x(2(t - 2))$

Plot of $x(3t + 2)$; Plot of $x(3t) + x(3t + 4)$

2.67 Same as problem 66.

2.68

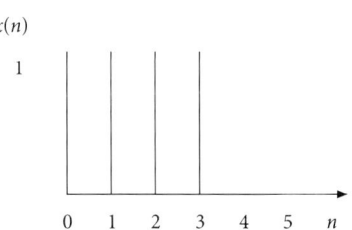

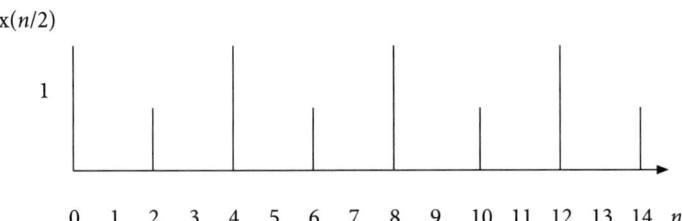

2.69

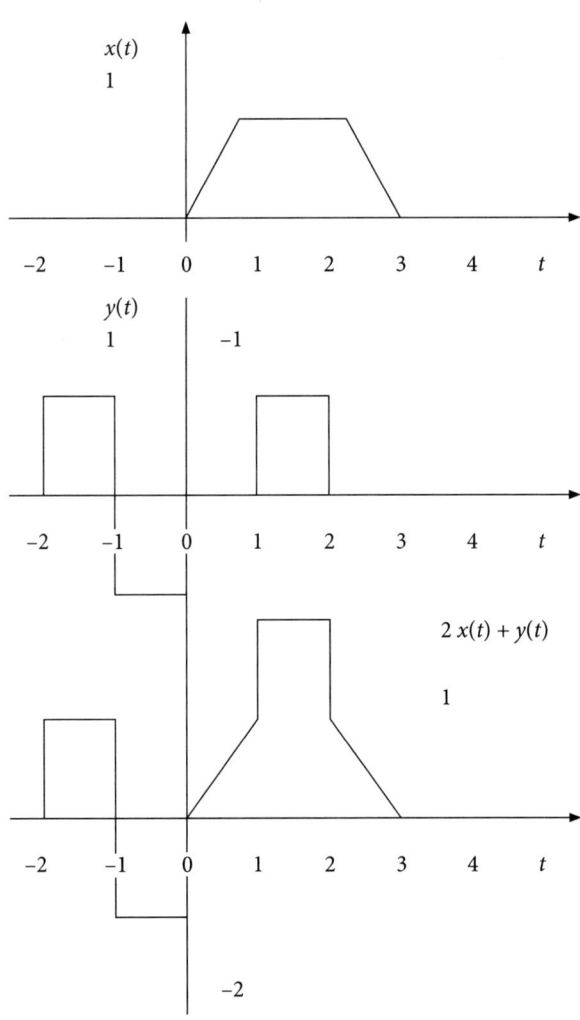

Plot of $x(t) + y(t)$

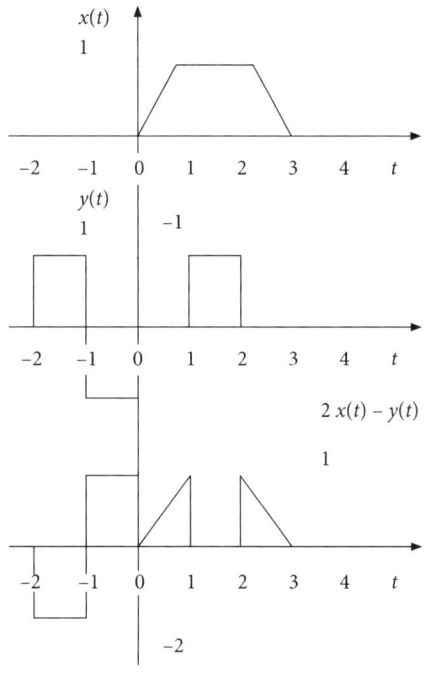

Plot of $x(t) - y(t)$

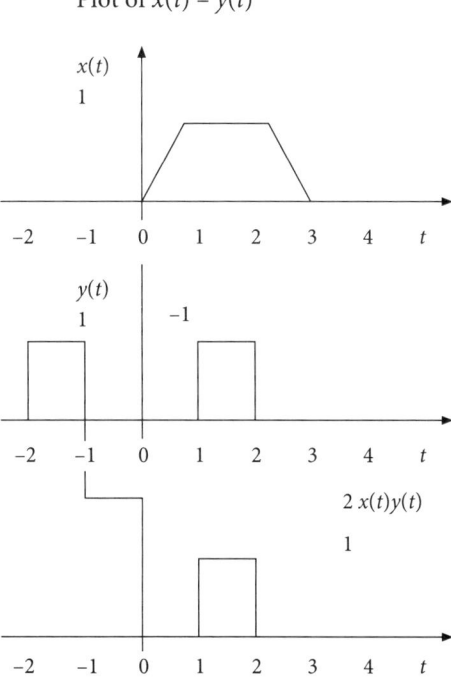

Plot of $x(t)$, $y(t)$ and $x(t)y(t)$

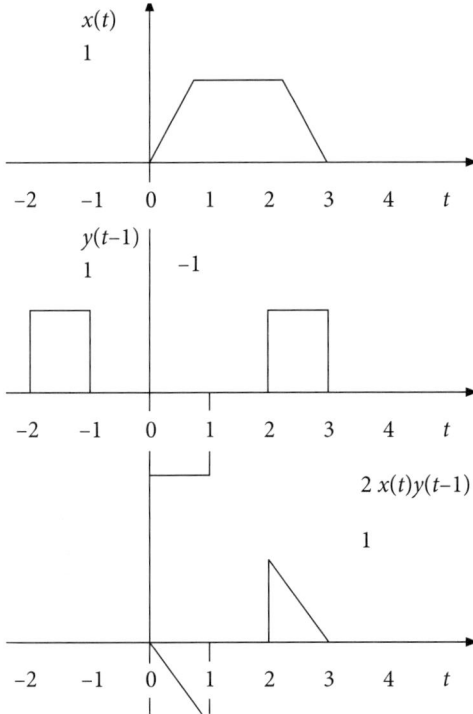

Plot of $x(t)$, $y(t-1)$, $x(t)y(t-1)$

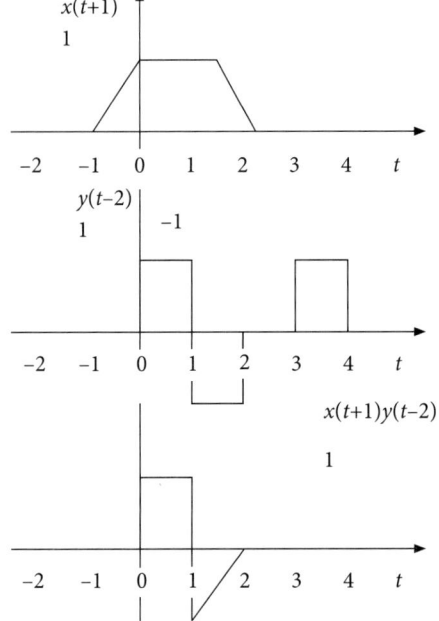

Plot of $x(t+1)y(t-2)$

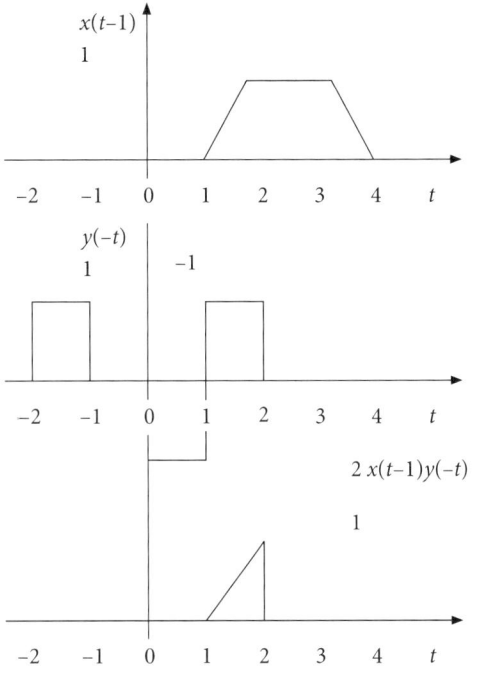

Plot of $x(t-1)y(-t)$

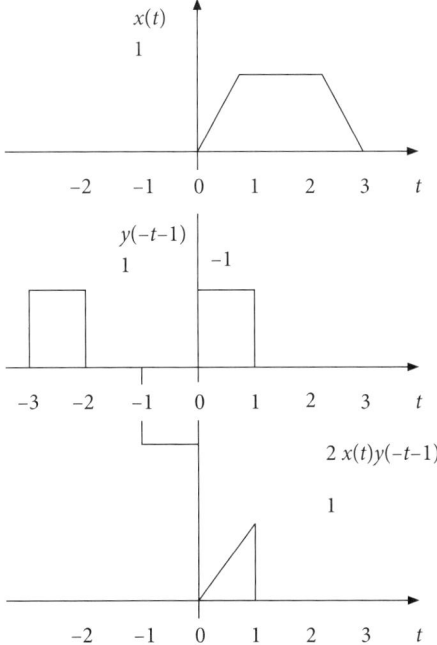

Plot of $x(2t)y(-t+2)$

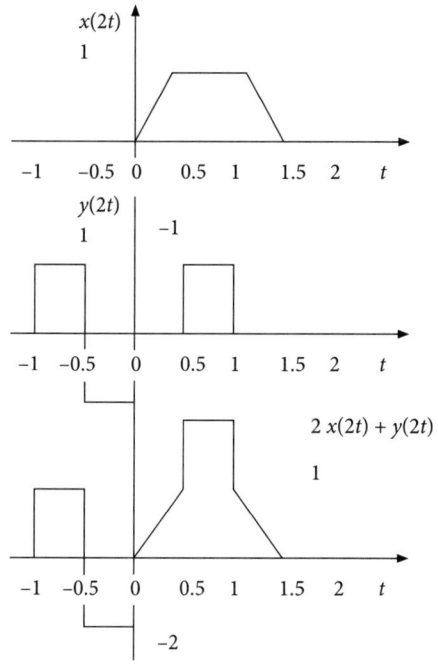

Plot of $x(2t)y(3t)$. (Note scale on x- axis)

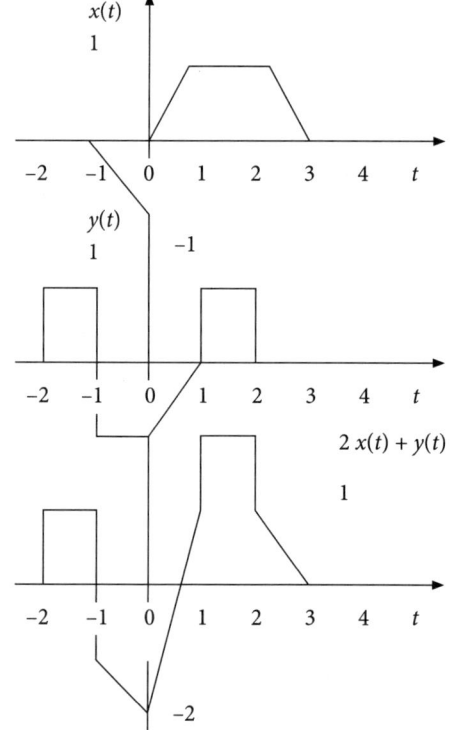

2.70

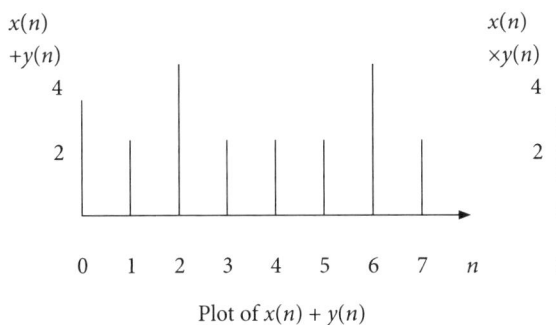
Plot of x(n) + y(n)

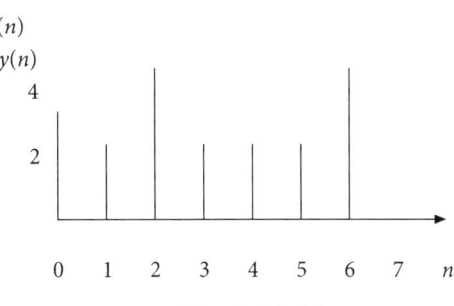
Plot of x(n)y(n)

2.71

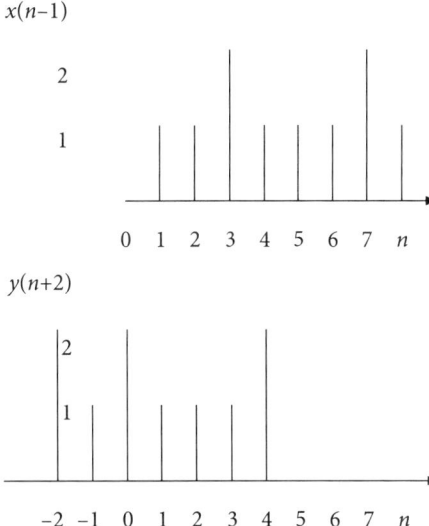

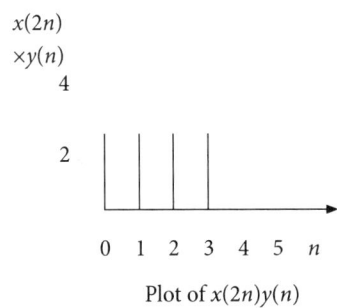
Plot of x(2n)y(n)

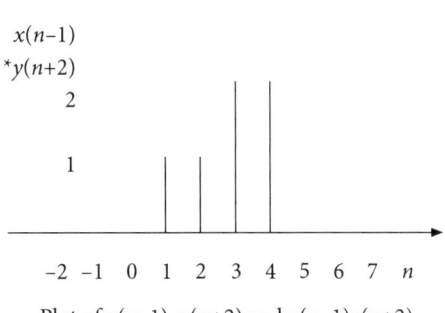
Plot of x(n−1), y(n+2) and x(n−1)y(n+2)

2.72

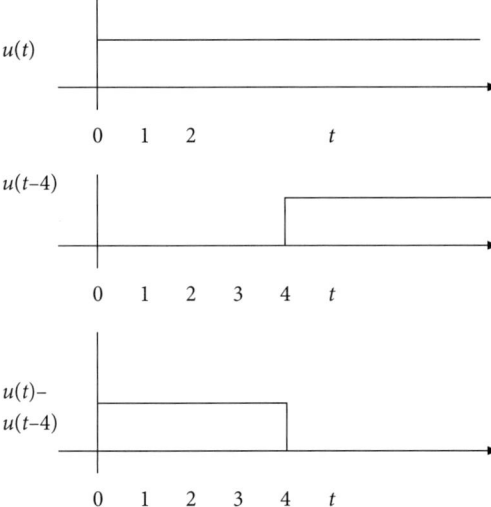

Plot of i) $x(t)$

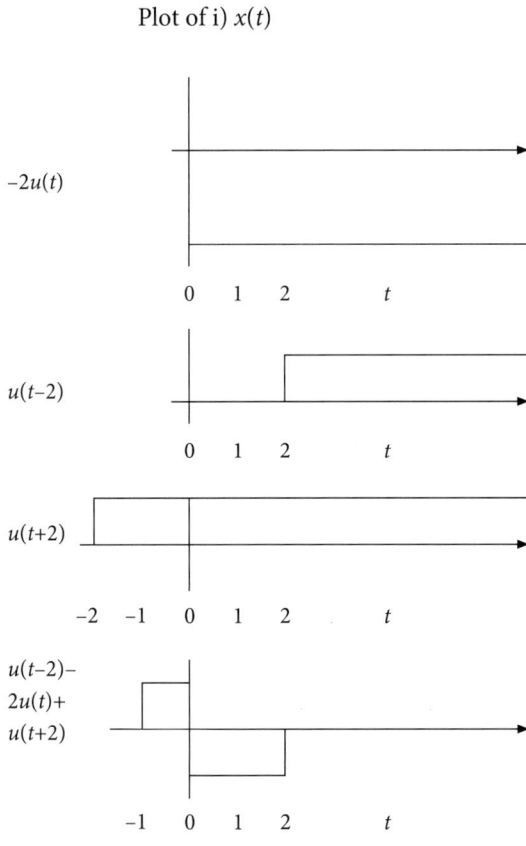

Pot of ii) $x(t)$

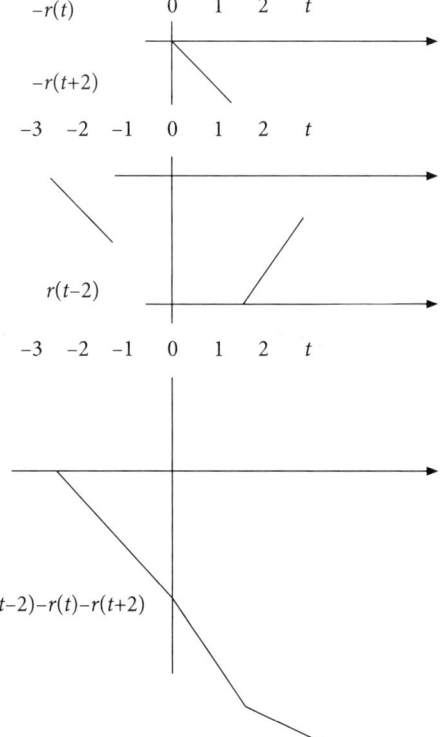

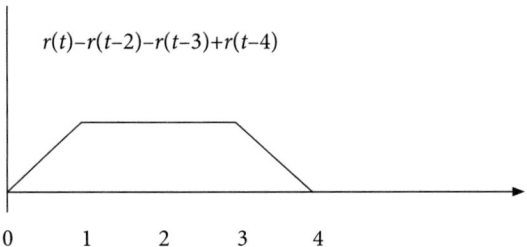

3

CT and DT Systems

Learning Objectives

- Properties of systems
- Linearity
- Time invariance
- Stability
- Causality
- Invertibility
- Memory
- LTI systems
- Systems as interconnected operators
- Series and parallel interconnection

This chapter concentrates on system definition and properties. The properties of systems such as linearity, time invariance, causality, invertibiity, memory and stability are discussed in detail. We have discussed sampling in chapter 1. The reader is already familiar with CT and DT signals. We will define properties of CT and DT systems simultaneously. Finally, we will see how to represent a system as an interconnection of operators and series/parallel; connection of systems.

3.1 Properties of CT and DT Systems – Linearity and Shift Invariance

We will discuss and explain properties of CT and DT systems in the following sections. We will study the property of linearity and Shift/time invariance in this section.

3.1.1 Linearity property

Any CT or DT system is said to be linear if it obeys two important properties. The first is *homogeneity* and the second is *additivity*.

Homogeneity property

The system is said to obey the property of homogeneity if the following condition holds:

- For CT and DT input signal of $x(t)$ or $x[n]$, respectively, if the output is given by $y(t)$ or $y[n]$, then if the input signal is scaled by a factor of k to get input signal of $kx(t)$ or $kx[n]$, then the output is also scaled by the same factor k i.e., output is $ky(t)$ or $ky[n]$, where k is any scaling factor. Let H represent the system operator, then

- if $y(t) = H[x(t)]$ then $ky(t) = H[kx(t)]$ and (3.1)

- if $y[n] = H[x[n]]$ then $ky[n] = H[kx[n]]$ (3.2)

Additivity Property (Superposition Property)

The system is said to obey additivity property if the following condition holds good.

- Let $y_1(t)$ and $y_1[n]$ be the output for the CT and DT input signal $x_1(t)$ and $x_1[n]$, respectively, and $y_2(t)$ and $y_2[n]$ be the output for input signal $x_2(t)$ and $x_2[n]$, respectively, for a CT and DT system. The property of additivity states the following:

  If the input to a system is the addition of two signals, then the output of the system is the addition of the respective outputs. Let H represent the system operator, then

- If $y_1(t) = H[x_1(t)]$ and $y_2(t) = H[x_2(t)]$, then

$$y_1(t) + y_2(t) = H[x\ (t) + x_2(t)] \qquad (3.3)$$

If $y_1[n] = H[x_1[n]]$ and $y_2[n] = H[x_2[n]]$, then

$$y_1[n] + y_2[n] = H[x\ [n] + x_2[n]] \qquad (3.4)$$

This indicates that the signals added at the input do not interfere with each other when they pass via a system.

Considering homogeneity and additivity holding good, we can write the following statement for CT and DT systems. This is a statement of superposition property. *All the linear systems obey the superposition property.*

- If $y_1(t) = H[x_1(t)]$ and $y_2(t) = H[x_2(t)]$, then

$$ay_1(t) + by_2(t) = H[ax_1(t) + bx_2(t)] \tag{3.5}$$

If $y_1[n] = H[x_1[nt]]$ and $y_2[n] = H[x_2[n]]$, then

$$ay_1[n] + by_2[n] = H[ax_1[n] + bx_2[n]] \tag{3.6}$$

Physical significance of linearity

If the system is linear, the transfer graph of output Vs input is a straight line graph passing through the origin. Linear systems have a meaning more than this! If the system is linear, the input signal can be suitably decomposed into component signals and the corresponding outputs for the component signals one at a time can be calculated by assuming all other inputs equal to zero. The component outputs can be scaled and added to generate the output of the system for the input signal. This procedure simplifies the computation of the system output for different input signals. This property is called as the *superposition property*.

We will consider some examples of CT and DT systems to illustrate the concept of linearity. We will first consider CT systems

Example 3.1

Check if the homogeneity property holds good for a system given by $y(t) = 5$? Find if the system is linear.

Solution

Note that for any input, the output is constant. We will have to check if the homogeneity and additivity property hold good. We will first check homogeneity property. If the input is doubled (scaled by a factor of 2), the output is not scaled by the same factor. It remains constant equal to 5. According to the property of homogeneity,

$$\text{if } y(t) = H[x(t)] \text{ then } 2y(t) = H[2 \times x(t)] \tag{3.7}$$

Here, the output remains constant even if the input is doubled. Hence, the system is not homogeneous.

We will now check for additivity.

If the input to the system is $x_1(t) + x_2(t)$, then the output must be

$$y_1(t) + y_2(t) = 5 + 5 = 10 \tag{3.8}$$

if the property of additivity holds good. The actual output is just equal to 5 whatever the applied input is. The additivity also fails.

The system is not homogeneous and not additive and hence, is not linear.

Example 3.2

Check if the system given by $y(t) = x(t) + 2$ is linear.

Solution

In the given system, for an input $x(t)$, the output is given by

$$y(t) = x(t) + 2 \tag{3.9}$$

Therefore for the input $kx(t)$, where k is any scaling factor, according to homogeneity property, the output would be

$$y(t) = k[x(t) + 2] \neq k[x(t)] + 2 \tag{3.10}$$

Hence, the system is not homogeneous.
We will now check for additivity.
If the input to the system is $x_1(t) + x_2(t)$, then output must be

$$y_1(t) + y_2(t) = x_1(t) + x_2(t) + c. \tag{3.11}$$

The actual output is equal to

$$y_1(t) + y_2(t) = x_1(t) + c + x_2(t) + c \tag{3.12}$$

The additivity property also fails.

Teaser The graph for both above systems is a straight line, yet they are non-linear systems! The above systems are said to have a bias. The systems with bias are incrementally linear. The incrementally linear system will have a graph of Δy Vs Δx as linear and it will pass through the origin.

Example 3.3

Check if the system given by $y(t) = x(t) + x(t-1)$ is linear.

Solution

Let the system be $y(t) = x(t) + x(t-1)$. Suppose the input is $kx(t)$, where k is any scaling factor. Then the output is

$$y(t) = kx(t) + kx(t-1) \tag{3.13}$$

To satisfy the homogeneity property the output should be $y(t) = k[x(t) + x(t-1)]$, which is the actual output. Hence, the system is homogeneous.
To satisfy the property of additivity, if the input to the system is $x_1(t) + x_2(t)$ then output must be

$$y_1(t) + y_2(t) = x_1(t) + x_1(t-1) + x_2(t) + x_2(t-1). \tag{3.14}$$

The actual output is equal to

$$y_1(t) + y_2(t) = x_1(t) + x_1(t-1) + x_2(t) + x_2(t-1) \quad (3.15)$$

The additivity property holds good. Hence, the system is linear.

Note The system obeys superposition property and hence, is linear!

Example 3.4
Find if the system given by $y(t) = 0.1 \times x(t)$ is linear?

Solution

Let the system be $y(t) = 0.1 \times x(t)$. Suppose the input is $kx(t)$, where k is any scaling factor. Then the output is

$$y(n) = 0.1k \times x(t) \quad (3.16)$$

To satisfy the homogeneity property the output should be

$$y(t) = k \times 0.1 \times x(t) = 0.1kx(t) \quad (3.17)$$

Hence, the system is homogeneous.

To satisfy the property of additivity, if the input to the system is $x_1(t) + x_2(t)$, then the output must be

$$y_1(t) + y_2(t) = 0.1 \times [x_1(t)] + 0.1 \times [x_2(t)]. \quad (3.18)$$

The actual output is equal to

$$y_1(t) + y_2(t) = 0.1 \times [x_1(t) + x_2(t)] \quad (3.19)$$

The additivity property also holds good. Hence, the system is linear.

Note The graph of the system is linear and it passes through the origin!
Let us solve some examples for DT systems.

Example 3.5
Does the homogeneity property hold for $y[n] = 2$? Is the system linear?

Solution

Note that for any input, the output is constant. We will have to check if the homogeneity and additivity property hold good. We will first check the homogeneity property. If the input is doubled (scaled by a factor of 2), the output is not scaled by the same factor. It remains constant equal to 2. Hence, the system is not homogeneous.

We will now check for additivity.
If the input to the system is $x_1[n] + x_2[n]$, then the output must be

$$y_1[n] + y_2[n] = 2 + 2 = 4 \qquad (3.20)$$

if the property of additivity holds good. The actual output is just equal to 2 whatever the applied input is. The additivity also fails.

The system is not homogeneous and not additive and hence, is not linear.

Example 3.6

Consider the following system $y[n] = x[n] + c$. Is the system homogeneous? Is it linear?

Solution

In the given system, for an input $x(n)$ the output is given by

$$y[n] = x[n] + c \qquad (3.21)$$

Therefore for the input $kx[n]$, where k is any scaling factor, the output would be

$$y[n] = k[x[n]] + c \qquad (3.22)$$

According to the condition of homogeneity property the output in this case should be k times the output when the input was $x[n]$, that is, the output should be

$$k[x[n] + c] \qquad (3.23)$$

which is not the case and hence, the system is not homogeneous.
We will now check for additivity.
If the input to the system is $x_1[n] + x_2[n]$, then the output must be

$$y_1[n] + y_2[n] = x_1[n] + x_2[n] + c. \qquad (3.24)$$

The actual output is just equal to

$$y_1[n] + y_2[n] = x_1[n] + c + x_2[n] + c \qquad (3.25)$$

The additivity property also fails.

✓ **Teaser** The graph for both the systems is a straight line, yet they are non-linear systems! The above systems are said to have a bias. The systems with bias are incrementally linear. The incrementally linear system will have a graph of Δy Vs Δx as linear and it will pass through the origin.

✓ **Things to remember** The system is a linear system if the output Vs input graph for a system is linear and it passes the through origin. If the graph is linear and it does not pass through the origin, it is an incrementally linear system, but not a linear system. A system with a non-linear transfer graph is linear if it obeys the superposition principle.

Example 3.7

Consider the following system $y[n] = |x[n]|$. Is the system linear?

Solution

Suppose the input is $kx[n]$, where k is any scaling factor, then the output is

$$y[n] = |kx[n]| \tag{3.26}$$

To satisfy the homogeneity property the output should be

$$y[n] = k|x[n]| \tag{3.27}$$

Hence, the system is not homogeneous. The system is not homogeneous, therefore it is non-linear.

To satisfy the property of additivity, if the input to the system is $x_1[n] + x_2[n]$, then the output must be

$$y_1[n] + y_2[n] = |x_1[n]| + |x_2[n]|. \tag{3.28}$$

The actual output is just equal to

$$y_1[n] + y_2[n] = |x_1[n] + x_2[n]|. \tag{3.29}$$

The additivity property also fails. Hence, the system is non-linear. The transfer graph of the system is shown in Fig. 3.1.

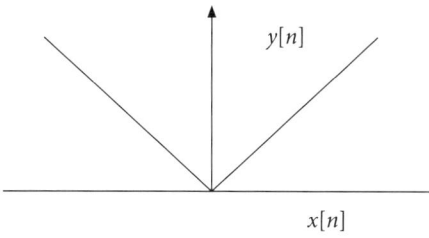

Fig. 3.1 The transfer graph for the DT system shown as a continuous graph

Let us now consider some examples of linear systems.

Example 3.8

Consider the following systems and verify the homogeneity and additivity property.

1. $y[n] = 0$
2. $y[n] = x[0]$
3. $y[n] = x[n-1]$
4. $y[n] = 2x[n] + 3x[n-1]$
5. $y[n] = 0.2x[n] - 0.1y[n-1]$

Solution

1. Let the system be $y[n] = 0$. Suppose the input is $kx[n]$, where k is any scaling factor, then the output is zero $= ky[n]$.

   To satisfy the homogeneity property the output should be $y[n] = k \times 0 = 0$. Hence, the system is homogeneous.

   To satisfy the property of additivity, if the input to the system is $x_1[n] + x_2[n]$, then the output must be

   $$y_1[n] + y_2[n] = 0 + 0 = 0. \qquad (3.30)$$

   The actual output is equal to

   $$y_1[n] + y_2[n] = 0. \qquad (3.31)$$

   The additivity property also holds good. Hence, the system is linear.

   Note The graph of the system is linear and it passes through the origin!

2. Let the system be $y[n] = x[0]$. Suppose the input is $kx[n]$, where k is any scaling factor, then the output is just $x[0]$.

   To satisfy the homogeneity property the output should be $y[n] = k \times x[0]$. Hence, the system is not homogeneous.

   To satisfy the property of additivity, if the input to the system is $x_1[n] + x_2[n]$, then the output must be

   $$y_1[n] + y_2[n] = x_1[0] + x_2[0]. \qquad (3.32)$$

   The actual output is equal to

   $$y_1[n] + y_2[n] = x_1[0] + x_2[0] \qquad (3.33)$$

   The additivity property holds good. Hence, the system is not linear.

Note The graph of the system is linear; however, it does not pass through the origin!

3. Let the system be $y[n] = x[n - 1]$. Suppose the input is $kx(n)$, where k is any scaling factor. Then the output is

$$y[n] = kx[n - 1] \tag{3.34}$$

To satisfy the homogeneity property the output should be $y[n] = k \times x[n - 1]$. Hence, the system is homogeneous.

To satisfy the property of additivity, if the input to the system is $x_1[n] + x_2[n]$, then the output must be

$$y_1[n] + y_2[n] = x_1[n-1] + x_2[n-1]. \tag{3.35}$$

The actual output is equal to

$$y_1[n] + y_2[n] = [x_1[n-1] + x_2[n-1]] \tag{3.36}$$

The additivity property holds good. Hence, the system is linear.

Note The graph of the system is linear and it does pass through the origin!

4. Let the system be $y([n] = 2x[n] + 3x[n - 1]$. Suppose the input is $kx[n]$, where k is any scaling factor, then the output is

$$y([n] = 2kx[n] + 3kx[n-1]. \tag{3.37}$$

To satisfy the homogeneity property the output should be

$$y([n] = 2kx[n] + 3kx[n-1] \tag{3.38}$$

Hence, the system is homogeneous.

To satisfy the property of additivity, if the input to the system is $x_1[n] + x_2[n]$, then the output must be

$$y_1[n] + y_2[n] = [2x_1[n] + 3x_1[n-1]] + [2x_2[n] + 3x_2[n-1]]. \tag{3.39}$$

The actual output is the same. The additivity property also holds good. Hence, the system is linear.

Note The system obeys superposition property and hence, is linear.

5. Let the system be $y[n] = 0.2x[n] - 0.1y[n - 1]$. Suppose the input is $kx[n]$, where k is any scaling factor, then the output is

$$y[n] = k \times (0.2x[n] - 0.1y[n-1]) \tag{3.40}$$

To satisfy the homogeneity property the output should be

$$y[n] = k \times 0.2x[n] - k \times 0.1y[n-1] \tag{3.41}$$

Hence, the system is homogeneous.

To satisfy the property of additivity, if the input to the system is $x_1[n] + x_2[n]$, then the output must be

$$y_1[n] + y_2[n] = [0.2x_1[n] - 0.1y_1[n-1]] + [0.2x_2[n] - 0.1y_2[n-1]]. \tag{3.42}$$

The actual output is the same. The additivity property also holds good. Hence, the system is linear.

Note The system obeys superposition property and hence, is linear.

Example 3.9

Consider the CT systems given by

1. $y(t) = \cos(x(t))$

2. $y(t) = \int_{-\infty}^{t/4} x(\tau) d\tau$

3. $y(t) = \dfrac{d}{dt} x(t)$

4. $y(t) = x(3 - t)$

5. $y(t) = x(t/5)$

Are the systems linear? If not, explain why.

Solution

1. Consider a system given by $y(t) = \cos(x(t))$

    Let the output for the inputs $x_1(t)$ and $x_2(t)$ be $y_1(t)$ and $y_2(t)$ given by $y_1(t) = \cos(x_1(t))$ and $y_2(t) = \cos(x_2(t))$, respectively.

    We will verify if the superposition property holds good. Let the input to the system be $ax_1(t) + bx_2(t)$, then the output will be

$$\cos(ax_1(t)) + \cos(bx_2(t)) \ne a\cos(x_1(t)) + b\cos(x_2(t)) \tag{3.43}$$

Hence, the system is not linear. We know that the cosine curve is non-linear! The system using any trigonometric function is non-linear!!

2. $y(t) = \int_{-\infty}^{t/4} x(\tau)d\tau$

Let output for the inputs $x_1(t)$ and $x_2(t)$ be $y_1(t)$ and $y_2(t)$ given by $y_1(t) = \int_{-\infty}^{t/4} x_1(\tau)d\tau$, $y_2(t) = \int_{-\infty}^{t/4} x_2(\tau)d\tau$, respectively.

We will verify if the superposition property holds good. Let the input be $ax_1(t) + bx_2(t)$, the output will be

$$ay_1(t) + by_2(t) = a\int_{-\infty}^{t/4} x_1(\tau)d\tau + b\int_{-\infty}^{t/4} x_2(\tau)d\tau = \int_{-\infty}^{t/4} [ax_1(\tau) + bx_2(\tau)]d\tau \quad (3.44)$$

Note The system obeys superposition property and hence, is linear.

3. $y(t) = \dfrac{d}{dt}x(t)$. Let us verify the superposition property.

Let output for the inputs $x_1(t)$ and $x_2(t)$ be $y_1(t)$ and $y_2(t)$ given by $y_1(t) = \dfrac{d}{dt}x_1(t)$, $y_2(t) = \dfrac{d}{dt}x_2(t)$, respectively.

We will verify if the superposition property holds good. Let the input be $ax_1(t) + bx_2(t)$, the output will be

$$ay_1(t) + by_2(t) = \dfrac{d}{dt}(ax_1(t) + bx_2(t)) = a\dfrac{d}{dt}x_1(t) + b\dfrac{d}{dt}x_2(t) \quad (3.45)$$

Note The system obeys superposition property and hence, is linear.

4. $y(t) = x(3 - t)$

Let output for the inputs $x_1(t)$ and $x_2(t)$ be $y_1(t)$ and $y_2(t)$ given by $y_1(t) = x_1(3 - t)$, $y_2(t) = x_2(3 - t)$, respectively.

We will verify if the superposition property holds good. Let the input be $ax_1(t) + bx_2(t)$, the output will be

$$ay_1(t) + by_2(t) = ax_1(3-t) + bx_2(3-t) \quad (3.46)$$

Note The system obeys superposition property and hence, is linear.

5. $y(t) = x(t/5)$

Let output for the inputs $x_1(t)$ and $x_2(t)$ be $y_1(t)$ and $y_2(t)$ given by $y_1(t) = x_1(t/5)$, $y_2(t) = x_2(t/5)$, respectively.

We will verify if the superposition property holds good. Let the input be $ax_1(t) + bx_2(t)$, the output will be

$$ay_1(t) + by_2(t) = ax_1(t/5) + bx_2(t/5) \qquad (3.47)$$

Note The system obeys superposition property and hence, is linear.

Example 3.10
Consider the DT systems given by

1. $y[n] = \sin(x[n])$

2. $y[n] = 3x[n]u[n]$

3. $y[n] = \log_{10}(|x[n]|)$

4. $y[n] = \sum_{k=-\infty}^{n} x[k+1]$

5. $y[n] = x[n] \sum_{k=-\infty}^{\infty} \delta[n-3k]$

6. $y[n] = x[n^3]$

Are the systems linear? If not, explain why.

Solution

1. $y[n] = \sin(x[n])$ Let output for the inputs $x_1[n]$ and $x_2[n]$ be $y_1[n]$ and $y_2[n]$ given by $y_1[n] = \sin(x_1[t])$ and $y_2[n] = \sin(x_2[n])$, respectively.

   We will verify if the superposition property holds good. Let the input be $ax_1[n] + bx_2[n]$, the output will be

   $$\sin(ax_1[n]) + \sin(bx_2[n]) \neq a\sin(x_1[n]) + b\sin(x_2[n]) \qquad (3.48)$$

   Hence, the system is not linear. We know that the cosine curve is non-linear!

   The system using any trigonometric function is non-linear!!

2. $y[n] = 3x[n]u[n]$
   Let output for the inputs $x_1[n]$ and $x_2[n]$ be $y_1[n]$ and $y_2[n]$ given by $y_1[n] = 3x_1[n]u[n]$ and $y_2[n] = 3x_2[n]u[n]$, respectively.

   We will verify if the superposition property holds good. Let the input be $ax_1[n] + bx_2[n]$, the output will be

   $$3ax_1[n]u[n] + 3bx_2[n]u[n] = [(a \times 3(x_1[n]) + b \times 3(x_2[n]))u[n] \qquad (3.49)$$

Hence, the system is linear.

Note The system obeys superposition property and hence is linear.

3. $y[n] = \log_{10}(|x[n]|)$

Let output for the inputs $x_1[n]$ and $x_2[n]$ be $y_1[n]$ and $y_2[n]$ given by $y_1[n] = \log_{10}(|x_1[n]|)$ and $y_2[n] = \log_{10}(|x_2[n]|)$, respectively.

We will verify if the superposition property holds good. Let the input be $ax_1[n] + bx_2[n]$, the output will be

$$a\log_{10} x_1[n] + b\log_{10} x_2[n] \neq \log_{10}[(a \times (x_1[n] + b \times (x_2[n]))] \quad (3.50)$$

Hence, the system is non-linear.

Note The system does not obey superposition property and the transfer curve of log function is not linear; hence, the system is non-linear.

4. $y[n] = \sum_{k=-\infty}^{n} x[k+1]$.

Let output for the inputs $x_1[n]$ and $x_2[n]$ be $y_1[n]$ and $y_2[n]$ given by $y_1[n] = \sum_{k=-\infty}^{n} x_1[k+1]$ and $y_2[n] = \sum_{k=-\infty}^{n} x_2[k+1]$, respectively.

We will verify if the superposition property holds good. Let the input be $ax_1[n] + bx_2[n]$, the output will be

$$a\sum_{k=-\infty}^{n} x_1[k+1] + b\sum_{k=-\infty}^{n} x_2[k+1] = \sum_{k=-\infty}^{n} ax_1[k+1] + bx_2[k+1] \quad (3.51)$$

Hence, the system is linear.

Note The system obeys superposition property and hence, is linear.

5. $y[n] = x[n] \sum_{k=-\infty}^{\infty} \delta[n-3k]$

Let output for the inputs $x_1[n]$ and $x_2[n]$ be $y_1[n]$ and $y_2[n]$ given by $y_1[n] = x_1[n]\sum_{k=-\infty}^{\infty} \delta[n-3k]$ and $y_2[n] = x_2[n]\sum_{k=-\infty}^{\infty} \delta[n-3k]$, respectively.

We will verify if the superposition property holds good. Let the input be $ax_1[n] + bx_2[n]$, the output will be

$$ax_1[n]\sum_{k=-\infty}^{\infty} \delta[n-3k] + bx_2[n]\sum_{k=-\infty}^{\infty} \delta[n-3k] = [ax_1[n] + bx_2[n]]\sum_{k=-\infty}^{\infty} \delta[n-3k] \quad (3.52)$$

Hence, the system is linear.

Note The system obeys superposition property and hence, is linear.

6. $y[n] = x[n^2]$

Let output for the inputs $x_1[n]$ and $x_2[n]$ be $y_1[n]$ and $y_2[n]$ given by $y_1[n] = x_1[n^2]$ and $y_2[n] = x_2[n^2]$, respectively.

We will verify if the superposition property holds good. Let the input be $ax_1[n] + bx_2[n]$, the output will be

$$ax_1[n^2] + bx_2[n^2] = ay_1[n] + by_2[n] \qquad (3.53)$$

Hence, the system is linear.

Note The system obeys superposition property and hence, is linear.

3.1.2 Time invariance / shift invariance property

Most signal-processing (DSP) techniques also require the system to be shift invariant even though it is not a requirement for linearity. The system is said to be time invariant or shift invariant if its input–output characteristics do not change with time. If $y(t)$ or $y[n]$ is the output for some input $x(t)$ or $x[n]$, respectively, for analog and digital systems, then for a shift invariant system, if the input is shifted by k time units / k samples to the right, that is, the input is $x(t - k)/x[n - k]$ then the output is $y(t - k)/y[n - k]$, that is, the output is also shifted by k time units/k samples to the right. The system that is linear and obeys the time invariance property is called a *linear time invariant (LTI) system*. We will refer to the LTI systems only unless stated otherwise.

Physical significance of time/shift invariance If the system is time/shift invariant, then the output of the system for any shifted input can be easily calculated by introducing same amount of shift in the output.

Physical significance of linear time/shift invariance If the system is LTI, one can characterize the system in terms of its impulse response. One may decompose any input signal into scaled and shifted delta functions and the output of the system can be obtained using the principle of superposition. The output of the system to a shifted delta function is just the shifted impulse response. The computation of the output to any input signal is greatly simplified.

We will illustrate the property of shift invariance with the help of CT and DT systems.

Example 3.11

(Non-linear and non shift invariant system) Is the following system linear and shift/time invariant? $y(t) = x(1)$.

Solution

We will verify the shift invariance property. Let the input be shifted by k time units, therefore the new input is $x(t - k)$. The output is still $x(1)$ and is not shifted by k time units. So, the system is not a shift variant system. We will now check for linearity. If the input is $kx(t)$, the output must be $kx(1)$. However, the actual output is just $x(1)$ and is constant. Similarly, if the input is $x_1(t) + x_2(t)$, still the output is

$$y(t) = x(1) \neq x(1) + x(1) \tag{3.54}$$

Teaser Verify that the system is non-linear by checking conditions for homogeneity and additivity. (The system is non-linear.)

Example 3.12

(Shift invariant, but non-linear system) Check if the following system is shift invariant and linear? $y(t) = [x(t)]^2$. Prove that the system is shift invariant, but non-linear.

Solution

We will verify the shift invariance property. Let the input be shifted by k time units, therefore the new input is $x(t - k)$. The output is $[x(t - k)]^2$ and is same as shifted by k time units i.e., $y(t - k)$. So, the system is a shift variant system. We will now check for linearity. If the input is $kx(t)$, the output must be $k[x(t)]^2$. However, the actual output is $[kx(t)]^2$ Similarly, if the input is $x_1(t) + x_2(t)$, the output is not

$$y(t) = [x_1(t)+x_2(t)]^2 \neq [x_1(t)]^2 + [x_2(t)]^2 \tag{3.55}$$

Teaser Verify that the system is non-linear by checking conditions for homogeneity and additivity. (The system is non-linear.)

Example 3.13

(Linear and shift invariant system) Consider the following system:

$$y(t) = x(t-2)$$

Prove that the system is both shift invariant and linear i.e., the system is LTI system.

Solution

We will verify the shift invariance property. Let the input be shifted by k units, that is the new input is $x(t - k)$. The output is

$$y(t-k) = x(t-2-k). \tag{3.56}$$

One can see that the output is also shifted by k units, therefore the system is a shift invariant system.

Let us now check the linearity of the system.

1. Homogeneity: Let the input be doubled. Then the output is also doubled. So, the system is homogeneous.
2. Additivity: We will check the additivity. Let the input be $a_1 x_1(t) + a_2 x_2(t)$
Then the output is

$$a_1 y_1(t) + a_2 y_2(t) = a_1 x_1(t-2) + a_2 x_2(t-2) \qquad (3.57)$$

Therefore the system is additive.

The system satisfies both the homogeneity and additivity properties, therefore the system is linear. It is also shift invariant. Hence, the system is an LTI system.

We will solve some examples for DT systems.

Example 3.14

(Non-linear and non-shift invariant) Is the following system shift variant? Is it linear? $y[n] = x[0]$

Solution

We will verify the shift invariance property. Let the input be shifted by k units; therefore, the new input is $x[n-k]$. The output is still $x[0]$ and is not shifted by k units. So, the system is not a shift variant system

Teaser Verify that the system is non-linear. (The system non-linear.) Verify homogeneity and additivity property.

Example 3.15

(Shift Invariant however, Non-Linear System)
Consider the following system: $y[n] = x[n]^2$
Prove that the system is shift invariant, but non-linear.

Solution

We will verify the shift invariance property. Let the input be shifted by k units, that is the new input be $x[n-k]$. Then the output is given by

$$y[n-k] = x[n-k]^2 \qquad (3.58)$$

One can see that the output is also shifted by k units. So the system is a shift invariant system.
Let us now check the linearity of the system. Let $x[n] = 3$, then $y[n] = x[n]^2 = 9$. If the input is now doubled, that is, if $x[n] = 6$, then the output is

$$y[n] = x[n]^2 = 36 \neq 2 \times 9. \qquad (3.59)$$

We see that the output is not doubled. So, the system is not homogeneous and hence, is not linear. The reader may also check for additivity property.

Example 3.16

(Linear and Shift Invariant System) Consider the following system:

$$y[n] = x[n-1]$$

Prove that the system is both shift invariant and linear i.e., the system is LTI system.

Solution

We will verify the shift invariance property. Let the input be shifted by k units, that is the new input is $x[n - k]$. The output is

$$y[n-k] = x[n-k-1]. \tag{3.60}$$

One can see that the output is also shifted by k units, therefore the system is a shift invariant system.

Let us now check the linearity of the system.

1. Homogeneity: Let the input be doubled. Then the output is also doubled. So, the system is homogeneous.
2. Additivity: We will check the additivity. Let the input be $a_1 x_1[n] + a_2 x_2[n]$
   Then the output is

$$a_1 y_1[n] + a_2 y_2[n] = a_1 x_1[n-1] + a_2 x_2[n-1] \tag{3.61}$$

Therefore, the system is additive.

Since the system satisfies both the homogeneity and additivity properties, therefore the system is linear. It is also shift invariant. Hence, the system is LTI system.

We will now prove the property of shift invariance for following CT systems.

Example 3.17

Consider the CT systems given by

1. $y(t) = \cos(x(t))$

2. $y(t) = \int_{-\infty}^{t/4} x(\tau) d\tau$

3. $y(t) = \dfrac{d}{dt} x(t)$

4. $y(t) = x(3-t)$

5. $y(t) = x(t/5)$

Are the systems shift/time invariant? If not, explain why.

Solution

1. Consider a system given by $y(t) = \cos(x(t))$

   Let the input be shifted by k units that is the new input is $x(t - \tau)$. The output is

   $$y(t-\tau) = \cos(x(t-\tau)) \tag{3.62}$$

   One can see that the output is also shifted by τ units, therefore the system is a shift invariant system.

2. $y(t) = \int_{-\infty}^{t/4} x(\tau)d\tau$ Let the input be shifted by k units that is the new input is $x(t - \tau)$. The output is

   $$y(t-\tau) = \int_{-\infty}^{(t-\tau)/4} x(\tau)d\tau. \tag{3.63}$$

   One can see that the output is also shifted by τ units, therefore the system is a shift invariant system.

3. $y(t) = \dfrac{d}{dt}x(t)$ Let us verify the superposition property.

   Let the input be shifted by k units that is the new input is $x(t - \tau)$. The output is

   $$y(t-\tau) = \dfrac{d}{dt}(x(t-\tau)). \tag{3.64}$$

   One can see that the output is also shifted by τ units, therefore the system is a shift invariant system.

4. $y(t) = x(3 - t)$

   Let the input be shifted by k units that is the new input is $x(t - \tau)$. The output is

   $$y(t-\tau) = x(3-(t-\tau)). \tag{3.65}$$

   One can see that the output is also shifted by τ units, therefore the system is a shift invariant system.

5. $y(t) = x(t/5)$

   Let the input be shifted by k units that is the new input is $x(t - \tau)$. The output is

   $$y(t - \tau) = x((t - \tau)/5). \qquad (3.66)$$

   One can see that the output is also shifted by τ units, therefore the system is a shift invariant system.

Example 3.18

Consider the DT systems given by

1. $y[n] = \sin(x[n])$

2. $y[n] = 3x[n]u[n]$

3. $y[n] = \log_{10}(|x[n]|)$

4. $y[n] = \sum_{k=-\infty}^{n} x[k+1]$

5. $y[n] = x[n] \sum_{k=-\infty}^{\infty} \delta[n - 3k]$

6. $y[n] = x[n^2]$

Are the systems shift/time invariant? If not, explain why.

Solution

1. $y[n] = \sin(x[n])$

   We will verify the shift invariance property. Let the input be shifted by k units, that is the new input is $x[n - k]$. The output is

   $$y[n - k] = \sin(x[n - k]) \qquad (3.67)$$

   One can see that the output is also shifted by k units, therefore the system is a shift invariant system.

2. $y[n] = 3x[n]u[n]$

   We will verify the shift invariance property. Let the input be shifted by k units, that is the new input is $x[n - k]$. The output is

   $$y[n - k] = 3x[n - k]u[n] \neq 3x[n - k]u[n - k] \qquad (3.68)$$

One can see that the output is not shifted by k units, therefore the system is not a shift invariant system.

3. $y[n] = \log_{10}(|x[n]|)$

We will verify the shift invariance property. Let the input be shifted by k units, that is the new input is $x[n - k]$. The output is

$$y[n-k] = \log_{10}(|x[n-k]|) \tag{3.69}$$

One can see that the output is also shifted by k units, therefore the system is a shift invariant system.

4. $y[n] = \sum_{k=-\infty}^{n} x[k+1]$.

We will verify the shift invariance property. Let the input be shifted by k units, that is the new input is $x[n - k]$. The output is

$$y[n-k] = \sum_{k=-\infty}^{n-k} x[k+1] \tag{3.70}$$

One can see that the output is also shifted by k units, therefore the system is a shift invariant system.

5. $y[n] = x[n] \sum_{k=-\infty}^{\infty} \delta[n - 3k]$

We will verify the shift invariance property. Let the input be shifted by k_1 units, that is the new input is $x[n - k_1]$. The output is

$$y[n-k_1] = x[n-k_1] \sum_{k=-\infty}^{\infty} \delta[n-3k] \neq x[n-k_1] \sum_{k=-\infty}^{\infty} \delta[n-3k-k_1] \tag{3.71}$$

One can see that the output is not shifted by k units, therefore the system is not a shift invariant system.

6. $y[n] = x[n^2]$ We will verify the shift invariance property. Let the input be shifted by k units, that is the new input is $x[n - k]$. The output is

$$y[n-k] = x[(n-k)^2] \tag{3.72}$$

One can see that the output is also shifted by k units, therefore the system is a shift invariant system.

Concept Check

- What is meaning of a linear system?
- What is the property of homogeneity?
- What is additivity property?
- What is the physical significance of linearity?
- When will you call the system as incrementally linear?
- Can a system with linear graph be non-linear?
- Can a system with non-linear transfer curve be linear?
- What is the physical significance of shift invariance?
- What is LTI system?

3.2 Properties of CT and DT Systems – Causality and Memory

We will study the properties of causality and memory in this section.

3.2.1 Causality property

A system is said to be causal if its present output depends only on past and present inputs and past outputs. Let us consider an example of a causal CT and DT system.

$$y(t) = a_0 x(t) + a_1 x(t-1) - b_1 y(t-1) \tag{3.73}$$

$$y[n] = a_0 x[n] + a_1 x[n-1] - b_1 y[n-1] \tag{3.74}$$

Here, the output at time instant 't' in case of CT system or 'n' in case of DT system depends on current input at time instant 't' or 'n' and past input and output at instant 't – 1' or 'n – 1' i.e., previous time sample. These systems are causal. Now consider CT and DT systems represented as

$$y(t) = a_0 x(t) + a_1 x(t+1) - b_1 y(t-1) \tag{3.75}$$

$$y[n] = a_0 x[n] + a_1 x[n+1] - b_1 y[n-1] \tag{3.76}$$

These are non-causal systems as the output of the system at current time instant 't' or 'n' depends on the current inputs and the next input at time instant 't + 1' and 'n + 1', respectively.

Physical significance of Causality Causal systems are practically realizable or implementable. The system of a human being is causal as we always keep on learning from the past inputs and past outputs of the system. The future inputs have no effect on our act at current or present time. The causal systems can be implemented in real time. The present and past inputs have meaning only for temporal systems where time is an independent variable. In case of spatial domain systems, present and past input has no meaning. Such non-causal systems can be implemented after grabbing the complete input data. Non-causal temporal systems can be implemented if some delay is tolerable. The system considers the future inputs for calculating the current outputs and can generate a bench mark for system performance. The real-time system performance may be compared with the ideal system implemented with some tolerable delay. Offline systems can always be implemented as non-causal systems.

We will consider some examples for CT systems first.

Example 3.19

Consider a system represented by

$$y(t) = \int_{\tau=-1}^{\tau=1} h(\tau)x(t-\tau)d\tau$$

Is the system causal? If not, explain why.

Solution

In this case, $h(t)$ exists for $\tau = -1$. The input for this value of time is $t + 1$.

The output of the system at time instant t depends on the next input, i.e., input at $t + 1$, therefore the system is non-causal.

Example 3.20

Consider a system given by $y(t) = x(t-1) + y(t-3)$ Is the system causal?

Solution

The output of the system at instant t depends on input at current time instant t and past input at $t = t - 1$ and past output at time instant $t = t - 3$, therefore the system is causal.

Example 3.21

Consider a system represented as $y(t) = x(1-t)$. Is it causal?

Solution

The system output at time instant t depends on time instant $1 - t$. consider $t = -2$, $1 - t$ becomes $1 + 2$ i.e., 3. So, the output at $t = -2$ time units depends on input at time instant $t = 3$. This is a future time instant. Hence, the system is non-causal.

We will now consider examples for DT systems.

Example 3.22

Consider a system represented by

$$y[n] = \sum_{k=-1}^{1} h[k]x[n-k]$$

Is the system causal? If not, explain why.

Solution

In this case, $h[n]$ exists for $n = -1$. From the given equation we have

$$y[n] = h[-1]x[n+1] + h[0]x[n] + h[1]x[n-1] \tag{3.77}$$

The output of the system at instant n depends on the next input, i.e., input at $n + 1$, therefore the system is non-causal.

Example 3.23

Consider the system represented by

$$y[n] = \sum_{k=-1}^{n+1} x[k]h[n-k]$$

Is the system causal? If not, explain why.

Solution

Expanding the given equation we have

$$y[n] = x[-1]h[n+1] + x[0]h[n] + \ldots + x[n+1]h[-1] \tag{3.78}$$

We see from Eq. (3.78) that the output of the system at n depends on the input at the next time instant, that is, at $n + 1$, if $h[n]$ exists for $n = -1$ and therefore the system becomes non-causal.

Example 3.24

Consider a system given by

$$y[n] = x[n] - x[n-1] + y[n-2]$$

Is the system causal?

Solution

The output of the system at instant n depends on current and past input and past output; therefore, the system is causal.

Example 3.25

Consider a system given by

$$y[n] = x[-n]$$

Is the system causal? If not, explain why.

Solution

The system is non-causal.

Reason To see why it is non-causal, put $n = -2$ in the given equation. We will get

$$y[-2] = x[2] \tag{3.79}$$

The output at $n = -2$ depends on the input at the next time instant, that is, $n = 2$; therefore, the system is non-causal;

Example 3.26

Consider the CT systems given by

1. $y(t) = \cos(x(t))$

2. $y(t) = \int_{-\infty}^{t/4} x(\tau)d\tau$

3. $y(t) = \dfrac{d}{dt} x(t)$

4. $y(t) = x(3 - t)$

5. $y(t) = x(t/5)$

Are the systems causal? If not, explain why.

Solution

1. $y(t) = \cos(x(t))$. Here, the system output at current instant of time 't' depends on only the current input at time 't'. Hence, the system is causal.

2. $y(t) = \int_{-\infty}^{t/4} x(\tau)d\tau$. Here, the system output is the integration of its previous inputs from minus infinity to $t/4$. Let us put $t = 4$, now we have to integrate the inputs up to $t = 1$. Let us put $t = -4$, now we have to integrate the inputs up to $t = -1$, which is the next time instant. Hence, the system is non-causal.

3. $y(t) = \dfrac{d}{dt} x(t)$. Here, the system output at current instant of time instant 't' depends on only the current input at 't'. Hence, the system is causal.

4. $y(t) = x(3 - t)$. Let us put $t = -1$. The output at time instant of -1 depends on input at $4 = 3 - (-1)$ i.e., the next time instant. Hence, the system is non-causal.

5. $y(t) = x(t/5)$. Let us put $t = -5$, the output at $t = -5$ depends on input at $t = -1$, which is the next time instant. Hence, the system is non-causal.

Example 3.27

Consider the DT systems given by

1. $y[n] = \sin(x[n])$

2. $y[n] = 3x[n]u[n]$

3. $y[n] = \log_{10}(|x[n]|)$

4. $y[n] = \sum\limits_{k=-\infty}^{n} x[k+1]$

5. $y[n] = x[n] \sum\limits_{k=-\infty}^{\infty} \delta[n - 3k]$

6. $y[n] = x[n^2]$

Are the systems causal? If not, explain why.

Solution

1. $y[n] = \sin(x[n])$. Here, the system output at current instant of time 'n' depends on only the current input at 'n'. Hence, the system is causal.

2. $y[n] = 3x[n]u[n]$.

   Let us first write the definition of $u[n]$.

   $u[n] = 1$ for $n \geq 0$

   $\quad\quad = 0$ otherwise

   Let us put $n = -1$, $u[-1] = 0$, The output at time '-1' is zero irrespective of the input $x[n]$. Let us put $n = +1$, $u[+1] = 1$, The output at time '1' is $3x[1]$ depend on current input $x[1]$. The system is causal.

3. $y[n] = \log_{10}(|x[n]|)$. Let us put $n = -1$, the output at $t = -1$ will depend on log of $x[-1]$, i.e., current time instant input. The system is causal.

4. $y[n] = \sum_{k=-\infty}^{n} x[k+1]$. Let us put $n = 1$, the output at instant 1 depends on inputs from minus infinity to $n = 2$ i.e., the next time instant. The system is non-causal.

5. $y[n] = x[n] \sum_{k=-\infty}^{\infty} \delta[n-3k]$. The output depends only on the current sample; if the delta function exists, then the system is causal.

6. $y[n] = x[n^2]$. Consider $n = 1$, the output at time instant 1 depends on input sample at time instant 1; however, the output at time instant 2 depends on the input at time instant 4. This makes the system non-causal.

3.2.2 Memory

The system is said to have memory or said to be dynamic if its current output depends on previous, future input or previous and future output signals. The system is said to be memoryless or instantaneous if its current output depends only on the current input. The extent of memory depends on the number of past inputs or outputs on which the current output depends.

We will consider examples of systems with memory.

Example 3.28

Consider a CT system given by

$$y(t) = \int_{-\infty}^{t} x(\tau) d\tau$$

Does the system have memory?

Solution

The current output of the system depends on past inputs. The system is with memory. The memory extends into infinite past inputs.

Example 3.29

Consider a CT system given by

$$y(t) = 4x(t)$$

Does the system have memory?

Solution

The current output of the system depends only on the current input. The system is without memory or is memoryless.

Example 3.30

Consider a DT system given by

$$y[n] = \frac{1}{2}[x[n] + x[n-1]]$$

Does the system have memory?

Solution

The current output of the system depends on current and one past input. The system is with memory. The memory extends into one past input.

Example 3.31

Consider a DT system given by

$$y[n] = 5 + 4x[n]$$

Does the system have memory?

Solution

The current output of the system depends only on the current input. The system is without memory or is memoryless.

Physical significance of memory The systems containing only the resistive network has no memory and is said to be memoryless or instantaneous systems. If the system contains active element like capacitor or inductor, the system memory will extend up to infinity. Such systems are said to be dynamic. Generally, we have to handle dynamic systems. We human beings always learn from the past. Our brain system has a memory that allows us to understand the new concepts.

Example 3.32

Consider the CT systems given by

1. $y(t) = \cos(x(t))$

2. $y(t) = \int_{-\infty}^{t/4} x(\tau) d\tau$

3. $y(t) = \dfrac{d}{dt} x(t)$

4. $y(t) = x(3 - t)$

5. $y(t) = x(t/5)$

Do the systems have memory? If yes, find the extent of memory.

Solution

1. $y(t) = \cos(x(t))$. Here, the system output at current instant of time 't' depends on only the current input at 't'. Hence, the system is memoryless.

2. $y(t) = \int_{-\infty}^{t/4} x(\tau)d\tau$. Here, the system output is the integration of its previous inputs from minus infinity to $t/4$. Hence, the system has memory and it extends from infinity to $t/4$.

3. $y(t) = \dfrac{d}{dt}x(t)$. Here, the system output at current instant of time instant 't' depends on only the current input at 't'. Hence, the system is memoryless.

4. $y(t) = x(3 - t)$. Let us put $t = 2$. The output at time instant of 2 depends on input at $t = 3 - (2) = 1$ i.e., the previous time instant. Hence, the system has memory. The extent of memory varies with the value of the current instant.

5. $y(t) = x(t/5)$. The output at the current instant depends on the input at previous instant $t/5$. The system has memory. The extent of memory is up to $t/5$ time.

Example 3.33

Consider the DT systems given by

1. $y[n] = \sin(x[n])$

2. $y[n] = 3x[n]u[n]$

3. $y[n] = \log_{10}(|x[n]|)$

4. $y[n] = \sum_{k=-\infty}^{n} x[k+1]$

5. $y[n] = x[n] \sum_{k=-\infty}^{\infty} \delta[n-3k]$

6. $y[n] = x[n^2]$

Do the systems have memory? If yes, find the extent of memory.

Solution:

1. $y(t) = \sin(x[n])$. Here, the system output at current instant of time 'n' depends on only the current input at 'n'. Hence, the system is memoryless.

2. $y[n] = 3x[n]u[n]$. Here, the system output at current instant of time 'n' depends on only the current input at 'n'. Hence, the system is memoryless.

3. $y[n] = \log_{10}(|x[n]|)$. Here, the system output at current instant of time 'n' depends on only the current input at 'n'. Hence, the system is memoryless.

4. $y[n] = \sum_{k=-\infty}^{n} x[k+1]$. Here, the system output at current instant of time 'n' depends on one future input, current input and past inputs form infinity to previous input. Hence, the system has memory.

5. $y[n] = x[n] \sum_{k=-\infty}^{\infty} \delta[n-3k]$. Here, the system output at current instant of time 'n' depends on current input if the delta function exists. Hence, the system is memoryless.

6. $y[n] = x[n^2]$. Consider $n = 1$, the output at time instant 1 depends on input sample at time instant 1; however, the output at time instant 2 depends on the input at time instant 4. The system has memory; however, it depends on future inputs and the extent of memory depends on the value of 'n' current sample.

Concept Check

- Explain the meaning of causality.
- What is physical significance of causality?
- When will you call the system as memoryless?
- What is the physical significance of system with memory?
- Can you realize a non-causal system?
- What are real-time systems?

3.3 Properties of CT and DT Systems – Invertibility and Stability

We will study the properties of invertibility and stability in this section.

3.3.1 Invertibility

If it is possible to recover the input of the system, then the system is said to be invertible. Consider the system represented by the operator H producing the output as $y(t)$ with input as $x(t)$. The idea of system invertbility is more clearly understood if one considers the new operator H^{-1} called as inverse operator with the so-called inverse system, as shown in Fig. 3.2. The H^{-1} is not the reciprocal of the operator H; however, it is the symbol used to indicate the inverse.

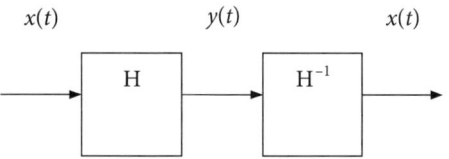

Fig. 3.2 Invertibility of systems

Note that there must be a one-to-one correspondence between the input and the output in order to have invertibility.

Physical significance of invertibility When we use a system for processing the signal, for example, for calculating some transform for removing the noise, the system is required to be invertible (inverse transform) to recover the signal back in time domain. If we process the signal to extract some features for recognition, then the system may be non-invertible as we are not recovering the signal. Invertible system also finds applications in communication field. For error-free transmission an equalizer is used at the input of the receiver that has inverse characteristics as that of the channel.

We will consider some examples of invertible systems.

Example 3.34

Consider a CT system given by

$$y(t) = \int_{-\infty}^{t} x(\tau) d\tau$$

Is the system invertible?

Solution

The output of the system can be differentiated to get the signal $x(t)$. Differentiator is the inverse system and the integrator is said to be the invertible system. We can obtain $x(t)$ by differentiating the output $y(t)$

$$\frac{d}{dt} y(t) = x(t) \tag{3.80}$$

Example 3.35

Consider a CT system given by

$$y(j\omega) = \int_{-\infty}^{\infty} x(\tau) e^{-j\omega \tau} d\tau$$

Is the system invertible?

Solution

The output of the system can be recognized as the FT of the input signal $x(t)$. Inverse FT is the inverse system and FT is said to invertible system. We can obtain $x(t)$ by taking Inverse FT of the output $y(t)$

$$x(t) = \int_{-\infty}^{\infty} y(j\omega) e^{j\omega\tau} d\omega \qquad (3.81)$$

Example 3.36

Consider a DT system given by

$$y[n] = 3x[n] + 4$$

Is the system invertible?

Solution

The input can be recovered from the output using the equation

$$x[n] = \frac{1}{3}[y[n] - 4] \qquad (3.82)$$

The system is invertible.

Example 3.37

Consider a DT system given by

$$X[K] = \sum_{n=0}^{N-1} x[n] e^{-j2\pi nk/N}$$

Is the system invertible?

Solution

The output of the system can be recognized as the DFT of the input signal $x[n]$. Inverse DFT is the inverse system and DFT is said to invertible system. We can obtain $x[n]$ by taking inverse DFT of the output $x[k]$.

$$x[n] = \sum_{k=0}^{N-1} X[k] e^{j2\pi nk/N} \qquad (3.83)$$

Example 3.38

Consider a CT system given by

$$y(t) = (x(t))^2$$

Is the system invertible?

Solution

The output of the system can be recognized as the square of the input signal $x(t)$. We can obtain $x(t)$ by taking the square root of the output.

$$x(t) = \pm\sqrt{y(t)} \qquad (3.84)$$

There is no one-to-one correspondence. Hence, the system is not invertible.

Example 3.39

Consider the CT systems given by

1. $y(t) = \cos(x(t))$

2. $y(t) = \int_{-\infty}^{t/4} x(\tau)d\tau$

3. $y(t) = \int_{t-3}^{t} x(\tau)d\tau$

4. $y(t) = \dfrac{d}{dt}x(t)$

5. $y(t) = x(3 - t)$

6. $y(t) = x(t/5)$

Are the systems invertible? If not, find why they are not invertible.

Solution

1. $y(t) = \cos(x(t))$

$$x(t) = \cos^{-1}(y(t)) \pm 2n\pi \qquad (3.85)$$

There is no one-to-one correspondence between $x(t)$ and $y(t)$. Hence, the system is non-invertible.

2. $y(t) = \int_{-\infty}^{t/4} x(\tau)d\tau$

$$x(t/4) = \dfrac{d}{dt}y(t) \qquad (3.86)$$

The system is invertible.

3. $y(t) = \int_{t-3}^{t} x(\tau)d\tau$

$$y(t) = \int_{-\infty}^{t} x(\tau)d\tau - \int_{-\infty}^{t-3} x(\tau)d\tau$$
(3.87)

$$x(t) - x(t-2) = \frac{d}{dt}y(t)$$

The system is invertible.

4. $y(t) = \dfrac{d}{dt}x(t)$

$$x(t) = \int_{-\infty}^{t} y(\tau)d\tau \tag{3.88}$$

There is one-to-one correspondence. The system is invertible.

5. $y(t) = x(3 - t)$

$$x(t) = y(t+3) \tag{3.89}$$

The system is invertible.

6. $y(t) = x(t/5)$

$$x(t) = y(5t) \tag{3.90}$$

The system is invertible.

Example 3.40

Consider the DT systems given by

1. $y[n] = \sin(x[n])$

2. $y[n] = 3x[n] - 5$

3. $y[n] = 3x[n]u[n]$

4. $y[n] = \log_{10}(|x[n]|)$

5. $y[n] = \displaystyle\sum_{k=-\infty}^{n} x[k+1]$

6. $y[n] = x[n^2]$

Are the systems invertible? If yes, find the reason for invertibility.

Solution

1. $y[n] = \sin(x[n])$

$$x[n] = \sin^{-1}(y[n]) \pm \frac{1}{2} \qquad (3.91)$$

There is no one-to-one correspondence. The system is non-invertible.

2. $y[n] = 3x[n] - 5$

$$x[n] = \frac{1}{3}(y[n] + 5) \qquad (3.92)$$

The system is invertible.

3. $y[n] = 3x[n]u[n]$

$$y[n] = \sum_{m=0}^{\infty} 3x[m]$$

$$x[n] = \frac{1}{3}y[n] - \sum_{m=1}^{\infty} 3x[m] \qquad (3.93)$$

The system is invertible.

4. $y[n] = \log_{10}(|x[n]|)$

$$x[n] = \pm 10^{y[n]} \qquad (3.94)$$

The system is non-invertible. There is no one-to-one correspondence.

5. $y[n] = \sum_{k=-\infty}^{n} x[k+1]$

$$x[n] = y[n] - x[n+1] - \sum_{k=-\infty}^{n-2} x[k+1] \qquad (3.95)$$

Here, the system is invertible.

6. $y[n] = x^2[n]$

$$x[n] = \pm\sqrt{y[n]} \qquad (3.96)$$

The system is non-invertible, as there is no one-to-one correspondence.

3.3.2 Stability

Stability is a notion that describes whether the system will be able to follow the input. A system is said to be unstable if its output is out of control or increases without bound. To define stability, we define the following types of responses for linear time invariant systems.

1. **Zero State Response** This is due to the input only. All initial conditions are zero.
2. **Zero Input Response** This is due to the initial conditions only. All inputs are zero.

The system response will be discussed in detail in chapter 4.

BIBO stability An arbitrary relaxed system (with zero initial conditions) is said to be bounded input bounded output (BIBO) stable if and only if its output is bounded for every bounded input.

if $|x(t)| < \infty$ then $|y(t)| < \infty$ for $-\infty \leq t \leq \infty$

if $|x[n]| < \infty$ then $|y[n]| < \infty$ for $-\infty \leq n \leq \infty$ \hfill (3.97)

For zero input stability, the magnitude of the output signal/sequence must be finite and it must asymptotically approach zero. If the output sequence is represented as $y[n]$, then

$|y(t)| < \infty; \lim_{t \to \infty} |y(t)| = 0$

$|y[n]| < \infty; \lim_{n \to \infty} |y[n]| = 0$ \hfill (3.98)

Physical significance of Stability The stable system is only of practical value. If the system is unstable, the extraneous noise or even thermal noise may allow the system to become unstable and the system output may become infinite. The characteristic roots of the system will basically decide the stability of the system. (Refer to section 4.2.)

We will consider some examples to illustrate the concepts.

Example 3.41

Consider a system given by $y(t) = x(t) + 3$. Determine if the system is stable.

Solution

Let the input signal be bounded. $|x(t)| < M$ for $-\infty \leq t \leq \infty$
The output can be written as

$|y(t)| = |x(t)| + 3 < M + 3$ \hfill (3.99)

The output of the system is also bounded. Hence, the system is stable.

Example 3.42

Consider a system given by $y(t) = \int_{-\infty}^{t} x(\tau)d\tau$

Solution

Let the input be bounded. $|x(t)| < M$ for $-\infty \leq t \leq \infty$

The output integrates the values of input from infinity to t. Hence, infinite values are used in the integration, which may lead to infinite output. The system cannot be guaranteed to be stable.

Example 3.43

Consider a system given by $y[n] = \sum_{m=0}^{10} x[m]$

Solution

Let the input be bounded. $|x(t)| < M$ for $-\infty \leq t \leq \infty$. The output sums the values of input from $m = 0$ to $m = 10$. Hence, only 11 values are used in the summation that results in the output of

$$|y(t)| < 11M \text{ for } -\infty \leq t \leq \infty \tag{3.100}$$

The system output is bounded and hence, the system is stable.

Example 3.44

Consider a system given by $y[n] = \sum_{m=-\infty}^{n} x[m]$

Solution

Let the input be bounded. The output sums the values of input from infinity to n. Hence, infinite values are used in the summation that may lead to infinite output. The system cannot be guaranteed to be stable.

Example 3.45

Consider the CT systems given by

1. $y(t) = \cos(x(t))$

2. $(t) = \int_{-\infty}^{t/4} x(\tau)d\tau$

3. $y(t) = \int_{t-3}^{t} x(\tau)d\tau$

4. $y(t) = \dfrac{d}{dt} x(t)$

5. $y(t) = x(3 - t)$

6. $y(t) = x(t/5)$

Are the systems stable? If not, find out why they are not stable.

Solution

1. $y(t) = \cos(x(t))$ Let the input be bounded.

$$|x(t)| < M \text{ for } -\infty \leq t \leq \infty \ |y(t)| = |\cos(x(t))| \leq 1 \qquad (3.101)$$

The cosine terms are always bounded and have a value less than 1. Hence, the system is stable.

2. $y(t) = \int_{-\infty}^{t/4} x(\tau)d\tau$ Let the input be bounded.

$$|x(t)| < M \text{ for } -\infty \leq t \leq \infty \ |y(t)| = |\int_{-\infty}^{t/4} x(\tau)d\tau| \qquad (3.102)$$

The output is the integration of values between minus infinity to $t/4$ i.e., infinite values. The output cannot be guaranteed to be bounded. Hence, the system is unstable.

3. $y(t) = \int_{t-3}^{t} x(\tau)d\tau$

Let the input be bounded.

$$|x(t)| < M \text{ for } -\infty \leq t \leq \infty, \ |y(t)| = |\int_{t-3}^{t} x(\tau)d\tau| \qquad (3.103)$$

The output is the integration of values between $t - 3$ to t i.e., a finite time interval of 3 time units. Hence, the system is stable.

4. $y(t) = \dfrac{d}{dt} x(t)$

Let the input be bounded.

$$|x(t)| < M \text{ for } -\infty \leq t \leq \infty, \ |y(t)| = |\dfrac{d}{dt}|x(t)|| \qquad (3.104)$$

The output is the differentiation of a finite quantity. Hence, the system is stable.

5. $y(t) = x(3 - t)$

Let the input be bounded.

$$|x(t)| < M \text{ for } -\infty \leq t \leq \infty, \ |y(t)| = |x(3-t)| < M \qquad (3.105)$$

Hence, the system is stable.

6. $y(t) = x(t/5)$

   Let the input be bounded.

   $$|x(t)| < M \text{ for } -\infty \leq t \leq \infty, \ |y(t)| = |x(t/5)| < M \qquad (3.106)$$

   Hence, the system is stable.

Example 3.46

Consider the DT systems given by

1. $y[n] = \sin(x[n])$

2. $y[n] = 3x[n] - 5$

3. $y[n] = 3x[n]u[n]$

4. $y[n] = \log_{10}(|x[n]|)$

5. $y[n] = \sum_{k=-\infty}^{n} x[k+1]$

6. $y[n] = x[n^2]$

Are the systems stable? If yes, find the reason for stability.

Solution

1. $y[n] = \sin(x[n])$

   Let the input be bounded.

   $$|x[n]| < M \text{ for } -\infty \leq n \leq \infty \ |y[n]| = |\sin(x[n])| \leq 1 \qquad (3.107)$$

   The sine terms are always bounded and have a value less than 1. Hence, the system is stable.

2. $y[n] = 3x[n] - 5$

   Let the input be bounded.

   $$|x[n]| < M \text{ for } -\infty \leq n \leq \infty, \ |y[n]| = |3|x[n]|| - 5 \leq 3M - 5 \qquad (3.108)$$

   The system output is bounded. Hence, the system is stable.

3. $y[n] = 3x[n]u[n]$

   Let the input be bounded.

$$|x[n]| < M \text{ for } -\infty \leq n \leq \infty, \ |y[n]| = |3|x[n]u[n]|| < M \qquad (3.109)$$

The system output is bounded. Hence, the system is stable.

4. $y[n] = \log_{10}(|x[n]|)$
   Let the input be bounded.

$$|x[n]| < M \text{ for } -\infty \leq n \leq \infty, \ |y[n]| = |\log_{10}|x[n]|| < M \qquad (3.110)$$

The system output is bounded. Hence, the system is stable.

5. $y[n] = \sum_{k=-\infty}^{n} x[k+1]$

   Let the input be bounded.

$$|x[n]| < M \text{ for } -\infty \leq n \leq \infty, \ |y[n]| = |\sum_{k=-\infty}^{n} x[k+1]| \to \infty \qquad (3.111)$$

The system output is the summation of infinite values. Hence, the system is unstable.

6. $y[n] = x^2[n]$
   Let the input be bounded.

$$|x[n]| < M \text{ for } -\infty \leq n \leq \infty, \ |y[n]| = |x^2[n]| < M^2 \qquad (3.112)$$

The system output is the square of a finite value. Hence, the system is stable.

Concept Check
- What is meaning of invertibility?
- Specify a situation when you need the invertibility property.
- What is the physical significance of invertibility?
- How will you confirm the stability of the system?
- What will happen to the output of the system if the system is unstable?
- What is the essential condition for stability of a system?
- What is BIBO stability?

3.4 System Representation as Interconnection of Operations

This section interprets the systems as interconnection of operations. Let us consider a simple example to illustrate the concept. When the input signal is transformed by some operator H to produce the output signal, the operator denotes the action of the system. Let the output of any system $y(t)$ or $y[n]$ be represented as

$$y(t) = H[x(t)]$$

$$y[n] = H[x[n]] \qquad (3.113)$$

Where, H is the operator that acts on the input signal $x(t)$ or $x[n]$ to produce the output $y(t)$ or $y[n]$. The operator H can be represented as interconnection of operators as follows.

Consider a system represented by the Eq. (3.114).

$$y(t) = \frac{1}{3}[x(t) + x(t-1) + x(t-2)]$$

$$\qquad (3.114)$$

$$y[n] = \frac{1}{3}[x[n] + x[n-1] + x[n-2]]$$

This equation represents a moving average system. Basically, the current output of the system is found by taking the average of current and previous two input samples. To obtain the next output sample, we have to take the average of next, current and the previous input sample. This is a sliding average and hence, called as the moving average system. Let us try to represent the system in the form of a block schematic, as shown in Fig. 3.1. Let us represent the time shifting operation by the operator S. Here, S represents a time shift by 1 time unit for CT systems and time shift by 1 sample for DT systems. Now, the operator H can be represented as

$$H = \frac{1}{3}[1 + S + S^2] \qquad \text{and}$$

$$y(t) = \frac{1}{3}[1 + S + S^2]x(t)$$

$$\qquad (3.115)$$

$$y[n] = \frac{1}{3}[1 + S + S^2]x[n]$$

The representation is clear from Fig. 3.3 drawn for a DT system. A similar representation will hold good for CT system as well.

We will go through further examples to illustrate the concept.

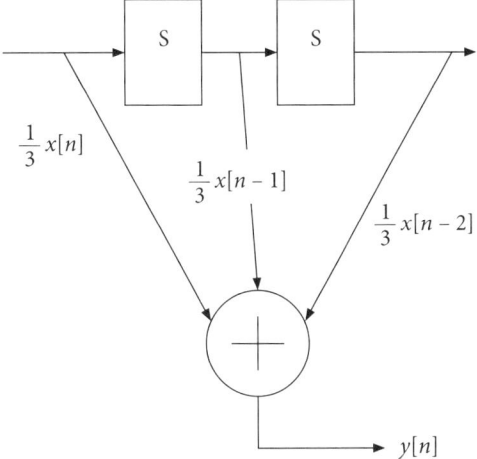

Fig. 3.3 Block schematic for a system

Example 3.47
Consider a CT system represented by Eq. (3.116), represent it as the interconnection of operators.

$$y(t) = [x(t) + \frac{1}{2}x(t-2) + \frac{1}{3}x(t-4) + \frac{1}{2}x(t-6)] \tag{3.116}$$

Solution
Let us try to represent the system in the form of a block schematic, as shown in Fig. 3.4. Let us represent the time shifting operation by the operator S, which represents a time shift by 2 time units. Now, the operator H can be represented as

$$H = \left[1 + \frac{1}{2}S + \frac{1}{3}S^2 + \frac{1}{2}S^3\right] \text{ and } y(t) = \left[1 + \frac{1}{2}S + \frac{1}{3}S^2 + \frac{1}{2}S^3\right]x(t) \tag{3.117}$$

The representation is clear from Fig. 3.4.

Example 3.48
Consider a CT system represented by the Eq. (3.118).

$$y(t) = \left[x(t) + \frac{1}{2}x(t-1) - \frac{1}{3}y(t-2)\right] \tag{3.118}$$

Represent it as the interconnection of operators.

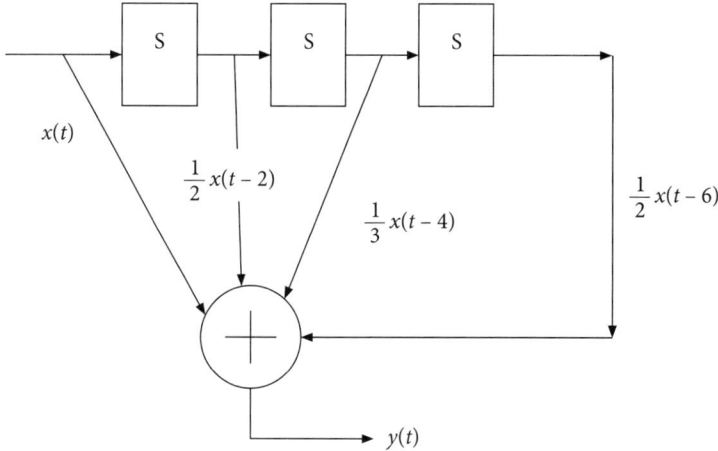

Fig. 3.4 Block schematic for Example 3.47

Solution

Let us try to represent the system in the form of a block schematic, as shown in Fig. 3.5. Let us represent the time shifting operation by the operator S that represents a time shift by 1 time unit. Now, the operator H can be represented as

$$H = \left[1 + \frac{1}{2}S\right] / \left[1 + \frac{1}{3}S^2\right] \quad \text{and} \quad y(t) = \left[1 + \frac{1}{2}S\right] / \left[1 + \frac{1}{3}S^2\right] x(t) \quad (3.119)$$

Note that we have to take the term containing $y(t - 2)$ on the left-hand side to combine it with $y(t)$. The representation is clear from Fig. 3.5.

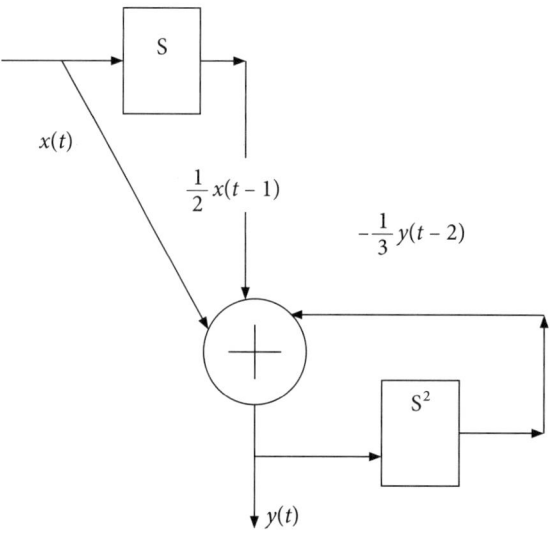

Fig. 3.5 Block schematic for Example 3.48

Example 3.49

Consider a DT system represented by Eq. (3.120), represent it as the interconnection of operators.

$$y[n] = [x[n] + \frac{1}{2}x[n-1] + \frac{1}{3}x[n-2] + \frac{1}{2}x[n-3]] \qquad (3.120)$$

Solution

Let us try to represent the system in the form of a block schematic, as shown in Fig. 3.6. Let us represent the time shifting operation by the operator S. Now, the operator H can be represented as

$$H = \left[1 + \frac{1}{2}S + \frac{1}{3}S^2 + \frac{1}{2}S^3\right] \text{ and } y[n] = \left[1 + \frac{1}{2}S + \frac{1}{3}S^2 + \frac{1}{2}S^3\right]x[n] \qquad (3.121)$$

The representation is clear from Fig. 3.6.

Example 3.50

Consider a system represented by the Eq. (3.122).

$$y[n] = [x[n] + \frac{1}{2}x[n-1] - \frac{1}{3}y[n-1]] \qquad (3.122)$$

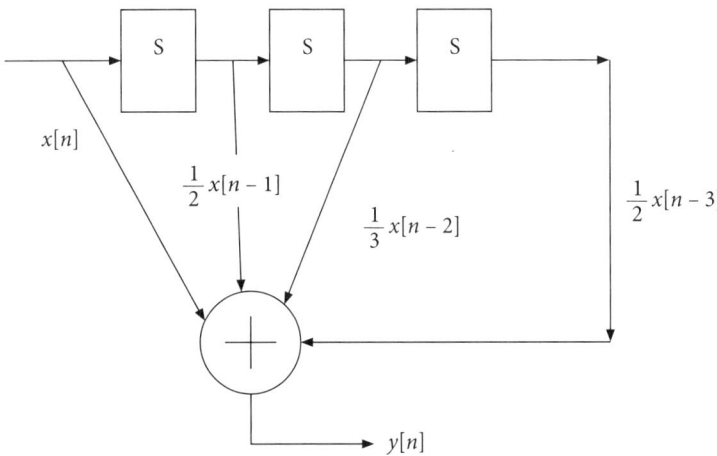

Fig. 3.6 Block schematic for Example 3.49

Represent it as the interconnection of operators.

Solution

Let us try to represent the system in the form of a block schematic, as shown in Fig. 3.7. Let us represent the time-shifting operation by the operator S. Now, the operator H can be represented as

$$H = \left[1+\frac{1}{2}S\right] / \left[1+\frac{1}{3}S\right] \text{ and } y[n] = \left[1+\frac{1}{2}S\right] / \left[1+\frac{1}{3}S\right] x[n] \qquad (3.123)$$

Note that we have to take the term containing $y(n-1)$ on the left-hand side to combine it with $y(n)$. The representation is clear from Fig. 3.7.

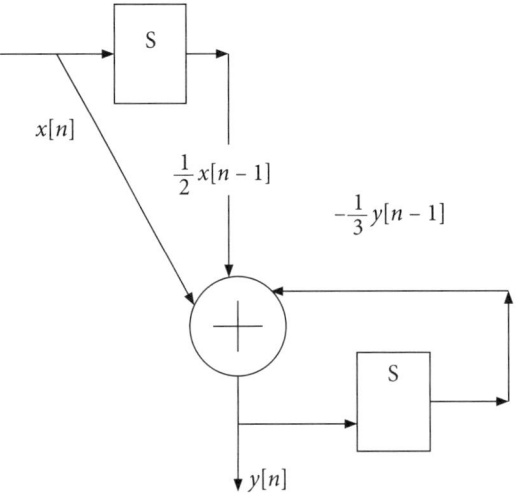

Fig. 3.7 Block schematic for Example 3.50

Concept Check

- What is a moving average system?
- How will you divide the system operator into number of operators such as time sifting operator?
- What is the advantage of using interconnection of operations?

3.5 Series and Parallel Interconnection of Systems

Let us understand the significance of series and parallel interconnections of the systems. We need to understand what will happen to the impulse response of the interconnected system. These interconnections have significance for LTI systems.

3.5.1 Series interconnection of systems

Consider the series connection of two linear and shift invariant systems, i.e., LTI CT systems, as shown in Fig. 3.8. Let the impulse responses of the two systems be specified as $h_1(t)$ and $h_2(t)$, respectively. The series connection is also termed as cascaded configuration. The output of the system is $y(t)$. Let $z(t)$ represent the output of the first system with impulse response $h_1(t)$. The output $z(t)$ can be written as

$$z(\tau) = x(\tau) * h_1(\tau)$$

$$z(\tau) = \int_{-\infty}^{\infty} x(\eta) h_1(\tau - \eta) d\eta$$

(3.124)

Here, the symbol * denotes the convolution operation. The meaning of convolution and the procedure to implement convolution in CT and DT domain will be discussed in detail in chapter 4.

The output of first system $z(t)$ is applied as input to the second system with impulse response $h_2(t)$. The output of the second system $y(t)$ will be given as

$$y(t) = z(t) * h_2(t)$$

$$y(t) = \int_{-\infty}^{\infty} z(\tau) h_2(t - \tau) d\tau$$

$$y(t) = \int_{-\infty}^{\infty} \int_{-\infty}^{\infty} x(\eta) h_1(\tau - \eta) h_2(t - \tau) d\eta d\tau$$

put $\tau - \eta = \upsilon$,

$$y(t) = \int_{-\infty}^{\infty} \int_{-\infty}^{\infty} x(\eta) h_1(\upsilon) h_2(t - \upsilon - \eta) d\eta d\upsilon$$

(3.125)

$$y(t) = \int_{-\infty}^{\infty} x(\eta) \int_{-\infty}^{\infty} h_1(\upsilon) h_2(t - \eta - \upsilon) d\upsilon$$

$$y(t) = \int_{-\infty}^{\infty} x(\eta) h(t - \eta) d\eta$$

$$h(t) = h_1(t) * h_2(t)$$

$$y(t) = x(t) * h(t) = \int_{-\infty}^{\infty} x(\eta) h(t - \eta) d\eta$$

We can conclude that when the two systems are cascaded or connected in series, the impulse response of the cascaded system is the convolution of impulse responses of the two systems.

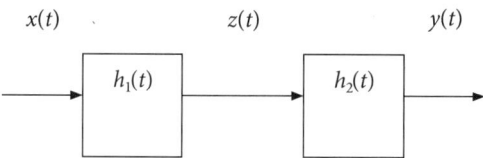

Fig. 3.8 Series interconnection of CT systems

In case of DT systems, same equations will hold good. When the two systems with impulse responses $h_1[n]$ and $h_2[n]$ are cascaded or connected in series, the impulse response of the cascaded system is the convolution of impulse responses of the two systems given by $h_1[n]*h_2[n]$.

Consider the series connection of two linear and shift invariant systems, i.e., LTI DT systems, as shown in Fig. 3.9. Let the impulse responses of the two systems be specified as $h_1[n]$ and $h_2[n]$, respectively. The series connection is also termed as cascaded configuration. The output of the system is $y[n]$. Let $z[n]$ represent the output of the first system with impulse response $h_1[n]$. The output $z[n]$ can be written as

$$z[n] = x[n] * h_1[n]$$

$$z[n] = \sum_{m=-\infty}^{n} x[m] h_1[n-m] \tag{3.126}$$

The output of first system $z[n]$ is applied as input to the second system with impulse response $h_2[n]$. The output of the second system $y[n]$ will be given as

$$y[n] = z[n] * h_2[n]$$

$$y[n] = \sum_{m=-\infty}^{n} z[m] h_2[n-m]$$

$$y[n] = \sum_{m=-\infty}^{n} \sum_{l=-\infty}^{m} x[l] h_1[m-l] h_2[n-m]$$

put $m - l = w$

$$y[n] = \sum_{w=-l-\infty}^{n-l} \sum_{l=-\infty}^{m} x[l]h_1[w]h_2[n-l-w] \tag{3.127}$$

$$y[n] = \sum_{l=-\infty}^{n} x[l]h[n-l]$$

$$h[n] = h_1[n] * h_2[n]$$

$$y[n] = x[n] * h[n]$$

We can conclude that when the two systems are cascaded or connected in series, the impulse response of the cascaded system is the convolution of impulse responses of the two systems.

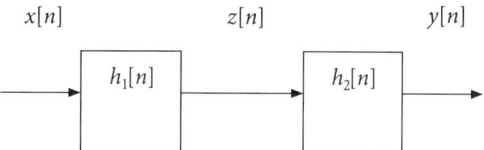

Fig. 3.9 Series interconnection of DT systems

3.5.2 Parallel interconnection of systems

Consider the parallel connection of two linear and shift invariant systems, i.e., LTI systems, as shown in Fig. 3.10. Let the impulse responses of the two systems be specified as $h_1(t)$ and $h_2(t)$, respectively.

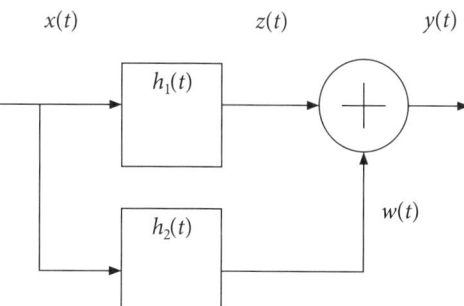

Fig. 3.10 Parallel interconnection of CT systems

The output of the system is $y(t)$. Let $z(t)$ represent the output of the first system with impulse response $h_1(t)$. Let $w(t)$ represent the output of the second system with impulse response $h_2(t)$. The output $z(t)$ can be written as

$$z(t) = x(t) * h_1(t)$$

$$z(t) = \int_{-\infty}^{\infty} x(\eta) h_1(t-\eta) d\eta \tag{3.128}$$

The symbol * is used to represent the convolution operation.
The output $w(t)$ can be written as

$$w(t) = x(t) * h_2(t)$$

$$w(t) = \int_{-\infty}^{\infty} x(\eta) h_2(t-\eta) d\eta \tag{3.129}$$

The output of the parallel interconnection $y(t)$ can be written as

$$y(t) = x(t) * h_1(t) + x(t) * h_2(t)$$

$$y(t) = \int_{-\infty}^{\infty} x(\eta) h_1(t-\eta) d\eta + \int_{-\infty}^{\infty} x(\eta) h_2(t-\eta) d\eta \tag{3.130}$$

$$y(t) = \int_{-\infty}^{\infty} x(\eta)[h_1(t-\eta) + h_2(t-\eta)] d\eta$$

Let the impulse response of the parallel configuration be written as $h(t)$ given by

$$h(t) = h_1(t) + h_2(t)$$

$$y(t) = \int_{-\infty}^{\infty} x(\eta)[h_1(t-\eta) + h_2(t-\eta)] d\eta \tag{3.131}$$

$$y(t) = \int_{-\infty}^{\infty} x(\eta) h(t-\eta) d\eta = x(t) * h(t)$$

We can conclude that when the two systems are connected in parallel, the impulse response of the parallel configuration system is the addition of impulse responses of the two systems.

In case of DT systems, same equations will hold good. When the two systems with impulse responses $h_1[n]$ and $h_2[n]$ are connected in parallel, the impulse response of the cascaded system is the addition of impulse responses of the two systems given by $h_1[n] + h_2[n]$.

Consider the parallel connection of two linear and shift invariant systems, i.e., LTI systems, as shown in Fig. 3.11. Let the impulse responses of the two systems be specified as $h_1[n]$ and $h_2[n]$, respectively.

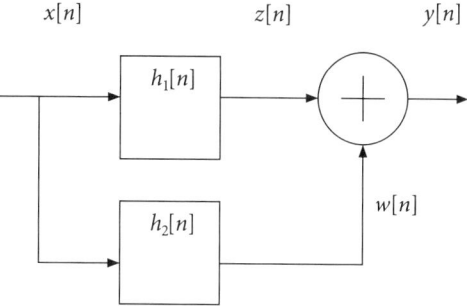

Fig. 3.11 Parallel interconnection of DT systems

The output of the system is $y[n]$. Let $z[n]$ represent the output of the first system with impulse response $h_1[n]$. Let $w[n]$ represent the output of the second system with impulse response $h_2[n]$. The output $z[n]$ can be written as

$$z[n] = x[n]*h_1[n]$$

$$z[n] = \sum_{m=-\infty}^{n} x[m]h_1[n-m] \tag{3.132}$$

The symbol * is used to represent the convolution operation.
The output $w[n]$ of the second system can be written as

$$w[n] = x[n]*h_2[n]$$

$$w[n] = \sum_{m=-\infty}^{n} x[m]h_2[n-m] \tag{3.133}$$

The output of the parallel interconnection $y[n]$ can be written as

$$y[n] = x[n]*h_1[n] + x[n]*h_2[n]$$

$$y[n] = \sum_{m=-\infty}^{n} x[m]h_1[n-m] + \sum_{m=-\infty}^{n} x[m]h_2[n-m] \tag{3.134}$$

$$y[n] = \sum_{m=-\infty}^{n} x[m]\{h_1[n-m] + h_2[n-m]\}$$

Let the impulse response of the parallel configuration be written as $h[n]$ given by

$$h[n] = h_1[n] + h_2[n] \tag{3.135}$$

$$y[n] = \sum_{m=-\infty}^{n} x[m]\{h_1[n-m] + h_2[n-m]\} \tag{3.136}$$

$$y[n] = \sum_{m=-\infty}^{n} x[m]h[n-m] = x[n] * h[n]$$

We can conclude that when the two systems are connected in parallel, the impulse response of the parallel configuration system is the addition of impulse responses of the two systems.

Let us go through some numerical problems to illustrate the concepts.

Example 3.51

Consider the configuration shown in Fig. 3.12 with impulse responses given by $h_1(t), h_2(t), h_3(t)$ for the systems, as shown. Find the impulse response of the overall configuration.

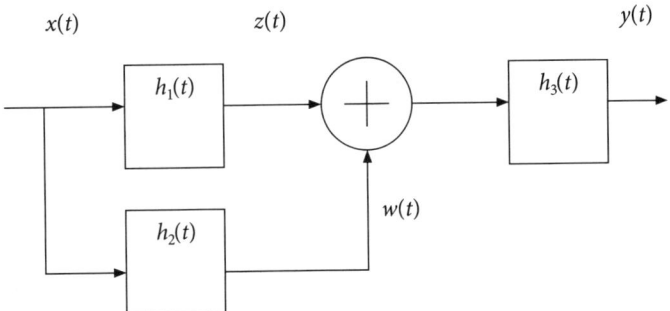

Fig. 3.12 Configuration of three systems for Example 3.51

Solution

Firstly we will find the impulse response of the parallel interconnection of two systems with impulse response $h_1(t)$ and $h_2(t)$. The impulse response of the parallel configuration is $h(t) = h_1(t) + h_2(t)$. The third system is connected in series with the parallel configuration. Hence, the impulse response of the series configuration will be the convolution of the two impulse responses. We find that the distributive law holds good.

$$h'(t) = [h_1(t) + h_2(t)] * h_3(t) = h_1(t) * h_3(t) + h_2(t) * h_3(t) \tag{3.137}$$

Example 3.52

Consider the configuration shown in Fig. 3.13 with impulse responses given by $h_1(t), h_2(t), h_3(t)$ for the systems, as shown in the figure below. Find the impulse response of the overall configuration.

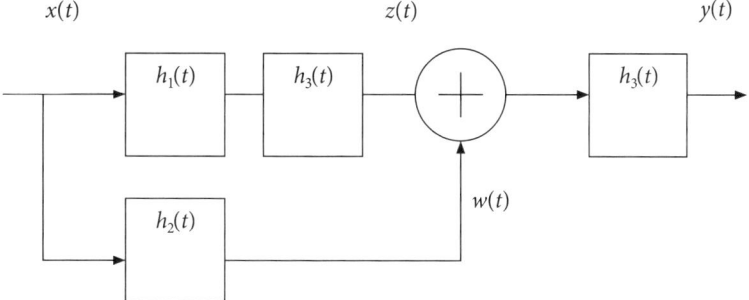

Fig. 3.13 Configuration of four systems for Example 3.52

Solution

Let us find the impulse response of the series system $h(t)$ giving output as $z(t)$

$$h(t) = h_1(t) * h_3(t). \tag{3.138}$$

This series configuration is in parallel with $h_2(t)$. The impulse response of the parallel configuration is

$$h'(t) = [h(t) + h_2(t)] = [h_1(t) * h_3(t)] + h_2(t) \tag{3.139}$$

The system with impulse response $h_3(t)$ is connected in series with this parallel configuration. The impulse response of the overall system is then given by

$$h''(t) = [h(t) + h_2(t)] * h_3(t) = \{[h_1(t) * h_3(t)] + h_2(t)\} * h_3(t) \tag{3.140}$$

Example 3.53

Consider the configuration shown in Fig. 3.14 with impulse responses given by $h_1(t)$, $h_2(t)$, $h_3(t)$ for the systems, as shown in the figure below. Find the impulse response of the overall configuration.

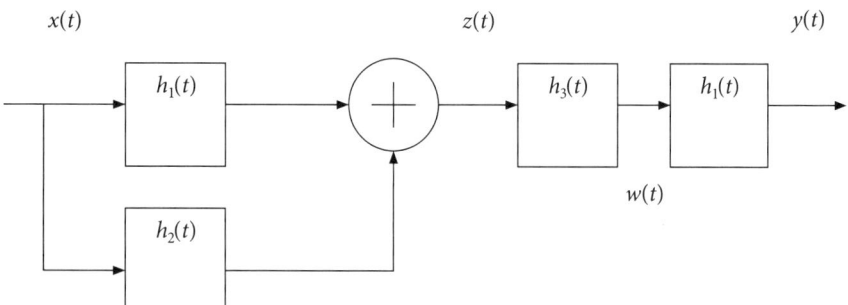

Fig. 3.14 Configuration of four systems for Example 3.53

Solution

Let us find the impulse response of the parallel system $h(t)$ giving output as $z(t)$

$$h(t) = h_1(t) + h_2(t). \qquad (3.141)$$

This parallel configuration is in series with $h_3(t)$. The impulse response of the series configuration is

$$h'(t) = [h_1(t) + h_2(t)] * h_3(t) \qquad (3.142)$$

The system with impulse response $h_1(t)$ is connected in series with this series configuration. The impulse response of the overall system is then given by

$$h''(t) = [h_1(t) + h_2(t)] * h_3(t) * h_1(t) \qquad (3.143)$$

Example 3.54

Consider the overall impulse response of the system given by $h''(t) = [h_1(t) * h_2(t)] + [h_3(t) * h_1(t)]$ for the three systems with impulse responses $h_1(t), h_2(t), h_3(t)$. Draw the configuration.

Solution

The overall impulse response for the configuration indicates that the series combination of $h_1(t)$ and $h_2(t)$ is connected in parallel with series combination of $h_3(t)$ and $h_1(t)$. It can be drawn as shown in Fig. 3.15.

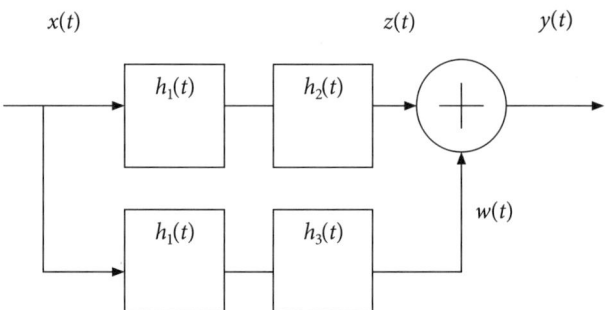

Fig. 3.15 Configuration of four systems for Example 3.54

We will now solve some examples for DT systems.

Example 3.55

Consider the configuration shown in Fig. 3.16 with impulse responses given by $h_1[t], h_2[t]$ and $h_3[t]$ in Fig. 3.16 for the systems, as shown in the figure below. Find the impulse response of the overall configuration.

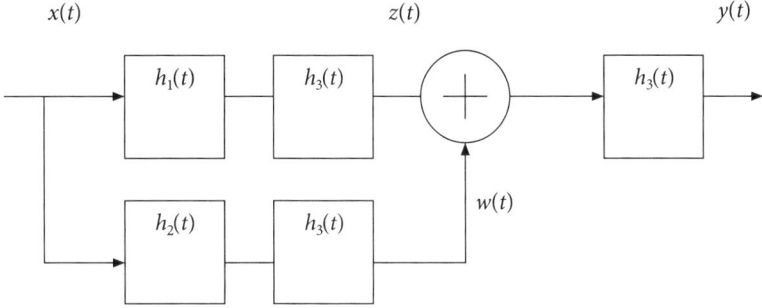

Fig. 3.16 Configuration of five systems for Example 3.55

Solution

Let us find the impulse response of the series system $h[n]$ giving output as $z[n]$

$$h[n] = h_1[n] * h_3[n]. \tag{3.144}$$

Let us find the impulse response of the series system $h'[n]$ giving output as $w[n]$

$$h'[n] = h_2[n] * h_3[n]. \tag{3.145}$$

These series configurations are in parallel. The impulse response of the parallel configuration $h''[n]$ is

$$h''[n] = [h[n] + h'(t)] = [h_1[n] * h_3[n]] + [h_2[n] * h_3[n]] \tag{3.146}$$

The system with impulse response $h_3[n]$ is connected in series with this parallel configuration. The impulse response of the overall system $h_{overall}$ is then given by

$$h[n]_{overall} = [h[n] + h'[n]] * h_3[n] = \{[h_1[n] * h_3[n]] + [h_2[n] * h_3[n]]\} * h_3(t) \tag{3.147}$$

Example 3.56

Consider the configuration shown in Fig. 3.17 with impulse responses given by $h_1[n]$, $h_2[n]$, $h_3[n]$ for the systems, as shown in the figure below. Find the impulse response of the overall configuration.

Solution

Let us find the impulse response of the parallel system $h[n]$ giving output as $z[n]$

$$h[n] = h_1[n] + h_2[n]. \tag{3.148}$$

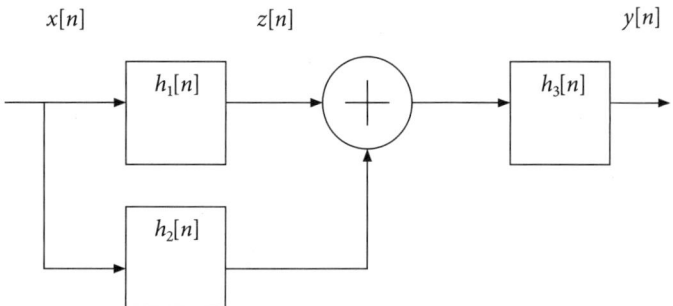

Fig. 3.17 Configuration of three systems for Example 3.56

This parallel configuration is in series with $h_3[n]$. The impulse response of the series configuration is

$$h'(t) = [h_1(t) + h_2(t)] * h_3(t) \qquad (3.149)$$

Example 3.57

Consider the overall impulse response of the system given by $h''[n] = \{[h_1[n]*h_2[n]]+[h_3[n]*h_1[n]]\}*h_1[n]$ for the three systems with impulse responses $h_1[n]$, $h_2[n]$, $h_3[n]$. Draw the configuration.

Solution

The overall impulse response for the configuration indicates that the series combination of $h_1[n]$ and $h_2[n]$ is connected in parallel with series combination of $h_3[n]$ and $h_1[n]$. This parallel configuration is in series with $h_1[n]$. It can be drawn as shown in Fig. 3.18.

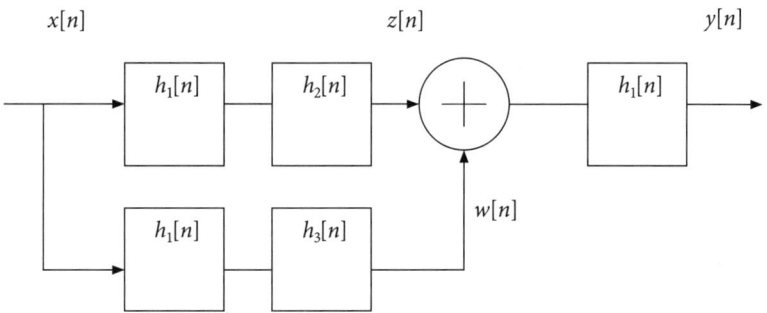

Fig. 3.18 Configuration of five systems for Example 3.57

Things to remember

If the two systems are connected in series, their impulse responses get convolved. If the two systems are connected in parallel, their impulse responses get added.

Concept Check

- What is the impulse response of the series of two systems?
- How will you find the impulse response of the parallel configuration?
- Does the rule for series and parallel interconnection hold good for DT systems?

Summary

In this chapter, we have described and explained the properties of DT systems.

- We have discussed the important properties of DT systems, namely, linearity and shift invariance. It was shown that a system is linear when it is homogeneous as well as additive. It was emphasized that if the transfer curve of the system is linear passing through the origin, then it is linear. Linearity has a meaning more than this. If the system obeys the principle of superposition i.e., additivity and homogeneity, then the system is linear. We then defined the property of time/shift invariance. If the system is linear, the input signal can be suitably decomposed into component signals and the corresponding outputs for the component signals one at a time can be calculated by assuming all other inputs equal to zero. The component outputs can be scaled and added to generate the output of the system for the input signal. This is exactly the property of superposition. The system is said to be time/shift invariant if the input to the system is shifted impulse $\delta(n - k)$, then it results in a shifted impulse response of $h(n - k)$. The linear and shift invariant system is termed as LTI system. If the system is LTI, one can characterize the system in terms of its impulse response. The calculation of the output for any given input in case of LTI system gets greatly simplified due to the principle of superposition.

- We further concentrated on causality and memory property of systems. We defined the property of causality for systems. The system is said to be causal if the present output of the system depends only on current and past input or output at previous instant. Causal systems are practically realizable or implementable. The system of a human being is causal as we always keep on learning from the past inputs and past outputs of the system. The future inputs have no effect on our act at current or present time. The present and past inputs have meaning only for temporal systems where time is an independent variable. In case of spatial domain systems, present and past input has no meaning. Non-causal temporal systems can be implemented if some delay is tolerable. We can generate a bench mark for system performance using non-causal systems. Offline systems can always be implemented as non-causal systems. The system is said to have memory or said to be dynamic if its current output depends on previous, future input or previous and future output signals. The system is said to be memoryless or instantaneous if its current output depends only on current input. Examples of memoryless systems are the systems containing only resistive elements.

- We further discussed invertibility and stability. If it is possible to recover the input of the system, then the system is said to be invertible. Invertible system also finds applications in communication field. For error-free transmission an equalizer is used at the input of the receiver that has inverse characteristics as that of the channel. Stability is a notion that describes whether the system will be able to follow the input. A system is said to be unstable if its output is out of control or increases without bound. An arbitrary relaxed system (with zero initial conditions) is said to be bounded input bounded output (BIBO) stable if and only if its output is bounded for every bounded input.

- The system can be described as an interconnection of operations. If the time delay is represented as S block, we can draw the block schematic for the time difference equations. We have discussed series and parallel interconnections of LTI systems and have shown that for series connections, the impulse responses of the individual systems get convolved and for parallel interconnections, the impulse responses of individual systems get added.

Multiple Choice Questions

1. The system is linear if
   (a) it is homogeneous
   (b) bit is additive
   (c) it is additive or homogeneous
   (d) it is additive and homogeneous

2. The system is causal when the current output sample depends on
   (a) current input sample
   (b) current or next and past input samples
   (c) current and/or past input samples and/or past output samples
   (d) next or past input samples or past output samples

3. The range of values of "a" for which the system with impulse response $h(n) = a^n u(n)$ is stable is
   (a) $|a| > 1$ (b) $|a| < 1$
   (c) $a > 0$ (d) $a < 0$

4. If the transfer graph for a system is linear and passes through origin
   (a) the system is nonlinear
   (b) the system s linear
   (c) the system may be additive
   (d) the system may be homogeneous

5. The system of human being is
   (a) non-causal (b) non-linear
   (c) causal (d) non-linear and non-causal

6. The system is said to be memoryless if
   (a) Only on a current input sample
   (b) current or next and past input samples
   (c) current and/or past input samples and/or past output samples
   (d) next or past input samples or past output samples
7. The following system is invertible
   (a) different transforms
   (b) all systems
   (c) if the equation exists to find signal values
   (d) non-linear system such as squaring device
8. The bench mark system can be designed using
   (a) causal systems (b) invertible systems
   (c) non-invertible systems (d) non-causal systems
9. The following systems are prone to noise
   (a) non-causal systems (b) causal systems
   (c) non stable systems (d) invertible systems
10. The system is BIBO stable if
    (a) the output is bonded for every bounded input
    (b) the output is always bounded
    (c) the output goes on increasing
    (d) the output decreases
11. The system is called as LTI if
    (a) the system is linear
    (b) the system is linear and time invariant
    (c) the system is time invariant
    (d) the system is additive
12. The system given by $y(t) = x(t) + 5$
    (a) is memoryless
    (b) is with memory
    (c) is unstable and without memory
    (d) is stable and with memory
13. The system given by $y[n] = x[3 - n]$ is
    (a) causal (b) anti causal
    (c) invertible and causal (d) causal and non-invertible
14. The system given by $y(t) = x^2(t)$ is
    (a) invertible (b) non-invertible
    (c) invertible with memory (d) invertible and memoryless

15. The system given by $y[n] = (n - 1)x[n]$ is
    (a) time invariant
    (b) time dependant
    (c) time bound
    (d) time variant
16. The system given by $y[n] = x[n - 1]\sin(n\omega)$ is
    (a) time invariant
    (b) LTI
    (c) Non-linear and time invariant
    (d) Time variant
17. Series interconnection of system results in
    (a) addition of the impulse responses
    (b) convolution of impulse responses
    (c) subtraction of impulse responses
    (d) multiplication of impulse responses
18. Parallel interconnection of systems results in
    (a) addition of the impulse responses
    (b) convolution of impulse responses
    (c) subtraction of impulse responses
    (d) multiplication of impulse responses

Review Questions

3.1 What is linearity? Define additivity and homogeneity. Is the transfer curve for a linear system always linear? Explain physical significance of linearity.

3.2 What is the time/shift invariance property of systems? Explain physical significance of shift invariance property.

3.3 Explain property of superposition.

3.4 When will you say that the system is memoryless? Give one example of a memoryless system.

3.5 Define causality for a system. Can we design and use a non-causal system? Is a causal system a requirement for spatial systems?

3.6 Explain the meaning of causality for a system of a human being.

3.7 What is invertibility? Can we use a non-invertible transform for processing a signal?

3.8 Explain the meaning of BIBO stability for a system.

3.9 Explain the physical significance of stability.

3.10 How will you interpret the system as interconnection of operators? Explain using a suitable example.

3.11 Find the impulse response for a series interconnected and parallel interconnected systems. Prove that the impulse response of the series interconnection of two LTI CT systems is a convolution of the two impulse responses.

3.12 Find the impulse response for a series interconnected and parallel interconnected systems. Prove that the impulse response of the series interconnection of two LTI DT systems is a convolution of the two impulse responses.

Problems

3.1 Is the system given by $y[n] = x[-n]$ a linear and shift invariant system?

3.2 Is the system given by $y(t) = x(t-2)$ a linear and shift invariant system?

3.3 Verify that the systems given by $y(n) = x[n]\cos(\omega n)$ and $y[n] = nx[n]$ are shift variant.

3.4 Check if the systems given by $y(t) = (t-1)x(t)$ and $y(t) = x(t)\cos(\omega t + \pi/4)$ are shift invariant?

3.5 Find if the following systems are time invariant.
  (a) $y[n] = x[n] - x[n-1]$
  (b) $y[n] = nx[n]$
  (c) $y[n] = x[1-n]$
  (d) $y[n] = x[n]\sin(\omega n)$
  (e) $y(t) = x(t) + x(t+1)$
  (f) $y(t) = t^2 x(t)$
  (g) $y(t) = x(4-t)$
  (h) $y(t) = x(t)\sin(t)$

3.6 Find if the following systems are linear.
  (a) $y[n] = (n+1)x[n]$
  (b) $y[n] = x[n^2]$
  (c) $y[n] = x^3[n]$
  (d) $y[n] = 2x[n] + 3$
  (e) $y(t) = (t+2)x(t)$
  (f) $y(t) = x^3(t)$
  (g) $y(t) = 3x(t) + 1$
  (h) $y(t) = \sin(t)x(t)$

3.7 Find if the following systems are causal.
  (a) $y[n] = 5x[n]$
  (b) $y[n] = \sum_{k=-\infty}^{n+1} x(k)$

(c) $y[n] = x[3 - n]$

(d) $y[n] = x[3n]$

(e) $y(t) = x(t^2)$

(f) $y(t) = x(5 - t)$

(g) $y(t) = x(2t - 2)$

(h) $y(t) = x(-2t)$

3.8 Find if the following systems are memoryless

(a) $y(t) = e^{-2}x(t)$

(b) $y(t) = \cos(x(t))$

(c) $y[n] = 5x[n] + 2x[n]u[n]$

(d) $y(t) = \int_{-\infty}^{t/3} x(\tau)d\tau$

(e) $y(t) = x(7 - 2t)$

(f) $y(t) = x(t/5)$

3.9 Find if the following systems are stable.

(a) $y(t) = \cos(x(t))$

(b) $y[n] = \log_{10}(|x[n]|)$

(c) $y[n] = \cos(2\pi x[n]) + x[n]$

(d) $y(t) = \dfrac{d}{dt}[e^{-t}x(t)]$

(e) $y(t) = x(t/3)$

(f) $y[n] = \sum\limits_{m=-\infty}^{n} x[m+3]$

(g) $y[n] = x[n] \sum\limits_{m=-\infty}^{\infty} \delta[n - 5m]$

3.10 Find if the following systems are invertible.

(a) $y[n] = \sum\limits_{m-\infty}^{n} x[m+3]$

(b) $y[n] = x[n-1] + 4$

(c) $y(t) = x^3(t)$

(d) $y(t) = x(t/9)$

(e) $y(t) = \sqrt{x(t)}$

(f) $y[n] = x[2n]$

3.11 Represent the following systems in terms of interconnection of operators

(a) $y(t) = x(t) + x(t-3) + y(t-6)$

(b) $y(t) = x(t-1) - y(t-2) - y(t-3)$

(c) $y[n] = x[n] + y[n-1] + y[n-2]$

(d) $y[n] = x[n-2] + y[n-2] - y[n-4]$

3.12 Find the overall impulse response for the interconnection of three systems.

(a)

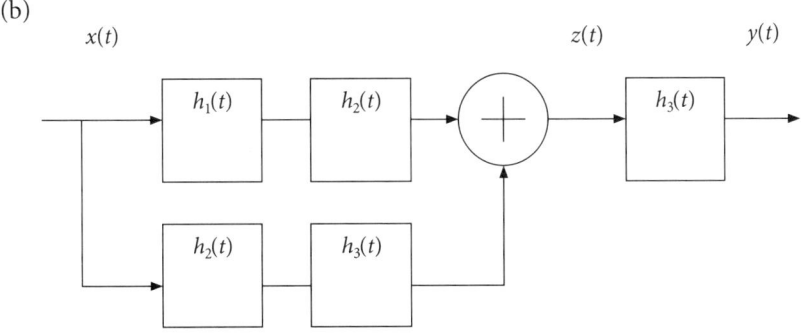

(b)

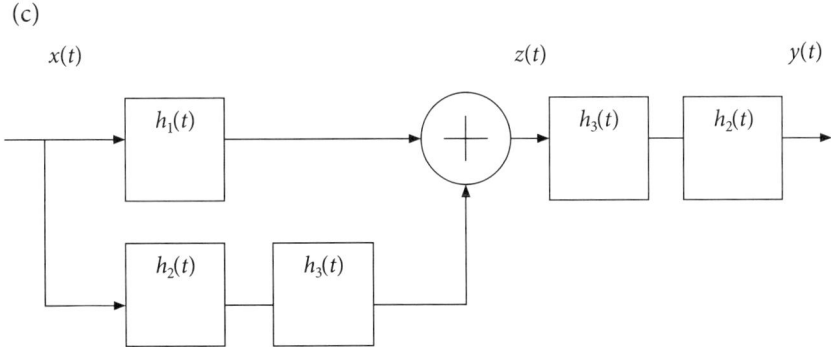

(c)

(d)

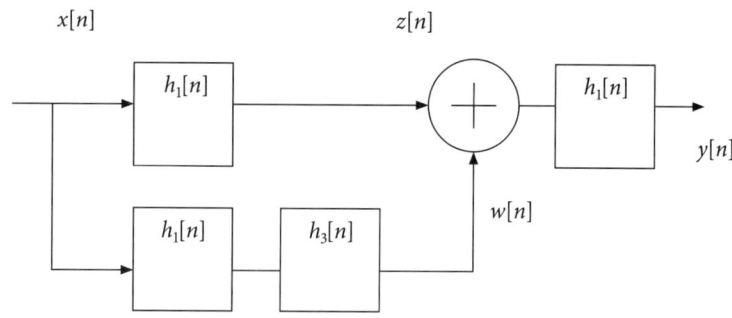

(e)

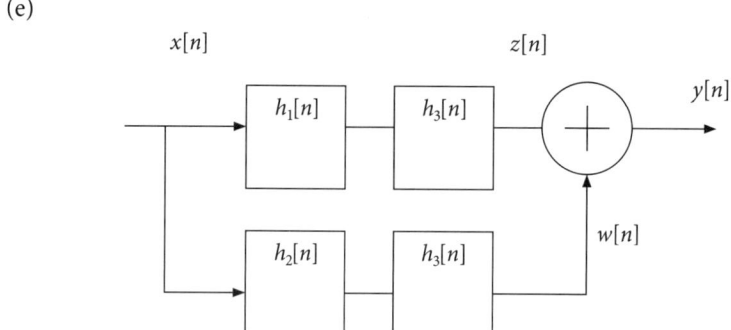

3.13 Find the possible interconnection for the following equation of the overall impulse response of the system.

(a) $h_{overall}[n] = \{[h_1[n] + h_2[n]] * [h_3[n] + h_1[n]]\} * h_1[n]$

(b) $h_{overall}[n] = \{[h_1[n] * h_2[n]] + [h_3[n] * h_1[n]]\} * [h_1[n] + h_2[n]]$

(c) $h_{overall}(t) = \{h_1(t) + [h_2(t) * h_3(t)]\} * h_3(t)$

Answers

Multiple Choice Questions

1 (a)	2 (c)	3 (b)	4 (b)	5 (c)
6 (a)	7 (a)	8 (d)	9 (c)	10 (a)
11 (b)	12 (a)	13 (b)	14 (b)	15 (d)
16 (d)	17 (b)	18 (a)		

Problems

3.1 Yes – Linear and shift invariant

3.2 Yes – linear and shift invariant

3.3 Yes, systems are time variant

3.4 The systems are time variant

3.5
| (a) Yes | (b) No | (c) Yes | (d) No | (e) Yes |
| (f) No | (g) Yes | (h) No | | |

3.6
| (a) Yes | (b) Yes | (c) No | (d) Yes | (e) Yes |
| (f) No | (g) No | (h) No | | |

3.7.
| (a) Yes | (b) No | (c) No | (d) No | (e) No |
| (f) No | (g) No | (h) No | | |

3.8
| (a) Yes | (b) Yes | (c) Yes | (d) No | (e) No |
| (f) No | | | | |

3.9
| (a) Yes | (b) Yes | (c) Yes | (d) Yes | (e) Yes |
| (f) No | (g) Yes | | | |

3.10
| (a) No | (b) Yes | (c) Yes | (d) Yes | (e) No |
| (f) No | | | | |

3.11 (1)

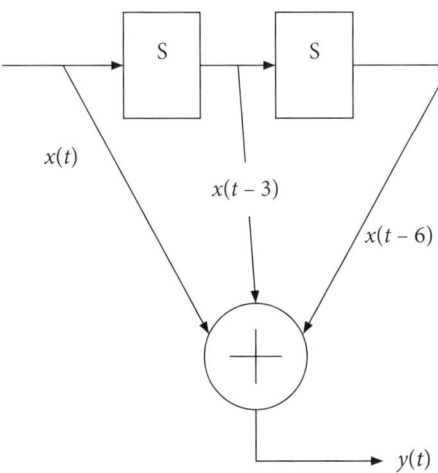

3.11 (2)

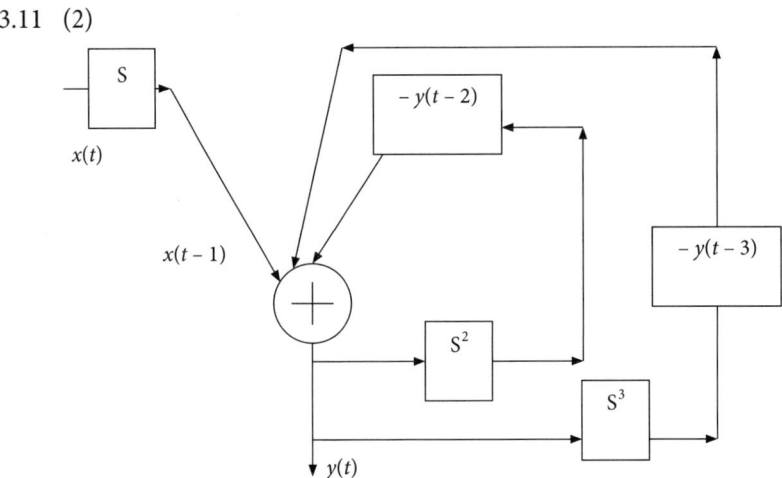

3.11 (3)

3.11 (4)

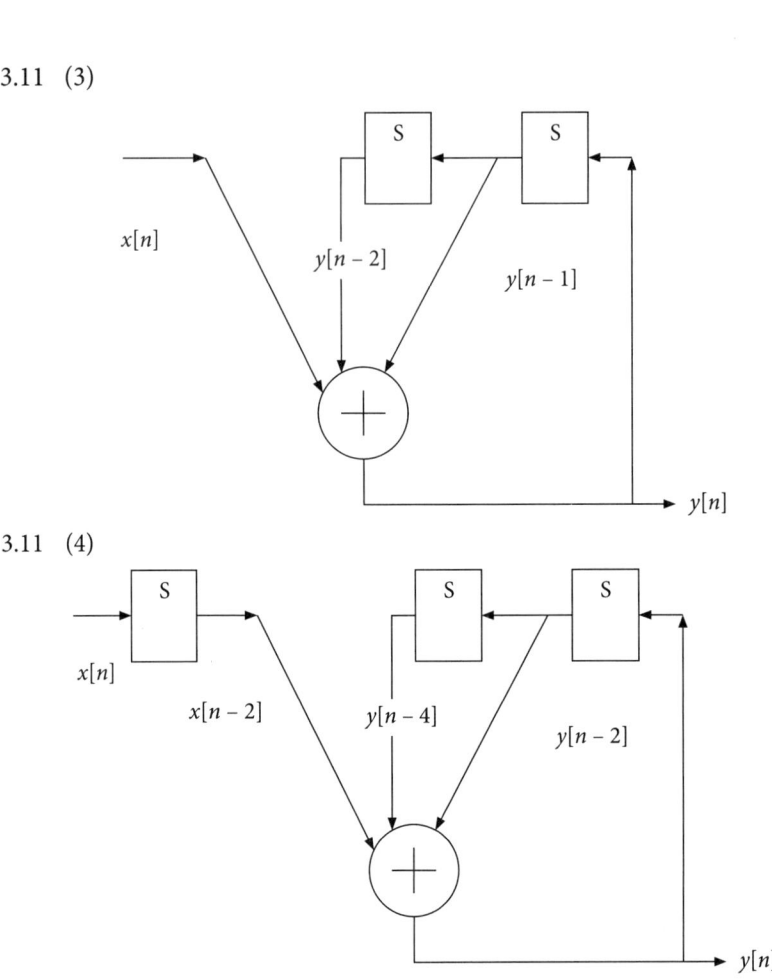

3.12 (a) $h_{overall}(t) = \{[h_1(t) + h_2(t)] * h_3(t)\}$

(b) $h_{overall}(t) = \{[h_1(t) * h_2(t)] + [h_2(t) * h_3(t)]\} * h_3(t)$

(c) $h_{overall}(t) = \{h_1(t) + [h_2(t) * h_3(t)]\} * h_3(t) * h_2(t)$

(d) $h_{overall}[n] = \{[h_1[n]] + [h_3[n] * h_1[n]]\} * h_1[n]$

(e) $h''[n] = \{[h_1[n] * h_2[n]] + [h_3[n] * h_1[n]]\}$

3.13 (a)

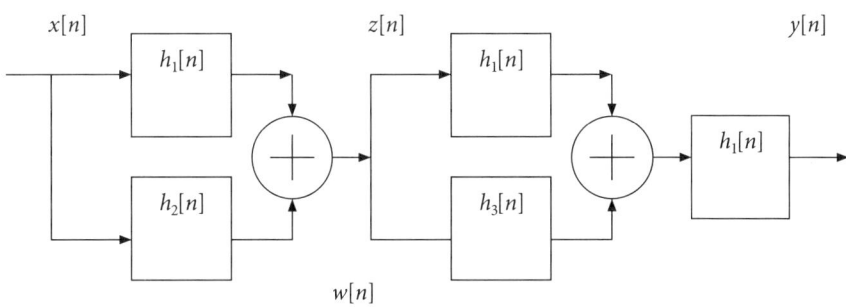

(b)

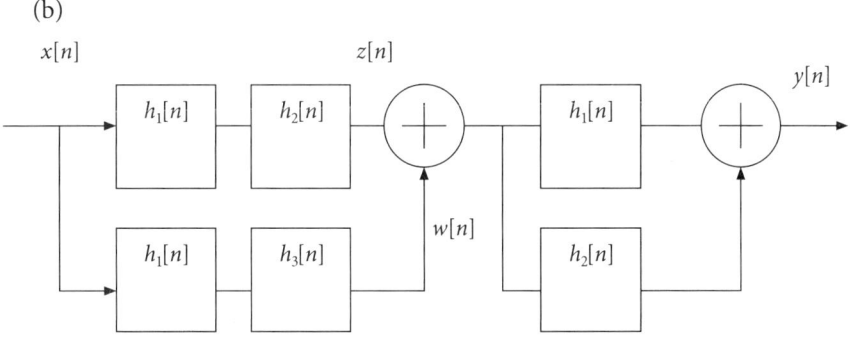

(c)

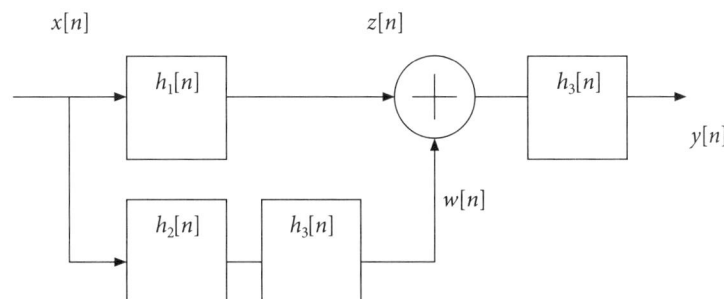

4

Time Domain Response of CT and DT LTI Systems

Learning Objectives

- Response of CT and DT LTI systems.
- Zero input response.
- System representation as impulse response.
- Representation of signal in terms of impulses.
- Convolution integral for CT LTI systems.
- Zero state response for CT and DT systems.
- Convolution sum for DT LTI systems.
- Causality, stability and memory for LTI systems in terms of impulse response.
- Series and parallel interconnections.
- Calculation of effective impulse response.

4.1 Response of CT Systems

Let us discuss the response of the system based on its differential equation. We will consider a system described by the differential equation and illustrate the response of the system due to initial conditions called as zero input response and the response due to externally applied input with all initial conditions zero namely zero state response. The total response of the system is then the addition of zero input response and the zero state response.

Consider a generalized differential equation for a system.

$$\frac{d^n y}{dt^n} + a_{n-1}\frac{d^{n-1} y}{dt^{n-1}} + \ldots\ldots + a_0 y = b_m \frac{d^m x}{dt^M} + b_{m-1}\frac{d^{m-1} x}{dt^{m-1}} + \ldots\ldots + b_0 x \qquad (4.1)$$

Let us denote $\frac{d}{dt} = D$, we may write the equation as

$$(D^n + a_{n-1}D^{n-1} + \ldots + a_0)y(t) = (b_m D^m + b_{m-1}D^{m-1} + \ldots + b_0)x(t) \quad (4.2)$$

$$(D^n + a_{n-1}D^{n-1} + \ldots + a_0)/(b_m D^m + b_{m-1}D^{m-1} + \ldots + b_0) = y(t)/x(t) \quad (4.3)$$

where $y(t)$ denotes the output of the system and $x(t)$ the input to the system.

$(D^n + a_1 D^{n-1} + \ldots + a_n)$ is referred to as a denominator polynomial in D of order n and $(b_m D^m + b_{m-1}D^{m-1} + \ldots + b_0)$ is termed as a numerator polynomial of order m. We will consider two cases for the ratio of degree of a numerator to the degree of a denominator namely m/n.

Case 1 $n < m$, i.e., the ratio is greater than 1. In this case, the system behaves as a differentiator of the order $m - n$. It will amplify high-frequency noise through differentiation. High-frequency signal will have high amplitude of differentiation output. The system will be more prone to noise.

Case 2 $n > m$, i.e., the ratio is less than 1. In this case, the system will be stable and it will not amplify noise.

We will assume that the degree of the numerator is less than the degree of the denominator polynomial i.e., case 2 holds good.

4.1.1 Zero input response

The denominator polynomial is called as the characteristic polynomial. When the externally applied input $x(t)$ is zero, we get the system equation as

$$(D^n + a_1 D^{n-1} + \ldots + a_n) = 0. \quad (4.4)$$

Let us assume the solution as $y(t) = ce^{-\lambda t}$. Putting this value of $y(t)$ in the polynomial, we get

$$(c\lambda^n e^{-\lambda t} + a_1 c\lambda^{n-1} e^{-\lambda t} + \ldots + a_n) = 0$$

$$c(\lambda^n + a_1 \lambda^{n-1} + \ldots + a_n)e^{-\lambda t} = 0 \quad (4.5)$$

This polynomial is same as the polynomial in D or with D replaced by λ. This equation is characteristic of the system and will decide the zero input response. Let us decompose the characteristic polynomial in 'n' number of roots to get the equation as

$$(\lambda - \lambda_1)(\lambda - \lambda_2)\ldots(\lambda - \lambda_n) = 0 \quad (4.6)$$

$\lambda_1, \lambda_2, \ldots, \lambda_n$ are roots of characteristic polynomial.

The roots of the characteristic polynomial are called as characteristic roots or eigen values or natural frequencies of the system. The characteristic modes or natural modes are denoted as $e^{-\lambda_1 t}, e^{-\lambda_2 t} \ldots$, etc. These are also termed as the eigen functions of the system.

We write the roots as exponential functions for two reasons.

1. If $(D^n + a_1 D^{n-1} + \ldots + a_n) = 0$, this indicates that all the n derivatives of the function are zero at all values of time. The nth derivative is $(-\lambda)^n e^{-\lambda t}$; if all n derivatives are to be zero, then this is true only for the exponential function.

2. The second reason is that in case of the LTI systems, if the input is exponential function, the output is also an exponential function. The other functions cannot have this claim. The exponential solutions are called as eigen functions.

The characteristic modes decide the behaviour of the system when the external input is zero. The response of the system when the external input is zero is termed as the zero input response. The characteristic modes will also play a major role in deciding the zero state response. The characteristic roots may be all distinct, may be repeated roots or may be complex roots.

We will consider a numerical problem to illustrate the concepts.

Example 4.1

Consider a simple second order system with characteristic equation given by $(D^2 + 3D + 2)y(t) = 0$. Find the zero input response if the initial conditions are $y(0) = 0$ and $Dy(0) = 5$.

Solution

$$(D^2 + 3D + 2) = 0 \Rightarrow (D+2)(D+1) = 0 \tag{4.7}$$

The roots are $D = -2$ and $D = -1$.

D is of the form $e^{-\lambda t}$ and e^{-t}. The solution can be written as

$$y(t) = c_1 e^{-2t} + c_2 e^{-t} \tag{4.8}$$

This is called as zero input response. We will now apply initial conditions to find the values of c_1 and c_2.

Put $t = 0$ in the solution to get

$$y(0) = c_1 + c_2 = 0 \tag{4.9}$$

Find the derivative of the solution and put $t = 0$ in the equation to get

$$Dy(t) = -2c_1 e^{-2t} - c_2 e^{-t} = 5 \tag{4.10}$$

put $t = 0$ to get $-2c_1 = c_2 + 5$

put $c_1 = -c_2$, $2c_2 = c_2 + 5$

$c_2 = 5$ and $c_1 = -5$

$y(t) = (5e^{-t} - 5e^{-2t})u(t)$

Zero state response The zero state response is obtained when the system is relaxed i.e. initial conditions are made zero and the external input is applied. Let $x(t) = 2e^{-3t}$ be applied as input to the system. We will discuss zero state response in the further section.

Concept Check
- What is characteristic polynomial?
- What is characteristic equation?
- What are eigen values?
- What are natural frequencies?
- What are characteristic modes?
- Why is eigen function selected as exponential function?
- What is zero input response?
- What is zero state response?

4.2 System Representation as Impulse Response

Let us first understand the meaning of impulse response of the system. As discussed in chapter 2, there are some standard inputs used for testing the response of the system. One such standard input is an impulse. The response of the system to the impulse function is termed as an *impulse response* of the system and is generally denoted as $h(t)$ or $h(n)$ in case of CT and DT systems, respectively. Impulse response goes by a different name in some other applications. If a system is functioning as a filter, then the impulse response is called the *filter Kernel*, the *convolution Kernel* or simply the *Kernel*. In image processing, impulse response is called the *point spread function*.

Any CT or DT system can be characterized by its impulse response. Any DT signal can be decomposed into summation of weighted and shifted unit impulses. Unit impulse function in CT domain is used as a standard input for testing of the system. When any CT input signal is applied, the output is obtained either as a convolution integral of the input signal with the impulse response of the system. Let us try to understand the concept of convolution.

Let us deal with CT systems first. Any input signal can be represented in terms of unit impulses. Ideally impulse function exists in time domain. Practically, generating the impulse function with infinite amplitude and zero width is impossible. Impulses can be realized as the pulses of very small duration and very high amplitude. Ideal impulse function cannot be realised in practice. Summarizing, we can say that the impulse function is used as a standard input to the LTI systems for the following reasons.

1. Any signal can be decomposed into a linear combination of scaled and shifted impulses.
2. LTI system is completely characterised by its impulse response.
3. The output is obtained by the convolution of input signal with the impulse response.

Consider any analog signal. We will try to decompose the input signal into closely spaced impulses. If we know the response of the system to the impulse function, the response of the system to any input can be found out as follows.

1. Decompose the input signal in terms number of impulse functions.
2. Find the response of the system for each of these impulses.
3. Integrate the responses to get the final response.

4.2.1 Representation of signals in terms of impulses

Let us first understand the decomposition of the signal in terms of impulses. The signal under consideration can be decomposed in two ways.

1. The signal can be decomposed into shifted and scaled impulses.
2. The signal can be decomposed into scaled exponential functions.

If the signal is decomposed using impulses, we will represent the system in terms of the impulse response and if the signal is decomposed using exponential functions, we will represent the system in terms of its frequency response. Decomposition of the signal using exponential functions and the frequency response will be discussed in chapter 5.

The representation of the system in terms of its impulse response is discussed in chapter 4 and representation of the system in terms of its frequency response will be discussed in chapter 5. Let us first understand the decomposition of the CT signal in scaled and translated unit impulses. Consider a CT signal $x(t)$ as shown in Fig. 4.10. Let us consider a pulsed or staircase approximation to $x(t)$. A small duration pulse is defined as $\delta_\Delta(t)$.

A CT signal $x(t)$ may be decomposed into pulses of small duration Δ and with different scaling factors equal to the height of the pulse, as shown in Fig. 4.1. The pulsed approximation to $x(t)$ can be written as

$$\bar{x}(t) = \sum_{k=-\infty}^{\infty} x(k\Delta)\delta_\Delta(t - k\Delta)\Delta \tag{4.11}$$

Here, $\delta_\Delta(t - k\Delta)$ represents a pulse at location $k\Delta$. It has a duration of Δ and amplitude of $x(k\Delta)$. In order to convert the pulse to unit amplitude delta pulse, it is multiplied by Δ. $\delta_\Delta(t-k\Delta)\Delta$ has unit amplitude. The staircase approximation to a CT signal $x(t)$ can be depicted in Fig. 4.1.

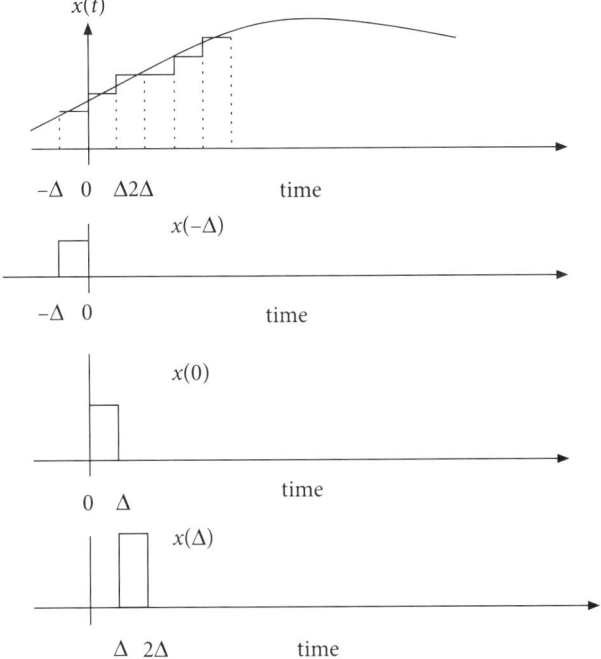

Fig. 4.1 Staircase approximation to a continuous time signal

The approximation is closer as the duration of the pulse approaches zero. As Δ approaches zero, the staircase approximation to $x(t)$ will change into the integral and can be written as

$$x(t) = \int_{-\infty}^{\infty} x(\tau)\delta(t-\tau)d\tau \qquad (4.12)$$

where $d\tau$ represents a small duration of the pulse tending to zero, $x(\tau)$ is the amplitude of the pulse and $\delta(t-\tau)$ represents a delta function at $t = \tau$. Using the property of shifting for the delta function, the integral is equal to $x(t)$.

We assume that the system is a linear time invariant (LTI) system. When the input signal is decomposed in terms of impulses, it is easy to find the system response to any input signal using the principle of superposition. Let the system be characterized by its impulse response $h(t)$.

The procedure to find the response of the system to the signal $x(t)$ using a graphical method can be stated as follows.

1. Decompose the signal $x(t)$ in terms of scaled and shifted unit impulses.
2. Find the response of the system $x(t)$ to unit impulse at $t = 0$.

3. The system being shift invariant, the response of the system to impulse at $t = t_1$ is $h(t - t_1)$.
4. Find the response of the system to all shifted impulses.
5. Finally add all the responses.

4.2.2 Calculation of impulse response of the system

Referring to section 4.1 the system equation must have a degree of numerator polynomial less than the degree of the denominator polynomial for noise considerations and for stability. The section 4.2 discusses about the decomposition of the signal into unit impulses. To find the response of the system to any input, we must know the impulse response of the system. The impulse response can be found out by applying an impulse at $t = 0$ and assuming zero initial conditions before the application of the impulse. The impulse is like a sudden shock and it generates the energy storages in the system after $t = 0$. It creates new initial conditions in the system. These newly created initial conditions will generate the system response even after the impulse is removed immediately after $t = 0$. The system response generated will naturally depend on the characteristic modes of the system. At the instant $t = 0$, the response will at the most be equal to the scaled impulse and it will exist only if the degree of the numerator is same as the degree of the denominator. The generalized equation for the impulse response can be written as

$$h(t) = a_0 \delta(t) + \text{Characteristic mode terms} \tag{4.13}$$

For the proof of the above equation the reader may refer to the chapter on Laplace Transform. If $m < n$, then $a_0 = 0$. We will illustrate the concepts with the help of the simple example.

Example 4.2
Determine the impulse response for a system given by the differential equation
$(D^2 + 3D + 2)y(t) = Dx(t)$.

Solution
Let us first evaluate the characteristic equation of the system and then evaluate the roots of the characteristic equation.
The characteristic equation is given by

$$(D^2 + 3D + 2) = 0 \Rightarrow (D+2)(D+1) = 0 \tag{4.14}$$

The roots are $D = -2$ and $D = -1$. The solution can be written as

$$y(t) = (c_1 e^{-2t} + c_2 e^{-t})u(t) \tag{4.15}$$

We need to find the values of constants. The derivative of $y(t)$ can be written as

$$\bar{y}(t) = -2c_1 e^{-2t} - c_2 e^{-t} \tag{4.16}$$

For any system with the denominator polynomial of order n, the initial conditions are given as follows.

$$y(0) = 0, Dy(0) = 0, \ldots\ldots D^{n-2}y(0) = 0 \ \& \ D^{n-1}y(0) = 1; \tag{4.17}$$

We will use the result without going into the proof of the result. The initial conditions for the system with denominator polynomial of order 2, the initial conditions will be translated as

$y(0) = 0$ and $\bar{y}(0) = 1$. Where $\bar{y}(0) = Dy(0)$ is first derivative of y.

Putting values of initial conditions in the above equations gives

$$y(0) = c_1 + c_2 = 0 \text{ and } \bar{y}(0) = -2c_1 - c_2 = 1 \tag{4.18}$$

Solving the equations leads to $c_1 = -c_2 = 1$

$$y(t) = (-e^{-2t} + e^{-t})u(t). \tag{4.19}$$

The second order term is zero in the numerator polynomial i.e., $m < n$. Hence, put $a_0 = 0$ in the impulse response equation. The solution is

$$h(t) = (-e^{-2t} + e^{-t})u(t) \tag{4.20}$$

which contains only the characteristic mode terms.
We will solve more problems to clarify the concepts further.

Example 4.3
Determine the impulse response for a system given by the differential equation $D(D+2)y(t) = (D+1)x(t)$.

Solution
Let us first evaluate the characteristic equation of the system and then evaluate the roots of the characteristic equation.
The characteristic equation is given by

$$D(D+2) = 0 \Rightarrow \text{roots are } D = -2 \text{ and } D = 0 \tag{4.21}$$

The solution can be written as

$$y(t) = c_1 e^{-2t} + c_2 \tag{4.22}$$

We need to find the values of constants. The derivative of $y(t)$ can be written as

$$\bar{y}(t) = -2c_1 e^{-2t} \tag{4.23}$$

for any system with the denominator polynomial of order n, the initial conditions are given as follows.

The initial conditions for the system with denominator polynomial of order 2, the initial conditions will be translated as

$y(0) = 0$ and $\bar{y}(0) = 1$. Where $\bar{y}(0) = Dy(0)$ is the first derivative of y.

Putting values of initial conditions in the above equations gives

$$y(0) = c_1 + c_2 = 0 \text{ and } \bar{y}(0) = -2c_1 = 1 \Rightarrow c_1 = -1/2 \tag{4.24}$$

Solving the equations leads to $c_1 = -c_2 = -1/2$

$$y(t) = \left(-\frac{1}{2}e^{-2t} + \frac{1}{2}\right)u(t). \tag{4.25}$$

The second order term is zero in the numerator polynomial i.e., $m < n$. Hence, put $a_0 = 0$ in the impulse response equation. The solution is

$$h(t) = \left(-\frac{1}{2}e^{-2t} + \frac{1}{2}\right)u(t) \tag{4.26}$$

which contains only the characteristic mode terms.

Example 4.4

Determine the impulse response for a system given by the differential equation $(D+2)y(t) = (D+1)x(t)$.

Solution

Let us first evaluate the characteristic equation of the system and then evaluate the roots of the characteristic equation.

The characteristic equation is given by

$$(D+2) = 0 \Rightarrow \text{roots are } D = -2 \tag{4.27}$$

The solution can be written as

$$y(t) = a_0 \delta(t) + c_1 e^{-2t} \tag{4.28}$$

We need to find the values of constants. The derivative of $y(t)$ can be written as a_0 is the value of the nth order term in the denominator polynomial and is equal to 1.

$$\overline{y}(t) = -2c_1 e^{-2t} \tag{4.29}$$

For any system with the denominator polynomial of order n, the initial conditions are given as follows.

The initial conditions for the system with denominator polynomial of order 2, the initial conditions will be translated as

$y(0) = 0$ and $\overline{y}(0) = 1$. Where $\overline{y}(0) = Dy(0)$ is first derivative of y.

Putting values of initial conditions in above equations gives

$$y(0) = a_0 + c_1 = 0 \text{ and } \overline{y}(0) = -2c_1 = 1 \Rightarrow c_1 = -1/2 \tag{4.30}$$

Solving the equations leads to $a_0 = 1/2$

$$y(t) = \left(-\frac{1}{2} e^{-2t} u(t) + \frac{1}{2} \delta(t) \right). \tag{4.31}$$

The order of the numerator is same as that of the denominator polynomial i.e., $m = n$. Hence, a_0 term exists in the impulse response equation. The solution is

$$h(t) = \left(-\frac{1}{2} e^{-2t} u(t) + \frac{1}{2} \delta(t) \right) \tag{4.32}$$

which contains the characteristic mode terms and the response due to the unit impulse at $t = 0$.

Example 4.5

Determine the impulse response for a system given by the differential equation $(D+2)y(t) = x(t)$.

Solution

Let us first evaluate the characteristic equation of the system and then evaluate the roots of the characteristic equation.

The characteristic equation is given by

$$(D+2) = 0 \Rightarrow \text{roots are } D = -2 \tag{4.33}$$

The solution can be written as

$$y(t) = a_0 \delta(t) + c_1 e^{-2t} \tag{4.34}$$

We need to find the values of constants.

As $m < n$, put $a_0 = 0$ in the impulse response equation.

For any system with the denominator polynomial of order n, the initial conditions are given as follows.

If the initial conditions for the system with denominator polynomial of order 1, the initial conditions will be translated as

$y(0) = 1.$

Putting values of initial conditions in above equations gives

$y(0) = c_1 = 1$

Solving the equations leads to

$$y(t) = (e^{-2t}u(t)). \qquad (4.35)$$

The solution is $h(t) = (e^{-2t}u(t))$, which contains only the characteristic mode terms.

Concept Check

- Why the impulse function is used as the standard input?
- Can you realize ideal impulse function?
- What is impulse response?
- How will you decompose the signal in terms of impulse function?
- How will you find the impulse response of the system?
- What is the nature of the impulse response?
- What are characteristic mode terms?

4.3 Convolution Integral for CT Systems

We will assume the system to be linear and time invariant i.e., LTI. Let us represent the impulse response of the CT system as $h(t)$ for the impulse at $t = 0$. We have implemented step 1 of the procedure in section 4.2.1. Let us implement steps 3, 4 and 5. The output of the system can be written as

$$y(t) = H\left[\int_{-\infty}^{\infty} x(\tau)\delta(t-\tau)d\tau\right] \qquad (4.36)$$

We can use the property of linearity to get

$$y(t) = \int_{-\infty}^{\infty} x(\tau)H[\delta(t-\tau)]d\tau \qquad (4.37)$$

The system is time invariant. So, the equation for the output becomes

$$y(t) = \int_{-\infty}^{\infty} x(\tau)h(t-\tau)d\tau \qquad (4.38)$$

The output can be seen as the weighted sum of the shifted impulse responses. The weights are the amplitude of the impulse at each time instant i.e., $x(\tau)$. The equation is termed as the convolution integral and is written as

$$y(t) = \int_{-\infty}^{\infty} x(\tau)h(t-\tau)d\tau = x(t)*h(t) = h(t)*x(t) \qquad (4.39)$$

Here, * denotes convolution operation.

The procedure for the calculation of the convolution integral can be stated as follows:

1. Plot a graph of $x(\tau)$ and $h(-\tau)$. To plot $h(-\tau)$, we have to reflect $h(\tau)$ about $\tau = 0$.
2. Start with time shift t large and negative.
3. Find $x(\tau) h(t - \tau)$ and $y(t)$.
4. Increase the time shift t until functional form of the product $x(\tau) h(t - \tau)$ changes.
5. Find $x(\tau) h(t - \tau)$ and $y(t)$ for the new interval of t.
6. Continue by increasing t in positive direction and each of new interval t find $y(t)$.

Let us illustrate this procedure by some numerical examples. If the impulse response of the system is known, then the response of the system to any external input can be easily calculated by convolving the impulse response with the input signal. This is called as convolution integral.

Example 4.6

Let $x(t) = e^{-2t}[u(t) - u(t-2)]$, $h(t) = e^{-t}u(t)$, Find $x(t)*h(t)$.

Solution

We start with step 1 i.e., drawing $x(\tau)$ and $h(-\tau)$.

Step 1 Let us draw both these waveforms. Plots of $x(\tau)$ and $h(t - \tau)$ for different intervals are shown in Fig. 4.2. We have to shift $h(t - \tau)$ slowly towards the right.

Step 2 Start with time shift t large and negative. Let t vary from minus infinity to zero. We find that until t crosses zero, there is no overlap between the two signals. So, the convolution integral has a value ranging from minus infinity to zero. At $t = 0$, the right edge of $h(-\tau)$ touches left edge of $x(\tau)$.

Step 3 Consider the second interval between $t = 0$ to 2.

$$x(\tau)h(t-\tau) = e^{-2\tau}e^{-(t-\tau)} \qquad (4.40)$$

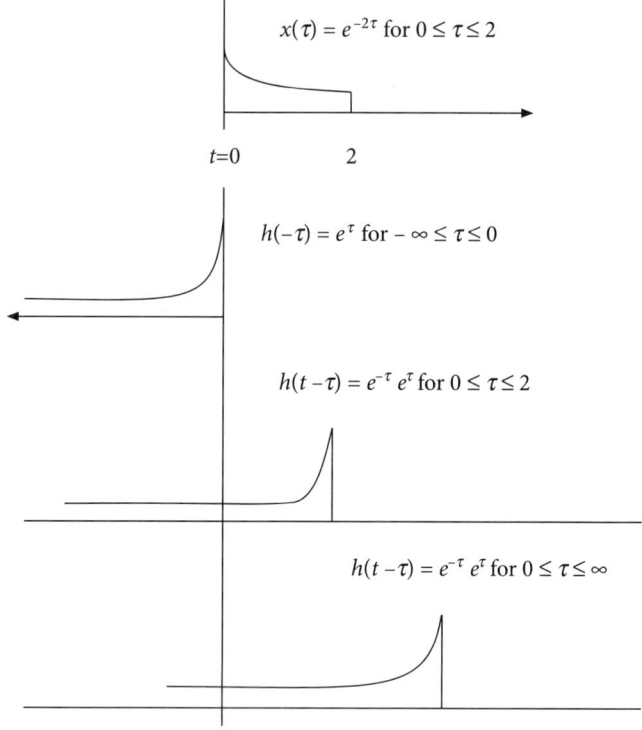

Fig. 4.2 Plots of $x(\tau)$ and $h(t - \tau)$ for various time intervals

The overlapping interval will be between 0 and t. The output can be calculated as

$$y(t) = \int_0^t x(\tau)h(t-\tau)d\tau = \int_0^t e^{-2\tau} e^{-(t-\tau)} d\tau$$

$$= \int_0^t e^{-t} e^{-\tau} d\tau = e^{-t}[-e^{-\tau}]_0^t = e^{-t}[-e^{-t} + 1] \qquad (4.41)$$

$$= e^{-t} - e^{-2t}$$

$$y(2) = e^{-2} - e^{-4}$$

Step 4 The third interval is between 2 and infinity. For $t \geq 2$, the overlapping interval will be from 0 to 2. The output is given by

$$y(t) = \int_0^2 x(\tau)h(t-\tau)d\tau = \int_0^2 e^{-2\tau} e^{-(t-\tau)} d\tau$$

$$= \int_0^2 e^{-t} e^{-\tau} d\tau = e^{-t}[-e^{-\tau}]_0^2 = e^{-t}[-e^{-2} + 1] \qquad (4.42)$$

$$= e^{-t}[1-e^{-2t}]$$

$$y(2) = e^{-2} - e^{-4}$$

The output of the system is plotted in Fig. 4.3.

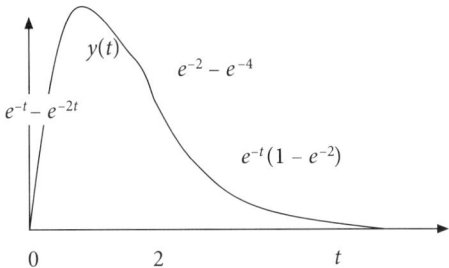

Fig. 4.3 The output of the system

The output can be specified as follows.

$$y(t) = 0 \text{ for } t < 0$$
$$= e^{-t} - e^{-2t} \text{ for } 0 \leq t \leq 2 \quad (4.43)$$
$$= e^{-t}(1 - e^{-2}) \text{ for } t > 2$$

The MATLAB program for plotting the convolution integral is given below. The output of MATLAB program is shown in Fig. 4.4.

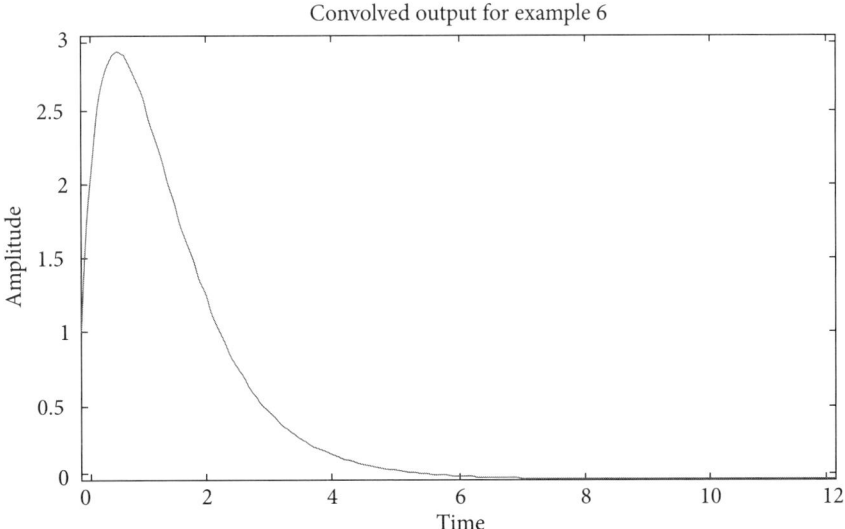

Fig. 4.4 Convolution output for Example 4.6

```
clear all;
t=0:0.1:2;
x=exp(-2*t);
t1=0:0.1:10;
y=exp(-t1);
z=conv(x,y);
n=0:0.1:12;
plot(n,z);title('convolved output for example 4.6');
xlabel('time');ylabel('amplitude');
```

Example 4.7

Let $x(t)=1-t$ for $0 \le t \le 1$, $h(t)=e^{-2t}u(t)$, find $x(t)*h(t)$.

Solution

We start with step 1 i.e., drawing $x(\tau)$ and $h(-\tau)$.

Step 1 Let us draw both these waveforms. Figure 4.5 shows plots of $x(\tau)$ and $h(-\tau)$.

Step 2 Start with time shift t large and negative. Let t vary from minus infinity to zero. We find that until t crosses zero, there is no overlap between the two signals. So, the convolution integral has a value ranging from minus infinity to zero. At $t = 0$, the right edge of $h(-\tau)$ touches the left edge of $x(\tau)$.

Step 3 Consider the second interval between $t = 0$ to 1.

$$x(\tau)h(t-\tau)=e^{-2\tau}(1-t+\tau). \tag{4.44}$$

The overlapping interval will be between 0 and t. The output can be calculated as

$$y(t) = \int_0^t x(\tau)h(t-\tau)d\tau = \int_0^t e^{-2\tau}(1-t+\tau)d\tau$$

$$= (1-t)\int_0^t e^{-2\tau}d\tau + \int_0^t \tau e^{-2\tau}d\tau = (1-t)\left[-e^{-2\tau}/2\right]_0^t$$

$$+ \left[\frac{1}{4}\left(e^{-2\tau}(-2\tau-1)\right)\right]_0^t$$

$$= \frac{1}{2}(1-t)\left[1-e^{-2t}\right]+\left[-\frac{1}{2}te^{-2t}-\frac{1}{4}e^{-2t}+\frac{1}{4}\right] \tag{4.45}$$

$$y(t)=\frac{1}{2}-\frac{t}{2}-\frac{1}{2}e^{-2t}+\frac{1}{2}te^{-2t}-\frac{1}{2}te^{-2t}-\frac{1}{4}e^{-2t}+\frac{1}{4}$$

$$=\frac{1}{4}\left[3-2t-3e^{-2t}\right] \text{ for } 0 \le t \le 1$$

Step 4 The third interval is between 1 and infinity. For $t \geq 1$, the overlapping interval will be $t - 1$ to t. The output is given by

$$y(t) = \int_{t-1}^{t} x(\tau)h(t-\tau)d\tau = \int_{t-1}^{t} e^{-2\tau}(1-t+\tau)d\tau$$

$$= (1-t)\int_{t-1}^{t} e^{-2\tau}d\tau + \int_{t-1}^{t} \tau e^{-2\tau}d\tau = (1-t)[e^{-2\tau}]_{t-1}^{t} + \left[\frac{1}{4}e^{-2\tau}(-2\tau-1)\right]_{t-1}^{t}$$

(4.46)

$$= \frac{1}{2}(1-t)\left(e^{-2(t-1)} - e^{-2t}\right) + \frac{1}{4}\left[e^{-2t}(-2t-1) - e^{-2(t-1)}(-2(t-1)-1)\right]$$

$$y(t) = \frac{1}{4}e^{-2(t-1)} - \frac{3}{4}e^{-2t} \text{ for } t > 1$$

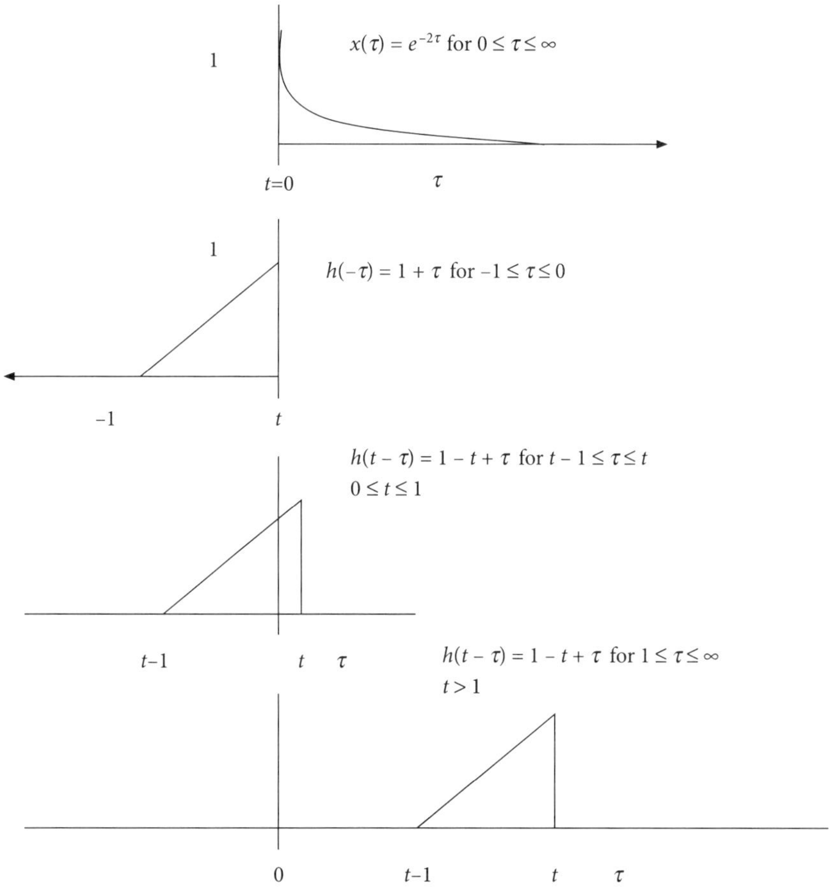

Fig. 4.5 Plot of $x(\tau)$ and $h(t - \tau)$ for different intervals

The output can be specified as follows.

$$y(t) = 0 \text{ for } t < 0$$

$$= \frac{1}{4}(3 - 2t - 3e^{-2t}) \text{ for } 0 \leq t \leq 1 \qquad (4.47)$$

$$= \frac{1}{4}e^{-2(t-1)} - \frac{3}{4}e^{-2t} \text{ for } t > 1$$

A MATLAB program to plot the convolution integral is given below. The convolution output is shown in Fig. 4.6.

```
clear all;
t=0:0.1:1;
x=1-t;
t1=0:0.1:10;
y=exp(-2*t1);
z=conv(x,y);
n=0:0.1:11;
plot(n,z);title('convolved output for example 4.7');
xlabel('time');ylabel('amplitude');
```

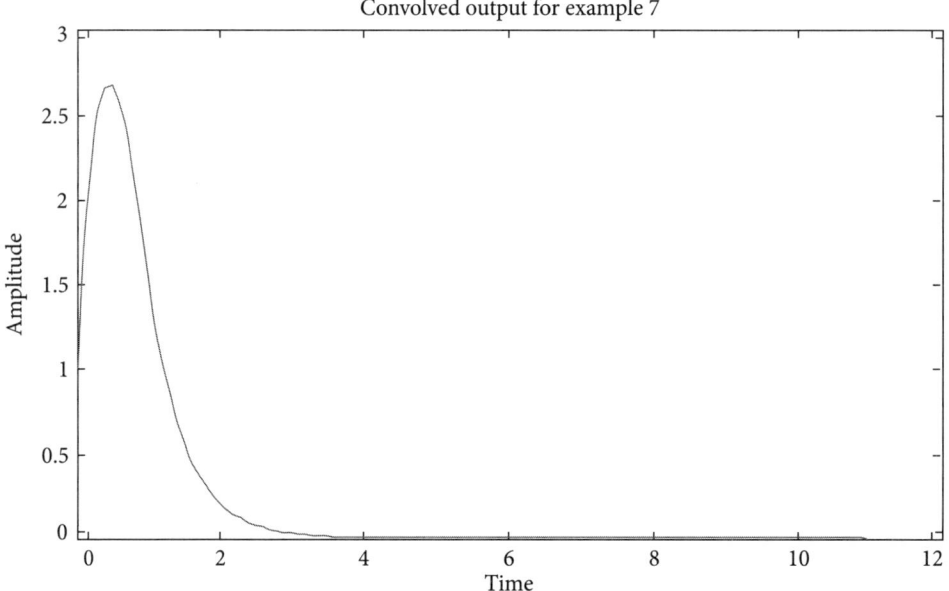

Fig. 4.6 Convolution output for Example 4.7

Example 4.8

Let $x(t) = 2$ for $1 \leq t \leq 2$, $h(t) = 1$ for $0 \leq t \leq 3$, , find $x(t)*h(t)$.

Solution

We start with step 1 i.e., drawing $x(\tau)$ and $h(-\tau)$.

Step 1 Let us draw both these waveforms. Figure 4.6 shows plots of $x(\tau)$ and $h(t - \tau)$ for different intervals.

Step 2 Start with time shift t large and negative. Let t vary from minus infinity to 1. We find that until t crosses 1, there is no overlap between the two signals. So, the convolution integral has a value of zero from minus infinity to zero. At $t = 1$, the right edge of $h(-\tau)$ touches the left edge of $x(\tau)$.

Step 3 Consider the second interval between $t = 1$ to 2. $x(\tau)h(t - \tau) = 2$.
The overlapping interval will be between 0 and t. The output can be calculated as

$$y(t) = \int_1^t 2 \times 1 d\tau = 2\tau \Big|_1^t = (2t - 2) \tag{4.48}$$

Step 4 Consider the second interval between $t = 2$ to 4. $x(\tau)h(t - \tau) = 2$.
The overlapping interval will be from 1 to 2. The output can be calculated as

$$y(t) = \int_1^2 2 \times 1 d\tau = 2\tau \Big|_1^2 = 2 \tag{4.49}$$

The output is constant, equal to 2.

Step 5 Consider the second interval between $t = 4$ to 5. $x(\tau)h(t - \tau) = 2$.
The overlapping interval will be from $t - 3$ to 2. The output can be calculated as

$$y(t) = \int_{t-3}^2 2 \times 1 d\tau = 2\tau \Big|_{t-3}^2 = 2(2 - t + 3) = 2(5 - t) \tag{4.50}$$

Step 6 Consider the second interval between $t = 5$ and infinity.
No overlapping interval will be there. The output is zero.
The overall output can be summarized as

$$y(t) = 0 \text{ for } t < 1$$

$$= 2t - 2 \text{ for } 1 \leq t \leq 2$$

$$= 2 \text{ for } 2 \leq t \leq 4 \tag{4.51}$$

$$= 2(5 - t) \text{ for } 4 \leq t \leq 5$$

$$= 0 \text{ for } t > 5$$

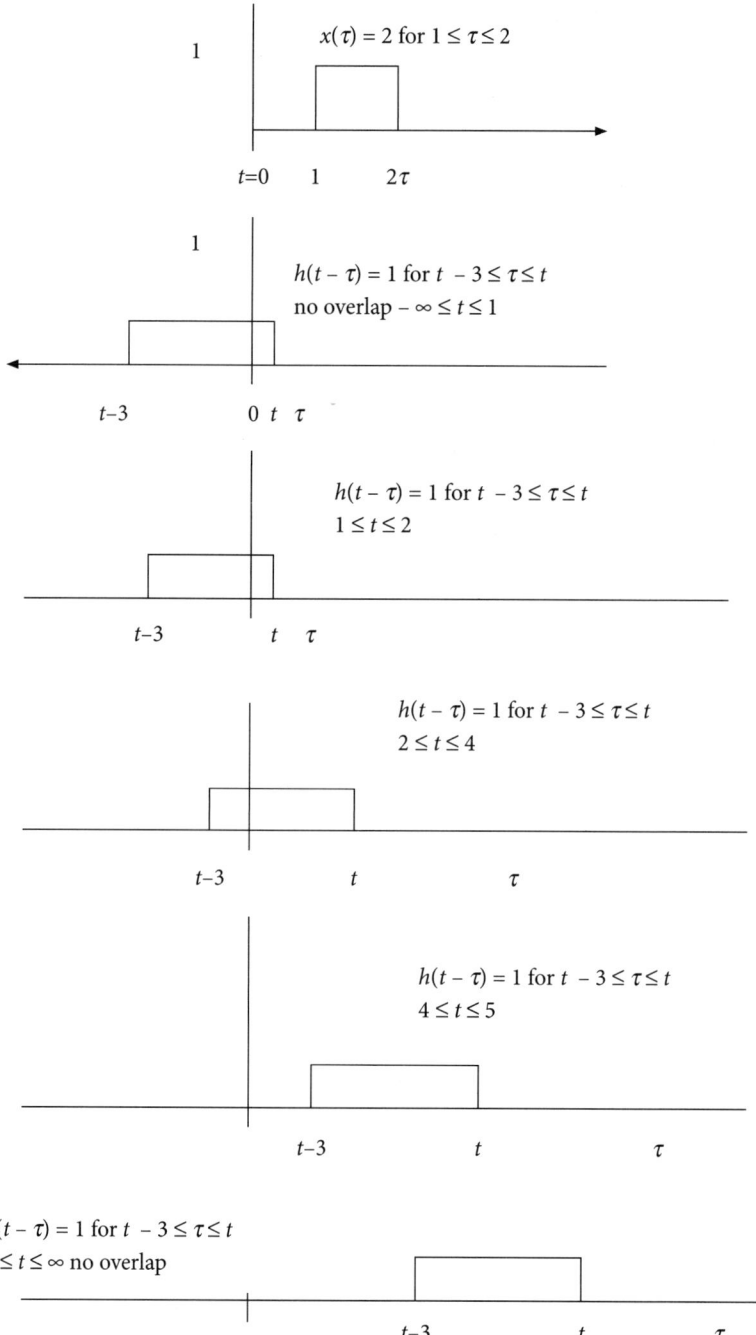

Fig. 4.7 Plots of $x(\tau)$ and $h(t - \tau)$ for different intervals

The output $y(t)$ is drawn in Fig. 4.8 below.

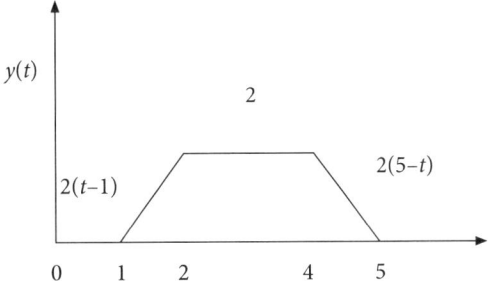

Fig. 4.8 Plot of the output of the system

A MATLAB program for finding the convolution integral is given below. The output is shown in Fig. 4.9.

```
clear all;
x=[0 0 0 0 0 0 0 0 0 0 2 2 2 2 2 2 2 2 2 2 2 0];
y=[1 1 1 1 1 1 1 1 1 1 1 1 1 1 1 1 1 1 1 1 1 1 1 1 1 1 1 1 1 1 0];
z=conv(x,y);
n=1:1:53;
plot(n,z);title('convolved output for example 4.8');
xlabel('time');ylabel('amplitude');
```

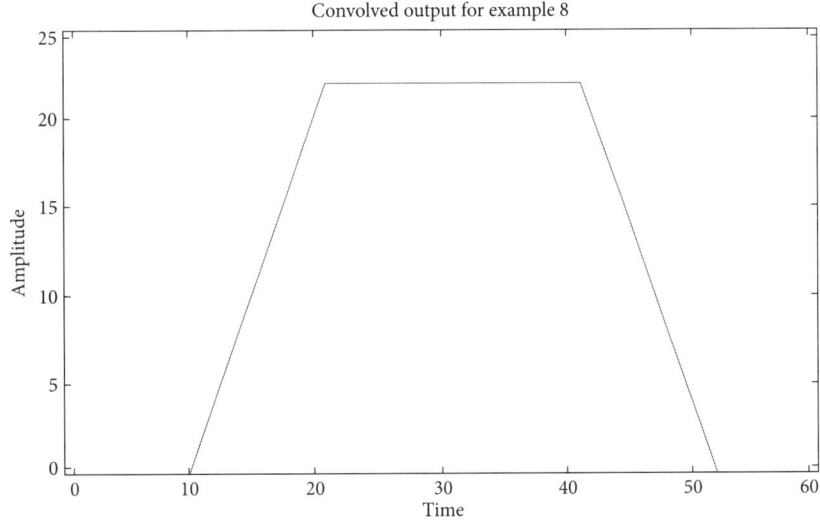

Fig. 4.9 Convolved output for Example 4.8

Example 4.9

Let $x(t) = 2$ for $0 \leq t \leq 2$, $h(t) = t$ for $0 \leq t \leq 3$, find $x(t)*h(t)$.

Solution

We start with step 1 i.e., drawing $x(\tau)$ and $h(-\tau)$.

Step 1 Let us draw both these waveforms. Figure 4.10 shows the plots of $x(\tau)$ and $h(t - \tau)$ for different intervals.

Step 2 Start with time shift t large and negative. Let t vary from minus infinity to zero. We find that until t crosses zero, there is no overlap between the two signals. So, the convolution integral has a value of zero from minus infinity to zero. At $t = 0$, the right edge of $h(-\tau)$ touches the left edge of $x(\tau)$.

Step 3 Consider the second interval between $t = 0$ to 2. $x(\tau)h(t - \tau) = 2(t - \tau)$. The overlapping interval will be between 0 to t. The output can be calculated as

$$y(t) = \int_0^t 2(t-\tau)d\tau = 2t\tau \Big|_0^t - \tau^2 \Big|_0^t = (2t^2 - t^2) = t^2, \qquad (4.52)$$

as shown in Fig. 4.11.

Step 4 Consider the second interval between $t = 2$ to 3. $x(\tau)h(t - \tau) = 2(t - \tau)$. The overlapping interval will be from 0 to 2. The output can be calculated as

$$y(t) = \int_0^2 2(t-\tau)d\tau = 2t\tau \Big|_0^2 - \tau^2 \Big|_1^2 = 4t - 4 \qquad (4.53)$$

The output is as shown in Fig. 4.11.

Step 5 Consider the second interval between $t = 3$ to 5. $x(\tau)h(t - \tau) = 2(t - \tau)$. The overlapping interval will be from $t - 3$ to 2. The output can be calculated as

$$y(t) = \int_{t-3}^2 2(t-\tau)d\tau = 2t\tau \Big|_{t-3}^2 - \tau^2 \Big|_{t-3}^2 = 2t(2-t+3) \\ - (4 - (t^2 - 6t + 9)) \qquad (4.54)$$

$$= 2t(5 - t) - (4 - t^2 + 6t - 9) = 10t - 2t^2 + t^2 - 6t + 5 = 4t - t^2 + 5,$$

as shown in Fig. 4.10.

Step 6 Consider the second interval between $t = 5$ to infinity. $x(\tau)h(t - \tau) = 2(t - \tau)$.

No overlapping interval will be there. The output is zero.
The overall output can be summarized as

$y(t) = 0$ for $t < 1$

$\quad = t^2$ for $0 \leq t \leq 2$

$\quad = 4t - 4$ for $2 \leq t \leq 3$ $\qquad (4.55)$

$\quad = 4t - t^2 + 5$ for $3 \leq t \leq 5$

$\quad = 0$ for $t > 5$

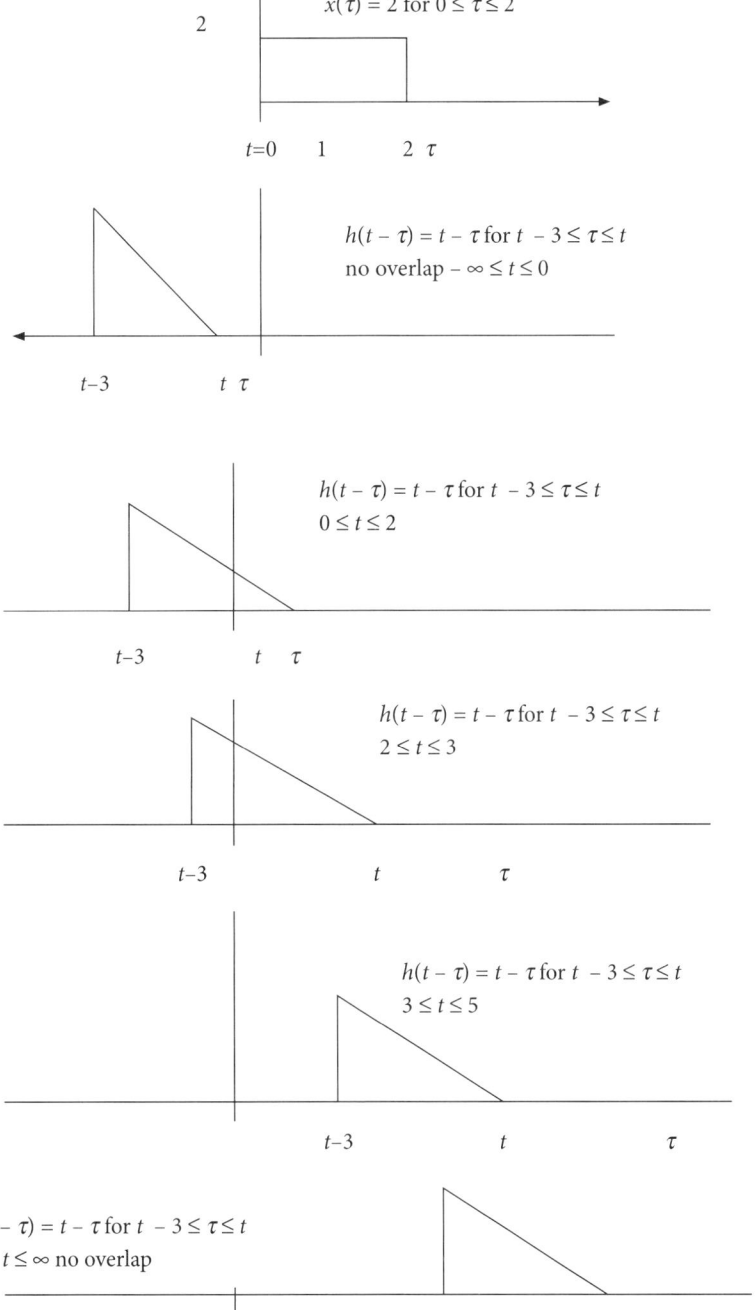

Fig. 4.10 Plots of $x(\tau)$ and $h(t - \tau)$ for different intervals

The overall output can be drawn as shown in Fig. 4.11 below.

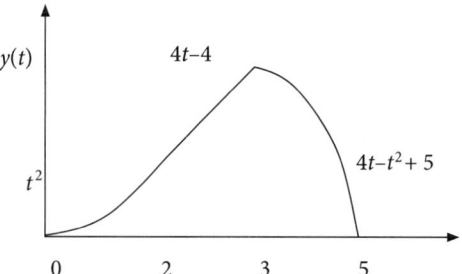

Fig. 4.11 Overall output of the system

A MATLAB program to evaluate the integral is given below. The output of the program is plotted in Fig. 4.12.

```
clear all;
clc;
t=0:0.1:2;
x=2*(t-t+1);
t1=0:0.1:3;
x1=t1;
x2=conv(x,x1);
t2=0:0.1:5;
plot(t2,x2);
title('convolution integral for eaxmple 4.9');xlabel('time');ylabel('amplitude');
```

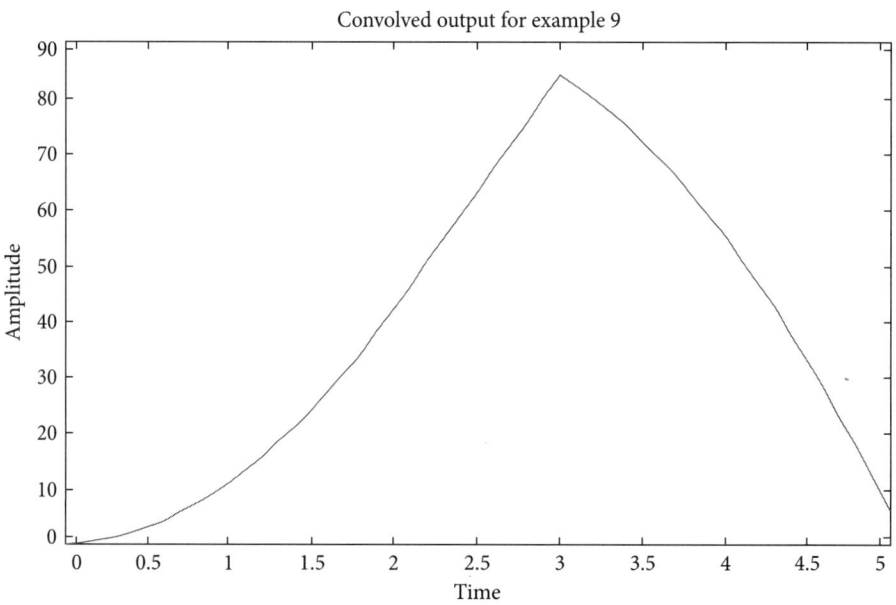

Fig. 4.12 Plot of convolution for Example 4.9

4.3.1 Zero state response

The zero state response can be defined as the response of the system for the external input assuming zero initial conditions. When the initial conditions are zero, the system is said to be relaxed. We have seen the procedure for the calculation of impulse response of the system in section 4.2.2. The computation of the convolution integral is studied in section 4.3.1. To find the response of the system for external input, we have to convolve the impulse response of the system with the input using the procedure studied in section 4.3.1. We will illustrate the calculation of zero state response with the help of a simple numerical problem.

Example 4.10

Consider a system given by $(D + 2)y(t) = x(t)$. The impulse response is calculated as $h(t) = (e^{-2t}u(t))$. Let the external input be applied as $x(t) = (e^{-t}u(t))$. Find the response of the system for the applied input.

Solution

We have to find the zero state response by convolving the impulse response with the externally applied input assuming all initial conditions as zero.

We start with step 1 i.e., drawing $x(\tau)$ and $h(-\tau)$.

Step 1 Let us draw both these waveforms. Figure 4.13 shows plots of $x(\tau)$ and $h(t - \tau)$ for different intervals.

Step 2 Start with time shift t large and negative. Let t vary from minus infinity to zero. We find that until t crosses zero, there is no overlap between the two signals. So, the convolution integral has a value of zero from minus infinity to zero. At $t = 0$, the right edge of $h(-\tau)$ touches the left edge of $x(\tau)$.

Step 3 Consider the second interval between $t = 0$ to ∞. $x(\tau)h(t - \tau) = e^{-2\tau} e^{-(t-\tau)}$.

The overlapping interval will be between 0 and t. The output can be calculated as

$$y(t) = \int_0^t x(\tau)h(t-\tau)d\tau = \int_0^t e^{-2\tau} e^{-(t-\tau)} d\tau$$

$$= \int_0^t e^{-t}e^{-\tau}d\tau = e^{-t}[-e^{-\tau}]_0^t = e^{-t}[-e^{-t} +1] \qquad (4.56)$$

$$= e^{-t} - e^{-2t}$$

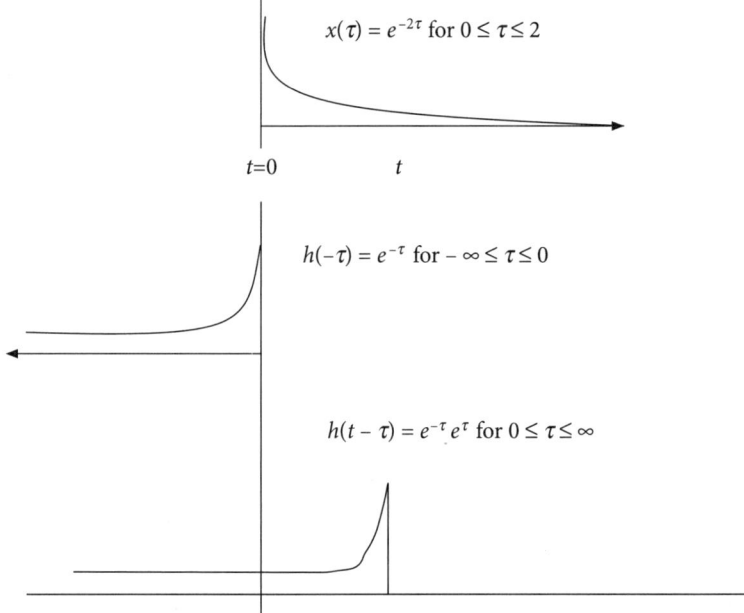

Fig. 4.13 Plots of $x(\tau)$ and $h(t - \tau)$ for various time intervals

The output of the system is plotted in Fig. 4.14.

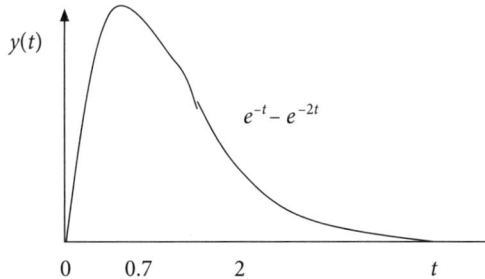

Fig. 4.14 Overall output of the system

The output can be specified as follows:

$$y(t) = 0 \text{ for } t < 0$$
$$= e^{-t} - e^{-2t} \text{ for } 0 \leq t \leq \infty \tag{4.57}$$
$$\text{or } y(t) = (e^{-t} - e^{-2t})u(t)$$

The output has a term due to external input and a term due to system characteristic equation.

A MATLAB program to find the convolved output is given below. The convolved output is shown in Fig. 4.15.

```
clear all;
clc;
t=0:0.1:10;
x=exp(-t);
t=0:0.1:10;
y=exp(-2*t);
z=conv(x,y);
plot(z);title('convolution integral for
example 4.10');
xlabel('time*10'); ylabel('amplitude');
```

Concept Check

- Explain the steps to find the convolution integral
- What is zero state response?
- What is the meaning of zero state?
- What are initial conditions?

4.4 Response of DT Systems

Let us discuss the response of the system based on its difference equation. DT systems are described in terms of the difference equation. We will consider a

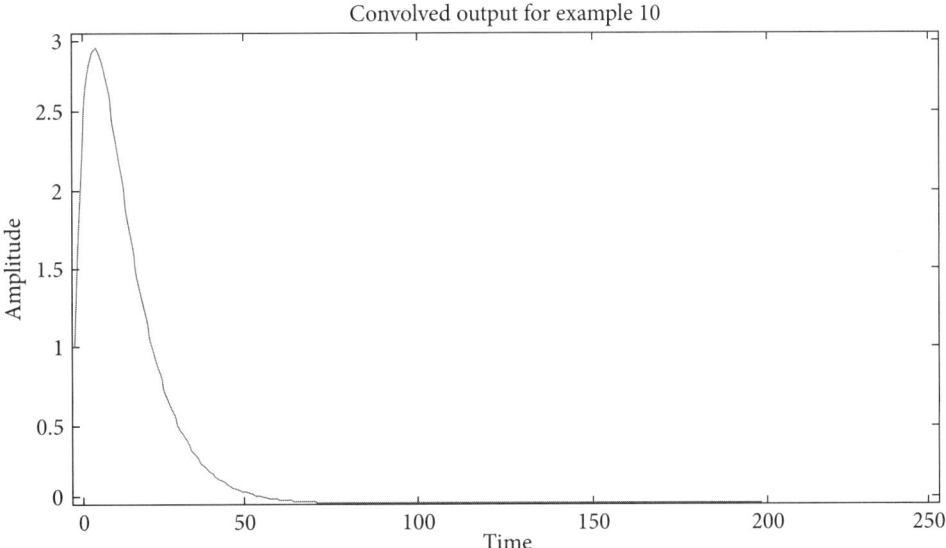

Fig. 4.15 Convolved output for Example 4.10

system described by the difference equation and illustrate the procedure for evaluating the response of the system. The reader may recall the meaning of zero input response and the zero state response. When the external input is zero, the response of the system is termed as zero input response. When the external input is applied and the initial conditions are zero i.e., the system is relaxed, the output of the system is termed as the zero state response. We may evaluate the zero input response and zero state response separately and then find the total response. The total response of the system is then found as the addition of zero input response and the zero state response.

Consider a generalized difference equation in advance form for a DT system given by the equation

$$y[k+n] + a_{n-1}y[k+n-1] + \ldots + a_0 y[k]$$

$$= b_m f[k+m] + b_{m-1} f[k+m-1] + \ldots + b_0 f[k] \tag{4.58}$$

For a causal system $m \leq n$, so that the output of the system does not depend on future inputs or future output samples. The coefficient of $y[k+n]$ is normalized to unity. We will introduce a delay of n samples for all the terms. The equation in the delay operator form can be written as

$$y[k] + a_{n-1}y[k-1] + \ldots + a_0 y[k-n]$$

$$= b_m f[k] + b_{m-1} f[k-1] + \ldots + b_0 f[k-m] \tag{4.59}$$

This represents the equation for a causal system if $f[-1], f[-2], \ldots f[-n] = 0$. We can use the iterative method to find the solution. Put $k = 0$ to find $y[0]$.

$$y[0] + a_{n-1}y[-1] + \ldots + a_0 y[-n]$$

$$= b_m f[0] + b_{m-1} f[-1] + \ldots + b_0 f[-m] \tag{4.60}$$

Putting values of samples for a causal system and knowing the initial conditions, we get

$$y[0] = -a_{n-1}y[-1] - \ldots + a_0 y[-n] + b_m f[0] \tag{4.61}$$

Similarly knowing $y[0]$, and initial conditions along with values of inputs $f[0]$ and $f[1]$, one can evaluate $y[1]$.

$$y[1] = -a_{n-1}y[0] - \ldots + a_0 y[-n+1] + b_m f[1] + b_{m-1} f[0] \tag{4.62}$$

This is called as the recursive solution.

Let us illustrate the concepts with the help of simple examples.

Example 4.11

Consider the difference equation

$$y[k] - 0.5y[k-1] = f[k]$$

$y[-1] = 8$ and $f[k] = k^2$

Find the solution using recursive procedure.

Solution

Put $k = 0$ to get

$$y[0] - 0.5y[-1] = 0, \; y[0] = 4$$

put $k = 1$, $y[1] - 0.5y[0] = f[1]$ \hfill (4.63)

$$y[1] = 2 + 1 = 3$$

Similarly, one can find $y[2]$ and so on.

It is often convenient to use operational notation to represent the time delay. We can write the same equation in operational form as

Let us denote $Df[k] = f[k + 1]$, we may write the generalized difference equation as

$$(D^n + a_{n-1}D^{n-1} + \ldots + a_0)y[k] = (b_m D^m + b_{m-1}D^{m-1} + \ldots + b_0)f[k] \quad (4.64)$$

$$(D^n + a_{n-1}D^{n-1} + \ldots + a_0)/(b_m D^m + b_{m-1}D^{m-1} + \ldots + b_0) = y[k]/f[k] \quad (4.65)$$

Where $y[k]$ denotes the kth output sample and $f[k]$ the kth input sample of the system.

$(D^n + a_{n-1}D^{n-1} + \ldots + a_0)$ is referred to as a denominator polynomial in D of order n and $(b_m D^m + b_{m-1}D^{m-1} + \ldots + b_0)$ is termed as a numerator polynomial of order m for a ratio of output to the input of the system. Let us consider two cases.

Case 1 $n < m$, i.e., the ratio is greater than 1. In this case, the system behaves as a differentiator of order $m - n$. It will amplify high-frequency noise through differentiation. High-frequency signal will have high amplitude of differentiation output. The system will be more prone to noise.

Case 2 $n > m$, i.e., the ratio is less than 1. In this case, the system will be stable and it will not amplify noise.

We will assume that the degree of the numerator is less than the degree of the denominator polynomial i.e., case 2 holds good.

4.4.1 Zero input response

The denominator polynomial is called as the characteristic polynomial. When the externally applied input $f[k]$ is zero, we get the system equation as
$(D^n + a_1 D^{n-1} + \ldots + a_n) = 0$. Let us assume the solution as $y[k] = cr^k$. Putting this value of $y[k]$ in the polynomial, we get

$$(cr^n + a_{n-1} cr^{n-1} + \ldots + a_0 cr^k) = 0$$

$$c(r^n + a_{n-1} r^{n-1} + \ldots + a_0) = 0 \tag{4.66}$$

This polynomial is same as the polynomial in D or with D replaced by r. This equation is characteristic of the system and will decide the zero input response. Let us decompose the characteristic polynomial in n number of roots to get the equation as

$$(r - r_1)(r - r_2) \ldots (r - r_n) = 0 \tag{4.67}$$

$r_1, r_2, \ldots r_n$ are roots of the charateristic polynomial.

The roots of the characteristic polynomial are called as characteristic roots or eigen values or natural frequencies of the system. The characteristic modes or the natural modes are denoted as r_i^k $i = 1, 2, \ldots$, etc. These are also termed as eigen functions of the system.

We write the roots as exponential functions for two reasons:

To get the linear combination of $y[k]$ to get the values of $y[k], y[k+1], \ldots y[k+n]$ etc. zero for all values of k, $y[k]$ and $y[n+k]$ must have the same form. Only an exponential function satisfies this condition.

The second reason is that in case of the LTI systems, if the input is an exponential function, the output is also an exponential function. The other functions cannot have this claim. The exponential solutions are called as eigen functions.

The characteristic modes decide the behaviour of the system when the external input is zero. The response of the system is termed as the zero input response. The characteristic modes will also play a major role in deciding the zero state response. The characteristic roots may be all distinct, may be repeated roots or may be complex roots.

We will consider a numerical problem to illustrate the concepts.

Example 4.12

Consider a simple second order system with characteristic equation given by

$$y[k+2] - 0.4 y[k+1] - 0.12 y[k] = 5 f[k+2] \text{ with } y[-1] = 0 \text{ and } y[-2] = 25$$

$$f[k] = 4^{-k} u[k]$$

Find the zero input response for the system.

Solution

The system equation is given in advance form so that we can write it in operational form. It can be written as $(D^2 - 0.4D - 0.12)\, y[k] = 5D^2 f[k]$

We will first write the characteristic equation by equating the denominator polynomial to zero when $f[k] = 0$. i.e.,

$$(D^2 - 0.4D - 0.12) = 0 \tag{4.68}$$

$$(D - 0.6)(D + 0.2) = 0$$

The solution is $D = 0.6$ and $D = -0.2$

The zero input response can be written as

$$y[k] = c_1 (0.6)^k + c_2 (-0.2)^k \tag{4.69}$$

We will now use the initial conditions to find the constants.

$$k = -1,\; y[-1] = c_1 (0.6)^{-1} + c_2 (-0.2)^{-1} = \frac{5}{3} c_1 - 5 c_2 = 0 \tag{4.70}$$

$$k = -2,\; y[-2] = c_1 (0.6)^{-2} + c_2 (-0.2)^{-2} = \frac{25}{9} c_1 + 25 c_2 = 25$$

$$k = -1,\; c_1 - 3 c_2 = 0 \tag{4.71}$$

$$k = -2,\; c_1 + 9 c_2 = 9$$

After solving the above, we get,

$$c_1 = 9/4,\; c_2 = 3/4$$

The solution can be written as

$$y[k] = \frac{9}{4}(0.6)^k + \frac{3}{4}(-0.2)^k \tag{4.72}$$

4.4.2 Impulse response of DT system

Let us now see how to find the impulse response of the system. To find the impulse response we have to put input as impulse i.e., $\delta[n]$ and find the output of the system. To find zero state response for a DT system we apply the input. The initial conditions will now be assumed to be zero and the input will be applied to find the zero state response. The input can be decomposed into linear combination of impulses and the responses to all impulses can be added to get the total response. We will illustrate this with the help of a simple numerical example.

Example 4.13

Consider a simple second order system with the characteristic equation given by

$$y[k+2] - 0.4y[k+1] - 0.12y[k] = 5f[k+2] \text{ with } y[-1] = 0 \text{ and } y[-2] = 25$$

$$f[k] = 4^{-k} u[k]$$

Find the impulse response of the system.

Solution

The system equation is given in advance form. We can write it by allowing a delay of samples as

$$y[k] - 0.4y[k-1] - 0.12y[k-2] = 5f[k] \tag{4.73}$$

We will put $f[k] = \delta[k]$ and $y[k] = h[k]$. The equation becomes

$$h[k] - 0.4h[k-1] - 0.12h[k-2] = 5\delta[k] \tag{4.74}$$

We will put initial conditions as zero.

$$h[-1] = h[-2] = \ldots h[-n] = 0 \text{ for causal system} \tag{4.75}$$

Put $k = 0$, $h[0] - 0.4h[-1] - 0.12h[-2] = 5$

$$h[0] = 5 \tag{4.76}$$

Put $k = 1$, $h[1] - 0.4h[0] - 0.12h[-1] = 5 \times 0$

$$h[1] - 0.4 \times 5 = 0 \Rightarrow h[1] = 2$$

To find closed form expression for $h[n]$, we proceed as follows.
The characteristic equation for the system is

$$(D^2 - 0.4D - 0.12) = 0$$

$$(D - 0.6)(D + 0.2) = 0 \tag{4.77}$$

The solution is
$D = 0.6$ and $D = -0.2$
The solution can be written as

$$h[k] = [c_1(0.6)^k + c_2(-0.2)^k]u[k] \qquad (4.78)$$

The solution is appended by $u[k]$ as it exists only for $k \geq 0$, because the system is causal.

We will now use initial conditions to find the constants.

$$k = 0, h[0] = c_1(0.6)^0 + c_2(-0.2)^0 = c_1 + c_2 = 5$$

$$k = 1, h[1] = c_1(0.6)^1 + c_2(-0.2)^1 = 0.6c_1 - 0.2c_2 = 2 \qquad (4.79)$$

After solving the above, we get

$$c_1 = 15/4, c_2 = 5/4$$

The solution can be written as

$$h[k] = \left[\frac{15}{4}(0.6)^k + \frac{5}{4}(-0.2)^k\right]u[k] \qquad (4.80)$$

4.4.3 Zero state response of DT system

We will illustrate the concept with the help of simple numerical example.

Example 4.14

Consider a simple second order system with impulse response and the input signal given by

$$h[k] = (0.3)^k u[k] \text{ and } f[k] = (0.8)^k u[k]$$

Find the zero state response of the system.

Solution

Here, we have to convolve the impulse response with applied input. The zero state response can be written as

$y[k] = \sum_{m=0}^{k} f[m]h[k-m]$ for a causal system. (Refer to section 4.8 for LTI causal system impulse response).

$$y[k] = \sum_{m=0}^{k}[(0.8)^m] \times (0.3)^{(k-m)}$$

$$y[k] = (0.3)^k \sum_{m=0}^{k}\left[\left(\frac{0.8}{0.3}\right)^m\right]$$

$$y[k] = (0.3)^k \left[\left\{\frac{(0.8)^{k+1} - (0.3)^{k+1}}{(0.3)^k(0.8-0.3)}\right\}\right] \qquad (4.81)$$

$$y[k] = 2[(0.8)^{k+1} - (0.3)^{k+1}]u[k]$$

Concept Check

- What is characteristic polynomial?
- What are characteristic functions?
- What are eigen values?
- What is the difference equation?
- What do you mean by the initial conditions?
- What is zero input response?
- What is zero state response?

4.5 Representation of DT Signals in Terms of Delta Functions

We will use the property of superposition for decomposition of a sequence using scaled and shifted delta functions. Let us first go through the superposition property of LTI systems.

Property of Superposition

For LTI systems, once the impulse response is known, the system response to any DT input signal can easily be analyzed using the principle of superposition.

Most of the DSP techniques work on the principle of divide and conquer. This technique allows us to break the input into small parts and find the output for every small part and finally combine the outputs for small parts using the principle of superposition. A system is said to obey the property of superposition if it obeys the properties of homogeneity and additivity and is then called a linear system. Superposition is a powerful tool for analyzing complicated input signals.

Figure 4.16 depicts the principle of superposition clearly. In this figure, for the input $k_1 x_1[n] + k_2 x_2[n]$ to a system the output is $k_1 y_1[n] + k_2 y_2[n]$.

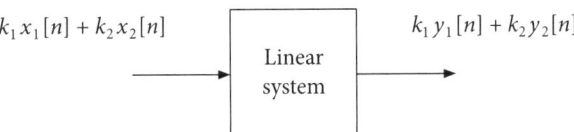

Fig. 4.16 Response of a linear system

Example 4.15

Consider a sequence given by

$$x[n] = \left\{ \begin{array}{c} 1, 2, -1, 2, 3 \\ \uparrow \end{array} \right\} \quad (4.82)$$

Decompose the sequence in linear combination of impulses and find the response to input $x[n]$.

Solution

The position of the arrow indicates a sample at $n = 0$. The sequence $x[n]$ can be decomposed using impulse decomposition. The decomposition can be written as

$$x[n] = \delta[n] + 2\delta[n-1] - \delta[n-2] + 2\delta[n-3] + 3\delta[n-4] \quad (4.83)$$

where $\delta[n]$ is a delta function defined as

$$\delta[n] = \left\{ \begin{array}{ll} 1 & \text{for } n=0 \\ 0 & \text{otherwise} \end{array} \right\} \quad (4.84)$$

Here, $\delta[n-1]$ represents a delta function shifted by one sample to the right and $\delta[n-2]$ represents a delta function shifted by two samples to the right and so on. Now, find the response of the system to delta function at every location and then reunite the result.

Using the properties of linearity and shift invariance the output of the system with impulse response $h[n]$ to input of $x[n]$ is given by

$$y[n] = h[n] + 2h[n-1] - h[n-2] + 2h[n-3] + 3h[n-4] \quad (4.85)$$

Here we have used the property of shift invariance, namely, the output of the system for $\delta[n-1]$ is $h[n-1]$ and so on. Reuniting the output for each delta function in this manner is called synthesis.

4.5.1 Convolution sum for DT systems

Consider an LTI system with an impulse response $h[n]$. When $\delta[n]$ is the input to the system, $h[n]$ is the output. Let the sequence $x[n]$ specified in Eq. (4.86) be the input to the system. The output is given by the convolution sum

$$y[n] = x[0]h[0] + x[1]h[n-1] + x[2]h[n-2] + x[3]h[n-3] + x[4]h[n-4] \quad (4.86)$$

$$y[n] = \sum_{k=0}^{4} x[k]h[n-k] \quad (4.87)$$

The output $y(n)$ is given by the convolution sum:

$$y[n] = x[n] * h[n] \quad (4.88)$$

Here, the symbol * denotes the convolution sum.

Input Side Algorithm for Convolution

We will consider the input side view for the convolution algorithm. Consider $x[n]$ as a $P = 4$ point sequence given by {1 2 2 1} and $h[n]$ as a $Q = 3$ point sequence given by {1 2 1}, as shown in Fig. 4.17. Let us convolve the two sequences.

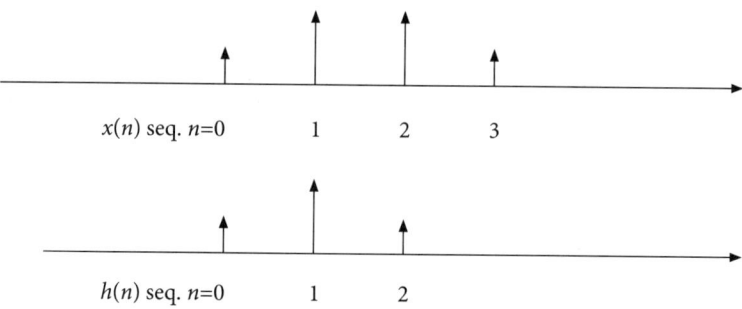

Fig. 4.17 Two sequences, $x[n]$ and $h[n]$

Decompose the input sequence in four delta functions (input sequence has four samples)

$$x[n] = \delta[n] + 2\delta[n-1] + 2\delta[n-2] + \delta[n-3] \quad (4.89)$$

The amplitude of delta function is multiplied by the sample value at that location.

Write $h[n]$ in terms of delta functions:

$$h[n] = \delta[n] + 2\delta[n-1] + \delta[n-2] \quad (4.90)$$

Find the output for each delta function in the input sequence. Let $y_1[n]$, $y_2[n]$, $y_3[n]$ and $y_4[n]$ denote the outputs for four delta functions $\delta[n]$, $\delta[n-1]$, $\delta[n-2]$ and $\delta[n-3]$, respectively, as shown in Fig. 4.13. The output of the system is the impulse response of the system, namely, $h[n]$

$$y_1[n] = \delta[n] + 2\delta[n-1] + \delta[n-2] \quad (4.91)$$

The output of the system to $2\delta[n-1]$ is the impulse response shifted by one sample, that is, $2h[n-1]$

$$y_2[n] = 2\delta[n-1] + 4\delta[n-2] + 2\delta[n-3] \tag{4.92}$$

The impulse response to $2\delta[n-2]$ is shifted by 2 samples by again using shift invariance property, that is, $h[n-2]$:

$$y_3[n] = 2\delta[n-2] + 4\delta[n-3] + 2\delta[n-4] \tag{4.93}$$

The impulse response for $\delta[n-3]$ is shifted by 3 samples using shift invariance property.

$$y_4[n] = \delta[n-3] + 2\delta[n-4] + \delta[n-5] \tag{4.94}$$

Add the responses for all delta functions to obtain the output $y[n]$.

$$y[n] = y_1[n] + y_2[n] + y_3[n] + y_4[n] \tag{4.95}$$

$$\begin{aligned} y[n] = &\delta[n] + 2\delta[n-1] + \delta[n-2] + 2\delta[n-1] + 4\delta[n-2] \\ &+ 2\delta[n-3] + 2\delta[n-2] + 4\delta[n-3] + 2\delta[n-4] \\ &+ \delta[n-3] + 2\delta[n-5] \end{aligned} \tag{4.96}$$

$$y[n] = \delta[n] + 4\delta[n-1] + 7\delta[n-2] + 7\delta[n-3] + 4\delta[n-4] + \delta[n-5] \tag{4.97}$$

The output $y(n)$ has $(4 + 3 - 1)$, that is, $(P + Q - 1) = 6$ samples, as expected. The procedure is depicted in Fig. 4.18.

	N=0	1	2	3	4	5

Input sequence $x(n)$ 1   2   2   1

Impulse response $h(n)$ 1

            2   1

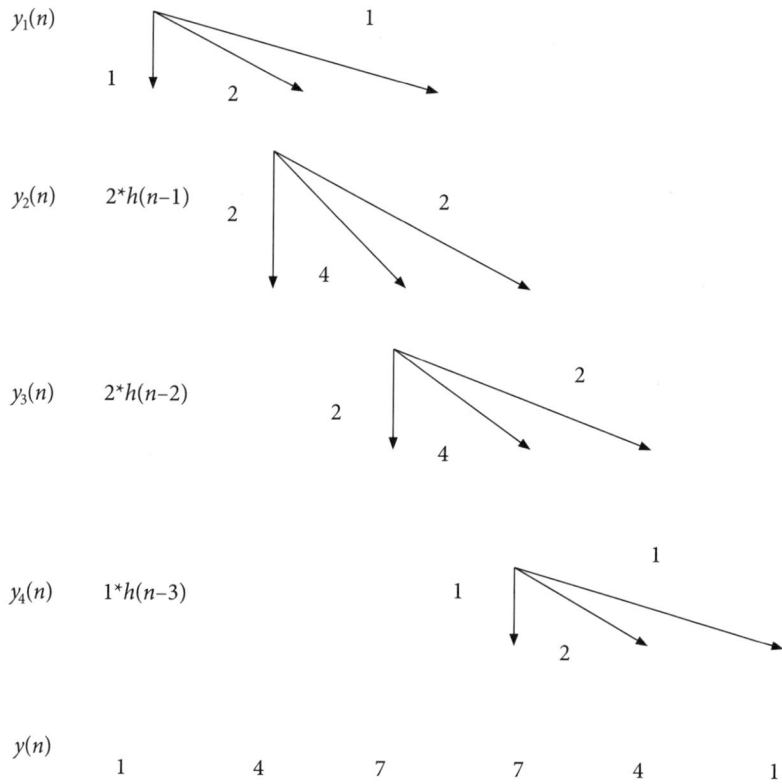

Fig. 4.18 Input side algorithm

Output side algorithm for convolution

This algorithm changes the viewpoint for the process of convolution. It finds out number of input signal components contributing the output point. Contribution at each time instant is shown in Fig. 19.

Step 1 Concentrate on the output at $n = 0$. The only input component contributing to this output point is at $n = 0$. The output is $1*1 = 1$, as shown in Fig. 4.9.

Step 2 The output at $n = 1$ is contributed from input at $n = 0$ and input at $n = 1$(red color lines). The output is $2*1 + 1*2 = 4$. The output is contributed by solid lines.

Step 3 The output at $n = 2$ is shown to be contributed by three green lines. We observe from the three dotted lines that it is as if the impulse response is flipped around to find the output at $n = 2$. The output is $1*1 + 2*2 + 2*1 = 7$.

Step 4 Similarly the output at $n = 3$ is contributed by the three dash dotted paths. The output is again 7.

Step 5 Calculate the output at $n = 3$ on similar lines. Output at $n = 3$ is obtained from the two long dashed lines $2*1 + 1*2 = 4$

Step 6 The output at $n = 4$ is $1*1 = 1$ from the single long dash dot dot line.

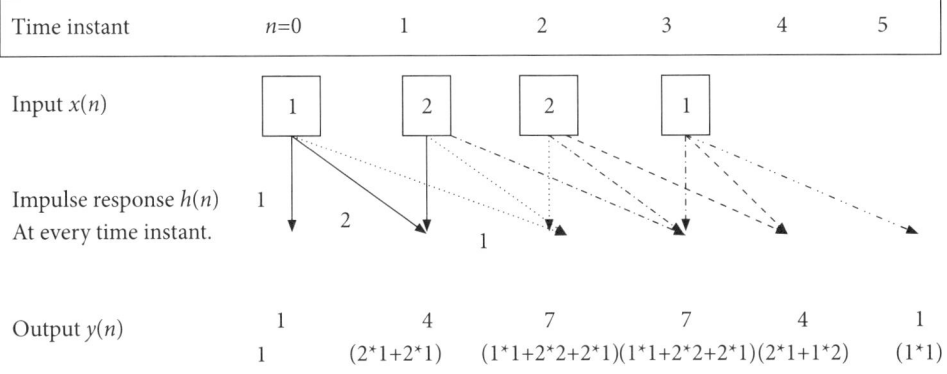

Fig. 4.19 Output side algorithm for convolution

Conventional method for computation of convolution of two sequences

Example 4.16

Consider the convolution of two sequences given by

$x[n] = \{1\ 2\ 2\ 1\}$

$h[n] = \{1\ 2\ 1\}$

This convolution is called linear convolution.

We will consider the formula for convolution. The output $y[n]$ is given by

$$y[n] = \sum_{m=0}^{P+Q-1} x[m]h[n-m] = \sum_{m=0}^{P+Q-1} h[m]x[n-m] \tag{4.98}$$

where P is the number of points in the sequence $x[n]$ and Q is the number of points in the sequence $h[n]$. The procedure for computation of the sum can be stated as follows.

Step 1 Invert one of the sequences, that is, calculate $x[-m]$ or $h[-m]$. Let us assume that we calculate $h[-m]$.

Step 2 Perform sample-by-sample multiplication, that is, $h[-m]x[m]$ for all m and accumulate the sum. This operation of sample-by-sample multiplication and summation is called multiply and accumulate operation.

Step 3 Shift the inverted sequence towards the right by one sample and again calculate the sum.

Step 4 Repeat step 3 until all $P + Q - 1$ number of samples are calculated.
In this example $P = 4$, $Q = 3$, so we need to calculate $P + Q - 1 = 6$ samples. The procedure is represented in Table 4.1.

Table 4.1 Conventional convolution procedure (Example 4.16)

Sample Number	−2	−1	0	1	2	3	4	5
$x[m]$			1	2	2	1		
$h[-m]$	1	2	1					
$h[1-m]$		1	2	1				
$h[2-m]$			1	2	1			
$h[3-m]$				1	2	1		
$h[4-m]$					1	2	1	
$h[5-m]$						1	2	1
$y[n]$			1	4	7	7	4	1

We see from Table 4.1 that for the calculation of $y[0]$, the last sample of $h[-m]$ overlaps with the first sample of $x[m]$; therefore, multiply and accumulate operation will result in output = 1. Successive shifting of $h[-m]$ towards the right and application of multiply and accumulate operation will generate the output as shown in Table 4.1.

The result is same as obtained on the output-side and input-side algorithms.

Let us solve some examples to illustrate convolution for DT systems.

Example 4.17

LTI system has the impulse response given by $h[n] = u[n] - u[n-8]$. If the input is $x[n] = u[n-2] - u[n-5]$, find the output of the system.

Solution

Let us first plot the two signals. Figure 4.20 shows the plots of $h[n]$, $x[n]$ and $x[-n]$. Let us find the output $y[n]$ sample by sample using conventional method by shifting $x[-n]$ towards the right, one sample at a time. Here, $h[n]$ has 8 samples, so $P = 8$. $x[n]$ has 3 samples, i.e., $Q = 3$. We have to find $P + Q - 1 = 10$ output samples.

We will follow the step by step procedure.

Step 1 Invert one of the sequences, that is, calculate $x[-m]$ or $h[-m]$. Let us assume that we calculate $x[-m]$.

Step 2 Perform sample-by-sample multiplication, that is, $x[-m]\,h[m]$ for all m and accumulate the sum. This operation of sample-by-sample multiplication and summation is called multiply and accumulate operation.

Step 3 Shift the inverted sequence towards the right by one sample and again calculate the sum.

Step 4 Repeat step 3 until all $P + Q - 1$ number of samples are calculated.

In this example $P = 8$, $Q = 3$, so we need to calculate $P + Q - 1 = 10$ samples.

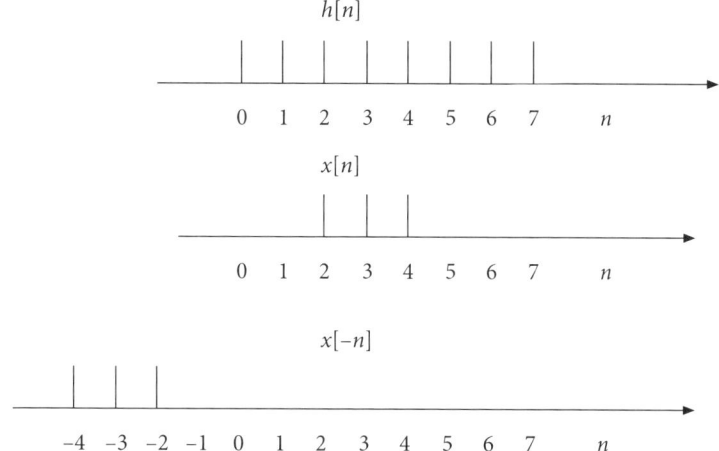

Fig. 4.20 Plots of $h[m]$, $x[m]$ and $x[-m]$

The procedure is represented in Table 4.2.

Table 4.2 Conventional convolution procedure (Example 4.17)

Sample Number	−4	−3	−2	−1	0	1	2	3	4
$h[m]$					1	1	1	1	1..
$x[-m]$	1	1	1						
$x[1-m]$		1	1	1					
$x[2-m]$			1	1	1				
$x[3-m]$				1	1	1			
$x[4-m]$					1	1	1		
$x[5-m]$						1	1	1	
$y[m]$							1	2	3...

We see from Table 4.2 that for the calculation of $y[2]$, last sample of $x[-m]$ overlaps with the first sample of $h[m]$; therefore, multiply and accumulate operation will result in output = 1. Successive shifting of $x[-m]$ towards the right and application of multiply and accumulate operation will generate the output as shown in Table 4.1. At $y[10]$, only two samples start overlapping. For $y[11]$, only one sample overlaps. The output is listed below.

$$y[0]=0,\ y[1]=0,\ y[2]=1,\ y[3]=2,\ y[4]=3,\ y[5]=3$$
(4.99)
$$y[6]=3,\ y[7]=3,\ y[8]=3,\ y[9]=3,\ y[10]=2,\ y[11]=1,\ y[12]=0$$

we can write $y[n] = \begin{cases} 0 \text{ for } n \leq 1 \text{ and } n \geq 12 \\ 1 \text{ for } n = 2 \text{ and } n = 11 \\ 2 \text{ for } n = 3 \text{ and } n = 10 \\ 3 \text{ for } 2 \leq n \leq 9 \end{cases}$

The plot of the output signal is shown in Fig. 4.21.

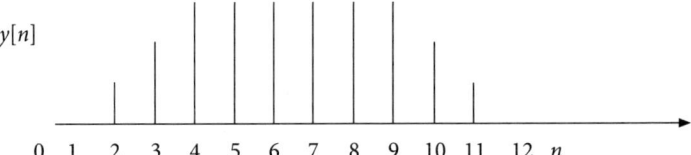

Fig. 4.21 Plot of output signal $y[n]$

Example 4.18

LTI system has the impulse response given by $h[n] = (0.2)^n \{u[n] - u[n-3]\}$. If the input is $x[n] = (0.3)^n u[n]$, find the output of the system.

Solution

Let us first plot the two signals. Figure 4.22 shows plots of $h[n]$, $x[n]$ and $x[-n]$. Let us find the output $y[n]$ sample by sample using conventional method by shifting $x[-n]$ towards the right, one sample at a time.

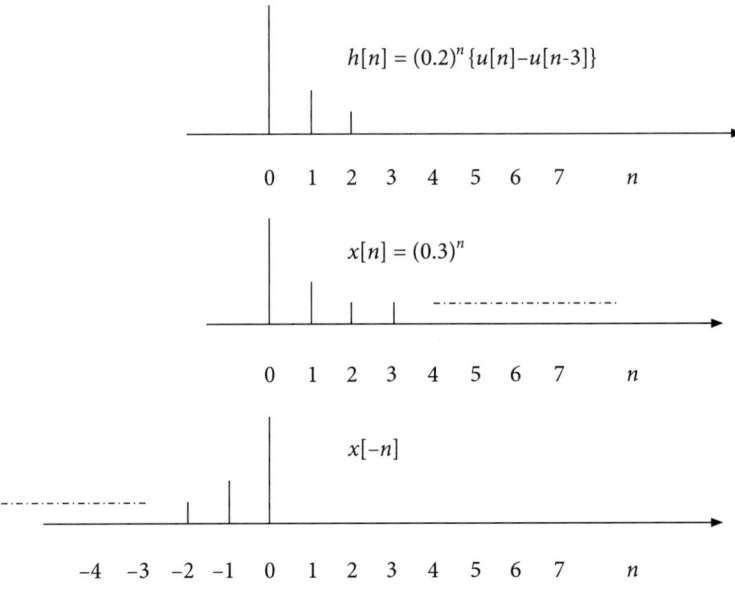

Fig. 4.22 Plots of $h[n]$, $x[n]$ and $x[-n]$

$y(0) = 1, y[1] = \{(0.2) + (0.3)\}, y[2] = \{(0.3)^2 + (0.3)(0.2) + (0.2)^2\},$

$y[3] = \{(0.3)^3 + (0.2)(0.3)^2 + (0.3)(0.2)^2\}...$

We can write $y[0] = 1, y[1] = 0.5$

$$y[n] = (0.3)^n + (0.2)(0.3)^{n-1} + (0.2)^2(0.3)^{n-2} \text{ for } n \geq 2 \qquad (4.100)$$

The plot of the output signal is shown in Fig. 4.23.

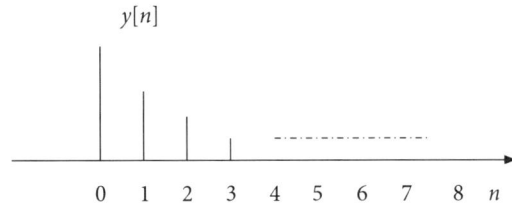

Fig. 4.23 Plot of output signal $y[n]$

Example 4.19

LTI system has the impulse response given by $h[n] = u[n-3]$. If the input is $x[n] = u[n]$, find the output of the system.

Solution

Let us first plot the two signals. Figure 4.24 shows plots of $h[m]$, $x[m]$ and $x[-m]$. Let us find the output $y[n]$ sample by sample using conventional method by shifting $x[-m]$ towards the right, one sample at a time.

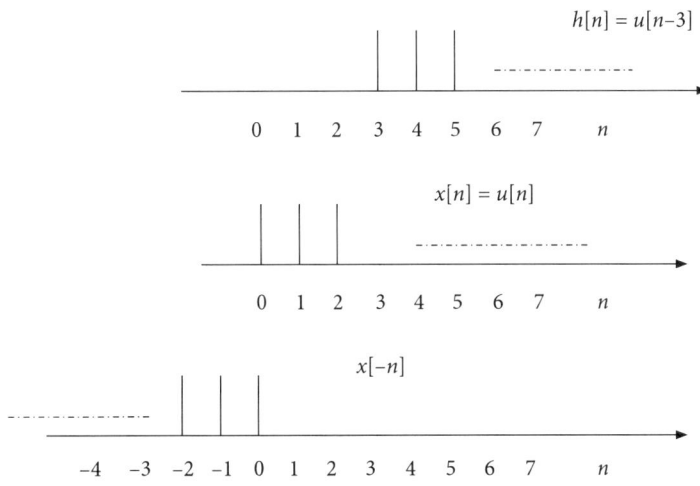

Fig. 4.24 Plots of $h[n]$, $x[n]$ and $x[-n]$

We will follow the step-by-step procedure.

Step 1 Invert one of the sequences, that is, calculate $x[-m]$ or $h[-m]$. Let us assume that we calculate $x[-m]$.

Step 2 Perform sample-by-sample multiplication, that is, $x[-m]\,h[m]$ for all m and accumulate the sum. This operation of sample-by-sample multiplication and summation is called multiply and accumulate operation.

Step 3 Shift the inverted sequence towards the right by one sample and again calculate the sum.

Step 4 Repeat step 3 until all $P + Q - 1$ number of samples are calculated.

In this example $P = 3$ to infinity, $Q = \infty$ so we need to calculate $P + Q - 1 = $ 1 to infinity samples.

The procedure is represented in Table 4.3.

Table 4.3 Conventional convolution procedure (Example 4.19)

Sample Number	-4	-3	-2	-1	0	1	2	3	4	5	6...
$h[m]$								1	1	1	1...1
$x[-m]$	...1	1	1	1	1						
$x[1-m]$		1	1	1	1	1					
$x[2-m]$			1	1	1	1	1				
$x[3-m]$				1	1	1	1	1			
$x[4-m]$					1	1	1	1			1
$x[5-m]$						1	1	1			
$y[m]$						1	2	3	4	5	6...

We see from Table 4.3 that for the calculation of $y[3]$, the last sample of $x[-m]$ overlaps with the first sample of $h[m]$; therefore, multiply and accumulate operation will result in output = 1. Successive shifting of $x[-m]$ towards the right and application of multiply and accumulate operation will generate increasing output, as shown in Table 4.3, as more and more samples overlap. The output is listed below.

$$y[0] = 0,\ y[1] = 0,\ y[2] = 0,\ y[3] = 1,\ y[4] = 2,\ y[5] = 3 \ldots\ldots \quad (4.101)$$

We can write $y[n] = \begin{cases} 0 \text{ for } 0 \leq n \leq 2 \\ (n-2) \text{ for all } n \geq 3 \end{cases}$

The plot of the output signal is shown in Fig. 4.25.

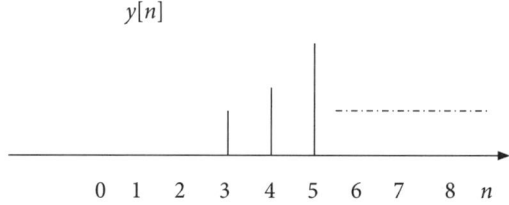

Fig. 4.25 Plot of output signal $y[n]$

Example 4.20

LTI system has the impulse response given by $h[n]=1$ for $n=0,-1,-2$ and if the input is $x[n]=\begin{cases}1 \text{ for } n=0,1\\ 2 \text{ for } n=2,3\end{cases}$, find the output of the system.

We will follow the step-by-step procedure.

Step 1 Invert one of the sequences, that is, calculate $x[-m]$ or $h[-m]$. Let us assume that we calculate $x[-m]$.

Step 2 Perform sample-by-sample multiplication, that is, $x[-m]\,h[m]$ for all m and accumulate the sum. This operation of sample-by-sample multiplication and summation is called multiply and accumulate operation.

Step 3 Shift the inverted sequence towards the right by one sample and again calculate the sum. Shift the inverted sequence towards the left by one sample and again calculate the sum until the output is non-zero.

Step 4 Repeat step 3 until all $P+Q-1$ number of samples are calculated.

In this example $P=3$, $Q=4$, so we need to calculate $P+Q-1=6$ samples. Note that here we have to shift the inverted sequence towards the left as well until the generated output is zero.

The plots of $x[m]$, $h[m]$ and $x[-m]$ are shown below in Fig. 4.26.

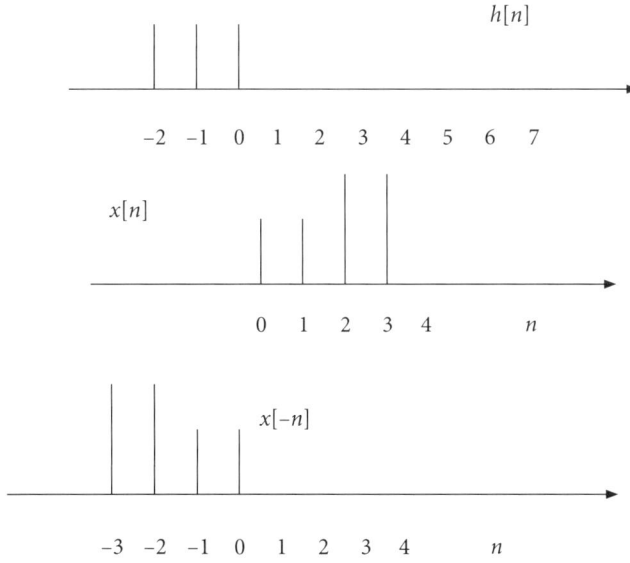

Fig. 4.26 Plot of $h[n]$, $x[n]$ and $x[-n]$

The procedure is represented in Table 4.4.

Table 4.4 Conventional convolution procedure (Example 4.20)

Sample Number	-5	-4	-3	-2	-1	0	1	2	3	4
h[m]				1	1	1				
x[-m]			2	2	1	1				
x[1 - m]				2	2	1	1			
x[2 - m]					2	2	1	1		
x[3 - m]						2	2	1	1	
x[1+ m]		2	2	1	1					
x[2 + m]	2	2	1	1						
y[m]				1	2	4	5	4	2	

We see from Table 4.4 that for the calculation of y[0], 3 samples of x[-m] overlap with samples of h[m]; therefore, multiply and accumulate operation will result in output = 4. Successive shifting of x[-m] towards the right and application of multiply and accumulate operation will generate the output y[1] = 5, y[2] = 4, y[3] = 2 etc., as shown in Table 4.4, as less and less samples overlap. To find y[-1], we have to shift x[-m] towards the left by one sample to get y[-1] = 2. Here, only 2 samples overlap. To find y[-2], shift x[1 - m] towards the left again by one sample to get y[-2] = 1, as shown in Table 4.4. The output is calculated and listed below.

$y[0] = 4, \ y[1] = 5, y[2] = 4, y[3] = 2, y[4] = 0, y[5] = 0......$

$y[-1] = 2, y[-2] = 1, y[-3] = 0,.....$

We can write $y[n] = \begin{cases} 0 \text{ for } n \leq -3 \text{ and } n \geq 4 \\ 1 \text{ for } n = -2 \\ 2 \text{ for } n = -1 \text{ and } n = 3 \\ 4 \text{ for } n = 0 \text{ and } n = 2 \\ 5 \text{ for } n = 1 \end{cases}$ (4.102)

The plot of the output signal is shown in Fig. 4.27.

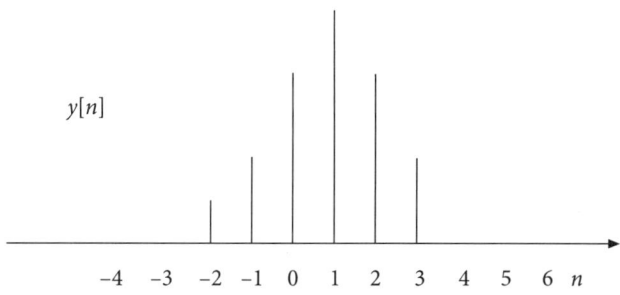

Fig. 4.27 Plot of output signal y[n]

4.5.2 Convolution using MATLAB

We will illustrate the use of MATLAB for solving convolution problems in Examples 4.17, 4.18, 4.19 and 4.20.

Let us consider Example 4.17. LTI system has the impulse response given by $h[n] = u[n] - u[n-8]$. If the input is $x[n] = u[n-2] - u[n-5]$, find the output of the system. Let us write MATLAB program for the same. MATLAB uses 'conv' command to execute convolution. The output of the convolution is shown in Fig. 4.28. Compare Fig. 4.28 with Fig. 4.21.

```
clear all;
x=[1,1,1,1,1,1,1,1];
h=[0,0,1,1,1,0,0,0,];
b=conv(x,h);
n=0:1:14;
stem(n,b);title('output of x(n)*h(n)');
xlabel('sample number');ylabel('Ampltude');
```

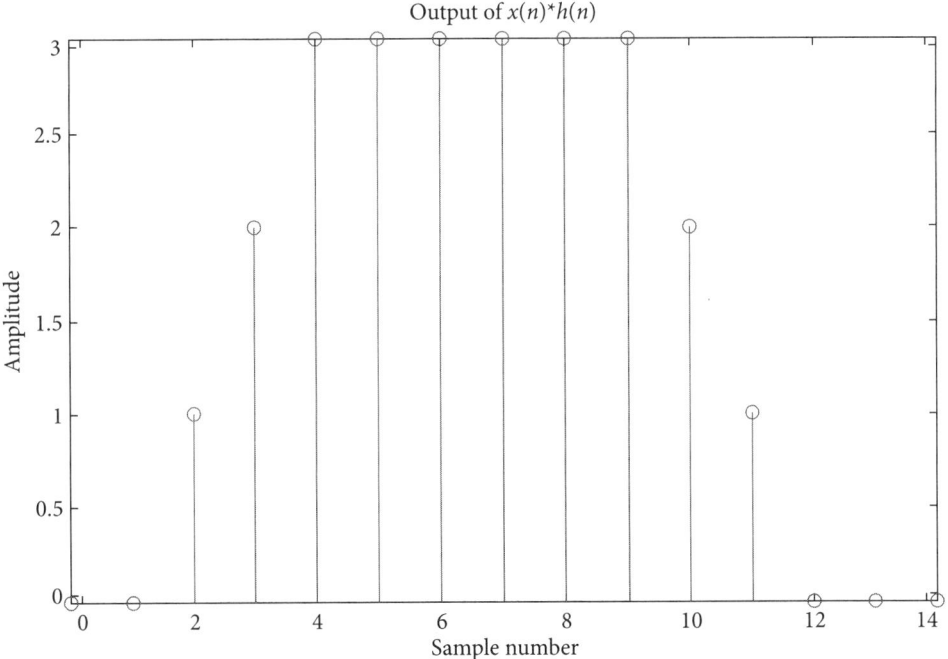

Fig. 4.28 Plot of convolved output for Example 4.17

Let us consider Example 4.18. LTI system has the impulse response given by $h[n] = (0.2)^n \{u[n] - u[n-3]\}$. If the input is $x[n] = (0.3)^n u[n]$, find the output of the system. Let us write MATLAB program for the same. The output of the convolution is shown in Fig. 4.29. Compare Fig. 4.29 with Fig. 4.23. Note that the index for the output starts at $n = 1$ rather than zero.

```
clear all;
x=[1,0.2,0.04,0,0];
h=[1,0.3,(0.3)^2,(0.3)^3,(0.3)^4,(0.3)^5,(0.3)^6,
0.3)^7,(0.3)^8,(0.3)^9,(0.3)^10];
b=conv(x,h);
disp(b);
stem(b);title('output of x(n)*h(n)');
xlabel('sample number');ylabel('Ampltude');
```

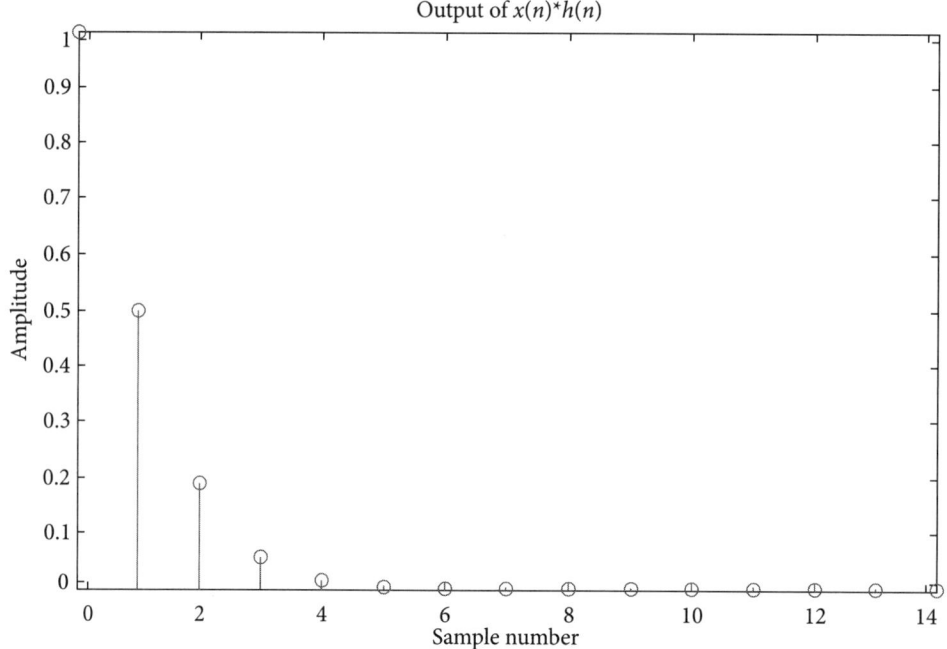

Fig. 4.29 Plot of convolved output for Example 4.18

First 9 output values displayed are
   Columns 1 through 9
   1.0000 0.5000 0.1900 0.0570 0.0171 0.0051 0.0015 0.0005 0.0001

Let us consider Example 4.19. LTI system has the impulse response given by $h[n] = u[n-3]$. If the input is $x[n] = u[n]$, find the output of the system. Let us write MATLAB program for the same. The output of the convolution is shown in Fig. 4.30. Compare Fig. 4.30 with Fig. 4.25. Here, we have taken only 10 samples of h and x sequence. Hence, the output decreases after sample number 10. Actually, it extends upto infinity. Hence, the actual output after sample number 10 will continuously increase as a function of $(n-2)$ and will tend to infinity.

```
clear all;
for i=4:10,
x(i)=1;
```

```
end
for i=1:10,
h(i)=1;
end
b=conv(x,h);
for i=1:10,
b1(i)=b(i);
end
n=0:1:9;
stem(n,b1);title('output of x(n)*h(n)');
xlabel('sample number');ylabel('Ampltude');
```

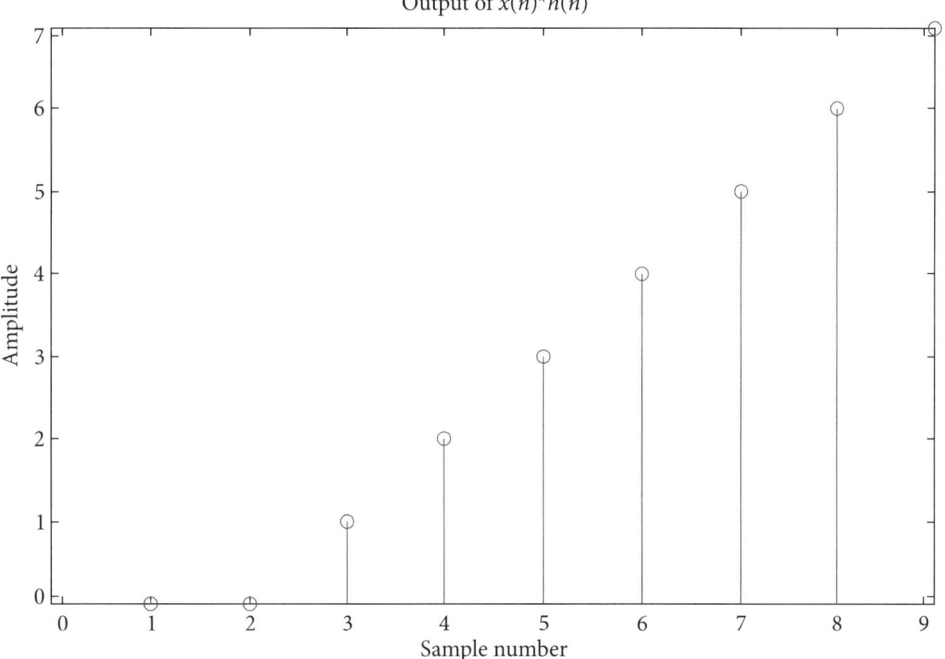

Fig. 4.30 Convolved output for Example 4.19

Let us consider Example 4.20. LTI system has the impulse response given by $h[n]$ and if the input is $x[n]$, as shown in Fig. 4.26, find the output of the system. The output of the convolution is shown in Fig. 4.31. Compare Fig. 4.31 with Fig. 4.27.

```
clear all;
n=-2:1:0;
x=[1,1,1];
m=0:1:3;
h=[1,1,2,2];
l=-2:1:3;
```

```
b=conv(x,h);
stem(l,b);title('output of x(n)*h(n)');
xlabel('sample number');ylabel('Ampltude');
```

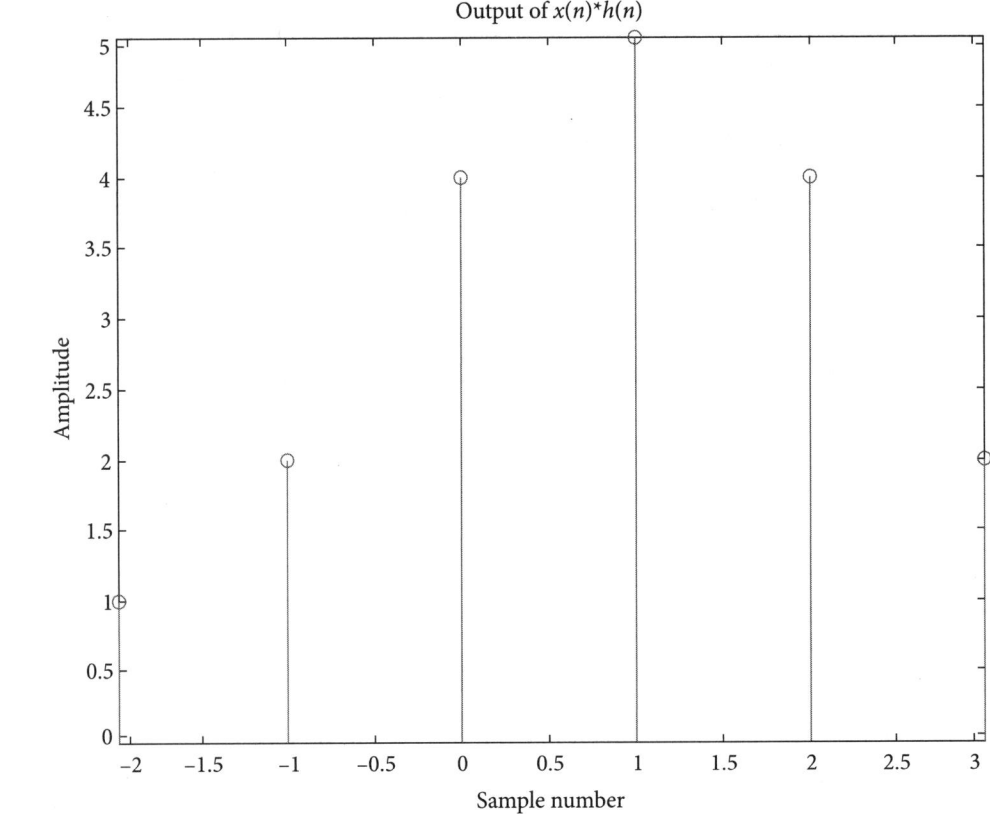

Fig. 4.31 Convolved output for Example 4.20

Concept Check
- How will you decompose any DT signal in terms of unit impulses?
- What is the property of superposition
- What is impulse response?
- What is a zero state response?
- What is convolution sum?
- Explain the steps for input side algorithm.
- What is output side algorithm?
- Explain steps for conventional convolution algorithm?
- How will you implement convolution using MATLAB commands?

4.6 Unit Step Response of CT and DT LTI Systems

We have studied the calculation of impulse response for CT and DT systems in the previous sections. We will discuss the step response in this section. To find the unit step response of the system, we have to convolve the impulse response of the system with the unit step applied as input. The output of the system is the desired step response. We will illustrate the concepts with the help of simple numerical examples. We will first consider the step response for a CT system.

Example 4.21

Find the step response of the CT LTI system with impulse response given by

$$h(t) = (e)^{-|t|}.$$

Solution

We need to find

$$h[n] * u[n] = (e)^{-|t|} * u(t) \tag{4.103}$$

Let us plot both the signals and $u(t - \tau)$. The plot is shown in Fig. 4.32.

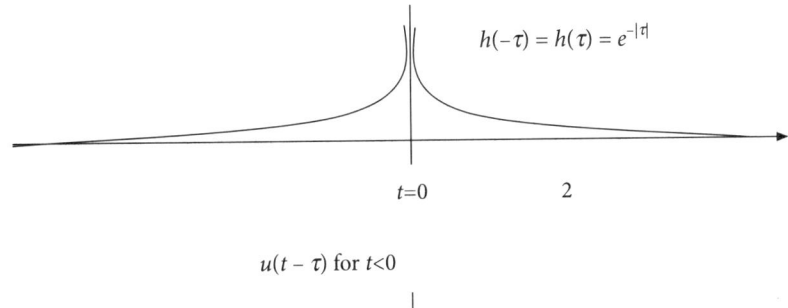

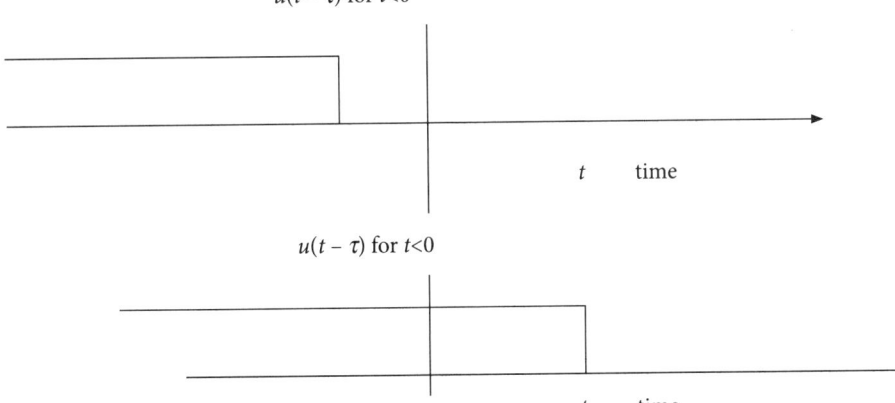

Fig. 4.32 Plot of $h(\tau)$, $u(\tau)$ and $u(t - \tau)$ for different intervals for Example 4.21

Consider the first interval for $-\infty < t < 0$. The convolution integral can be written as

$$h(t)*u(t) = (e)^{-|t|}*u(t) = \int_{-\infty}^{t} e^{-\tau}d\tau = -e^{-\tau}\Big\downarrow_{-\infty}^{t} = -e^{-t} - 0 = -e^{-t} \quad (4.104)$$

For interval $0 < t < \infty$, the convolution integral can be written as

$$h(t)*u(t) = (e)^{-|t|}*u(t) = \int_{0}^{t} e^{-\tau}d\tau = -e^{-\tau}\Big\downarrow_{0}^{t} = -e^{-t} - (-1) = 1 - e^{-t} \quad (4.105)$$

The signal has infinite values. Hence, the MATLAB program output is not shown.

Example 4.22

Find the step response of the CT LTI system with the impulse response given by

$$h(t) = u(t) - u(t-2).$$

Solution

We need to find

$$h(t)*u(t) = [u(t) - u(t-2)]*u(t) \quad (4.106)$$

Let us plot both the signals and $u(t-\tau)$. The plot is shown in Fig. 4.33.

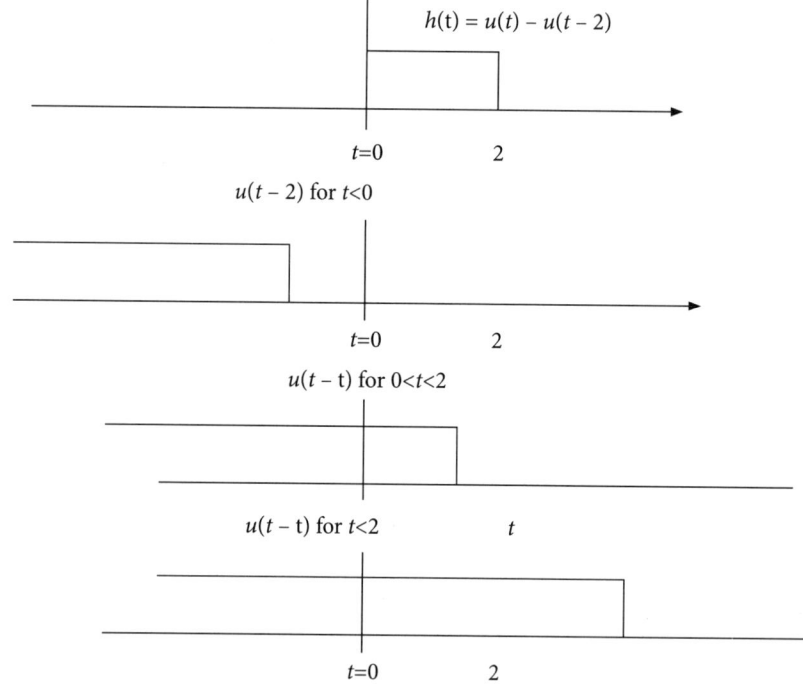

Fig. 4.33 Plot of $h(\tau)$, $u(\tau)$ and $u(t-\tau)$ for different intervals for Example 4.22

Consider first interval for $-\infty < t < 0$. The convolution integral is zero as there is no overlap.

Consider the interval $0 < t < 2$. The convolution integral can be written as

$$h[n] * u[n] = [u(t) - u(t-2)] * u(t) = \int_0^t d\tau = \tau \Big|_0^t = t \qquad (4.107)$$

For interval $2 < t < \infty$, the convolution integral can be written as

$$h[n] * u[n] = [u(t) - u(t-2)] * u(t) = \int_0^2 d\tau = \tau \Big|_0^2 = 2 \qquad (4.108)$$

Example 4.23

Find the step response of the CT LTI system with the impulse response given by

$$h(t) = \delta(t) - \delta(t-2).$$

Solution

We need to find

$$h(t) * u(t) = [\delta(t) - \delta(t-2)] * u(t) \qquad (4.109)$$

Let us plot both the signals and $u(t - \tau)$. The plot is shown in Fig. 4.34.

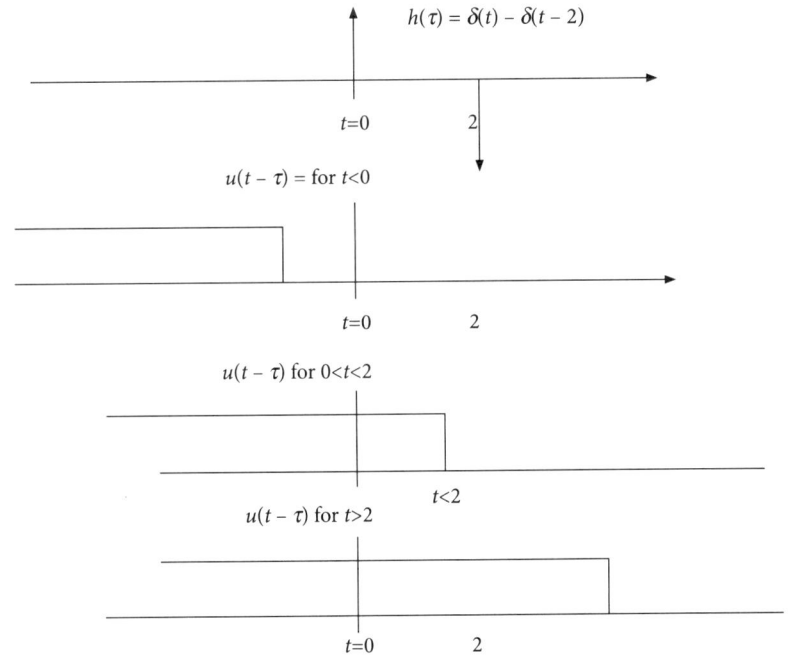

Fig. 4.34 Plot of $h(\tau)$, $u(\tau)$ and $u(t - \tau)$ for different intervals for Example 4.23

Consider first interval for $-\infty < t < 0$. The convolution integral is zero as there is no overlap.

Consider the interval $-\infty < t < 2$. The convolution integral can be written as

$$h(t) * u(t) = [\delta(t) - \delta(t-2)] * u(t) = \int_{0-}^{t} \delta(t)d\tau = 1 \qquad (4.110)$$

For interval $2 < t < \infty$, the convolution integral can be written as

$$h(t) * u(t) = [\delta(t) - \delta(t-2)] * u(t)$$

$$= \int_{-\infty}^{2} \delta(\tau)d\tau - \int_{2-}^{\infty} \delta(\tau - 2)d\tau = 1 - 1 = 0 \qquad (4.111)$$

Example 4.24

Find the step response of the CT LTI system with impulse response given by

$$h(t) = tu(t).$$

Solution

We need to find

$$h(t) * u(t) = [tu(t)] * u(t) \qquad (4.112)$$

Let us plot both the signals and $u(t - \tau)$. The plot is shown in Fig. 4.35.

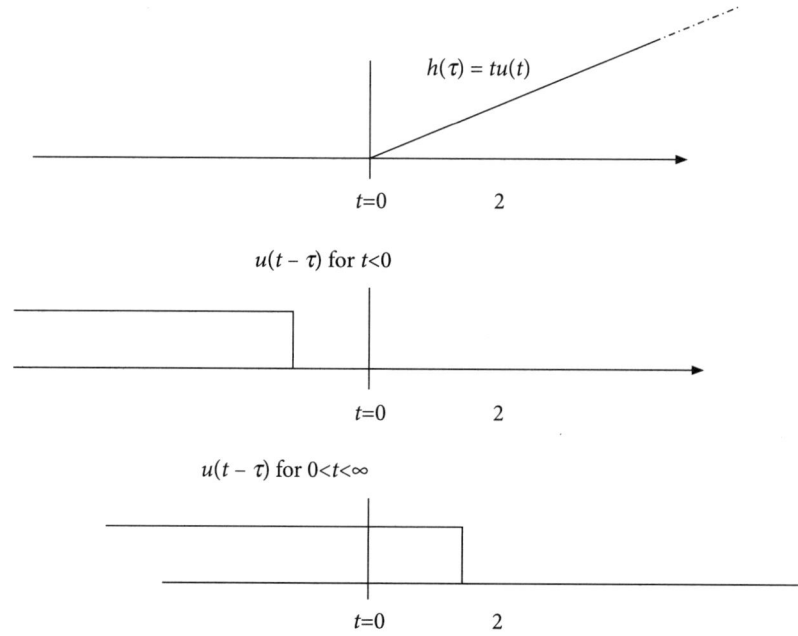

Fig. 4.35 Plot of $h(\tau)$, $u(\tau)$ and $u(t - \tau)$ for different intervals for Example 4.24

Consider first interval for $-\infty < t < 0$. The convolution integral is zero as there is no overlap.

Consider the interval $0 < t < \infty$. The convolution integral can be written as

$$h(t) * u(t) = [tu(t)] * u(t) = \int_0^t \tau \, d\tau = \frac{\tau^2}{2} \Big|_0^t = \frac{t^2}{2} \tag{4.113}$$

Example 4.25

Find the step response of the DT LTI system with impulse response given by

$$h[n] = \left(\frac{1}{2}\right)^n u[n].$$

Solution

We need to convolve the impulse response of the system with the step function and find

$$h[n] * u[n] = \left(\frac{1}{2}\right)^n u[n] * u[n] \tag{4.114}$$

Let us plot both the signals. We will invert one of the signals, for example, $u[n]$ and find $u[-n]$. The reader may recall that this is step 1 in the procedure for conventional convolution of DT signals. We will follow the step-by-step procedure.

Step 1 Invert one of the sequences, that is, calculate $x[-m]$ or $h[-m]$. Let us assume that we calculate $x[-m]$.

Step 2 Perform sample-by-sample multiplication, that is, $x[-m] h[m]$ for all m and accumulate the sum. This operation of sample-by-sample multiplication and summation is called multiply and accumulate operation.

Step 3 Shift the inverted sequence towards the right by one sample and again calculate the sum. Shift the inverted sequence towards the left by one sample and again calculate the sum until the output is non-zero.

Step 4 Repeat step 3 until all $P + Q - 1$ number of samples are calculated.

In this example $P = $ infinity, $Q = $ infinity, so we need to calculate $P + Q - 1 = $ infinite samples. Note that here we have to shift the inverted sequence towards the left as well until the generated output is zero.

The plot is shown in Fig. 4.36.

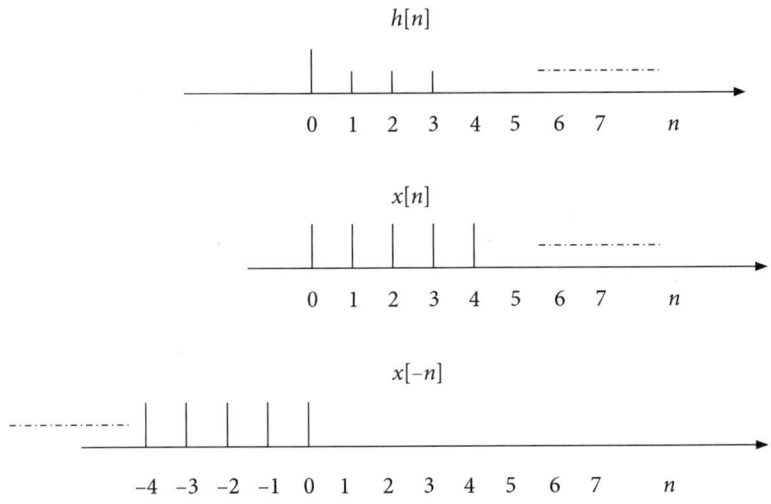

Fig. 4.36 Plot of $h[n]$, $x[n]$ and $x[-n]$

Table 4.5 Conventional convolution procedure (Example 4.25)

Sample Number	...,-5	-4	-3	-2	-1	0	1	2	3	4...
$h[m]$						1	½	¼	1/8	1/16
$x[-m]$	...1	1	1	1	1	1				
$x[1-m]$	...1	1	1	1	1	1	1			
$x[2-m]$	...1	1	1	1	1	1	1	1		
$x[3-m]$	...1	1	1	1	1	1	1	1	1	
$x[4-m]$	...1	1	1	1	1	1	1	1	1	1
$y[m]$						1	1 + ½	1 + ½ + ¼	...	..

We see from Table 4.5 that for the calculation of $y[0]$, only one sample of $x[-m]$ overlaps with the samples of $h[m]$; therefore, multiply and accumulate operation will result in output = 1. Successive shifting of $x[-m]$ towards the right and application of multiply and accumulate operation will generate output $y[1] = 1 + 1/2$, $y[2] = 1 + 1/2 + 1/4$, $y[3] = 1 + 1/2 + 1/4 + 1/8$, etc., as shown in Table 4.5, as more and more samples overlap. The output is calculated and listed below.

The output of convolution is given by

$$y[0]=1, y[1]=1+1/2, y[2]=1+1/2+1/4,$$

In general

$$y[n]=1+1/2+1/4+\ldots\left(\frac{1}{2}\right)^n = \frac{1-\left(\frac{1}{2}\right)^{n+1}}{1-\frac{1}{2}} = 2(1-(1/2)^{n+1}) \qquad (4.115)$$

Example 4.26

Find the step response of the DT LTI system with impulse response given by

$$h[n] = \delta[n] - \delta[n-1].$$

Solution

We need to find

$$h[n] * u[n] = \{\delta[n] - \delta[n-1]\} * u[n] \tag{4.116}$$

Let us plot both the signals and $u[-n]$. The plot is shown in Fig. 4.37.

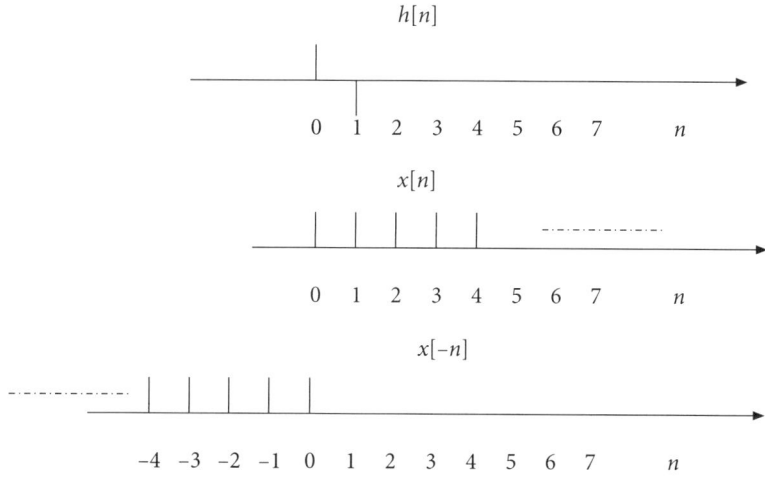

Fig. 4.37 Plot of $h[n]$, $x[n]$ and $x[-n]$

The output of convolution is given by

$$y[0] = 1, y[1] = 0, y[2] = 0, \ldots$$

$$y[n] = \delta[n] \tag{4.117}$$

Example 4.27

Find the step response of the DT LTI system with impulse response given by

$$h[n] = (-1)^n \left[u[n+2] - u[n-1] \right].$$

Solution

We need to find

$$h[n] * u[n] = (-1)^n \{u[n+2] - u[n-1]\} * u[n] \tag{4.118}$$

Let us plot both the signals and $u[-n]$. The plot is shown in Fig. 4.38.

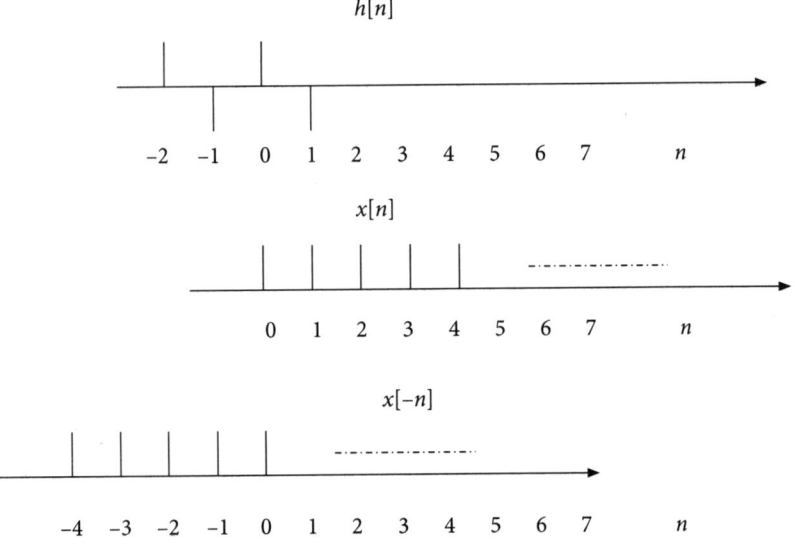

Fig. 4.38 Plot of $h[n]$, $x[n]$ and $x[-n]$

The output of convolution is given by

$$y[-1] = 0, y[-2] = 1$$

$$y[0] = 1, y[1] = 0, y[2] = 0, \ldots$$

$$y[n] = \delta[n+2] + \delta[n] \tag{4.119}$$

Example 4.28

Find the step response of the DT LTI system with impulse response given by

$$h[n] = u[n].$$

Solution

We need to find

$$h[n] * u[n] = u[n] * u[n] \tag{4.120}$$

Let us plot both the signals and $u[-n]$. The plot is shown in Fig. 4.39.
The output of convolution is given by

$$y[0] = 1, y[1] = 2, y[2] = 3, \ldots$$

$$y[n] = n \tag{4.121}$$

Concept Check

- What is a step response?
- Find $u[n] - u[n-1]$.
- Find $u[n-1] * \delta[n+1]$

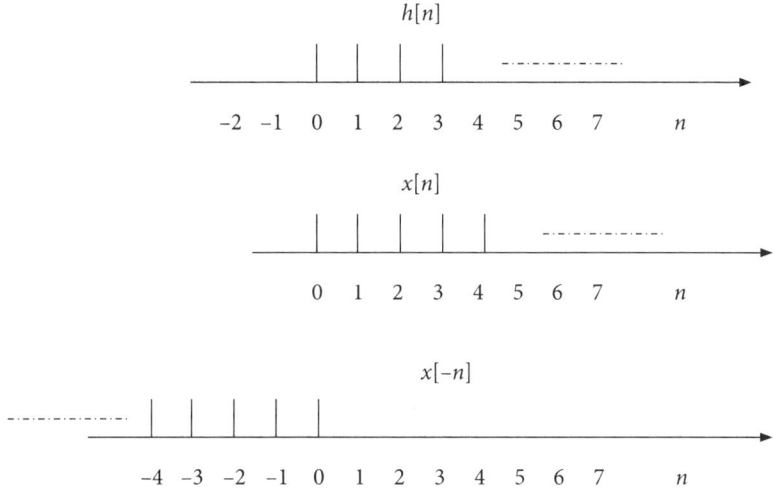

Fig. 4.39 Plot of $h[n]$, $x[n]$ and $x[-n]$

4.7 Properties of LTI DT Systems

Let us discuss the properties like memory, causality and stability for CT and DT LTI systems. We will study how the properties translate for LTI systems.

4.7.1 Memory property of CT and DT LTI systems

The system is said to have memory or said to be dynamic if its current output depends on previous, future input or previous and future output signals. The system is said to be memoryless or instantaneous if its current output depends only on the current input. Let the system impulse response of the DT system be $h[n]$. Let the input to the system be $x[n]$ and the output be $y[n]$. The output of the system can be written as the convolution sum in case of DT system as

$$y[n] = \sum_{m=-\infty}^{\infty} h[m]x[n-m] \qquad (4.122)$$

In order that the system is memoryless, $y[n]$ should depend only on $x[n]$ i.e., $x[n-m]$ must be zero for all non-zero m. This condition will be true if

$h[n] = 0$ for non-zero m or $h[m]$ should exist only when $m = 0$. The condition can be translated as

$$h[n] = c\delta[n] \text{ for } DT \text{ systems} \tag{4.123}$$

The memoryless DT system just scales the unit impulse.
Similarly for a CT system

$$y(t) = \int_{-\infty}^{\infty} h(\tau)x(t-\tau)d\tau \tag{4.124}$$

In order that the system is memoryless, $y(t)$ should depend on only $x(t)$ i.e., $x(t - \tau)$ must be zero for all non-zero τ. This condition will be true if $h(\tau)$ is zero all non-zero τ. This condition can be translated as

$$h(t) = c\delta(t) \text{ for } CT \text{ systems} \tag{4.125}$$

The memoryless CT system just scales the unit impulse. All memoryless systems perform scalar multiplication on the input.

Example 4.29
Check if the following CT systems are memoryless.

1. $h(t) = e^{-|t|}$

2. $h(t) = e^{3t}u(t-1)$

3. $h(t) = u(t+2) - 2u(t-1)$

4. $h(t) = 5\delta(t)$

5. $h(t) = \sin(\pi t)u(t)$

Solution
We have to check if the impulse response of the system is a delta function.
  i. $h(t) = e^{-|t|}$. The impulse response is not a delta function; hence, the system is with memory.
  ii. $h(t) = e^{3t}u(t - 1)$. The impulse response is not a delta function; hence, the system is with memory.
  iii. $h(t) = u(t + 2) - 2u(t - 1)$. The impulse response is not a delta function; hence, the system is with memory.
  iv. $h(t) = 5\delta(t)$. The impulse response is a scaled delta function; hence, the system is memoryless.

v. $h(t) = \sin(\pi t)u(t)$. The impulse response is not a delta function; hence, the system is with memory.

Example 4.30

Check if the following DT systems are memoryless.

1. $h[n] = 2^n u[-n]$

2. $h[n] = e^{3n} u(n-2)$

3. $h[n] = \cos\left(\dfrac{1}{4}\pi n\right)[u([n]) - u[n-2]]$

4. $h[n] = 5u[n] - 5u[n-1]$

5. $h[n] = \sin(3\pi n)u[n]$

6. $h[n] = \delta[n] + \cos(\pi n)$

Solution

We have to check if the impulse response of the system is a delta function.

i. $h[n] = 2^n u[-n]$. The impulse response is not a delta function; hence, the system is with memory.

ii. $h[n] = e^{3n} u[n-2]$. The impulse response is not a delta function; hence, the system is with memory.

iii. $h[n] = \cos\left(\dfrac{1}{4}\pi n\right)[u[n] - u[n-2]]$. The impulse response is not a delta function; hence, the system is with memory.

iv. $h[n] = 5u[n] - 5u[n-1] = 5\delta[n]$. The impulse response is a scaled delta function; hence, the system is memoryless.

v. $h[n] = \sin(3\pi n)u[n]$. The impulse response is not a delta function; hence, the system is with memory.

vi. $h[n] = \delta[n] + \cos(\pi n)$. The impulse response is not a delta function; hence, the system is with memory.

4.7.2 Condition of causality for CT and DT LTI systems

Let us now consider an LTI system. An LTI system is causal or realizable if its current output depends on past and current inputs and past outputs. To put it formally, a DT LTI system is causal if its impulse response $h[n]$ is zero for all $n < 0$. If $x[n]$ is the input to a system, then the output $y[n]$ is given by

$$y[n] = \sum_{k=0}^{\infty} h[k]x[n-k] \qquad (4.126)$$

If, for example, $h[-1]$ exists, then $y[n]$ will depend on $x[n + 1]$, which is the next input, and the system will be non-causal.
or

$$y[n] = \sum_{k=-\infty}^{n} x[k]h[n-k] \qquad (4.127)$$

as $y[n] = x[n] * h[n] = h[n] * x[n]$ \qquad (4.128)

As the system is an LTI system, the response of the system to input at '$n - k$' is $h[n - k]$ due to property of time invariance. If we put $h[n] = 0$ for $n < 0$, then $h[n-k] = 0$ for $k > n$

Hence, the limits for k will be from $-\infty$ to n.

We will now consider the causality for CT LTI systems. A CT LTI system is causal if its impulse response $h(t)$ is zero for all $t < 0$. If $x(t)$ is the input to a system then the output $y(t)$ is given by

$$y(t) = \int_{0}^{\infty} h(\tau)x(t-\tau)d\tau \qquad (4.129)$$

If, for example, $h(t)$ exists for $t < 0$ i.e., $h(-1)$, then $y(t)$ will depend on $x(t + 1)$, which is the input at next time instant, and the system will be non-causal.
or

$$y(t) = \int_{-\infty}^{t} x(\tau)h(t-\tau)d\tau \qquad (4.130)$$

as $y(t) = x(t) * h(t) = h(t) * x(t)$ \qquad (4.131)

As the system is an LTI system, the response of the system to input at '$t - \tau$' is $h(t - \tau)$ due property of time invariance. If we put

$$h(t) = 0 \text{ for } t < 0 \text{ then } h(t-\tau) = 0 \text{ for } \tau > t \qquad (4.132)$$

Hence, the limits for τ will be from $-\infty$ to t.

To see why the impulse response of the system $h(t)$ or $h[n]$ must be zero for all $t < 0$ or $n < 0$ for causality, we will consider the following example.

Example 4.31
Consider the system represented by

$$y[n] = \sum_{k=-1}^{2} x[k]h[n-k]$$

Is the system causal? If not, explain why.

Solution

$$y[n] = x[-1]h[n+1] + x[0]h[n] + x[1]h[n-1] + x[2]h[n-2] \qquad (4.133)$$

The current output depends on next two input sample. Hence, the system is non-causal. Hence, if $h[n]$ is non-zero for $n < 0$, we can conclude that the system is non-causal.

Example 4.32
Check if the following CT systems are causal.

1. $h(t) = e^{-|t|}$

2. $h(t) = e^{3t}u(t-1)$

3. $h(t) = u(t+2) - 2u(t-1)$

4. $h(t) = 5\delta(t)$

5. $h(t) = \sin(\pi t)u(t)$

Solution

We have to check if the impulse response of the system is zero for all $n < 0$.

i. $h(t) = e^{-|t|}$. The impulse response is not zero for all $t < 0$; hence, the system is non-causal.

ii. $h(t) = e^{3t} u(t-1)$. The impulse response exists for all $t \geq 1$ as $u(t-1) = 1$ for all $t \geq 1$; hence, the system is causal.

iii. $h(t) = u(t+2) - 2u(t-1)$. The function $h(t)$ is equal to 1 for $-2 \leq t \leq 1$. The impulse response exists for negative vales of t up to $t = -2$. Hence, the system is non-causal.

iv. $h(t) = 5\delta(t)$. The impulse response is a scaled delta function and is zero for all negative values of t; hence, the system is causal.

v. $h(t) = \sin(\pi t)u(t)$. The impulse response is zero for all negative values of t; hence, the system is causal.

Example 4.33

Check if the following DT systems are causal.

1. $h[n] = 2^n u[-n]$

2. $h[n] = e^{3n} u(n-2)$

3. $h[n] = \cos\left(\frac{1}{4}\pi n\right)[u([n]) - u[n-2]]$

4. $h[n] = 5u[n] - 5u[n-1]$

5. $h[n] = \sin(3\pi n)u[n]$

6. $h[n] = \delta[n] + \cos(\pi n)$

Solution

We have to check if the impulse response of the system is zero for all negative values of n.

i. $h[n] = 2^n\, u[-n]$. The impulse response is existing for all negative values of n, hence, the system is non-causal.

ii. $h[n] = e^{3n}u[n-2]$. The impulse response is zero all negative values of n as it exists for $n \geq 2$; hence, the system is causal.

iii. $h[n] = \cos\left(\frac{1}{4}\pi n\right)[u[n] - u[n-2]]$. The impulse response exists for $0 \leq n \leq 2$ and is zero for all negative value of n; hence, the system is causal.

iv. $h[n] = 5u[n] - 5u[n-1] = 5\delta[n]$. The impulse response is a scaled delta function and is zero for all negative values of n; hence, the system is causal.

v. $h[n] = \sin(3\pi n)u[n]$. The impulse response exists for $n \geq 0$ and is zero for all negative values of n; hence, the system is causal.

vi. $h[n] = \delta[n] + \cos(\pi n)$. The impulse response exists for $n = 0$ and for all values of n; hence, the system is non-causal.

4.7.3 Stability for CT and DT LTI systems

Theorem

*CT or DT LTI system is BIBO stable **if and only if** its impulse response is absolutely summable.*

Let us first prove this theorem for CT LTI systems.

Proof

Let the input to the system $x(t)$ be bounded, then there exists a finite constant M such that $|x(t)| \leq M$.

The output $y(t)$ is given by the convolution sum

$$y(t) = \int_{-\infty}^{\infty} h(\tau) x(t-\tau) d\tau \tag{4.134}$$

Therefore, absolute values on both sides of Eq. (4.134)

$$|y(t)| = \left| \int_{-\infty}^{\infty} h(\tau) x(t-\tau) d\tau \right| \tag{4.135}$$

As the absolute value of sum is always less than or equal to the sum of the absolute values, therefore, we have

$$|y(t)| = \left| \int_{-\infty}^{\infty} h(\tau) x(t-\tau) \right| \leq \int_{-\infty}^{\infty} |h(\tau)| |x(t-\tau)| d\tau \tag{4.136}$$

Substitute upper bound for $x(t - \tau)$, we get

$$|y(t)| = \left| \int_{-\infty}^{\infty} h(\tau) x(t-\tau) \right| \leq M \int_{-\infty}^{\infty} |h(\tau)| d\tau \tag{4.137}$$

Equation (4.137) indicates that $y(t)$ is bounded if the impulse response of the system is absolutely summable. We have proved the '*if*' part of the theorem.

Let the sum be represented as S.

It is now required to be shown that if the impulse response is unbounded, then there exists some input $x(t)$ for which the output is unbounded.

Consider the input given by

$$x(t) = \begin{cases} \dfrac{h'(-t)}{|h(-t)|} & \text{for } h(t) \neq 0 \\ 0 & \text{otherwise} \end{cases} \tag{4.138}$$

where $h'(t)$ represents the complex conjugate of $h(t)$.

Substitute $t = 0$ in Eq. (4.137) with $x(t)$ given by Eq. (4.138). We get

$$y(0) = \int_{-\infty}^{\infty} h(t) x(0-\tau) d\tau \tag{4.139}$$

$$y(0) = \int_{-\infty}^{\infty} h(t) \frac{h'(\tau)}{|h(\tau)|} \tag{4.140}$$

$$y(0) = \int_{-\infty}^{\infty} \frac{|h(\tau)|^2}{|h(\tau)|} = \int_{-\infty}^{\infty} |h(\tau)| = S = \infty \tag{4.141}$$

The system is found to produce infinite output at t = 0 for above $x(t)$ if its impulse response is not absolutely summable. We have proved the *only if* part of the stability statement!

Let us now prove the theorem for DT systems as well.

Proof

Let the input to the system $x[n]$ be bounded, then there exists a finite constant M such that $|x[n]| \leq M$.

The output $y[n]$ is given by the convolution sum

$$y[n] = \sum_{k=-\infty}^{\infty} h[k]x[n-k] \tag{4.142}$$

Therefore, absolute values on both sides of Eq. (4.126)

$$|y([n]| = \left| \sum_{k=-\infty}^{\infty} h[k]x([n-k] \right| \tag{4.143}$$

As the absolute value of sum is always less than or equal to the sum of the absolute values, therefore, we have

$$|y[n]| = \left| \sum_{k=-\infty}^{\infty} h[k]x[n-k] \right| \leq \sum_{k=-\infty}^{\infty} |h[k]||x[n-k]| \tag{4.144}$$

Substitute upper bound for $x[n-k]$, we get

$$|y[n]| \leq \sum_{k=-\infty}^{\infty} |h[k]||x[n-k]| \leq M \sum_{k=-\infty}^{\infty} |h[k]| \tag{4.145}$$

Equation (4.145) indicates that $y[n]$ is bounded if the impulse response of the system is absolutely summable. We have proved the '*if*' part of the theorem.

Let the sum be represented as S.

It is now required to show that if the impulse response is unbounded, then there exists some input $x[n]$ for which the output is unbounded.

Consider the input given by

$$x[n] = \begin{cases} \dfrac{h'[-n]}{|h[-n]|} & \text{for } h[n] \neq 0 \\ 0 & \text{otherwise} \end{cases} \quad (4.146)$$

where $h'[n]$ represents the complex conjugate of $h[n]$.

Substitute $n = 0$ in Eq. (4.145) with $x[n]$ given by Eq. (4.146). We get

$$y[0] = \sum_{k=-\infty}^{\infty} h[k]x[0-k] \quad (4.147)$$

$$y[0] = \sum_{k=-\infty}^{\infty} h[k]\frac{h'[k]}{|h[k]|} \quad (4.148)$$

$$y[0] = \sum_{k=-\infty}^{\infty} \frac{|h[k]|^2}{|h[k]|} = \sum_{k=-\infty}^{\infty} |h[k]| = S = \infty \quad (4.149)$$

The system is found to produce infinite output at $n = 0$ for above $x(n)$ if its impulse response is not absolutely summable. We have proved the *only if* part of the stability statement!

Example 4.34

Find the range of values of "a" for which the system with impulse response $h[n]$ is stable. The impulse response is given by

$$h[n] = a^n u[n]$$

Solution

We have to prove that the impulse response is absolutely summable. That is

$$\sum_{k=-\infty}^{\infty} h[k] \text{ is finite} \quad (4.150)$$

The impulse response $h[n]$ is a right-handed sequence or is causal. The limit for k will be between zero and infinity. That is

$$\sum_{k=0}^{\infty} a^k = 1 + a + a^2 + a^3 + \cdots + a^\infty \quad (4.151)$$

This infinite geometric series will converge only if $|a| < 1$. The system is then stable for $|a| < 1$.

Example 4.35

Check if the following CT systems are stable

1. $h(t) = e^{-|t|}$

2. $h(t) = e^{3t}u(t-1)$

3. $h(t) = u(t+2) - 2u(t-1)$

4. $h(t) = 5\delta(t)$

5. $h(t) = \sin(\pi t)u(t)$

Solution

We have to prove that the impulse response is absolutely summable. That is

$$\int_{-\infty}^{\infty} h(t)dt \text{ is finite.}$$

i. $h(t) = e^{-|t|}$,

$$\int_{-\infty}^{\infty} h(t)dt = \int_{-\infty}^{0} e^{t}dt + \int_{0}^{\infty} e^{-t}dt = 2\int_{0}^{\infty} e^{-t}dt = -2e^{-t}\Big|_{0}^{\infty} = -2[0-1] = 2 \quad (4.152)$$

is finite.

The impulse response is absolutely summable; hence, the system is stable.

ii. $h(t) = e^{3t}u(t-1)$.

$$\int_{-\infty}^{\infty} h(t)dt = \int_{1}^{\infty} e^{3t}dt = \frac{1}{3}e^{3t}\Big|_{1}^{\infty} = \frac{1}{3}[\infty - e^{3}] \to \infty \quad (4.153)$$

The impulse response is not absolutely summable; hence, the system is not stable.

iii. $h(t) = u(t+2) - 2u(t-1)$.

$$\int_{-\infty}^{\infty} h(t)dt = \int_{-2}^{1} dt + \int_{1}^{\infty} -1 dt = 3 - \infty \to -\infty \quad (4.154)$$

The impulse response is not absolutely summable; hence, the system is not stable.

iv. $h(t) = 5\delta(t)$.

$$\int_{-\infty}^{\infty} h(t)dt = 5\int_{-\infty}^{\infty} \delta(t)dt = 5\times 1 = 5 \text{ is finite.} \tag{4.155}$$

The impulse response is absolutely summable; hence, the system is stable.

v. $h(t) = \sin(\pi t)u(t)$.

$$\int_0^{\infty} h(t)dt = \int_0^{\infty} \sin(\pi t)dt = \lim_{N\to\infty} \frac{N}{\pi}\left[\cos(\pi t)\downarrow_0^T\right] \tag{4.156}$$

$$= \lim_{N\to\infty} \frac{N}{\pi}[1-(-1)] = \infty$$

T represents the period of the cos function. The impulse response is not absolutely summable; hence, the system is not stable.

Example 4.36

Check if the following DT systems are stable.

1. $h[n] = 2^n u[-n]$
2. $h[n] = e^{3n} u(n-2)$
3. $h[n] = \cos\left(\frac{1}{4}\pi n\right)[u([n]) - u[n-2]]$
4. $h[n] = 5u[n] - 5u[n-1]$
5. $h[n] = \sin(3\pi n)u[n]$
6. $h[n] = \delta[n] + \cos(\delta n)$

Solution

We have to prove that the impulse response is absolutely summable. That is

$$\sum_{-\infty}^{\infty} h[n] = \text{finite}$$

i. $h[n] = 2^n u[-n]$,

$$\sum_{-\infty}^{\infty} h[n] = \sum_{-\infty}^{0} 2^n = 1 + \frac{1}{2} + \frac{1}{4} + \dots = \frac{1}{1-\frac{1}{2}} = 2 = \text{finite.} \tag{4.157}$$

The impulse response is absolutely summable; hence, the system is stable.

ii. $h[n] = e^{3n}u[n-2]$

$$\sum_{-\infty}^{\infty} h[n] = \sum_{2}^{\infty} e^{3n} = 1 + e^3 + e^6 + \ldots \to \text{infinity} \qquad (4.158)$$

The impulse response is not absolutely summable; hence, the system is not stable.

iii. $h[n] = \cos\left(\frac{1}{4}\pi n\right)[u[n] - u[n-2]]$

$$\sum_{0}^{1} h[n] = \sum_{0}^{1} \cos\left(\frac{1}{4}\pi n\right) = 1 + \cos\left(\frac{1}{4}\pi\right) = 1 + \frac{1}{\sqrt{2}} \text{ is finite.} \qquad (4.159)$$

The impulse response is absolutely summable; hence, the system is stable

iv. $h[n] = 5u[n] - 5u[n-1] = 5\delta[n]$.

$$5\sum_{-\infty}^{\infty} \delta[n] = 5 \text{ is finite.} \qquad (4.160)$$

The impulse response is absolutely summable; hence, the system is stable

v. $h[n] = \sin(3\pi n)u[n]$.

$$\sum_{0}^{\infty} h[n] = \sum_{0}^{\infty} \sin(3\pi n) = \to \text{infinity} \qquad (4.161)$$

The impulse response is not absolutely summable; hence, the system is unstable.

vi. $h[n] = \delta[n] + \cos(\pi n)$.

$$\sum_{0}^{\infty} h[n] = \delta[n] + \sum_{0}^{\infty} \cos(\pi n) = \to \text{infinity} \qquad (4.162)$$

The impulse response is not absolutely summable; hence, the system is unstable.

Concept Check

- What is property of memory for LTI systems?
- When will you say that the system is memoryless?
- What is causality?
- State property of causality in terms of impulse response.
- What is stablit?
- What is BIBO stability?
- State property of stability in terms of impulse response.

4.8 Series and Parallel Interconnection of Systems

We have studied series and parallel interconnection of systems in section 3.5. We will go through the calculation of impulse response of the overall system in case of series and parallel interconnection of systems as we have studied convolution integral and convolution sum for CT and DT systems in this chapter. We will illustrate the concepts with the help of simple numerical calculations.

Example 4.37

Consider the interconnections of the systems, as shown in Fig. 4.40. Let the impulse responses be specified as

$h_1(t) = u(t)$

$h_2(t) = \delta(t)$

$h_3(t) = \delta(t - 1)$

Find the response of the overall interconnection.

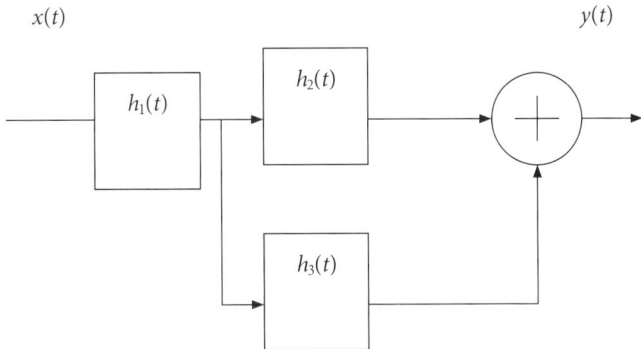

Fig. 4.40 Interconnection of systems for Example 4.37

Solution

The impulse response of the overall interconnection can be written as

$$h_1(t) * \{h_2(t) + h_3(t)\}$$

$$= \{u(t)\} * \{\delta(t) + \delta(t-1)\}$$

(4.163)

Figure 4.41 shows a plot of two signals and the output convolved signal.

Consider first interval for $-\infty < t < 0$. The convolution integral is zero, as there is no overlap.

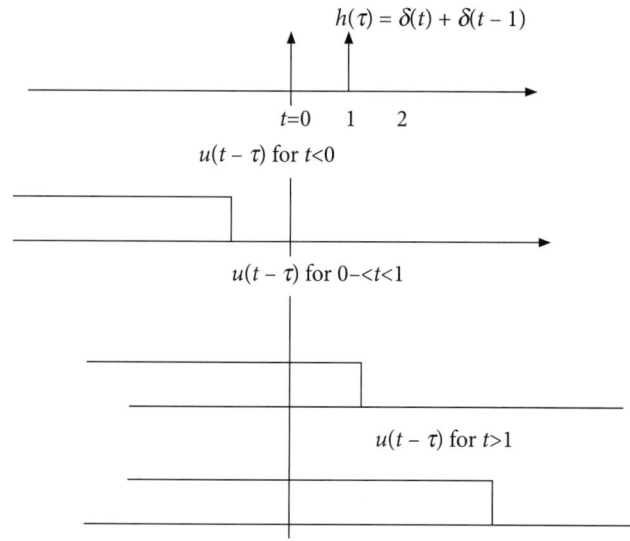

Fig. 4.41 Plot of $h(\tau)$, $u(\tau)$ and $u(t - \tau)$ for different intervals for Example 4.37

Consider the interval $-\infty < t < 1$. The convolution integral can be written as

$$h(t) * u(t) = [\delta(t) + \delta(t-1)] * u(t) = \int_{0-}^{t} \delta(t)d\tau = 1 \qquad (4.164)$$

For interval $1 + < t < \infty$, the convolution integral can be written as

$$h(t) * u(t) = [\delta(t) + \delta(t-1)] * u(t) = \int_{-\infty}^{1-} \delta(\tau)d\tau + \int_{1-}^{\infty} \delta(\tau-1)d\tau \qquad (4.165)$$

$$= 1 + 1 = 2$$

Example 4.38

Consider the interconnections of the systems, as shown in Fig. 4.42. Let the impulse responses be specified as

$$h_1[n] = u[n]$$

$$h_2[n] = u[n+3] - u[n]$$

$$h_3[n] = \delta[n-3]$$

$$h_4[n] = \left(\frac{1}{3}\right)^n u[n]$$

Find the response of the overall interconnection.

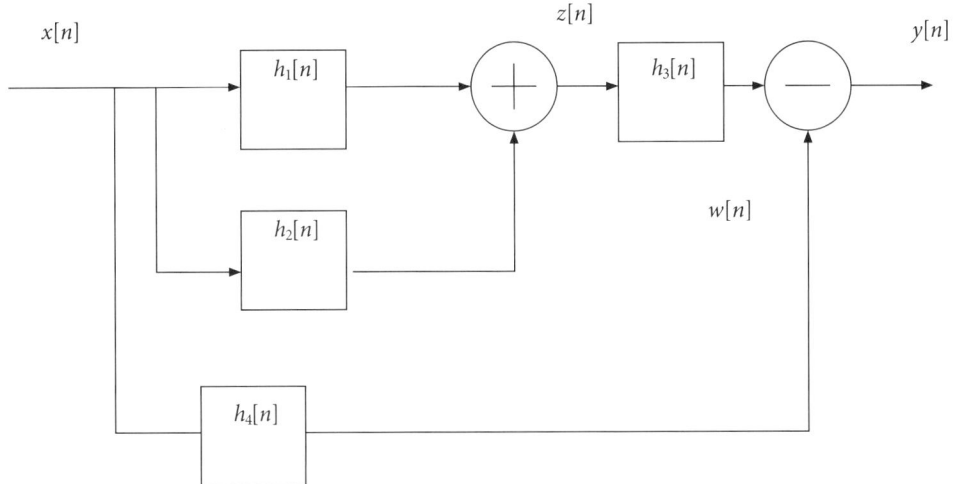

Fig. 4.42 Interconnection of systems for Example 4.38

Solution

The impulse response of the overall interconnection can be written as

$$\{(h_1[n]+h_2[n])*h_3[n]\}-h_4[n]$$

$$\{(u[n]+u[n+3]-u[n])*\delta[n-3]\}-\left(\frac{1}{3}\right)^n u[n]$$

$$=u[n+3]*\delta[n-3]-\left(\frac{1}{3}\right)^n u[n] \quad (4.166)$$

$$=u[n]-\left(\frac{1}{3}\right)^n u[n]$$

$$=[1-\left(\frac{1}{3}\right)^n]u[n]$$

Note that $u[n+3]*\delta[n-3]=u[n]$, as is clear from Fig. 4.43.

Example 4.39

Consider the interconnections of the systems, as shown in Fig. 4.44. Let the impulse responses be specified as

$$h_1[n]=\left(\frac{1}{2}\right)^n [u[n+2]-u[n-2]]$$

$$h_2[n]=\delta[n]$$

$$h_3[n]=u[n-1]$$

Find the response of the overall interconnection.

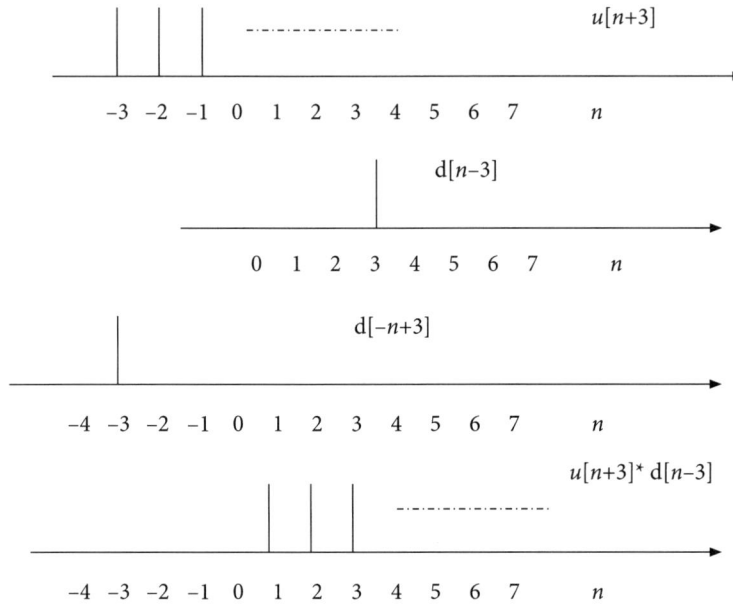

Fig. 4.43 Convolution of $u[n + 3]$ with $\delta[n - 3]$

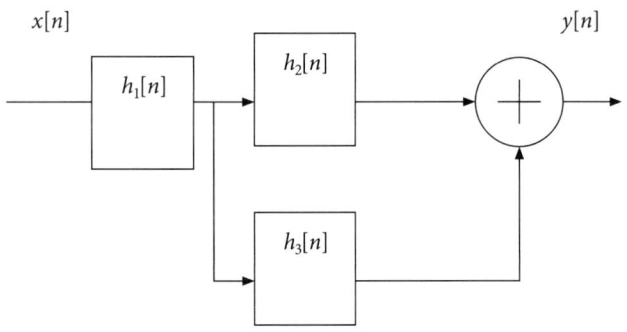

Fig. 4.44 Interconnection of systems for Example 4.39

Solution

The impulse response of the overall interconnection can be written as

$$h_1[n] * \{h_2[n] + h_3[n]\}$$

$$= \left(\frac{1}{2}\right)^n \{u[n+2] - u[n-2]\} * \{\delta[n] + u[n-1]\}$$

(4.167)

$$= \left(\frac{1}{2}\right)^n \{u[n+2] - u[n-2]\} * u[n]$$

Figure 4.45 shows a plot of $h_1[n]$, $u[-n]$ and output $y[n]$.

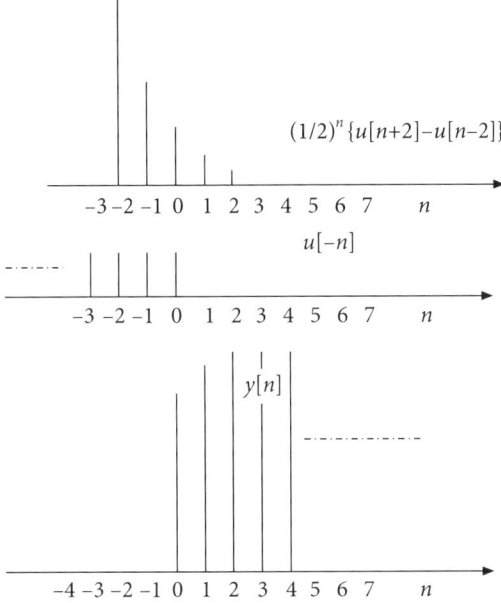

Fig. 4.45 Plot of $h_1[n]$, $u[-n]$ and output $y[n]$

The output $y[n]$ can be calculated as

$$y[0] = [4+2+1] = 7, \ y[1] = [4+2+1+1/2] = 7.5,$$
(4.168)
$$y[2] = [4+2+1+1/2+1/4] = 7.75, \ y[3] = 7.75, \ y[4] = 7.75, \ldots\ldots$$

Concept Check

- Can you find the impulse response for a series interconnection of systems?
- Can you find the impulse response of the parallel interconnection of systems?
- What is series configuration?
- What is parallel configuration?

Summary

We have introduced the concept of solution of the differential equation and difference equation.

- We have discussed the important properties of CT system equation namely the differential equation. We have explained the meaning of characteristic equation, which is the denominator polynomial. We write the solution of the characteristic equation as $y(t) = ce^{-\lambda t}$. The roots of the characteristic equation namely $e^{-\lambda_1 t}$, $e^{-\lambda_2 t}$ $e^{-\lambda_1 t}, e^{-\lambda_2 t}$ are the eigen values or characteristic modes or the natural modes of the system. The reason for selecting the exponential terms as solution is two-fold. If $(D^n + a_1 D^{n-1} + \ldots + a_n) = 0$, this indicates that all the n derivatives of the function are zero at all values of time. The nth derivative is $(-\lambda)^n e^{-\lambda t}$, if all n derivatives are to be zero, then this is true only for the exponential function. Secondly, in case of the LTI systems, if the input is exponential function, the output is also exponential function. No other function can have this claim. The exponential solutions called as eigen functions.

- We then discussed the concept of zero input response with initial conditions specified and showed that the response will contain the characteristic mode terms. When external input is zero, the response of the system with initial conditions applied is the zero input response. The zero state response was described as the response of the system to external input when the initial conditions are zero. Several examples are solved to illustrate the concepts. The evaluation of impulse response is described and explained with the help of a simple example. To find the impulse response, we have to apply impulse as an input to the system. Impulse response is also known as filter Kernel, convolution Kernel or simply Kernel. Practically, generating the impulse function with infinite amplitude and zero width is impossible. Ideal impulse function cannot be realised in practical Kernel. The impulse function is used as a standard input to the LTI systems for three reasons. Any signal can be decomposed into a linear combination of scaled and shifted impulses. Secondly, LTI system is completely characterised by its impulse response, and the output is obtained by adding the individual output responses. Numbers of numerical problems are solved for the calculation of convolution of system impulse response with the input signal. To find the zero state response, find the impulse response and convolve the input with the impulse response.

- We further concentrated on DT systems. DT systems are described by the difference equation. The concepts of zero input response and zero state response are explained for DT systems as well. Here, we have to find the characteristic mode terms for the difference equation. The roots of the characteristic equation are again assumed as exponential functions for the same reason. The calculation of zero input response and zero state response is explained with simple numerical examples. The procedure for convolution sum is explained with three different algorithms namely input-side algorithm, output-side algorithm and conventional method. Several examples are solved to further illustrate the process of convolution. The problems are also solved using MATLAB programs. The convolution outputs can be compared for hand calculations and using MATLAB.

- We considered the unit step response as the next section. It is illustrated using number of solved problems both for CT and DT systems. The properties of

CT and DT LTI systems like memory, causality and stability are explained related to the impulse response of the system. The condition for DT system to be memoryless is translated in terms of impulse response as $h[n] = c\delta[n]$ for DT systems and $h[t] = c\delta[t]$ for CT systems. The system is causal if h[n] is zero for all n < 0 and in case of CT system, the system is causal if $h(t) = 0$ for all $t < 0$. The stability condition is named as BIBO stability. DT or CT system is said to be stable if the impulse response is absolutely summable. Several problems are solved to illustrate the properties of LTI systems. The series and parallel interconnection of systems were explained in chapter 3. This chapter explains the calculation of effective impulse response if the systems are interconnected using series or parallel connection.

Multiple Choice Questions

1. Characteristic modes decide the system response when
    (a) external input is zero
    (b) no initial conditions are assumed
    (c) initial conditions are zero
    (d) impulse is applied as input

2. Zero input response refers to
    (a) Input response of the system
    (b) Response of the system when input is zero
    (c) Response of the system when initial conditions are zero.
    (d) Response of the system to any external input.

3. The roots of the characteristic polynomial are called as
    (a) eigen values   (b) polynomial roots
    (c) special roots   (d) solution for equation

4. Impulse function is used as a standard input to the LTI systems
    (a) because impulse function is ideal
    (b) because impulse function cannot be realized
    (c) because impulse function is easy to understand
    (d) because LTI system can be characterized by impulse response

5. To find the step response
    (a) convolve the impulse response with $u(n)$
    (b) convolve the impulse response with $\delta[n]$
    (c) multiply the input with the impulse response
    (d) multiply the impulse with system response

6. Input-side algorithm finds the contribution of each delta function in the input
   (a) to the output
   (b) to each output sample
   (c) to all output samples
   (d) to the respective output samples
7. Output-side algorithm finds out number of input signal components
   (a) contributing to each output sample
   (b) used to find the output samples
   (c) used for calculation of all output samples
   (d) contributing to next output sample
8. Convolution of 2 DT sequences of lengths P and Q results in convolved sequence of length
   (a) $P + Q$
   (b) $P + Q + 1$
   (c) $P + Q - 1$
   (d) $P - Q$
9. The system is causal when the impulse response $h(n)$ satisfies the condition
   (a) $h(n) = 0$ for all $n < 0$
   (b) $h(n) = 0$ for all n except $n = 0$
   (c) $h(n)$ exists for all $n < 0$
   (d) $h(n)$ is non-zero for all n
10. $u[n+3] * \delta[n-3] =$
    (a) $u[n + 3]$
    (b) $\delta[n + 3]$
    (c) $\delta[n]$
    (d) $u[n]$
11. The LTI system is stable if its impulse response is
    (a) absolutely summable
    (b) exists
    (c) is infinite
    (d) is not convergent
12. If the LTI system has impulse response given by $\delta[n]$, then
    (a) the system is with memory
    (b) the system is memoryless
    (c) the system is linear
    (d) the system may be homogeneous
13. Series interconnection of system results in
    (a) addition of the impulse responses
    (b) convolution of impulse responses
    (c) subtraction of impulse responses
    (d) multiplication of impulse responses
14. Parallel interconnection of systems results in
    (a) addition of the impulse responses
    (b) convolution of impulse responses

(c) subtraction of impulse responses

(d) multiplication of impulse responses

Review Questions

4.1 What is a characteristic equation? What are characteristic mode terms?

4.2 What are eigen values? Why the solutions of the characteristic equations are assumed as exponential functions?

4.3 What is a zero input response? Explain the procedure to find the zero input response for CT and DT systems.

4.4 What is a zero state response? How will you find the response to any input signal?

4.5 How will you calculate the impulse response of the system? How will you find the step response for the system?

4.6 Explain input-side algorithm for calculation of convolution of DT signals.

4.7 Explain output-side algorithm for calculation of convolution of DT signals.

4.8 Explain conventional algorithm for calculation of convolution of DT signals.

4.9 Explain the procedure to find the convolution integral.

4.10 Explain the property of memory in terms of impulse response of the CT and DT system.

4.11 Explain the property of causality in terms of impulse response of the CT and DT system.

4.12 Explain the property of stability in terms of impulse response of the CT and DT system.

4.13 Find the impulse response for a series interconnected and parallel interconnected systems. Prove that the impulse response of the series interconnection of two LTI systems is a convolution of the two impulse responses.

Problems

4.1 Consider a simple second order system with characteristic equation given by $(D^2 + 5D + 6)y(t) = 0$. Find the zero input response if the initial conditions are $y(0) = 0$ and $Dy(0) = 5$.

4.2 Determine the impulse response for a system given by the differential equation $(D^2 + 5D + 6)y(t) = Dx(t)$.

4.3 Determine the impulse response for a system given by the differential equation $D(D + 3)y(t) = (D + 2)x(t)$.

4.4 Determine the impulse response for a system given by the differential equation $(D + 3)y(t) = (D + 1)x(t)$.

4.5 Determine the impulse response for a system given by the differential equation $(D + 1)y(t) = x(t)$.

4.6 Let $x(t) = e^{-2t}[u(t) - u(t - 3)]$, $h(t) = e^{-t} u(t)$, find $x(t)*h(t)$.

4.7 Let $x(t) = 1 - t$ for $0 \le t \le 1$, $h(t) = e^{-t} u(t)$, find $x(t)*h(t)$.

4.8 Let $x(t) = 2$ for $1 \le t \le 2$, $h(t) = 1$ for $0 \le t \le 4$, find $x(t)*h(t)$.

4.9 Let $x(t) = 1$ for $0 \le t \le 2$, $h(t) = t$ for $0 \le t \le 3$, find $x(t)*h(t)$.

4.10 Consider a system given by $(D + 1)y(t) = x(t)$. The impulse response is calculated as $h(t) = (e^{-t} u(t))$. Let the external input be applied as $x(t) = (e^{-t} u(t))$. Find the response of the system for the applied input.

4.11 Consider the difference equation

$$y[k] - 0.4y[k-1] = f[k]$$

$$y[-1] = 8 \text{ and } f[k] = k^2$$

Find the solution using recursive procedure.

4.12 Find the zero input response for the system. Consider a simple second order system with characteristic equation given by

$$y[k+2] - 0.2y[k+1] - 0.15y[k] = 5f[k+2] \text{ with } y[-1] = 0 \text{ and } y[-2] = 25$$

$$f[k] = 2^{-k} u[k]$$

4.13 Consider a simple second order system with characteristic equation given by

$$y[k+2] - 0.2y[k+1] - 0.15y[k] = 5f[k+2] \text{ with } y[-1] = 0 \text{ and } y[-2] = 5$$

$$f[k] = 2^{-k} u[k]$$

Find the impulse response of the system.

4.14 Consider a simple second order system with impulse response and the input signal given by $h[k] = (0.2)^k u[k]$ and $f[k] = (0.3)^k u[k]$. Find the zero state response.

4.15 Use input-side algorithm for convolution to convolve the two sequences $x[n] = [1\ 2\ 1\ 1]$ and $h[n] = [1\ 2\ 2]$.

4.16 Use output-side algorithm to convolve the same two sequences. Use conventional method to convolve the two sequences.

4.17 LTI system has the impulse response given by $h[n] = u[n] - u[n-7]$. If the input is $x[n] = u[n-2] - u[n-4]$, find the output of the system.

4.18 LTI system has the impulse response given by $h[n] = (0.2)^n \{u[n] - u[n-2]\}$. If the input is $x[n] = (0.4)^n u[n]$, find the output of the system.

4.19 LTI system has the impulse response given by $h[n] = u[n - 2]$. If the input is $x[n] = u[n]$, find the output of the system.

4.20 LTI system has the impulse response given by $h[n] = 1$ for $n = 0, -1$ and if the input is $x[n] = \begin{cases} 1 \text{ for } n = 0,1 \\ 3 \text{ for } n = 2,3 \end{cases}$, find the output of the system.

4.21 Solve problems 4.17, 4.18, 4.19 and 4.20 using MATLAB programs.

4.22 Find the step response of the LTI system with impulse response given by $h(t) = (e)^{-t} u(t)$.

4.23 Find the step response of the LTI system with impulse response given by $h(t) = u(t) - u(t - 1)$.

4.24 Find the step response of the LTI system with impulse response given by $h(t) = \delta(t) - \delta(t - 3)$.

4.25 Find the step response of the LTI system with impulse response given by $h(t) = tu(t - 1)$.

4.26 Find the step response of the LTI system with impulse response given by $h[n] = \left(\dfrac{1}{3}\right)^n u[n]$.

4.27 Find the step response of the LTI system with impulse response given by $h[n] = \delta[n] - [n - 2]$.

4.28 Find the step response of the LTI system with impulse response given by $h[n] = (-1)^n [u[n + 1] - u[n - 1]]$.

4.29 Find the step response of the LTI system with impulse response given by $h[n] = u[n - 1]$.

4.30 Check if the following CT systems are memoryless, causal and stable.

$h(t) = e^{-t}$

$h(t) = e^{2t} u(t - 2)$

$h(t) = u(t + 4) - 2u(t - 2)$

$h(t) = 2\delta(t)$

$h(t) = \sin(5\pi t) u(t)$

4.31 Check if the following DT systems are memoryless, causal and stable.

$h[n] = 5^n u[-n + 1]$

$h[n] = e^{4n} u(n - 1)$

$h[n] = \cos\left(\dfrac{1}{4}\pi n\right)[u([n+1] - u[n - 3]]$

$h[n] = 2u[n] - 2u[n-1]$

$h[n] = \sin(7\pi n)u[n]$

$h[n] = \delta[n] + \cos(2\pi n)$

4.32 Consider the system represented by

$$y[n] = \sum_{k=-2}^{1} x[k]h[n-k]$$

Is the system causal? If not, explain why.

4.33 Find if the system with impulse response $h[n]$ is stable. The impulse response is given by

$$h[n] = \left(\frac{1}{2}\right)^n u[n]$$

4.34 Consider the interconnections of the systems, as shown in Fig. 1. Let the impulse responses be specified as

$h_1(t) = u(t)$

$h_2(t) = \delta(t)$

$h_3(t) = \delta(t-2)$

Find the response of the overall interconnection.

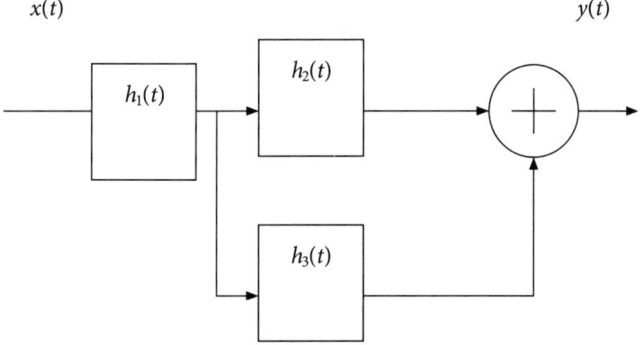

4.35 Consider the interconnections of the systems, as shown in Fig. 2. Let the impulse responses be specified as

$h_1[n] = u[n]$

$h_2[n] = u[n+2] - u[n]$

$h_3[n] = \delta[n-2]$

$$h_4[n] = \left(\frac{1}{2}\right)^n u[n]$$

Find the response of the overall interconnection.

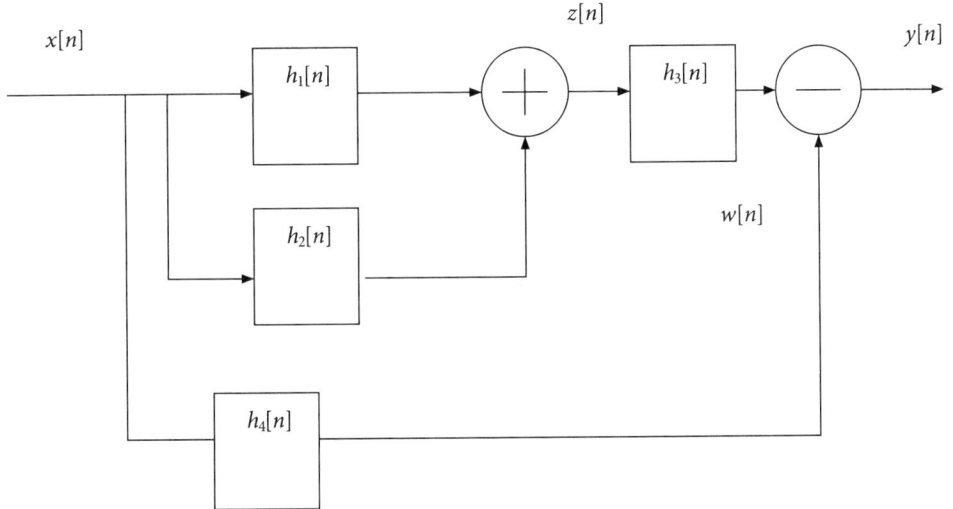

4.36 Consider the interconnections of the systems, as shown in Figure 3. Let the impulse responses be specified as

$$h_1[n] = \left(\frac{1}{2}\right)^n [u[n+1] - u[n-1]]$$

$$h_2[n] = \delta[n]$$

$$h_3[n] = u[n-1]$$

Find the response of the overall interconnection.

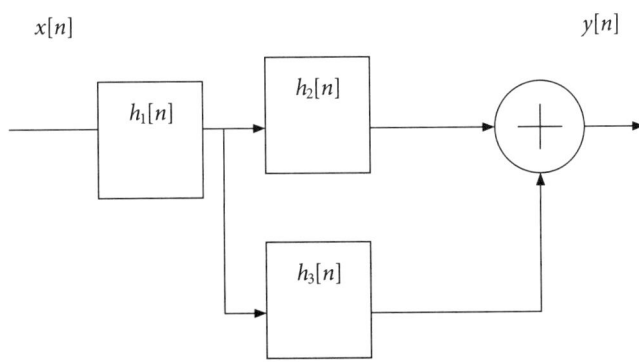

Answers

Multiple Choice Questions

1 (a)	2 (b)	3 (a)	4 (d)	5 (a)
6 (d)	7 (a)	8 (c)	9 (a)	10 (d)
11 (a)	12 (b)	13 (b)	14 (a)	

Problems

4.1 $y(t) = (5e^{-t} - 5e^{-2t})u(t)$

4.2 $h(t) = (-e^{-3t} + e^{-2t})u(t)$

4.3 $h(t) = \left(-\dfrac{1}{3}e^{-2t} + \dfrac{1}{3}\right)u(t)$

4.4 $h(t) = \left(-\dfrac{1}{3}e^{-2t}u(t) + \dfrac{1}{3}\delta(t)\right)$

4.5 $h(t) = (e^{-t}u(t))$

4.6 $y(t) = 0$ for $t < 0$
$\quad\quad = e^{-t} - e^{-3t}$ for $0 \leq t \leq 3$
$\quad\quad = e^{-t}(1 - e^{-3})$ for $t < 3$

4.7 $y(t) = 0$ for $t < 0$
$\quad\quad = (2 - t - 2e^{-t})$ for $0 \leq t \leq 1$
$\quad\quad = e^{-(t-1)} - 2e^{-t}$ for $t < 1$

4.8 $y(t) = 0$ for $t < 0$
$\quad\quad = 2(t - 2)$ for $1 \leq t \leq 2$
$\quad\quad = 2$ for $2 \leq t \leq 5$
$\quad\quad = 2(6 - t)$ for $5 \leq t \leq 6$
$\quad\quad = 0$ for $t > 6$

4.9 $y(t) = 0$ for $t < 0$
$\quad\quad = t^2 / 2)$ for $0 \leq t \leq 2$
$\quad\quad = 2t - 2$ for $2 \leq t \leq 3$

$= 2t - t^2/2 + 5/2$ for $3 \le t \le 5$

$= 0$ for $t > 5$

4.10 $y(t) = te^{-t}$

4.11 $y[0] = 3.2, y[1] = 2.28$

4.12 $y[k] = \dfrac{15}{32}(0.5)^k + \dfrac{9}{32}(-0.2)^k$

4.13 $h[k] = \left[\dfrac{35}{8}(0.6)^k + \dfrac{5}{8}(-0.2)^k\right]u[k]$

4.14 $y[k] = 10[(0.3)^{k+1} - (0.2)^{k+1}]u[k]$

4.15 $y[n] = [1\ 4\ 6\ 5\ 3\ 1]$

4.16 $y[n] = [1\ 4\ 6\ 5\ 3\ 1]$

4.17 $y[n] = \begin{cases} 0 \text{ for } n \le 1 \text{ and } n \ge 10 \\ 1 \text{ for } n = 2 \text{ and } n = 9 \\ 2 \text{ for } 3 \le n \le 8 \end{cases}$

4.18 $y[0] = 1, y[1] = 0.6,$
$y[n] = (0.2)(0.4)^{n-1} + (0.4)^n$ for $n \ge 2$

4.19 $y[n] = \begin{cases} 0 \text{ for } 0 \le n \le 2 \\ (n-1) \text{ for all } n \ge 2 \end{cases}$

4.20 $y[0] = 2, y[1] = 4, y[2] = 6, y[3] = 3, y[4] = 0, y[5] = 0\ldots$
$y[-1] = 1, y[-2] = 0, y[-3] = 0,\ldots$

we can write $y[n] = \begin{cases} 0 \text{ for } n \le -2 \text{ and } n \ge 4 \\ 1 \text{ for } n = -1 \\ 2 \text{ for } n = 0 \\ 4 \text{ for } n = 1 \\ 6 \text{ for } n = 2 \\ 3 \text{ for } n = 3 \end{cases}$

4.21 Same as 4.17, 4.18, 4.19, 4.20.

For Example 4.18, First 11 values displayed are

Columns 1 through 11

1.0000   0.6000   0.2400   0.0960   0.0384   0.0154   0.0061   0.0025   0.0010
0.0004   0.0002

4.22 $y(t) = (1 - e^{-t})u(t)$

4.23 $y(t) = \begin{cases} t \text{ for } 0 < t \le 1 \\ 1 \text{ for } t > 1 \end{cases}$

4.24 $y(t) = \begin{cases} 1 \text{ for } t < 3 \\ 0 \text{ for } t > 3 \end{cases}$

4.25 $y(t) = \left\{ \dfrac{t^2 - 1}{2} \text{ for } t > 1 \right\}$

4.26 $y[n] = \dfrac{3}{2}(1 - (1/3)^{n+1})$

4.27 $y[n] = \delta[n] + \delta[n-1]$

4.28 $y[n] = \delta[n] - 1$

4.29 $y[n] = n - 1$

4.30 i) ii) iii) with memory iv) memoryless v) with memory,
   iii) non-causal i) ii) iv) v) causal.
   i) iv) stable, ii) iii) v) non stable

4.31 i) ii) iii) v) vi) with memory, iv) memoryless,
   i) iii) vi) non-causal, ii) iv) v) causal,
   i) iii) iv) stable, ii) v) vi) unstable

4.32 The current output depends on next input sample. Hence, the system is non-causal.

4.33 Yes, stable.

4.34 $y[n] = 1$ for $-\infty < t < 2$
   $= 2$ for $2 < t < \infty$

4.35 $y[n] = \left[1 - \left(\dfrac{1}{2} \right)^n \right] u[n]$

4.36 $y[0] = [2 + 1] = 3, y[1] = [2 + 1 + 1/2] = 3.5,$
   $y[2] = 3.5, y[3] = 3.5, y[4] = 3.5, \ldots\ldots$

5

Fourier Series Representation of Periodic Signals

Learning Objectives

- Vector representation of signals
- Concept of orthogonality
- Orthogonal basis functions
- Trigonometric Fourier series expansion of periodic signals
- Exponential Fourier series representation of CT periodic signals (CTFS)
- Why exponential functions?
- Properties of FS
- Recovery of signal from FS
- Gibbs phenomenon
- FS representation of DT periodic signals

This chapter deals with the Fourier series representation of periodic CT signals. Chapter 4 dealt with the representation of signals in terms of impulses where the impulse function is treated as a basic unit. Any signal can be represented as a linear combination of scaled and shifted impulses. This chapter uses the decomposition of a signal in terms of sines and cosines or exponential signals and represents the signal as a vector with each coefficient of the vector representing the amount of similarity of the signal with the basic functions namely sines and cosines or exponential functions. This representation is termed as Fourier series representation after the scientist named Fourier who claimed that every signal can be represented as a linear combination of sinusoidal functions.

Firstly, we will study the representation of signals in terms of vectors, which makes it easy to understand the concept of orthogonality. The use of exponential series representation is elaborated. The trigonometric series representation and exponential representations are compared. We will then classify the signals as periodic and aperiodic. Periodic signals will be represented in terms of Fourier series representation. Aperiodic signals will be represented in terms of Fourier transform representation. Fourier series representation will be illustrated for CT and DT signals.

5.1 Signal Representation in Terms as Sinusoids

We have discussed the decomposition of any CT signal in terms of impulses and decomposition of any DT signal in terms of unit impulse function in chapter 4. Unit impulse is considered as a basic unit because the LTI systems can be characterized by impulse response. We can also decompose any signal in terms of sinusoids or in terms of complex exponentials. Here, the sinusoids or exponential functions are used as basic units. The decomposition of a signal as a linear combination of scaled sinusoids or complex exponentials is discussed further.

Why we select a sinusoid or complex exponentials?

The reason is when a sinusoid or complex exponential is given as input to any LTI system, the output is the same sinusoid or the same complex exponential except the change in amplitude and/or the change of phase. No other signal can have this claim. We will first prove that the sinusoids of different frequencies are orthogonal to each other and complex exponentials of different frequencies are also orthogonal to each other. The orthogonal set of functions forms the set of basis functions that are then used for unique representation of any CT signal.

5.1.1 Orthogonality property

We will first define orthogonality. The two signals are said to be orthogonal over the time interval $t = [t, t']$ if

$$\int_t^{t'} x(t)y^*(t)dt = \int_t^{t'} x^*(t)y(t)dt = 0 \tag{5.1}$$

Where * denotes the complex conjugate if the signal is complex. If the two signals also satisfy the condition that

$$\int_t^{t'} x(t)x^*(t)dt = \int_t^{t'} y^*(t)y(t)dt = 1 \tag{5.2}$$

They are said to be orthonormal.

Let us first prove that two sinusoidal functions or two cosine functions of different frequencies are orthogonal to each other. We will prove this with the help of simple numerical examples.

Example 5.1
Prove that $\cos(\omega t)$ and $\cos(2\omega t)$ are orthogonal to each other.

Solution
Let T represent the period of the cosine function with an angular frequency of ω. The period for a cosine function with an angular frequency of 3ω will be $T/3$ and so on. We have to prove that the dot product of the two functions is zero. We know the integration of cosine function over one period or multiple periods is zero.

$$\int_0^T \cos(\omega t)\cos(2\omega t)dt = \frac{1}{2}\int_0^T [\cos(3\omega t) + \cos(\omega t)]dt$$

$$= \frac{1}{2}\int_0^T \cos(3\omega t)dt + \frac{1}{2}\int_0^T \cos(\omega t)dt \qquad (5.3)$$

$$= \frac{1}{2}\left[\frac{1}{3\omega}\sin(3\omega t)\right]_0^T + \frac{1}{2}\left[\frac{1}{\omega}\sin(\omega t)\right]_0^T$$

$$= 0$$

Example 5.2
Prove that $\cos(2\omega t)$ and $\cos(3\omega t)$ are orthogonal to each other.

Solution
Let T represent the period of the cosine function with an angular frequency of ω. The period for a cosine function with an angular frequency of 3ω will be $T/3$ and the period for a cosine function with an angular frequency of 2ω will be $T/2$ so on. We have to prove that the dot product of the two functions is zero. We know that the integration of cosine function over one period or multiple periods is zero.

$$\int_0^T \cos(2\omega t)\cos(3\omega t)dt = \frac{1}{2}\int_0^T [\cos(5\omega t) + \cos(\omega t)]dt$$

$$= \frac{1}{2}\int_0^T \cos(5\omega t)dt + \frac{1}{2}\int_0^T \cos(\omega t)dt \qquad (5.4)$$

$$= \frac{1}{2}\left[\frac{1}{5\omega}\sin(5\omega t)\right]_0^T + \frac{1}{2}\left[\frac{1}{\omega}\sin(\omega t)\right]_0^T$$

$$= 0$$

Example 5.3
Prove that $\cos(\omega t)$ and $\cos(3\omega t)$ are orthogonal to each other.

Solution
Let T represent the period of the cosine function with an angular frequency of ω. The period for a cosine function with an angular frequency of 3ω will be $T/3$

and the period for a cosine function with an angular frequency of 2ω will be $T/2$ so on. We have to prove that the dot product of the two functions is zero. We know the integration of cosine function over one period or multiple periods is zero.

$$\int_0^T \cos(\omega t)\cos(3\omega t)dt = \frac{1}{2}\int_0^T [\cos(4\omega t) + \cos(2\omega t)]dt$$

$$= \frac{1}{2}\int_0^T \cos(4\omega t)dt + \frac{1}{2}\int_0^T \cos(2\omega t)dt \quad (5.5)$$

$$= \frac{1}{2}\left[\frac{1}{4\omega}\sin(4\omega t)\right]_0^T + \frac{1}{2}\left[\frac{1}{2\omega}\sin(2\omega t)\right]_0^T$$

$$= 0$$

Example 5.4

Prove that $\sin(\omega t)$ and $\sin(2\omega t)$ are orthogonal to each other.

Solution

Let T represent the period of the sine function with an angular frequency of ω. The period for a sine function with an angular frequency of 3ω will be $T/3$ and the period for a sine function with an angular frequency of 2ω will be $T/2$, and so on. We have to prove that the dot product of the two functions is zero. We know that the integration of cosine function over one period or multiple periods is zero.

$$\int_0^T \sin(\omega t)\sin(2\omega t)dt = \frac{1}{2}\int_0^T [\cos(\omega t) - \cos(3\omega t)]dt$$

$$= \frac{1}{2}\int_0^T \cos(\omega t)dt - \frac{1}{2}\int_0^T \cos(3\omega t)dt \quad (5.6)$$

$$= \frac{1}{2}\left[\frac{1}{\omega}\sin(\omega t)\right]_0^T - \frac{1}{2}\left[\frac{1}{3\omega}\sin(3\omega t)\right]_0^T$$

$$= 0$$

Example 5.5

Prove that $\sin(\omega t)$ and $\cos(2\omega t)$ are orthogonal to each other.

Solution

Let T represent the period of the sine function with an angular frequency of ω. The period for a sine function with an angular frequency of 3ω will be $T/3$ and the period for a sine function with an angular frequency of 2ω will be $T/2$ so on. We have to prove that the dot product of the two functions is zero. We know the integration of sine function over one period or multiple periods is zero.

$$\int_0^T \sin(\omega t)\cos(2\omega t)dt = \frac{1}{2}\int_0^T [\sin(\omega t) + \sin(3\omega t)]dt$$

$$= \frac{1}{2}\int_0^T \sin(\omega t)dt + \frac{1}{2}\int_0^T \sin(3\omega t)dt \tag{5.7}$$

$$= \frac{1}{2}\left[\frac{1}{\omega}\cos(\omega t)\right]_0^T + \frac{1}{2}\left[\frac{1}{3\omega}\cos(3\omega t)\right]_0^T$$

$$= 0$$

We can conclude that $\sin(\omega t)$ and $\sin(2\omega t)$ are orthogonal. Sine functions of all frequencies that are multiples of ω i.e., 3ω, 4ω ... are all orthogonal to each other. Similarly, $\cos(\omega t)$ and $\cos(2\omega t)$ are orthogonal. Cosine functions of all frequencies that are multiples of ω i.e., 3ω, 4ω ... are all orthogonal to each other. Also, $\cos(\omega t)$ and $\sin(2\omega t)$ are orthogonal. Cosine functions of all frequencies that are multiples of ω i.e., 3ω, 4ω ... are all orthogonal to each other and orthogonal to all sine functions of all frequencies that are multiples of ω i.e., 3ω, 4ω ......

Example 5.6

We can now show that the set consisting of {1, $\sin(m\omega t)$, $\sin(n\omega t)$, $\cos(m\omega t)$, $\cos(n\omega t)$ for positive integer values of m and n forms an orthogonal set over one period T.

Solution

We will decompose the problem in 4 parts. In part 1, we will show that

$$\int_0^T \cos(m\omega t)\cos(n\omega t)dt = \begin{cases} 0 \text{ for } m \neq n \\ \dfrac{T}{2} \text{ for } m = n \end{cases} \tag{5.8}$$

$$\int_0^T \cos(m\omega t)\cos(n\omega t)dt = \frac{1}{2}\int_0^T [\cos(m-n)\omega t + \cos(m+n)\omega t]dt$$

$$= \begin{cases} \dfrac{\sin(m-n)\omega t}{2(m-n)\omega} + \dfrac{\sin(m+n)\omega t}{2(m+n)\omega} \text{ for } m \neq n \\ \dfrac{t}{2} + \dfrac{\sin(2m\omega t)}{4m\omega} \text{ for } m = n \end{cases}_0^T = \begin{cases} 0 \text{ for } m \neq n \\ \dfrac{T}{2} \text{ for } m = n \end{cases} \tag{5.9}$$

In part 2, we will prove that

$$\int_0^T \sin(m\omega t)\sin(n\omega t)dt = \begin{cases} 0 \text{ for } m \neq n \\ \dfrac{T}{2} \text{ for } m = n \end{cases} \tag{5.10}$$

$$\int_0^T \sin(m\omega t)\sin(n\omega t)dt = \frac{1}{2}\int_0^T [\cos(m-n)\omega t - \cos(m+n)\omega t]dt$$

$$\left\{\begin{array}{l}\dfrac{\sin(m-n)\omega t}{2(m-n)\omega} - \dfrac{\sin(m+n)\omega t}{2(m+n)\omega} \text{ for } m \neq n \\ \dfrac{t}{2} - \dfrac{\sin(2m\omega t)}{4m\omega} \text{ for } m = n\end{array}\right\}_0^T = \left\{\begin{array}{l}0 \text{ for } m \neq n \\ \dfrac{T}{2} \text{ for } m = n\end{array}\right\} \quad (5.11)$$

In part 3, we will prove that

$$\int_0^T \sin(m\omega t)\cos(n\omega t)dt = \left\{\begin{array}{l}0 \text{ for } m \neq n \\ \dfrac{T}{2} \text{ for } m = n\end{array}\right\} \quad (5.12)$$

$$\int_0^T \sin(m\omega t)\cos(n\omega t)dt = \frac{1}{2}\int_0^T [\sin(m-n)\omega t + \sin(m+n)\omega t]dt$$

$$\left\{\begin{array}{l}-\dfrac{\cos(m-n)\omega t}{2(m-n)\omega} - \dfrac{\cos(m+n)\omega t}{2(m+n)\omega} \text{ for } m \neq n \\ \dfrac{t}{2} - \dfrac{\cos(2m\omega t)}{4m\omega} \text{ for } m = n\end{array}\right\}_0^T = \left\{\begin{array}{l}0 \text{ for } m \neq n \\ \dfrac{T}{2} \text{ for } m = n\end{array}\right\} \quad (5.13)$$

In part 4, we will prove that

$$\int_0^T 1 \times \cos(n\omega t)dt = 0 \text{ for all } n \text{ and } \int_0^T 1 \times \sin(m\omega t)dt = 0 \text{ for all } m$$

$$\int_0^T 1 \times \cos(n\omega t)dt = \left[\frac{1}{n\omega}\sin(n\omega t)\right]_0^T = 0 \text{ for all } n \text{ and} \quad (5.14)$$

$$\int_0^T 1 \times \sin(m\omega t)dt = \left[-\frac{1}{m\omega}\cos(m\omega t)\right]_0^T = 0 \text{ for all } m$$

We can also prove that the complex exponentials of frequencies that are multiples of ω i.e., $2\omega, 3\omega, 4\omega \ldots$ are all orthogonal to each other.

Example 5.7

Prove that $\exp(jm\omega t)$ and $\exp(jn\omega t)$ are orthogonal to each other.

Solution

Let T represent the period of the exponential function with an angular frequency of ω. The period for a exponential function with an angular frequency of 3ω will be $T/3$ and so on. We have to prove that the dot product of the two functions is zero. We know the integration of cosine function over one period or multiple periods is zero.

$$\int_0^T \exp(jm\omega t)\exp^*(jn\omega t)dt = \int_0^T [\exp(j(m-n)\omega t)]dt$$

$$= \begin{cases} T \text{ for } m=n \\ \exp(j(m-n)\omega t)/j(m-n)\omega \text{ for } m \neq n \end{cases}\Big|_0^T$$

$$= \begin{cases} T \text{ for } m=n \\ [\cos(m-n)\omega t - j\sin(m-n)\omega t]/j(m-n)\omega \end{cases}\Big|_0^T \quad (5.15)$$

$$= \begin{cases} T \text{ for } m=n \\ 0 \text{ for } m \neq n \end{cases}$$

We have proved that the set of sinusoids and complex exponentials form two sets of orthogonal functions or what are called as basis functions. Let us try to understand the definition of basis functions.

5.1.2 Basis functions

Basis functions are orthogonal functions that are used to find the similarity of the signal with. When they are combined in a linear manner it will result in a signal. In mathematics, particularly numerical analysis, a basis function is an element of the basis for a function space. It can also be defined as a set of functions that can be linearly combined to form a more general set of functions. We will consider a simple example to illustrate the concept. Consider representation of point in a space. We use orthogonal basis as x axis and y axis. These orthogonal basis functions can represent any point in a plane with unique coordinates x and y. The point P is represented as doublet (x, y) where x represents the projection of a vector joining a point to origin and y represents the projection of the vector on y axis, as shown in Fig. 5.1. Here, we try to represent a point P using two orthogonal basis functions namely x and y coordinates as a vector given by $X\bar{i} + Y\bar{j}$. The two axis namely x-axis and y-axis are the elements of the basis for a two dimensional space. The point P is represented as a linear combination of the basis functions namely unit vector along x-axis and unit vector along y-axis. The x-axis and y-axis are orthogonal i.e., perpendicular to each other namely their dot product is zero.

There is a great analogy between signals and vectors. We know that we can express any vector as a linear combination of two weighted unit vectors one in x and other in y direction for a planner representation and as a linear combination of three weighted unit vectors in case of a 3-D representation, we express any vector in 3-D space as a linear combination of 3 weighted unit

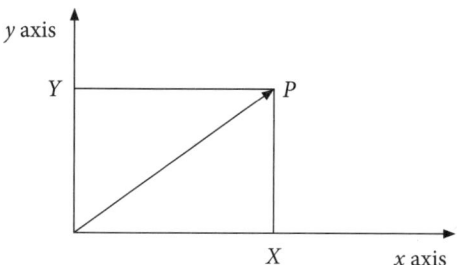

Fig. 5.1 Representation of point in space

vectors namely the unit vectors in *x*, *y* and *z* directions. Similarly, we can express any signal as a linear combination of weighted complex exponential functions that form a complete orthogonal set of basis functions. We will consider signals as vectors. We will represent any signal as a linear combination of these basis functions namely sinusoids and co-sinusoids. The alternative representation of the signal is in terms of the complex sinusoids i.e., complex exponential functions. Consider any CT signal $x(t)$. We will express the signal as a weighted sum of complex exponential functions given by

$$x(t) = \sum_{k=1}^{M} a_k e^{j\omega_k t} \tag{5.16}$$

Let us illustrate the concepts with a simple numerical example.

Example 5.8

Consider any LTI system and apply the complex exponential signal as input to any LTI system. Let the LTI system have a real-valued impulse response $h(t)$. Prove that the output of the system is complex exponential again with a change of magnitude and phase.

Solution

Let, the output be represented as $y(t)$ given by

$$y(t) = \int_{-\infty}^{\infty} h(\tau) x(t-\tau) d\tau = a_k e^{j\omega_k t} \int_{-\infty}^{\infty} h(\tau) e^{-j\omega_k \tau} d\tau$$

We define $H(\omega) = \int_{-\infty}^{\infty} h(\tau) e^{-j\omega_k \tau} d\tau$ is a complex-valued constant

$$y(t) = a_k e^{j\omega_k t} H(\omega) \tag{5.17}$$

We can express

$$H(\omega) = A e^{j\varphi} \tag{5.18}$$

$$y(t) = Aa_k e^{j(\omega_k t + \varphi)} \quad (5.19)$$

The expression for $y(t)$ proves that LTI system only changes the amplitude and phase of the exponential function. Similarly, it can be easily proved that when a sinusoid is given as input to the LTI system, the output is again the same sinusoid with change of magnitude and phase. Let us consider one numerical example to illustrate the concepts.

Example 5.9

Consider LTI system with real-valued impulse response given by $h(t) = e^{-2t}u(t)$. Let the input to the system be $x(t) = \sin(2t)$. Find the output of the system.

Solution

Let us evaluate the output.

$$y(t) = \int_0^\infty h(\tau)x(t-\tau)dt = \int_0^\infty e^{-2\tau} \operatorname{Im} g[e^{-2j(t-\tau)}]d\tau$$

$$y(t) = \operatorname{Im} g[e^{-2jt}]\int_0^\infty e^{-2\tau} \operatorname{Im} g[e^{2j\tau}]d\tau$$

$$\text{Define } H(\omega) = \int_0^\infty e^{-2\tau} \operatorname{Im} g[e^{2j\tau}]d\tau = Ae^{j\vartheta} \quad (5.20)$$

$$y(t) = \operatorname{Im} g[e^{-2jt}]H(\omega)$$

$$y(t) = A \operatorname{Im} g[e^{-2jt - j\vartheta}]$$

$$y(t) = A \sin(2t + \vartheta)$$

The output is again a sine function with a change of phase.

Concept Check

- Why the sine wave or exponential function is used for decomposition of the signal?
- What is orthogonality?
- What are basis functions?
- Are $\sin(3\omega t)$ & $\cos(4\omega t)$ orthogonal to each other?

5.2 FS Representation of Periodic CT Signals

Any periodic signal can be represented in frequency domain as a Fourier series. **Let us first understand why Fourier series?** Fourier series representation is a representation of a signal as a sum of scaled fundamental and its scaled harmonics. When we say that a signal is periodic with a period of T or a fundamental frequency of '$f = 1/T$', the signal contains fundamental frequency component 'f' and its harmonics like '$2f$', '$3f$', etc. or it may contain only the harmonics of some fundamental frequency with fundamental frequency being absent. We have to find out the amplitude of the fundamental and amplitudes of the harmonic frequencies present in the signal for writing a Fourier series for a signal. These amplitudes will then be called as the Fourier series coefficients. The Fourier series coefficients will represent the degree of matching of the signal with the corresponding harmonic component. In short, Fourier series expansion of a signal $x_p(t)$ is equivalent to resolving a signal into sum of sine and cosine terms. This expansion can be written as

$$x_p(t) = a_0 + \sum_{n=1}^{\infty} \left[a_n \cos\left(\frac{2\pi nt}{T}\right) + b_n \sin\left(\frac{2\pi nt}{T}\right) \right] \tag{5.21}$$

Where a_n = amplitude of cos term for n^{th} harmonic and b_n = amplitude of sin term n^{th} harmonic with T - period

The terms $\cos\left(\frac{2\pi nt}{T}\right)$ and $\sin\left(\frac{2\pi nt}{T}\right)$ are called as basis functions and as we have understood, they are orthogonal. We have proved in section 5.1.1 that each of the cosine and sine term for all harmonics is orthogonal to every other term i.e., their dot product is zero. The periodic signals usually encountered in practice are found to satisfy Dirichlet's conditions necessary for Fourier series representation to hold good. Dirichlet's conditions can be stated as follows:

1. The function $x_p(t)$ is single valued within the period.
2. It has at the most finite number of discontinuities.
3. It has finite number of maxima and minima.
4. The function is absolutely integrable.

Fourier series representation is used for periodic signals due to following reasons:
1. Fourier series decomposes the signal into linear combination of scaled sines and cosines.
2. Sines and cosines are orthogonal functions and hence the decomposition is unique.
3. The periodic functions encountered in practice satisfy the Dirichlet's conditions.

5.2.1 Evaluation of fourier coefficients of trigonometric FS

The coefficients in Eq. (5.21) namely a_0, a_n, b_n are termed as Fourier coefficients of the trigonometric Fourier Series representation. Let us see how to evaluate these coefficients. To find value of a_0, we have integrate both sides of Eq. (5.14) over one period of $x(t)$. Integrating we get

$$\int_t^{t+T} x_p(t)dt = a_0 \int_t^{t+T} dt + \sum_{n=1}^{\infty} a_n \left[a_n \int_t^{t+T} \cos\left(\frac{2\pi nt}{T}\right) dt + b_n \int_t^{t+T} \sin\left(\frac{2\pi nt}{T}\right) dt \right] \quad (5.22)$$

Integration of the sine and cosine terms over one period is always zero as the total area under sine or cosine function for one period is zero. Hence, we can simplify the Eq. (5.22) as

$$\int_t^{t+T} x_p(t)dt = a_0 \int_t^{t+T} dt = a_0 T$$

$$a_0 = \frac{1}{T} \int_t^{t+T} x_p(t)dt \quad (5.23)$$

To evaluate a_n, b_n, we have to use the results stated in Eqs (5.8), (5.10) and (5.12). Let us multiply Eq. (5.21) by $\cos\left(\frac{2\pi mt}{T}\right)$ or $\cos(m\omega t)$ and integrate over one period.

$$\int_t^{t+T} x_p(t)\cos(n\omega t)dt = a_0 \int_t^{t+T} \cos(n\omega t)dt + \sum_{n=1}^{\infty} [a_n \int_t^{t+T} \cos(n\omega t)\cos(m\omega t)dt \\ + b_n \int_t^{t+T} \sin(n\omega t)\cos(m\omega t)dt] \quad (5.24)$$

Using the result of Eqs (5.8), (5.10) and (5.12), we can write Eq. (5.24) as

$$\int_t^{t+T} x_p(t)\cos(n\omega t)dt = a_n T/2$$

$$a_n = \frac{2}{T} \int_t^{t+T} x_p(t)\cos(n\omega t)dt \quad (5.25)$$

Let us now multiply Eq. (5.21) by $\sin\left(\frac{2\pi mt}{T}\right)$ or $\sin(m\omega t)$ and integrate over one period.

$$\int_t^{t+T} x_p(t)\sin(n\omega t)dt = a_0 \int_t^{t+T} \sin(n\omega t)dt + \sum_{n=1}^{\infty} [a_n \int_t^{t+T} \cos(n\omega t)\sin(m\omega t)dt \\ + b_n \int_t^{t+T} \sin(n\omega t)\sin(m\omega t)dt] \quad (5.26)$$

Using the result of Eqs (5.8), (5.10) and (5.12), we can write Eq. (5.26) as

$$\int_{t}^{t+T} x_p(t)\sin(n\omega t)dt = b_n T/2$$

$$b_n = \frac{2}{T}\int_{t}^{t+T} x_p(t)\sin(n\omega t)dt$$

(5.27)

5.2.2 Exponential FS representation of periodic CT signals

Generally instead of writing Fourier series expansion in terms of cosine and sine terms, the expansion is written in terms of exponential functions using the trigonometric identities namely

$$\cos\left(\frac{2\pi nt}{T}\right) = \frac{1}{2}\left[\exp\left(\frac{j2\pi nt}{T}\right) + \exp\left(\frac{-j2\pi nt}{T}\right)\right]$$

(5.28)

$$\sin\left(\frac{2\pi nt}{T}\right) = \frac{1}{2j}\left[\exp\left(\frac{j2\pi nt}{T}\right) - \exp\left(\frac{-j2\pi nt}{T}\right)\right]$$

(5.29)

The representation of the signal in terms of exponential series can be written as

$$x_P(t) = \left\{a_0 + \frac{1}{2}\sum_{n=1}^{\infty}\left[(a_n - jb_n)\exp\left(\frac{j2\pi nt}{T}\right) + (a_n + jb_n)\exp\left(\frac{-j2\pi nt}{T}\right)\right]\right\}$$

(5.30)

Let us define

$$c_n = \begin{cases} \frac{1}{2}[a_n - jb_n] & n > 0 \\ a_0 & n = 0 \\ \frac{1}{2}[a_n + jb_n] & n < 0 \end{cases}$$

(5.31)

The signal can now be written as

$$x_P(t) = \sum_{n=-\infty}^{\infty} c_n \exp\left(\frac{j2\pi nt}{T}\right)$$

(5.32)

The exponential series coefficients can be calculated using the formula

$$c_n = \frac{1}{T}\int_{-T/2}^{T/2} x_P(t)\exp\left(\frac{-j2\pi nt}{T}\right)dt \quad n = 0, \pm 1, \pm 2......$$

(5.33)

To find the 0th coefficient, we will put n=0 in equation 5.33 to get

$$c_0 = \frac{1}{T} \int_{-T/2}^{T/2} x_p(t)dt \qquad (5.34)$$

The exponential representation is easy to handle. It also simplifies the mathematical calculations. The presence of negative frequencies appearing in exponential series representation is a by-product. It has no physical significance. Conventionally we write Fourier series using exponential terms with positive as well as negative frequencies and the graph representing frequency domain representation, which is a graph of amplitude or magnitude with respect to frequency, has both positive and negative frequencies represented. The representation of negative frequency side is just the replica or mirror image of the representation of positive frequency side.

Thus the frequency domain representation of a periodic analog signal in terms of exponential Fourier series is a spectrum consisting of discrete components corresponding to fundamental and harmonic terms. It is a discrete spectrum.

Fourier for the first time presented a paper on the decomposition of any signal using complex exponentials. Any periodic signal can be decomposed as a linear combination of weighted exponentials or weighted sine and cosine functions. Consider any periodic signal $x(t)$. We can write the FS representation as

$$x(t) = \sum_{k=-\infty}^{\infty} X[k]e^{jk\omega_0 t} \qquad (5.35)$$

Where $x[k]$ are FS coefficients. Here, $\omega_0 = 2\pi/T$, $n = k$ and $x[k] = c_n$. $x[k]$ can be evaluated from $x(t)$ as

$$X[k] = \frac{1}{T} \int_{(T)} x(t)e^{-jk\omega_0 t} dt$$

$$X[0] = \frac{1}{T} \int_{(T)} x(t)dt \qquad (5.36)$$

The fundamental period of $x(t)$ is represented as T and $\omega_0 = 2\pi/T$. We can say that $x(t)$ and $x[k]$ are FS pairs. The FS representation is also called as frequency domain representation. Each FS coefficient $x[k]$ is associated with each complex sinusoid of a different angular frequency $k\omega_0$. We will solve some numerical examples to illustrate the evaluation of FS coefficients.

Example 5.10

Determine the FS representation for the signal given as $x(t) = 5\cos\left(\frac{\pi}{3}t + \frac{\pi}{6}\right)$.

Solution

Let us first determine the fundamental period of $x(t)$.

$$\omega_0 = \frac{\pi}{3} = \frac{2\pi}{6} = \frac{2\pi}{T} \text{ Comparing we can find } T = 6. \tag{5.37}$$

Let us write $x(t)$ as a linear sum of weighted exponentials.

$$x(t) = \sum_{k=-\infty}^{\infty} X[k] e^{jk\left(\frac{\pi}{3}\right)t} \tag{5.38}$$

$x(t)$ is already given in terms of a cosine function. So, let us write it in terms of exponentials and pull the coefficients.

$$x(t) = 5\cos\left(\frac{\pi}{3}t + \frac{\pi}{6}\right)$$

$$= 5 \frac{e^{j\left(\frac{\pi}{3}\right)t + j\frac{\pi}{6}} + e^{-j\left(\frac{\pi}{3}\right)t - j\frac{\pi}{6}}}{2} \tag{5.39}$$

$$= \frac{5}{2} e^{j\frac{\pi}{6}} e^{j\left(\frac{\pi}{3}\right)t} + \frac{5}{2} e^{-j\frac{\pi}{6}} e^{-j\left(\frac{\pi}{3}\right)t}$$

Referring to Eq. (5.15), we can identify that $k = 0$ will give the constant term and $k = 1$ will be the coefficient of the fundamental frequency.

$$X[k] = \begin{cases} \frac{5}{2} e^{-j\left(\frac{\pi}{6}\right)} & \text{for } k = -1 \\ \frac{5}{2} e^{j\left(\frac{\pi}{6}\right)} & \text{for } k = 1 \\ 0 & \text{otherwise} \end{cases} \tag{5.40}$$

Let us plot the magnitude and phase of $x[k]$. It is shown in Fig. 5.2.

Example 5.11

Determine the FS representation for the signal given as $x(t) = 2\sin(2\pi t - 3) + \sin(6\pi t)$.

Solution

Let us first determine the fundamental period of $x(t)$.

For the first term, $\omega_0 = \frac{2\pi}{1} = \frac{2\pi}{T}$. After comparing we can find that $T = 1$.

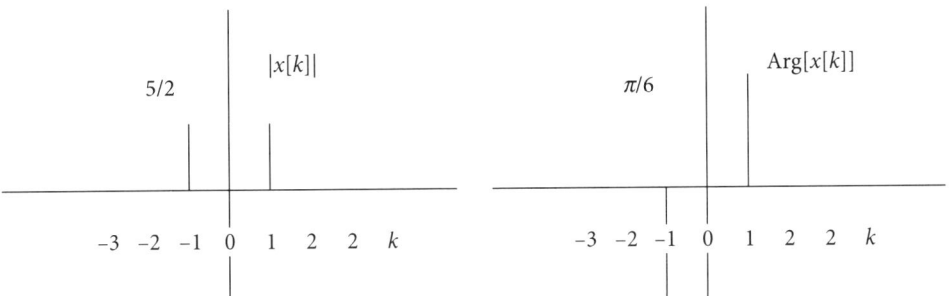

Fig. 5.2 Magnitude and phase plot of $x[k]$

For the second term, $\omega_0 = \dfrac{6\pi}{1} = \dfrac{2\pi}{1/3} = \dfrac{2\pi}{T}$. After comparing we can find that $T = 1/3$.

Hence, the fundamental period is the larger of the two and is equal to 1. Let us write $x(t)$ as a linear sum of weighted exponentials.

$$x(t) = \sum_{k=-\infty}^{\infty} X[k] e^{jk(2\pi)t} \qquad (5.41)$$

is already given in terms of a cosine function. So, let us write it in terms of exponentials and pull the coefficients.

$$x(t) = 2\sin(2\pi t - 3) + \sin(6\pi t)$$

$$= 2\dfrac{1}{2j}\left[e^{j(2\pi t - 3)} - e^{-j(2\pi t - 3)}\right] + \dfrac{1}{2j}\left[e^{j6\pi t} - e^{-j6\pi t}\right] \qquad (5.42)$$

$$= \dfrac{1}{j}\left[e^{-3j}e^{j2\pi t} - e^{3j}e^{-j2\pi t}\right] + \dfrac{1}{2j}\left[e^{j6\pi t} - e^{-j6\pi t}\right]$$

Referring to Eq. (5.15), we can identify that $k = 0$ will give the constant term and $k = 1$ will be the coefficient of the fundamental frequency and $k = 3$ will be the coefficient of third harmonic.

$$X[k] = \begin{cases} \dfrac{1}{j}e^{-3j} = -je^{-3j} & \text{for } k = 1 \\ -\dfrac{1}{j}e^{3j} = je^{3j} & \text{for } k = -1 \\ \dfrac{1}{2j} = \dfrac{-j}{2} & \text{for } k = 3 \\ -\dfrac{1}{2j} = \dfrac{j}{2} & \text{for } k = -3 \\ 0 & \text{otherise} \end{cases} \qquad (5.43)$$

Let us plot the magnitude and phase of x[k]. It is shown in Fig. 5.3.

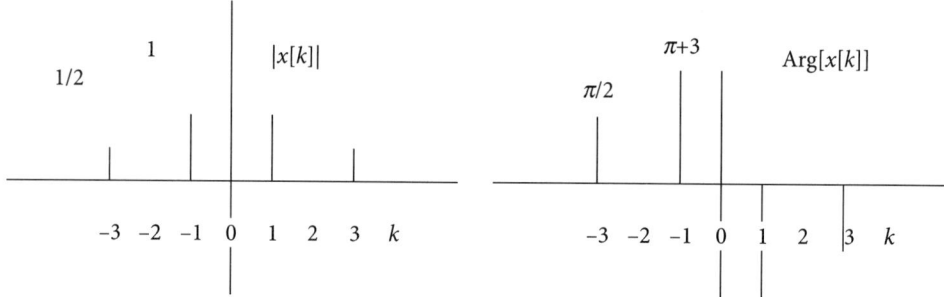

Fig. 5.3 Magnitude and phase plot of x[k]

Example 5.12

Determine the FS representation for the signal given as $x(t) = 2\sin(4\pi t) + \cos(3\pi t)$.

Solution

Let us first determine the fundamental period of $x(t)$.

For the first term, $\omega_0 = \dfrac{2\pi}{1/2} = \dfrac{2\pi}{T}$. After comparing we can find that $T=1/2$.

For the second term, $\omega_0 = \dfrac{3\pi}{1} = \dfrac{2\pi}{2/3} = \dfrac{2\pi}{T}$. After comparing we can find that $T=2/3$.

Hence, the fundamental period is the larger of the two and is equal to $T = 1/2$ and fundamental frequency is 2 Hz. The first term has a frequency of 2, which is the 4th harmonic of ½, and the second term has a frequency of 3/2, which is the third harmonic of ½.

Let us write $x(t)$ as a linear sum of weighted exponentials.

$$x(t) = \sum_{k=-\infty}^{\infty} X[k] e^{jk(\pi)t} \qquad (5.44)$$

is already given in terms of a cosine function. So, let us write it in terms of exponentials and pull the coefficients.

$$x(t) = 2\sin(4\pi t) + \cos(3\pi t)$$

$$= 2\dfrac{1}{2j}\left[e^{j4\pi t} - e^{-j(4\pi t)}\right] + \dfrac{1}{2}\left[e^{j3\pi t} + e^{-j3\pi t}\right] \qquad (5.45)$$

$$= \dfrac{1}{j}\left[e^{j4\pi t} - e^{-j4\pi t}\right] + \dfrac{1}{2}\left[e^{j3\pi t} + e^{-j3\pi t}\right]$$

$$X[k] = \begin{cases} \dfrac{1}{j} = -j & \text{for } k = 4 \\ -\dfrac{1}{j} = j & \text{for } k = -4 \\ \dfrac{1}{2} & \text{for } k = 3 \\ \dfrac{1}{2} & \text{for } k = -3 \\ 0 & \text{otherwise} \end{cases} \qquad (5.46)$$

Let us plot the magnitude and phase of $x[k]$. It is shown in Fig. 5.4.

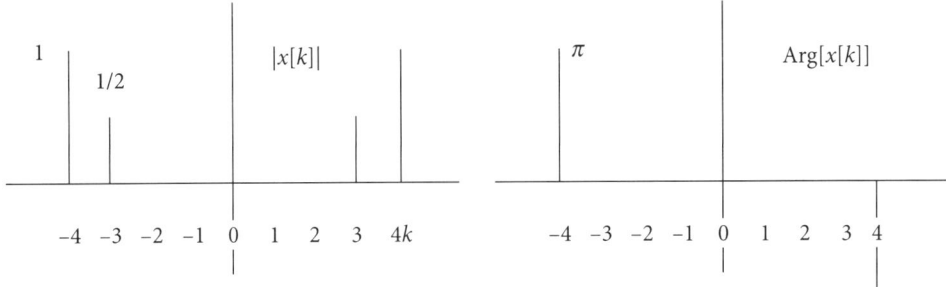

Fig. 5.4 Magnitude and phase plot of $x[k]$

Concept Check

- Why is FS representation used for periodic signals?
- State Dirichlet's conditions.
- Write the formula to find the DC component in the signal.
- State the formula for evaluation of sine and cosine function coefficient.
- What is the advantage of using exponential series representation?
- What is the significance of negative frequency?
- Write the formula for calculation of exponential series coefficients.

5.3 Application of Fourier Series Representation

Communication systems use a technique called as pulse modulation. In this case the carrier is a pulse train. Let us study the representation of a periodic pulse train using exponential Fourier series as a discrete spectrum. This enables us to draw the spectrum of the modulated signal.

Example 5.13

Consider a pulse train of rectangular pulses of duration T and period T_0, as shown in Fig. 5.5. Find FS representation.

Solution

The representation of the signal for one period can be written as

$$x_p(t) = \begin{cases} A & \text{for } -\dfrac{T}{2} \le t \le \dfrac{T}{2} \\ 0 & \text{otherwise} \end{cases} \quad (5.47)$$

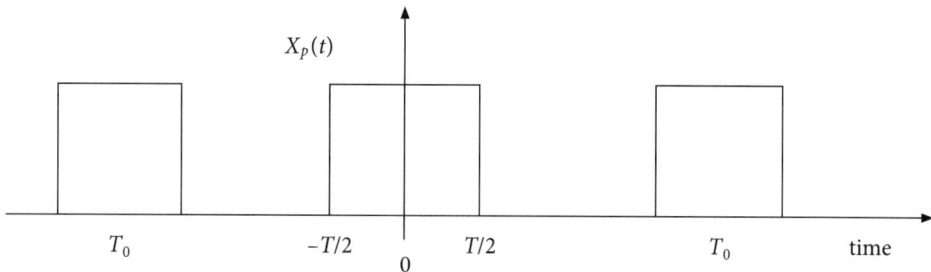

Fig. 5.5 Periodic train of rectangular pulses

The signal is an even signal. It is said to have even symmetry if $x(t) = x(-t)$. If we fold the signal with respect to the vertical axis, the signal on left of vertical axis exactly falls on to the signal on right of the vertical axis.

We will use the formula specified in Eq. (5.36) to find the Fourier series coefficients. The plot of c_n coefficients is a sync function. The sync function magnitude plot for T/T_0 equal to 0.1 is plotted in Fig. 5.6, which indicates that the sync function will have first zero crossing for $n = 10$. Referring to Eq. (5.48), we can verify that if T/T_0 is 0.1, the sync function will have first zero crossing for $n = 10$, as shown in Fig. 5.6 i.e., when the angle is equal to π and when T/T_0 is 0.05, the sync function will have first zero crossing for $n = 20$, as shown in Fig. 5.7.

$$c_n = \frac{1}{T_0} \int_{-T_0/2}^{T_0/2} x_p(t) \exp\left(\frac{-j2\pi nt}{T_0}\right) dt \quad n = 0, \pm 1, \pm 2 \ldots$$

$$= \frac{1}{T_0} \int_{-T/2}^{T/2} A \exp\left(\frac{-j2\pi nt}{T_0}\right) dt \quad (5.48)$$

$$= \frac{A}{n\pi} \sin\left(\frac{n\pi T}{T_0}\right) = \frac{AT}{T_0} \text{sinc}\left(\frac{nT}{T_0}\right)$$

where $\text{sinc}(x) = \sin(\pi x)/\pi x$

A MATLAB program to plot the Fourier series coefficients using Eq. (5.48) is given below. The number of coefficients plotted is 81. Actually the sync function extends from minus infinity to plus infinity. The centre coefficient with index 41 is actually the coefficient at $n = 0$. The coefficients plotted for $n = 1$ to 40 are with negative index. The index of the coefficients is to be understood as $n = -40$ to 40. The reader can easily verify that the second zero crossing occurs for T/T_0 equal to 0.05 is at $n = 60$, which is actually for $n = 20$, and the first zero crossing occurs for T/T_0 equal to 0.1 is at $n = 50$, which is actually for $n = 10$ in Fig. 5.7. The phase plot is shown in Fig. 5.8.

The zero crossings will occur when the angle is multiples of π. The spacing between the lines for T/T_0 equal to 0.1 is at π/10 and for T/T_0 equal to 0.2 is at π/5. We can conclude that

1. If we assume that T is kept constant, when T/T_0 is reduced, actually T_0 increases and the spacing between the lines is reduced.
2. If we assume that T_0 is kept constant, when T/T_0 is reduced, actually T reduces and the first zero crossing occurs farther away i.e., at $n = 10$.

```
clear all;
T0=4;
T=0.4;
A=1;

for n=1:40,
    c(n+41)=1/(n*pi)*sin(n*pi*T/T0);
    if n<10,
        p(n+41)=0;
    else
    if (n>10&&n<20),
        p(n+41)=-180;
    else
        if (n>20&&n<30),
        p(n+41)=0;
        else
            p(n+41)=-180;
        end
        end
    end
end
c(41)=0.1;
p(41)=0;
disp(p);
for n=1:40,
    c(n)=c(82-n);
  if n<10,
      p(n)=180;
```

```
        else
            if (n>10&&n<20),
                p(n)=0;
                else
                if (n>20&&n<30),
                p(n)=180;
                else
                    p(n)=0;
                end
                end
            end
        end
s=-40:1:40;
stem(s,abs(c));
title('plot of magnitude of discrete spectrum');
xlabel('coefficient number');
ylabel('amplitude');
figure;
stem(s,p);
title('plot of phase of discrete spectrum');
xlabel('coefficient number');
ylabel('amplitude');
```

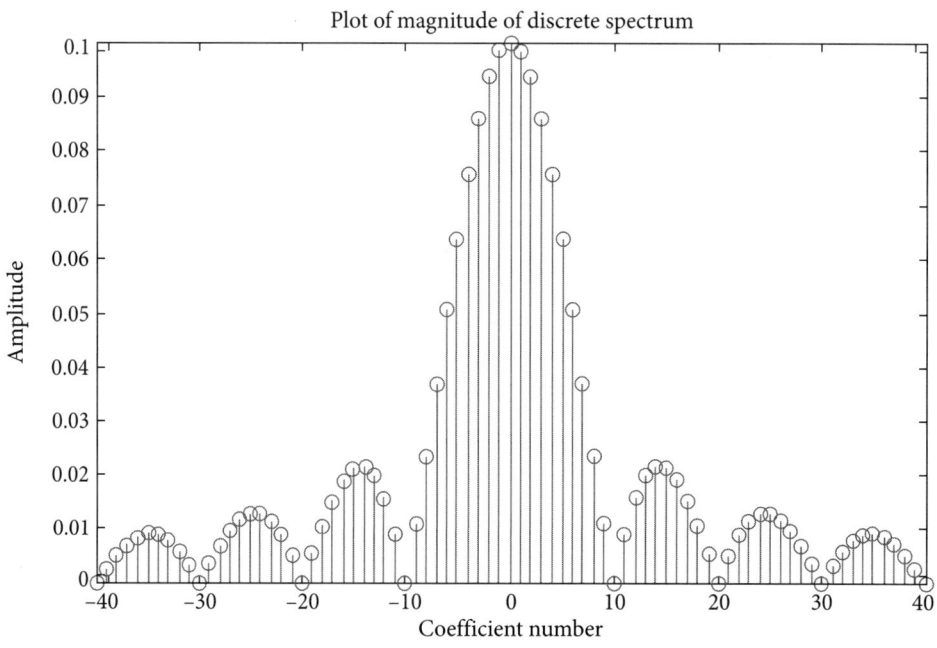

Fig. 5.6 Sync function plot for T/T_0 equal to 0.1

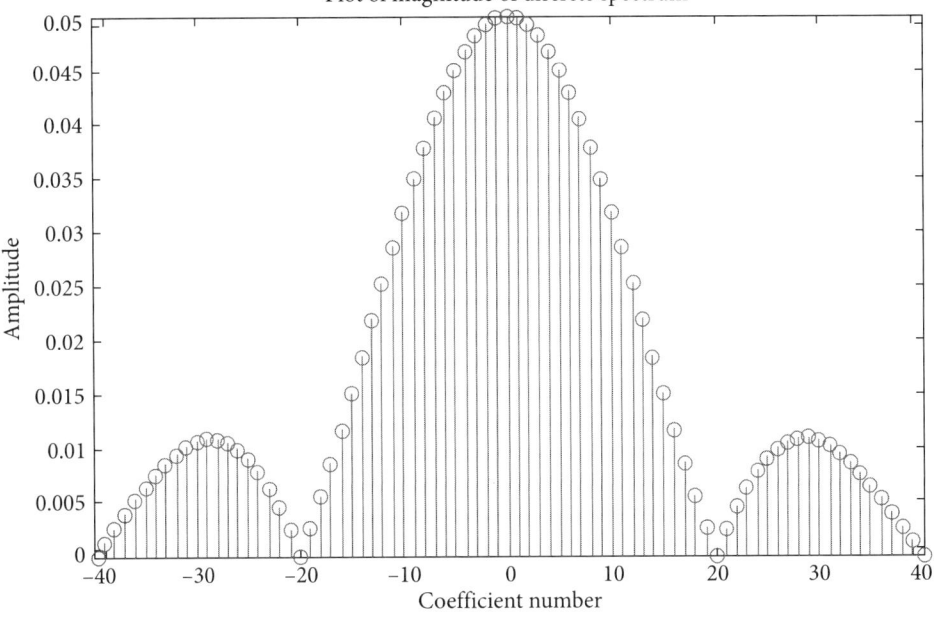

Fig. 5.7 Sync function plot for T/T_0 equal to 0.05

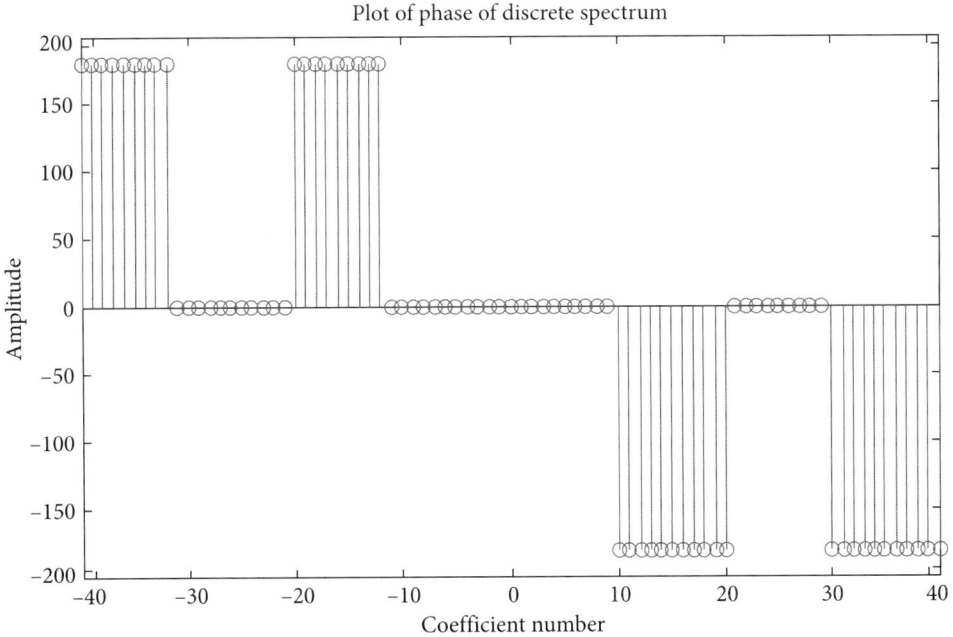

Fig. 5.8 Phase plot for T_0 equal to 0.1

We can note the following points from the above exercise.
1. The line spacing in the spectrum is decided by the period T_0.
2. The envelope of the amplitude spectrum is determined by the ratio T/T_0.
3. The zero crossings occur at frequencies that are multiple of $1/T$.
4. The phase spectrum has values between $0°$ and $\pm 180°$.

We will use the property of even symmetry to find Trigonometric FS. Let us first find

$$a_0 = \frac{1}{T_0}\int_{-T/2}^{T/2} A\,dt = \frac{A}{T}t\Big\downarrow_{-T/2}^{T/2} = AT/T_0 \tag{5.49}$$

$$a_n = \frac{2}{T_0}\int_{t}^{t+T} x_p(t)\cos(n\omega t)\,dt$$

$$= \frac{4}{T_0}\int_{0}^{T/2} A\cos(n\omega t)\,dt = \frac{4A\sin(n\omega t)}{T_0 n\omega}\Big\downarrow_{0}^{T/2} = \frac{4A\sin(n\omega T/2)}{T_0 n\omega} \tag{5.50}$$

$$= \frac{2AT\sin(n\omega T/2)}{T_0 n\omega T/2}$$

$$b_n = \frac{2}{T_0}\int_{t}^{t+T} x_p(t)\sin(n\omega t)\,dt$$

$$= \frac{2}{T_0}\left[\int_{0}^{T/2} A\sin(n\omega t)\,dt + \int_{-T/2}^{0} A\sin(n\omega t)\,dt\right] \tag{5.51}$$

$$= \frac{2}{T_0}\left[\int_{0}^{T/2} A\sin(n\omega t)\,dt - \int_{0}^{T/2} A\sin(n\omega t)\,dt\right] = 0$$

The FS can be written as

$$x(t) = \frac{AT}{T_0} + \sum_{n=-\infty, n\neq 0}^{\infty} \frac{4A\sin(n\omega T/2)}{T_0 n\omega}\cos(n\omega t) \tag{5.52}$$

No sine terms exist for the waveform having even symmetry.

Example 5.14
Determine the FS representation for the square wave in Fig. 5.9.

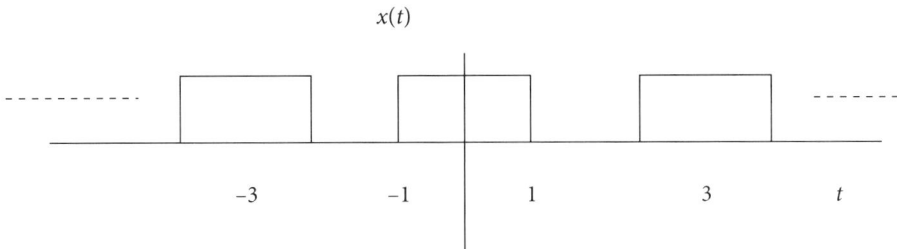

Fig. 5.9 Square wave for Example 5.14

Solution

Step 1 Let us first find the period of the wave. The wave repeats after $T = 3$ seconds. The period is T.

$$\omega_0 = \frac{2\pi}{3} \tag{5.53}$$

Step 2 Let us write $x(t)$ as a linear sum of weighted exponentials.

$$x(t) = \sum_{k=-\infty}^{\infty} X[k]e^{jk(2\pi/3)t} \tag{5.54}$$

$X(t)$ can be found using the formula

$$X[k] = \frac{1}{T}\int_{(T)} x(t)e^{-jk(2\pi/3)t}\,dt \tag{5.55}$$

$$X[k] = \frac{1}{T}\int_{-T_1}^{T_1} e^{-jk(2\pi/3)t}\,dt$$

$$= \frac{1}{3}\left[\frac{1}{-jk(2\pi/3)}e^{-jk(2\pi/3)t}\right]_{-1}^{1}$$

$$= \frac{1}{3}\left[\frac{1}{-jk(2\pi/3)}\{e^{-jk(2\pi/3)} - e^{jk(2\pi/3)}\}\right] \tag{5.56}$$

$$= \frac{j}{2\pi k}[e^{-jk(2\pi/3)} - e^{jk(2\pi/3)}]$$

$$= -\frac{1}{\pi k}\sin(2\pi k/3) \text{ for } k \neq 0$$

Step 3 Find $X[0]$.

$$X[0] = \frac{1}{3}\int_{-1}^{1} x(t)dt = \frac{1}{3}t\Big|_{-1}^{1} \tag{5.57}$$
$$= \frac{2}{3}$$

Step 4 The exponential Fourier series can be written as

$$x(t) = \frac{2}{3} + \sum_{k=-\infty, k\neq 0}^{\infty} \frac{1}{2j\pi k}[e^{-j(2\pi/3)k} - e^{j(2\pi/3)k}] \tag{5.58}$$

Let us find trigonometric FS.

$$a_0 = X[0] = \frac{2}{3}$$

$$a_n = \frac{2}{T}\int_{t}^{t+T} x_p(t)\cos(n\omega t)dt$$

$$= \frac{2}{3}\int_{-1}^{1} \cos(n(2\pi/3)t)dt$$

$$= \frac{2}{3}[\sin(n(2\pi/3)t)]\Big|_{-1}^{1} \tag{5.59}$$

$$= \frac{2}{3}\Big[\sin((2\pi/3)nt) - (-\sin((2\pi/3)nt))\Big]\Big/2\pi n/3 \text{ for } n$$

$$= \frac{2}{3}\frac{6}{2\pi n}\sin((2\pi/3)n) = \frac{2}{\pi n}\frac{1}{2} = \frac{1}{\pi n}$$

$$b_n = \frac{2}{T}\int_{t}^{t+T} x_p(t)\sin(n\omega t)dt$$

$$= \frac{2}{3}\int_{-1}^{1} \sin(n(2\pi/3)t)dt$$
$$\tag{5.60}$$

$$= \frac{2}{3}[\cos(n(2\pi/3)t)]/2\pi n/3\Big|_{-1}^{1}$$

$$= \frac{2}{\pi n}[0] = 0 \text{ for } n$$

The FS can be written as

$$x(t) = \frac{2}{3} + \sum_{n=-\infty, n\neq 0}^{\infty} \frac{1}{\pi n} \cos(n(2\pi/3)t) \qquad (5.61)$$

Let us convert trigonometric FS coefficients to exponential FS coefficients.

$$c_n = \frac{1}{2}[a_n - jb_n], \ c_{-n} = \frac{1}{2}[a_n + jb_n], \ c_0 = a_0$$

$$c_n = \frac{1}{2}\left[\frac{-j}{\pi k}\right] = \frac{-j}{2\pi k} \qquad (5.62)$$

$$c_{-n} = \frac{j}{2\pi k}$$

No sine terms exist as the waveform has even symmetry.

We will consider the following example again with even symmetry.

Example 5.15

Determine the FS representation for the signal given by $x(t) = |\sin(2\pi t)|$. The periodic wave is shown in Fig. 5.10.

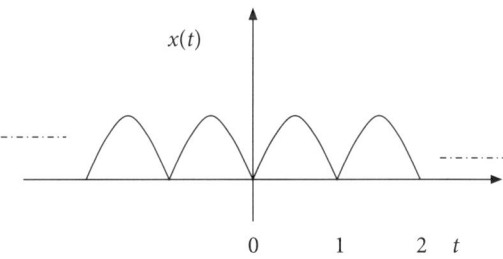

Fig. 5.10 The signal wave for Example 5.15

Solution

Step 1 Let us first find the period of the wave. The wave repeats after time period of 1 second. $\omega = \frac{2\pi}{1}$, period of the rectified sine wave is 2π second.

Step 2 Let us now find the equation for the sine wave between 0 and 1 second. This is a half part of the sine wave with a period of 2 seconds. So, the angular frequency of the sine wave $\omega_0 = \pi = \frac{2\pi}{2}$, and the signal between 0 and 1 seconds can be written as $x(t) = \sin(\pi t)$.

Step 3 Let us write $x(t)$ as a linear sum of weighted exponentials.

$$x(t) = \sum_{k=-\infty}^{\infty} X[k] e^{jk(2\pi)t} \qquad (5.63)$$

We can find $X[k]$ using the formula

$$X[k] = \frac{1}{T}\int_{(T)} x(t)e^{-jk\omega t}\,dt \qquad (5.64)$$

Step 4 Use formula in Eq. (5.36) to find $X[k]$.

$$X[k] = \frac{1}{1}\int_0^1 \sin(\pi t)e^{-jk(2\pi)t}\,dt$$

$$= \frac{1}{2j}\left[\frac{e^{j\pi(1-2k)t}}{(1-2k)j\pi} + \frac{e^{-j\pi(1+2k)t}}{(1+2k)j\pi}\right]_0^1$$

$$= -\frac{1}{2}\left[\frac{e^{-j\pi(1-2k)}-1}{(1-2k)\pi} + \frac{e^{j\pi(1+2k)}-1}{(1+2k)\pi}\right] \qquad (5.65)$$

$$= -\frac{1}{2}\left[\frac{-2(1+2k)-2(1-2k)}{(1-4k^2)\pi}\right]$$

$$= -\frac{-2-2k-2+4k}{2(1-4k^2)\pi} = \frac{2}{(1-4k^2)\pi}$$

Note that $e^{-j\pi} = e^{j\pi} = (-1)$ and $e^{-j\pi(1-2k)} = e^{j\pi(1+2k)} = (-1)$ for even and odd values of k.

Step 5 Evaluate $X[0]$ by integrating the signal over the period.

$$X[0] = \frac{1}{T}\int_0^1 x(t)\,dt = \frac{1}{1}\int_0^1 \sin(\pi t)\,dt$$

$$= -\cos(\pi t)/\pi \Big\downarrow_0^1 = \frac{2}{\pi} \qquad (5.66)$$

The magnitude response can be written as

$$X[k] = \begin{cases} \dfrac{2}{(1-4k^2)\pi} & \text{for all } k \neq 0 \\ \dfrac{2}{\pi} & \text{for } k=0 \end{cases} \qquad (5.67)$$

Step 6 The exponential Fourier series can be written as

$$x(t) = \frac{2}{\pi} + \sum_{k=-\infty, k\neq 0}^{\infty} \frac{2}{(1-4k^2)\pi} e^{j2\pi kt} \qquad (5.68)$$

Let us find trigonometric FS.

$$a_0 = \frac{1}{1}\int_0^1 \sin(\pi t)dt = -\cos(\pi t)/\pi \Big|_0^1 = -\frac{(-1)-1}{\pi} = \frac{2}{\pi} \qquad (5.69)$$

$$a_n = \frac{2}{T}\int_t^{t+T} x_p(t)\cos(n\omega t)dt$$

$$= \frac{2}{1}\int_0^1 \sin(\pi t)\cos(2n\pi t)dt = \int_0^1 [\sin(2n+1)\pi t + \sin(1-2n)\pi t]dt \qquad (5.70)$$

$$= [-\cos(2n+1)\pi t/(2n+1)\pi - \cos(1-2n)\pi t/(1-2n)\pi]\Big|_0^1$$

$$= -\frac{(-1)^{2n+1}-1}{(2n+1)\pi} - \frac{(-1)^{1-2n}-1}{(1-2n)\pi} = \frac{2(1-2n)+2(1+2n)}{(1-4n^2)\pi} = \frac{4}{(1-4n^2)\pi} \text{ for } n$$

$$b_n = \frac{2}{T_0}\int_t^{t+T} x_p(t)\sin(n\omega t)dt$$

$$= \frac{2}{1}\left[\int_0^1 \sin(\pi t)\sin(2n\pi t)dt\right] = \int_0^1 [\cos(1-2n)\pi t - \cos(1+2n)\pi t]dt \qquad (5.71)$$

$$= \left[-\frac{\sin(1+2n)\pi t}{(1+2n)\pi} + \frac{\sin(1-2n)\pi t}{(1-2n)\pi}\right]\Big|_0^1$$

$$= 0 \text{ for all } n$$

The FS can be written as

$$x(t) = \frac{2}{\pi} + \sum_{n=-\infty, n\neq 0}^{\infty} \frac{4}{(1-4n^2)\pi}\cos(2n\pi t) \qquad (5.72)$$

Let us convert trigonometric FS coefficients to exponential FS coefficients.

$$c_n = \frac{1}{2}[a_n - jb_n], \; c_{-n} = \frac{1}{2}[a_n + jb_n], \; c_0 = a_0$$

$$c_n = \frac{1}{2}\left[\frac{4}{(1-4k^2)\pi}\right] = \frac{2}{(1-4k^2)\pi} \qquad (5.73)$$

$$c_{-n} = \frac{2}{(1-4k^2)\pi},$$

No sine terms exist as the waveform has even symmetry.

Let us write a MATLAB program to plot the spectrum for the signal.

```
clear all;
t=-5:0.1:5;
x=abs(sin(pi*t));
plot(t,x);title('plot of rectified sine wave');
xlabel('time');ylabel('amplitude');
for k=1:21,
y(k)=2/((1-4*(k-11).*(k-11))*pi);
end
figure;
k1=-10:1:10;
stem(k1,y);title('plot of spectrum of the signal');
xlabel('frequency index');ylabel('amplitude');
```

Figure 5.11 shows the plot of the signal and Fig. 5.12 shows the plot of the magnitude spectrum for the FS representation of the signal.

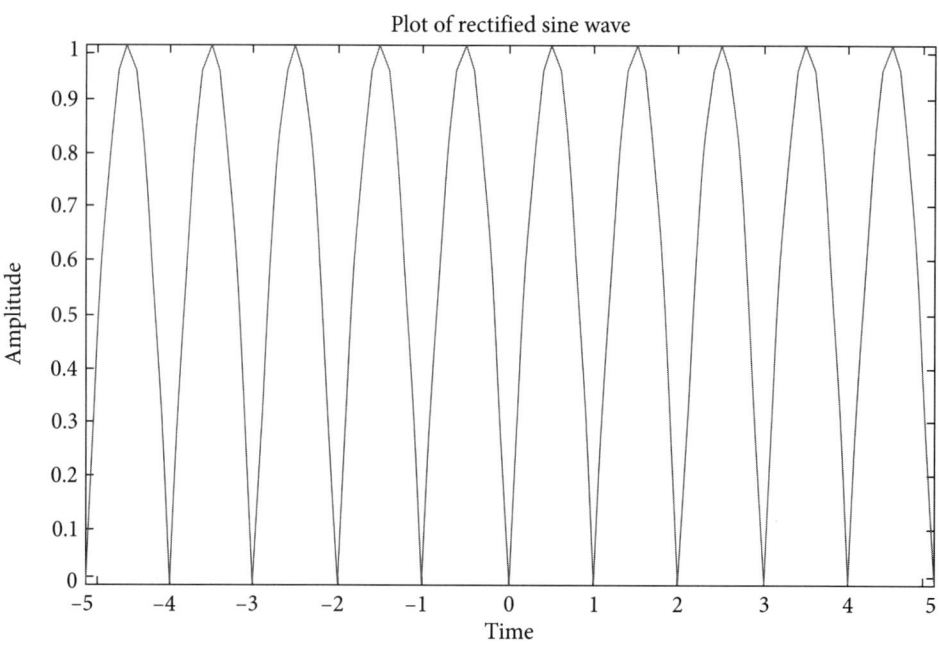

Fig. 5.11 Plot of the full wave rectified sine wave

The following example is having no symmetry. It will have sine and cosine terms.

Example 5.16

Determine the FS representation for the signal with the periodic wave, as shown in Fig. 5.13.

Solution

Step 1 Let us first find the period of the wave. The wave repeats after time period of 2 seconds. $\omega = \pi = \frac{2\pi}{2}$, period of the rectified sine wave is π second.

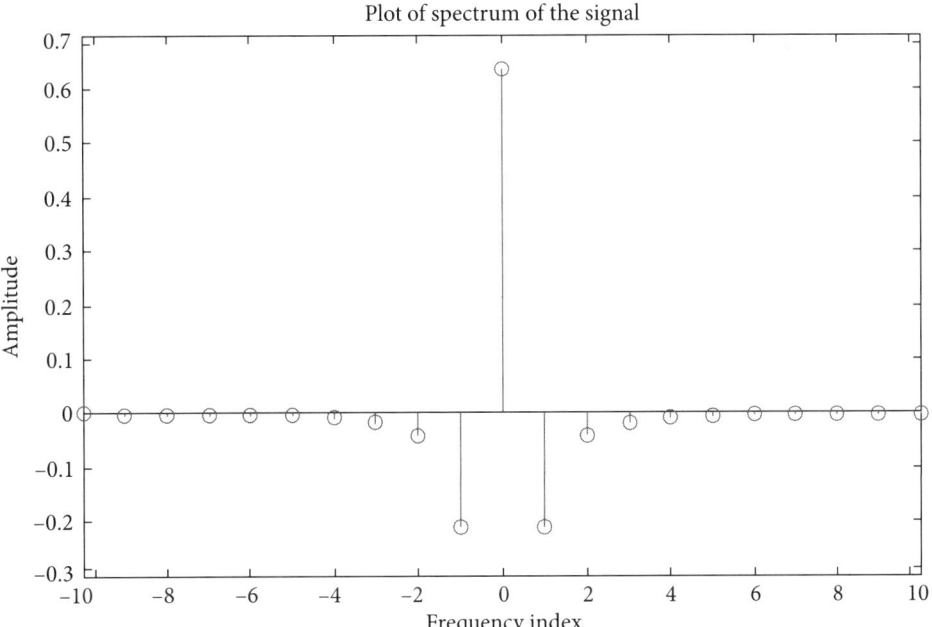

Fig. 5.12 Plot of the FS spectrum for the signal

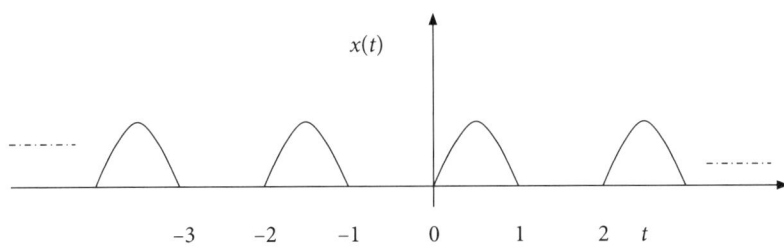

Fig. 5.13 The signal wave for Example 5.16

Step 2 Let us now find the equation for the sine wave between 0 to 1 second. This is a half part of the sine wave with a period of 2 seconds. So, the angular frequency of the sine wave $\omega_0 = \pi = \frac{2\pi}{2}$, and the signal between 0 to 1 seconds can be written as $x(t) = \sin(\pi t)$.

Step 3 Let us write $x(t)$ as a linear sum of weighted exponentials.

$$x(t) = \sum_{k=-\infty}^{\infty} X[k] e^{jk(\pi)t} \qquad (5.74)$$

We can find $X[k]$ using the formula

$$X[k] = \frac{1}{T}\int_{(T)} x(t)e^{-jk\omega t}\,dt \qquad (5.75)$$

Step 4 Use formula for FS to find $X[k]$.

$$X[k] = \frac{1}{2}\int_0^1 \sin(\pi t)e^{-jk(\pi)t}\,dt$$

$$= \frac{1}{4j}\left[\frac{e^{j\pi(1-k)t}}{(1-k)j\pi} + \frac{e^{-j\pi(1+k)t}}{(1+k)j\pi}\right]_0^1$$

$$= -\frac{1}{4}\left[\frac{e^{j\pi(1-k)} - 1}{(1-k)\pi} + \frac{e^{-j\pi(1+k)} - 1}{(1+k)\pi}\right]$$

$$= -\frac{1}{4}\left[\frac{-2(1+k) - 2(1-k)}{(1-k^2)\pi}\right] \text{ for } k \text{ even and } k \neq 1$$

$$= -\frac{1}{4}\left[\frac{-2 - 2k - 2 + 2k}{(1-k^2)\pi}\right] \qquad (5.76)$$

$$= \frac{1}{(1-k^2)\pi}$$

for $k=1$, $X[1] = \frac{1}{2}\int_0^1 \sin(\pi t)e^{j\pi t}\,dt = \frac{1}{4j}\left[\frac{t}{1} + \frac{e^{-2j\pi t}}{2\pi j}\right]_0^1$

$$= \frac{1}{4j}\left[1 + \frac{1-1}{2\pi j}\right] = \frac{1}{4j}$$

for k odd, $X[k] = 0$

Note that $e^{-j\pi} = e^{j\pi} = (-1)$ and $e^{-j\pi(1-k)} = e^{j\pi(1+k)} = (-1)$ for even values of k and $e^{-j\pi(1-k)} = e^{j\pi(1+k)} = (1)$ for odd values of k.

Step 5 Evaluate $X[0]$ by integrating the signal over the period.

$$X[0] = \frac{1}{T}\int_0^1 x(t)\,dt = \frac{1}{2}\int_0^1 \sin(\pi t)\,dt$$

$$= -\cos(\pi t)/2\pi \Big|_0^1 = \frac{1}{\pi} \qquad (5.77)$$

The magnitude response can be written as

$$X[k] = \begin{cases} \dfrac{1}{(1-k^2)\pi} & \text{for all } k \neq 0 \text{ and } k \text{ even} \\ \dfrac{1}{\pi} & \text{for } k = 0 \end{cases} \qquad (5.78)$$

Step 6 The exponential Fourier series can be written as

$$x(t) = \frac{1}{\pi} + \sum_{k=-\infty, k\neq 0}^{\infty} \frac{1}{(1-k^2)\pi} e^{j\pi kt} \qquad (5.79)$$

Let us find sine and cosine series using formula for a_n and b_n.

$$a_0 = \frac{1}{2}\int_0^1 \sin(\pi t)dt = -\cos(\pi t)/\pi \Big|_0^1 = -\frac{(-1)-1}{\pi} = \frac{2}{\pi} \qquad (5.80)$$

$$a_n = \frac{2}{T}\int_{(T)} x(t)\cos(nwt)dt$$

$$= \frac{2}{2}\left[\int_0^1 \sin(\pi t)\cos(n\pi t)dt = \frac{1}{2}\int_0^1 \left[\sin(1+n)\pi t + \sin(1-n)\pi t\right]dt\right.$$

$$= \frac{1}{2}\left[-\cos(1+n)\pi t/(1+n)\pi \Big|_0^1 - \cos(1-n)\pi t/(1-n)\pi \Big|_0^1\right]$$

$$= -\frac{1}{2(1+n)\pi}\left[\text{con}(1+n)\pi - 1\right] - \frac{1}{2(1-n)\pi}\left[\cos(1-n)\pi - 1\right] \text{ for } k \text{ even} \qquad (5.81)$$

$$= -\frac{1}{2(1+n)\pi}[-2] - \frac{1}{2(1-n)\pi}[-2] = \frac{2[1-n+1+n]}{2(1-n^2)\pi} = \frac{2}{(1-n^2)\pi}$$

for k odd $a_n = 0$

$$a_1 = \frac{2}{2}\int_0^1 \sin(\pi t)\cos(\pi t)dt = \frac{1}{2}\int_0^1 \sin(2\pi t)dt = -\frac{1}{2}\left[\cos(2\pi t)/2\pi\right]\Big|_0^1$$
$$= 0$$

$$b_n = \frac{2}{T}\int_{(T)} x(t)\sin(wt)dt$$

$$= \frac{2}{2}\left[\int_0^1 \sin(\pi t)\sin(n\pi t)dt = \frac{1}{2}\int_0^1 \left[\cos(n-1)\pi t - \cos((1-n)\pi t\right]dt\right.$$

$$= \frac{1}{2}\left[\frac{\sin(n-1)\pi t}{(n-1)\pi} - \frac{\sin(1+n)\pi t}{(1+n)\pi}\right]\Big\downarrow_0^1 = 0 \text{ for all } n \text{ except } n=1 \quad (5.82)$$

$$= \frac{1}{2} \text{ for } n=1 \, (\sin(n-1)\pi / (n-1)\pi \to 1)$$

$$x(t) = a_0 + \sum a_n \cos(n\omega t) + \sum b_n \sin(n\omega t)$$

$$x(t) = \frac{2}{\pi} + \frac{1}{2}\sin(\pi t) + \sum_{k=even}^{\infty}\left[\frac{2}{(1-k^2)\pi}\right]\cos(k\pi t) \text{ for } k \text{ odd}$$

(5.83)

Let us derive the exponential series from the trigonometric series.

$$c_n = \frac{1}{2}[a_n - jb_n], \, c_{-n} = \frac{1}{2}[a_n + jb_n], \, c_0 = a_0 = \frac{2}{\pi}$$

$$c_n = \frac{1}{2}\left[\frac{2}{(1-k^2)\pi}\right] = \frac{1}{(1-k^2)\pi} \quad (5.84)$$

$$c_{-n} = \frac{1}{(1-k^2)\pi}, \, c_1 = \frac{-j}{4} = \frac{1}{4j}, \, c_{-1} = -\frac{1}{4j}$$

Let us write a MATLAB program to plot the Fourier Spectrum. Spectrum plot is shown in Fig. 5.14.

```
clear all;
for k=1:21,
y(k)=1/((1-(k-11).*(k-11))*pi);
end
y(11)=1/4;
disp(y);
k1=-10:1:10;
stem(k1,y);title('plot of spectrum of the signal');
xlabel('frequency index');ylabel('amplitude');
plot(z1);title('Phase plot of FT of exponential sign
al');xlabel('frequency');ylabel('angle');
```

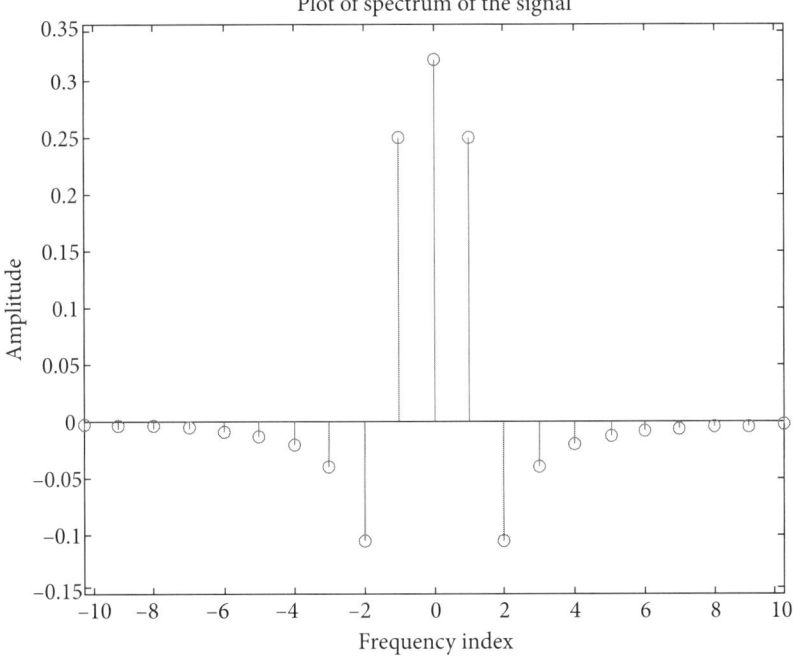

Fig. 5.14 Plot of spectrum for a half wave rectified sine wave

The following example shows even symmetry for the signal between −1 and 1. It will have only cosine terms.

Example 5.17

Determine the FS representation using exponential series for the signal with periodic wave, as shown in Fig. 5.15. Find FS representation using sine and cosine series.

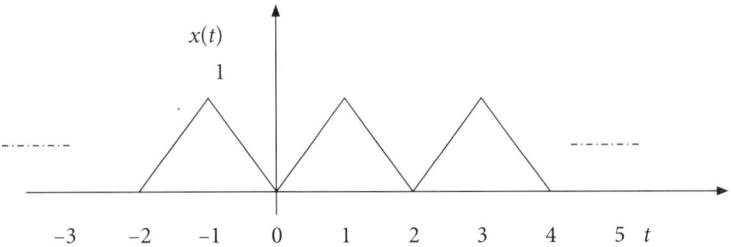

Fig. 5.15 The signal wave for Example 5.17

Solution

Step 1 Let us first find the period of the wave. The wave repeats after 2 seconds. The period is π. $\omega = \dfrac{2\pi}{2} = \pi$. The given waveform has even symmetry $x(t) = x(-t)$.

Step 2 Find equations of the straight lines between −1 to 0 and 0 to 1. The straight line from 0 to 1 can be written as t and that from −1 to 0 can be written as $-t$. For $x(t)$ we will put the equation of the straight line for that corresponding interval.

Step 3 Find $X[k]$. $X[k] = \dfrac{1}{T}\displaystyle\int_{(T)} x(t)e^{-jk\omega t}\,dt$

$$X[k] = \frac{1}{2}\left[\int_0^1 t e^{-jk(\pi)t}\,dt + \int_{-1}^0 -t e^{-jk\pi t}\,dt\right]$$

$$= \frac{1}{2}\left\{\left[\frac{1}{-jk\pi}t e^{-jk\pi t}\Big\downarrow_0^1 - \frac{1}{k^2\pi^2}e^{-jk\pi t}\Big\downarrow_0^1\right] + \int_0^1 t e^{jk\pi t}\,dt\right\}$$

$$= \frac{1}{2}\left\{\left[\frac{e^{-jk\pi}}{-jk\pi} - \frac{[e^{-jk\pi}-1]}{k^2\pi^2}\right] + \left[\frac{1}{jk\pi}t e^{jk\pi t}\Big\downarrow_0^1 - \frac{1}{k^2\pi^2}e^{jk\pi}\Big\downarrow_0^1\right]\right\}$$

$$= \frac{1}{2}\left[\frac{1}{-jk\pi}((-1)^k) - \frac{(-1)^k - 1}{k^2\pi^2}\right] + \left[\frac{1}{jk\pi}(-1)^k - \frac{[e^{jk\pi}-1]}{k^2\pi^2}\right] \quad (5.85)$$

$$= \frac{1}{2}\left[\frac{-(-1)^k + 1 - (-1)^k + 1}{k^2\pi^2}\right]$$

$$= \frac{2 - 2(-1)^k}{2k^2\pi^2} = \frac{2}{k^2\pi^2} \text{ for } k \text{ odd}$$

$= 0$ for k even

The magnitude response can be written as

$$X[k] = \begin{cases} \dfrac{1}{2} & \text{for } k \text{ even} \\ \dfrac{1}{2} + \dfrac{2}{k^2\pi^2} & \text{for } k \text{ odd} \end{cases} \quad (5.86)$$

Step 5 Evaluate $X[0]$ by integrating the signal over the period.

$$X[0] = \frac{1}{T}\int_0^1 x(t)\,dt = \frac{2}{2}\left[\int_0^1 t\,dt\right]$$

$$(5.87)$$

$$= \left[t^2/2\,\Big\downarrow_0^1\right] = \frac{1}{2}$$

Step 6 The exponential Fourier series can be written as

$$x(t) = \frac{1}{2} \text{ for } k \text{ even}$$

(5.88)

$$x(t) = \frac{1}{2} + \sum_{k=-\infty, k\neq 0}^{\infty} \left[\frac{2}{k^2\pi^2}\right] e^{jk\pi t} \text{ for } k \text{ odd}$$

Let us find sine and cosine series using formula for a_n and b_n.

$$a_0 = \frac{2}{2}\int_0^1 t\,dt = t^2/2 \downarrow_0^1 = \frac{1}{2}$$

(5.89)

$$a_n = \frac{2}{T}\int_{(T)} x(t)\cos(nwt)\,dt$$

$$= \frac{4}{2}\int_0^1 t\cos(n\pi t)\,dt = 2\left[t\sin(n\pi t)/n\pi \downarrow_0^1 - \cos(n\pi t)/n^2\pi^2 \downarrow_0^1\right]$$

(5.90)

$$= 2\left[(0-0) - \frac{(-1)^n - 1}{n^2\pi^2}\right] = \frac{4}{n^2\pi^2} \text{ for } n \text{ odd and } = 0 \text{ for } n \text{ even}$$

$$b_n = \frac{2}{T}\int_{(T)} x(t)\sin(nwt)\,dt$$

$$= \frac{2}{2}\left[\int_0^1 x(t)\sin(n\pi t)\,dt + \int_{-1}^0 x(-t)\sin(n\pi t)\,dt\right]$$

(5.91)

$$= \left[\int_0^1 t\sin(n\pi t)\,dt - \int_0^1 t\sin(n\pi t)\right] = 0$$

$$x(t) = x(-t)$$

$$x(t) = a_0 + \sum a_n \cos(n\omega t) + \sum b_n \sin(n\omega t)$$

(5.92)

$$x(t) = \frac{1}{2} + \sum_{k=-\infty, k\neq 0}^{\infty} \left[\frac{4}{k^2\pi^2}\right]\cos(k\pi t) \text{ for } k \text{ odd}$$

We can see that it has all cosine terms.

Let us derive the exponential series from the trigonometric series.

$$c_n = \frac{1}{2}[a_n - jb_n], c_{-n} = \frac{1}{2}[a_n + jb_n], c_0 = a_0 = \frac{1}{2}$$

$$c_n = \frac{1}{2}\left[\frac{4}{k^2\pi^2}\right] = \frac{2}{k^2\pi^2} \tag{5.93}$$

Note that when the function has even symmetry, the FS contains only cosine terms.

Let us consider the signal with odd symmetry. We will show that it will have only odd terms i.e., sine terms. The following signal satisfies the condition $x(-t) = -x(t)$ for signal between −1 to 1.

Example 5.18

Find the trigonometric FS and exponential FS for the periodic signal shown in Fig. 5.16.

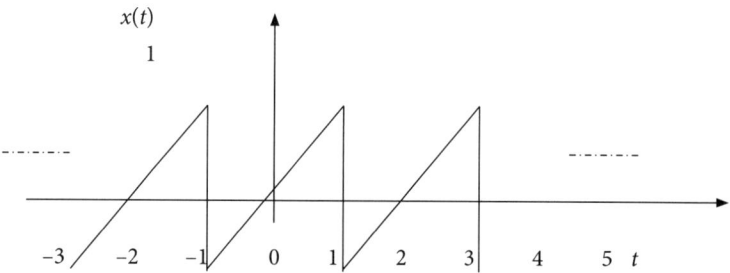

Fig. 5.16 Signal for Example 5.18

Solution

Step 1 Let us first find the period of the wave. The wave repeats after 2 seconds. The period is π. $\omega = \frac{2\pi}{2} = \pi$. The given waveform has odd symmetry. Hence, it will have only sine terms.

Step 2 Find equations of the straight lines between −1 to 0 and 0 to 1. The straight line from 0 to 1 can be written as t and that from −1 to 0 can also be written as t. For $x(t)$ we will put the equation of the straight line for that corresponding interval.

Step 3 Find $X[k]$. $X[k] = \frac{1}{T}\int_{(T)} x(t)e^{-jk\omega t} dt$

$$X[k] = \frac{1}{2}\left[\int_{-1}^{1} te^{-jk(\pi)t} dt\right]$$

$$= \frac{1}{2}\left\{\left[\frac{1}{-jk\pi}te^{-jk\pi t}\downarrow_{-1}^{1} - \frac{1}{k^2\pi^2}e^{-jk\pi t}\downarrow_{-1}^{1}\right]\right\}$$

$$= \frac{1}{2}\left\{\left[\frac{e^{-jk\pi} + e^{jk\pi}}{-jk\pi} - \frac{[e^{-jk\pi} - e^{jk\pi}]}{k^2\pi^2}\right]\right\} \qquad (5.94)$$

$$= \frac{1}{2}\left[\frac{(-1)^k + (-1)^k}{-jk\pi} - \frac{(-1)^k - (-1)^k}{k^2\pi^2}\right]$$

$$= \left[\frac{(-1)^k}{-jk\pi}\right] \text{ for all } k$$

The magnitude response can be written as

$$X[k] = \left\{-\frac{(-1)^k}{jk\pi} \text{ for all } k\right\} \qquad (5.95)$$

Step 5 Evaluate $X[0]$ by integrating the signal over the period.

$$X[0] = \frac{1}{T}\int_{-1}^{1} x(t)dt = \frac{1}{2}\left[\int_{-1}^{1} t\, dt\right]$$

$$= \frac{1}{2}\left[t^2/2\downarrow_{-1}^{1}\right] = 0 \qquad (5.96)$$

Step 6 The exponential Fourier series can be written as

$$x(t) = \sum_{k=-\infty, k\neq 0}^{\infty}\left[-\frac{2(-1)^k}{jk\pi}\right]e^{jk\pi t} \text{ for all } k \qquad (5.97)$$

Let us find sine and cosine series using formula for a_n and b_n.

$$a_0 = \frac{1}{2}\int_{-1}^{1} t\, dt = \frac{1}{2}t^2/2\downarrow_{-1}^{1} = 0 \qquad (5.98)$$

$$a_n = \frac{2}{T}\int_{(T)} x(t)\cos(nwt)dt$$

$$= \frac{2}{2}\int_{-1}^{1} t\cos(n\pi t)dt = \left[t\sin(n\pi t)/n\pi \downarrow_{-1}^{1} -\cos(n\pi t)/n^2\pi^2 \downarrow_{-1}^{1}\right] \quad (5.99)$$

$$= (0-0) - \frac{(-1)^n - (-1)^n}{n^2\pi^2} = 0 \text{ for all } n$$

$$b_n = \frac{2}{T}\int_{(T)} x(t)\sin(nwt)dt$$

$$= \frac{2}{2}\int_{-1}^{1} t\sin(n\pi t)dt = \left[-t\cos(n\pi t)/n\pi \downarrow_{-1}^{1} -\sin(n\pi t)/n^2\pi^2 \downarrow_{-1}^{1}\right] \quad (5.100)$$

$$= -((-1)^n + (-1)^n)/n\pi = -2(-1)^n/n\pi \text{ for all } n$$

$$x(t) = \sum_{k=1,}^{\infty}\left[-\frac{2(-1)^k}{k\pi}\right]\sin(k\pi t) \text{ for all } k \quad (5.101)$$

We can see that it has all sine terms.
Let us derive the exponential series from the trigonometric series.

$$c_n = \frac{1}{2}[a_n - jb_n], \; c_{-n} = \frac{1}{2}[a_n + jb_n], \; c_0 = a_0 = \frac{1}{2}$$

$$c_n = \frac{1}{2}[-2j(-1)^n/n\pi] = \frac{j(-1)^n}{n\pi} \quad (5.102)$$

$$c_{-n} = -\frac{j(-1)^n}{n\pi},$$

Note that when the function has odd symmetry, the FS contains only the sine terms.

Example 5.19

Find the trigonometric FS and exponential FS for the periodic signal shown in Fig. 5.17.

Solution

Step 1 Let us first find the period of the wave. The wave repeats after 2 seconds. The period is π. $\omega = \frac{2\pi}{2} = \pi$. The given waveform has odd symmetry.

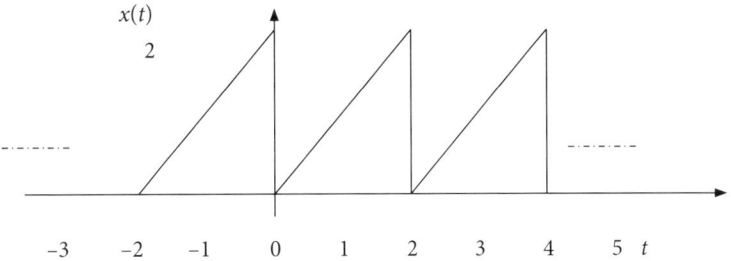

Fig. 5.17 Signal for Example 5.19

Hence, it will have only sine terms. If we remove the DC term and observe the signal between 0 to 2, we will confirm odd symmetry.

Step 2 Find equations of the straight line from 0 to 2. The straight line from 0 to 2 can be written as t. For $x(t)$ we will put the equation of the straight line for that corresponding interval.

Step 3 Find $X[k]$. $X[k] = \dfrac{1}{T}\int_{(T)} x(t) e^{-jk\omega t} dt$

$$X[k] = \frac{1}{2}\left[\int_0^2 t e^{-jk(\pi)t} dt\right]$$

$$= \frac{1}{2}\left\{\left[\frac{1}{-jk\pi} t e^{-jk\pi t}\Big\downarrow_0^2 - \frac{1}{k^2\pi^2} e^{-jk\pi t}\Big\downarrow_0^2\right]\right\}$$

$$= \frac{1}{2}\left\{\left[\frac{2e^{-j2k\pi}}{-jk\pi} - \frac{[e^{-j2k\pi} - 1]}{k^2\pi^2}\right]\right\} \qquad (5.103)$$

$$= \frac{1}{2}\left[\frac{2(1)^{2k}}{-jk\pi} - \frac{(1)^{2k} - 1}{k^2\pi^2}\right]$$

$$= \left[\frac{1}{-jk\pi}\right] \text{for all } k$$

The magnitude response can be written as

$$X[k] = \left\{-\frac{1}{jk\pi} \text{for all } k\right\} \qquad (5.104)$$

Step 5 Evaluate $X[0]$ by integrating the signal over the period.

$$X[0] = \frac{1}{T}\int_0^2 x(t)dt = \frac{1}{2}\left[\int_0^2 t\,dt\right]$$

(5.105)

$$= \frac{1}{2}\left[t^2/2 \Big\downarrow_0^2\right] = 1$$

Step 6 The exponential Fourier series can be written as

$$x(t) = \sum_{k=-\infty, k\neq 0}^{\infty}\left[-\frac{1}{jk\pi}\right]e^{jk\pi t} \text{ for all } k \quad (5.106)$$

Let us find sine and cosine series using formula for a_n and b_n.

$$a_n = \frac{2}{T}\int_{(T)} x(t)\cos(nwt)dt$$

$$= \frac{2}{2}\int_0^2 t\cos(n\pi t)dt = \left[t\sin(n\pi t)/n\pi \Big\downarrow_0^2 - \cos(n\pi t)/n^2\pi^2 \Big\downarrow_0^2\right] \quad (5.107)$$

$$= (0-0) - \frac{(1)^n - (1)^n}{n^2\pi^2} = 0 \text{ for all } n$$

$$b_n = \frac{2}{T}\int_{(T)} x(t)\sin(nwt)dt$$

$$= \frac{2}{2}\int_0^2 t\sin(n\pi t)dt = \left[-t\cos(n\pi t)/n\pi \Big\downarrow_0^2 - \sin(n\pi t)/n^2\pi^2 \Big\downarrow_0^2\right] \quad (5.108)$$

$$= -2(1)^n/n\pi = -2/n\pi \text{ for all } n$$

$$x(t) = \sum_{k=1}^{\infty}\left[-\frac{2}{k\pi}\right]\sin(k\pi t) \text{ for all } k \quad (5.109)$$

Let us derive the exponential series from the trigonometric series.

$$c_n = \frac{1}{2}[a_n - jb_n], c_{-n} = \frac{1}{2}[a_n + jb_n], c_0 = a_0 = 1$$

$$c_n = \frac{1}{2}[-2j/n\pi] = \frac{-j}{n\pi}$$

$$c_{-n} = \frac{j}{n\pi},$$

(5.110)

Note that when the function has odd symmetry, the FS contains only the sine terms.

Let us consider a signal with half wave symmetry given by $x(t) = -x(t \pm T/2)$. Consider a function shown in Fig. 5.18. It is neither purely even nor purely odd. It has half wave symmetry. Such functions contain only odd harmonics in FS. As can be seen the function has DC term equal to zero.

Example 5.20

Determine the FS representation for the signal with periodic wave, as shown in Fig. 5.18.

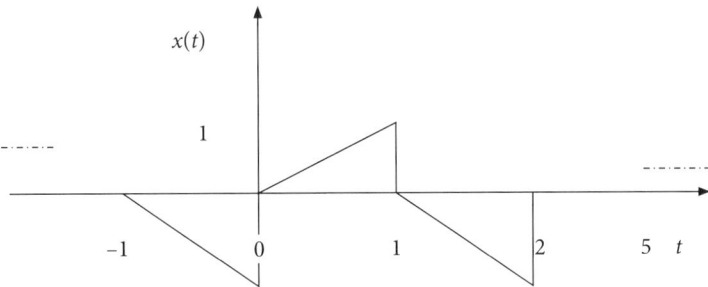

Fig. 5.18 The signal wave for Example 5.20

Solution

Step 1 Let us first find the period of the wave. The wave repeats after 2 seconds. The period is π.

$$\omega_0 = \frac{2\pi}{2} = \pi \tag{5.111}$$

Step 2 Find equations of the straight lines between 0 to 1 and 1 to 2. The straight line from 0 to 1 can be written as t and that from 1 to 2 can be written as $1 - t$.

Step 3 We can find $X[k]$

$$X[k] = \frac{1}{2}\left[\int_0^1 t e^{-jk\pi t} dt + \int_1^2 (1-t) e^{-jk\pi t} dt\right]$$

$$= \frac{1}{2}\left\{\left[\frac{1}{-jk\pi} t e^{-jk\pi t} \Big|_0^1 - \frac{1}{k^2 \pi^2} e^{-jk\pi t} \Big|_0^1\right] \right. \tag{5.112}$$

$$\left. - \left[\frac{1}{-jk\pi} t e^{-jk\pi t} \Big|_1^2 - \frac{1}{k^2 \pi^2} e^{-jk\pi t} \Big|_1^2\right]\left[+\frac{e^{-jk\pi}}{-jk\pi} \Big|_1^2\right]\right\}$$

$$= \frac{1}{2}\left[\frac{e^{-jk\pi}}{-jk\pi} - \frac{1}{k^2\pi^2}\left[e^{-jk\pi} - 1\right]\right]$$

$$-\left[\frac{2e^{-j2k\pi} - e^{-jk\pi}}{-jk\pi} - \frac{e^{-j2k\pi} - e^{-jk\pi}}{k^2\pi^2}\right] + \left[\frac{e^{-j2k\pi} - e^{-jk\pi}}{-jk\pi}\right]$$

$$= \frac{1}{2}\left[\frac{(-1)^k - 1}{-jk\pi}\right] - \frac{1}{2}\left[\frac{(-1)^k - 1 - 1 + (-1)^k}{k^2\pi^2}\right]$$

$$= -\frac{j}{k\pi} + \frac{2}{k^2\pi^2} \text{ for odd } k$$

Step 5 Evaluate $X[0]$ by integrating the signal over the period.

$$X[0] = \frac{1}{T}\int_0^2 x(t)dt = \frac{1}{2}\left[\int_0^1 t\,dt + \int_1^2 (1-t)dt\right] \tag{5.113}$$

$$= \frac{1}{2}\left[t^2/2 \Big|_0^1 + 1 + (-t^2/2)\Big|_1^2\right] = \frac{1}{2}\left[1/2 + 1 - 3/2\right] = 0$$

Step 6 The exponential Fourier series can be written as

$$x(t) = \sum_{k=-\infty, k\neq 0}^{\infty}\left[-\frac{j}{k\pi} + \frac{2}{k^2\pi^2}\right]e^{jk\pi t} \text{ for all odd } k \tag{5.114}$$

Let us find sine and cosine series using formula for a_n and b_n. We use the property of half wave symmetry.

$$a_n = \frac{4}{T}\int_{(T/2)} x(t)\cos(nwt)dt$$

$$= \frac{4}{2}\int_0^1 t\cos(n\pi t)dt = 2\left[t\sin(n\pi t)/n\pi \Big|_0^1 - \cos(n\pi t)/n^2\pi^2 \Big|_0^1\right] \tag{5.115}$$

$$= 2\left[(0-0) - \frac{(-1)^n - (1)^n}{n^2\pi^2}\right] = \frac{-4}{n^2\pi^2} \text{ for all odd } n$$

$$= 0 \text{ for even } n$$

$$b_n = \frac{4}{T}\int_{(T/2)} x(t)\sin(nwt)dt$$

$$= \frac{4}{2}\int_0^1 t\sin(n\pi t)dt = 2\left[-t\cos(n\pi t)/n\pi \downarrow_0^1 - \sin(n\pi t)/n^2\pi^2 \downarrow_0^1\right] \quad (5.116)$$

$$= -2\left[(-1)^n/n\pi = -2/n\pi \text{ for all } n\right.$$

$$x(t) = \sum_{k=1,}^{\infty}\left[-\frac{2}{k\pi}\right]\sin(k\pi t) - \frac{4}{k^2\pi^2}\cos(k\pi t) \text{ for all odd } k \quad (5.117)$$

Let us derive the exponential series from the trigonometric series.

$$c_n = \frac{1}{2}[a_n - jb_n], c_{-n} = \frac{1}{2}[a_n + jb_n], c_0 = a_0 = 0$$

$$c_n = \left[-j/n\pi - 2/k^2\pi^2\right] \quad (5.118)$$

$$c_{-n} = \left[-j/n\pi + 2/k^2\pi^2\right]$$

Example 5.21
Determine the FS representation for the signal with periodic wave, which is shown in Fig. 5.20.

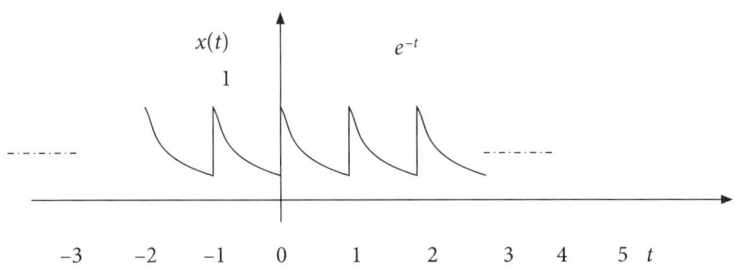

Fig. 5.19 Plot of signal for Example 5.21

Solution
Step 1 Let us first find the period of the wave. The wave repeats after time period of 1 seconds. $\omega = \pi = \frac{2\pi}{1}$, period of the wave is 2π second.
Step 2 The equation for the wave is $x(t) = e^{-t}$.
Step 3 Let us write $x(t)$ as a linear sum of weighted exponentials.

$$x(t) = \sum_{k=-\infty}^{\infty} X[k]e^{jk(2\pi)t} \qquad (5.119)$$

We can find $X[k]$ using the formula

$$X[k] = \frac{1}{T}\int_{(T)} x(t)e^{-jk\omega t}dt \qquad (5.120)$$

Step 4 Use formula in Eq. (5.52) to find $X[k]$.

$$X[k] = \frac{1}{1}\int_0^1 e^{-t}e^{-jk(2\pi)t}dt = \int_0^1 e^{-(1+j2\pi k)t}dt$$

$$= \left[\frac{e^{-(1+j2\pi k)t}}{-(1+j2\pi k)}\right]_0^1$$

$$= \left[\frac{e^{-(1+j2\pi k)}-1}{-(1+j2\pi k)}\right] \qquad (5.121)$$

$$= \left[\frac{1-e^{-1}e^{-j2\pi k}}{(1+j2\pi k)}\right] = \frac{1-e^{-1}}{1+j2\pi k} \text{ for all } k$$

Note that $e^{-j2\pi} = 1, e^{-j2\pi k} = 1$ for all k.

Step 5 Evaluate $X[0]$ by integrating the signal over the period.

$$X[0] = \frac{1}{T}\int_0^1 x(t)dt = \frac{1}{1}\int_0^1 e^{-t}dt$$

$$= \frac{e^{-t}}{-1}\bigg|_0^1 = 1-e^{-1} \qquad (5.122)$$

The magnitude response can be written as

$$X[k] = \left\{\frac{1-e^{-1}}{(1+j2\pi k)} \text{ for all } k\right\} \qquad (5.123)$$

Step 6 The exponential Fourier series can be written as

$$x(t) = \sum_{k=-\infty}^{\infty} \frac{1-e^{-1}}{(1+j2\pi k)}e^{j2\pi kt} \qquad (5.124)$$

Let us write a MATLAB program to find the Fourier line spectrum. Figure 5.20 shows the line spectrum.

```
clear all;
for k=1:21,
y(k)=abs((1-exp(-1))/((1+1j*2*pi*(k-11))));
end
disp(y);
k1=-10:1:10;
stem(k1,y);title('plot of spectrum of the signal');
xlabel('frequency index');ylabel('amplitude');
```

Note that the signal has odd symmetry and contains only sine terms.

Note that for all examples solved above, even though the signal is periodic in time domain its Fourier series representation is aperiodic. We will consider a periodic train of impulses, which is again a periodic signal, and find its FS representation.

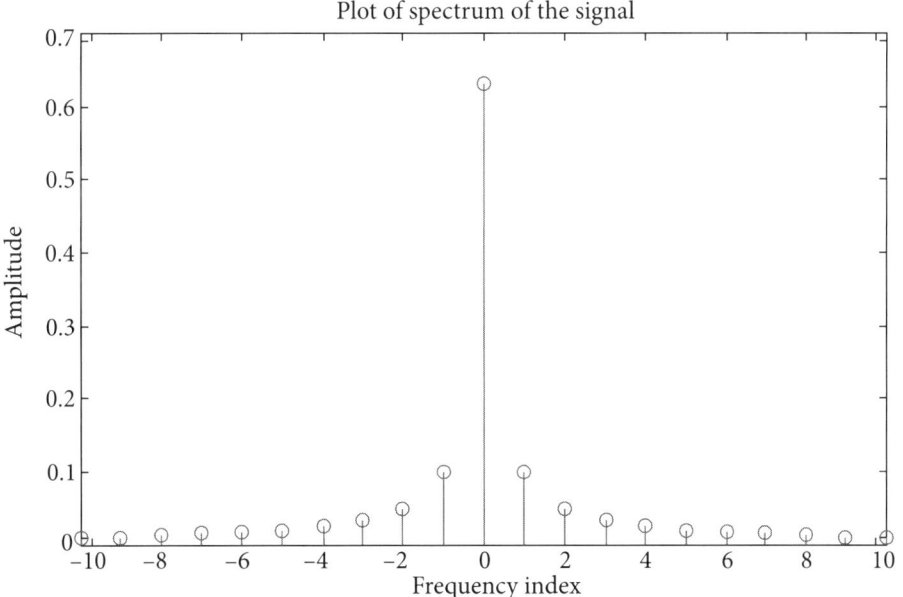

Fig. 5.20 Plot of line spectrum for the signal

Example 5.22

Consider a train of pulses as shown in Fig. 5.21. Find FS representation for this periodic signal.

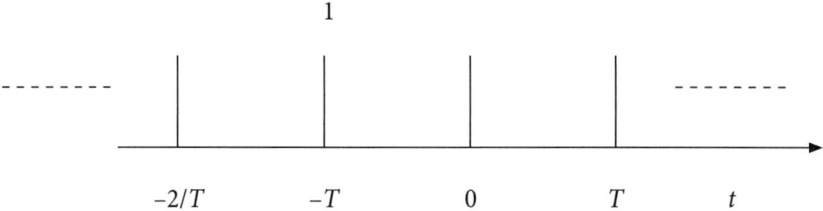

Fig. 5.21 Signal for Example 5.22

Solution

We will use the formula for exponential FS. The signal has a period of T. Consider the time interval between $-T/2$ to $T/2$.

$$X[k] = \frac{1}{T}\int_{(T)} x(t)e^{-jk\omega t}\,dt = \frac{1}{T}\int_{-T/2}^{T/2}\delta(t)e^{-jk2\pi t/T}\,dt$$

put $t = 0$ equation

$$X[k] = c_k = \frac{1}{T},\ \delta(t) = \begin{cases} 1 & \text{for } t=0 \\ 0 & \text{otherwise} \end{cases} \tag{5.125}$$

$$x(t) = \sum_{k=-\infty}^{\infty} \frac{1}{T} e^{jk\omega t}$$

The FS is again a train of impulses with a separation of $1/T$, as shown in Fig. 5.22.

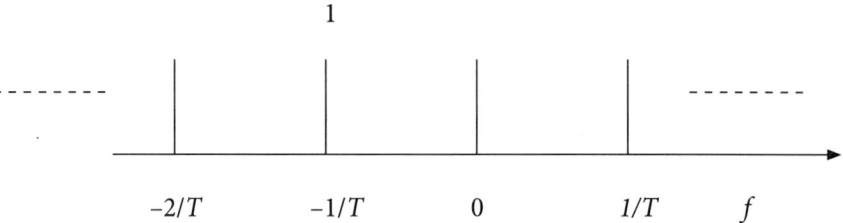

Fig. 5.22 Plot of spectrum for the train of impulse in example

Example 5.23

Consider a train of pulses given by $x(t) = \sum_{k=-\infty}^{\infty}\delta\!\left(t-\frac{1}{2}T\right) + \sum_{k=-\infty}^{\infty}\delta\!\left(t-\frac{3}{2}T\right)$, as shown in Fig. 5.23. Find FS representation for this periodic signal.

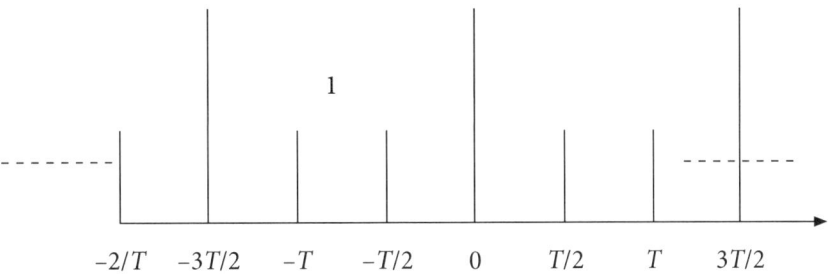

Fig. 5.23 Signal for Example 5.23

Solution

We will use the formula for exponential FS. The signal has a period of $3T/2$. Consider the time interval between $-5T/4$ to $5T/4$. It will contain 3 impulses one at $t = -T/2$, second at $t = 0$ and third at $t = T/2$.

$$X[k] = \frac{1}{T}\int_{(T)} x(t)e^{-jk\omega t}dt = \frac{2}{3}\int_{-4T/4}^{5T/4}[\delta\left(t+\frac{1}{2}\right) + 2\delta(t) + \delta\left(t-\frac{1}{2}\right)]e^{-jk4\pi t/3}dt$$

$$X[k] = c_k = \left(\frac{2}{3}e^{jk2\pi t/3} + 4/3 + \frac{2}{3}e^{-jk2\pi t/3}\right), \delta(t) = \begin{cases} 1 \text{ for } t=0 \\ 0 \text{ otherwise} \end{cases} \quad (5.126)$$

$$X[k] = \frac{4}{3} + \frac{4}{3}\cos(k2\pi/3)$$

The FS is again a train of impulses with a separation of 2/3. Let us write a MATLAB program to plot the line spectrum for the signal. The line spectrum is plotted in Fig. 5.24.

```
clear all;
for k=1:20,
    x(k)=(4/3)+(4/3)*cos((2*(k-1)*pi)/(3));
end
k1=-9:1:10;
stem(k1,x);title('spectrum for signal');
xlabel('Fourier Series coefficient
number');ylabel('amplitude');
```

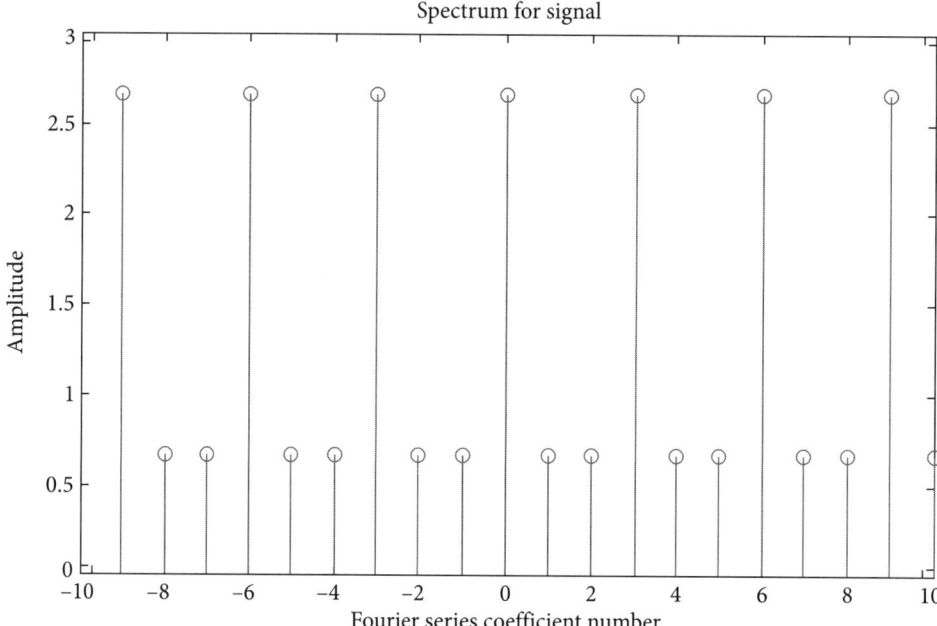

Fig. 5.24 Line spectrum for Example 5.23

Note that for the above examples, the signal is periodic but is discrete in time domain. The FS representation is periodic in this case. We will study FS for DT signals in section 5.6.

Concept Check

- State the application of FS representation.
- Draw the line spectrum for the periodic pulse train.
- What will happen when the period of the pulse tin increases?
- What is the effect of reduction of the pulse width?
- Define even signal? State the characteristic of the FS for the even signal.
- Define odd signal? Will the FS contain only sine terms?
- What is half wave symmetry?
- Define FS for the periodic train of impulses.

5.4 Properties of Fourier Series for CT Signals

We will state and prove the following properties of CT FS namely linearity time shifting, time reversal, time scaling, time differentiation, time integration, convolution, modulation, complex conjugate property and Parseval's relation.

1. **Property of Linearity**

   If

   $$x_1(t) \leftrightarrow C_n \text{ and } x_2(t) \leftrightarrow D_n$$
   $$\text{then } ax_1(t) + bx_2(t) \leftrightarrow aC_n + bD_n \tag{5.127}$$

   Proof

   Using the definition of FS, we can write

   $$FS[ax_1(t) + bx_2(t)] = \frac{1}{T}\int_t^{t+T}[ax_1(t) + bx_2(t)]e^{-jn\omega t}dt$$

   $$= a\left[\frac{1}{T}\int_t^{t+T}x_1(t)e^{-jn\omega t}dt\right] + b\left[\frac{1}{T}\int_t^{t+T}x_2(t)e^{-jn\omega t}dt\right] \tag{5.128}$$

   $$= aC_n + bD_n$$

   Physical significance of property of linearity If the signal can be decomposed into linear combination of component signals. The FS representation can be found for each signal and the FS can be added to get the FS for the linear combination. This simplifies the mathematical computations.

2. **Time Shifting Property**

   The time shifting property states that if

   $$x(t) \leftrightarrow C_n,$$
   $$\text{then } x(t - t_0) \leftrightarrow e^{-jn\omega t_0}C_n \tag{5.129}$$

   Proof

   Using the definition of FS, we can write

   $$x(t-t_0) = \sum_{n=-\infty}^{\infty} C_n e^{jn\omega(t-t_0)} = \sum_{n=-\infty}^{\infty}(C_n e^{-jn\omega t_0})e^{jn\omega t}$$

   $$= FS^{-1}(C_n e^{-jn\omega t_0}) \tag{5.130}$$

   Physical significance of property of Time shifting Consider two vectors with one time-shifted with respect to the other. If FS is evaluated for both, the FS will be identical in magnitude. The time-shifted signal will have a FS with phase shift of $e^{-j\vartheta}$ where $\vartheta = n\omega t_0$. This is because the magnitude of $e^{-j\vartheta}$ is one. This fact can be used for template matching when two templates are same but just shifted in time.

3. **Time Reversal Property**

The time reversal property states that if

$$x(t) \leftrightarrow C_n, \qquad (5.131)$$

then $x(-t) \leftrightarrow C_{-n}$

Proof

Using the definition of FS, we can write

$$x(-t) = \sum_{n=-\infty}^{\infty} C_n e^{jn\omega(-t)} = \sum_{n=-\infty}^{\infty} (C_n e^{j(-n)\omega t}) \qquad (5.132)$$

$$= FS^{-1}(C_{-n})$$

Physical significance of property of Time reversal The FS coefficients for a time-reversed signal are C_{-n}. This means the FS coefficients will have same magnitude but the phase angle is negative of the phase angle for FS coefficients of original signal. The FS coefficients of the time-reversed signal will be a complex conjugate of the FS coefficient of the original signal.

4. **Time Scaling Property**

The time scaling property states that if

$$x(t) \leftrightarrow C_n, \qquad (5.133)$$

then $x(\alpha t) \leftrightarrow C_n \ \omega \to \alpha\omega$

Proof

Using the definition of FS, we can write

$$x(t) = \sum_{n=-\infty}^{\infty} C_n e^{jn\omega t},$$

$$x(\alpha t) = \sum_{n'=\infty}^{-\infty} C_{n'} e^{jn\alpha\omega t} \qquad (5.134)$$

$$= FS^{-1}(C_n) \text{ with } \omega \to \alpha\omega$$

Physical significance of property of Time scaling FS coefficients of the time-scaled signal with a scaling factor α will have the same magnitude, and the phase is scaled by the factor α.

5. **Time Differentiation Property**

   The property of time differentiation states that if

   $$x(t) \leftrightarrow C_n, \qquad (5.135)$$

   then $\dfrac{d}{dt}x(t) \leftrightarrow jn\omega C_n$

Proof

Using the definition of FS, we can write

$$x(t) = \sum_{n=-\infty}^{\infty} C_n e^{jn\omega t}, \text{ differentiate both sides wrt } t,$$

$$\frac{d}{dt}x(t) = jn\omega \sum_{n'=\infty}^{-\infty} C_n . e^{jn\omega t} \qquad (5.136)$$

$$= FS^{-1}(C_n jn\omega)$$

Physical significance of property of Time differentiation When the signal is differentiated in time domain, its FS coefficients plane rotates by the angle of $n\omega$.

6. **Time Integration Property**

   The property of time integration states that if

   $$x(t) \leftrightarrow C_n \qquad (5.137)$$

   then $\int_{-\infty}^{t} x(\tau)d\tau \leftrightarrow C_n / jn\omega$ if $(C_0 = 0)$

Proof

Using the definition of FS, we can write

$$x(t) = \sum_{n=-\infty}^{\infty} C_n e^{jn\omega t}, \text{ integrate both sides wrt } t,$$

$$\int_{-\infty}^{t} x(\tau)d\tau = \sum_{n=-\infty}^{\infty} \int_{-\infty}^{t} C_n e^{jn\omega\tau} d\tau = \sum_{n=-\infty}^{\infty} C_n e^{jn\omega\tau} / jn\omega \downarrow_{-\infty}^{t} \qquad (5.138)$$

$$= \sum_{n=-\infty}^{\infty} C_n e^{jn\omega t} / jn\omega = FS^{-1}(C_n / jn\omega), \text{ with } C_0 = 0$$

Physical significance of property of Time integration When the signal is integrated in time domain, its FS coefficients plane rotates by angle of $-1/n\omega$.

7. **Convolution Property**

   The convolution property states that if

   $$x_1(t) \leftrightarrow C_n \text{ and } x_2(t) \leftrightarrow D_n, \qquad (5.139)$$

   then $x_1(t) * x_2(t) \leftrightarrow TC_n D_n$

   Proof

   Using the definition of FS, we can write

   $$FS[x_1(t) * x_2(t)] = \frac{1}{T} \int_t^{t+T} [(x_1(t) * x_2(t)] e^{-jn\omega t} dt$$

   $$= \left[\frac{1}{T} \int_t^{t+T} \left[\int_t^{t+T} x_1(\tau) x_2(t-\tau) d\tau \right] e^{-jn\omega t} dt \right] \text{put } t - \tau = t' \qquad (5.140)$$

   $$= T \left(\frac{1}{T} \int_t^{t+T} x_1(\tau) e^{-jn\omega \tau} d\tau \right) \left(\frac{1}{T} \int_t^{t+T} x_2(t') e^{-jn\omega t'} dt' \right)$$

   $$= TC_n D_n$$

   Physical significance of convolution property When the two signals are convolved in time domain, the FS coefficients get multiplied. Convolution is computationally costly. When the signal passes via the LTI system, the signal gets convolved with the impulse response of the system. To find the FS representation for the convolved output, one may find FS for the input signal and the impulse response and multiply the two FS.

8. **Modulation Property**

   The modulation property states that If

   $$x_1(t) \leftrightarrow C_n \text{ and } x_2(t) \leftrightarrow D_n, \qquad (5.141)$$

   then $x_1(t) \times x_2(t) \leftrightarrow \sum_{m=-\infty}^{\infty} C_m D_{n-m}$

Proof

Using the definition of FS, we can write

$$FS[x_1(t) \times x_2(t)] = \frac{1}{T}\int_{t}^{t+T}[(x_1(t) \times x_2(t)]e^{-jn\omega t}dt$$

$$= \left[\frac{1}{T}\int_{t}^{t+T}x_1(t)\left[\sum_{m=-\infty}^{\infty}D_m e^{jm\omega t}\right]e^{-jn\omega t}dt\right] \quad (5.142)$$

$$= \left(\frac{1}{T}\int_{t}^{t+T}x_1(t)\sum_{m=-\infty}^{\infty}D_m e^{-j(n-m)\omega \tau}dt\right)$$

Interchange the order of integration and summation

$$= \sum_{m=-\infty}^{\infty}D_m\left[\frac{1}{T}\int_{t}^{t+T}x_1(t)e^{-j(n-m)\omega t}\right]$$

$$= \sum_{m=-\infty}^{\infty}D_m C_{n-m} \; or = \sum_{m=-\infty}^{\infty}C_m D_{n-m} \;\; (C*D = D*C)$$

Physical significance of modulation property When the carrier signal multiplies with the message signal to get $\cos(2\pi f_c t)\cos(2\pi f_m t)$ we get double side band suppressed carrier modulated signal. This property states the property for FS for the multiplied signal, hence it is called modulation property. The FS for the modulated signal is the convolution of the FS of the component signals.

9. **Conjugate Symmetry Property**

The conjugate symmetry property states that if

$x(t) \leftrightarrow C_n$ for complex valued $x(t)$

$$x*(t) \leftrightarrow C^*_{-n} \quad (5.143)$$

and $C_{-n} = C^*_n$ for real $x(t)$

Proof

Using the definition of FS, we can write

$$x(t) = \sum_{n=-\infty}^{\infty} C_n e^{jn\omega t},$$

$$FS[x^*(t)] = \frac{1}{T}\int_t^{t+T} x^*(t)e^{-jn\omega t}dt = \left\{\frac{1}{T}\int_t^{t+T} x(t)e^{jn\omega t}\right\}^*$$

(5.144)

$$= \left\{\frac{1}{T}\int_t^{t+T} x(t)e^{-j(-n)\omega t}dt\right\}^* = \{C_{-n}\}^*$$

$$x^*(t) \leftrightarrow C_{-n}^*$$

Physical significance of the conjugate symmetry property If the signal is real, FS coefficients for the signal show the conjugate symmetry. The positive coefficients are complex conjugates of the negative coefficients. Refer to Figs 5.14 and 5.20 to confirm conjugate property.

10. Parseval's Relation

Parseval's relation states that

$x_1(t) \leftrightarrow C_n$ and $x_2(t) \leftrightarrow D_n$ for complex valued $x(t)$,

$$\text{then } \frac{1}{T}\int_t^{t+T} x_1(t)x_2^*(t)dt = \sum_{n=-\infty}^{\infty} C_n D_n^*$$

(5.145)

Parseval's identity states that

$x(t) \leftrightarrow C_n$ for complex valued $x(t)$,

$$\text{then } \frac{1}{T}\int_t^{t+T} |x(t)|^2 dt = \sum_{n=-\infty}^{\infty} |C_n|^2$$

(5.146)

Proof To prove Parseval's relation, we start with LHS

$$\frac{1}{T}\int_t^{t+T} x_1(t)x_2^*(t)dt = \frac{1}{T}\int_t^{t+T} \left\{\sum_{n=-\infty}^{\infty} C_n e^{jn\omega t}\right\} x_2^*(t)dt$$

(5.147)

Interchanging order of integration and summation

$$= \sum_{n=-\infty}^{\infty} C_n \left\{\frac{1}{T}\int_t^{t+T} x_2(t)e^{-jn\omega t}dt\right\}^* = \sum_{n=-\infty}^{\infty} C_n D_n^*$$

To prove Parseval's identity,

$$\text{LHS} = \frac{1}{T}\int_{t}^{t+T} x(t)x^*(t)dt = \sum_{n=-\infty}^{\infty} C_n C_n^* = \sum_{n=-\infty}^{\infty} |C_n|^2 \qquad (5.148)$$

Physical significance of Parseval's identity Parseval's identity states that the energy of the signal in time domain is the same as the energy of the FS coefficients. This simply means, when the signal is transformed in frequency domain, its energy is preserved.

To illustrate the significance of the above properties, we will consider numerical examples to illustrate the use of different properties.

Example 5.24

Determine the FS representation for the shifted signal given by $x(t) = |\sin(2\pi t)|$ by ½ second. The shifted periodic wave is shown in Fig. 5.25.

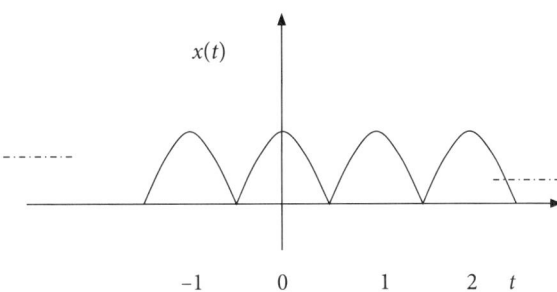

Fig. 5.25 The time shifted signal wave for Example 5.24

Solution

We will use the result of FS and will use the property of time shifting to find FS for the shifted signal.

FS for the equation $x(t) = |\sin(2\pi t)|$ is already solved in Example 5.15. We will borrow the result.

$$X[k] = \begin{cases} \dfrac{2}{(1-4k^2)\pi} & \text{for all } k \neq 0 \\ \dfrac{2}{\pi} & \text{for } k = 0 \end{cases} \qquad (5.149)$$

We will now use the property of time shifting, which states that

if $x(t) \leftrightarrow C_n$

then $x(t - t_0) \leftrightarrow e^{-jn\omega t_0} C_n$ \qquad (5.150)

Here, the time shift is of $t_0 = \frac{1}{2}$ second. The FS for the shifted signal can be written as

$$X[k] = \begin{cases} \dfrac{2}{(1-4k^2)\pi} e^{-j\omega k/2} & \text{for all } k \neq 0 \\ \dfrac{2}{\pi} & \text{for } k = 0 \end{cases} \quad (5.151)$$

Example 5.25

Determine the FS representation for the time-scaled signal given by $x(t) = |\sin(2\pi t)|$ by a factor of 2. The time scaled periodic wave is shown in Fig. 5.26. The period of the signal is reduced to ½.

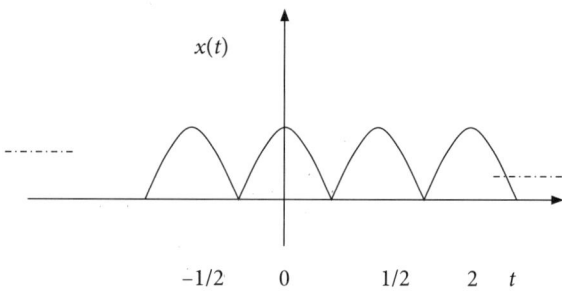

Fig. 5.26 The time shifted signal wave for Example 5.25

Solution

We will use the result of FS and will use the property of time scaling to find FS for the time scaled signal.

FS for the equation $x(t) = |\sin(2\pi t)|$ is already solved in Example 5.15. We will borrow the result.

$$X[k] = \begin{cases} \dfrac{2}{(1-4k^2)\pi} & \text{for all } k \neq 0 \\ \dfrac{2}{\pi} & \text{for } k = 0 \end{cases} \quad (5.152)$$

We will now use the property of time scaling, which states that

$x(t) \leftrightarrow C_n,$

then $x(\alpha t) \leftrightarrow C_n \;\; \omega \to \alpha\omega$ \quad (5.153)

Here, the time scaling is by a factor of ½. The FS for the time scaled signal can be written as

$$X[k] = \begin{cases} \dfrac{2}{(1-4k^2)\pi} \text{ with } \omega = \omega/2 \text{ for all } k \neq 0 \\ \dfrac{2}{\pi} \text{ for } k = 0 \end{cases} \quad (5.154)$$

The exponential Fourier series can be written as

$$x(t) = \frac{2}{\pi} + \sum_{k=-\infty, k \neq 0}^{\infty} \frac{2}{(1-4k^2)\pi} e^{j\pi kt} \quad (5.155)$$

Example 5.26

For the signal shown in Fig. 5.27, find FS using the property of time differentiation.

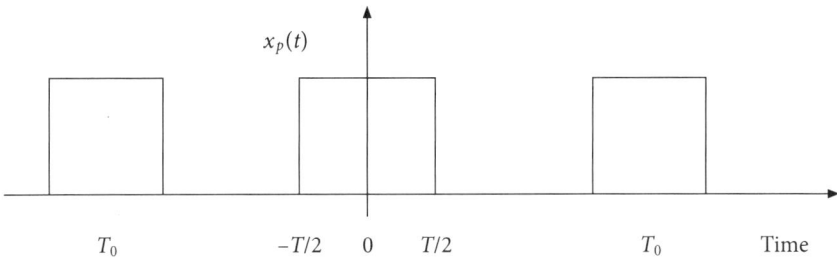

Fig. 5.27 Plot of signal for Example 5.26

Solution

Let us first differentiate the signal. The differentiated signal can be written as $y(t) = \dfrac{d}{dt}(x_p(t))$ shown in Fig. 5.28.

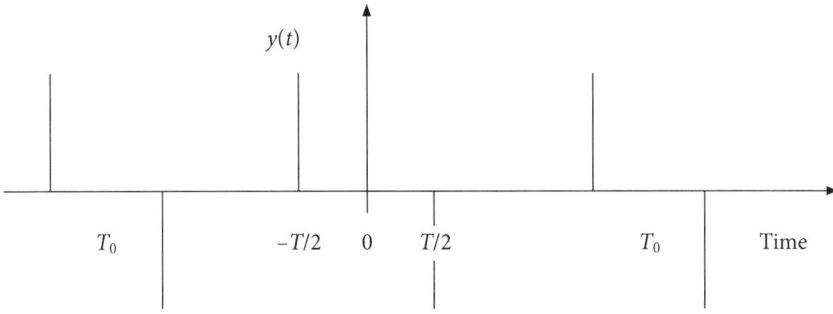

Fig. 5.28 Plot of differentiated signal

Step 1 Let us find the fundamental angular frequency $\omega = \dfrac{2\pi}{T_0}$

Step 2 FS for the signal $y(t)$ is given by

$$D_n = \frac{1}{T_0} \int_{-T/2}^{T/2} y(t) e^{-jn\omega t} dt$$

$$= \frac{1}{T_0} \int_{-T/2}^{T/2} [\delta(t+T/2) - \delta(t-T/2)] e^{-jn\omega t} dt \qquad (5.156)$$

$$= -\frac{1}{T_0} \left[e^{jn\omega T/2} - e^{-jn\omega T/2} \right]$$

$$= -\frac{2j}{T_0} \sin(n\omega T/2) = -\frac{2j}{T_0} \sin\left(\frac{2\pi n T}{2T_0} \right)$$

Step 3 Use the property of differentiation in time domain.

$D_n = jn\omega C_n$ where C_n is FS of $x_P(t)$

$$C_n = \frac{D_n}{jn\omega} = -\frac{2jT_0}{jn2\pi T_0} \sin\left(\frac{2\pi n T}{2T_0}\right) = -\frac{1}{n\pi} \sin\left(\frac{n\pi T}{T_0}\right) \qquad (5.157)$$

Example 5.27

Find the trigonometric FS for the signal shown in Fig. 5.29 with an even symmetry and prove the following identity.

$$\frac{\pi^2}{8} = 1 + \frac{1}{9} + \frac{1}{25} + \ldots\ldots$$

Solution

The signal has even symmetry, so FS will have only cosine terms. Let us find sine and cosine series using formula for a_n and b_n.

$$a_0 = \frac{1}{4} \int_{-2}^{2} x(t) dt = \frac{1}{4} \left[\int_{-2}^{0} (3t+3) dt + \int_{0}^{2} (3-3t) dt \right]$$

$$= \frac{1}{4} \left[(3t^2/2 + 3t) \Big|_{-2}^{0} + (3t - 3t^2/2) \Big|_{0}^{2} \right] \qquad (5.158)$$

$$= \frac{1}{4} [-6 + 6 + 6 - 6] = 0$$

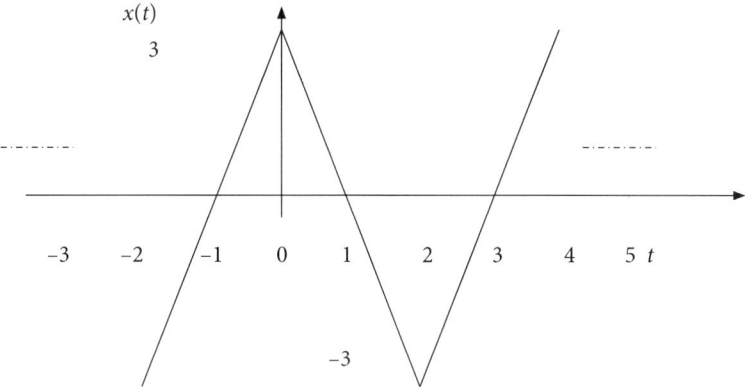

Fig. 5.29 Signal plot for Example 5.27

$\omega = 2\pi / 4 = \pi / 2$

$$a_n = \frac{4}{T} \int_{(T)} x(t)\cos(nwt)dt$$

$$= \frac{4}{4}\left[\int_{-2}^{0} (3t+3)\cos(n\pi t/2)dt\right]$$

$$= 3\left[t\sin(n\pi/2)/n\pi/2 \downarrow_{-2}^{0} - \cos(n\pi/2)/n^2\pi^2/4 \downarrow_{-2}^{0}\right] \quad (5.159)$$

$$+ 3\sin(n\pi/2)/n\pi/2 \downarrow_{-2}^{0}$$

$$= 3(0 + 4[-(-1)^n + 1]/n^2\pi^2 + 0)$$

$$= 12[-(-1)^n + 1]/n^2\pi^2 = 24/n^2\pi^2 \text{ for odd } n$$

$$b_n = 0 \quad (5.160)$$

$$x(t) = \sum_{k=1,}^{\infty}\left[\frac{24}{k^2\pi^2}\right]\cos(k\pi t/2) \text{ for all odd } k$$

$$(5.161)$$

$$= \frac{24}{\pi^2}\left[\cos(0.5\pi t) + \frac{1}{9}\cos(1.5\pi t) + \frac{1}{25}\cos(2.5\pi t) + \ldots\right]$$

Put $t = 2$ in above equation, at $t = 2$, $x(t) = -3$

$$-3 = \frac{24}{\pi^2}\left[(-1) + \frac{1}{9}(-1) + \frac{1}{25}(-1) + \ldots\right]$$

$$\frac{\pi^2}{8} = 1 + \frac{1}{9} + \frac{1}{25} + \ldots$$

This proves the identity.

Concept Check

- What is linearity property of FS?
- Define time scaling, time reversal and time shifting property of FS. State its significance.
- What is significance of property of differentiation in time domain?
- State integration property.
- What are modulation and convolution properties? State the physical significance.
- Explain the meaning of Parseval's identity.
- Explain complex conjugate property of FS.

5.5 Recovery of CT Signal from FS

The signal can be recovered from the FS by taking Inverse FS. The concept is better illustrated using some simple numerical examples.

Example 5.28

Determine the time domain signal using its magnitude and phase spectrum given in Fig. 5.30.

Solution

The exponential series can be written as

$$x(t) = \sum_{n=2\&3} C_n e^{-jn\omega t} = e^{j\pi/4}e^{-j2\omega t} + 2e^{-j\pi/8}e^{-j3\omega t} + e^{-j\pi/4}e^{j2\omega t} + 2e^{j\pi/8}e^{j3\omega t} \quad (5.162)$$

$$= 2\cos(2\omega t - \pi/4) + 4\cos(3\omega t + \pi/8)$$

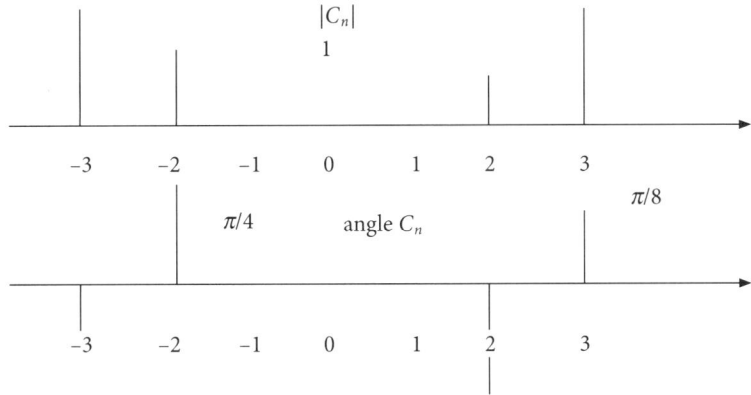

Fig. 5.30 Plot of magnitude and phase spectrum for Example 5.28

Example 5.29

Find the time domain signal with FS coefficients given below and with $\omega = \pi$.
$C_n = j\delta(n-1) - j\delta(n+1) + 2\delta(n-2) + 2\delta(n+2)$

Solution

The signal $x(t)$ can be written as

$$x(t) = \sum_n C_n e^{jn\omega t} = \sum_n C_n e^{jn\pi t}$$

$$x(t) = je^{j\pi t} - je^{-j\pi t} + 2e^{j2\pi t} + 2e^{-j2\pi t} \quad (5.163)$$

$$= -2\sin(\pi t) + 4\cos(2\pi t)$$

5.5.1 Gibbs phenomenon

In the above two examples, the FS coefficients are finite in number. Hence, we can recover the analog signal very easily. This will not be the case if the FS coefficients are infinite in number. We cannot process infinite number of coefficients due to finite processing power of our computer. Gibbs phenomenon cannot be eliminated as we can never use infinite coefficients in the recovery process. The recovered signal using finite number of coefficients will not be the exact square wave signal we started with but the recovered signal will show oscillations near the edge of the square wave. These oscillations near the edge are called as Gibbs phenomenon. When we use finite number of FS coefficients, we can say that we are truncating the infinite FS. This truncation leads to Gibbs phenomenon. Let us consider one example to illustrate the concepts.

Consider the signal given by

$$x_p(t) = \begin{pmatrix} A \text{ for } -\dfrac{T}{2} \leq t \leq \dfrac{T}{2} \\ 0 \text{ otherwise} \end{pmatrix} \qquad (5.164)$$

The spectrum of the signal can be written as

$$C_n = \dfrac{AT}{T_0} \operatorname{sinc}\left(\dfrac{nT}{T_0}\right) \qquad (5.165)$$

where $\operatorname{sinc}(x) = \sin(\pi x)/\pi x$

Let us consider the Fourier series coefficients, for example, 50, 100 and 200 so on, and see the effect of the number of coefficients on the recovered signal.

We will write a MATLAB program to recover the signal. We have used number of coefficients equal to 50 in the program below. The reader is required to change value of N to 100 and 200 and see its effect on the recovered signal. The plot of FS coefficients for $N = 50$ is shown in Fig. 5.31 and the recovered signal is shown in Fig. 5.32. The recovered signal after changing value of N to 100 is shown in Fig. 5.33.

```
clear all;
T0=4;
T=0.4;
A=1;
N=50;
for n=1:N,
    c(n+N+1)=1/(n*pi)*sin(n*pi*T/T0);
end
c(N+1)=0.1;

for n=1:N,
    c(n)=c((2*N+2)-n);
 end
stem(c);
title('plot of magnitude of discrete spectrum');
xlabel('coefficient number');
ylabel('amplitude');
figure;
t=-N:1:N;
stem(t,c);
x=ifft(c);
x1=ifftshift(x);
plot(abs(x1));
```

```
title('plot of recovered signal');
xlabel('sample number');
ylabel('amplitude');
```

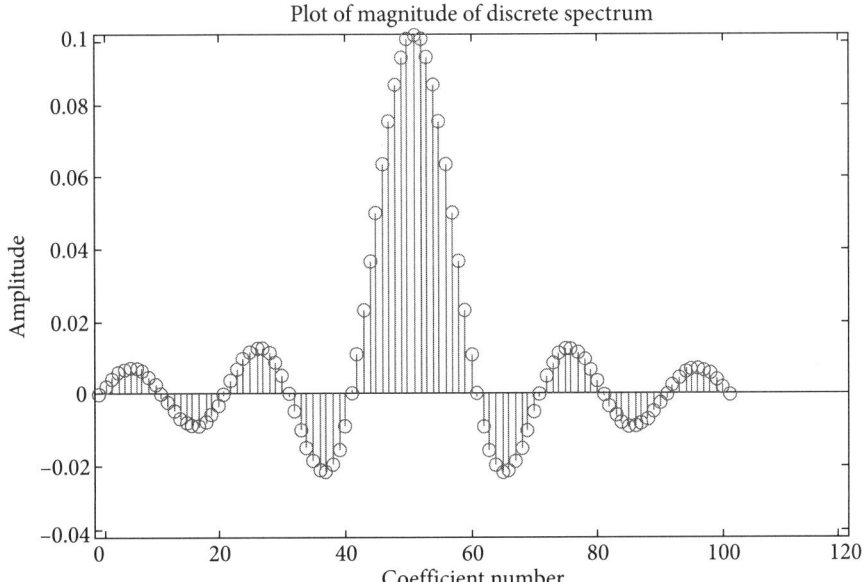

Fig. 5.31 Plot of FS coefficients for $N = 50$

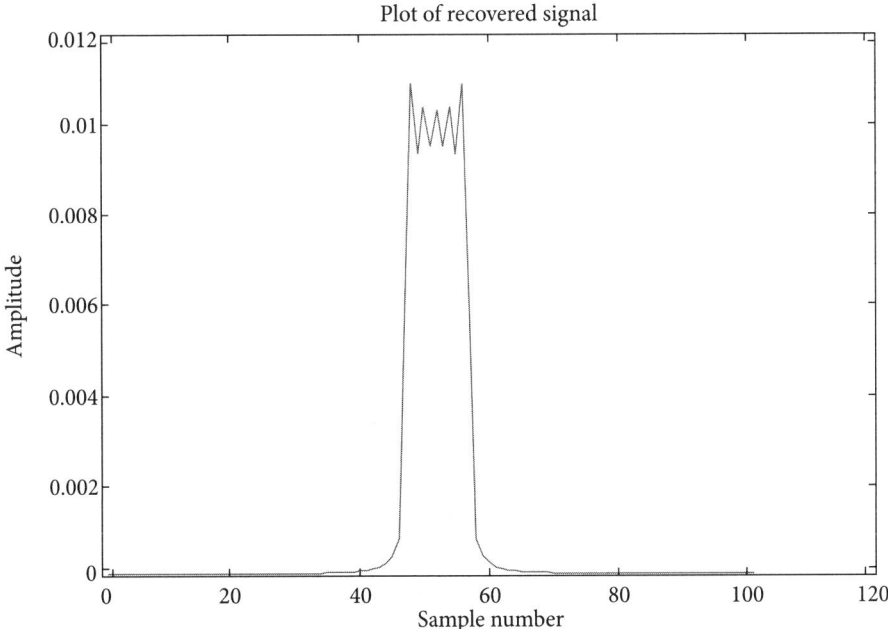

Fig. 5.32 Plot of recovered signal for $N = 50$

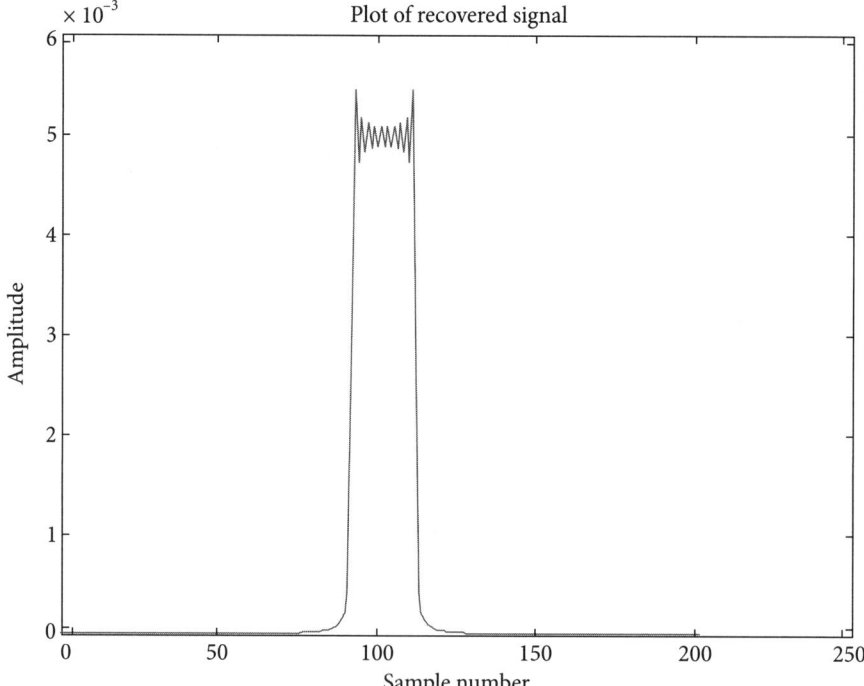

Fig. 5.33 Plot of recovered signal for $N = 100$

We can see from Figs 5.32 and 5.33 that as we use more number of FS coefficients in the recovery process the recovered signal matches more closely towards the original square wave. It can be observed that the amount of overshoot remains the same even after increasing the value of N. The peak overshoot is seen to shift towards the point of discontinuity.

Concept Check

- State the formula for recovery of the signal from FS.
- What will happen if the FS contains infinite terms?
- What is the effect of using only finite number of FS coefficients in the recovery process?
- What is Gibbs phenomenon?
- State the property of the overshoot in case of recovery of the square wave signal.

5.6 FS Representation of DT Periodic Signals

Let us define the orthogonality for DT periodic signals. Two DT periodic signals are said to be orthogonal if their inner product defined as the sum of values in their product is zero. Let us consider the complex sinusoids with frequencies separated by integer multiple of fundamental frequency, for example, $e^{jk\Omega_0 n}$ and $e^{jm\Omega_0 n}$. We will prove their inner product is zero.

$$I_{m,k} = \sum_{n=0}^{n-1} e^{jk\Omega_0 n} e^{-jm\Omega_0 n} = \sum_{n=0}^{N-1} e^{j(k-m)\Omega_0 n} = \begin{cases} N \text{ if } k = m \\ \dfrac{1-e^{-jk2\pi}}{1-e^{-jk\Omega_0}} = 0 \text{ for } k \neq m \end{cases} \quad (5.166)$$

note $e^{-jk2\pi} = 1$

We can conclude that the complex sinusoids with frequencies separated by integer multiple of fundamental frequency are orthogonal to each other.

Discrete Time Fourier Series (DTFS) for periodic DT signals can be written as

$$\bar{x}[n] = \sum_{k=(N)} X[k] e^{jk\Omega_0 n} \quad \text{where } \Omega_0 = \frac{2\pi}{N} \text{ is fundamental frequency}$$

$X[k]$ are FS coefficients

$$X[k] = \frac{1}{N} \sum_{n=(N)} x[n] e^{-jk\Omega_0 n} \quad (5.167)$$

$X[n]$ is DT periodic signal with period equal to N.

We will go through some numerical examples to illustrate the concepts.

Example 5.30

Find the DTFS coefficients for the signal given by $x[n] = \cos\left(\dfrac{\pi}{8} n + \theta\right)$

Solution

Step 1 Find the fundamental period and fundamental frequency.

$$x[n] = \cos\left(\frac{\pi}{8} n + \theta\right) = \cos\left(\frac{2\pi n}{16} + \theta\right) \quad (5.168)$$

Here, fundamental period is $N = 16$

The fundamental frequency is 1/16

Step 2 We will write the signal in terms of exponentials.

$$x[n] = \cos\left(\frac{\pi}{8} n + \theta\right) = \frac{1}{2}\left[e^{j2\pi n/16} e^{j\theta} + e^{-j2\pi n/16} e^{-j\theta}\right] \quad (5.169)$$

The fundamental period is 16. The DTFS will consist of 16 coefficients varying from $k = -7$ to 8.

$$x[n] = \sum_{k=-7}^{8} X[k]e^{jk(\pi/8)n}$$

$$X[k] = \begin{cases} \dfrac{1}{2}e^{-j\theta} & \text{for } k=-1 \\ \dfrac{1}{2}e^{j\theta} & \text{for } k=1 \\ 0 & \text{otherwise} \end{cases} \qquad (5.170)$$

$X[k]$ is also periodic with period of 16. The magnitude and phase plot is shown in Fig. 5.34.

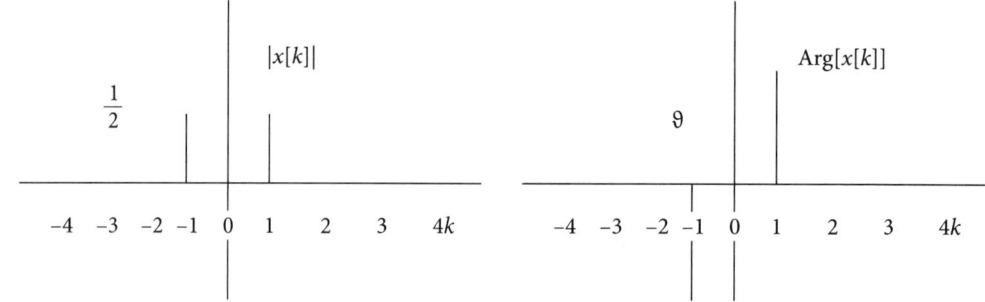

Fig. 5.34 Magnitude and phase plot for the FS in Example 5.30

Example 5.31

Find the DTFS coefficients for the signal given by $x[n] = 2 + \sin\left(\dfrac{\pi}{10}n + \dfrac{\pi}{2}\right)$

Solution

Step 1 Find the fundamental period and fundamental frequency.

$$x[n] = 2 + \sin\left(\frac{\pi}{10}n + \frac{\pi}{2}\right) = 2 + \sin\left(\frac{2\pi n}{20} + \frac{\pi}{2}\right) \qquad (5.171)$$

Here, fundamental period is $N = 20$

The fundamental frequency is $1/20$

Step 2 We will write the signal in terms of exponentials.

$$x[n] = 2 + \sin\left(\frac{\pi}{10}n + \frac{\pi}{2}\right) = 2 + \frac{1}{2j}\left[e^{j2\pi n/20}e^{j\pi/2} - e^{-j2\pi n/20}e^{-j\pi/2}\right] \qquad (5.172)$$

The fundamental period is 20. The DTFS will consist of 20 coefficients varying from $k = -9$ to 10.

$$x[n] = \sum_{k=-9}^{10} X[k]e^{jk(\pi/10)n}$$

$$X[k] = \begin{cases} -\dfrac{1}{2j}e^{-j\pi/2} \text{ for } k=-1 \\ \dfrac{1}{2j}e^{j\pi/2} \text{ for } k=1 \\ 2 \text{ for } k=0 \\ 0 \text{ otherwise} \end{cases} \qquad (5.173)$$

$X[k]$ is also periodic with period of 20. The magnitude and phase plot of FS is shown in Fig. 5.34.

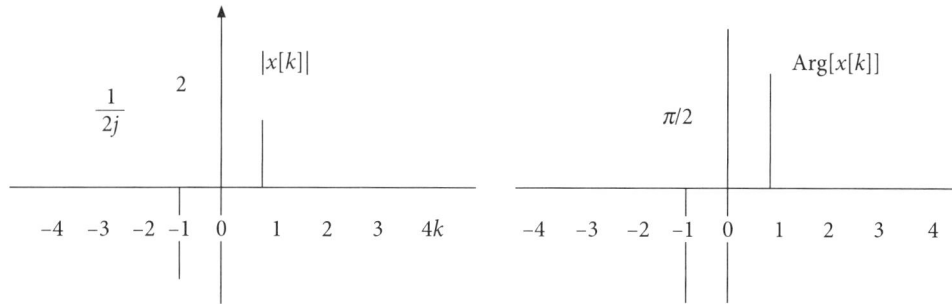

Fig. 5.35 Magnitude and phase plot for the FS in Example 5.31

Example 5.32

Find the DTFS coefficients for the DT periodic signal shown in Fig. 5.36.

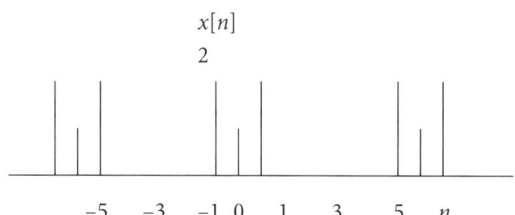

Fig. 5.36 Periodic DT signal

Solution

Step 1 Find the fundamental period and fundamental frequency. The signal is periodic with a period of 6 samples from −3 to 3. We will use the formula for $x[k]$.

$$X[k] = \frac{1}{N}\sum_{n=-2}^{3} x[n]e^{-jkn2\pi/6} = \frac{1}{6}\left[2e^{j\pi k/3} + 2e^{-j\pi k/3} + 1\right]$$

$$X[k] = \frac{1}{6} + \frac{2}{3}\cos(\pi k/3)$$

(5.174)

$$X[k] = \begin{cases} \frac{1}{3}e^{jk\pi/3} & \text{for } 1 \leq k \leq 3 \\ \frac{1}{3}e^{-jk\pi/3} & \text{for } -2 \leq k \leq -1 \\ \frac{1}{6} & \text{for } k = 0 \end{cases}$$

(5.175)

Note from the examples solved above for DT periodic signals that FS for DT periodic signals is also periodic. DTFS has the same period as that of DT signal in terms of number of samples. We can summarize this fact in Table 5.1.

Table 5.1 FS for CT and DT periodic signals

Time domain signal	If periodic?	FS representation
CT	Periodic	Aperiodic (CTFS)
DT	Periodic	Periodic (DTFS)

Summary

We have introduced the concept of representation of signals in terms of linear combination of sinusoids and discussed Fourier series representation for CT periodic signals in detail.

- We started with the introduction of decomposition of the signal in terms of sinusoids or exponential functions. The reason for selection of these functions as basic functions is explained. The concept of orthogonality of basic functions is emphasized for unique representation of signals. The basic functions selected are called as basis functions. Examples are solved to show the orthogonal condition for sinusoids and co-sinusoids of different frequencies. The example of x-axis and y-axis for the representation of point in a plane is illustrated. It is shown that the sinusoids and exponential functions when given as input to LTI system result in the output of the same frequency except for the change in amplitude and phase.

- The formula for computation of Fourier series representation is given and the concept is illustrated with numerical problems. It is shown that when sinusoidal signal or exponential signal is given as input to the LTI system, the output is again a sinusoidal signal or exponential signal of same frequency with a change in amplitude and phase.

- Fourier series representation is a representation of a signal as a sum of scaled fundamental and its scaled harmonics. The amplitude of the fundamental and amplitudes of the harmonic frequencies must be found for writing a Fourier series for a signal. These amplitudes are called as the Fourier series coefficients. The Fourier series coefficient for a particular harmonic represents the degree of matching of the signal with the corresponding harmonic component. Fourier series expansion of a periodic signal is equivalent to resolving a signal into sum of scaled sine and cosine terms of the harmonics. FS representation requires that the signals satisfy the Dirichlet's conditions namely the signal must be single-valued with finite discontinuities or with finite minima and maxima. The signal must also be absolutely integrable. The exponential representation is preferred as it is easy to handle. It also simplifies the mathematical calculations. The presence of negative frequencies appearing in exponential series representation is a by-product. The negative frequencies have no physical significance. FS representation gives a discrete spectrum.

- Numerical examples are solved to illustrate the evaluation of FS coefficients. The magnitude and phase spectrum is drawn. Applications of FS are in the area of communication where the signal is modulated using a pulse train. The representation of a periodic pulse train using exponential Fourier series is illustrated as a discrete spectrum. This can be used to find the spectrum of the modulated signal. The line spectrum for a square wave pulse indicates that the zero crossings will occur at multiple of 1/pulse width. The spacing between the lines will increase when the period of the square wave reduces. The first zero crossing occurs farther away when the width of the pulse reduces. Numerical examples are solved to illustrate the evaluation of FS using trigonometric and exponential signals for periodic signals of different types namely with even symmetry, with odd symmetry and with half wave symmetry. Signals with even symmetry will have no sine terms. Signals with odd symmetry will have no cosine terms. Signals with half wave symmetry will have only odd harmonics.

- The properties of FS namely linearity, time shifting, scaling, reversal are explained with their physical significance. Properties like time differentiation, time integration, modulation and convolution with Parseval's identity are also explained with physical significance for each. Numerical examples illustrating the property of time shifting, scaling and differentiation are solved. Recovery of the signal from FS is explained with examples. When the FS contains infinite coefficients, the practical recovery process does not use infinite coefficients, but uses only finite terms. The effect of using only finite terms in the recovery gives rise to Gibbs phenomenon. The example of recovery of square wave is explained using finite terms. It is shown that there are oscillations and the overshoot is independent of the number of terms used. As the number of terms increases, the overshoot shifts towards the edges. The oscillations near the edges are called as Gibbs phenomenon.

- FS representation of the DT periodic signal is also discussed. Two DT periodic signals are said to be orthogonal if their inner product defined as the sum of values in their product is zero. The complex sinusoids with frequencies separated by integer multiple of fundamental frequency are orthogonal to each other. These sinusoids are used as the basis function for FS representation. The FS representation of periodic DT signals is illustrated with numerical examples. FS for CT periodic signals is aperiodic, whereas FS for DT periodic signals is periodic with same period as that of DT signal.

Multiple Choice Questions

1. Fourier series representation is used for
   - (a) Periodic signals
   - (b) Aperiodic signals
   - (c) Pseudo-periodic signals
   - (d) All signals

2. Sinusoidal function is used for signal decomposition because
   - (a) Response of the LTI system to sinusoid is a sinusoid with same frequency
   - (b) Response of the system is stable
   - (c) Response of LTI system is non-linear
   - (d) Response of LTI system to sinusoid is zero

3. Two signals are said to be orthogonal if
   - (a) the cross product of two signals is zero
   - (b) the dot product of two signals is zero
   - (c) the projection of signal on other signal is maximum
   - (d) the two signals are not related to each other

4. Basis functions must be orthogonal so that
   - (a) the signal representation is uniform
   - (b) the signal representation is unique
   - (c) the signal representation is zero
   - (d) the signal representation is concise

5. Dirichlet's condition requires that
   - (a) the signal is absolute
   - (b) the signal is non integrable
   - (c) the signal is absolutely integrable
   - (d) the signal integration is zero

6. The exponential FS is used because
   - (a) it is a complex series
   - (b) it is having magnitude equal to 1
   - (c) it is giving negative frequency representation
   - (d) it simplifies the mathematics

7. FS coefficients for the signal $x(t) = 5\cos\left(\dfrac{\pi}{2}t\right)$ are
   (a) $X[k] = 5/2$ for $k = \pm 1$
   (b) $X[k] = 5$ for $k = \pm 1$
   (c) $X[k] = 2/5$ for $k = \pm 1$
   (d) $X[k] = 10$ for $k = \pm 1$

8. If the signal has even symmetry, its FS contains only
   (a) sine terms
   (b) sine and cosine terms
   (c) cosine terms
   (d) tan terms

9. If the signal has odd symmetry, its FS contains only
   (a) sine terms
   (b) sine and cosine terms
   (c) cosine terms
   (d) tan terms

10. If the signal has half wave symmetry, then FS contains
    (a) even harmonics
    (b) odd harmonics
    (c) all harmonic terms
    (d) sine terms

11. FS representation of the train of impulses with spacing of T is
    (a) a train of impulses with spacing equal to $2T$
    (b) a train of impulses with spacing equal to $T/2$
    (c) a train of impulses with spacing equal to $1/T$
    (d) a train of impulses wit spacing equal to $2/T$

12. Time shifting property of FS states that
    (a) $x(t) \leftrightarrow C_n$
    then $x(t - t_0) \leftrightarrow e^{-jn\omega t_0} C_n$
    (b) $x(t) \leftrightarrow C_n$
    then $x(t - t_0) \leftrightarrow e^{jn\omega t_0} C_n$
    (c) $x(t) \leftrightarrow C_n$
    then $x(t - t_0) \leftrightarrow e^{-j\omega t_0} C_n$
    (d) $x(t) \leftrightarrow C_n$
    then $x(t - t_0) \leftrightarrow e^{-jn} C_n$

13. Time differentiation of a square wave will result in
    (a) impulses at regular interval
    (b) impulses at the edge locations
    (c) pulses at edge locations
    (d) positive impulse for rising edge and negative impulse for trailing edge

14. Time shifted signals will have
    (a) same magnitude response of FS
    (b) same phase response of FS
    (c) same response of FS
    (d) shifted response of FS

15. FS coefficients of the time scaled signal with scaling factor of α will have
    (a) Same magnitude response and phase is scaled by the same factor α
    (b) Magnitude and phase are scaled by same factor α

(c) Magnitude response is scaled by factor of α

(d) Magnitude and phase are scaled by factor of $1/\alpha$

16. Modulation property of FS states that the FS of modulated signal is
    (a) convolution of FS of message signal and carrier
    (b) multiplication of FS of message signal and carrier
    (c) addition of FS of message signal and carrier
    (d) linear combination of scaled FS of signal and carrier

17. Convolution property states that the FS of convolved signal is
    (a) convolution of FS of two signals
    (b) multiplication of FS of two signals
    (c) addition of FS of two signals
    (d) linear combination of scaled FS of two signals

18. Recovery of the square wave signal using truncated infinite FS results in
    (a) Large oscillations in the recovered signal
    (b) Gibbs phenomenon
    (c) Distortion in the recovered signal
    (d) Small oscillations in the flat portion of the square wave

19. Gibbs phenomenon reduces when
    (a) large number of FS coefficients are used in the recovery
    (b) less number of FS coefficients are used in the recovery
    (c) all infinite FS coefficients are used
    (d) few FS coefficients are used in the recovery

20. FS representation of a periodic DT signal is
    (a) aperiodic
    (b) periodic
    (c) continuous
    (d) discrete and aperiodic

Review Questions

5.1 How will you classify the signals as periodic or aperiodic? What is FS representation of the periodic signal?

5.2 Can you call the signal as a vector? What are basis functions? Why they should be orthogonal?

5.3 Prove that sine and cosine functions of fundamental and the harmonics are orthogonal to each other.

5.4 Why are the exponential functions used in place of sine and cosine functions?

5.5 Why at all use sinusoidal or exponential functions as basis functions?

5.6 State Dirichlet's conditions?

5.7 Why is the FS representation preferred for the decomposition of the signal?

5.8 Can you convert trigonometric FS to exponential FS and vice versa?

5.9 If the signal has some type of symmetry, can you reduce the number of computations by drawing certain conclusions?

5.10 State properties of FS related to time shifting. Explain the physical significance of each property.

5.11 State the convolution and modulation property of FS. What are its applications?

5.12 State Parseval's Identity and explain its significance.

5.13 What is Gibbs phenomenon? What is the cause of this effect? Can you eliminate this effect? How will you reduce it?

Problems

5.1 Prove that $\cos(5\omega t)$ and $\cos(6\omega t)$ are orthogonal to each other.

5.2 Prove that $\exp(j3\omega t)$ and $\exp(j7\omega t)$ are orthogonal to each other.

5.3 Determine the FS representation for the signal given as $x(t) = 3\cos\left(\dfrac{\pi}{3}t + \dfrac{\pi}{2}\right)$.

5.4 Determine the FS representation for the signal given as $x(t) = 4\cos(2\pi t + 2) + \sin(4\pi t)$.

5.5 Determine the FS representation for the signal given as $x(t) = \cos(4\pi t) + \cos(6\pi t)$.

5.6 Consider a pulse train of rectangular pulses of duration T and period T_0, as shown in the figure below. Find FS representation.

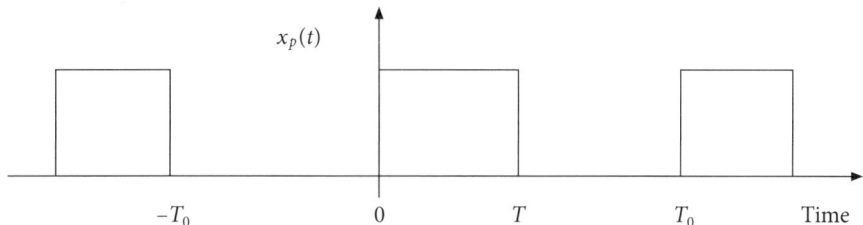

5.7 Determine the FS representation for the signal given by $x(t) = |\sin(2t)|$. The periodic wave is shown in the figure below.

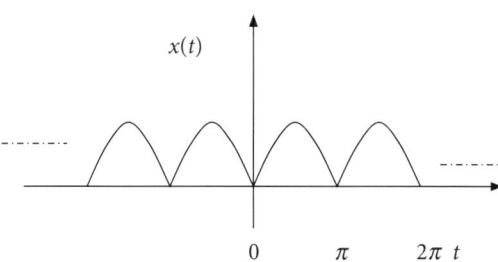

5.8 Determine the Trigonometric series and exponential series representation for the following signals.

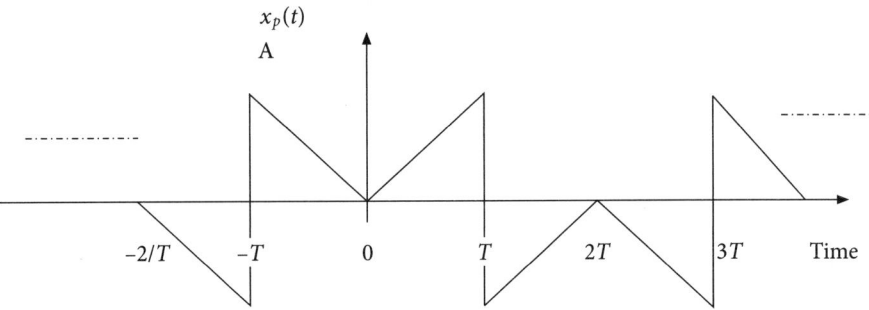

5.9 Find the exponential and trigonometric Fourier series for the signal given below.

$$x(t) = \begin{cases} 0 & \text{for } -1 \leq t < 0 \\ 1 - 0.5\sin(\pi t) & \text{for } 0 \leq t < 1 \end{cases}$$

5.10 Find if the following signals satisfy the Dirichlet conditions.

$x(t) = 2\tan(\pi t)$

$x(t) = \sin(0.5\pi / t)$ for $0 < t < 1$ and signal repeats with period of 1

5.11 Consider a train of pulses as shown in the figure below. Find FS representation for this periodic signal.

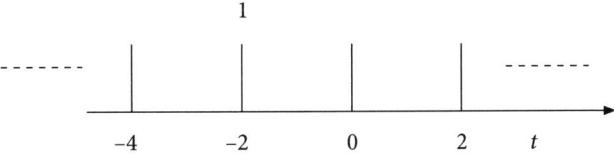

5.12 Use the Fourier series representation for problem 5.6 and find Fourier series representation for the following signal. Use the property of time shifting.

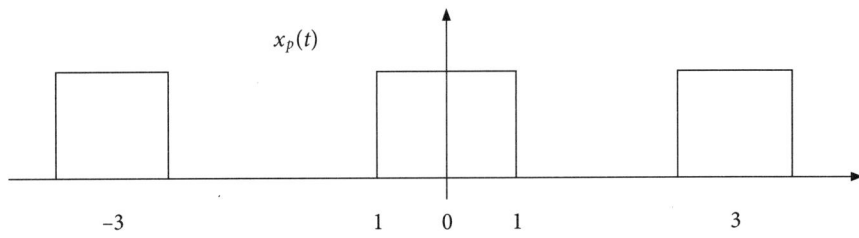

5.13 Use the property of differentiation in time to find FS representation for the signal shown in figure for Example 4.12 above.

5.14 Find the time domain signal with FS coefficients given below and with $\omega = \pi$. $C_n = j\delta(n-2) - j\delta(n+2) + 4\delta(n-3) + 4\delta(n+3)$

5.15 Determine the time domain signal using its magnitude and phase spectrum given in the figure below.

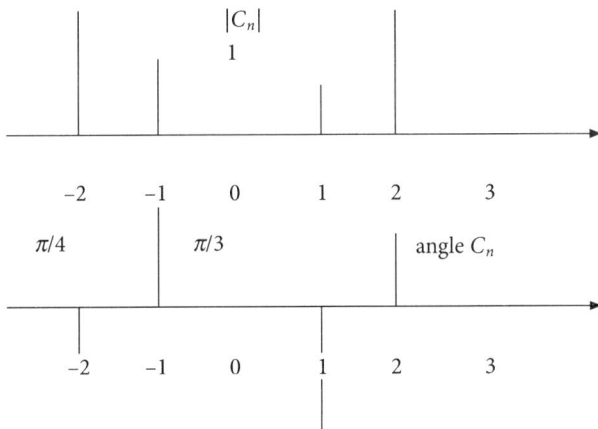

5.16 Find the DTFS coefficients for the signal given by $x[n] = 3\sin\left(\dfrac{\pi}{4}n + \beta\right)$

5.17 Find the DTFS coefficients for the signal given by $x[n] = 1/2 + \cos\left(\dfrac{\pi}{5}n + \dfrac{\pi}{4}\right)$

5.18 Find the DTFS coefficients for the DT periodic signal shown in the figure below.

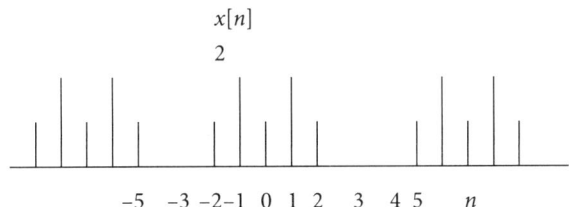

Answers

Multiple Choice Questions

1 (a)	2 (a)	3 (b)	4 (b)	5 (c)
6 (d)	7 (a)	8 (c)	9 (a)	10 (b)
11 (c)	12 (a)	13 (d)	14 (a)	15 (a)
16 (a)	17 (b)	18 (b)	19 (a)	20 (b)

Problems

5.1 --

5.2 --

5.3 $X[k] = \begin{cases} \dfrac{3}{2} e^{-j\left(\frac{\pi}{2}\right)} & \text{for } k = -1 \\ \dfrac{3}{2} e^{j\left(\frac{\pi}{2}\right)} & \text{for } k = 1 \\ 0 & \text{otherwise} \end{cases}$

5.4 $X[k] = \begin{cases} 2e^{2j} & \text{for } k = 1 \\ 2e^{-2j} & \text{for } k = -1 \\ 2/j = -2j & \text{for } k = 2 \\ -\dfrac{2}{j} = 2j & \text{for } k = -2 \\ 0 & \text{otherwise} \end{cases}$

5.5. $X[k] = \begin{cases} 1/2 & \text{for } k = 2 \\ 1/2 & \text{for } k = -2 \\ 1/2 & \text{for } k = 3 \\ 1/2 & \text{for } k = -3 \\ 0 & \text{otherwise} \end{cases}$

5.6 $c_n = \dfrac{A}{2n\pi} \left[\exp\left(-\dfrac{j2\pi nT}{T_0}\right) - 1 \right]$

or $c_n = -j\dfrac{2A}{n\pi}$ for odd n

5.7 $X[k] = \dfrac{2j}{(1-4k^2)\pi}$ for all k

5.8   The waveform has quarter wave symmetry.

$$X[k] = -\frac{2}{k\pi} + \frac{2}{k^2\pi^2} \text{ for } k \text{ odd}$$

$$= \frac{2}{k\pi} \text{ for } k \text{ even}$$

$$a_n = 0, a_1 = 0 \text{ and } b_n = \frac{4}{n\pi}$$

5.9 $x(t) = \frac{1}{2} + \frac{1}{2\pi} + \sum_{k=-\infty, k\neq 0}^{\infty} \left[\frac{1}{jk\pi} + \frac{-k}{2(k^2-1)\pi} \right] e^{jk\pi t}$ for k odd

5.10 Signal 1 is not absolutely integrable and signal 2 has infinite number of extrema points, so both the signals do not satisfy Dirichlet conditions.

5.11 $x(t) = \sum_{k=-\infty}^{\infty} \frac{1}{2} e^{jk\omega t}$

5.12 $X[k] = \frac{-j}{2\pi k} \left[e^{-jk(2\pi/3)} - e^{jk(2\pi/3)} \right]$

$$= -\frac{1}{\pi k} \sin(2\pi k/3) \text{ for } k \neq 0$$

5.13 $C_n = \frac{D_n}{jn\omega} = \frac{2jT_0}{jn2\pi T_0} \sin\left(\frac{2\pi nT}{2T_0}\right) = -\frac{1}{n\pi} \sin\left(\frac{2n\pi}{3}\right)$ put $T/2 = 1$ and $T_0 = 3$

5.14 $x(t) = -2\sin(2\pi t) + 8\cos(3\pi t)$

5.15 $x(t) = 2\cos(2\omega t - \pi/4) + 4\cos(\omega t + \pi/3)$

5.16 $X[k] = \begin{cases} -\frac{1}{2j} e^{-j\beta} & \text{for } k = -1 \\ \frac{1}{2j} e^{j\beta} & \text{for } k = 1 \\ 0 & \text{otherwise} \end{cases}$

5.17 $X[k] = \begin{cases} \dfrac{1}{2}e^{-j\pi/4} & \text{for } k = -1 \\ \dfrac{1}{2}e^{j\pi/4} & \text{for } k = 1 \\ \dfrac{1}{2} & \text{for } k = 0 \\ 0 & \text{otherwise} \end{cases}$

5.18 $X[k] = \begin{cases} \dfrac{1}{3}e^{jk\pi/3} + \dfrac{1}{6}e^{2jk\pi/3} & \text{for } 1 \leq k \leq 3 \\ \dfrac{1}{3}e^{-jk/3} + \dfrac{1}{6}e^{-2jk\pi/3} & \text{for } -2 \leq k \leq -1 \\ \dfrac{1}{6} & \text{for } k = 0 \end{cases}$

6

Fourier Transform Representation of Aperiodic Signals

Learning Objectives

- Fourier transform of aperiodic CT signal
- Fourier transform for some special signals
- Properties of FT
- Applications of FT
- Use of Dirac delta function
- Evaluation of FT of periodic signals using Dirac delta function
- Fourier transform of aperiodic DT signals
- Fourier transform of periodic DT signals
- Response of LTI system using FT

The chapter deals with Fourier transform (FT) representation for continuous time and discrete time aperiodic signals. The FT of CT signals is called as CTFT and that for DT signals is called as DTFT. The Fourier transform is defined and is evaluated for all standard aperiodic signals such as exponential signal, rectangular pulse, triangular pulse, etc. The IFT of some standard signals such as sinc function is also discussed. The use of Dirac Delta function is explained for evaluation of FT for periodic signals. FT of DT signals are illustrated with some numerical examples. Properties of FT and DTFT are emphasized with their physical significance. The numerical examples for the calculation of FT using FT properties are illustrated. The calculation of response of LTI system to the input signal is simplified using FT.

6.1 Fourier Transform Representation of Aperiodic CT Signals

If the signal $x(t)$ is aperiodic, a similar representation for a signal in frequency domain can be developed in terms of Fourier Transform using exponential signals as basis function. The signal $x_p(t)$ a periodic signal can be generated by repeating $x(t)$ after a period of T_0. We can now define $x(t)$ as

$$x(t) = \lim_{T_0 \to \infty} x_p(t) \tag{6.1}$$

The Fourier series representation can be written for the periodic signal $x_p(t)$. Equation (5.38) becomes

$$x(t) = \lim_{T_0 \to \infty} \sum_{n=-\infty}^{\infty} C_n \exp\left(\frac{j2\pi nt}{T_0}\right) \tag{6.2}$$

The coefficients can be written as

$$C_n = \lim_{T_0 \to \infty} \frac{1}{T_0} \int_{-T_0/2}^{T_0/2} x_p(t) \exp\left(\frac{-j2\pi nt}{T_0}\right) dt \quad n = 0, \pm 1, \pm 2 \ldots \tag{6.3}$$

As the period tends to infinity, the spacing between the spectral lines, which is equal to $1/T_0$, will tend to zero and the spectrum will be a continuous spectrum. Now, the summation will turn into the integral. The coefficients will represent a frequency point on the frequency axis represented by $X(f)$. The Eqs (6.2) and (6.3) can be written as

$$x(t) = \int_{-\infty}^{\infty} X(f) \exp(j2\pi f t) df \text{ or } x(t) = \frac{1}{2\pi} \int_{-\infty}^{\infty} X(j\omega) e^{j\omega t} d\omega \tag{6.4}$$

where $f = \dfrac{n}{T_0}$ as $T_0 \to \infty$

$$X(f) = \frac{1}{T_0} \int_{-\infty}^{\infty} x(t) \exp(-j2\pi f t) dt$$

$$X(j\omega) = \int_{-\infty}^{\infty} x(t) e^{-j\omega t} dt \tag{6.5}$$

The Eqs (6.4) and (6.5) represent the inverse Fourier Transform and Fourier Transform of the signal. FT of continuous time aperiodic signal can be denoted as Continuous Time Fourier Transform (CTFT) i.e., Fourier Transform representation of a continuous time signal namely the analog signal that is aperiodic. CT aperiodic signal is represented as a superposition of complex sinusoids involving continuum of frequencies ranging from $-\infty$ to ∞. The signal $x(t)$ and its FT are said to be transform pairs and are denoted as $x(t) \leftrightarrow X(j\omega)$. The integrals in Eqs (6.4) and (6.5) may not always converge for all $x(t)$ and $X(j\omega)$. Without going into the details of analysis of convergence, we will simply state the condition for convergence. Let us define

$$\bar{x}(t) = \frac{1}{2\pi} \int_{-\infty}^{\infty} X(j\omega) e^{j\omega t} d\omega \qquad (6.6)$$

The condition of convergence means the signal recovered after inverse FT must approach the actual signal $x(t)$. We may reduce this condition as the mean squared error (MSE) between the two approaches zero.

$$MSE = \int_{-\infty}^{\infty} |x(t) - \bar{x}(t)|^2 \, dt \to 0 \qquad (6.7)$$

This condition will be true only if the function $x(t)$ is square integrable i.e., if

$$\int_{-\infty}^{\infty} |x(t)|^2 \, dt < \infty \qquad (6.8)$$

This simply means the difference has zero energy and it does not guarantee the convergence at all points in time. The point-wise convergence can be confirmed if the signal satisfies the Dirichlet conditions stated below.

1. The signal is absolutely integrable. i.e., the Eq. (6.8) holds good.
2. The signal $x(t)$ has finite number of maxima, minima and discontinuities in the finite range for which the signal exists or is defined. The signal shown in Fig. 6.1 has finite number of maxima, minima and discontinuities.

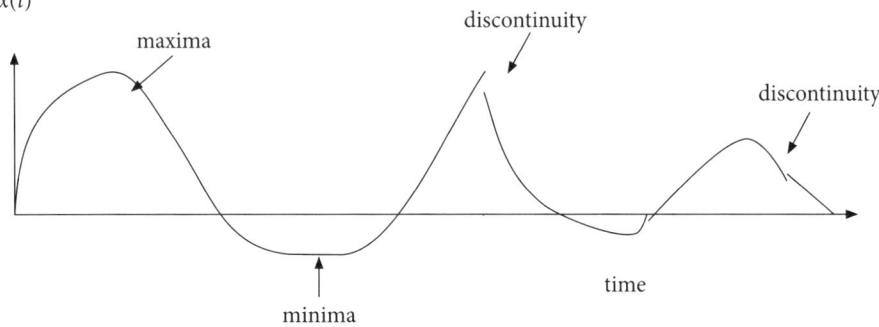

Fig. 6.1 Signal with finite number of maxima, minima and discontinuities

3. The size of the discontinuities is finite i.e., the signal must not tend to infinity at the point of discontinuity. We can observe that the signal $x(t)$ in Fig. 6.1 is not tending to infinity at the point of discontinuity.

FT of CT aperiodic signal is a continuous function of frequency and is also aperiodic. If we consider the signals encountered in practice, the conditions 2 and 3 are satisfied. However, sometimes the signal may not be square or absolutely integrable, for example, the unit step signal. It is possible to define FT for some of such cases using impulses as will be explained in section 6.2.1. Here, FT does not converge. However, we can still make use of FT for simplifying the mathematical calculations.

The reader is advised to closely observe Eqs (6.5) for FT and (6.6) for IFT. It can be observed that there is an additional factor of $1/2\pi$ in IFT and the sign of exponential function is reversed in IFT as compared to FT. Otherwise, the equations have a similarity. We will discuss it further in the property of duality. We will solve some numerical problems to illustrate the calculation of FT.

Example 6.1

Determine the FT representation for the signal given as $x(t) = e^{-3t}u(t)$.

Solution

The exponential signal $x(t)$ exists between zero to infinity and is termed as a right-handed signal. We can use Eq. (6.5) to find the FT.

$$X(j\omega) = \int_{-\infty}^{\infty} e^{-3t}u(t)e^{-j\omega t}dt = \int_{0}^{\infty} e^{-(3+j\omega)t}dt$$

$$X(j\omega) = -\frac{1}{3+j\omega}e^{-(3+j\omega)t}\Big\downarrow_{0}^{\infty} \qquad (6.9)$$

$$= \frac{1}{3+j\omega}$$

6.1.1 Evaluation of magnitude and phase response using hand calculations

We define the magnitude response of signal characterized by its FT. FT expression is a complex function. Magnitude response defines the magnitude of FT (a complex number) at different values of angular frequency ω. Phase response defines the phase angle of FT (a complex number) at different values of angular frequency ω. Let us first understand the physical significance of the Magnitude and Phase response plots. Magnitude response plot signifies the magnitudes of different angular frequencies in a signal. Phase response plot signifies the phase angle of different angular frequencies in a signal.

We can write the magnitude and phase spectrum as

$$|X(j\omega)| = \left|\frac{1}{3+j\omega} \times \frac{3-j\omega}{3-j\omega}\right| = \left|\frac{3-j\omega}{9+\omega^2}\right|$$

$$= sqaue\ root(real\ part^2 + img\ part^2)$$

$$= square\ root\left(\frac{3}{9+\omega^2}\right)^2 + \left(\frac{j\omega}{9+\omega^2}\right)^2 \tag{6.10}$$

$$= \sqrt{\frac{9+\omega^2}{(9+\omega^2)^2}} = \frac{1}{\sqrt{9+\omega^2}}$$

To evaluate magnitude response, we have to put different values of ω in Eq. (6.10). We have to find the magnitude of FT by finding a square root of sum of real part square and imaginary part square, as the equation for FT is a complex quantity. Let us put values of ω as 1, 2, etc. in the Eq. (6.10).

if $\omega = 0, |X(j\omega)| = \frac{1}{\sqrt{9}} = \frac{1}{3} = 0.3333$

if $\omega = \pm 1, |X(j\omega)| = \frac{1}{\sqrt{9+1}} = 0.3162$

if $\omega = \pm 2, |X(j\omega)| = \frac{1}{\sqrt{9+4}} = 0.2774$

$$\tag{6.11}$$

if $\omega = \pm 3, |X(j\omega)| = \frac{1}{\sqrt{9+9}} = 0.2357$

if $\omega = \pm 4, |X(j\omega)| = \frac{1}{\sqrt{9+16}} = 0.2$

if $\omega = \pm 5, |X(j\omega)| = \frac{1}{\sqrt{9+25}} = 0.1715,\ etc.$

Note We can see that the magnitude of FT decreases as the angular frequency increases.

Owing to symmetry property, the magnitude response on negative ω axis is symmetric with respect to the magnitude response on positive ω axis. FT represents the frequency contents of the signal with its magnitude and phase. Let us understand the physical significance of this symmetry, if any.

Physical significance of the symmetry in magnitude response The evaluation of FT uses the exponential representation. A sinusoid or a co-sinusoid of any angular frequency is represented as

$$\sin(\omega t) = \frac{1}{2j}\left[e^{j\omega t} - e^{-j\omega t}\right]$$

(6.12)

$$\cos(\omega t) = \frac{1}{2}\left[e^{j\omega t} + e^{-j\omega t}\right]$$

The representation uses the range of both positive as well as negative angular frequencies. The frequency as determined by number of oscillations per second is definitely a positive quantity. We have introduced negative frequencies for mathematical convenience. The exponential terms simplify the mathematical calculations. This negative frequency information is just a replication of the positive frequency information. It carries no additional meaning. It has no physical significance. It is just appearing as a by-product because we use exponential signal representation.

$$\arg(X(j\omega)) = \arctan\frac{\text{img part}}{\text{real part}} = -\arctan\left(\frac{\omega}{3}\right)$$

(6.13)

To evaluate phase response, we have to put different values of ω in Eq. (6.13). Let us put values of ω as 1, 2, etc. in the equation. The Eq. (6.13) indicates that the phase response for negative frequencies is anti-symmetric with respect to the response for positive frequencies.

if $\omega = 0$, $\arg(X(j\omega)) = -\arctan(0) = 0.00$

if $\omega = 1$, $\arg(X(j\omega)) = -\arctan(1/3) = -0.3218$

if $\omega = 2$, $\arg(X(j\omega)) = -\arctan(2/3) = -0.5880$

if $\omega = 3$, $\arg(X(j\omega)) = -\arctan(3/3) = -0.7854$

if $\omega = 4$, $\arg(X(j\omega)) = -\arctan(4/3) = -0.9273$

if $\omega = 5$, $\arg(X(j\omega)) = -\arctan(5/3) = -1.0304$ (6.14)

if $\omega = -1$, $\arg(X(j\omega)) = -\arctan(1/3) = 0.3218$

if $\omega = -2$, $\arg(X(j\omega)) = -\arctan(2/3) = 0.5880$

if $\omega = -3$, $\arg(X(j\omega)) = -\arctan(3/3) = 0.7854$

if $\omega = -4$, $\arg(X(j\omega)) = -\arctan(4/3) = 0.9273$

if $\omega = -5$, $\arg(X(j\omega)) = -\arctan(5/3) = 1.0304$, etc.

We see that as the value of ω increases, the angle value decreases. The angle value is negative for positive values of ω. The phase (angle) value is positive for negative values of ω and it increases as the negative value of ω increases.

Let us now write a MATLAB program to plot the magnitude and phase response and verify the result of hand calculations. Fig. 6.2 shows the plot of the signal and Fig. 6.3 shows magnitude and phase plot of the signal. The result of magnitude value and phase value may be verified by the reader using the following MATLAB program and display the result of z and z_1 in the command window.

```
clear all;
t=0:0.1:40;
x=exp(-3*t);
plot(t,x);title('plot of exponential signal');xlabel
('time');ylabel('amplitude');
for i=1:20,
    y(i)=abs(1/sqrt((9+(i)*(i))));
end;
z(21)=1/3;
for i=1:20,
    z(i+21)=y(i);
end
for i=1:20,
    z(i)=y(21-i);
end
figure;
subplot(2,1,1);
s=-20:1:20;
plot(s,z);title('Magnitude plot of Fourier Transform
of exponential signal');xlabel('frequency');ylabel
('amplitude');
subplot(2,1,2);
for i=1:20,
    y1(i)=angle(1/(3+1j*i));
end;
z1(21)=0.0;
for i=1:20,
    z1(i+21)=y1(i);
end
for i=1:20,
    z1(i)=-y1(21-i);
end;
plot(s,z1);title('Phase plot of FT of exponential si
gnal');xlabel('frequency');ylabel('angle');
```

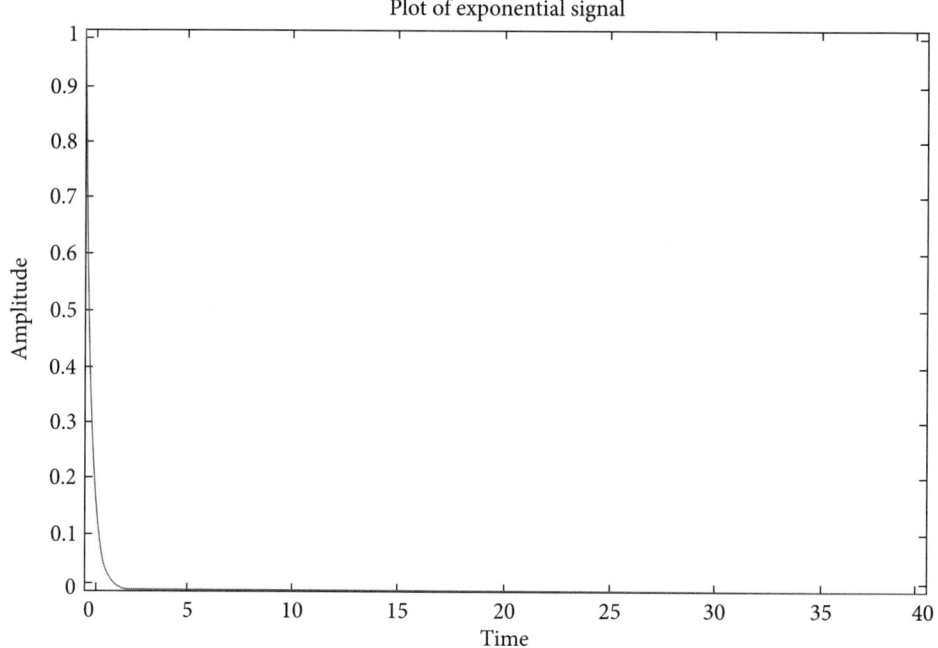

Fig. 6.2 Plot of exponential signal

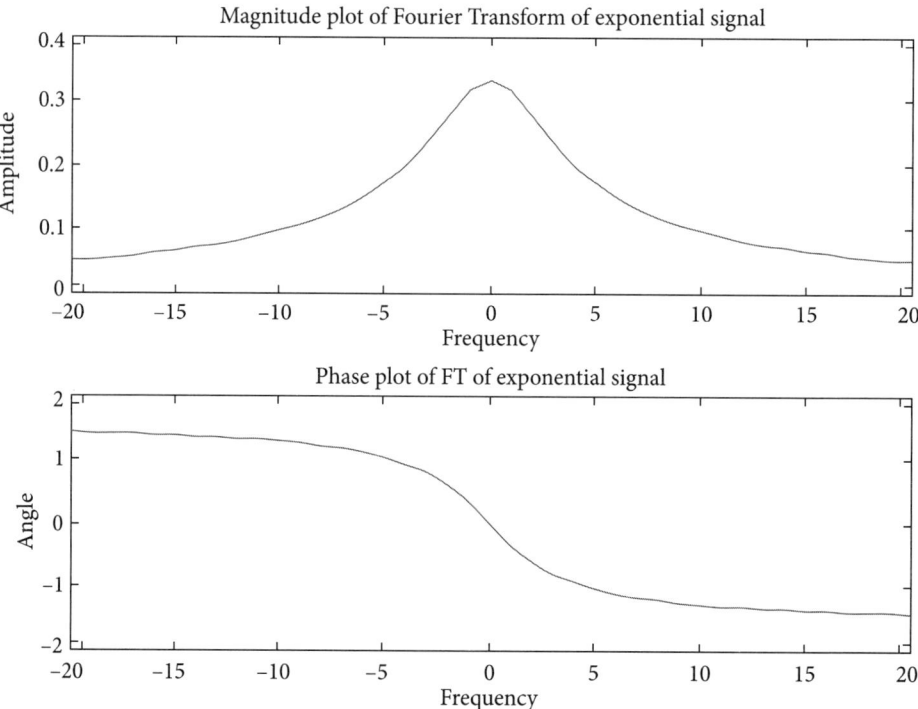

Fig. 6.3 Plot of magnitude and phase of FT of the exponential signal

Example 6.2

Determine the FT representation for the signal given as $x(t) = e^{-3t}u(-t)$.

Solution

The signal exists from minus infinity to zero and is termed as the left-handed signal. We can use Eq. (6.5) to find the FT.

$$X(j\omega) = \int_{-\infty}^{\infty} e^{3t}u(-t)e^{-j\omega t}dt = \int_{-\infty}^{0} e^{(3-j\omega)t}dt = \int_{0}^{\infty} e^{-(3-j\omega)t}dt$$

$$X(j\omega) = -\frac{1}{3-j\omega}e^{-(3-j\omega)t}\Big|_{0}^{\infty} \qquad (6.15)$$

$$= \frac{1}{3-j\omega}$$

We can write the magnitude and phase spectrum as

$$|X(j\omega)| = \left|\frac{1}{3-j\omega} \times \frac{3+j\omega}{3+j\omega}\right| = \left|\frac{3+j\omega}{9+\omega^2}\right|$$

$$= \text{square root}(\text{real part}^2 + \text{img part}^2)$$

$$= \text{square root}\left(\frac{3}{9+\omega^2}\right)^2 + \left(\frac{j\omega}{9+\omega^2}\right)^2 \qquad (6.16)$$

$$= \sqrt{\frac{9+\omega^2}{(9+\omega^2)^2}} = \frac{1}{\sqrt{9+\omega^2}}$$

$$\arg(X(j\omega)) = \arctan\frac{\text{img part}}{\text{real part}} = -\arctan\left(-\frac{\omega}{3}\right) \qquad (6.17)$$

To evaluate magnitude response, we have to put different values of ω in Eq. (6.16). Let us put values of ω as 1, 2, etc. in the equation.

if $\omega = 0, |X(j\omega)| = \frac{1}{\sqrt{9}} = 0.3333$

if $\omega = \pm 1, |X(j\omega)| = \frac{1}{\sqrt{9+1}} = 0.3162$

$$\text{if } \omega = \pm 2, |X(j\omega)| = \frac{1}{\sqrt{9+4}} = 0.2774$$

$$\text{if } \omega = \pm 3, |X(j\omega)| = \frac{1}{\sqrt{9+9}} = 0.2357$$

(6.18)

$$\text{if } \omega = \pm 4, |X(j\omega)| = \frac{1}{\sqrt{9+16}} = 0.2$$

$$\text{if } \omega = \pm 5, |X(j\omega)| = \frac{1}{\sqrt{9+25}} = 0.1715, \text{ etc.}$$

Note We can see that the magnitude of FT decreases as the angular frequency increases. Owing to symmetry property, the magnitude response on negative ω axis is symmetric with the magnitude response on positive ω axis. The magnitude response for Example 6.2 is the same as that for Example 6.1. Note that the signal in Example 6.2 is the time-reversed signal of Example 6.1. The reader is requested to refer to the time reversal property of FT in section 6.5.

To evaluate phase response, we have to put different values of ω in Eq. (6.17). Let us put values of ω as 1, 2, etc. in the equation.

The phase response of FT distinguishes the two signals. Phase response for Example 6.1 is negative of the phase response for Example 6.2. Phase response is responsible for distinguishing the causal and anti-causal signal.

$$\begin{aligned}
&\text{if } \omega = 0, \ \arg(X(j\omega)) = -\arctan(0) \quad = 0.00 \\
&\text{if } \omega = 1, \ \arg(X(j\omega)) = -\arctan(-1/3) = 0.3218 \\
&\text{if } \omega = 2, \ \arg(X(j\omega)) = -\arctan(-2/3) = 0.5880 \\
&\text{if } \omega = 3, \ \arg(X(j\omega)) = -\arctan(-3/3) = 0.7854 \\
&\text{if } \omega = 4, \ \arg(X(j\omega)) = -\arctan(-4/3) = 0.9273 \\
&\text{if } \omega = 5, \ \arg(X(j\omega)) = -\arctan(-5/3) = 1.0304 \\
&\text{if } \omega = -1, \ \arg(X(j\omega)) = -\arctan(-1/3) = -0.3218 \\
&\text{if } \omega = -2, \ \arg(X(j\omega)) = -\arctan(-2/3) = -0.5880 \\
&\text{if } \omega = -3, \ \arg(X(j\omega)) = -\arctan(-3/3) = -0.7854 \\
&\text{if } \omega = -4, \ \arg(X(j\omega)) = -\arctan(-4/3) = -0.9273 \\
&\text{if } \omega = -5, \ \arg(X(j\omega)) = -\arctan(-5/3) = -1.0304, \text{ etc.}
\end{aligned}$$

(6.19)

A MATLAB program is written below for this example. The plot of signal shown in Figs 6.4 and 6.5 shows the magnitude and phase response of the FT of the signal. We can observe that the phase is only reversed and the magnitude response is unaltered.

```
clear all;
t=-40:0.1:0;
x=exp(3*t);
plot(t,x);title('plot of exponential signal');xlabel
('time');ylabel('amplitude');
for i=1:20,
    y(i)=abs(1/sqrt((9+(i)*(i))));
end;
z(21)=1/3;
for i=1:20,
    z(i+21)=y(i);
end
for i=1:20,
    z(i)=y(21-i);
end
figure;
subplot(2,1,1);
s=-20:1:20;
plot(s,z);title('Magnitude plot of Fourier Transform
of exponential signal');xlabel('frequency');ylabel(
'amplitude');
subplot(2,1,2);
for i=1:20,
    y1(i)=angle(1/(3-1j*i));
end;
z1(21)=0.0;
for i=1:20,
    z1(i+21)=y1(i);
end
for i=1:20,
    z1(i)=-y1(21-i);
end;
plot(s,z1);title('Phase plot of FT of exponential si
gnal');xlabel('frequency');ylabel('angle');
```

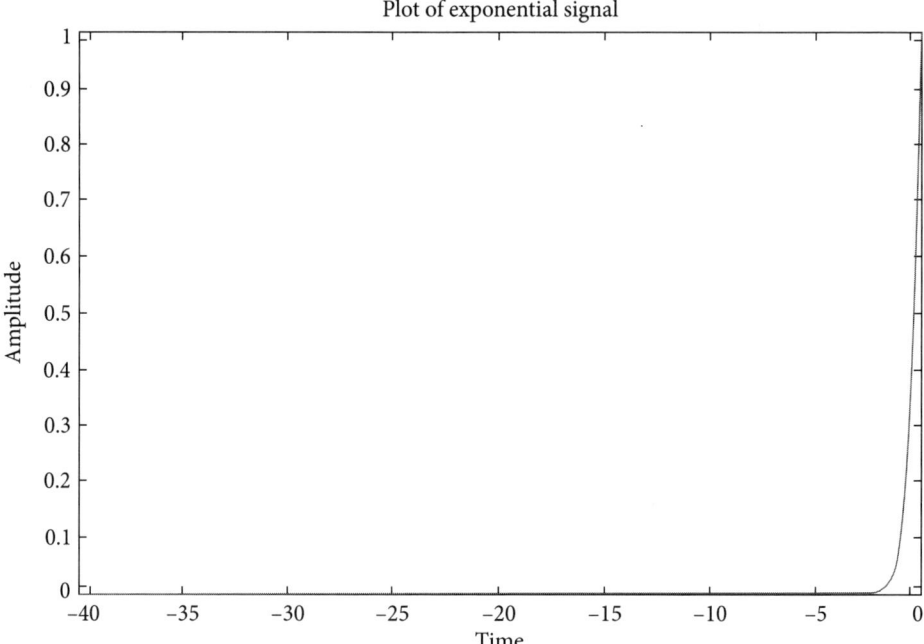

Fig. 6.4 Plot of exponential signal for Example 6.2

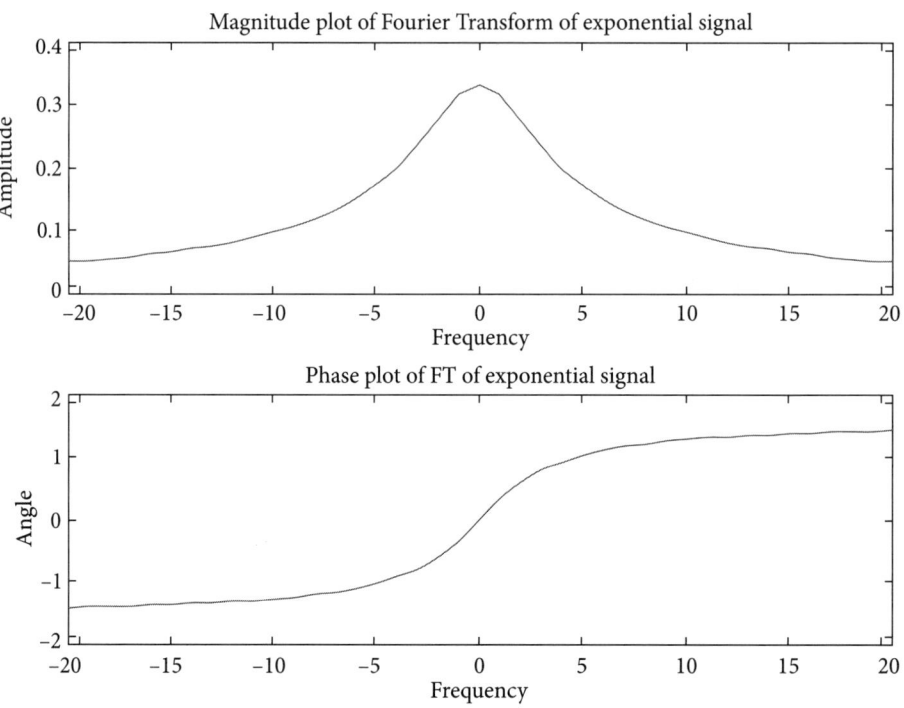

Fig. 6.5 Plot of magnitude and phase response of FT of exponential signal

Example 6.3

Determine the FT representation for the double-sided exponential signal given as $x(t) = e^{-3|t|}$ for all t.

Solution

We can use Eq. (6.5) to find the FT.

$$X(j\omega) = \int_{-\infty}^{\infty} e^{-3|t|} e^{-j\omega t} dt = \int_{-\infty}^{0} e^{(3-j\omega)t} dt + \int_{0}^{\infty} e^{-(3+j\omega)t} dt$$

$$= \frac{1}{(3-j\omega)} e^{(3-j\omega)t} \Big\downarrow_{-\infty}^{0} - \frac{1}{(3+j\omega)} e^{-(3+j\omega)t} \Big\downarrow_{0}^{\infty}$$

$$X(j\omega) = \frac{1}{3-j\omega} + \frac{1}{3+j\omega} \qquad (6.20)$$

$$= \frac{3+j\omega+3-j\omega}{9+\omega^2} = \frac{6}{9+\omega^2} \text{ is real}$$

Note– The phase is zero for all frequencies

A MATLAB program to plot the signal and its magnitude response is given below. Note that the phase values for all frequencies are zero. The phase plot is hence, not shown. Figs 6.6 and 6.7 show the plot of a signal and its magnitude plot of FT.

```
clear all;
t=-40:0.1:40;
x=exp(-3*abs(t));
plot(t,x);title('plot of exponential signal');xlabel
('time');ylabel('amplitude');
for i=1:20,
    y(i)=abs(6/(9+(i)*(i)));
end;
z(21)=6/9;
for i=1:20,
    z(i+21)=y(i);
end
for i=1:20,
    z(i)=y(21-i);
end
figure;
s=-20:1:20;
plot(s,z);title('Magnitude plot of Fourier Transform
of exponential signal'); xlabel('frequency');ylabel
('amplitude');
```

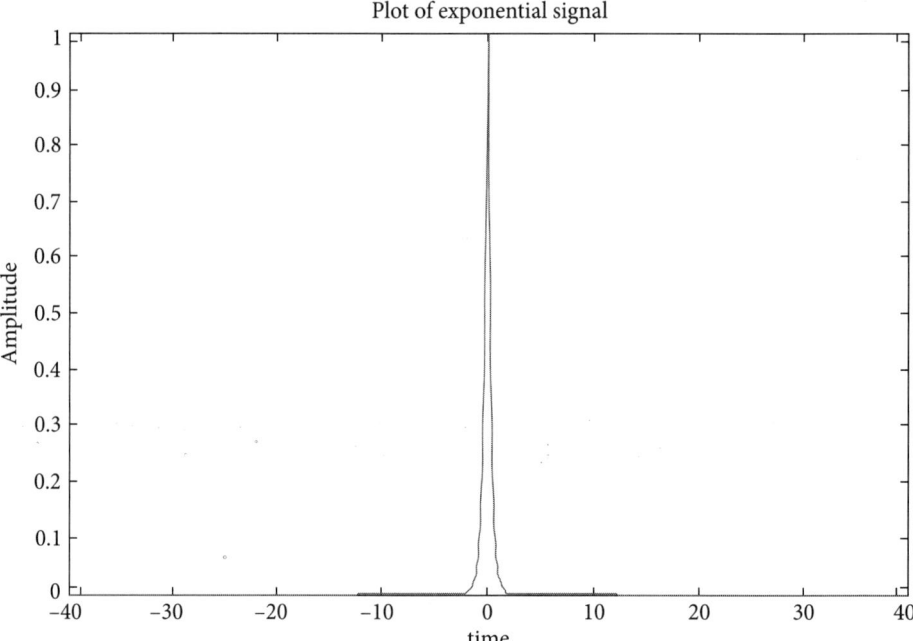

Fig. 6.6 Plot of the signal

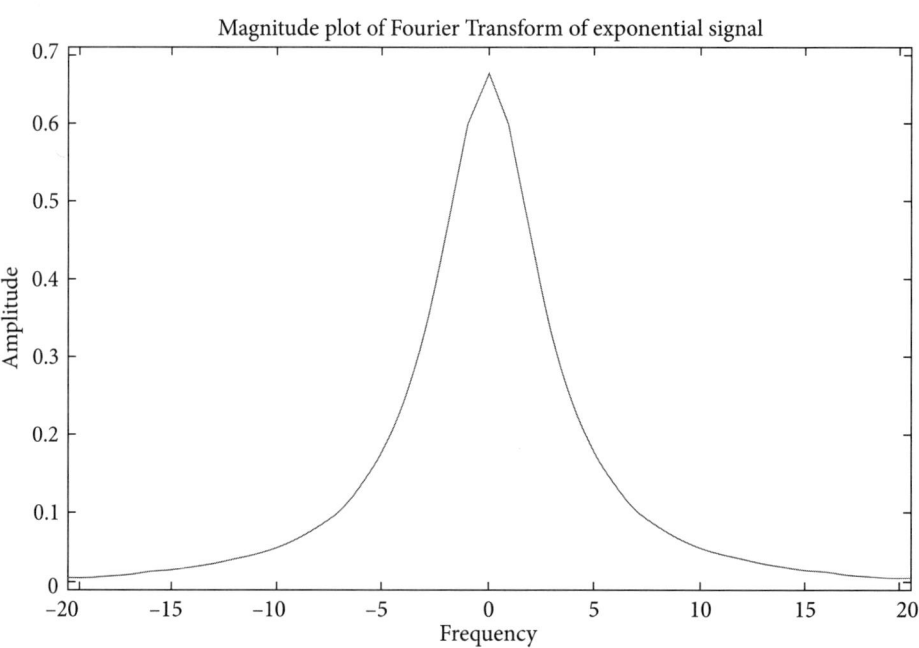

Fig. 6.7 Magnitude plot of FT of the signal

Concept Check

- Write the equations for FT and IFT of aperiodic CT signal and find the similarity and differences in the two equations.
- Write Dirichlet conditions for defining the FT of the signal.
- Can we define a FT for a signal that does not satisfy the Dirichlet conditions?
- Check if CTFT of CT aperiodic signal is also continuous and aperiodic.
- How will you find the magnitude response of the FT $X(j\omega)$ that is a complex number?
- Comment on the symmetric nature of the magnitude response.
- How will you find the phase response of FT?
- Comment on the nature of phase response of FT of the exponential function.
- When the signal is a double-sided exponential, what is the nature of phase response and the magnitude response of FT?

6.2 Fourier Transform of Some Standard CT Signals

Communication systems use a technique called as pulse code modulation (PCM) and then binary phase shift keying. We are required to find the spectrum of the rectangular pulse in PCM signal and the spectrum of BPSK signal. After reading this section the reader will understand the nature of FT of the carrier, the nature of FT of the rectangular pulse and its variation with the width of the pulse. This will give the reader the insight of finding FT of the BPSK signal, which is a multiplication of rectangular pulse with the carrier. The modulation property of FT tells us that multiplication of two signals in the time domain results in the convolution of the transforms of the two signals.

Let us study the representation of aperiodic pulse using Fourier Transform.

Example 6.4

Consider a rectangular pulse of duration T and amplitude A, as shown in Fig. 6.8. Find its FT.

Solution

The rectangular pulse in Fig. 6.7 can be mathematically defined as

$$\text{rect}(t) = \begin{cases} A & -\dfrac{T}{2} \leq t \leq \dfrac{T}{2} \\ 0 & |t| > \dfrac{T}{2} \end{cases} \qquad (6.21)$$

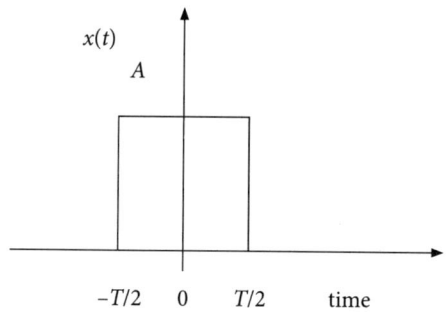

Fig. 6.8 A rectangular pulse of duration T and amplitude A

A rectangular pulse of duration T and amplitude A can be written as

$$x(t) = A\,\text{rect}(t/T) \tag{6.22}$$

A Fourier Transform of this rectangular pulse can be written as

$$X(\omega) = \int_{-T/2}^{T/2} A\exp(-j\omega t)\,dt$$

$$= -\frac{1}{j\omega}Ae^{-j\omega t}\Big\downarrow_{-T/2}^{T/2} = -\frac{A1}{j\omega}[e^{-j\omega T/2} - e^{j\omega T/2}]$$

$$= \frac{2jA\sin(\omega T/2)}{j\omega} = \frac{2A\sin(\omega T/2)}{\omega} \tag{6.23}$$

$$= \frac{2A}{\omega}\sin(\omega T/2) = AT\frac{\sin(2\pi fT/2)}{2\pi fT/2} = AT\,\text{sinc}(fT/2)$$

We thus have a FT pair:

$$A\,\text{rect}(t/T) \Leftrightarrow AT\,\text{sinc}(fT/2) \tag{6.24}$$

The nature of the amplitude spectrum of the rectangular pulse is shown in Fig. 6.9.

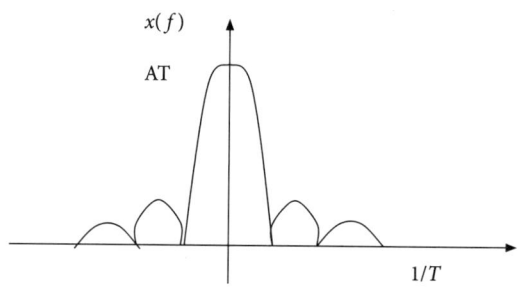

Fig. 6.9 Nature of amplitude spectrum of Rectangular pulse

The zero crossings occur in the spectrum at multiples of $1/T$. As the pulse duration decreases, the zero crossings are separated and the spectrum spreads with the decrease in the peak amplitude. This fact is used in the spread spectrum technique to spread the spectrum of the BPSK signal for application in communication engineering.

Let us write a MATLAB program to plot the continuous spectrum. We have used a value of $T = 0.1$ and $T = 0.2$. The value of A is kept constant, equal to 1. The index 'n' represents the frequency. The plots for $T = 0.2$ and $T = 0.1$ are shown in Figs 6.10 and 6.11. We can observe that as T decreases from 0.2 to 0.1, the spacing between the zero crossings increases and the peak amplitude decreases. This is the important finding. The zero crossing for $T = 0.2$ occurs after an interval of $1/T$ i.e., 5 and for $T = 0.1$, zero crossing occurs after an interval of $1/T$ i.e., 10 samples. The frequency domain picture shows a main peak at the centre with certain width, which is called as the main lobe, and the other similar lobes on either side of are called as side lobes. We can say that as the width of the rectangular pulse in time domain increases, the width of the main lobe in frequency domain decreases and vice versa. This indicates that if the signal gets concentrated in the time domain, it is more spread out in the frequency domain and vice versa. In other words, we say that the time bandwidth product remains constant and has a lower bound. This fact is summarized as a principle called Heisenberg uncertainty principle written as $\delta t \times \delta f \geq \frac{1}{2}$. This equation also indicates that if we want to get good resolution in time domain, we will have to sacrifice for resolution in frequency domain.

```
clear all;
T=0.2;
A=1;
N=50;
for n=1:N,
     c(n+N+1)=A*T*sin(n*pi*T)/(n*pi*T);
 end
c(N+1)=T;
%p(41)=0;
%disp(p);
for n=1:N,
     c(n)=c(2*N+2-n);
end
plot(abs(c));
title('plot of magnitude of continuous spectrum for T=0.2');
xlabel('frequency');
ylabel('amplitude');
```

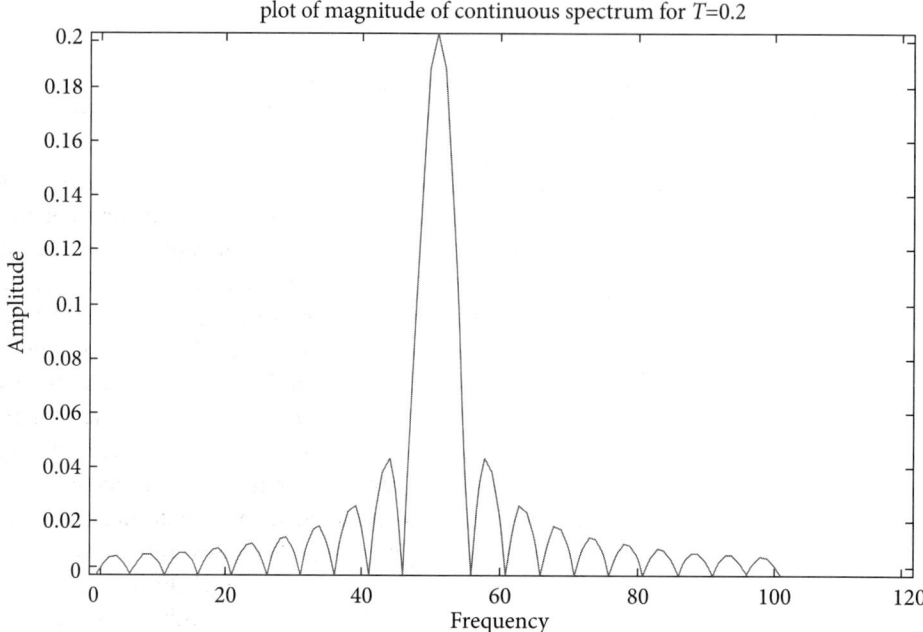

Fig. 6.10 Plot of magnitude response for FT of rectangular pulse-duration 0.2 sec

Note Width of main lobe in frequency domain is 10 units.

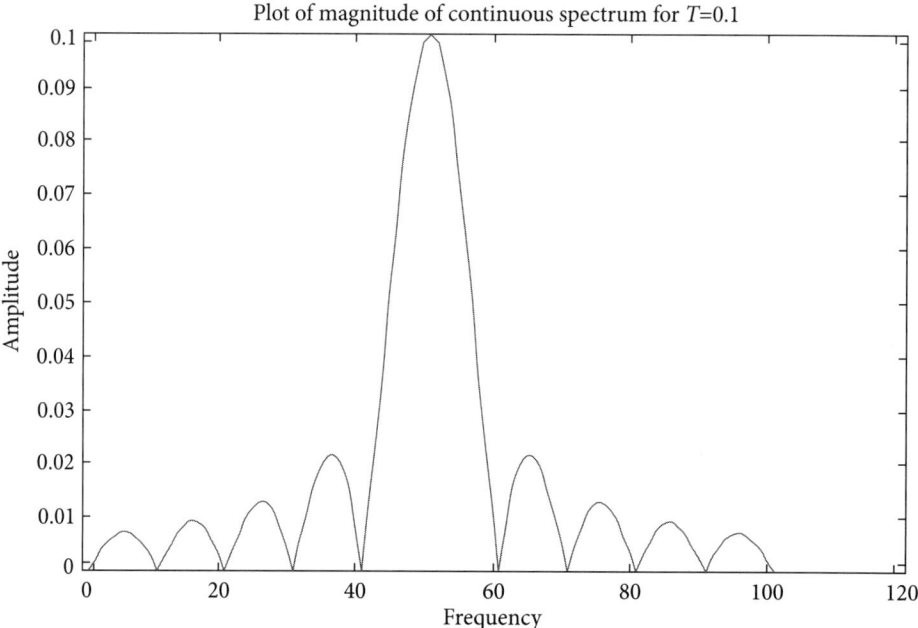

Fig. 6.11 Plot of magnitude response for FT of rectangular pulse -duration 0.1 sec

Note Width of the main lobe in the frequency domain is 20 units. The main lobe is 2 times wider when the width of the rectangular pulse is made half.

Example 6.5

Find the IFT of

$$X(j\omega) = \begin{cases} 1 & -w \leq \omega \leq w \\ 0 & |\omega| > w \end{cases}, \text{ as shown in Fig. 6.12.}$$

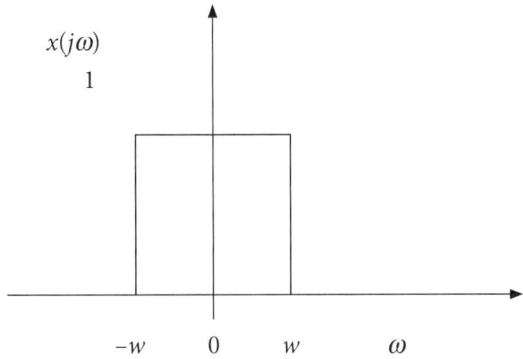

Fig. 6.12 A rectangular frequency domain signal of width w and amplitude 1

Solution

To find IFT, we will use the equation for IFT

$$x(t) = \frac{1}{2\pi} \int_{-\infty}^{\infty} X(j\omega) e^{j\omega t} d\omega$$

$$x(t) = \frac{1}{2\pi} \int_{-w}^{w} e^{j\omega t} d\omega$$

$$x(t) = \frac{1}{2\pi} \int_{-w}^{w} e^{j\omega t} d\omega = \frac{1}{2\pi} \left[\frac{1}{jt} e^{j\omega t} \Big\downarrow_{-w}^{w} \right] \quad (6.25)$$

$$= \frac{1}{2\pi} \left[\frac{1}{jt} \{ e^{jwt} - e^{jwt} \} \right]$$

$$= \frac{1}{\pi t} [\sin(wt)]$$

This is a sinc function. Let us write a MATLAB program for this example. Figure 6.13 shows the magnitude response plot of time domain response for the width of the rectangular pulse in frequency domain equal to 0.2 and the magnitude plot for time domain is shown in Fig. 6.14 for the frequency domain width of 0.1 seconds. We can see that the width of the main lobe of the time domain response decreases as the width of the rectangular pulse in frequency domain increases.

```
clear all;
w=0.2;
A=1;
N=50;
for n=1:N,
    X(n+N+1)=(1/pi)*sin(n*w)/(n);
 end
X(N+1)=w/pi;
for n=1:N,
    X(n)=X(2*N+2-n);
end
plot(abs(X));
title('plot of magnitude of IFT of frequency domain rectangular pulse of width w=0.2');
xlabel('time');
ylabel('amplitude');
```

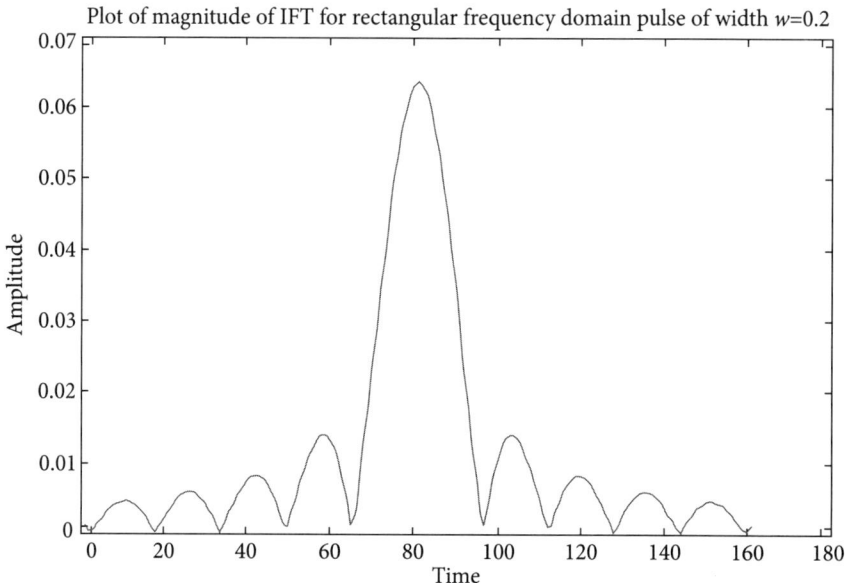

Fig. 6.13 Magnitude plot of the response for frequency domain width = 0.2
(time domain main lobe has width = 96−80 = 16 time units)

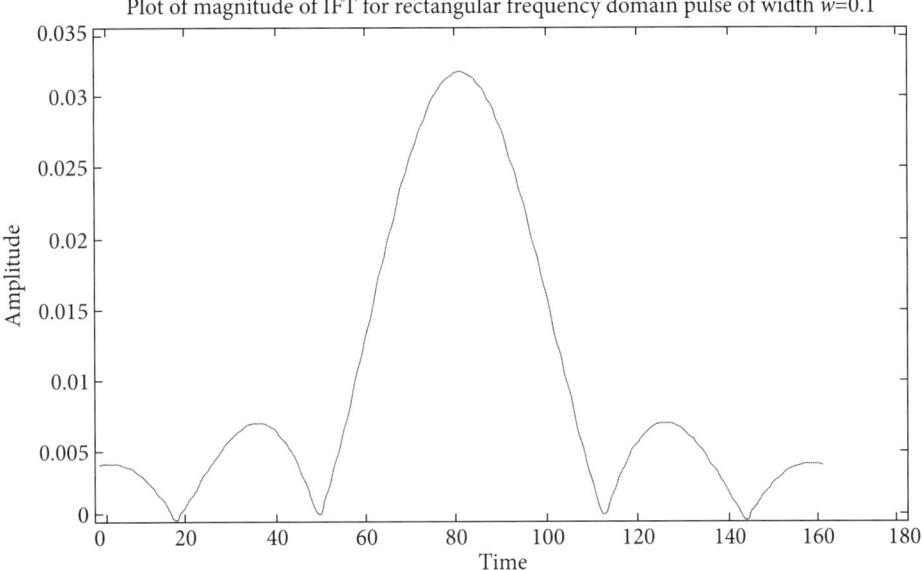

Plot of magnitude of IFT for rectangular frequency domain pulse of width w=0.1

Fig. 6.14 Magnitude plot of the response for frequency domain width = 0.1

(time domain main lobe has width = 112–80 = 32 time units)

Note that a time domain rectangular pulse has a FT as a sinc function and a frequency domain rectangular pulse also has IFT as a sinc function. Hence, if the function is a rectangular function in one domain, it is a sinc function in the other domain. If the function is localized in one domain, it is well spread out in the other domain. When the width of the rectangular pulse increases in one domain time/frequency, then the width of main lobe in the other domain frequency/time decreases and vice versa. This is exactly stated in Heisenberg's uncertainty principle. If we interchange the word time with frequency we get the other property. This phenomenon is called as duality property.

Example 6.6

Find FT of the aperiodic signal given by

$$x(t) = \begin{cases} t \text{ for } |t| \le 1 \\ 0 \text{ for } t > 1 \end{cases}$$

Solution

Let us plot the signal first. The plot of the signal is shown in Fig. 6.15.

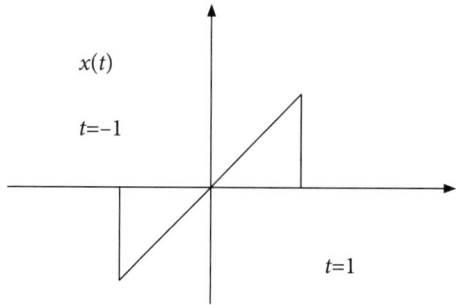

Fig. 6.15 Plot of $x(t)$ for Example 6.6

FT of $x(t)$ can be written as

$$X(j\omega) = \int_{-\infty}^{\infty} x(t)e^{-j\omega t}\,dt = \int_{-1}^{1} te^{-j\omega t}\,dt$$

$$= t\int_{-1}^{1} e^{-j\omega t}\,dt - \int_{-1}^{1}\int_{-1}^{1} e^{-j\omega t}\,dt$$

$$= \frac{te^{-j\omega t}}{-j\omega}\bigg|_{-1}^{1} - \frac{e^{-j\omega t}}{-\omega^2}\bigg|_{-1}^{1} \qquad (6.26)$$

$$= \frac{e^{-j\omega} + e^{j\omega}}{-j\omega} + \frac{e^{-j\omega} - e^{j\omega}}{\omega^2}$$

$$= \frac{j2}{\omega}\cos(\omega) - \frac{j2}{\omega^2}\sin(\omega)$$

Note that, we have to use integration by parts to solve the problem.

Example 6.7

Find FT of the aperiodic triangular pulse signal given by

$$x(t) = \begin{cases} 1 + t/T & \text{for } -T < t \le 0 \\ 1 - t/T & \text{for } 0 < t < T \end{cases}$$

Solution

Let us plot the signal first. The plot of the signal is shown in Fig. 6.16.

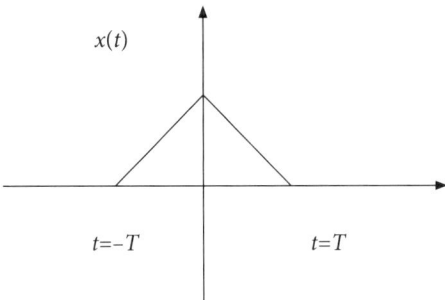

Fig. 6.16 Plot of $x(t)$ for Example 6.7

FT of $x(t)$ can be written as

$$X(j\omega) = \int_{-\infty}^{\infty} x(t)e^{-j\omega t}dt = \int_{-1}^{1} x(t)e^{-j\omega t}dt$$

$$= \int_{-T}^{0}(1+t/T)e^{-j\omega t}dt + \int_{0}^{T}(1-t/T)e^{-j\omega t}dt$$

$$= \frac{e^{-j\omega t}}{-j\omega}\Big\downarrow_{-T}^{0} + \frac{1}{T}\left[\frac{te^{-j\omega t}}{-j\omega}\Big\downarrow_{-T}^{0} + \frac{e^{-j\omega t}}{\omega^2}\Big\downarrow_{-T}^{0}\right] + \frac{e^{-j\omega t}}{-j\omega}\Big\downarrow_{0}^{T} \qquad (6.27)$$

$$-\frac{1}{T}\left[\frac{te^{-j\omega t}}{-j\omega}\Big\downarrow_{0}^{T} + \frac{e^{-j\omega t}}{\omega^2}\Big\downarrow_{0}^{T}\right] +$$

$$= \left[\frac{2}{T\omega^2}(1 - \cos\omega T)\right]$$

Note that we have to use integration by parts to solve the problem.

Example 6.8
Find FT of aperiodic signal given by $x(t) = e^{-2t}u(t-1)$

Solution
FT can be written as

$$X(j\omega) = \int_{-\infty}^{\infty} x(t)e^{-j\omega t}dt = \int_{1}^{\infty} e^{-2t}e^{-j\omega t}dt$$

$$= \int_{1}^{\infty} e^{-(2+j\omega)t}dt$$

$$= \frac{e^{-(2+j\omega)t}}{-(2+j\omega)} \Big\downarrow_1^\infty$$

$$= \frac{1}{-(2+j\omega)} - \frac{e^{-(2+j\omega)}}{-(2+j\omega)} \qquad (6.28)$$

$$= \frac{e^{-(2+j\omega)} - 1}{2+j\omega}$$

Example 6.9

Find FT of an aperiodic signal given by $x(t) = e^{-2t}\cos(\pi t)u(t)$. This is a decaying sinusoid.

Solution

FT can be written as

$$X(j\omega) = \int_{-\infty}^{\infty} x(t)e^{-j\omega t}\,dt = \int_0^\infty e^{-2t}\cos(\pi t)e^{-j\omega t}\,dt$$

$$= \frac{1}{2}\int_0^\infty e^{-(2+j\omega)t}(e^{j\pi t} + e^{-j\pi t})\,dt$$

$$= \frac{1}{2}\Big[\int_0^\infty e^{-(2+j(\omega-\pi))t}\,dt + \int_0^\infty e^{-(2+j(\omega+\pi))t}\,dt\Big] \qquad (6.29)$$

$$= \frac{1}{2}\Big[\frac{e^{-(2+j(\omega-\pi))t}}{-(2+j(\omega-\pi))}\Big\downarrow_0^\infty + \frac{e^{-(2+j(\omega+\pi))t}}{-(2+j(\omega+\pi))}\Big\downarrow_0^\infty\Big]$$

$$= \frac{1}{2}\Big[\frac{1}{2+j(\omega-\pi)} + \frac{1}{2+j(\omega+\pi)}\Big]$$

Example 6.10

Find FT of an aperiodic signal given by $x(t) = (1 - t^2)$ for $|t| < 1$.

Solution

FT can be written as

$$X(j\omega) = \int_{-\infty}^{\infty} x(t)e^{-j\omega t}dt = \int_{-1}^{1}(1-t^2)e^{-j\omega t}dt$$

$$= \frac{e^{-j\omega t}}{-j\omega}\Big\downarrow_{-1}^{1} - \left[t^2\frac{e^{-j\omega t}}{-j\omega}\Big\downarrow_{-1}^{1} - \int_{-1}^{1}2t\int_{-1}^{1}e^{-j\omega t}dt\right]$$

$$= \frac{e^{-j\omega}-e^{j\omega}}{-j\omega} - \frac{e^{-j\omega}-e^{j\omega}}{-j\omega} + \int_{-1}^{1}2t\frac{e^{-j\omega t}}{-j\omega}dt$$

$$= \frac{1}{-j\omega}\left[2t\frac{e^{-j\omega t}}{-j\omega}\Big\downarrow_{-1}^{1} - 2\frac{e^{-j\omega t}}{-\omega^2}\Big\downarrow_{-1}^{1}\right] \qquad (6.30)$$

$$= \left[\frac{2(e^{-j\omega}+e^{j\omega})}{-\omega^2} + 2\frac{e^{-j\omega}-e^{j\omega}}{-j\omega^3}\right]$$

$$= \left[\frac{4\cos(\omega)}{-\omega^2} + \frac{4j\sin(\omega)}{j\omega^3}\right]$$

$$= \frac{4\cos(\omega)}{-\omega^2} + \frac{4\sin(\omega)}{\omega^3}$$

Example 6.11
Find FT of an aperiodic signal given by $x(t) = e^{(1+2t)}u(-t+2)$.

Solution

FT can be written as

$$X(j\omega) = \int_{-\infty}^{\infty} x(t)e^{-j\omega t}dt = \int_{-\infty}^{2} e^{1+2t}e^{-j\omega t}dt$$

$$= \int_{-\infty}^{2} e^{1+(2-j\omega)t}dt$$

$$= e\int_{-\infty}^{2} e^{(2-j\omega)t}dt$$

$$= e\frac{1}{2-j\omega}e^{(2-j\omega)t}\Big\downarrow_{-\infty}^{2}]$$

$$=\frac{e^{5-j2\omega}}{2-j\omega} \qquad (6.31)$$

Example 6.12
Find FT of the signum function.

Solution
The signum function is defined as

sgn(t) = 1 for t > 0 and

= −1 for t < 0

Let us express the signum function as the limiting case of two-sided exponential function.

$$\text{sgn}(t) = \lim_{a \to 0}[e^{-at}u(t) - e^{at}u(-t)]$$

$$X(j\omega) = \lim_{a \to 0}\Big[\int_{-\infty}^{\infty}e^{-a|t|}e^{-j\omega t}dt = -\int_{-\infty}^{0}e^{at}e^{-j\omega t}dt + \int_{0}^{\infty}e^{-at}e^{-j\omega t}dt$$

$$= \lim_{a \to 0}\left[-\frac{1}{(a-j\omega)}e^{(a-j\omega)t}\Big\downarrow_{-\infty}^{0} - \frac{1}{(a+j\omega)}e^{-(a+j\omega)t}\Big\downarrow_{0}^{\infty}\right]$$

(6.32)

$$X(j\omega) = \lim_{a \to 0}\left[-\frac{1}{a-j\omega} + \frac{1}{a+j\omega}\right]$$

$$= \frac{1}{j\omega} + \frac{1}{j\omega} = \frac{2}{j\omega}$$

The signum function and its magnitude plot are shown in Fig. 6.17.

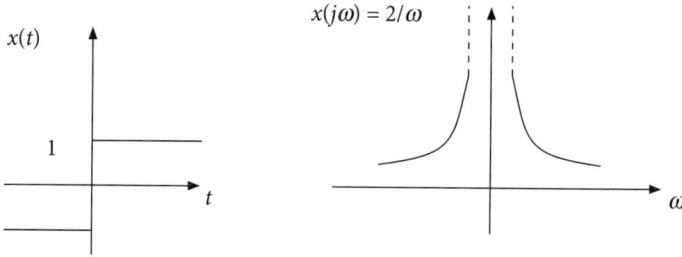

Fig. 6.17 Plot of signum function and its FT

Example 6.13

Find FT of the unit step function.

Solution

The unit step function can be expressed in terms of the signum function given by

$$u(t) = \frac{1}{2}[1 + \text{sgn}(t)] \tag{6.33}$$

We will now take the FT of the unit step.

$$X(j\omega) = \frac{1}{2} FT[1 + \text{sgn}(t)]$$

$$= \frac{1}{2}\left[2\pi\delta(\omega) + \frac{2}{j\omega}\right] = \pi\delta(\omega) + \frac{1}{j\omega} \tag{6.34}$$

Figure 6.18 shows the step function and FT of the step function.

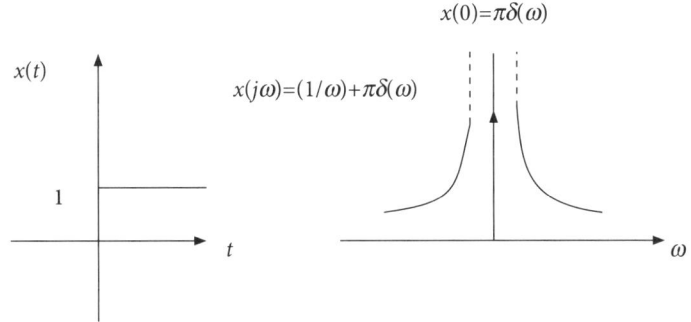

Fig. 6.18 Unit step function and it's FT

6.2.1 Use of dirac delta function

The theory of Fourier Transform is applicable to signals that satisfy Dirichlet's conditions. This means the function must be integrable and these conditions include all energy signals. It is required to extend the theory of Fourier Transform to all power signals as well. It is also desirable to combine Fourier series and Fourier Transform and treat Fourier series as a special case of Fourier Transform. Both these requirements can be fulfilled if we can make a proper use of Dirac delta function or unit impulse function.

Example 6.14

Let us recall the definition of Dirac delta function. It is defined as follows.

$$\delta(t) = 0 \quad \text{for } t \neq 0$$

$$\int_{-\infty}^{\infty} \delta(t)dt = 1 \tag{6.35}$$

We will now find the Fourier Transform of delta function.

Solution

$$FT[\delta(t)] = \int_{-\infty}^{\infty} \delta(t)\exp(-j2\pi ft)dt \tag{6.36}$$

$$= 1 \quad [(\exp(-j2\pi ft) = 1 \text{ at } t = 0]$$

FT of Dirac delta function and delta function is shown in Fig. 6.19.

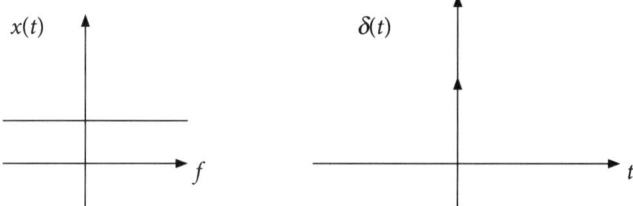

Fig. 6.19 FT of delta function and delta function plots

This equation states that the Fourier Transform of Dirac delta function extends uniformly over entire frequency interval from $-\infty$ to ∞. We will now consider the applications of Dirac delta function.

6.2.2 Applications of dirac delta function

Example 6.15

Find FT of D.C. signal

Solution

We will use the duality property of Fourier Transform to find the Fourier Transform of D.C. signal. The D.C. signal can be presented as

$$x(t) = 1 \text{ for all } t$$

$$x(t) \Leftrightarrow 2\pi\delta(\omega) \quad (6.37)$$

This also gives a useful relation

$$\int_{-\infty}^{\infty} \exp(-j2\pi ft)dt = \delta(\omega) \quad (6.38)$$

A DC signal can be viewed as a double-sided exponential signal with the limit that 'a' tends to zero. $1 = \lim_{a \to 0} 1 \times e^{-a|t|}$. We will use the result for the double-sided exponential to get the FT of the constant.

$$X(j\omega) = \lim_{a \to 0} \int_{-\infty}^{\infty} e^{-a|t|} e^{-j\omega t} dt = \int_{-\infty}^{0} e^{(a-j\omega)t} dt + \int_{0}^{\infty} e^{-(a+j\omega)t} dt$$

$$= \lim_{a \to 0} \frac{1}{(a-j\omega)} e^{(a-j\omega)t} \Big\downarrow_{-\infty}^{0} - \frac{1}{(a+j\omega)} e^{-(a+j\omega)t} \Big\downarrow_{0}^{\infty}$$

$$X(j\omega) = \lim_{a \to 0} \frac{1}{a-j\omega} + \frac{1}{a+j\omega}$$

$$= \lim_{a \to 0} \frac{a+j\omega+a-j\omega}{a^2+\omega^2} = \lim_{a \to 0} \frac{2a}{a^2+\omega^2} = 0 \text{ for } \omega \neq 0 \quad (6.39)$$

for $\omega = 0$, $X(j\omega) = A\delta(\omega)$ where

$$A = \int_{-\infty}^{\infty} \frac{2a}{a^2+\omega^2} d\omega = 2a \int_{-\infty}^{\infty} \frac{1}{a^2+\omega^2} d\omega = 2a \left[\frac{1}{a} \tan^{-1}\left(\frac{\omega}{a}\right) \right] \Big\downarrow_{-\infty}^{\infty}$$

$$= 2\left[\frac{\pi}{2} + \frac{\pi}{2}\right] = 2\pi$$

$$X(j\omega) = 2\pi\delta(\omega)$$

The Fourier Transform of D.C. signal is a Dirac delta function in frequency domain that exists only at $\omega = 0$. The D.C. signal and its Fourier Transform are shown in Fig. 6.20.

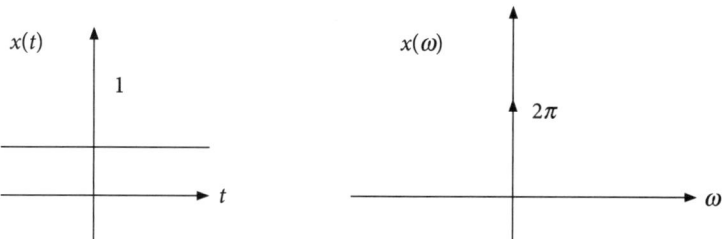

Fig. 6.20 Plot of DC signal and its FT

Example 6.16

Find the FT of the complex Exponential signal.

Solution

Let us find Fourier Transform of exponential signal. Let the signal be given by

$$x(t) = \exp(j2\pi f_c t) \text{ for all } t \tag{6.40}$$

To find the Fourier transform, we will use the result for the Dirac delta function and will use the frequency shifting property of Fourier Transform namely

if $x(t) \Leftrightarrow X(f)$

$$\exp(j2\pi f_0 t) x(t) \Leftrightarrow X(f - f_0) \tag{6.41}$$

$$X(j\omega) = 2\pi\delta(\omega) \downarrow_{\omega = \omega - \omega_c} = 2\pi\delta(\omega - \omega_c)$$

Let x(t) be a D. C. signal, we know it transforms to

$$1 \Leftrightarrow 2\pi\delta(\omega) \tag{6.42}$$

We will multiply the D.C. signal by the complex exponential to get the signal

$$x(t) = 1 \times \exp(j2\pi f_c t) \text{ for all } t \tag{6.43}$$

Now, the Fourier Transform of the signal can be found by using the frequency shifting property of Fourier Transform.

If $1 \Leftrightarrow 2\pi\delta(\omega)$

$$\exp(j2\pi f_c t) \times 1 \Leftrightarrow 2\pi\delta(\omega - \omega_c) \tag{6.44}$$

We note that FT of exponential signal is a Dirac delta function in frequency domain at $f = f_c$, as shown in Fig. 6.21.

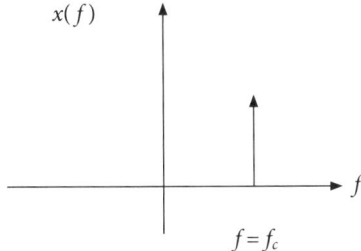

Fig. 6.21 FT of complex exponential signal

Example 6.17

Find FT of Sine wave signal:

Solution

Let us find the Fourier Transform of a sine function by representing it in terms of exponentiation function. Let the signal be given by

$$x(t) = \sin(2\pi f_c t) = \frac{1}{2j}[\exp(j2\pi f_c t) - \exp(-j2\pi f_c t)] \text{ for all } t \quad (6.45)$$

To find the Fourier transform, we will use the result for the Dirac delta function and will use the frequency shifting property of Fourier Transform.

Let $x(t)$ be a D. C. signal, we know it transforms to $1 \Leftrightarrow 2\pi\delta(\omega)$.

We will multiply the D.C. signal by the sine signal in terms of complex exponential to get the signal

$$x(t) = 1 \times \frac{1}{2j}[\exp(j2\pi f_c t) - \exp(-j2\pi f_c t)] \text{ for all } t \quad (6.46)$$

Now, the Fourier Transform of the signal can be found by using the frequency shifting property of Fourier Transform.

if $1 \Leftrightarrow 2\pi\delta(\omega)$

$$1 \times \frac{1}{2j}\left[\exp(j2\pi f_c t) - \exp(-j2\pi f_c t)\right] \Leftrightarrow \frac{\pi}{j}[\delta(\omega - \omega_c) - \delta(\omega + \omega_c)] \quad (6.47)$$

We note that it is a sum of Dirac delta function in frequency domain at $f = f_c$ with unity amplitude and a Dirac delta function at $f = -f_c$ with minus unity amplitude, as shown in Fig. 6.22.

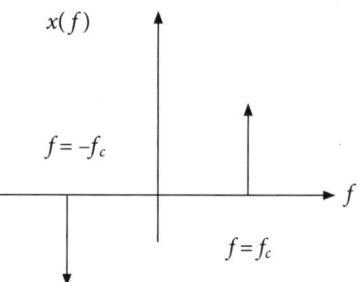

Fig. 6.22 FT of sine wave function

Example 6.18

Find the FT of a Cosine signal

Let us find the Fourier Transform of a cosine function by representing it in terms of exponentiation function. Let the signal be given by

$$x(t) = \cos(2\pi f_c t) = \frac{1}{2}[\exp(j2\pi f_c t) + \exp(-j2\pi f_c t)] \text{ for all } t \qquad (6.48)$$

To find the Fourier transform, we will use the result for the Dirac delta function and will use the frequency shifting property of Fourier Transform.

Let $x(t)$ be a D. C. signal, we know it transforms to $1 \Leftrightarrow 2\pi\delta(\omega)$.

We will multiply the D.C. signal by the sine signal in terms of complex exponential to get the signal

$$x(t) = 1 \times \frac{1}{2}[\exp(j2\pi f_c t) + \exp(-j2\pi f_c t)] \text{ for all } t \qquad (6.49)$$

Now, the Fourier Transform of the signal can be found by using the frequency shifting property of Fourier Transform.

if $1 \Leftrightarrow 2\pi\delta(f)$

$$1 \times \frac{1}{2}[\exp(j2\pi f_c t) + \exp(-j2\pi f_c t)] \Leftrightarrow \pi[\delta(\omega - \omega_c) + \delta(\omega + \omega_c)] \qquad (6.50)$$

We note that it is sum of Dirac delta function in frequency domain at $f = f_c$ with unity amplitude and a Dirac delta function at $f = -f_c$ with unity amplitude, as shown in Fig. 6.23.

The above examples indicate that if the function is spread out in time domain, it is concentrated in the other domain and vice versa. This result is useful for finding the spectrum of the carrier wave used in communication systems.

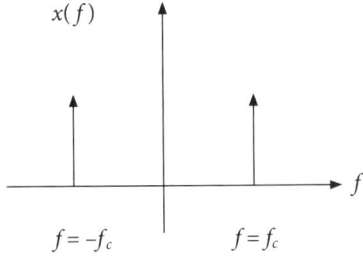

Fig. 6.23 FT of a cosine function

We can extend this to find the spectrum of the amplitude modulated wave, frequency modulated wave, etc. Note that we use these basic results in amplitude modulation to study the spectrum of amplitude modulated wave, spectrum of double side band suppressed carrier wave, single side band wave, etc.

Concept Check

- What is the nature of the FT of a rectangular pulse?
- What will be the change in the sinc function if the width of the rectangular pulse is increased?
- What is Heisenberg uncertainty principle?
- What is the nature of FT of a sinc function?
- What is a duality property related to a rectangular function and its FT?
- What is the FT of the delta function?
- How will you find FT of a sine wave using frequency shifting property and delta function?

6.3 Fourier Transforms of Periodic CT Signals

We have seen in section 3.3 that periodic signals can be represented in terms of Fourier series in frequency domain. Fourier Transform representation is used for aperiodic signals. We will now see how to write Fourier Transform representation of periodic signals. We will make use of delta functions to represent the transform.

Example 6.19
Consider a periodic signal $x_p(t)$ of period T_0. Find the FT of the periodic signal.

Solution
We know that we can represent the periodic signal using exponential Fourier series given by Eq. (5.21) and the coefficients of the series can be calculated using Eq. (5.22). The equations are reproduced here for easy reference.

$$x_P(t) = \sum_{n=-\infty}^{\infty} c_n \exp\left(\frac{j2\pi nt}{T_0}\right) \qquad (6.51)$$

$$c_n = \frac{1}{T_0} \int_{-T_0/2}^{T_0/2} x_P(t) \exp\left(\frac{-j2\pi nt}{T_0}\right) dt \quad n = 0, \pm 1, \pm 2 \ldots \ldots \qquad$$

Let us consider a signal $x(t)$, which is a pulse signal, and it is the same as $x_p(t)$ over one period of it and is zero elsewhere, given as

$$x(t) = \begin{cases} x_P(t) & -T_0/2 \leq t \leq T_0/2 \\ 0 & \text{elsewhere} \end{cases} \qquad (6.52)$$

We can represent $x_p(t)$ in terms of $x(t)$ as an infinite summation given by

$$x_P(t) = \sum_{m=-\infty}^{\infty} x(t - mT_0) \qquad (6.53)$$

Where $x(t)$ can be considered as a generating function. It is Fourier transformable as it is an energy signal. The Fourier transform coefficients can be written as

$$C_n = \frac{1}{T_0} \int_{-\infty}^{\infty} x(t) \exp(-j2\pi nT/T_0) dt$$

$$= \frac{1}{T_0} X\left(\frac{n}{T_0}\right) \quad \text{where } X\left(\frac{n}{T_0}\right) \text{ is FT of } x(t) \text{ evaluated at } n/T_0 \qquad (6.54)$$

We can write Eq. (6.48) as

$$x_P(t) = \sum_{m=-\infty}^{\infty} x(t - mT_0) = \frac{1}{T_0} \sum_{m=-\infty}^{\infty} X\left(\frac{n}{T_0}\right) \exp(j2\pi nT/T_0) \qquad (6.55)$$

We know that using a frequency shifting property FT, we can find the FT of complex exponential as

$$\exp(j2\pi f_c t) \Leftrightarrow \delta(f - f_c)$$

Hence, $x_P(t) = \sum_{m=-\infty}^{\infty} x(t - mT_0) \Leftrightarrow \frac{1}{T_0} \sum_{m=-\infty}^{\infty} X\left(\frac{n}{T_0}\right) \delta\left(f - \frac{n}{T_0}\right) \qquad (6.56)$

This relation means that FT of a periodic signal is a set of delta functions occurring at integer multiples of the fundamental frequency $1/T_0$, including the origin, and each delta function is weighted by a factor equal to the value of FT coefficient at n/T_0.

Example 6.20

Use the above result to find FT of an ideal sampling function shown in Fig. 6.20.

Solution

An ideal sampling function is an infinite sequence of uniformly spaced delta functions. The ideal sampling function can be written as

$$\delta_T(t) = \sum_{m=-\infty}^{\infty} \delta(t - mT_0) \tag{6.57}$$

We can recognize the generating function for an ideal sampling function as the delta function $\delta(t)$ with FT of $X\left(\dfrac{n}{T_0}\right) = 1$ for all n. We can now write FT of the ideal sampling function using Eq. (6.52) as

$$\delta_T(t) = \sum_{m=-\infty}^{\infty} \delta(t - mT_0) \Leftrightarrow \frac{1}{T_0} \sum_{m=-\infty}^{\infty} \delta\left(f - \frac{n}{T_0}\right) \tag{6.58}$$

Equation (6.53) states that the FT of a periodic train of delta functions spaced T_0 apart, consists of another set of delta functions weighted by a factor of $1/T_0$ and spaced by frequency equal to $1/T_0$ apart. The ideal sampling function and its FT is depicted in Fig. 6.24.

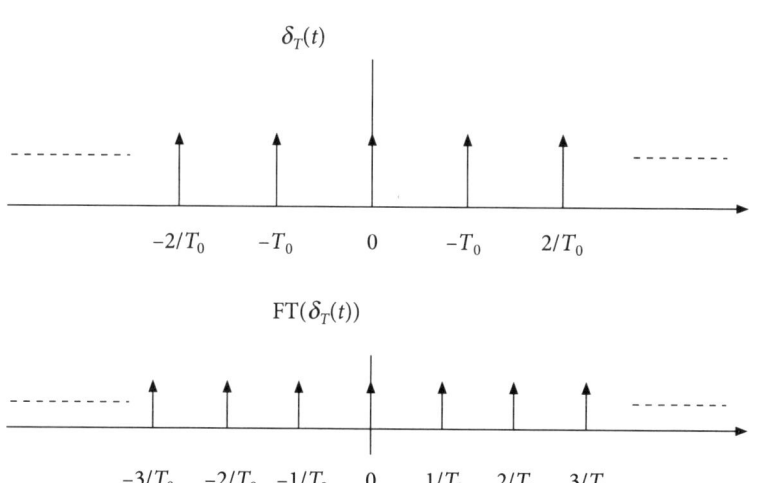

Fig. 6.24 A periodic pulse train and its FT

Concept Check

- Comment on the FT of periodic signal.
- What is the nature of the FT of a train of impulses?

6.4 Inverse Fourier Transform

The inverse FT is obtained by directly using the equation for inverse FT. Let us consider one numerical problem to illustrate the concepts.

Example 6.21

Find the inverse FT of

$$X(j\omega) = \begin{cases} 2\cos(\omega) & \text{for } |\omega| \leq \pi \\ 0 & \text{for } |\omega| > \pi \end{cases}$$

Solution

To find IFT, we will use the equation for IFT.

$$x(t) = \frac{1}{2\pi} \int_{-\infty}^{\infty} X(j\omega) e^{j\omega t} d\omega$$

$$x(t) = \frac{1}{2\pi} \int_{-\pi}^{\pi} 2\cos(\omega) e^{j\omega t} d\omega$$

$$x(t) = \frac{1}{2\pi} \int_{-\pi}^{\pi} (e^{j\omega} + e^{-j\omega}) e^{j\omega t} d\omega = \frac{1}{2\pi} \left[\int_{-\pi}^{\pi} e^{j\omega(t+1)} d\omega + \int_{-\pi}^{\pi} e^{j\omega(t-1)} d\omega \right]$$

$$= \frac{1}{2\pi} \left[\frac{1}{j(t+1)} e^{j\omega(t+1)} \Big|_{-\pi}^{\pi} + \frac{1}{j(t-1)} e^{j\omega(t-1)} \Big|_{-\pi}^{\pi} \right]$$

$$= \frac{1}{2\pi j} \left[\frac{1}{t+1} \{e^{j\pi(t+1)} - e^{-j\pi(t+1)}\} + \frac{1}{t-1} \{e^{j\pi(t-1)} - e^{-j\pi(t-1)}\} \right]$$

$$= \frac{1}{\pi} \left[\frac{1}{t+1} \sin(\pi(t+1)) + \frac{1}{t-1} \sin(\pi(t-1)) \right] \tag{6.59}$$

If the FT equation is in the form of a polynomial in $j\omega$ in numerator and a polynomial in $j\omega$ in the denominator, then a normally partial fraction expansion is required to be used for decomposing the polynomial of higher degree in component factors. The process of decomposing the equation of higher degree into addition of component factors is called as partial fraction expansion (PFE). The FT equation is called as the transfer function of the system. Each factor of the transfer function is then inverse Fourier transformed by inspection. To find IFT by inspection, we make use of the result of some standard FT pairs. The property of linearity of FT is made use of in this case. To illustrate the concept of partial fraction expansion, we will consider some numerical examples.

Example 6.22

Find inverse FT using partial fraction expansion.

$$X(j\omega) = \frac{1}{(j\omega)^2 + 5j\omega + 6}$$

Solution

Step 1 We will first decompose the denominator in two factors.

$$X(j\omega) = \frac{1}{(j\omega)^2 + 5j\omega + 6} = \frac{1}{(j\omega + 3)(j\omega + 2)} \quad (6.60)$$

Step 2 Decompose the transfer function in component functions using partial fraction expansion.

$$X(j\omega) = \frac{1}{(j\omega + 3)(j\omega + 2)} = \frac{k_1}{j\omega + 3} + \frac{k_2}{j\omega + 2}$$

find k_1 and k_2

$$k_1 = \frac{1}{j\omega + 2}\bigg\downarrow_{j\omega = -3} = -1$$

$$k_2 = \frac{1}{j\omega + 3}\bigg\downarrow_{j\omega = -2} = 1$$

$$X(j\omega) = \frac{1}{j\omega + 2} - \frac{1}{j\omega + 3} \quad (6.61)$$

Step 3 Find the IFT of each component term

$$X(j\omega) = \frac{1}{j\omega + 2} - \frac{1}{j\omega + 3} \quad (6.62)$$

$$x(t) = (e^{-2t} - e^{-3t})u(t)$$

We know that $\text{IFT}\left(\dfrac{1}{j\omega+1}\right) = e^{-t}u(t)$

The above result used is a standard FT pair.

Example 6.23
Find inverse FT using partial fraction expansion.

$$X(j\omega) = \dfrac{j\omega+3}{(j\omega+4)^2}$$

Solution

Step 1 We will first decompose the denominator in two factors.

$$X(j\omega) = \dfrac{j\omega+4}{(j\omega+4)^2} \tag{6.63}$$

Step 2 Decompose the transfer function in component functions using partial fraction expansion.

$$X(j\omega) = \dfrac{j\omega+3}{(j\omega+4)^2} = \dfrac{k_1}{(j\omega+4)^2} + \dfrac{k_2}{j\omega+4}$$

Find k_1 and k_2

$$k_1 = \dfrac{j\omega+3}{(j\omega+4)^2}(j\omega+4)^2 \Big\downarrow_{j\omega=-4} = -1 \tag{6.64}$$

$$k_2 = \dfrac{d}{d\omega}\left(\dfrac{j\omega+3}{(j\omega+4)^2}(j\omega+4)^2\right)\Big\downarrow_{j\omega=-4} = 1$$

$$X(j\omega) = -\dfrac{1}{(j\omega+4)^2} + \dfrac{1}{j\omega+4}$$

Step 3 Find the IFT of each component term.

$$X(j\omega) = -\dfrac{1}{(j\omega+4)^2} + \dfrac{1}{j\omega+4}$$

$$x(t) = (-te^{-4t} + e^{-4t})u(t)$$

we know that $\text{IFT}\left(\dfrac{1}{j\omega+1}\right) = e^{-t}u(t)$

(6.65)

and $\text{IFT}\left(\dfrac{1}{j\omega+1}\right)^2 = te^{-t}u(t)$

Example 6.24
Find the inverse FT using partial fraction expansion.

$$X(j\omega) = \dfrac{4}{-\omega^2 + 4j\omega + 3}$$

Solution
Step 1 We will first decompose the denominator in two factors.

$$X(j\omega) = \dfrac{4}{-\omega^2 + 4j\omega + 3} = \dfrac{4}{(j\omega+3)(j\omega+1)} \quad (6.66)$$

Step 2 Decompose the transfer function in component functions using partial fraction expansion.

$$X(j\omega) = \dfrac{4}{(j\omega+3)(j\omega+1)} = \dfrac{k_1}{j\omega+3} + \dfrac{k_2}{j\omega+1} \quad (6.67)$$

Find k_1 and k_2

$$k_1 = \dfrac{1}{j\omega+1} \Bigg\downarrow_{j\omega=-3} = -1/2$$

$$k_2 = \dfrac{1}{j\omega+3} \Bigg\downarrow_{j\omega=-1} = 1/2$$

$$X(j\omega) = \dfrac{1}{2}\left[\dfrac{1}{j\omega+1} - \dfrac{1}{j\omega+3}\right]$$

Step 3 Find the IFT of each component term.

$$X(j\omega) = \dfrac{1}{2}\left[\dfrac{1}{j\omega+1} - \dfrac{1}{j\omega+3}\right]$$

$$x(t) = \frac{1}{2}(e^{-t} - e^{-3t})u(t)$$

(6.68)

We know that $\text{IFT}\left(\dfrac{1}{j\omega+1}\right) = e^{-t}u(t)$

The above result used is a standard FT pair.

Example 6.25
Find inverse FT using partial fraction expansion.

$$X(j\omega) = \frac{-(j\omega)^2 - 4j\omega - 6}{[(j\omega)^2 + 3j\omega + 2](j\omega + 4)}$$

Solution
Step 1 We will first decompose the denominator in two factors.

$$X(j\omega) = \frac{-(j\omega)^2 - 4j\omega - 6}{[(j\omega)^2 + 3j\omega + 2](j\omega + 4)} = \frac{-(j\omega)^2 - 4j\omega - 6}{(j\omega + 2)(j\omega + 1)(j\omega + 4)}$$

(6.69)

Step 2 Decompose the transfer function in component functions using partial fraction expansion.

$$X(j\omega) = \frac{-(j\omega)^2 - 4j\omega - 6}{(j\omega + 2)(j\omega + 1)(j\omega + 4)} = \frac{k_1}{j\omega + 2} + \frac{k_2}{j\omega + 1} + \frac{k_3}{j\omega + 4}$$

Find k_1, k_2 and k_3

$$k_1 = \frac{-(j\omega)^2 - 4j\omega - 6}{(j\omega + 1)(j\omega + 4)}\bigg|_{j\omega = -2} = \frac{-4 + 8 - 6}{(-1)(2)} = \frac{-2}{-2} = 1$$

$$k_2 = \frac{-(j\omega)^2 - 4j\omega - 6}{(j\omega + 2)(j\omega + 4)}\bigg|_{j\omega = -1} = \frac{-1 + 4 - 6}{(1)(3)} = \frac{-3}{3} = -1$$

$$k_3 = \frac{-(j\omega)^2 - 4j\omega - 6}{(j\omega + 2)(j\omega + 1)}\bigg|_{j\omega = -4} = \frac{-16 + 16 - 6}{(-2)(-3)} = \frac{-6}{6} = -1$$

(6.70)

$$X(j\omega) = \left[\frac{1}{j\omega + 2} - \frac{1}{j\omega + 1} - \frac{1}{j\omega + 4}\right]$$

Step 3 Find the IFT of each component term.

$$X(j\omega) = \left[\frac{1}{j\omega+2} - \frac{1}{j\omega+1} - \frac{1}{j\omega+4} \right]$$

$$x(t) = (e^{-2t} - e^{-t} - e^{-4t})u(t) \tag{6.71}$$

We know that $\text{IFT}\left(\dfrac{1}{j\omega+1}\right) = e^{-t}u(t)$

The above result used is a standard FT pair.

Concept Check

- Write the equation for IFT.
- Compare the IFT equation with that of FT.
- What is partial fraction expansion?
- When PFE is used for calculation of IFT, which property of FT is used?
- How will you find IFT by inspection?

6.5 FT of Aperiodic DT Signals (DTFT)

Consider a band-limited aperiodic analog signal (energy signal) $x(t)$. Let it be suitably sampled as per the sampling theorem to generate a DT signal $x(n)$. In the process of sampling, the signal $x(t)$ is multiplied by an impulse train with spacing between the impulses as $T = 1/Fs$ where Fs stands for the sampling frequency, as shown in Fig. 6.25 (refer to Fig. 1.3 in chapter 1 on sampling).

When we look at the Fourier domain picture of the $x(n)$, we see that it is periodic in the spectral domain, as shown in Fig. 6.26. (Refer to Fig. 1.4 in chapter 1 on sampling). The resulting spectrum of $x(n)$ i.e., DTFT is a convolution of the spectrum of $x(t)$ and the spectrum of the train of impulse. The reader is encouraged to refer to the modulation property of FT in section 6.6, which states that the multiplication of two signals in time domain is a convolution of the FTs of the two signals. Note that it is customary to show the spectrum of the signal with a bandwidth of W as a triangular representation. It does not mean that that the respective frequencies have decreasing amplitude. The x axis is either the frequency or the normalized angular frequency given by $\omega = 2\pi \dfrac{F}{F_s}$, when $F = Fs$, the normalized angular frequency is equal to 2π, as shown in Fig. 6.26. We see that the continuous spectrum is periodic with a period equal to 2π. It is given by Eq. (6.72).

$$X(e^{j\Omega}) = \sum_{n=-\infty}^{\infty} x[n]e^{-j\Omega n} \qquad (6.72)$$

The signal $x[n]$ can be recovered from DTFT using the formula

$$x[n] = \frac{1}{2\pi} \int_{-\infty}^{\infty} X(e^{j\Omega})e^{j\Omega n} d\Omega \qquad (6.73)$$

The DTFT being periodic has all the information in one period.

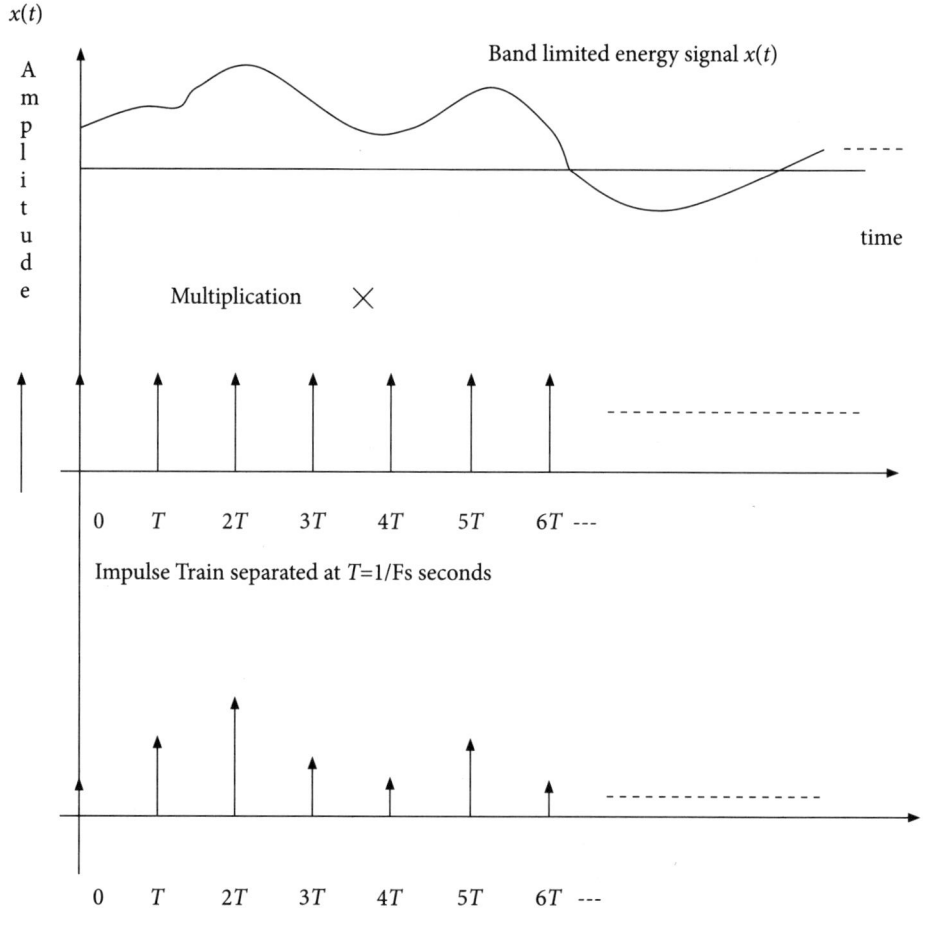

Fig. 6.25 Sampled signal obtained by multiplication of CT signal with a train of impulse

Example 6.26

Find the FT of $x[n] = \begin{cases} 1 \text{ for } 0 \le n \le 3 \\ 0 \text{ otherwise} \end{cases}$

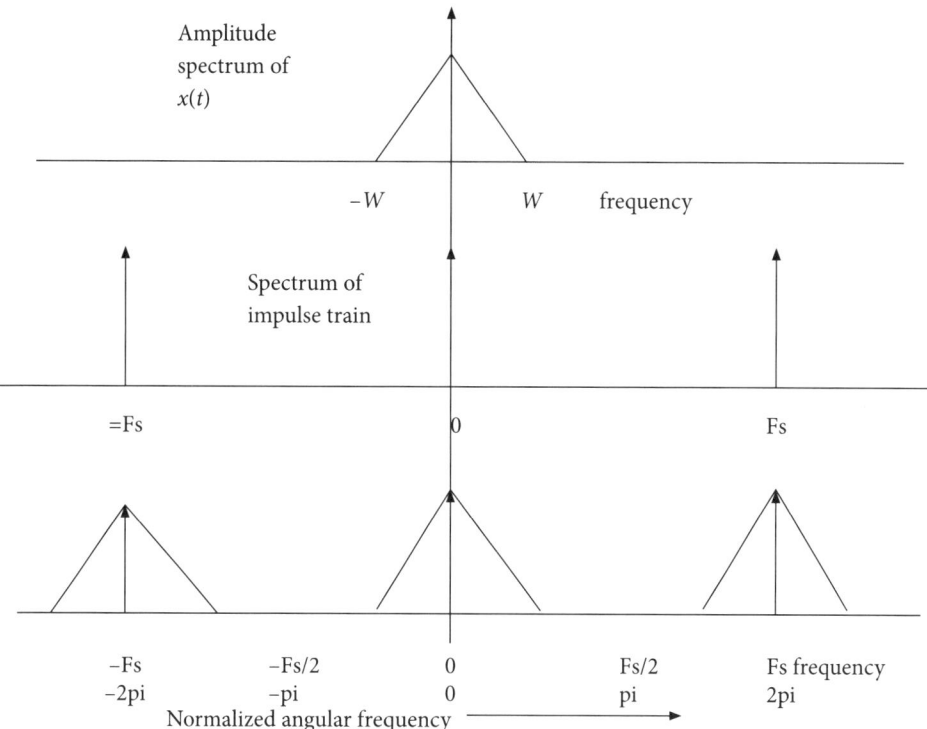

Fig. 6.26 Replicated continuous spectrum of x[n] (DTFT) [obtained as a result of convolution of FTs of signal and train of impulse]

Solution

Let us use the definition of DTFT.

$$X(e^{j\omega n}) = \sum_{n=-\infty}^{\infty} x[n]e^{-j\omega n} = \sum_{n=0}^{3} x[n]e^{-j\omega n} = 1 + e^{-j\omega} + e^{-2j\omega} + e^{-3j\omega} \quad (1)$$

Multiply both sides by e^{-jw} and subtract from the first equation.

$$e^{-j\omega}X(e^{-j\omega n}) = e^{-j\omega} + e^{-2j\omega} + e^{-3j\omega} + e^{-4j\omega} \quad (2)$$

(1)–(2)

$$(1-e^{-j\omega})X(e^{j\omega n}) = (1-e^{-4j\omega}) \tag{6.74}$$

$$X(e^{j\omega n}) = \frac{1-e^{-4j\omega}}{1-e^{-j\omega}}. \text{ This is called as closed form expression}$$

$$X(e^{j\omega n}) = (1 + \cos(\omega) + \cos(2\omega) + \cos(3\omega)) - j(\sin(\omega) + \sin(2\omega) + \sin(3\omega))$$

To find the magnitude response using hand calculations, we have to put different values of ω in the equation and find real and imaginary parts. Magnitude and phase can be calculated using rectangular to polar conversion. Let us write a MATLAB program to plot magnitude and phase response. We will use abs command and angle command to find magnitude and phase. Figures 6.27 and 6.28 show the magnitude and phase response. We can note that the magnitude response and phase response, both are periodic with a period equal to 2pi. The magnitude response is a periodic sinc function.

```
clear all;
w=0:0.1:20;
w1=w/pi;
x=(1+cos(w)+cos(2*w)+cos(3*w))-
1j*(sin(w)+sin(2*w)+sin(3*w));
plot(w1,abs(x));
title('magnitude response of FT');
xlabel('angular frequency as multiple of pi');
ylabel('magnitude');
figure;
plot(w1,angle(x));
title('phase response of FT');
xlabel('anglular frequency as multiple of pi');
ylabel('phase value');
```

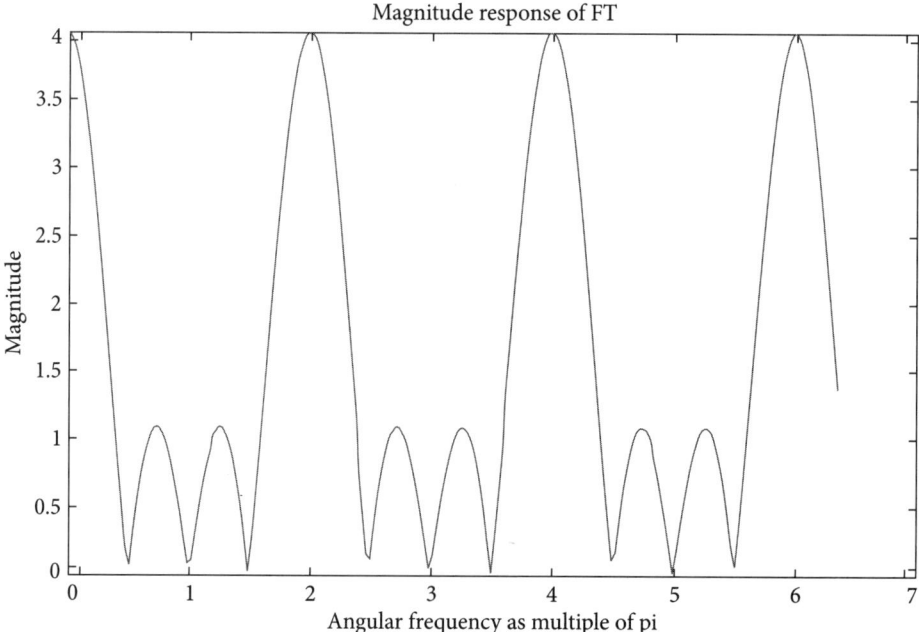

Fig. 6.27 Magnitude response with period 2π

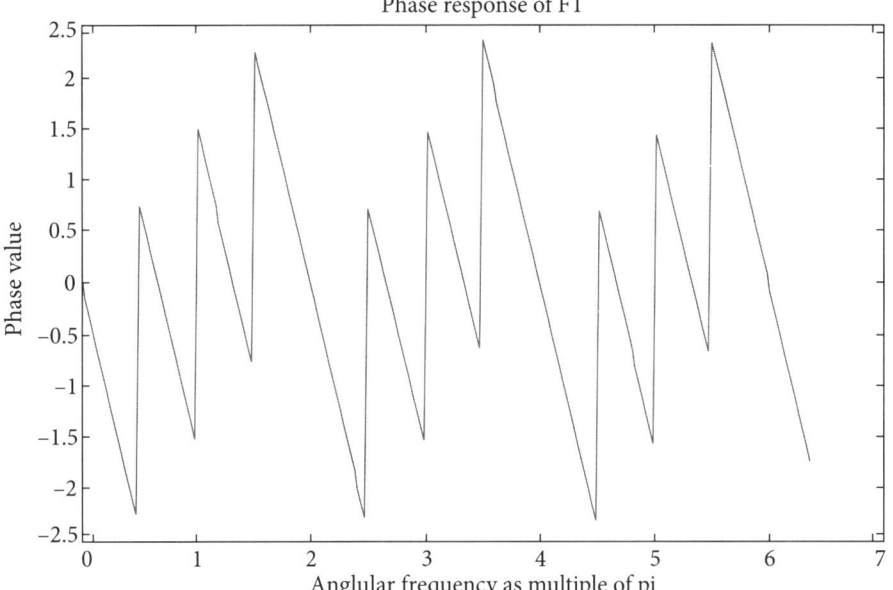

Fig. 6.28 Phase response with period 2π

Example 6.27

Find the FT of $x[n] = \begin{cases} 1/3 \text{ for } 0 \leq n \leq 2 \\ 0 \text{ otherwise} \end{cases}$

Solution

Using the definition of DTFT.

$$X(e^{j\Omega n}) = \sum_{n=-\infty}^{\infty} x[n]e^{-j\Omega n} = \sum_{n=0}^{2} x[n]e^{-j\Omega n} = \frac{1}{3}\left[1 + e^{-j\Omega} + e^{-2j\Omega}\right]$$

(6.75)

$$= \frac{1}{3}(1 + \cos(\Omega) + \cos(2\Omega)) - \frac{1}{3}j(\sin(\Omega) + \sin(2\Omega))$$

To find the magnitude response using hand calculations, we have to put different values of ω in the equation and find real and imaginary parts. Magnitude and phase can be calculated using rectangular to polar conversion. Let us write a MATLAB program to plot the magnitude and phase response. We will use abs command and angle command to find magnitude and phase. Figures 6.29 and 6.30 show the magnitude and phase response. We can note that the magnitude response and phase response, both are periodic with a period equal to 2pi.

```
clear all;
w=0:0.1:20;
w1=w/pi;
```

```
x=1/3*(1+cos(w)+cos(2*w))-1j/3*(sin(w)+sin(2*w));
plot(w1,abs(x));
title('magnitude response of FT');
xlabel('angular    frequency    as    multiple    of    pi');
ylabel('magnitude');
figure;
plot(w1,angle(x));
title('phase response of FT');
xlabel('anglular    frequency    as    multiple    of    pi');
ylabel('phase value');
```

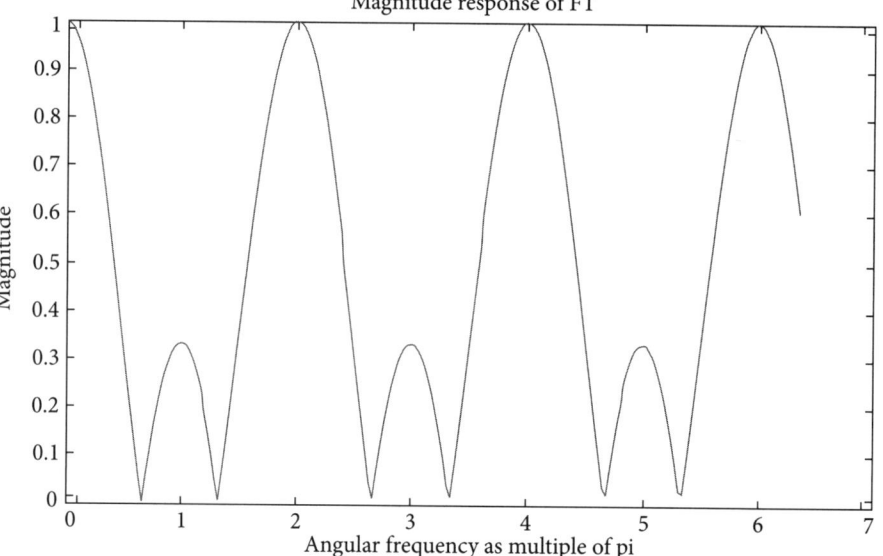

Fig. 6.29 Magnitude response with period 2π

Example 6.28

Find inverse DTFT of $X(e^{j\Omega}) = \begin{cases} 1 & |\Omega| \leq w \\ 0 & w < |\Omega| < \pi \end{cases}$.

Solution

As DTFT is periodic with a period of 2*pi, it is specified only between –pi to +pi. We will use the formula for inverse DTFT to find the signal $x[n]$.

$$x[n] = \frac{1}{2\pi}\int_{-\infty}^{\infty} X(e^{j\Omega})e^{j\omega n}d\omega = \frac{1}{2\pi}\int_{-w}^{w} e^{j\Omega n}d\Omega = \frac{1}{2\pi nj}[e^{jwn} - e^{-jwn}]$$

(6.76)

$$= \frac{\sin(wn)}{n\pi} \text{ for } n \neq 0 \text{ and } x[n] = \frac{\sin(wn)}{wn}\frac{w}{\pi} \to \frac{w}{\pi} \text{ as } n \to 0$$

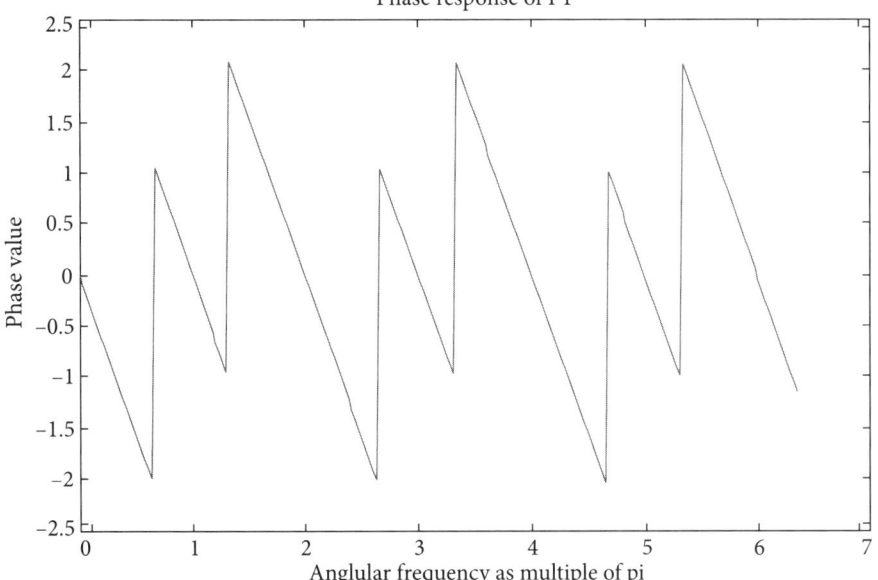

Fig. 6.30 Phase response with period 2π

Figure 6.31 shows the plot of the signal and the DTFT of the signal. A MATLAB program to plot the inverse DTFT of the periodic rectangular pulse in frequency domain is given below. Inverse DTFT is a DT signal that is a sinc function (aperiodic signal).

```
clear all;
w=0.2;
A=1;
N=50;
for n=1:N,
    X(n+N+1)=(1/(n*pi))*sin(n*w);
  end
X(N+1)=w/pi;
for n=1:N,
    X(n)=X(2*N+2-n);
end
n1=-50:1:50;
stem(n1,X);
title('plot of magnitude of Inverse
DTFT of frequency domain rectangular
pulse of width w=0.2');
xlabel('time');ylabel('amplitude');
```

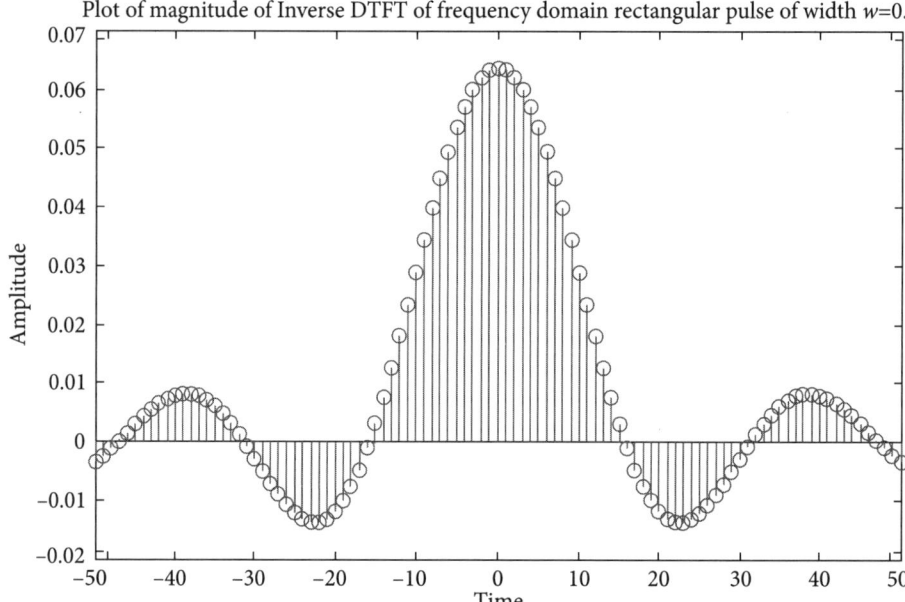

Fig. 6.31 Inverse DTFT of a periodic rectangular pulse

Note When the DT signal is aperiodic, the DTFT is continuous and periodic in frequency domain with a period of pi and when a signal is continuous and periodic in frequency domain, its inverse DTFT is aperiodic and discrete.

6.5.1 FT of standard aperiodic DT signals

We will evaluate the DTFT of some standard DT signals such as unit impulse, exponential sequence. We will also find the inverse DTFT of train of impulse in the frequency domain.

Example 6.29

Find DTFT of the unit impulse in time domain.

Solution

We know that the unit impulse is an aperiodic DT signal. Let us find DTFT using the formula in Eq. (6.73).

$$X(e^{j\Omega}) = \sum_{n=-\infty}^{\infty} \delta[n]e^{-j\Omega n} = 1 \tag{6.77}$$

The plot of the signal and DTFT is shown in Fig. 6.32.

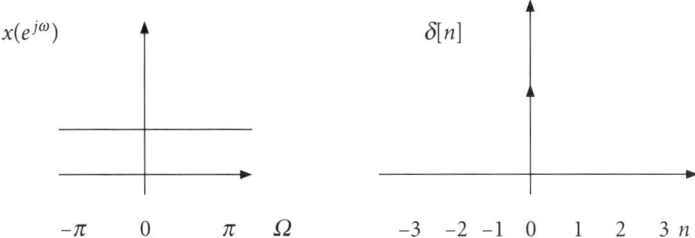

Fig. 6.32 Plot of DTFT and DT signal unit impulse

Example 6.30

Find the inverse DTFT of $X(e^{j\Omega}) = \delta(\Omega)$, for $-\pi < \Omega < \pi$.

Solution

We will use the formula for inverse DTFT.

$$x[n] = \frac{1}{2\pi}\int_{-\infty}^{\infty} X(e^{j\Omega})e^{j\Omega n}d\omega = \frac{1}{2\pi}\int_{-\pi}^{\pi}\delta(\Omega)e^{j\Omega n}d\Omega$$

$$= \frac{1}{2\pi}$$

(6.78)

We have used the shifting property of impulse function. We can define DTFT as a periodic function given by $X(e^{j\Omega}) = \sum_{k=-\infty}^{\infty} \delta(\Omega - k2\pi)$. The plot of the signal and its DTFT are shown in Fig. 6.33.

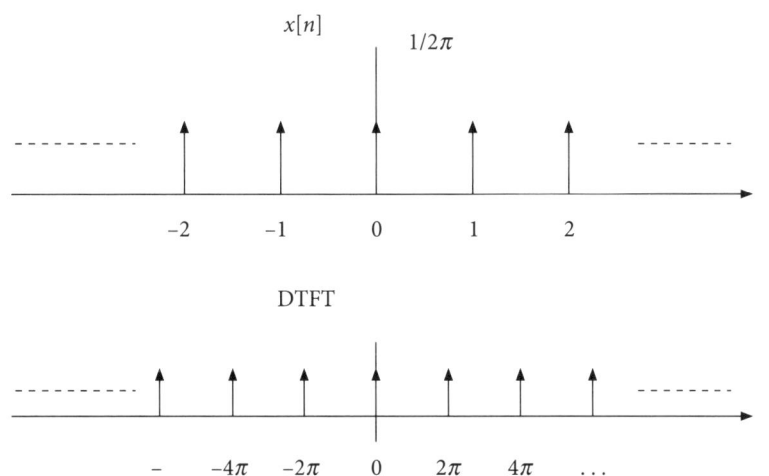

Fig. 6.33 Plot of signal and its DTFT

Example 6.31

Find the DTFT of the exponential sequence in time domain given by $x[n] = a^n u[n]$ with $|a|<1$

Solution

We know that the exponential sequence is an aperiodic DT signal. Let us find DTFT using the formula in Eq. (6.73).

The exponential sequence can be written as

$$x[n] = a^n u[n] \text{ with } |a|<1 \tag{6.79}$$

$$X(e^{j\Omega}) = \sum_{n=0}^{\infty} a^n e^{-j\Omega n} = \sum_{n=0}^{\infty} (ae^{-j\Omega})^n$$

$$= \frac{1}{1-ae^{-j\Omega}} \tag{6.80}$$

The limits are from 0 to infinity as the sequence is appended by $u[n]$.

Note that if a is real, we can find the magnitude response by finding the square root of the sum of real part square and imaginary part square.

$$|X(e^{j\Omega})| = \frac{1}{1-ae^{-j\Omega}} = \frac{1}{[(1-a\cos(\Omega))^2 + a^2 \sin^2(\Omega)]^{1/2}}$$

$$\arg(X(j\Omega)) = -\arctan\left(\frac{a\sin(\Omega)}{1-a\cos(\Omega)}\right) \tag{6.81}$$

Let us take the value of $a = 0.1$ and plot the magnitude and phase response.

```
clear all;
w=0:1:20*pi;
w1=w/pi;
x=1./sqrt((1-(0.5)*cos(w1)).*(1-(0.5)*cos(w1))+(0.5
)*(0.5)*(sin(w1).*sin(w1)));
plot(w1,x);
title('magnitude response of FT');
xlabel('angular     frequency    in     radians');
ylabel('magnitude');
figure;
%a=-atan(((0.5)*sin(w1))/(1-(0.5)*cos(w1)));
a=-angle(1./(1-(0.5)*exp(1j*w1)));
plot(w1,a);
title('phase response of FT');
xlabel('anglular frequency in radians'); ylabel('phase
```

value');

Note that we have used the negative sign before the angle as per the formula in Eq. (6.67). Figures 6.34 and 6.35 show the plot of magnitude response and the plot of phase response.

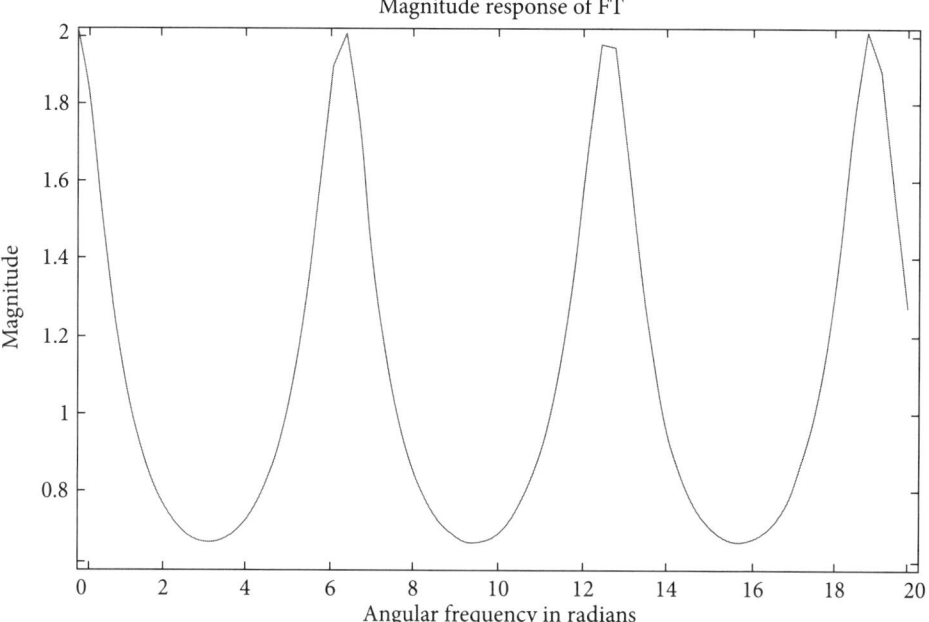

Fig. 6.34 Plot of magnitude response

Example 6.32

Find the DTFT of

$$x[n] = 2^n \text{ for } 0 \leq n \leq 7$$
$$= 0 \text{ otherwise}$$

Solution

We will use the formula for DTFT.

$$X(e^{j\Omega}) = \sum_{n=0}^{7} 2^n e^{-j\Omega n} = \sum_{n=0}^{7} (2e^{-j\Omega})^n$$

$$= 1 + 2e^{-j\Omega} + 2^2 e^{-j2\Omega} + \ldots + 2^7 e^{-j7\Omega} \tag{6.82}$$

Multiply both sides by $2e^{-j\Omega}$

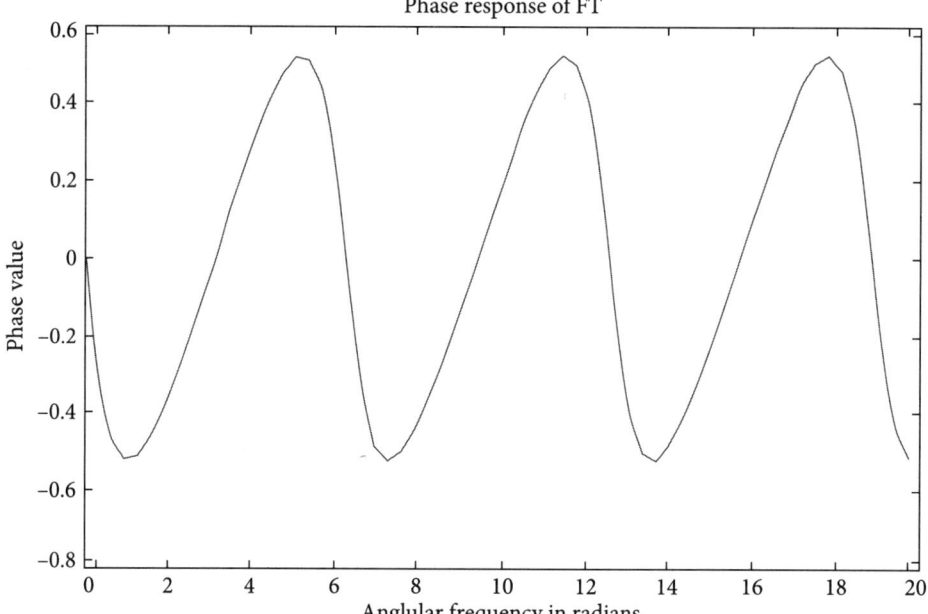

Fig. 6.35 Plot of phase response

$$2e^{-j\Omega}X(e^{j\Omega}) = 2e^{-j\Omega} + 2^2 e^{-j2\Omega} + 2^3 e^{-j3\Omega} + \ldots\ldots + 2^8 e^{-j8\Omega}$$

Subtract the second equation from the first.

$$(1 - 2e^{-j\Omega})X(e^{j\Omega}) = 1 - 2^7 e^{-j7\Omega}$$

$$X(e^{j\Omega}) = \frac{1 - 2^8 e^{-j8\Omega}}{1 - 2e^{-j\Omega}}$$

A MATLAB program to plot magnitude and phase response of DTFT is given below. Figures 6.36 and 6.37 show the magnitude and phase response, respectively. It can be observed that the magnitude and phase response is periodic with a period of 2pi (from −3.14(−pi) to 3.14(pi)).

```
clear all;
clc;
w=-20:0.2:20;
x=1+2*exp(-1j*w)+4*exp(-1j*2*w)+8*exp
(-1j*3*w)+16*exp(-1j*4*w)+32*exp(-1j*5*w)+64*exp
(-1j*6*w)+128*exp(-1j*7*w);
plot(w,abs(x));title('plot of magnitude response');
xlabel('angular frequency');ylabel('amplitude');
```

```
figure;
plot(w,angle(x));title('plot of phase response');
xlabel('angular frequency');ylabel('phase value');
```

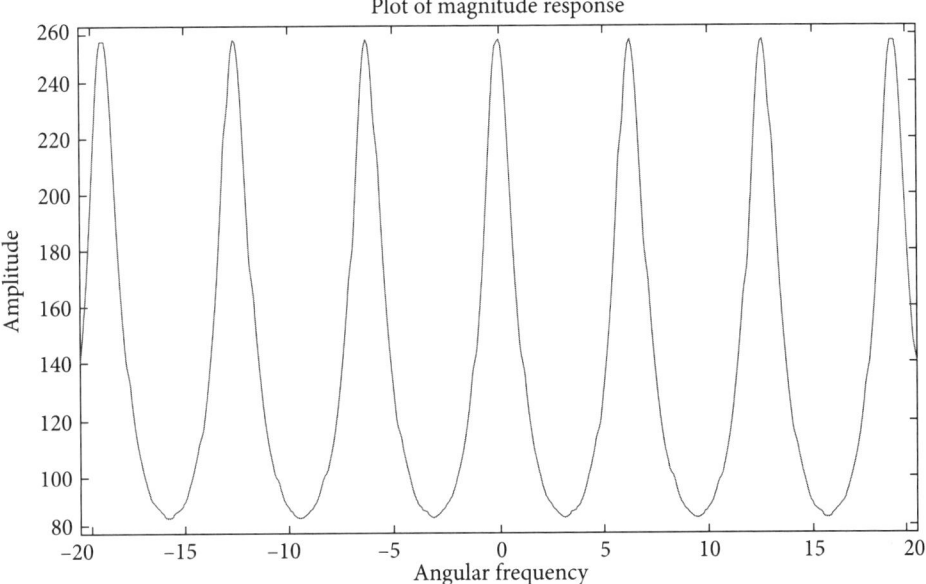

Fig. 6.36 Magnitude response

Example 6.33

Find DTFT of the signal $x[n] = a^{|n|}$ for $-1 < a < 1$

Solution

We will use the formula for DTFT.

$$x[n] = a^n; n \geq 0$$
$$= a^{-n}; n < 0 \quad \text{i.e., from } -\infty \text{ to } -1$$

$$X(e^{j\Omega}) = \sum_{n=-\infty}^{-1}(a^{-n}e^{-j\Omega n}) + \sum_{n=0}^{\infty} a^n e^{-j\Omega n} = \sum_{n=1}^{\infty}(ae^{j\Omega})^n + \sum_{n=0}^{\infty}(ae^{-j\Omega})^n$$

(6.83)

$$= \sum_{n=0}^{\infty}(ae^{j\Omega})^n - 1 + \sum_{n=0}^{\infty}(ae^{-j\Omega})^n \quad (\text{note}: (ae^{j\Omega})^0 = 1)$$

We will use the closed-form expression for infinite series.

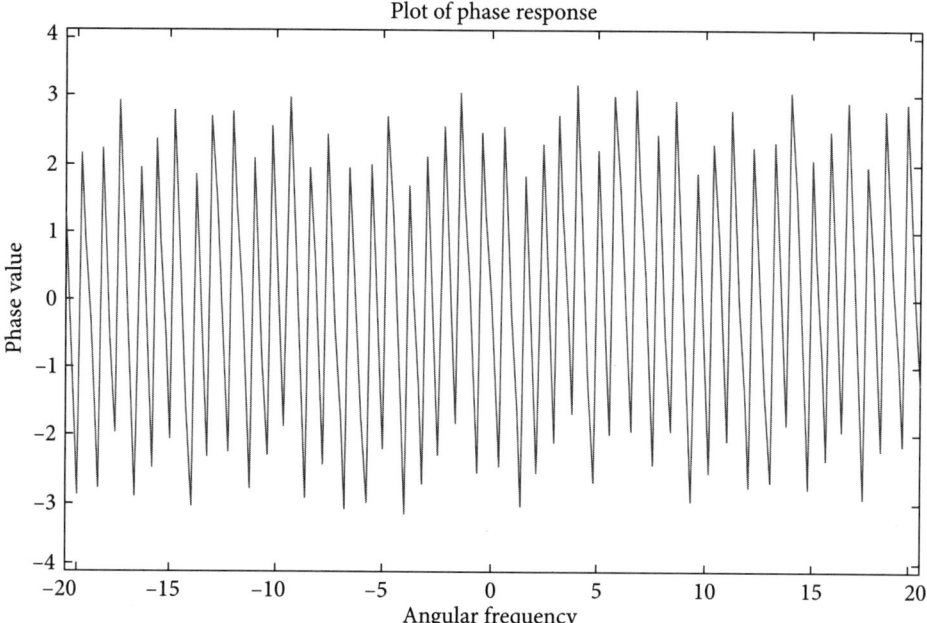

Fig. 6.37 Phase response

$$X(e^{j\Omega}) = \frac{1}{1-ae^{j\Omega}} - 1 + \frac{1}{1-ae^{-j\Omega}}$$

$$= \frac{1-ae^{-j\Omega}-1-a^2+ae^{-j\Omega}+ae^{j\Omega}+1-ae^{j\Omega}}{(1-ae^{j\Omega})(1-ae^{-j\Omega})}$$

$$= \frac{1-a^2}{1-2a\cos\Omega+a^2}$$

Example 6.34

Find DTFT of the signal $x[n] = \frac{1}{2}\left[\left(\frac{1}{2}\right)^n + \left(\frac{1}{4}\right)^n\right]u[n]$.

Solution

We will use the formula for DTFT.

$$X(e^{j\Omega}) = \frac{1}{2}\left[\sum_{n=0}^{\infty}\left(\left(\frac{1}{2}\right)^n e^{-j\Omega n}\right) + \sum_{n=0}^{\infty}\left(\frac{1}{4}\right)^n e^{-j\Omega n}\right]$$

$$= \frac{1}{2}\left[\sum_{n=0}^{\infty}\left(\frac{1}{2}e^{-j\Omega}\right)^n + \sum_{n=0}^{\infty}\left(\frac{1}{4}e^{-j\Omega}\right)^n\right]$$

$$X(e^{j\Omega}) = \frac{1}{2}\left[\frac{1}{1-\frac{1}{2}e^{-j\Omega}} + \frac{1}{1-\frac{1}{4}e^{-j\Omega}}\right] \quad (6.84)$$

$$= \frac{1}{2}\left[\frac{1-\frac{1}{4}e^{-j\Omega} + 1 - \frac{1}{2}e^{j\Omega}}{\left(1-\frac{1}{2}e^{-j\Omega}\right)\left(1-\frac{1}{4}e^{-j\Omega}\right)}\right]$$

$$= \frac{1}{2}\left[\frac{2-\frac{3}{4}}{1-\frac{3}{4}e^{-j\Omega}+\frac{1}{8}e^{-2j\Omega}}\right] = \frac{1-\frac{3}{8}}{1-\frac{3}{4}e^{-j\Omega}+\frac{1}{8}e^{-2j\Omega}}$$

Example 6.35

Find the inverse DTFT of $X(e^{j\Omega}) = 3\cos(4\Omega)$.

Solution

We will use the formula for inverse DTFT.

$$X(e^{j\Omega}) = 0 + \frac{3}{2}e^{j4\Omega} + \frac{3}{2}e^{-j4\Omega} \quad (6.85)$$

$$X(e^{j\Omega}) = \sum_{n=-\infty}^{\infty} x[n]e^{-j\Omega n} = x[-4]e^{j4\Omega} + x[0] + x[4]e^{-j4\Omega}$$

Comparing two equations

$$x[-4] = \frac{3}{2}, x[0] = 0; x[4] = \frac{3}{2}$$

We can obtain the same result using the direct formula for inverse DTFT.

$$x[n] = \frac{1}{2\pi}\int_{-\pi}^{\pi} X(e^{j\Omega})e^{j\Omega n}d\omega = \frac{3}{4\pi}\int_{-\pi}^{\pi}[e^{j4\Omega}e^{j\Omega n} + e^{-j4\Omega}e^{j\Omega n}]d\Omega$$

$$= \frac{3}{4\pi}[\int_{-\pi}^{\pi} e^{j(4+n)\Omega}d\Omega + \int_{-\pi}^{\pi} e^{j(n-4)\Omega}d\Omega]$$

$$= \frac{3}{4\pi}\left[\frac{e^{j(4+n)\Omega}}{j(4+n)} + \frac{e^{j(n-4)\Omega}}{j(n-4)}\right]_{-\pi}^{\pi} = \frac{3}{4\pi}\left[\frac{2j\sin(4+n)\pi}{j(4+n)} + \frac{2j\sin(n-4)\pi}{j(n-4)}\right] \quad (6.86)$$

$$= \frac{3}{4\pi}[2\pi] \text{ for } n = \pm 4$$

$$= \frac{3}{2} \text{ for } n = \pm 4$$

and $= 0$ otherwise

Example 6.36
Find the inverse DTFT of $X(e^{j\Omega}) = \cos(2\Omega) + j\sin(2\Omega)$.

Solution
We will use the formula for inverse DTFT.

$$X(e^{j\Omega}) = \cos(2\Omega) + j\sin(2\Omega) = e^{j2\Omega} \quad (6.87)$$

$$X(e^{j\Omega}) = 0 + e^{j2\Omega}$$

$$\quad (6.88)$$

$$X(e^{j\Omega}) = \sum_{n=-\infty}^{\infty} x[n]e^{-j\Omega n} = x[-2]e^{j2\Omega}$$

Comparing two equations

$x[-2] = 1$, $x[0] = 0$; $x[n] = 0$, otherwise

We can use the direct formula for inverse DTFT to find the DT signal.

$$x[n] = \frac{1}{2\pi} \int_{-\infty}^{\infty} X(e^{j\Omega}) e^{j\Omega n} d\Omega$$

$$= \frac{1}{2\pi} \int_{-\pi}^{\pi} e^{j2\Omega} e^{j\Omega n} d\Omega$$

$$= \frac{1}{2\pi j(n+2)} \left[e^{j(n+2)\Omega} \right]_{-\pi}^{\pi} \qquad (6.89)$$

$$= \frac{1}{2\pi j(n+2)} \left[e^{j(n+2)\pi} - e^{-j(n+2)\pi} \right]$$

$$= \frac{\sin(n+2)\pi}{(n+2)\pi}$$

$$= 1 \text{ for } n = -2$$

$$= 0 \text{ otherwise}$$

Example 6.37
Find the inverse DTFT of

$$X(e^{j\Omega}) = e^{-j2.5\Omega}; |\Omega| \le 1$$
$$= 0; 1 < \Omega < \pi$$

Solution
We will use the formula for inverse DTFT.

$$x[n] = \frac{1}{2\pi} \int_{-\pi}^{\pi} X(e^{j\Omega}) e^{j\Omega n} d\Omega$$

$$= \frac{1}{2\pi} \int_{-1}^{1} e^{-2.5j\Omega} e^{j\Omega n} d\Omega$$

$$\qquad (6.90)$$

$$= \frac{1}{2\pi} [\int_{-1}^{1} e^{(jn-2.5j)\Omega} d\Omega$$

$$= \frac{1}{2\pi} \left\{ \left[\frac{e^{j(n-2.5)\Omega}}{j(n-2.5)} \right]_{-1}^{1} \right\}$$

$$= \frac{1}{2\pi} \left[\frac{2j\sin(n-2.5)}{j(n-2.5)} \right] = \frac{\sin(n-2.5)}{\pi(n-2.5)} \text{ for all } n$$

We can write a MATLAB program to calculate $x[n]$ for different values of n and plot it. Inverse DTFT is plotted in Fig. 6.38. The signal has a centre of symmetry around 2.5.

```
clear all;
clc;
for n=1:1:21,
    x(n)=sin((n-11)-2.5)/(pi*((n-11)-2.5));
end
n1=-10:1:10;
stem(n1,x);
title('plot of the signal');xlabel('sample number');
ylabel('amplitude');
```

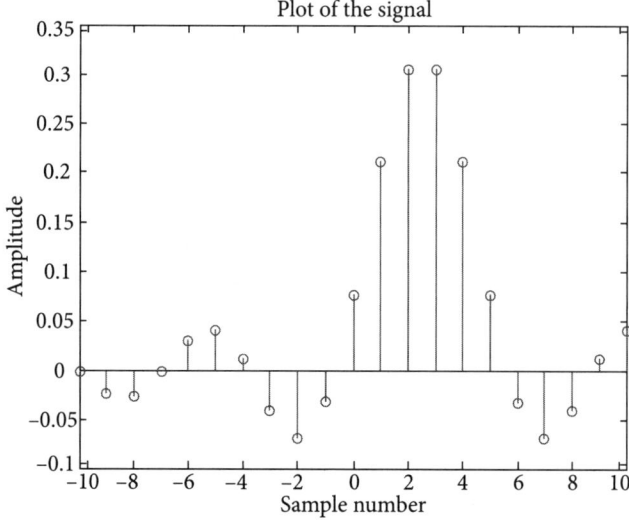

Fig. 6.38 Plot of inverse DTFT for Example 6.37

Example 6.38

Find the inverse DTFT of $X(e^{j\Omega}) = \begin{cases} e^{\Omega} & \text{for } -\pi < \Omega \leq 0 \\ e^{-\Omega} & \text{for } 0 < \Omega \leq \pi \end{cases}$.

Solution

We will use the formula for inverse DTFT.

$$x[n] = \frac{1}{2\pi}\int_{-\infty}^{\infty} X(e^{j\Omega})e^{j\Omega n} d\Omega$$

$$= \frac{1}{2\pi}\int_{-\pi}^{0} e^{\Omega} e^{j\Omega n} d\Omega + \frac{1}{2\pi}\int_{0}^{\pi} e^{-\Omega} e^{j\Omega n} d\Omega$$

$$= \frac{1}{2\pi}\left[\int_{-\pi}^{0} e^{(jn+1)\Omega} d\Omega + \int_{0}^{\pi} e^{(jn-1)\Omega} d\Omega\right] \qquad (6.91)$$

$$= \frac{1}{2\pi}\left\{\left[\frac{e^{(jn+1)\Omega}}{jn+1}\right]_{-\pi}^{0} + \left[\frac{e^{(jn-1)\Omega}}{jn-1}\right]_{0}^{\pi}\right\}$$

$$= \frac{1}{2\pi}\left[\frac{1-e^{-(jn+1)\pi}}{jn+1} + \frac{e^{(jn-1)\pi}-1}{jn-1}\right]$$

Example 6.39

Find the inverse DTFT of $X(e^{j\Omega}) = \begin{cases} \sin(-\Omega) \text{ for } -\pi < \Omega \leq 0 \\ \sin(\Omega) \text{ for } 0 < \Omega \leq \pi \end{cases}$.

Solution

We will use the formula for inverse DTFT.

$$X(e^{j\Omega}) = \frac{1}{2j}[0 + e^{j\Omega} - e^{-j\Omega}]$$

$$(6.92)$$

$$X(e^{j\Omega}) = \sum_{n=-\infty}^{\infty} x[n]e^{-j\Omega n} = x[-1]e^{j\Omega} + x[1]e^{-j\Omega} = \frac{1}{2j}e^{j\Omega} - \frac{1}{2j}e^{-j\Omega}$$

Comparing two equations

$$x[-1] = 1/2j, x[1] = -1/2j; x[n] = 0, \text{ otherwise}$$

$$x[n] = \frac{1}{2\pi} \int_{-\pi}^{\pi} X(e^{j\Omega}) e^{j\Omega n} d\omega$$

$$= \frac{1}{2\pi} \int_{-\pi}^{0} \sin(-\Omega) e^{j\Omega n} d\Omega + \frac{1}{2\pi} \int_{0}^{\pi} \sin(\Omega) e^{j\Omega n} d\Omega$$

$$= \frac{1}{4\pi j}[-\int_{-\pi}^{0}[e^{j(n+1)\Omega} + e^{j(n-1)\Omega}]d\Omega + \int_{0}^{\pi}[e^{j(n+1)\Omega} + e^{j(n-1)\Omega}]d\Omega$$

$$= \frac{1}{4\pi j}\left\{\left[-\frac{e^{j(n+1)\Omega}}{j(n+1)}\right]_{-\pi}^{0} - \left[\frac{e^{j(n-1)\Omega}}{j(n-1)}\right]_{-\pi}^{0} + \left[\frac{e^{j(n+1)\Omega}}{j(n+1)}\right]_{0}^{\pi} + \left[\frac{e^{j(n-1)\Omega}}{j(n-1)}\right]_{0}^{\pi}\right\}$$

$$= \frac{1}{4\pi j}\left[-\frac{1-e^{-j(n+1)\pi}}{j(n+1)} - \frac{1-e^{-j(n-1)\pi}}{j(n-1)} + \frac{e^{j(n+1)\pi}-1}{j(n+1)} + \frac{e^{j(n-1)\pi}-1}{j(n-1)}\right] \quad (6.93)$$

$$= \frac{1}{4\pi j}\left[\frac{e^{j(n+1)\pi}-e^{-j(n+1)\pi}}{j(n+1)} + \frac{e^{j(n-1)\pi}-e^{-j(n-1)\pi}}{j(n-1)}\right]$$

$$= \frac{1}{4\pi j}\left[\frac{2j\sin(n+1)\pi}{j(n+1)} + \frac{2j\sin(n-1)\pi}{j(n-1)}\right]$$

$$= \frac{1}{2\pi j}\left[\frac{\sin(n+1)\pi}{(n+1)} + \frac{\sin(n-1)\pi}{(n-1)}\right]$$

$$= 0 \text{ for } n \neq -1 \ \& \ n \neq 1$$

$$= \frac{1}{2j} \text{ for } n = \pm 1$$

Example 6.40

Find the inverse DTFT of $X(e^{j\Omega}) = \begin{cases} 1 \text{ for } -\pi/2 < \Omega \leq 0 \\ e^{-j\pi} \text{ for } 0 < \Omega \leq \pi/2 \end{cases}$.

Solution

We will use the formula for inverse DTFT.

$$x[n] = \frac{1}{2\pi}\int_{-\infty}^{\infty} X(e^{j\Omega})e^{j\Omega n}d\omega$$

$$= \frac{1}{2\pi}\int_{-\pi/2}^{0} e^{j\Omega n}d\Omega + \frac{1}{2\pi}\int_{0}^{\pi/2} e^{-j\pi}e^{j\Omega n}d\Omega$$

$$= \frac{1}{2\pi}\left\{\left[\frac{e^{j\Omega n}}{jn}\right]_{-\pi/2}^{0} - \left[\frac{e^{j\Omega n}}{jn}\right]_{0}^{\pi/2}\right\} \text{Note}: e^{-j\pi} = \cos\pi - j\sin\pi = -1 \quad (6.94)$$

$$= \frac{1}{2\pi}\left[\frac{1-e^{-jn\pi/2}}{jn} - \frac{e^{jn\pi/2}-1}{jn}\right] = \frac{1}{2\pi}\left[\frac{-e^{jn\pi/2} - e^{-jn\pi/2}+2}{jn}\right]$$

$$= \frac{2-2\cos(n\pi/2)}{2\pi jn} = \frac{(1-\cos(n\pi/2))}{j\pi n} \text{ for } n \ne 0$$

A MATLAB program to plot the time response is given below. Figure 6.39 shows the plot of time domain signal.

```
clear all;
clc;
for n=1:1:20,
x(n)=(1-cos(n*pi/2))/(n*pi);
end
stem(x);title('signal plot');xlabel('sample number');
ylabel('amplitude');
```

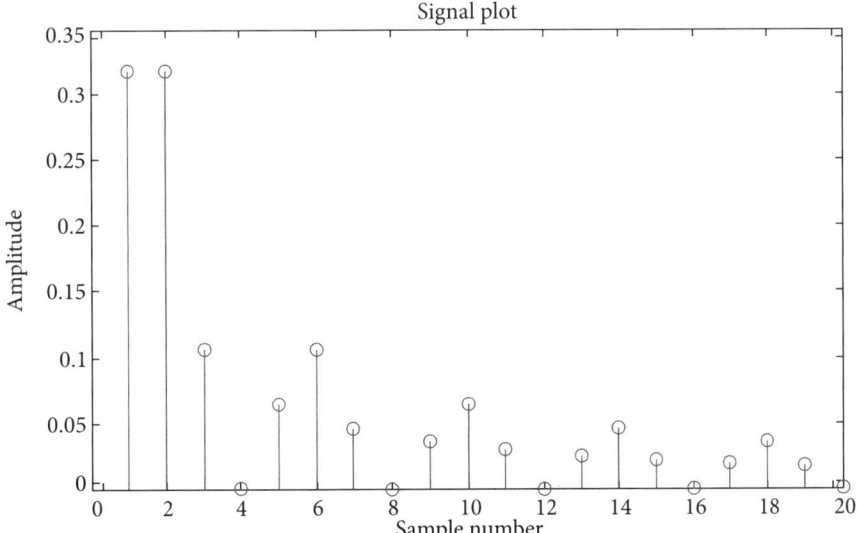

Fig. 6.39 Time domain response plot

Example 6.41

Find the inverse DTFT of $X(e^{j\Omega}) = \left\{\dfrac{1}{2}(1+\cos\Omega)\right\}$.

Solution

$$X(e^{j\Omega}) = \frac{1}{2}\left[1 + \frac{1}{2}e^{j\Omega} + \frac{1}{2}e^{-j\Omega}\right] \quad (6.95)$$

$$X(e^{j\Omega}) = \sum_{n=-\infty}^{\infty} x[n]e^{-j\Omega n} = x[-1]e^{j\Omega} + x[0] + x[1]e^{-j\Omega} = \frac{1}{4}e^{j\Omega} + 1 + \frac{1}{4}e^{-j\Omega}$$

Comparing two equations

$$x[-1] = 1/4, x[1] = 1/4j; x[n] = 1$$

Concept Check

- What is DTFT?
- Explain the reason for DTFT to be periodic.
- What is the period of DTFT?
- How will you calculate the magnitude and phase response using hand calculations?
- What will be inverse DTFT of a rectangular periodic pulse in frequency domain?
- What is DTFT of a unit impulse function?
- What is the inverse DTFT of a periodic impulse function in frequency domain?
- What will be the inverse DTFT of a periodic function like cosine?

6.6 Properties of FT and DTFT

We will study the properties of FT and DTFT. We will state and prove the following properties linearity, time shifting, time reversal, time scaling, time differentiation, time integration, Frequency shifting, frequency differentiation, convolution, frequency convolution, modulation, conjugation property, duality and Parseval's relation.

1. **Property of Linearity**

   If

   $$x_1(t) \leftrightarrow X_1(j\omega) \text{ and } x_2(t) \leftrightarrow X_2(j\omega),$$

   then $ax_1(t) + bx_2(t) \; aX_1(j\omega) + bX_2(j\omega)$ \hfill (6.96)

Proof

Using the definition of FT, we can write

$$FT[ax_1(t)+bx_2(t)] = \int_{-\infty}^{\infty}[ax_1(t)+bx_2(t)]e^{-j\omega t}dt$$

$$= a\left[\int_{-\infty}^{\infty} x_1(t)e^{-j\omega t}dt\right] + b\left[\int_{-\infty}^{\infty} x_2(t)e^{-j\omega t}dt\right] \quad (6.97)$$

$$= aX_1(j\omega) + bX_2(j\omega)$$

Similarly, linearity also holds for the FT of DT signals.
If

$x_1[n] \leftrightarrow X_1(e^{j\Omega})$ and $x_2[n] \leftrightarrow X_2(e^{j\Omega})$,

then $ax_1[n] + bx_2[n]\ aX_1(e^{j\Omega}) + bX_2(e^{j\Omega})$ \hfill (6.98)

Physical significance of property of linearity If the signal can be decomposed into linear combination of two or more component signals, then the FT of the signal is the linear combination of the component signals. This simplifies the mathematical computations. Similarly, if the signal can be decomposed into linear combination of two or more component signals in FT domain, then IFT of the signal is the linear combination of the component IFT signals. This property is used in the calculation of IFT using partial fraction expansion.

2. **Time Shifting Property**

    The time shifting property states that if

    $x_1(t) \leftrightarrow X(j\omega)$,

    then $x(t-t_0) \leftrightarrow e^{-j\omega t_0} X(j\omega)$ \hfill (6.99)

    Proof

    Using the definition of FT, we can write

    $$FT(x(t-t_0)) = \int_{-\infty}^{\infty} x(t-t_0)e^{-j\omega t}dt \text{ (put } t-t_0 = \tau)$$

    $$= \int_{-\infty}^{\infty} x(\tau)e^{-j\omega(\tau+t_0)}dt$$

    $$= e^{-j\omega t_0} \int_{-\infty}^{\infty} x(\tau)e^{-j\omega\tau}$$

    $$= e^{-j\omega t_0} X(j\omega)$$

    (6.100)

Similarly, for DTFT the property can be translated as

If

$x[n] \leftrightarrow X(e^{j\Omega})$,

then $x[n - n_0] \leftrightarrow e^{-j\Omega n_0} X(e^{j\Omega})$ \hfill (6.101)

Physical significance of property of time shifting Consider two signals with one time shifted with respect to the other. If FT is evaluated for both, the FT will be identical in magnitude. The time-shifted signal will have a FT with a phase shift of $e^{-j\vartheta}$ where $\vartheta = \omega t_0$. The magnitude of $e^{-j\vartheta}$ is one and hence, the magnitude response for a signal and its shifted version will be identical. This fact can be used for template matching when two templates are same, but are just shifted in time.

3. **Time Reversal Property**

The time reversal property states that if

$x(t) \leftrightarrow X(j\omega)$,

then $x(-t) \leftrightarrow X(-j\omega)$ \hfill (6.102)

Proof

Using the definition of FT, we can write

$$X(j\omega) = \int_{-\infty}^{\infty} x(t) e^{-j\omega t} dt \text{ (put } -\tau = t)$$

$$= \int_{-\infty}^{\infty} x(\tau) e^{-j\omega(-\tau)} d\tau = FT(x(-t)) \hfill (6.103)$$

For DTFT, the property can be stated as if

$x[n] \leftrightarrow X(e^{j\Omega})$,

then $x[-n] \leftrightarrow X(e^{-j\Omega})$ \hfill (6.104)

Physical significance of property of Time reversal The FT coefficients for a time-reversed signal are $X(-j\omega)$. This means the FT coefficients will have same magnitude, but the phase angle is negative of the phase angle for FT coefficients of the original signal. The FT coefficients of the time-reversed signal will be a complex conjugate of the FT coefficient of the original signal.

4. **Time Scaling Property**

The time scaling property states that if

$$x(t) \leftrightarrow X(j\omega),$$

then $x(at) \leftrightarrow \dfrac{1}{|a|} X(j\omega/a)$ \hfill (6.105)

Proof

Using the definition of FT, we can write

$$X(j\omega) = \int_{-\infty}^{\infty} x(t) e^{-j\omega t} dt$$

$$FT(x(at)) = \int_{-\infty}^{\infty} x(at) e^{-j\omega t} dt \tag{6.106}$$

Put $at = \tau$

$$= \int_{-\infty}^{\infty} x(\tau) e^{-j\omega \tau/a} (d\tau/a)$$

$$= \frac{1}{a} X\left(\frac{j\omega}{a}\right)$$

It can also be proved that if the scaling factor is negative, then $x(at) \leftrightarrow -\dfrac{1}{a} X(j\omega/a)$. Hence, in general for both positive and negative values of a, $x(at) \leftrightarrow \dfrac{1}{|a|} X(j\omega/a)$

Similarly, time scaling property for DTFT states that if

$$x[n] \leftrightarrow X(e^{j\Omega}),$$

then $x[an] \leftrightarrow \dfrac{1}{|a|} X(e^{j\Omega/a})$ \hfill (6.107)

Physical significance of property of Time scaling FT of the time-scaled signal with scaling factor a will have both magnitude and phase inverse scaled by the factor a. Consider the example of perceiving higher pitch of recorded sound when played at a higher speed. When the recorded sound is played again with higher speed, the time domain signal is compressed. This expands the frequency domain signal, which is perceived as a higher pitch.

5. **Time Differentiation Property**

    The property of time differentiation states that if

    $$x(t) \leftrightarrow X(j\omega),$$

    then $\dfrac{d}{dt} x(t) \leftrightarrow j\omega X(j\omega)$ \hfill (6.108)

    Proof

    Using the definition of FT, we can write

    $$X(j\omega) = \int_{-\infty}^{\infty} x(t) e^{-j\omega t} dt$$

    Differentiating both sides

    $$\frac{d}{dt}(x(t)) = \frac{d}{dt}\left[\int_{-\infty}^{\infty} x(t) e^{-j\omega t} dt\right] = \int_{-\infty}^{\infty} \frac{d}{dt}(x(t) e^{-j\omega t}) dt$$

    $$= \int_{-\infty}^{\infty} e^{-j\omega t} \frac{d}{dt}(x(t)) dt \quad \text{(use formula for } \int uv = u\int v - \int du \int v$$
    \hfill (6.109)

    $$= e^{-j\omega t} x(t) \Big\downarrow_{-\infty}^{\infty} + \int_{-\infty}^{\infty} j\omega e^{-j\omega t} x(t) dt \quad (x(-\infty) = 0 \text{ and } e^{-\infty} = 0)$$

    $$= j\omega X(j\omega)$$

    The properties of differentiation and integration apply only for continuous function and hence, are discussed for CT signals.

    Physical significance of property of Time differentiation When the signal is differentiated in time domain, its FT is multiplied by $j\omega$. For higher frequencies, the multiplying factor will also be high. Hence, after time differentiation, the higher frequencies are enhanced in frequency domain. Low frequencies will be reduced. The DC component will be destroyed when the signal is differentiated. The average value of the signal is reduced to zero when the signal is differentiated in time domain. FT of the differentiated signal will be zero at ω equal to zero.

6. **Time Integration Property**

    The property of time integration states that if

    $$x(t) \leftrightarrow X(j\omega),$$

    then $\int_{-\infty}^{t} x(\tau) d\tau \leftrightarrow X(j\omega) / j\omega$ \hfill (6.110)

Proof

Using the definition of FS, we can write

$$x(t) \leftrightarrow X(j\omega)$$

$$\frac{d}{dt}\left[\int_{-\infty}^{t} x(\tau)d\tau\right] = x(t)$$

so, $FT\{\frac{d}{dt}\left[\int_{-\infty}^{t} x(\tau)d\tau\right]\} = FT\{x(t)\}$ (6.111)

$$j\omega FT\left\{\int_{-\infty}^{t} x(\tau)d\tau\right\} = FT(x(t))$$

$$\int_{-\infty}^{t} x(\tau)d\tau \leftrightarrow X(j\omega)/j\omega$$

Physical significance of property of time integration When the signal is integrated in time domain, its FT gets multiplied in time by $-1/j\omega$. This factor is indeterminate at $\omega = 0$. If the signal has a zero DC component, then $X(j0) = 0$, and the integration equation exists and applies only to signals with zero DC value. Integration is basically a smoothing operation and hence, it smoothes the signal. The smoothing in time deemphasizes the high frequencies as is indicated by a factor of ω in denominator.

7. **Conjugation Property**

The conjugation property of FT states that

if $x(t) \leftrightarrow X(j\omega)$,

then $x^*(t) \leftrightarrow X^*(-j\omega)$ (6.112)

Proof: using definition of FT

$$X(j\omega) = \int_{-\infty}^{\infty} x(t)e^{-j\omega t} dt$$

$$FT\{x^*(t)\} = \int_{-\infty}^{\infty} x^*(t)e^{-j\omega t} dt$$

$$= \left[\int_{-\infty}^{\infty} x(t)e^{-(j(-\omega t))} dt\right]^*$$ (6.113)

(here it is FT equation, except ω is replaced by $-\omega$)

$$= [X(-j\omega)]^* = X^*(-j\omega)$$

Physical significance of conjugation property FT of the conjugate of the signal appears on negative frequency axis and is a conjugate of FT of the original signal. If the signal is real, FT coefficients for the signal show the conjugate symmetry. The positive coefficients are complex conjugates of the negative coefficients.

8. **Frequency Shifting Property**

The frequency shifting property of FT states that

if $x(t) \leftrightarrow X(j\omega)$,

then $e^{j\omega_0 t} x(t) \leftrightarrow X(j(\omega - \omega_0))$ (6.114)

Proof
Using definition of FT

$$X(j\omega) = \int_{-\infty}^{\infty} x(t) e^{-j\omega t} dt$$

$$FT\{e^{j\omega_0 t} x(t)\} = \int_{-\infty}^{\infty} x(t) e^{-j\omega t} e^{j\omega_0 t} dt = \int_{-\infty}^{\infty} x(t) e^{-j(\omega - \omega_0)t} dt \quad (6.115)$$

Here it is FT equation except ω is replaced by $\omega - \omega_0$

$$= [X(j(\omega - \omega_0))]$$

Similarly, for DTFT, the frequency shifting property states that

if $x[n] \leftrightarrow X(e^{j\Omega})$,

then $e^{j\Omega_0 n} x[n] \leftrightarrow X(e^{j(\Omega - \Omega_0)n})$ (6.117)

Physical significance of frequency shifting property When the FT of the signal is shifted in frequency domain by ω_0, the inverse transformation leads to a phase shifting of the signal by the same angular frequency ω_0. The resulting time domain signal is complex and its FT does not exhibit the symmetry property. Frequency shifting property is a dual of time shifting property. A shift in one domain leads to multiplication by a complex sinusoid in the other domain.

9. Frequency Differentiation Property

The frequency differentiation property of FT states that

if $x(t) \leftrightarrow X(j\omega)$,

$$\text{then } tx(t) \leftrightarrow j\frac{d}{d\omega}[X(j(\omega))] \tag{6.117}$$

Proof

Using the definition of FT

$$X(j\omega) = \int_{-\infty}^{\infty} x(t)e^{-j\omega t}dt$$

$$\frac{d}{d\omega}X(j\omega) = \frac{d}{d\omega}\left[\int_{-\infty}^{\infty} x(t)e^{-j\omega t}dt\right] = \int_{-\infty}^{\infty} x(t)\left[\frac{d}{d\omega}e^{-j\omega t}\right]dt$$

$$= \int_{-\infty}^{\infty} -jtx(t)e^{-j\omega t}dt \quad \left[-j = \frac{-j \times j}{j} = \frac{1}{j}\right] \tag{6.118}$$

$$j\frac{d}{d\omega}X(j\omega) = \int_{-\infty}^{\infty} tx(t)e^{-j\omega t}dt = FT(tx(t))$$

$$\frac{d}{d\omega}X(j\omega) = FT(-jtx(t))$$

For DTFT, the equation can be translated as

if $x[n] \leftrightarrow X(e^{j\Omega})$,

$$\text{then } nx[n] \leftrightarrow j\frac{d}{d\Omega}[X(e^{j\Omega})] \tag{6.119}$$

Physical significance of frequency differentiation property If we take differentiation of the FT of the signal and take its IFT, it will result in a signal that is multiplied by $(-t)$ time value for the signal.

10. Convolution Property

The convolution property states that

If

$$x_1(t) \leftrightarrow X_1(j\omega) \text{ and } x_2(t) \leftrightarrow X_2(j\omega),$$

$$\tag{6.120}$$

then $x_1(t) * x_2(t) \leftrightarrow X_1(j\omega)X_2(j\omega)$

Proof

Using the definition of FT, we can write

$$X_1(j\omega) = \int_{-\infty}^{\infty} x_1(t)e^{-j\omega t} dt$$

$$X_2(j\omega) = \int_{-\infty}^{\infty} x_2(t)e^{-j\omega t} dt$$

$$FT[x_1(t) * x_2(t)] = \int_{-\infty}^{\infty} [(x_1(t) * x_2(t)]e^{-j\omega t} dt$$

$$= \left[\int_{-\infty}^{\infty} \left[\int_{-\infty}^{\infty} x_1(\tau)x_2(t-\tau)d\tau\right] e^{-j\omega t} dt\right] \text{put } t-\tau = t' \quad (6.121)$$

$$= \left[\int_{-\infty}^{\infty} x_1(\tau)e^{-j\omega \tau} d\tau\right]\left[\int_{-\infty}^{\infty} x_2(t')e^{-j\omega t'} dt'\right]$$

$$= X_1(j\omega)X_2(j\omega)$$

The convolution property for DTFT states that

if $x_1[n] \leftrightarrow X_1(e^{j\Omega})$ and $x_2[n] \leftrightarrow X_2(e^{j\Omega})$,

then $x_1[n] * x_2[n] \leftrightarrow X_1(e^{j\Omega})X_2(e^{j\Omega})$ \quad (6.122)

Physical significance of convolution property When the two signals are convolved in time domain, their FTs get multiplied. Convolution is computationally costly. The convolution can be performed in transform domain using a multiplication operation that is less costly. Another important use of FT is to find the response of the system to a given input. When the signal passes via the LTI system, the signal gets convolved with the impulse response of the system. To find the FT representation for the convolved output, one may find FT for the input signal and FT of the impulse response and multiply the two FTs. The convolution in one domain transforms to modulation in the other domain. (Modulation refers to the multiplication of the two signals.) Both convolution and modulation properties are due to the fact that complex sinusoids are eigen functions of LTI system.

11. Frequency Convolution or Modulation Property

Frequency convolution property is also called as the modulation property. The frequency convolution or the modulation property states that if

$x_1(t) \leftrightarrow X_1(j\omega)$ and $x_2(t) \leftrightarrow X_2(j\omega)$,

then $x_1(t) \times x_2(t) \leftrightarrow X_1(j\omega) * X_2(j\omega)$ \hfill (6.123)

Proof

Using the definition of FT, we can write

$$FT[x_1(t) \times x_2(t)] = \int_{-\infty}^{\infty} [(x_1(t) \times x_2(t))]e^{-j\omega t} dt$$

$$= [\int_{-\infty}^{\infty} x_1(t)[\frac{1}{2\pi} \int_{-\infty}^{\infty} X_2(j\lambda)e^{j\lambda t} d\lambda]e^{-j\omega t} dt$$

$$= (\int_{-\infty}^{\infty} x_1(t)[\frac{1}{2\pi} \int_{-\infty}^{\infty} X_2(j\lambda)e^{-j(\omega-\lambda)t} d\lambda dt) \hfill (6.124)$$

Interchange the order of integration

$$= \frac{1}{2\pi} \int_{-\infty}^{\infty} X_2(j\lambda)(\int_{-\infty}^{\infty} x_1(t)e^{-j(\omega-\lambda)t} dt)d\lambda$$

$$= \frac{1}{2\pi} \int_{-\infty}^{\infty} X_1(j\lambda)X_2(j(\omega-\lambda))d\lambda$$

$$= X_1(j\omega) * X_2(j\omega)$$

The frequency convolution or modulation property for DTFT can be written as
   if

$x_1[n] \leftrightarrow X_1(e^{j\Omega})$ and $x_2[n] \leftrightarrow X_2(e^{j\Omega})$,

then $x_1[n] \times x_2[n] \leftrightarrow X_1(e^{j\Omega})(*)X_2(e^{j\Omega})$ \hfill (6.125)

(*)–denotes circular convolution as DTFT is periodic

Physical significance of modulation property When the carrier signal multiplies with the message signal to get $\cos(2\pi f_c t)\cos(2\pi f_m t)$, we get double side band suppressed carrier modulated signal. This property states the property of FT for the multiplied signal (multiplication of two sinusoids results in modulation); hence, it is called modulation property. The FT for the modulated signal is the convolution of the FTs of the component signals. We can understand the effect of signal truncation using the modulation property. When the signal is truncated, it is multiplied by a rectangular window in time domain. The result of multiplication of two signals in time domain is a convolution in the frequency domain. Hence, truncation of the signal in time domain results in convolution of signal spectrum and a sinc function (FT of a rectangular window). The convolution in frequency domain gives rise to a ripple near the band edge of the signal spectrum. This is termed as Gibbs phenomenon.

12. Duality Property

Throughout this chapter, we have observed symmetry between frequency domain and time domain representations of the signals. The reader may refer to the following examples.

1. A rectangular pulse in time/frequency domain results in a sinc function in frequency/time domain.
2. An impulse in time/frequency domain corresponds to a constant in frequency/time domain.
3. Convolution in time/frequency domain corresponds to multiplication in frequency/time domain.

The above examples indicate that we may interchange the words time and frequency. This interchangeability property is called as duality.

Consider the definitions of FT and IFT.

$$X(j\omega) = \int_{-\infty}^{\infty} x(t)e^{-j\omega t}\,dt$$

$$\bar{x}(t) = \frac{1}{2\pi}\int_{-\infty}^{\infty} X(j\omega)e^{j\omega t}\,d\omega \tag{6.126}$$

If one observes the above equation carefully, it can be concluded that

if $x(t) \leftrightarrow X(j\omega)$,

then $2\pi x(-\omega) \leftrightarrow X(jt)$, (6.127)

as indicated in Fig. 6.40.

The duality of rectangular pulses and sinc function is clear from Examples 6.4 and 6.5.

13. Time Bandwidth Product

Consider the example of a rectangular wave pulse. It was observed that the FT of a square wave pulse is a sinc function. Referring to Example 6.4 in section 6.2, the reader can recollect that if the width of the rectangular

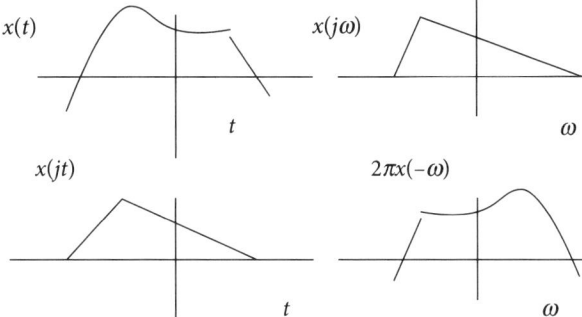

Fig. 6.40 FT duality property

pulse increases, the width of the main lobe of the sinc function decreases and vice versa. If the time extent of the signal increases, the frequency extent decreases. The property can be stated in other words as the product of the time extent and the frequency extent (time bandwidth product) remains constant.

Let us explain the concept of bandwidth for a sinc function for example. If the FT is a sinc function, the signal extends up to infinity in frequency domain. Hence, it is difficult to define the bandwidth. The bandwidth is actually referring to the significant content of the signal for positive frequencies. Considering the word significant, the bandwidth of the sinc function is defined as the positive width of the main lobe. Other commonly used definition of bandwidth is the frequency extent where the gain falls to 0.707 of its maximum value.

The time bandwidth product for any signal is lower bounded, as shown in the equation

$$T_d B_w \geq 1/2 \qquad (6.128)$$

This lower bound indicates that the signal duration and bandwidth cannot be simultaneously decreased. The signal satisfying the equation with equality is Gaussian pulse. Equation (6.128) is also called as uncertainty principle. The same result holds good for all Fourier representations.

14. **Parseval's Relation**

    Parseval's relation states that

    $x(t) \leftrightarrow X(j\omega)$ for complex valued $x(t)$,

    then $\int_{-\infty}^{\infty} |x(t)|^2 \, dt = \dfrac{1}{2\pi} \int_{-\infty}^{\infty} |X(j\omega)|^2 d\omega \qquad (6.129)$

Proof

To prove Parseval's relation, we start with LHS

$$|x(t)|^2 = x(t)x^*(t)$$

$$x(t) = \frac{1}{2\pi}\int_{-\infty}^{\infty} X(j\omega)e^{j\omega t}d\omega$$

$$x^*(t) = \frac{1}{2\pi}\int_{-\infty}^{\infty} X^*(j\omega)e^{-j\omega t}d\omega$$

$$\int_{-\infty}^{\infty} x(t)x^*(t)dt = \int_{-\infty}^{\infty} x(t)\left[\frac{1}{2\pi}\int_{-\infty}^{\infty} X^*(j\omega)e^{-j\omega t}d\omega\right]dt \qquad (6.130)$$

$$= \frac{1}{2\pi}\int_{-\infty}^{\infty} X^*(j\omega)\left[\int_{-\infty}^{\infty} x(t)e^{-j\omega t}dt\right]d\omega$$

$$= \frac{1}{2\pi}\int_{-\infty}^{\infty} X^*(j\omega)X(j\omega)d\omega$$

$$= \frac{1}{2\pi}\int_{-\infty}^{\infty} |X(j\omega)|^2 d\omega$$

Physical significance of Parseval's identity Parseval's identity states that the energy of the signal in time domain is the same as the energy of the FT coefficients. This simply means, when the signal is transformed in frequency domain, its energy is preserved. $|X(j\omega)|^2$ represents the energy spectral density of the signal.

Concept Check

- What is the significance of linearity property? How is it useful for finding inverse DTFT using partial fraction expansion method?
- Explain the time shifting and time scaling property of FT. What is its significance?
- What is the effect of shift in frequency on the time domain signal?
- What happens to a frequency domain signal when a signal is differentiated in time domain?
- State differentiation in frequency property.
- What is the modulation property?

- If two signals are multiplied in time domain, what is the effect on frequency domain signal?
- What is duality property?
- Explain the relation between time extent and frequency extent of the signal?
- What is time bandwidth product? What is the lower bound on this product?
- Explain the physical significance of the Parseval's identity.

6.7 FT and DTFT of Signals using FT/DTFT Properties

To illustrate the significance of the above properties, we will consider numerical examples to illustrate the use of different properties.

Example 6.42

Find inverse FT using partial fraction expansion. Use the property of linearity.

$$X(j\omega) = \frac{5(j\omega+1)}{j\omega[(j\omega)^2 + 7j\omega + 10]}$$

Solution

Step 1 We will first decompose the denominator in two factors.

$$X(j\omega) = \frac{5(j\omega+1)}{j\omega[(j\omega)^2 + 7j\omega + 10]} = \frac{5(j\omega+1)}{j\omega(j\omega+5)(j\omega+2)} \tag{6.131}$$

Step 2 Decompose the transfer function in component functions using partial fraction expansion.

$$X(j\omega) = \frac{k_1}{j\omega} + \frac{k_2}{j\omega+5} + \frac{k_3}{j\omega+2}$$

find k_1, k_2 and k_3

$$k_1 = \frac{5(j\omega+1)}{(j\omega+2)(j\omega+5)} \bigg\vert_{j\omega=0} = \frac{5}{10} = 1/2$$

$$\tag{6.132}$$

$$k_2 = \frac{5(j\omega+1)}{j\omega(j\omega+2)} \bigg\vert_{j\omega=-5} = \frac{5(-4)}{-5(-3)} = -4/3$$

$$k_3 = \frac{5(j\omega+1)}{j\omega(j\omega+5)}\bigg\downarrow_{j\omega=-2} = \frac{5(-1)}{-2(3)} = 5/6$$

$$X(j\omega) = \frac{1/2}{j\omega} - \frac{4/3}{j\omega+5} + \frac{5/6}{j\omega+2}$$

Step 3 Find IFT of each component term

$$X(j\omega) = \frac{1/2}{j\omega} - \frac{4/3}{j\omega+5} + \frac{5/6}{j\omega+2}$$

using property of linearity

$$x(t) = \frac{1}{2}u(t) - \frac{4}{3}e^{-5t}u(t) + \frac{5}{6}e^{-2t}u(t)$$

(6.133)

$$x(t) = \left(\frac{1}{2} - \frac{4}{3}e^{-5t} + \frac{5}{6}e^{-2t}\right)u(t)$$

Example 6.43

Find the FT of $x(t) = \cos(\omega_0 t)u(t)$ using the property of frequency shifting.

Solution

Let us first find FT of $u(t)$.

$$X(j\omega) = \int_{-\infty}^{\infty} u(t)e^{-j\omega t}dt = \int_0^{\infty} e^{-j\omega t}dt = -\frac{e^{-j\omega t}}{j\omega}\bigg\downarrow_0^{\infty} = \frac{1}{j\omega}$$

(6.134)

We will now use frequency shifting property of FT to find FT of $x(t) = \cos(\omega_0 t)u(t)$. Frequency shifting property of FT states that

if $x(t) \leftrightarrow X(j\omega)$,

then $e^{j\omega_0 t}x(t) \leftrightarrow X(j(\omega-\omega_0))$

$$\cos(\omega_0 t)u(t) = \frac{1}{2}\left[e^{j\omega_0 t} + e^{-j\omega_0 t}\right]u(t) \leftrightarrow \frac{1}{2}\left[\frac{1}{j(\omega-\omega_0)} + \frac{1}{j(\omega+\omega_0)}\right]$$

$$= \frac{1}{2}\left[\frac{2j\omega}{((j\omega)^2+\omega_0^2)}\right] = \frac{j\omega}{((j\omega)^2+\omega_0^2)}$$

(6.135)

Example 6.44

Find the FT of $x(t) = e^{-2jt}\cos(\omega_0 t)u(t)$ using the property of frequency shifting.

Solution

Frequency shifting property of FT states that

if $x(t) \leftrightarrow X(j\omega)$,

then $e^{j\omega_0 t}x(t) \leftrightarrow X(j(\omega - \omega_0))$ \hfill (6.136)

So, we have evaluated FT of $x(t) = \cos(\omega_0 t)u(t)$. We will then use frequency shifting property to find FT of the signal $x(t)$.

$$\cos(\omega_0 t)u(t) = \frac{1}{2}\left[e^{j\omega_0 t} + e^{-j\omega_0 t}\right]u(t) \leftrightarrow \frac{1}{2}\left[\frac{1}{j(\omega - \omega_0)} + \frac{1}{j(\omega + \omega_0)}\right]$$

$$e^{-2jt}\cos(\omega_0 t)u(t) \leftrightarrow \frac{1}{2}\left[\frac{1}{j(\omega - \omega_0 - 2)} + \frac{1}{j(\omega - \omega_0 - 2)}\right] \quad (6.137)$$

Example 6.45

Find the FT of $x(t) = \cos(2\pi f_0 (t-3))u(t)$ using the property of time shifting.

Solution

The time shifting property states that if

$x(t) \leftrightarrow X(j\omega)$,

then $x(t - t_0) \leftrightarrow e^{-j\omega t_0}X(j\omega)$

We know that

$$\cos(2\pi f_0 t)u(t) \leftrightarrow \frac{j\omega}{(j\omega)^2 + \omega_0^2}$$

$$\cos(2\pi f_0 (t-3))u(t) \leftrightarrow \frac{j\omega e^{-j3\omega}}{(j\omega)^2 + \omega_0^2} \quad (6.138)$$

Example 6.46

Use time shifting property to find the FT of the rectangular pulse shown in Fig. 6.41 below.

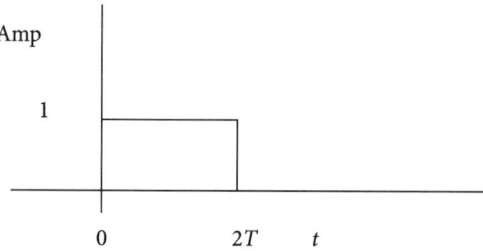

Fig. 6.41 Signal for Example 6.42

Solution

We can note that $y(t) = x(t - T)$

We can use time shifting property to find $Y(j\omega)$

$$X(j\omega) = \frac{2}{\omega}\sin(\omega T)$$

$$Z(j\omega) = e^{-j\omega T}\frac{2}{\omega}\sin(\omega T) \qquad (6.139)$$

Example 6.47

Use frequency shifting property to find inverse DTFT of $Z(j\omega) = \dfrac{1}{1 - 3e^{j(\Omega + \pi/4)}}$

Solution

We know the following result

$$a^n u[n] \leftrightarrow \frac{1}{1 - ae^{j\Omega}} \text{ so, } 3^n u[n] \leftrightarrow \frac{1}{1 - 3e^{j\Omega}} \qquad (6.140)$$

We will now use the property of frequency shifting.

$$3^n u[n] \leftrightarrow \frac{1}{1 - 3e^{j\Omega}} \Rightarrow 3^n e^{-j\pi n/4} u[n] \leftrightarrow \frac{1}{1 - 3e^{j(\Omega + \pi/4)}} \qquad (6.141)$$

Example 6.48

Use frequency differentiation property to find FT of $x(t) = te^{-2t}u(t)$.

Solution

The frequency differentiation property states that if

if $x(t) \leftrightarrow X(j\omega)$,

$$\text{then } tx(t) \leftrightarrow j\frac{d}{d\omega}[X(j(\omega))] \qquad (6.142)$$

We know that $x(t) = u(t) \leftrightarrow \dfrac{1}{j\omega}$

We will use frequency shifting property to find FT of $x(t) = e^{-2t}u(t)$
Frequency shifting property states that

if $x(t) \leftrightarrow X(j\omega)$,

then $e^{j\omega_0 t} x(t) \leftrightarrow X(j(\omega - \omega_0))$

So, $e^{-2t}u(t) \leftrightarrow \dfrac{1}{j\omega + 2}$ \hfill (6.143)

We will now use frequency differentiation property to find the FT of $x(t) = te^{-2t}u(t)$.

if $u(t) \leftrightarrow \dfrac{1}{j\omega}$

So, $e^{-2t}u(t) \leftrightarrow \dfrac{1}{j\omega + 2}$

$te^{-2t}u(t) \leftrightarrow j\dfrac{d}{d\omega}\left[\dfrac{1}{j\omega + 2}\right] = -j\dfrac{j}{(j\omega + 2)^2} = \dfrac{1}{(j\omega + 2)^2}$ \hfill (6.144)

Example 6.49

Use time differentiation property to find FT of $\dfrac{d}{dt}(e^{-at}u(t))$.

Solution

We know that $e^{-at}u(t) \leftrightarrow \dfrac{1}{j\omega + a}$. We will now use time differentiation property

$e^{-at}u(t) \leftrightarrow \dfrac{1}{j\omega + a}$,

then $\dfrac{d}{dt}(e^{-at}u(t)) \leftrightarrow \dfrac{j\omega}{j\omega + a}$ \hfill (6.145)

Example 6.50

Use differentiation in frequency, time scaling property to find the IFT of $X(j\omega) = j\dfrac{d}{d\omega}\left(\dfrac{e^{2j\omega}}{5 + j\omega}\right)$.

Solution

We know that, using time shifting property

$$x(t) \leftrightarrow X(j\omega),$$

then $x(t-t_0) \leftrightarrow e^{-j\omega t_0} X(j\omega)$ (6.146)

$$e^{-at}u(t) \leftrightarrow \frac{1}{j\omega+a}$$

So, $e^{-5t}u(t) \leftrightarrow \dfrac{1}{j\omega+5}$ (6.147)

we will use time shifting property

$$e^{-5(t+2)}u(t) \leftrightarrow \frac{e^{2j\omega}}{j\omega+5}$$

We will now use frequency differentiation property which states that

if $x(t) \leftrightarrow X(j\omega),$

then $tx(t) \leftrightarrow j\dfrac{d}{d\omega}[X(j(\omega))]$ (6.148)

$$te^{-5(t+2)}u(t) \leftrightarrow j\frac{d}{d\omega}\left(\frac{e^{2j\omega}}{5+j\omega}\right)$$ (6.149)

Example 6.51

Use the result of FT for a rectangular pulse of amplitude 1 between −1 and 1, and find the FT of the scaled rectangular pulse of amplitude 1 between −3 and 3.

Solution

We know that the FT of a rectangular pulse of width $2T$ (between $-T$ to T) is given by

$$X(j\omega) = \frac{2}{\omega}\sin(\omega T) \text{ put } T=1$$

$$X(j\omega) = \frac{2}{\omega}\sin(\omega)$$ (6.150)

To find the FT of a scaled rectangular pulse with a scaling factor of 1/3, we will use the property of scaling for FT to get

$$y(t) = x(t/3)$$

$Y(j\omega) = 3X(3j\omega)$, replace ω by 3ω in Eq. 6.139

$$Y(j\omega) = \frac{6}{3\omega}\sin(3\omega) = \frac{2}{\omega}\sin(3\omega) \qquad (6.151)$$

Example 6.52

Find DTFT of signal $x[n]$ and use the result of the property of scaling to find DTFT of $y[n]$ shown in Fig. 6.42 below.

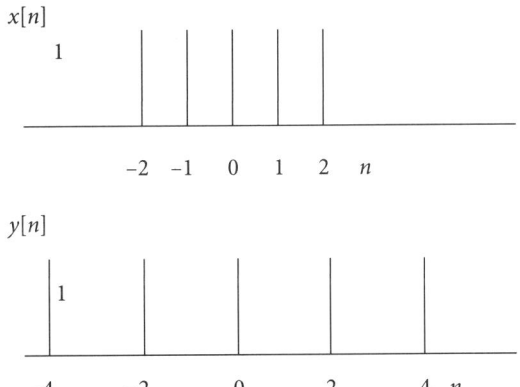

Fig. 6.42 The signal $x[n]$ and its scaled version $y[n]$

Solution:

$$X(e^{j\Omega}) = \sum_{n=-M}^{M} x[n]e^{-j\Omega n} = \sum_{n=-2}^{2} x[n]e^{-j\Omega n} = e^{j\Omega} + e^{2j\Omega} + 1 + e^{-j\Omega} + e^{-2j\Omega}$$

$$X(e^{j\Omega}) = 1 + 2\cos(\Omega) + 2\cos(2\Omega) \text{ put } m = M + n$$

It can also be written as

$$X(e^{j\Omega}) = \sum_{m=0}^{2M} x[m]e^{-j\Omega(m-M)} = e^{j\Omega M}\sum_{m=0}^{2M} e^{-j\Omega m}$$

$$X(e^{j\Omega}) = e^{j\Omega M}[1 + e^{-j\Omega} + e^{-j2\Omega} .... + e^{-j2M\Omega}]$$

$$(1-e^{-j\Omega})X(e^{j\Omega}) = e^{j\Omega M}(1-e^{-j(2M+1)\Omega})$$

$$X(e^{j\Omega}) = e^{j\Omega M}\frac{1-e^{-j(2M+1)\Omega}}{1-e^{-j\Omega}}$$

$$= e^{j\Omega M}\frac{e^{-j(2M+1)\Omega/2}e^{j(2M+1)\Omega/2} - e^{-j(2M+1)\Omega/2}e^{-j(2M+1)\Omega/2}}{e^{-j\Omega/2}e^{j\Omega/2} - e^{-j\Omega/2}e^{-j\Omega/2}}$$

$$= e^{jM\Omega}\frac{e^{-j(2M+1)\Omega/2}(2j\sin((2M+1)\Omega/2))}{e^{-j\Omega/2}(2j\sin(\Omega/2))}$$

(6.152)

$$X(e^{j\Omega}) = \frac{\sin((2M+1)\Omega/2)}{\sin(\Omega/2)} \text{ for } \Omega \neq 0, \pm 2\pi, \pm 4\pi..., \text{ etc.}$$

$$= (2M+1) \text{ for } \Omega = 0, \pm 2\pi, \pm 4\pi..., \text{ etc.}$$

We will now use the property of scaling to obtain DTFT of $y[n]$. Here, the scaling factor is ½.

$$x[n] \leftrightarrow 1 + 2\cos(\Omega) + 2\cos(2\Omega)$$

$$\text{then } y[n] = x[n/2] \leftrightarrow \frac{1}{|a|}X(j\Omega/a) = 2(1+2\cos(2\Omega)+2\cos(4\Omega)) \quad (6.153)$$

$$Y(e^{j\Omega}) = 2 + 4\cos(2\Omega) + 4\cos(4\Omega)$$

Example 6.53

Use the property of differentiation in time and differentiation in frequency to find the FT of a Gaussian pulse given by $g(t) = \frac{1}{\sqrt{2\pi}}e^{-t^2/2}$.

Solution

We will differentiate $g(t)$ with respect to t to find

$$g(t) \leftrightarrow G(j\omega),$$

$$\text{then } \frac{d}{dt}g(t) = -tg(t) \leftrightarrow j\omega G(j\omega) \quad (6.154)$$

We will now use the property of differentiation in frequency to get

if $g(t) \leftrightarrow G(j\omega)$,

then $tg(t) \leftrightarrow j\dfrac{d}{d\omega}[G(j(\omega))]$

$$-tg(t) \leftrightarrow \frac{1}{j}\frac{d}{d\omega}[G(j\omega)] \tag{6.155}$$

Equate the right-hand sides of Eqs (6.153) and (6.154), to get

$$j\omega G(j\omega) = \frac{1}{j}\frac{d}{d\omega}[G(j\omega)]$$

$$\frac{d}{d\omega}[G(j\omega)] = -\omega G(j\omega) \tag{6.156}$$

This differential equation has the general solution of the form $G(j\omega) = Ae^{-\omega^2/2}$ where A is a constant.

$$\text{for } \omega = 0, G(j0) = A = \int_{-\infty}^{\infty}\frac{1}{\sqrt{2\pi}}e^{-t^2/2}dt = 1 \tag{6.157}$$

This is the initial value theorem.

So, a Gaussian pulse has FT of the same form and is given by

$$g(t) = \frac{1}{\sqrt{2\pi}}e^{-t^2/2} \leftrightarrow e^{-\omega^2/2} \tag{6.158}$$

Example 6.54

Show that the DTFT of $x[n] = ne^{-j(\pi/4)}a^{n-2}u[n-2]$ is $X(e^{j\Omega}) = j\dfrac{d}{d\Omega}\left\{\dfrac{e^{-2j(\Omega-\pi/4)}}{1-ae^{-j(\Omega-\pi/4)}}\right\}$

Solution

We will use the result

$$a^n u[n] \leftrightarrow \frac{1}{1-ae^{-j\Omega}} \tag{6.159}$$

We will use time shifting property,

$$x[n] = a^n u[n] \leftrightarrow X(e^{j\Omega}) = \frac{1}{1-ae^{-j\Omega}},$$

then $x[n-2] = a^{n-2} u[n-2] \leftrightarrow e^{-j2\Omega}/(1-ae^{-j\Omega})$ \hfill (6.160)

We will use frequency shifting property

$$\text{if } a^{n-2} x[n-2] \leftrightarrow \frac{e^{-j2\Omega}}{1-ae^{-j\Omega}},$$

then $e^{-j\pi/4} a^{n-2} x[n-2] \leftrightarrow \dfrac{e^{-j2(\Omega-\pi/4)}}{1-ae^{-j(\Omega-\pi/4)}}$ \hfill (6.161)

We will use frequency differentiation property to get

$$\text{if } e^{-j\pi/4} a^{n-2} x[n-2] \leftrightarrow \frac{e^{-2j(\Omega-\pi/4)}}{1-ae^{-j(\Omega-\pi/4)}},$$

then $n e^{-j\pi/4} a^{n-2} x[n-2] \leftrightarrow j \dfrac{d}{d\Omega}\left[\dfrac{e^{-2j(\Omega-\pi/4)}}{1-ae^{-j(\Omega-\pi/4)}}\right]$ \hfill (6.162)

Example 6.55

Use the property of convolution in frequency or property of modulation to find the inverse DTFT of $X(e^{j\Omega}) = \dfrac{e^{-3j\Omega}}{1+\frac{1}{2}e^{-j\Omega}} * \dfrac{\sin(13\Omega/2)}{\sin(\Omega/2)}$

Solution

Property of modulation states that

$$x_1[n] \leftrightarrow X_1(e^{j\Omega}) \text{ and } x_2[n] \leftrightarrow X_2(e^{j\Omega})$$

then $x_1[n] \times x_2[n] \leftrightarrow X_1(e^{j\Omega})(*)X_2(e^{j\Omega})$ \hfill (6.163)

(*)– denote circular convolution as DTFT is periodic

We will find the inverse DTFT of the two individual terms that are convolved.
We will use the result

$$a^n u[n] \leftrightarrow \frac{1}{1-ae^{-j\Omega}}$$

$$\left(-\frac{1}{2}\right)^n u[n] \leftrightarrow \frac{1}{1+\frac{1}{2}e^{-j\Omega}} \qquad (6.164)$$

We will use time shifting property

$$\left(-\frac{1}{2}\right)^n u[n] \leftrightarrow \frac{1}{1+\frac{1}{2}e^{-j\Omega}},$$

$$\text{then}\left(-\frac{1}{2}\right)^{n-3} u[n-3] \leftrightarrow e^{-j3\Omega} / \left(1+\frac{1}{2}e^{-j\Omega}\right) \qquad (6.165)$$

We will use the result of 6.140, which is repeated here for ready reference. We will now find inverse DTFT of the second term.

$$x[n] \leftrightarrow X(e^{j\Omega}) = \frac{\sin((2M+1)\Omega/2)}{\sin(\Omega/2)} \quad \text{for} \quad \Omega \neq 0, \pm 2\pi, \pm 4\pi..., \text{ etc.}$$

$$= (2M+1) \quad \text{for} \quad \Omega = 0, \pm 2\pi, \pm 4\pi..., \text{ etc.}$$

Here, $x[n]$ is a rectangular wave sequence between $-M$ to $+M$

$$\frac{\sin(13\Omega/2)}{\sin(\Omega/2)} \leftrightarrow y[n] \text{ a rectangular wave sequence between } -6 \text{ to } +6$$

$$\text{So, } y[n] = \begin{cases} 1 \text{ for } -6 \leq n \leq 6 \\ 0 \text{ otherwise} \end{cases} \qquad (6.166)$$

The inverse DTFT of the given convolved signal is now the multiplication of their inverse DTFTs.

$$\left(-\frac{1}{2}\right)^{n-3} u[n-3] \leftrightarrow e^{-j3\Omega} / \left(1+\frac{1}{2}e^{-j\Omega}\right)$$

$$y[n] \leftrightarrow \frac{\sin(13\Omega/2)}{\sin(\Omega/2)} \qquad (6.167)$$

The inverse DTFT of the convolution of two transforms is

$$\left(-\frac{1}{2}\right)^{n-3} u[n-3] \times \{u[n+6] - u[n-7]\}$$

$$= \left(-\frac{1}{2}\right)^{n-3} \{u[n-3] - u[n-7]\}$$

Example 6.56

Use the property of duality to find FT of $x(t) = \dfrac{1}{1+jt}$.

Solution

We know the following result.

$$x(t) = e^{-t} u(t) \leftrightarrow \frac{1}{1+j\omega} \tag{6.168}$$

Replace ω by t. We get

$$x(jt) = \frac{1}{1+jt} \leftrightarrow 2\pi x(-\omega) = 2\pi e^{\omega} u(-\omega) \tag{6.169}$$

Example 6.57

Use Parseval's property to evaluate the sum $= \sum\limits_{n=-\infty}^{\infty} \dfrac{\sin^2(4n)}{\pi^2 n^2}$.

Solution

Consider the DT signal

$$x[n] = \frac{\sin(4n)}{\pi n} \leftrightarrow X(e^{j\Omega}) = \begin{cases} 1 \text{ for } |\Omega| \le 4 \\ 0 \text{ for } 4 < |\Omega| < \pi \end{cases} \tag{6.170}$$

We can write

$$\text{sum} = \sum_{n=\infty}^{\infty} \frac{\sin^2(4n)}{\pi^2 n^2} = \sum_{-\infty}^{\infty} \{x[n]\}^2 \tag{6.171}$$

Using Parseval's theorem

$$\text{sum} = \frac{1}{2\pi} \int_{-4}^{4} |X(e^{j\Omega})|^2 \, d\Omega = \frac{1}{2\pi} \Omega \downarrow_{-4}^{4} = 4/\pi$$

Concept Check

- Which property of FT have you used when you found the inverse FT using partial fraction expansion?
- How will you make use of frequency shifting property to find FT of a cosine function?
- Which property of FT can be used to find the FT of an exponentially damped sinusoid?
- Which properties of FT are useful for finding the FT of a Gaussian pulse?
- How will you find the inverse DTFT of the two convolved signals in frequency domain?

6.8 Analysis of LTI System using FT and DTFT

The FT and DTFT can be used for the analysis of LTI systems in two ways.
1. The system performance can be easily tested and analyzed in the frequency domain. The impulse response of the system is known in time domain. The impulse response signal is aperiodic. Depending upon if it is a CT or a DT signal, we can take its FT or DTFT. The FT or DTFT can be analyzed to find the magnitude and phase response of the system for different frequencies.
2. The response of any LTI system to a given input is evaluated as a convolution of the input signal with the impulse response of the LTI system in time domain. The convolution is computationally costly. The convolution property of FT helps in this regard. The convolution in time domain is equivalent to a multiplication in FT domain. Multiplication is less costly. One may first find FTs of the signal and impulse response of the LTI system, multiply the two FTs and then take inverse FT of the result. This gives the result of convolution of the signal and impulse response of the LTI system in time domain. Let us understand the procedure for finding frequency response and output of the system to any given input by considering some numerical examples.

Example 6.58

Let the impulse response of the system be given by $h(t) = e^{-2t}u(t-2)$ and the input to the system be $x(t) = e^{-4t}u(t)$. Find the output of the system.

Solution

Let us first find the FT of the input signal $x(t)$.

$$e^{-at}u(t) \leftrightarrow \frac{1}{a+j\omega}$$

$$x(t) = e^{-4t}u(t) \leftrightarrow \frac{1}{4+j\omega} \qquad (6.172)$$

We will now find the FT of the impulse response.

$$e^{-at}u(t) \leftrightarrow \frac{1}{a+j\omega}$$

$$e^{-2t}u(t) \leftrightarrow \frac{1}{2+j\omega} \tag{6.173}$$

$$e^{-2t}u(t-2) = e^{-4}e^{-2(t-2)}u(t-2) \leftrightarrow \frac{e^{-4}e^{-2j\omega}}{2+j\omega}$$

We have used time shifting property of FT
Let us multiply the two transforms.

$$\frac{1}{4+j\omega} \times \frac{e^{-4}e^{-2j\omega}}{2+j\omega} = \frac{k_1}{4+j\omega} + \frac{k_2}{2+j\omega}$$

We have used partial fraction expansion

$$k_1 = \frac{e^{-4}e^{-2j\omega}}{2+j\omega}\bigg\downarrow_{j\omega=-4} = \frac{e^{-4}e^{8}}{-2} = -\frac{e^4}{2}$$

$$k_2 = \frac{e^{-4}e^{-2j\omega}}{4+j\omega}\bigg\downarrow_{j\omega=-2} = \frac{e^{-4}e^{4}}{2} = \frac{1}{2} \tag{6.174}$$

Taking inverse FT of both terms

$$\text{the output} = -\frac{e^4}{2}e^{-4t}u(t) + \frac{1}{2}e^{-2t}u(t)$$

Example 6.59
The output of a system in response to an input $x(t) = e^{-3t}u(t)$ is $y(t) = e^{-t}u(t)$. Find the frequency response and impulse response of the system.

Solution
We will find the FT of the input as well as the output.

$$e^{-at}u(t) \leftrightarrow \frac{1}{a+j\omega}$$

$$x(t) = e^{-3t}u(t) \leftrightarrow \frac{1}{3+j\omega}$$

$$y(t) = e^{-t}u(t) \leftrightarrow \frac{1}{1+j\omega}$$

$$H(j\omega) = \frac{Y(j\omega)}{X(j\omega)} = \frac{3+j\omega}{1+j\omega} = 1 + \frac{2}{1+j\omega} \qquad (6.175)$$

The frequency response can be evaluated by putting different values of ω in the equation. Let us now find the inverse FT to get the impulse response of the system.

$$e^{-at}u(t) \leftrightarrow \frac{1}{a+j\omega}$$

$$H(j\omega) = \frac{Y(j\omega)}{X(j\omega)} = \frac{3+j\omega}{1+j\omega} = 1 + \frac{2}{1+j\omega} \qquad (6.176)$$

$$h(t) = \delta(t) + 2e^{-t}u(t)$$

Example 6.60

If the system is described by the differential equation given by
$\frac{d^2}{dt^2}y(t) + 5\frac{d}{dt}y(t) + 6y(t) = 2\frac{d}{dt}x(t) + x(t)$, find the frequency response and the input response of the system.

Solution

To find the frequency response, we will put

$$\frac{d^2}{dt^2} = (j\omega)^2, \frac{d}{dt} = j\omega \text{ in the differential equation}$$

$$\frac{d^2}{dt^2}y(t) + 5\frac{d}{dt}y(t) + 6y(t) = 2\frac{d}{dt}x(t) + x(t) \text{ to get}$$

$$(j\omega)^2 Y(j\omega) + 5(j\omega)Y(j\omega) + 6Y(j\omega) = 2(j\omega)X(j\omega) + X(j\omega) \qquad (6.177)$$

$$H(j\omega) = \frac{Y(j\omega)}{X(j\omega)} = \frac{2j\omega+1}{(j\omega)^2 + 5j\omega + 6} = \frac{k_1}{j\omega+3} + \frac{k_2}{j\omega+2}$$

$$k_1 = \frac{2j\omega+1}{j\omega+2}\bigg\downarrow_{j\omega=-3} = \frac{-6+1}{-3+2} = 5$$

$$k_2 = \frac{2j\omega+1}{j\omega+3}\bigg\downarrow_{j\omega=-2} = \frac{-4+1}{-2+3} = -3$$

$$H(j\omega) = \frac{5}{j\omega+3} - \frac{3}{j\omega+2}$$

$$h(t) = 5e^{-3t}u(t) - 3e^{-2t}u(t)$$

Example 6.61

If the impulse response of the system is given by $x(t) = \frac{4}{\pi^2 t^2}\sin^2(2t)$, find the frequency response.

Solution

We will use the modulation property of FT to get the FT of $x(t)$.

$$x(t) = \frac{4}{\pi^2 t^2}\sin^2(2t) = \left(\frac{2}{\pi t}\sin(2t)\right)^2 \quad (6.178)$$

We know the result

$$\frac{\sin(wt)}{\pi t} \leftrightarrow X(j\omega) = \begin{cases} 1 \text{ for } -w \leq \omega \leq w \\ 0 \text{ for } |\omega| > w \end{cases}$$

$$2\frac{\sin(2t)}{\pi t} = x_1(t) \leftrightarrow X(j\omega) = \begin{cases} 2 \text{ for } -2 \leq \omega \leq 2 \\ 0 \text{ for } |\omega| > 2 \end{cases}$$

$$x(t) \leftrightarrow X(j\omega) * X(j\omega) \quad (6.179)$$

We need to convolve the two rectangular window functions

The transform $X(j\omega)$ is plotted in Fig. 6.43. The same signal is convolved with itself to get the result of convolution, as shown in Fig. 6.44 below.

Note from the examples solved above for DT aperiodic signals that FT for DT aperiodic signals is periodic. DTFT has the period of 2*pi. We can summarize this fact in Table 6.1. Representations that are continuous in one domain either time or frequency are non-periodic in other domains. Representations that are discrete in one domain are periodic in the other domain, as listed in Table 6.2.

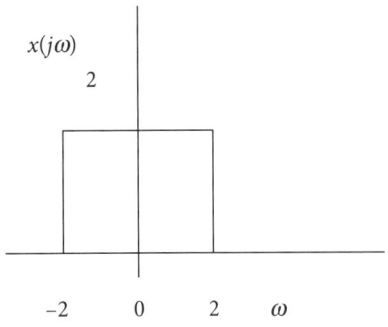

Fig. 6.43 Signal $X(j\omega)$

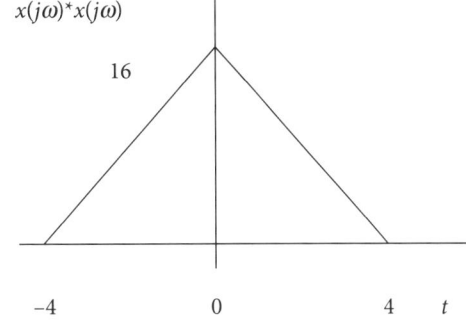

Fig. 6.44 Result of convolution of $X(j\omega)$ with itself

Table 6.1 FT for CT and DT periodic signals

Time domain signal	If periodic?	FS representation
CT	aperiodic	Aperiodic (FT)
DT	aperiodic	Periodic (DTFT)

Table 6.2 FS and FT periodicity properties

Time domain property	Frequency domain property
Continuous	Aperiodic (FT)
Discrete	Periodic (DTFT)
Periodic	Discrete (FS)
Non periodic	Continuous (DTFS)

Concept Check

- How will you analyze the frequency response of the system using FT?
- Which property of FT is useful for finding the output of the system to any given input?
- State the modulation property of FT. How is it useful for finding the FT of a multiplied signal?

Summary

In this chapter, we have described the FT of aperiodic CT signals and DTFT of aperiodic DT signals. The important concepts are explained.

- We have started with the Fourier representation of periodic signals and used the limit as the period tends to infinity to get the FT representation of aperiodic signals. The integrals of the FT equation may not always converge. The convergence can be confirmed if the signal satisfies Dirichlet conditions. Even if the signal is not square integrable, FT can still be defined using impulses. The procedure for the evaluation of magnitude and phase response using hand calculations is explained. MATLAB programs are given to confirm the results for magnitude and phase response of exponential signals. FT of a CT aperiodic signal is a continuous function of frequency and is aperiodic. The FT of left-handed and right-handed exponential signal have the same magnitude response. The phase values are reversed.

- The significance of FT is explained for the evaluation of the spectrum of BPSK in communication. FT of a rectangular pulse for different pulse widths is evaluated and is shown to be a sinc function. Effect of the pulse width variation on width of the main lobe of the sinc function is discussed. As the width of the rectangular pulse increases, the width of the main lobe of the sinc function decreases and vice versa. The extent of the signal in time multiplied by the extent of the signal in frequency domain is constant and has a lower bound. This is to state that time bandwidth product remains constant. This is also called as Heisenberg's uncertainty principle. It was also shown that IFT of a rectangular pulse in frequency domain is a sinc function in time domain. This is called as the property of duality. The FT of some standard signals is evaluated including the delta function, a constant, unit step, signum function, etc. The use of Dirac delta function for the evaluation of FT for complex exponential, sine function, cosine function, etc. is illustrated.

- FT of the periodic pulse train with spacing between the pulses of T is shown to be again a pulse train of spacing of $1/T$. The inverse FT calculations are shown using a direct formula for IFT. The IFT computation using partial fraction expansion is illustrated with several numerical examples. FT of DT signals are then discussed. The DT signal is generated by multiplication of a train of impulses and the signal. The multiplication in time domain is equivalently a convolution in transform domain. In FT domain, there is a convolution of the spectrum of the signal and the FT of a train of impulses, which is again a train of impulses. This results in DTFT as the replicated continuous spectrum. The DTFT of some standard aperiodic DT signals is evaluated. The procedure for calculation of inverse DTFT using a direct formula for DTFT is illustrated using numerical examples. DTFT of the periodic DT signal such as impulse train is shown to be the impulse train function.

- The properties of FT and DTFT are discussed with their physical significance. The property of linearity is used in inverse FT calculations. We have explained time shifting, time reversal, time scaling and time differentiation and integration properties of FT. The physical significance of all these properties is discussed. Time shifting property states that time shifting of a signal results in FT with same magnitude with phase changed by the angle ωt_0, where t_0 is the amount of shift. Time-reversed signals will also have the same magnitude response, but the sign of phase angle is reversed. Time scaling of a signal results in the scaling of both magnitude and phase angle of its FT. Time differentiation operation of a signal will remove the DC component of the FT and will multiply the magnitude of all frequencies by the frequency value resulting in the enhancement of high frequencies in FT domain. Time integration of the signal is basically a smoothing operation and hence, it deemphasizes the high frequencies in FT domain. Similar properties also apply to DTFT except the differentiation and integration.

- Conjugation property, Frequency shifting, frequency scaling, frequency differentiation properties are then discussed. FT of a real signal exhibits complex conjugate property. Frequency shifting by ω_0 in FT domain results in signal with phase change by same angular frequency. A shift in the signal in one domain results in the multiplication by the complex exponential in the other domain. Frequency differentiation results in multiplying the same by 't'. Further the properties of modulation and convolution are discussed. These properties indicate that multiplication in one domain results in the convolution in the other domain. This is a duality property. Duality is also indicated in FT and IFT calculations. FT of a rectangular pulse is a sinc function and vice versa. The Parseval's theorem is then explained. Several numerical examples are solved based on all the properties discussed.

- Output of the LTI system to any given input can be found by convolving the impulse response of the LTI system with the input signal. FT helps in this regard. Instead of using convolution, one may simplify the computations using FT. Find FTs of both the input signal and impulse response of LTI system, multiply the two and take inverse FT of the product. FT is useful for analyzing the frequency response of the system. Several numerical examples are solved to illustrate the concepts.

Multiple Choice Questions

1. A left-handed exponential signal extends from
   (a) Zero to infinity
   (b) Minus infinity to zero
   (c) Minus infinity to plus infinity
   (d) Zero to some positive integer

2. Dirichlet conditions state that
   (a) Signal must be square integrable
   (b) Signal must be finite
   (c) Signal must have infinite discontinuities
   (d) Signal must always be positive

3. Fourier transform of left-handed and right-handed exponential signal has
   (a) Same magnitude and phase response
   (b) Same phase response, but magnitude response reversed

(c) Same magnitude response, but phase valued reversed
(d) Different magnitude and phase response

4. Fourier transform of rectangular pulse in time domain is
   (a) a sinc function
   (b) a train of impulse
   (c) a modified sinc function
   (d) a rectangular window

5. As the width of the rectangular pulse in time domain increases
   (a) Width of main lobe of a sinc function in frequency domain also increases
   (b) Width of main lobe of a sinc function in frequency domain decreases
   (c) Width of main lobe of a sinc function in frequency domain
   (d) Width of main lobe of a sinc function in frequency domain reduces to zero

6. FT of a delta function in time domain is
   (a) A constant value 1 for all frequencies
   (b) A constant value 2*pi for all frequencies
   (c) A delta function existing at frequency of zero
   (d) A series of delta functions

7. FT of a signum function has magnitude value of
   (a) $1/j\omega$
   (b) $2/\omega$
   (c) $1/\omega^2$
   (d) 2ω

8. FT of a cosine function of frequency f_c is
   (a) a set of delta functions of unit magnitude at f_c and $2f_c$
   (b) a set of delta functions of unit magnitude at $-f_c$ and $2f_c$
   (c) a set of delta functions of unit magnitude at f_c, zero and $-f_c$
   (d) a set of delta functions of unit magnitude at f_c and $-f_c$

9. FT of a unit step function is
   (a) $\pi\delta(\omega) + 1/j\omega$
   (b) $\pi\delta(\omega) + 2/j\omega$
   (c) $2\pi\delta(\omega) + 1/j\omega$
   (d) $2\pi\delta(\omega) + 2/j\omega$

10. FT of DC is
    (a) $\pi\delta(\omega)$
    (b) $4\pi\delta(\omega)$
    (c) $2\delta(\omega)$
    (d) $2\pi\delta(\omega)$

11. Heisenberg uncertainty principle states that
    (a) $\Delta t \times \Delta f \geq 1/2$
    (b) $\Delta t \times \Delta f \geq 1$
    (c) $\Delta t \times \Delta f \leq 1/2$
    (d) $\Delta t \times \Delta f \geq 1/2$

12. When we find IFT using partial fraction expansion, we are using
    (a) Non-linearity property of FT
    (b) Convolution property of FT

(c) Multiplication property of FT
  (d) linearity property of FT
13. Time shifting of a signal results in FT with
    (a) same magnitude and phase response
    (b) same magnitude but, different phase response
    (c) same magnitude response and phase changed by factor of ωt_0 for each frequency
    (d) Different magnitude and phase responses
14. Time scaling of a signal results in FT with
    (a) magnitude and phase response changed by the same scale
    (b) Different magnitude response but, phase response unchanged
    (c) Same magnitude response and phase changed by same scale
    (d) Same phase response and magnitude changed by same scale
15. Time differentiation of a signal results in FT with
    (a) low frequencies enhanced
    (b) high frequencies enhanced
    (c) all frequencies enhanced
    (d) high frequencies deemphasized
16. Time differentiation of the signal results in FT with
    (a) low DC value
    (b) high DC value
    (c) same DC value
    (d) zero DC value
17. Frequency differentiation in FT domain results in IFT with
    (a) low frequencies deemphasized
    (b) high frequencies deemphasized
    (c) all frequencies deemphasized equally
    (d) no frequencies deemphasized
18. FT of a real signal shows
    (a) complex symmetry
    (b) complex conjugation property
    (c) phase symmetry property
    (d) Magnitude and phase symmetry
19. FT of a rectangular function is a sinc function. If the width of rectangular pulse increases, then
    (a) width of main lobe of sinc function decreases
    (b) width of main lobe remains constant

(c) width of main lobe of sinc function increases
(d) width of main lobe changes erratically

20. Convolution of the signals in time domain results in
    (a) Addition of the signals in transform domain
    (b) multiplication of the signals in transform domain
    (c) multiplication of transform of one signal with other signal
    (d) convolution of the two transforms

21. Modulation property is also termed as
    (a) Frequency convolution property
    (b) convolution property
    (c) additive property
    (d) linearity property

22. To find the output of the LTI system to any input is found by
    (a) Multiplying the input with the impulse response of the system
    (b) Convolving the input with the impulse response of the system
    (c) Convolving the FTs of the input and impulse response of the system
    (d) Multiplying the input with FT of the impulse response of the system

23. Parseval's theorem states that
    (a) $\int_{-\infty}^{\infty}|x(t)|^2\,dt = \int_{-\infty}^{\infty}|X(j\omega)|^2\,d\omega$
    (b) $\int_{-\infty}^{\infty}|x(t)|^2\,dt = \frac{1}{2}\int_{-\infty}^{\infty}|X(j\omega)|^2\,d\omega$
    (c) $\int_{-\infty}^{\infty}|x(t)|^2\,dt = \frac{1}{2\pi}\int_{-\pi}^{\pi}|X(j\omega)|^2\,d\omega$
    (d) $\int_{-\infty}^{\infty}|x(t)|^2\,dt = \frac{1}{2\pi}\int_{-\infty}^{\infty}|X(j\omega)|^2\,d\omega$

24. If $H(j\omega) = 1 + \dfrac{2}{1+j\omega}$, then $h(t)$ is given by
    (a) $h(t) = \delta(t) + 2e^{-t}u(t)$
    (b) $h(t) = \delta(t) + e^{-2t}u(t)$
    (c) $h(t) = u(t) + 2e^{-t}u(t)$
    (d) $h(t) = \delta(t) + 2e^{-2t}u(t)$

25. If the signal is given by $x(t) = e^{-at}u(t)$, its FT is given by
    (a) $\dfrac{1}{-a+j\omega}$
    (b) $\dfrac{1}{a+j\omega}$
    (c) $\dfrac{a}{a+j\omega}$
    (d) $\dfrac{1}{a-j\omega}$

26. Time scaling property of DTFT states that

   (a) if $x[n] \leftrightarrow X(e^{j\Omega})$ then $x[an] \leftrightarrow \dfrac{1}{|a|} X(e^{j\Omega})$

   (b) if $x[n] \leftrightarrow X(e^{j\Omega})$ then $x[an] \leftrightarrow X(e^{j\Omega/a})$

   (c) if $x[n] \leftrightarrow X(e^{j\Omega})$ then $x[an] \leftrightarrow \dfrac{1}{|a|} X(e^{j\Omega/a})$

   (d) if $x[n] \leftrightarrow X(e^{j\Omega})$ then $x[an] \leftrightarrow a\, X(e^{j\Omega/a})$

27. DTFT of a DT signal is periodic with a period of
   (a) 2pi
   (b) Pi
   (c) 4pi
   (d) Pi/2

28. FT of the Gaussian pulse is given by

   (a) $g(t) = \dfrac{1}{\sqrt{2\pi}} e^{-t^2/2} \leftrightarrow \dfrac{1}{2} e^{-\omega^2/2}$

   (b) $g(t) = \dfrac{1}{\sqrt{2\pi}} e^{-t^2/2} \leftrightarrow e^{-\omega^2/2}$

   (c) $g(t) = \dfrac{1}{\sqrt{2\pi}} e^{-t^2/2} \leftrightarrow \dfrac{1}{\sqrt{2\pi}} e^{-\omega^2/2}$

   (d) $g(t) = \dfrac{1}{\sqrt{2\pi}} e^{-t^2/2} \leftrightarrow \dfrac{1}{\pi} e^{-\omega^2/2}$

29. Consider a DT signal $x[n] = \dfrac{\sin(2n)}{\pi n}$, DTFT is given by

   (a) $X(e^{j\Omega}) = \begin{cases} 1 & \text{for } |\Omega| \leq 4 \\ 0 & \text{for } 4 < |\Omega| < \pi \end{cases}$

   (b) $X(e^{j\Omega}) = \begin{cases} 1 & \text{for } |\Omega| \leq 2 \\ 0 & \text{for } 2 < |\Omega| < \pi \end{cases}$

   (c) $X(e^{j\Omega}) = \begin{cases} 2 & \text{for } |\Omega| \leq 2 \\ 0 & \text{for } 2 < |\Omega| < \pi \end{cases}$

   (d) $X(e^{j\Omega}) = \begin{cases} 1 & \text{for } |\Omega| \leq \pi/2 \\ 0 & \text{for } \pi/2 < |\Omega| < \pi \end{cases}$

30. FT of $x(t) = \cos(\omega_0 t) u(t)$ is

   (a) $\dfrac{1}{2}\left[\dfrac{1}{j(\omega-\omega_0)} - \dfrac{1}{j(\omega+\omega_0)}\right]$

   (b) $\dfrac{1}{2}\left[\dfrac{1}{j(\omega-\omega_0)} + \dfrac{1}{j(\omega+\omega_0)}\right]$

   (c) $\dfrac{1}{2}\left[\dfrac{1}{j(\omega+\omega_0)} - \dfrac{1}{j(\omega-\omega_0)}\right]$

   (d) $\dfrac{1}{2}\left[\dfrac{1}{j(\omega+\omega_0)} + \dfrac{1}{j(\omega-2\omega_0)}\right]$

Review Questions

6.1 Define FT and IFT for aperiodic CT signals. State the Dirichlet conditions. How will you define FT for a signal if it is not square integrable?

6.2 Define the magnitude and phase response of FT of a signal.

6.3 Explain the procedure for magnitude and phase response using hand calculations.

6.4 Explain the difference between the phase responses of right-handed and left-handed exponential signals.

6.5 Discuss the effect of width of the rectangular pulse in time domain on the width of the main of the sinc function infrequency domain.

6.6 Explain Heisenberg's uncertainty principle.

6.7 Explain the use of Dirac delta function for finding the FT of exponential function, sine function, cosine function, etc.

6.8 Explain the procedure of IFT calculation using partial fraction expansion.

6.9 Explain why DTFT is periodic with a period of pi.

6.10 Explain the procedure for obtaining the DTFT in the form of closed from expression.

6.11 Explain how the DT signal is obtained?

6.12 Explain the DTFT of the train of impulses.

6.13 What is linearity property of FT? What are its applications?

6.14 Explain time shifting, time reversal and time scaling property of FT and DTFT along with the physical significance of each property.

6.15 Explain time differentiation and time integration property of FT. Does it have any significance for DTFT?

6.16 Explain frequency shifting and frequency differentiation property of FT and DTFT.

6.17 Explain modulation and convolution properties of FT and DTFT with the applications for both.

6.18 What is Parseval's theorem? What is its significance?

6.19 How will you use FT to analyze the LTI system? Explain with the help of suitable example.

6.20 How will you do frequency analysis of the signal? How will you analyze frequency response of the LTI system.

Problems

6.1 Determine the FT representation for the signal given as $x(t) = [2e^{-2t} + 3e^{-4t}]\, u(t)$. Find Magnitude response and phase response using hand calculations for 5 different values of angular frequency. Write a MATLAB program to find magnitude and phase response.

6.2 Consider a rectangular pulse of duration 0.4 sec and amplitude 2, as shown in the figure below. Find its FT. Write a MATLAB program to find the magnitude and phase response. The rectangular pulse in Figure can be mathematically defined as

$$\text{rect}(t) = \begin{cases} 2 & -0.2 \leq t \leq 0.2 \\ 0 & |t| > 0.2 \end{cases}$$

6.3 Find the IFT of

$$X(j\omega) = \begin{cases} 1 & -0.1 \leq \omega \leq 0.1 \\ 0 & |\omega| > 0.1 \end{cases}, \text{ as shown in the figure below. Write a}$$

MATLAB program to find IFT.

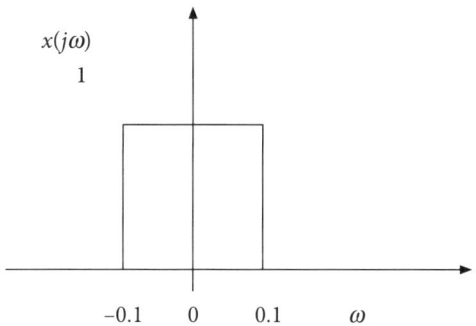

6.4 Find FT of the aperiodic signal given by

$$x(t) = \begin{cases} 2t & \text{for } 0 \leq t \leq 1 \\ 0 & \text{otherwise} \end{cases}$$

6.5. Find FT of the aperiodic triangular signal given by

$$x(t) = \begin{cases} 1 - t/2 & \text{for } 0 \leq t \leq 2 \\ 0 & \text{otherwise} \end{cases}$$

6.6 Find FT of aperiodic signal given by $x(t) = e^{-3t}u(t-3)$.
6.7 Find FT of aperiodic signal given by $x(t) = e^{-t}\cos(3\pi t)u(t)$.
6.8 Find FT of aperiodic signal given by $x(t) = t$ for $|t| < 1$.
6.9 Find FT of aperiodic signal given by $x(t) = e^{2t}u(-t+1)$.
6.10 Find FT of $x(t) = \delta(t) + u(t)$.

6.11 Find FT of the signal $x(t) = [3\cos(2\pi t) + 2\sin(3\pi t)]$ for all t using Dirac delta function.

6.12 FT of ideal sampling function with sampling interval of 4 seconds.

6.13 Find Inverse FT of

$$X(j\omega) = \begin{cases} 4\cos(3\omega) & \text{for } |\omega| \leq \pi \\ 0 & \text{for } |\omega| > \pi \end{cases}$$

6.14 Find the inverse FT using partial fraction expansion.

$$X(j\omega) = \frac{1}{(j\omega)^2 + 7j\omega + 10}$$

6.15 Find the inverse FT using partial fraction expansion.

$$X(j\omega) = \frac{j\omega + 1}{(j\omega + 2)^2}$$

6.16 Find the inverse FT using partial fraction expansion.

$$X(j\omega) = \frac{1}{-\omega^2 + 3j\omega + 2}$$

6.17 Find the inverse FT using partial fraction expansion.

$$X(j\omega) = \frac{-(j\omega)^2 - 3j\omega - 3}{[(j\omega)^2 + 3\omega + 2](j\omega + 3)}$$

6.18 Find the DTFT of $x[n] = \begin{cases} 1 & \text{for } 0 \leq n \leq 2 \\ -1 & \text{for } -2 \leq n \leq -1 \\ 0 & \text{otherwise} \end{cases}$

6.19 Find the DTFT of $x[n] = \begin{cases} 1/2 & \text{for } 0 \leq n \leq 3 \\ 0 & \text{otherwise} \end{cases}$

6.20 Find inverse DTFT of $X(e^{j\omega}) = \begin{cases} 1 & |\Omega| \leq \pi/2 \\ 0 & \pi/2 < |\Omega| < \pi \end{cases}$.

6.21 Find the DTFT of $x[n] = \delta[n] + \delta[n-1]$.

622 Find inverse DTFT of $X(e^{j\omega}) = \delta(\Omega) + \delta(\Omega - \pi/2)$, for $-\pi < \Omega < \pi$.

6.23 Find the DTFT of the exponential sequence $x[n] = \left(\frac{1}{3}\right)^n u[n] + \left(\frac{1}{4}\right)^n u[n]$.

6.24 Find the inverse DTFT of $X(e^{j\Omega}) = 2\cos(3\Omega)$.

6.25 Find the inverse DTFT of $X(e^{j\Omega}) = \cos(\Omega/2) + j\sin(\Omega/2)$.

6.26 Find the inverse DTFT of $X(e^{j\Omega}) = \begin{cases} e^{\Omega/2} & \text{for } -\pi < \Omega \leq 0 \\ e^{-\Omega/2} & \text{for } 0 < \Omega \leq \pi \end{cases}$.

6.27 Find the inverse DTFT of $X(e^{j\Omega}) = \begin{cases} -\sin(3\Omega) & \text{for } -\pi < \Omega \leq 0 \\ \sin(3\Omega) & \text{for } 0 < \Omega \leq \pi \end{cases}$.

6.28 Find the inverse FT using partial fraction expansion. Use the property of linearity.

$$X(j\omega) = \frac{2(j\omega + 3)}{j\omega[(j\omega)^2 + 3j\omega + 2]}$$

6.29 Find the FT of $x(t) = \cos(3\omega_0 t)u(t)$ using the property of frequency shifting.

6.30 Find the FT of $x(t) = e^{3jt}\sin(\omega_0 t)u(t)$ using the property of frequency shifting.

6.31 Find the FT of $x(t) = \sin(2\pi f_0(t-5))u(t)$ using the property of time shifting.

6.32 Use time shifting property to find the FT of the rectangular pulse shown in the figure below.

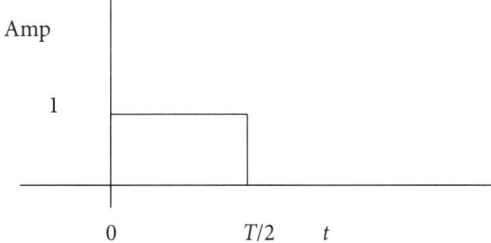

6.33 Use frequency shifting property to find inverse DTFT of

$$Z(j\omega) = \frac{1}{1 + 5e^{j(\Omega + \pi/3)}}.$$

6.34 Use frequency differentiation property to find FT of $x(t) = t\cos(10\pi t)u(t)$.

6.35 Use time differentiation property to find FT of $\dfrac{d}{dt}(\cos(at)u(t))$.

6.36 Use differentiation in frequency, time scaling property to find the IFT of

$$X(j\omega) = j\frac{d}{d\omega}\left(\frac{e^{-3j\omega}}{2 + j\omega}\right).$$

6.37 Use the result of FT for a rectangular pulse of amplitude 1 between −1 and 1 and find the FT of the scaled rectangular pulse of amplitude 2 between −1/2 and 1/2.

6.38 Find DTFT of signal $x[n]$ shown in Fig. 1 below and use the result of the property of scaling to find DTFT of $y[n]$ shown in Fig. 2 below.

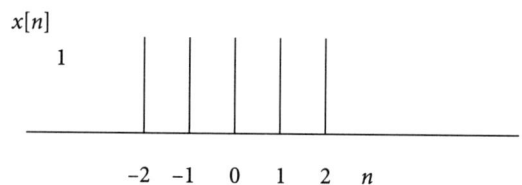

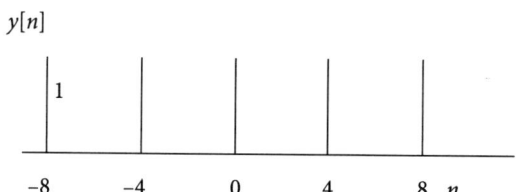

Answers

Multiple Choice Questions

1(b)	2(a)	3(c)	4(a)	5(b)
6(a)	7(b)	8(d)	9(a)	10(d)
11(a)	12(d)	13(c)	14(a)	15(b)
16(d)	17(b)	18 (b)	19 (a)	20 (b)
21 (a)	22 (b)	23 (d)	24 (a)	25 (b)
26 (c)	27 (a)	28 (b)	29(b)	30(a)

Problems

6.1 $X(j\omega) = \dfrac{2}{2+j\omega} + \dfrac{3}{4+j\omega}$, the magnitude of the first 5 frequency points is 1.75

1.6220   1.3779   1.1547   0.9775   0.8399

6.2 $X(\omega) = \dfrac{4\sin(0.2\omega)}{\omega}$

6.3 $x(t) = \dfrac{1}{\pi t}[\sin(0.1t)]$

6.4 $X(j\omega) = \dfrac{j2}{\omega}e^{-j\omega} - 2 + 2\dfrac{e^{-j\omega}}{\omega^2}$

6.5 $X(j\omega) = \dfrac{1-2e^{-2j\omega}}{j\omega} - \dfrac{e^{-2j\omega}-1}{2\omega^2}$

6.6 $X(j\omega) = \dfrac{e^{-(3+j\omega)} - 1}{3 + j\omega}$

6.7 $X(j\omega) = \dfrac{1}{2}\left[\dfrac{1}{1 + j(\omega - 3\pi)} + \dfrac{1}{1 + j(\omega + 3\pi)}\right]$

6.8 $X(j\omega) = \dfrac{2\sin\omega}{\omega} + \dfrac{4\cos\omega}{j\omega} - \dfrac{4\sin\omega}{\omega^3}$

6.9 $X(j\omega) = \dfrac{e^{2-j\omega}}{2 - j\omega}$

6.10 $FT[\delta(t) + u(t)] = 1 + \dfrac{1}{j\omega} + \pi\delta(\omega)$

6.11 $X(j\omega) = \dfrac{3}{2}[2\pi\delta(\omega - \omega_1) + 2\pi\delta(\omega + \omega_1)] - j[2\pi\delta(\omega - \omega_2) - 2\pi\delta(\omega - \omega_2)]$
where $\omega_1 = 2\pi$ & $\omega_2 = 3\pi$

6.12 $X(f) = \dfrac{1}{T_0}\sum_{m=-\infty}^{\infty}\delta\left(f - \dfrac{m}{T_0}\right) = \dfrac{1}{4}\sum_{m=-\infty}^{\infty}\delta\left(f - \dfrac{m}{4}\right)$

6.13 $x(t) = \dfrac{2}{\pi}\left[\dfrac{1}{t+3}\sin(\pi(t+3)) + \dfrac{1}{t-3}\sin(\pi(t-3))\right]$

6.14 $x(t) = \dfrac{1}{3}(e^{-2t} - e^{-5t})u(t)$

6.15 $x(t) = (-te^{-2t} + e^{-2t})u(t)$

6.16 $x(t) = (e^{-t} - e^{-2t})u(t)$

6.17 $x(t) = \left(e^{-2t} - \dfrac{1}{2}e^{-t} - \dfrac{3}{2}e^{-3t}\right)u(t)$

6.18 $X(e^{j\omega n}) = 1 - 2j\sin(\omega) - 2j\sin(2\omega)$

6.19 $X(e^{j\omega n}) = \dfrac{1}{2}(1 + \cos(\omega) + \cos(2\omega) + \cos(3\omega)) - \dfrac{1}{2}j(\sin(\omega) + \sin(2\omega) + \sin(3\omega))$

6.20
$x[n] = \dfrac{\sin(\pi n/2)}{n\pi}$ for $n \neq 0$ and
$x[n] = \dfrac{\sin(\pi n/2)}{\pi n/2}\dfrac{\pi/2}{\pi} \to \dfrac{1}{2}$ as $n \to 0$

6.21 $X(e^{j\omega}) = 1 + e^{-j\omega}$

6.22 $x[n] = \dfrac{1}{2\pi} + \dfrac{1}{2\pi}e^{j\pi n/2}$

6.23 $X(e^{j\Omega}) = \dfrac{1}{1-\dfrac{1}{3}e^{-j\Omega}} + \dfrac{1}{1-\dfrac{1}{4}e^{-j\Omega}}$

6.24 $x[n] = 2$ for $n = \pm 3$
and $= 0$ otherwise

6.25 $x[n] = 1$ for $n = -1/2$, $= 0$ otherwise

6.26 $x[n] = \dfrac{1}{2\pi}\left[\dfrac{1-e^{-(jn+1/2)\pi}}{jn+1/2} + \dfrac{e^{(jn-1/2)\pi}-1}{jn-1/2}\right]$

6.27 $x[n] = 0$ for $n \neq -3$ and $n \neq 3$
$= \dfrac{1}{2j}$ for $n = \pm 3$

6.28 $x(t) = (3 - 4e^{-t} - e^{-2t})u(t)$

6.29 $\cos(3\omega_0 t)u(t) \leftrightarrow \dfrac{1}{2}\left[\dfrac{1}{j(\omega-3\omega_0)} - \dfrac{1}{j(\omega+3\omega_0)}\right]$

6.30 $e^{3jt}\sin(\omega_0 t)u(t) \leftrightarrow \dfrac{1}{2j}\left[\dfrac{1}{j(\omega-\omega_0+3)} - \dfrac{1}{j(\omega+\omega_0+3)}\right]$

6.31 $\sin(2\pi f_0(t-5))u(t) \leftrightarrow \dfrac{\omega_0 e^{-j5\omega}}{(j\omega)^2 + \omega_0^2}$

6.32 $X(j\omega) = e^{-j\omega T/4}\dfrac{2}{\omega}\sin(\omega T/2)$

6.33 $(-5)^n e^{-j\pi n/3}u[n] \leftrightarrow \dfrac{1}{1+5e^{j(\Omega+\pi/3)}}$

6.34 $t\cos(10\pi t)u(t) \leftrightarrow \left[-\dfrac{1}{2}\left(\dfrac{1}{(\omega-10\pi)^2} + \dfrac{1}{(\omega+10\pi)^2}\right)\right]$

6.35 $\dfrac{d}{dt}(\cos(at)u(t)) \leftrightarrow \dfrac{1}{2}\left[\dfrac{j\omega}{j(\omega-a)} + \dfrac{j\omega}{j(\omega+a)}\right]$

6.36 $te^{-2(t-3)}u(t) \leftrightarrow j\dfrac{d}{d\omega}\left(\dfrac{e^{-3j\omega}}{2+j\omega}\right)$

6.37 $Y(j\omega) = \dfrac{8}{\omega}\sin(\omega/2)$

6.38 $Y(e^{j\Omega}) = 4 + 8\cos(4\Omega) + 8\cos(8\Omega)$

7

Laplace Transform

> **Learning Objectives**
> - Bilateral and unilateral Laplace transform
> - Complex frequency
> - Significance of Laplace transform
> - Properties of LT
> - Relationship between LT and FT
> - LT examples
> - Inverse LT
> - Stability of system in Laplace domain

We will discuss the unilateral and bilateral Laplace Transform (LT) and its significance in analyzing the systems. The significance of complex frequency will be discussed. The relation between LT and FT will be explained. Different examples based on calculation of LT and inverse LT will be solved. LT of some standard signals will be evaluated. LT is very useful for analyzing the stability of the system. Stability of the system in Laplace domain will be explained.

7.1 Definition of Laplace Transform

Laplace transform was first proposed by Laplace (year 1980). This is the operator that transforms the signal in time domain in to a signal in a complex frequency domain called as 'S' domain. The complex frequency domain will be denoted by S and the complex frequency variable will be denoted by 's'. Let us understand the significance of Laplace transform. The reader must be familiar with complex numbers. The complex frequency S can be likewise defined as $s = \sigma + j\omega$, where σ is the real part of s and $j\omega$ is the imaginary part of s. The complex numbers are defined by mathematicians and are the mathematical abstractions useful for the analysis of signals and systems. It simplifies the mathematics. Similarly, the complex frequency plane is also the mathematical abstraction useful for the simplification of mathematics. Laplace transform does not have any physical significance. We can just say that ω stands for the real frequency and Laplace transform transforms the signal from time domain to some kind of frequency domain.

Physical significance of Laplace transform Laplace transform has no physical significance except that it transforms the time domain signal to a complex frequency domain. It is useful to simply the mathematical computations and it can be used for the easy analysis of signals and systems. The stability of the system is directly revealed when the transfer function of the system is known in Laplace domain. LT is used for solving differential equations.

Definition of LT LT of signal $x(t)$ can be defined as

$$X(s) = \int_{-\infty}^{\infty} x(t)e^{-st}dt \tag{7.1}$$

Here, the signal $x(t)$ is a both-sided signal existing from $-\infty$ to $+\infty$ and LT is termed as bilateral LT. Unilateral LT is a special case of LT where the signal $x(t)$ is restricted to be causal i.e., it exists between 0 and ∞. Unilateral LT is a special case of LT.

We define inverse LT as

$$x(t) = \frac{1}{2\pi j} \int_{\sigma-j\infty}^{\sigma+j\infty} X(s)e^{st}ds \tag{7.2}$$

The inverse LT cannot be evaluated directly, as it requires contour integration. Inverse LT is usually obtained by inspection.

The integral defined for LT may not always converge. The region of values of σ for which LT converges is called as region of convergence (ROC). ROC is shown in the complex frequency plane called as S plane. S plane has σ on x-axis called real axis and $j\Omega$ on the y-axis called as imaginary axis, as shown in Fig. 7.1. The significance of ROC is LT for the signal does not exist for all σ, but it is convergent over certain range of values of σ for which $|x(t)e^{-\sigma t}|$ is absolutely integrable.

LT converts a time domain signal into a complex frequency plane. We can say that it is the representation of continuous time signal in terms of complex exponentials. Laplace transform possesses a number of properties useful for signal and system analysis. Many of the properties are similar to FT.

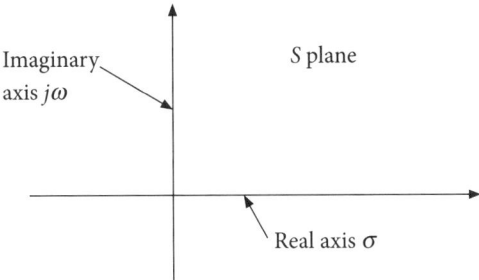

Fig. 7.1 S plane

1. Continuous time complex exponentials are eigen functions of LTI systems.
2. Convolution of two time domain signals results in multiplication of their LTs.
3. LT of the impulse response of the system is called as transfer function of the system in S domain.
4. LT converts a time domain differential equation into a simple algebraic equation in s.

LT comes in two varieties. One is a unilateral transform that is used for the analysis of differential equations with initial conditions specified. The second is bilateral transform, which is used for the analysis of the system with respect to its stability, causality and frequency response. LT can also be used for the analysis of signals that are not absolutely integrable such as impulse response of an unstable system.

Let us first prove that the complex exponential namely e^{st} is an eigen function of the LTI system. Let us apply e^{st} as input to the LTI system with impulse response $h(t)$. The system output $y(t)$ can be written as

$$y(t) = H\{e^{st}\}$$

$$= x(t) * h(t)$$

$$= \int_{-\infty}^{\infty} h(\tau) x(t-\tau) d\tau \qquad (7.3)$$

$$= \int_{-\infty}^{\infty} h(\tau) e^{s(t-\tau)} d\tau$$

$$= e^{st} \int_{-\infty}^{\infty} h(\tau) e^{-s\tau} d\tau$$

let us define

$$H(s) = \int_{-\infty}^{\infty} h(\tau)e^{-s\tau}d\tau$$

$$y(t) = e^{st}H(s)$$

The action of LTI system on the complex exponential is its multiplication with $H(s)$. Hence, the complex exponential is an eigen function of $H(s)$ and $H(s)$ is an eigen value.

Extended transform LT uses complex exponentials as basis functions namely e^{-st}. FT uses $e^{-j\omega t}$ as basis functions whose frequencies are restricted to the imaginary axis in a complex plane. These basis functions are not capable of synthesizing rising exponential signals. This basic problem is solved by using e^{-st} as basis function in LT where the frequency variable is generalized to $s = \sigma + j\omega$. This generalization permits us to synthesize the exponentially growing signals. FT is evaluated on the imaginary axis, whereas LT is evaluated on $s = \sigma + j\omega$ line, such as shown in Fig. 7.2, that σ lies in the region of convergence.

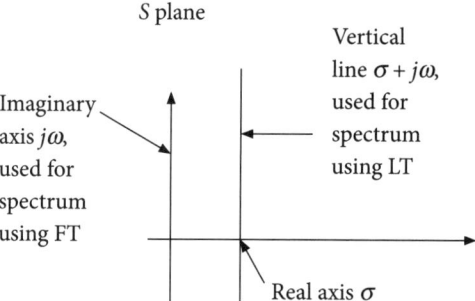

Fig. 7.2 Spectrum evaluation using FT and LT

Let us try to find the relation between LT and FT.

$$X(s) = \int_{-\infty}^{\infty} x(t)e^{-(\sigma + j\omega)t}dt \tag{7.4}$$

$$X(s) = \int_{-\infty}^{\infty} [x(t)e^{-\sigma t}]e^{-j\omega t}dt$$

This indicates that $X(s)$ i.e., LT of $x(t)$ is a FT of $x(t)e^{-\sigma t}$. Hence, we can say that LT of $x(t)$ converges if

$$\int_{-\infty}^{\infty} |x(t)e^{-\sigma t}|dt < \infty \tag{7.5}$$

$$X(j\omega) = X(s)\downarrow_{s=j\omega} \qquad (7.6)$$

LT can be converted to FT by substituting $s = j\omega$ i.e., $\sigma = 0$.

The range of values of σ for which LT converges is called as the region of convergence (ROC). Note that even though $x(t)$ is not absolutely integrable, by limiting ourselves to certain range of σ, we can ensure that Eq. (7.5) holds good.

The concepts can be better explained using one numerical example.

Example 7.1

Find LT and region of convergence for the signal $x(t) = e^t u(t)$.

Solution

The signal $x(t)$ can be plotted to see that the signal is divergent and is not absolutely integrable. The plot of the signal is shown in Fig. 7.3. The FT of the signal does not exist.

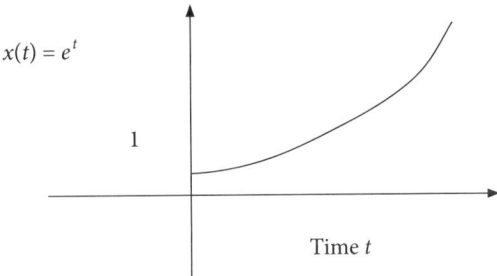

Fig. 7.3 Plot of signal $x(t)$

Now, let us find the LT of the signal.

$$X(s) = \int_0^\infty e^t e^{-st} dt = \int_0^\infty e^{(1-\sigma)t} e^{-j\omega t} dt$$

$$= \left[e^{[(1-\sigma)-j\omega]t}\right] / [(1-\sigma) - j\omega]\downarrow_0^\infty \quad \text{if } \sigma > 1 \qquad (7.7)$$

$$= \frac{1}{(1-\sigma) - j\omega}$$

The signal $x(t) = e^{-\sigma t} e^t u(t)$ is absolutely integrable if $\sigma > 1$. So, LT exists with ROC as $\sigma > 1$, as shown in Fig. 7.4.

Note that ROC will be the area of the S plane on the right or on the left of the vertical line with a pole on the vertical line intersecting the x-axis.

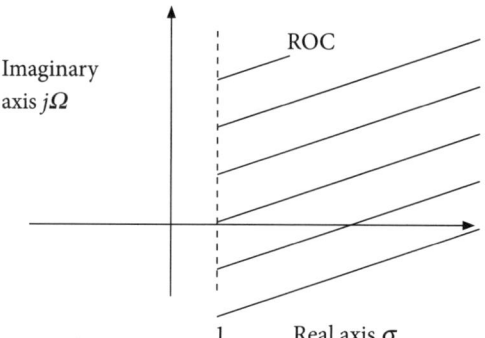

Fig. 7.4 ROC for Example 7.1

The comparison of FT and LT is listed in Table 7.1 below.

Table 7.1 Comparison of FT and LT

S. No.	FT	LT
1	Uses $e^{-j\omega t}$ as basis function, frequencies are restricted to the imaginary axis in a complex plane	Uses $e^{-(\sigma + j\omega)t} = e^{-st}$ as basis function, frequencies are the complex frequencies in a complex plane
2	FT exists only if $|x(t)|$ is absolutely integrable	LT exists only if $|x(t)e^{-\sigma t}|$ is absolutely integrable
3	No ROC specified	ROC is specified as a range of σ for which $|x(t)e^{-\sigma t}|$ is absolutely integrable
4	For some signals such as $x(t) = e^t$ i.e., growing exponentials, FT does not exist	For a signal such as $x(t) = e^t$, $|x(t)e^{-\sigma t}|$ is integrable for range $\sigma > 1$, and LT exists.
5	Rarely used for solving differential equations, as for many $x(t)$ FT is not convergent	Widely used for solving differential equations
6	FT is an LT of the same signal $x(t)$ evaluated by putting $s = j\omega$	LT of $x(t)$ is a FT of $x(t)e^{-\sigma t}$
7	Cannot be used for analysis of unstable or even marginally stable systems	Can be used for the analysis of unstable systems.

Concept Check

- What is the S domain?
- What is the significance of complex frequency plane?
- What is the physical significance of LT?
- Define LT.
- What is a bilateral LT? What is a unilateral LT?
- What is the method to find inverse LT?
- State properties of LT.

- Define eigen function and eigen value LTI system in S domain.
- Define basis functions for FT and LT.
- State the relation between FT and LT.
- State the ROC for existence of LT of $x(t) = e^t u(t)$.
- Can we use LT for analysis of unstable systems?

7.2 Laplace Transform of Some Standard Functions

Let us go through the LT of some standard signals.

Example 7.2

Find the LT of the **impulse** function $x(t) = \delta(t)$.

Solution

We will use the definition of LT.

$$X(s) = \int_{-\infty}^{\infty} \delta(t) e^{-st} dt = e^{-st} \Big\downarrow_{t=0} = 1 \quad \text{for all } s \tag{7.8}$$

ROC is the entire S plane

$\delta(t) \leftrightarrow 1$ for all s

There is no pole for $X(s)$. Hence, ROC is the entire S plane.

Example 7.3

Find the LT of the **unit step** function $x(t) = u(t)$.

Solution

We will use the definition of LT.

$$X(s) = \int_{-\infty}^{\infty} u(t) e^{-st} dt = \int_{0}^{\infty} e^{-st} dt = \frac{e^{-st}}{-s} \Big\downarrow_{0}^{\infty} = \frac{1}{s} \text{ for all Re}(s) > 0 \tag{7.9}$$

ROC is the entire right half S plane excluding $s = 0$

Similarly, LT of $x(t) = ku(t)$ is $\dfrac{k}{s}$ for $\text{Re}(s) > 0$

The LT $X(s)$ has a pole at $s = 0$. To find the pole we have to equate the denominator polynomial to zero and find its roots. Plot of the ROC and the pole is shown in Fig. 7.5.

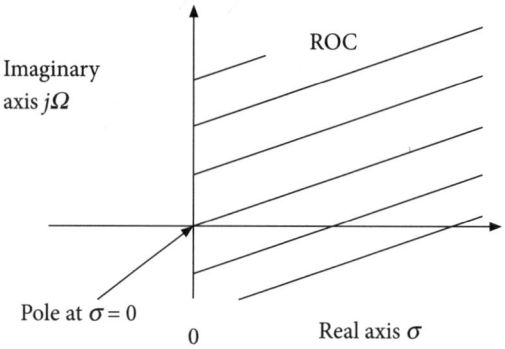

Fig. 7.5 Plot of ROC and a pole for $X(s)$ in Example 7.3

Example 7.4

Find the LT of the **ramp** function $x(t) = tu(t)$.

Solution

We will use the definition of LT.

$$X(s) = \int_{-\infty}^{\infty} u(t)e^{-st}dt = \int_{0}^{\infty} te^{-st}dt = \frac{te^{-st}}{-s}\Big|_{0}^{\infty} - \int_{0}^{\infty} \frac{e^{-st}}{-s}dt$$

$$= 0 - \frac{e^{-st}}{s^2}\Big|_{0}^{\infty} = \frac{1}{s^2} \text{ for all } \mathrm{Re}(s) > 0$$

(7.10)

ROC is the entire right half S plane excluding $s = 0$

There is a double pole at $s = 0$. Plot of ROC and pole is shown in Fig. 7.6.

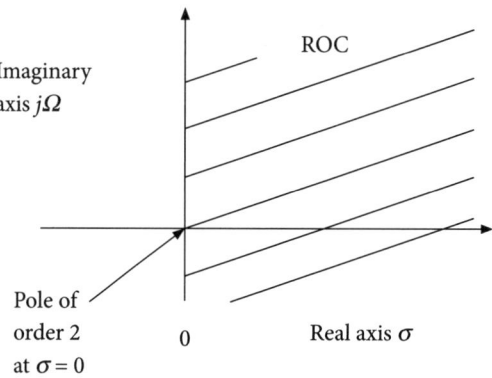

Fig. 7.6 Plot of ROC and pole for Example 7.4

Example 7.5
Find the LT of the **parabolic** function $x(t) = t^2 u(t)$.

Solution

We will use the definition of LT.

$$X(s) = \int_{-\infty}^{\infty} t^2 u(t) e^{-st} dt = \int_0^{\infty} t^2 e^{-st} dt = \frac{t^2 e^{-st}}{-s} \Big\downarrow_0^{\infty} - \int_0^{\infty} \frac{e^{-st} \times 2t}{-s} dt$$

$$= 0 + \frac{2}{s} \int_0^{\infty} t e^{-st} dt = \left\{ \frac{2}{s} \left[t \frac{e^{-st}}{-s} \right] \Big\downarrow_0^{\infty} - \int_0^{\infty} \frac{e^{-st}}{-s} dt \right\}$$

$$= \frac{2}{s}[0 + \frac{1}{s} \int_0^{\infty} e^{-st} dt = \frac{2}{s} \left[\frac{e^{-st}}{-s} \right] \Big\downarrow_0^{\infty}$$

$$= \frac{2}{s^3} \text{ for all } \mathrm{Re}(s) > 0 \tag{7.11}$$

ROC is the entire right half S plane excluding $s = 0$
There is a pole of order 3 at $s = 0$. Plot of ROC and pole is shown in Fig. 7.7.

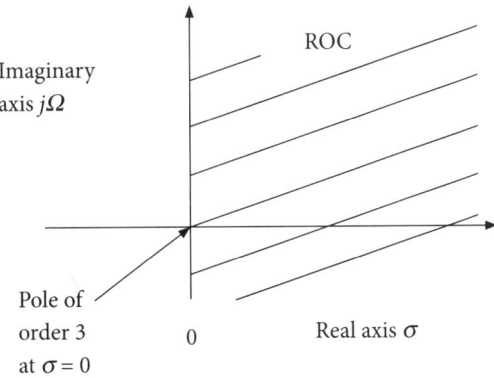

Fig. 7.7 Plot of ROC and pole for Example 7.5

Example 7.6
Find the LT of the **real exponential** function $x(t) = e^{at} u(t)$ and $x(t) = e^{-at} u(t)$.

Solution

We will use the definition of LT.

$$X(s) = \int_{-\infty}^{\infty} e^{at} u(t) e^{-st} dt = \int_{0}^{\infty} e^{at} e^{-st} dt = -\frac{e^{-(s-a)t}}{s-a} \Big\downarrow_{0}^{\infty}$$

$$= \frac{1}{s-a} \text{ for all } \text{Re}(s) > a \tag{7.12}$$

ROC is the right half S plane with Re(s) > a
ROC and a pole are shown plotted in Fig. 7.8.

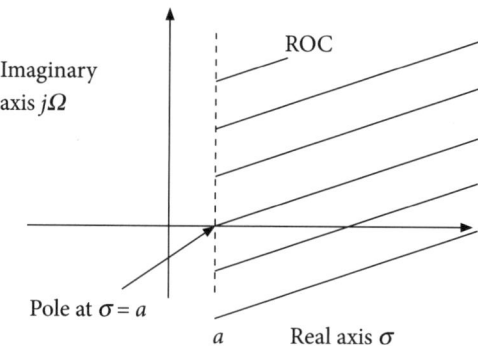

Fig. 7.8 ROC and pole for LT of $x(t) = e^{at}u(t)$

Let us find LT of the other function.

$$X(s) = \int_{-\infty}^{\infty} e^{-at} u(t) e^{-st} dt = \int_{0}^{\infty} e^{-at} e^{-st} dt = -\frac{e^{-(s+a)t}}{s+a} \Big\downarrow_{0}^{\infty}$$

$$= \frac{1}{s+a} \text{ for all } \text{Re}(s) > a \tag{7.13}$$

ROC is the right half S plane with Re(s) > −a
ROC is plotted in Fig. 7.9.

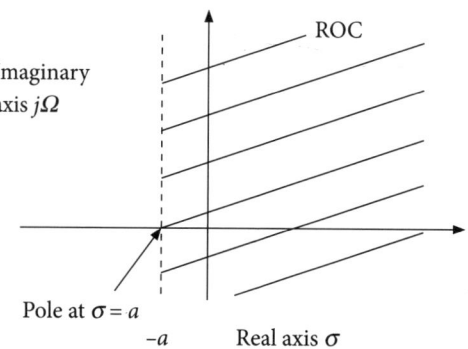

Fig. 7.9 ROC of the function $x(t) = e^{-at}u(t)$ and pole for LT function

Example 7.7

Find the LT of the **complex exponential** function $x(t) = e^{j\omega_0 t} u(t)$ and $x(t) = e^{-j\omega_0 t} u(t)$.

Solution

We will use the definition of LT.

$$X(s) = \int_{-\infty}^{\infty} e^{j\omega_0 t} u(t) e^{-st} dt = \int_{0}^{\infty} e^{j\omega_0 t} e^{-st} dt = -\frac{e^{-(s-j\omega_0)t}}{s - j\omega_0} \Big\downarrow_0^{\infty}$$

(7.14)

$$= \frac{1}{s - j\omega_0} \text{ for all } \text{Re}(s) > 0$$

ROC is the right half S plane with Re(s) > 0

The function has a pole at $s = j\omega_0$, which lies on the vertical line passing through zero i.e., the imaginary axis.

Note $s = \sigma + j\omega = 0 + j\omega_0$. We will equate the real parts on both sides and imaginary parts on both sides. We find that the pole is on the imaginary axis at $j\omega = j\omega_0$ and the ROC is the plane on the right side of the imaginary axis. The pole and the ROC are plotted in Fig. 7.10.

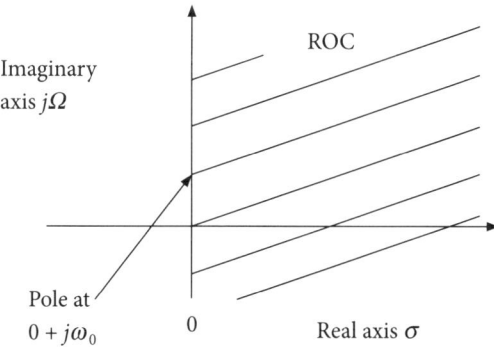

Fig. 7.10 Plot of ROC and a pole for $X(s)$ for Example 7.7

Let us find the LT of the other function.

$$X(s) = \int_{-\infty}^{\infty} e^{-j\omega_0 t} u(t) e^{-st} dt = \int_{0}^{\infty} e^{-j\omega_0 t} e^{-st} dt = -\frac{e^{-(s+j\omega_0)t}}{s + j\omega_0} \Big\downarrow_0^{\infty}$$

(7.15)

$$= \frac{1}{s + j\omega_0} \text{ for all } \text{Re}(s) > 0$$

ROC is the right half S plane with Re(s) > 0

The function has a pole at $s = -j\omega_0$, which lies on the vertical line passing through zero.

The pole and the ROC are plotted in Fig. 7.11.

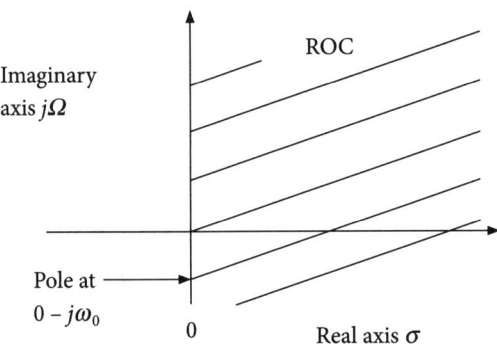

Fig. 7.11 Plot of ROC and a pole for $X(s)$ for Example 7.7

Note Both the functions have same ROC but different LTs.

Example 7.8

Find the LT of the sine and cosine function $x(t) = \cos(\omega_0 t)u(t)$ and $x(t) = \sin(\omega_0 t)u(t)$.

Solution

We will use the definition of LT.

$$X(s) = \int_{-\infty}^{\infty} \cos(\omega_0 t)u(t)e^{-st}dt = \int_0^{\infty} \cos(\omega_0 t)e^{-st}dt$$

$$= \int_0^{\infty} \frac{e^{j\omega_0 t} + e^{-j\omega_0 t}}{2} e^{-st}dt = \frac{1}{2}\left\{\left[\frac{e^{-[s-j\omega_0]t}}{-[(s-j\omega_0)]}\right]\downarrow_0^{\infty} + \left[\frac{e^{-[s+j\omega_0]t}}{-[s+j\omega_0]}\right]\downarrow_0^{\infty}\right\} \quad (7.16)$$

$$= \frac{1}{2}\left[\frac{1}{(s-j\omega_0)} + \frac{1}{(s+j\omega_0)}\right] \text{ for all } \operatorname{Re}(s) > 0$$

ROC is the right half S plane with $\operatorname{Re}(s) > 0$

$$X(s) = \frac{s}{(s^2 + \omega_0^2)}$$

We will find the LT of the other function.

$$X(s) = \int_{-\infty}^{\infty} \sin(\omega_0 t) u(t) e^{-st} dt = \int_{0}^{\infty} \sin(\omega_0 t) e^{-st} dt$$

$$= \int_{0}^{\infty} \frac{e^{j\omega_0 t} - e^{-j\omega_0 t}}{2j} e^{-st} dt = \frac{1}{2j} \left[\frac{e^{-(s-j\omega_0)t}}{-(s-j\omega_0)} \right]\Big\downarrow_0^\infty - \left[\frac{e^{-(s+j\omega_0)t}}{-(s+j\omega_0)} \right]\Big\downarrow_0^\infty \quad (7.17)$$

$$= \frac{1}{2j} \left[\frac{1}{(s-j\omega_0)} - \frac{1}{(s+j\omega_0)} \right] \text{ for all } \operatorname{Re}(s) > 0$$

ROC is the right half S plane with Re(s)

$$X(s) = \frac{1}{2j} \left[\frac{s + j\omega_0 - s + j\omega_0}{(s^2 + \omega_0^2)} \right] = \left[\frac{\omega_0}{(s^2 + \omega_0^2)} \right]$$

Example 7.8

Find LT of the **damped sine and cosine** function $x(t) = e^{-at} \sin(\omega_0 t) u(t)$ and $x(t) = e^{-at} \cos(\omega_0 t) u(t)$.

Solution

We will use the definition of LT.

$$X(s) = \int_{-\infty}^{\infty} e^{-at} \sin(\omega_0 t) u(t) e^{-st} dt = \int_{0}^{\infty} e^{-at} \sin(\omega_0 t) e^{-st} dt$$

$$= \int_{0}^{\infty} e^{-at} \frac{e^{j\omega_0 t} - e^{-j\omega_0 t}}{2j} e^{-st} dt = \frac{1}{2j} \left\{ \left[\frac{e^{-[(s+a)-j\omega_0]t}}{-[(s+a)-j\omega_0]} \right]\Big\downarrow_0^\infty \right.$$

$$\left. - \left[\frac{e^{-[(s+a)+j\omega_0]t}}{-[(s+a)+j\omega_0]} \right]\Big\downarrow_0^\infty \right\} \quad (7.18)$$

$$= \frac{1}{2j} \left[\frac{1}{(s+a)-j\omega_0} - \frac{1}{(s+a)+j\omega_0} \right] \text{ for all } \operatorname{Re}(s) > -a$$

ROC is the right half S plane with Re(s) > − a

$$X(s) = \frac{1}{2j} \left[\frac{s + a + j\omega_0 - s - a + j\omega_0}{(s+a)^2 + \omega_0^2} \right] = \left[\frac{\omega_0}{(s+a)^2 + \omega_0^2} \right]$$

There are two poles. One at $s = -a + j\omega_0$ and the other at $s = -a - j\omega_0$. The ROC and the pole plot are shown in Fig. 7.12.

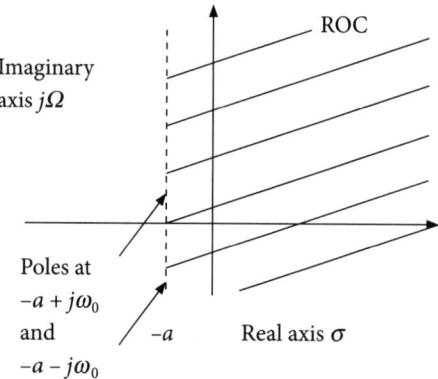

Fig. 7.12 Plot of ROC and poles for LT of damped sine function

Let us find the LT of the other function.

$$X(s) = \int_{-\infty}^{\infty} e^{-at} \cos(\omega_0 t) u(t) e^{-st} dt = \int_{0}^{\infty} e^{-at} \cos(\omega_0 t) e^{-st} dt$$

$$= \int_{0}^{\infty} e^{-at} \frac{e^{j\omega_0 t} + e^{-j\omega_0 t}}{2} e^{-st} dt = \frac{1}{2} \left\{ \left[\frac{e^{-[(s+a)-j\omega_0]t}}{-[(s+a) - j\omega_0]} \right] \downarrow_0^{\infty} \right.$$

$$\left. + \left[\frac{e^{-[(s+a)+j\omega_0]t}}{-[(s+a) + j\omega_0]} \right] \downarrow_0^{\infty} \right\}$$ (7.19)

$$= \frac{1}{2} \left[\frac{1}{(s+a) - j\omega_0} + \frac{1}{(s+a) + j\omega_0} \right] \text{ for all } \operatorname{Re}(s) > -a$$

ROC is the right half S plane with $\operatorname{Re}(s) > -a$

$$X(s) = \frac{(s+a)}{(s+a)^2 + \omega_0^2}$$

The ROC and the pole plot are shown in Fig. 7.13.

Note that ROCs and the pole zero plots for damped sin and cosine functions are the same but they have different LTs.

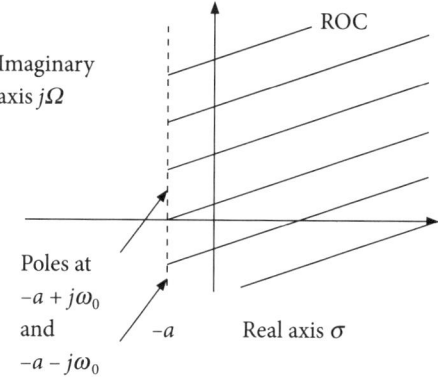

Fig. 7.13 Plot of ROC and poles for LT of damped cosine function

Concept Check

- What is LT of delta function? What is the ROC?
- What is LT of the unit step function? What is the ROC?
- Compare the LTs of the ramp and parabolic function. Compare their ROCs.
- Write the expression for the damped sine and cosine function.
- State the two functions with same ROC, same pole locations but different LTs.
- State the two functions with same ROC, different pole locations and different LTs.
- State a function with a double pole at $s = 0$.
- State a function with a triple pole at $s = 0$.

7.3 Properties of LT

We will state and explain different properties of LT that include linearity, time shifting, time scaling, time reversal, time differentiation, time integration, differentiation in S domain, frequency shifting, time convolution, modulation, conjugate property, Parseval's theorem, initial value theorem, and final value theorem. Let us consider the properties one by one.

1. **Property of Linearity**

    If

    $x_1(t) \leftrightarrow X_1(s)$ and $x_2(t) \leftrightarrow X_2(s)$,

    (7.20)

    then $ax_1(t) + bx_2(t) \leftrightarrow aX_1(s) + bX_2(s)$

Proof

Using the definition of LT, we can write

$$LT[ax_1(t)+bx_2(t)] = \int_{-\infty}^{\infty}[ax_1(t)+bx_2(t)]e^{-st}dt$$

$$= a\left[\int_{-\infty}^{\infty}x_1(t)e^{-st}dt\right] + b\left[\int_{-\infty}^{\infty}x_2(t)e^{-st}dt\right] \quad (7.21)$$

$$= aX_1(s) + bX_2(s)$$

Physical significance of property of linearity If the signal can be decomposed into linear combination of two or more component signals, then the LT of the signal is the linear combination of the component signals. This simplifies the mathematical computations. Similarly, if the signal can be decomposed into linear combination of two or more component signals in LT domain, then the inverse LT (ILT) of the signal is the linear combination of the component ILT signals. This property is used in calculation of ILT using partial fraction expansion.

2. **Time Shifting Property**

The time shifting property states that if

$x(t) \leftrightarrow X(s)$,

$$\quad (7.22)$$

then $x(t-t_0) \leftrightarrow e^{-st_0}X(s)$

Proof

Using the definition of LT, we can write

$$LT(x(t-t_0)) = \int_{-\infty}^{\infty}x(t-t_0)e^{-st}dt \text{ (put } t-t_0 = \tau)$$

$$= \int_{-\infty}^{\infty}x(\tau)e^{-s(\tau+t_0)}dt$$

$$\quad (7.23)$$

$$= e^{-st_0}\int_{-\infty}^{\infty}x(\tau)e^{-s\tau}$$

$$= e^{-st_0}X(s)$$

Physical significance of property of Time shifting A shift of t_0 in time corresponds to multiplication by the complex exponential e^{-st_0}. This property is usually applied to causal signals with positive time shift t_0. The restriction on the time shift arises because the time shift, if not restricted, may move the signal to negative time and the transform will no longer be unilateral.

3. **Time Reversal Property**

The time shifting property states that if

$$x(t) \leftrightarrow X(s),$$

then $x(-t) \leftrightarrow X(-s)$ (7.24)

Proof

Using the definition of LT, we can write

$$LT(x(-t)) = \int_{-\infty}^{\infty} x(-t)e^{-st} dt \text{ (put } -t = \tau)$$

$$= \int_{-\infty}^{\infty} x(\tau)e^{s\tau}(-d\tau)$$

$$= \int_{-\infty}^{\infty} x(\tau)e^{-(-s\tau)} d\tau$$

(7.25)

$$= X(-s)$$

Physical significance of property of time reversal A reversal of the signal in time domain corresponds to reversal of the complex variable s.

4. **Time Scaling Property**

The time scaling property states that if

$$x(t) \leftrightarrow X(s),$$

then $x(at) \leftrightarrow \dfrac{1}{|a|} X(s/a)$ (7.26)

Proof

Using the definition of LT, we can write

$$X(s) = \int_{-\infty}^{\infty} x(t)e^{-st} dt$$

$$LT(x(at)) = \int_{-\infty}^{\infty} x(at)e^{-st} dt$$

(7.27)

Put at $= \tau$

$$= \int_{-\infty}^{\infty} x(\tau) e^{-s\tau/a} (d\tau/a)$$

$$= \frac{1}{a} X\left(\frac{s}{a}\right)$$

Physical significance of property of time scaling Scaling in time introduces an inverse scaling in s.

5. **S domain shift Property**

   The property of s domain shift states that if

   $x(t) \leftrightarrow X(s)$, \hfill (7.28)

   then $e^{s_0 t} x(t) \leftrightarrow X(s - s_0)$

   Proof

   Using the definition of LT, we can write

   $$X(s) = \int_{-\infty}^{\infty} x(t) e^{-st} dt, \quad \text{multiply the signal by } e^{s_0 t} \hfill (7.29)$$

   $$\int_{-\infty}^{\infty} x(t) e^{s_0 t} e^{-st} dt] = \int_{-\infty}^{\infty} x(t) e^{-(s-s_0)t} dt$$

   $$= X(s - s_0)$$

   Physical significance of property of S domain shift Multiplication by the complex exponential in time introduces a shift in the complex frequency domain.

6. **Time Differentiation Property**

   The property of time differentiation states that if

   $x(t) \leftrightarrow X(s)$, \hfill (7.30)

   then $\dfrac{d}{dt} x(t) \leftrightarrow sX(s) - x(0^+)$

   Proof

   Using the definition of LT for unilateral signal, we can write

   $$X(s) = \int_{-\infty}^{\infty} x(t) e^{-st} dt \hfill (7.31)$$

Differentiating both sides

$$\frac{d}{dt}(x(t)) = \frac{d}{dt}\left[\int_{-\infty}^{\infty} x(t)e^{-st}dt\right] = \int_{-\infty}^{\infty} \frac{d}{dt}\left(x(t)e^{-st}\right)dt$$

$$= \int_{-\infty}^{\infty} e^{-st} \frac{d}{dt}(x(t))dt \quad \text{(use formula for } \int uv = u\int v - \int du \int v\text{)}$$

$$= e^{-st}x(t)\Big\downarrow_{0^+}^{\infty} + s\int_{0^+}^{\infty} se^{-st}x(t)dt \quad (X(S)\text{ exists so } x(t)e^{-st}\Big\downarrow_{t\to\infty} = 0)$$

$$= sX(s) - x(0^+)$$

Physical significance of property of time differentiation When the signal is differentiated in time domain, its LT is multiplied by s.

7. **Time Integration Property**

The property of time integration states that if

$x(t) \leftrightarrow X(s)$,

then $\int_{-\infty}^{t} x(\tau)d\tau \leftrightarrow X(s)/s + x^{-1}(0^+)/s$ \hfill (7.32)

where $x^{-1}(0^+) = \int_{-\infty}^{0^+} x(\tau)d\tau$

Proof

Using the definition of LS, we can write

$x(t) \leftrightarrow X(s)$

$$\frac{d}{dt}\left[\int_{-\infty}^{t} x(\tau)d\tau\right] = x(t)$$

so, $LT\left\{\frac{d}{dt}\left[\int_{-\infty}^{t} x(\tau)d\tau\right]\right\} = LT\{x(t)\}$ \hfill (7.33)

$sLT\left\{\int_{-\infty}^{0^+} x(\tau)d\tau + \int_{0}^{t} x(\tau)d\tau\right\} = LT(x(t))$

$$\int_{-\infty}^{t} x(\tau)d\tau \leftrightarrow X(s)/s - \frac{\int_{-\infty}^{0^+} x(\tau)d\tau}{s}$$

Physical significance of property of time integration When the signal is integrated in time domain, its LT gets multiplied in S domain by $1/s$. This factor is indeterminate at $s = 0$. Integration is basically a smoothing operation and hence it smoothes the signal. The smoothing in time deemphasizes the high frequencies, as is indicated by a factor of s in denominator.

8. **Conjugation property and conjugate symmetry property**

   The conjugation property of LT states that

   if $x(t) \leftrightarrow X(s)$,

   (7.34)

   then $x^*(t) \leftrightarrow X^*(s^*)$

   Proof

   Using definition of LT

   $$X(s) = \int_{-\infty}^{\infty} x(t)e^{-st}dt \tag{7.35}$$

   $$LT\{x^*(t)\} = \int_{-\infty}^{\infty} x^*(t)e^{-st}dt$$

   $$= \left[\int_{-\infty}^{\infty} x(t)e^{-(s^*t)}dt\right]^*$$

   $$= [X(s^*)]^* = X^*(s^*)$$

   Conjugate Symmetry property states that for real $x(t)$,

   $X(s) = X^*(s^*)$ (7.36)

   Proof

   $$X(s^*) = \int_{-\infty}^{\infty} x(t)e^{-(s^*)t}dt \tag{7.37}$$

   Take conjugate on both sides.

$$X^*(s^*) = \left[\int_{-\infty}^{\infty} x(t)e^{-(s^*)t}dt\right]^*$$

$$= \int_{-\infty}^{\infty} x(t)e^{-(s^*)^*t}dt = \int_{-\infty}^{\infty} x(t)e^{-st}dt$$

$$= X(s)$$

Physical significance of conjugation property LT of the conjugate of the signal is same as LT of the signal, if the signal is real.

9. **Frequency shifting property**

   The frequency shifting property of LT states that

   if $x(t) \leftrightarrow X(S)$,

   (7.38)

   then $e^{\pm at}x(t) \leftrightarrow X(s \mp a)$

   Proof

   Using definition of LT

   $$X(S) = \int_{-\infty}^{\infty} x(t)e^{-st}dt \quad (7.39)$$

   $$LT\{e^{\pm at}x(t)\} = \int_{-\infty}^{\infty} x(t)e^{-st}e^{\pm at}dt = \int_{-\infty}^{\infty} x(t)e^{-(s \mp a)t}dt$$

   $$= \text{Here it is } LT \text{ equation except } s \text{ is replaced by } s \mp a$$

   $$= X(s \mp a)$$

Physical significance of frequency shifting property When the LT of the signal is shifted in complex frequency domain by $\mp a$, the inverse transformation leads to the multiplication of the signal by the exponential $e^{\pm at}$.

10. **Frequency differentiation property**

    The frequency differentiation property of LT states that

    if $x(t) \leftrightarrow X(S)$,

    (7.40)

    then $tx(t) \leftrightarrow -\dfrac{d}{ds}[X(s)]$

Proof

Using definition of LT

$$X(S) = \int_{-\infty}^{\infty} x(t)e^{-st}\,dt$$

$$\frac{d}{ds}X(S) = \frac{d}{ds}\left[\int_{-\infty}^{\infty} x(t)e^{-st}\,dt\right] = \int_{-\infty}^{\infty} x(t)\left[\frac{d}{ds}e^{-st}\right]dt$$

$$= \int_{-\infty}^{\infty} -t \times x(t)e^{-st}\,dt \qquad (7.41)$$

$$-\frac{d}{ds}X(S) = \int_{-\infty}^{\infty} tx(t)e^{-st}\,dt = LT(tx(t))$$

$$-\frac{d}{ds}X(S) = LT(tx(t))$$

Physical significance of frequency differentiation property If we take differentiation of the LT of the signal and take its ILT, it will result in a signal that is multiplied by $(-t)$ time value for the signal.

11. **Convolution property**

The convolution property states that if

$x_1(t) \leftrightarrow X_1(s)$ and $x_2(t) \leftrightarrow X_2(s)$,

then $x_1(t) * x_2(t) \leftrightarrow X_1(s)X_2(s)$ $\qquad (7.42)$

Proof

Using the definition of LT, we can write

$$X_1(s) = \int_{-\infty}^{\infty} x_1(t)e^{-st}\,dt$$

$$X_2(s) = \int_{-\infty}^{\infty} x_2(t)e^{-st}\,dt \qquad (7.43)$$

Physical significance of convolution property When the two signals are convolved in time domain, their LTs get multiplied. Convolution is computationally costly. The convolution can be performed in transform domain using a multiplication operation that is less costly. To find the LT representation for the convolved output, one may find the LT for the one signal

$$LT[x_1(t) * x_2(t)] = \int_{-\infty}^{\infty} [(x_1(t) * x_2(t)] e^{-st} dt$$

$$= \left[\int_{-\infty}^{\infty} \left[\int_{-\infty}^{\infty} x_1(\tau) x_2(t-\tau) d\tau \right] e^{-st} dt \right] \quad \text{put } t - \tau = t'$$

$$= \left[\int_{-\infty}^{\infty} x_1(\tau) e^{-s\tau} d\tau \right] \left[\int_{-\infty}^{\infty} x_2(t') e^{-st'} dt' \right]$$

$$= X_1(s) X_2(s)$$

and LT of the other signal and multiply the two LTs. The convolution in one domain transforms to modulation in the other domain. (Modulation refers to the multiplication of the two signals.) Both convolution and modulation properties are due to the fact that complex exponentials are eigen functions of LTI system.

12. Frequency convolution or modulation property

Frequency convolution property is also called as the modulation property. The frequency convolution or the modulation property states that if

$$x_1(t) \leftrightarrow X_1(s) \quad \text{and} \quad x_2(t) \leftrightarrow X_2(s),$$

then $\quad x_1(t) \times x_2(t) \leftrightarrow X_1(s) * X_2(s)$

(7.44)

Proof

Using the definition of LT, we can write

$$LT[x_1(t) \times x_2(t)] = \int_{-\infty}^{\infty} [(x_1(t) \times x_2(t)] e^{-st} dt$$

$$= \left[\int_{-\infty}^{\infty} x_1(t) \left[\frac{1}{2\pi j} \int_{\sigma - j\infty}^{\sigma + j\infty} X_2(\lambda) e^{\lambda t} d\lambda \right] e^{-st} dt \right] \quad (7.45)$$

$$= \left(\int_{-\infty}^{\infty} x_1(t) \frac{1}{2\pi j} \int_{\sigma - j\infty}^{\sigma + j\infty} X_2(\lambda) e^{-(s-\lambda)t} d\lambda dt \right)$$

Interchange the order of integration

$$= \frac{1}{2\pi j} \int_{\sigma-j\infty}^{\sigma+j\infty} X_2(\lambda) \left(\int_{\sigma-j\infty}^{\sigma+j\infty} x_1(t) e^{-(s-\lambda)t} dt \right) d\lambda$$

$$= \frac{1}{2\pi j} \int_{\sigma-j\infty}^{\sigma+j\infty} X_2(\lambda) X_1(s-\lambda) d\lambda$$

$$= X_1(s) * X_2(s)$$

Physical significance of modulation property When the carrier signal multiplies with the message signal to get $\cos(2\pi f_c t)\cos(2\pi f_m t)$, we get double side band suppressed carrier modulated signal. This property states the property of LT for the multiplied signal (multiplication of two sinusoids results in modulation), hence it is called modulation property. The LT for the modulated signal is the convolution of the LTs of the component signals.

13. **Parseval's relation**

Parseval's relation states that

$x(t) \leftrightarrow X(s)$ for complex valued $x(t)$,

$$\text{then } \int_{-\infty}^{\infty} |x(t)|^2 \, dt = \frac{1}{2\pi j} \int_{\sigma-j\infty}^{\sigma+j\infty} |X(s)|^2 ds \tag{7.46}$$

Proof

To prove Parseval's relation, we start with LHS

$$|x(t)|^2 = x(t) x^*(t)$$

$$x(t) = \frac{1}{2\pi j} \int_{\sigma-j\infty}^{\sigma+j\infty} X(s) e^{st} ds \tag{7.47}$$

$$x^*(t) = \frac{1}{2\pi j} \int_{\sigma-j\infty}^{\sigma+j\infty} X^*(s) e^{-st} ds$$

$$\int_{-\infty}^{\infty} x(t) x^*(t) dt = \int_{\sigma-j\infty}^{\sigma+j\infty} x(t) \left[\frac{1}{2\pi j} \int_{\sigma-j\infty}^{\sigma+j\infty} X^*(s) e^{-st} ds \right] dt$$

$$= \frac{1}{2\pi j} \int_{\sigma-j\infty}^{\sigma+j\infty} X^*(s) \left[\int_{-\infty}^{\infty} x(t) e^{-st} dt \right] ds$$

$$= \frac{1}{2\pi j} \int_{\sigma-j\infty}^{\sigma+j\infty} X^*(s)X(s)ds$$

$$= \frac{1}{2\pi j} \int_{\sigma-j\infty}^{\sigma+j\infty} |X(s)|^2\, ds$$

Physical significance of parseval's identity   Parseval's identity states that the energy of the signal in time domain is same as the energy of the LT coefficients. This simply means, when the signal is transformed in any other domain, its energy is preserved. $|X(s)|^2$ represents the energy of the signal in S domain.

14. **Initial value theorem**

    The initial value theorem allows us to find the initial value of the signal without finding ILT. It states that

    if $x(t) \leftrightarrow X(S)$,

    (7.48)

    then $\lim_{t \to 0} x(t) = \lim_{s \to \infty} sX(s)$

    Proof

    Using time differentiation property, we can write

    $x(t) \leftrightarrow X(s)$,

    then $\dfrac{d}{dt}x(t) \leftrightarrow sX(s) - x(0^+)$

    Taking $\lim_{s \to \infty}$,

    $$LT\left(\frac{d}{dt}x(t)\right) = \lim_{s \to \infty}\left[\int_{-\infty}^{\infty} \frac{d}{dt}x(t)e^{-st}dt \leftrightarrow \lim_{s \to \infty}\left[sX(s) - x(0^-)\right]\right] \quad (7.49)$$

    $$0 = \lim_{s \to \infty} sX(s) - x(0^-)$$

    $$x(0) = \lim_{s \to \infty} sX(s)$$

    Physical significance of the initial value theorem   Initial value of the signal can be found from its LT by taking the limit of $sX(s)$ as s tends to infinity. We need not take ILT.

15. **Final value theorem**

    The final value theorem allows us to find the final value of the signal directly from its LT without finding ILT. It states that if $x(t) \leftrightarrow X(S)$,

$$\text{then } \lim_{t \to \infty} x(t) = \lim_{s \to 0} sX(s) \tag{7.50}$$

Proof

Using time differentiation property, we can write

$x(t) \leftrightarrow X(s)$,

$$\text{then } \frac{d}{dt} x(t) \leftrightarrow sX(s) - x(0^+) \tag{7.51}$$

Taking $\lim_{s \to 0}$,

$$LT\left(\frac{d}{dt} x(t)\right) = \lim_{s \to 0} \left[\int_0^\infty \frac{d}{dt} x(t) e^{-st} dt \leftrightarrow \lim_{s \to 0} [sX(s) - x(0^-)]\right]$$

$$\int_0^\infty \frac{d}{dt} x(t) dt = \lim_{s \to 0} [sX(s) - x(0^-)]$$

$$x(t) \Big\downarrow_0^\infty = \lim_{s \to 0} [sX(s) - x(0^-)]$$

$$x(\infty) - x(0) = \lim_{s \to 0} sX(s) - x(0)$$

$$\lim_{t \to \infty} x(t) = \lim_{s \to 0} sX(s)$$

Physical significance of the final value theorem The final value of the signal can be found from its LT by taking the limit of $sX(s)$ as s tends to zero. We need not take ILT. To apply final value theorem, we have to cancel poles and zeros of $sX(s)$ before applying the limit. All the poles of $X(s)$ must lie on the left-hand side of the S plane i.e., the system must be stable.

16. **Time periodicity property**

Let us consider a causal periodic signal with period equal to T, which satisfies the condition $x(t) = x(t + nT)$ for all $t > 0$ and $n = 0, 1, 2,\ldots$. We will use the definition of LT.

$$X(S) = \int_0^\infty x(t)e^{-st}\,dt$$

$$= \int_0^T x(t)e^{-st}\,dt + \int_T^{2T} x(t)e^{-st}\,dt + \ldots + \int_{nT}^{(n+1)T} x(t)e^{-st}\,dt$$

(7.52)

$$= \int_0^T x(t)e^{-st}\,dt + e^{-sT}\int_0^T x(t+T)e^{-st}\,dt + \ldots$$

$x(t)$ is periodic

$$= \int_0^T x(t)e^{-st}\,dt + e^{-sT}\int_0^T x(t)e^{-st}\,dt + e^{-2sT}\int_0^T x(t)e^{-st}\,dt + \ldots$$

$$= [1 + e^{-sT} + e^{-2sT} + \ldots]\int_0^T x(t)e^{-st}\,dt$$

$$X(S) = \frac{1}{1-e^{-sT}}\int_0^T x(t)e^{-st}\,dt = \frac{1}{1-e^{-sT}}[\text{transform of one period}]$$

the statement given below,

Physical Significance of Periodicity Property LT of a periodic signal can be found by taking LT of one period and dividing it by $1-e^{(-sT)}$, where T is a period of the signal.

Concept Check

- What is the significance of linearity property? How will you use it to find ILT using partial fraction expansion method?
- Write the statement for frequency shifting property of LT.
- What is the physical significance of conjugate symmetry property?
- Why frequency convolution property is called as the modulation property?
- What is the application for the convolution property?
- State the S domain shift property.
- What is the physical significance of Parseval's theorem?
- How will you find the initial value of the signal at $t = 0$?
- How will you find the value of the signal at $t = \infty$?
- Write the formula for LT of the periodic signals.
- Explain the time shifting and time scaling property of LT. What is its significance?

7.4 Solved Examples on LT

We will consider right-handed, left-handed and both-sided signals and find the LT of the signals and their ROCs.

Example 7.9

Find the LT of the right-handed signal $x(t) = e^{-at}u(t)$ and a left-handed signal $x(t) = -e^{-at}u(-t)$.

Solution

We will use the definition of LT.

$$X(S) = \int_{-\infty}^{\infty} e^{-at}u(t)e^{-st}dt = \int_{0}^{\infty} e^{-at}e^{-st}dt = -\frac{e^{-(s+a)t}}{s+a}\bigg\downarrow_{0}^{\infty} \quad (7.53)$$

$$= \frac{1}{s+a} \quad \text{for all} \quad s+a>0 \quad \text{i.e.,} \quad \text{Re}(s) > -a$$

ROC is the right half S plane with Re(s) > −a

Note   if $\text{Re}(a) = \sigma > -a, (\sigma + a) > 0, e^{-(s+a)\infty} \to 0$
if $\text{Re}(a) = \sigma < -a, (\sigma + a) < 0, e^{-(s+a)\infty} \to \infty$

ROC and pole are plotted in Fig. 7.14.

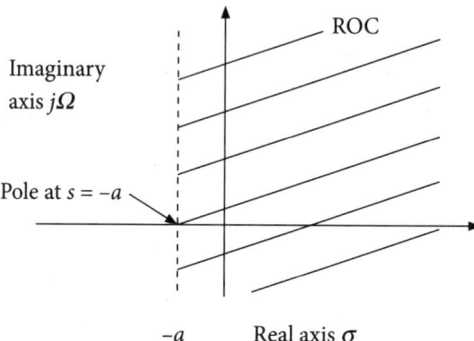

Fig. 7.14 ROC of $x(t) = e^{-at}u(t)$

We will use the definition of LT to find the LT of $x(t) = -e^{-at}u(-t)$.

$$X(S) = \int_{-\infty}^{\infty} -e^{-at}u(-t)e^{-st}dt = -\int_{-\infty}^{0} e^{-at}e^{-st}dt = \frac{e^{-(s+a)t}}{s+a}\bigg\downarrow_{-\infty}^{0} \quad (7.54)$$

$$= \frac{1}{s+a} \quad \text{for all} \quad \text{Re}(s) < -a$$

ROC is the left half S plane with Re(s) < −a
   Note: when $s > -a$, the integral $\to \infty$
   ROC for $x(t) = -e^{-at}u(-t)$ and pole are plotted in Fig. 7.15.
   Note that both the signals have the same LT but the ROCs are different. In order to recover the signal from LT, one needs to specify the ROC as well.

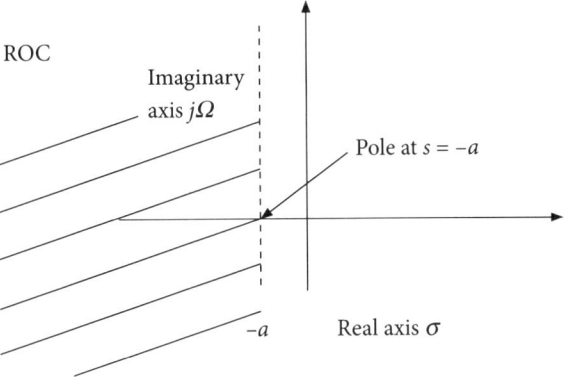

Fig. 7.15 ROC of $x(t) = -e^{-at}u(-t)$

Example 7.10

Find the LT of the both-sided signal $x(t) = e^{-at}u(t) + e^{-bt}u(-t)$. Find ROC.

Solution

We will use the definition of LT.

$$X(S) = \int_{-\infty}^{\infty} e^{-at}u(t)e^{-st}dt + \int_{-\infty}^{\infty} e^{-bt}u(-t)e^{-st}dt$$

$$= \int_{0}^{\infty} e^{-at}e^{-st}dt + \int_{-\infty}^{0} e^{-bt}e^{-st}dt$$

(7.55)

$$= -\frac{e^{-(s+a)t}}{s+a}\Big|_0^\infty - \frac{e^{-(s+b)t}}{s+b}\Big|_{-\infty}^0$$

$$= \frac{1}{s+a} - \frac{1}{s+b} = \frac{s+b-s-a}{s^2+(a+b)s+ab} = \frac{b-a}{s^2+(a+b)s+ab}$$

The first term converges for all $s + a > 0$ i.e., $\operatorname{Re}(s) > -a$
ROC is the right half S plane with $\operatorname{Re}(s) > -a$

Note : if $\operatorname{Re}(a) = \sigma > -a, (\sigma + a) > 0, e^{-(s+a)\infty} \to 0$

if $\operatorname{Re}(a) = \sigma < -a, (\sigma + a) < 0, e^{-(s+a)\infty} \to \infty$

Note : if $\operatorname{Re}(a) = \sigma > -a, (\sigma + a) > 0, e^{-(s+a)\infty} \to 0$

if $\operatorname{Re}(a) = \sigma < -a, (\sigma + a) < 0, e^{-(s+a)\infty} \to \infty$

The second term converges for all $-s - b > 0$ i.e., $\operatorname{Re}(s) < -b$
ROC is left half S plane with $\operatorname{Re}(s) < -b$

The common area of convergence exists from −a to −b provided a > b. ROC is plotted in Fig. 7.16. ROC does not exist if a < b. Here, the ROC for two terms does not intersect. The situation is shown in Fig. 7.17. Note that ROC is a strip parallel to the imaginary axis.

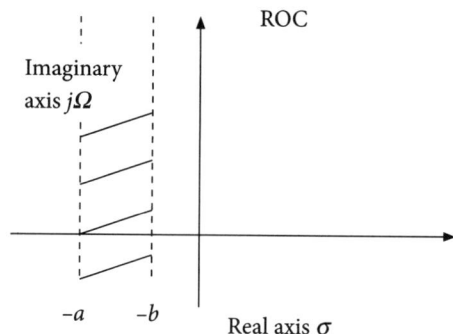

Fig. 7.16 ROC of $x(t) = e^{-at}u(t) + e^{-bt}u(-t)$ with $a > b$

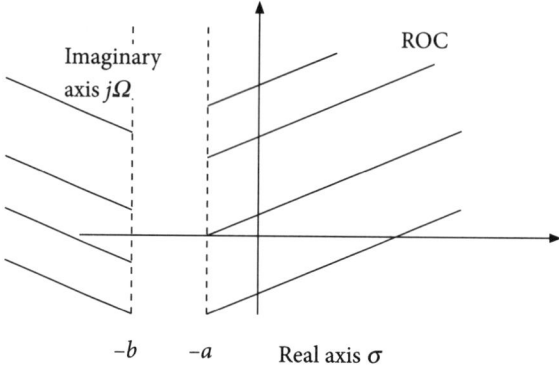

Fig. 7.17 ROC of $x(t) = e^{-at}u(t) + e^{-bt}u(-t)$ with $a < b$ (No ROC exists)

Example 7.11

Find the LT of the signal $x(t) = u(t - 4)$. Find ROC.

Solution

We will use the definition of LT.

$$X(S) = \int_{-\infty}^{\infty} u(t-4)e^{-st}\,dt = \int_{4}^{\infty} e^{-st}\,dt = \frac{e^{-st}}{-s}\Big|_{4}^{\infty}$$

(7.56)

$$= \frac{e^{-4s}}{s} \text{ for all } \mathrm{Re}(s) > 0$$

ROC is the right half S plane with Re(s) > 0
  Note when s < 0, the integral → ∞
  There is a pole at s = 0.
  Note the LT function has a pole at s = 0 and a zero at s = ∞

Example 7.12
Find the LT of $x(t) = e^{3t}u(t-3)$ and find ROC.

Solution
We will use the definition of LT.

$$X(S) = \int_{-\infty}^{\infty} e^{3t} u(t-3) e^{-st} dt = \int_{3}^{\infty} e^{3t} e^{-st} dt = \frac{e^{-(s-3)t}}{-(s-3)} \Big\downarrow_{3}^{\infty}$$

(7.57)

$$= \frac{e^{-3(s-3)}}{(s-3)} \text{ for all Re}(s) > 3$$

ROC is the right half S plane with Re(s) > 3
Note When s < 3, the integral → ∞
  There is a pole at s = 0. Pole and ROC are plotted in Fig. 7.18.

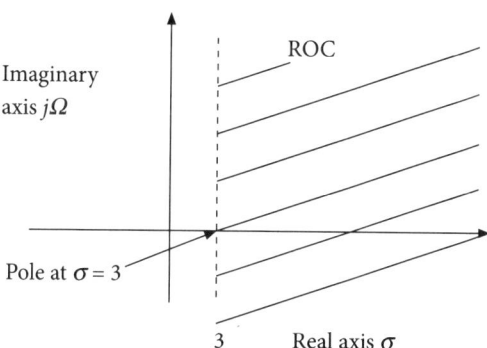

Fig. 7.18 Plot of ROC and pole of the function for Example 7.12

Note The LT function has a pole at s = 3.

Example 7.13
Find the LT of $x(t) = \delta(t-2)$ and find ROC.

Solution
We will use the definition of LT.

$$X(S) = \int_{-\infty}^{\infty} \delta(t-2)e^{-st}dt = e^{-st}\downarrow_{t=2}$$

(7.58)

$$= e^{-2s} \text{ for all } s$$

ROC is the right half S plane with Re(s) > 0
Note When s < 0, the integral → ∞
There is a pole at s = 0. Pole and ROC are plotted in Fig. 7.19. The function has a zero at s = ∞.

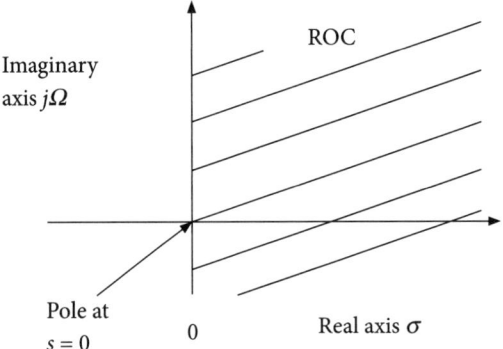

Fig. 7.19 Plot of ROC and pole of the function for Example 7.13

Note The LT function has a pole at s = 0.

Example 7.14
Find LT of $x(t) = \cos(2t)u(t-3)$ and find ROC.

Solution
We will use the definition of LT.

$$X(S) = \int_{-\infty}^{\infty} \cos(2t)u(t-3)e^{-st}dt = \frac{1}{2}\int_{3}^{\infty}[e^{2t}+e^{-2t}]e^{-st}dt$$

$$= \frac{1}{2}\left[\frac{e^{-(s-2)t}}{-(s-2)}\downarrow_{3}^{\infty} + \frac{e^{-(s+2)t}}{-(s+2)}\downarrow_{3}^{\infty}\right]$$

(7.59)

$$= \frac{1}{2}\left[\frac{e^{-3(s-2)}}{(s-2)} + \frac{e^{-3(s+2)}}{(s+2)}\right]$$

= ROC is the right half S place with Re(s) > 2 for first term and
ROC is right half S plane with Re(s) > −2 for the second term.

The common area of convergence is the right half S plane with Re(s) > 2.

There is a pole at $s = 2$ and $s = -2$. Pole and ROC are shown plotted in Fig. 7.20.

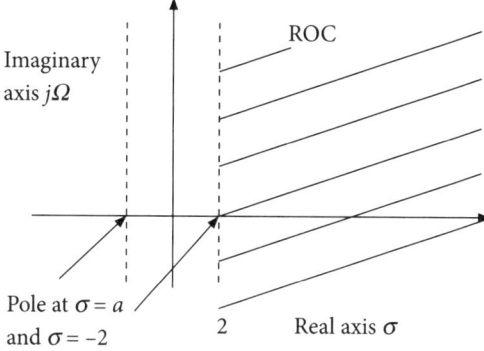

Fig. 7.20 Plot of ROC and pole of the function for Example 7.14

Note The LT function has a pole at $s = 2$ and $s = -2$.

Example 7.15

Find the LT of $x(t) = e^{-4|t|}$ and find ROC.

Solution

We will use the definition of LT.

$$X(S) = \int_{-\infty}^{\infty} e^{-4|t|} e^{-st} dt$$

$$= \int_{0}^{\infty} e^{-4t} e^{-st} dt + \int_{-\infty}^{0} e^{4t} e^{-st} dt$$

$$= -\frac{e^{-(s+4)t}}{s+4} \Big\downarrow_0^{\infty} - \frac{e^{-(s-4)t}}{s-4} \Big\downarrow_{-\infty}^0 \qquad (7.60)$$

$$= \frac{1}{s+4} - \frac{1}{s-4} = \frac{8}{s^2 - 16}$$

The first term converges for all $s + 4 > 0$ i.e., $\mathrm{Re}(s) > -4$
ROC is the right half S plane with $\mathrm{Re}(s) > -4$
Note: if $\mathrm{Re}(a) = \sigma > -4, (\sigma + 4) > 0, e^{-(s+4)\infty} \to 0$

If $\mathrm{Re}(a) = \sigma < -4, (\sigma + 4) < 0, e^{-(s+4)\infty} \to \infty$,

the second term converges for all $-s + 4 > 0$ i.e., $\mathrm{Re}(s) < 4$
ROC is left half S plane with $\mathrm{Re}(s) < 4$
ROC $-4 < \mathrm{Re}(s) < 4$

The plot of ROC and LT is depicted in Fig. 7.21.

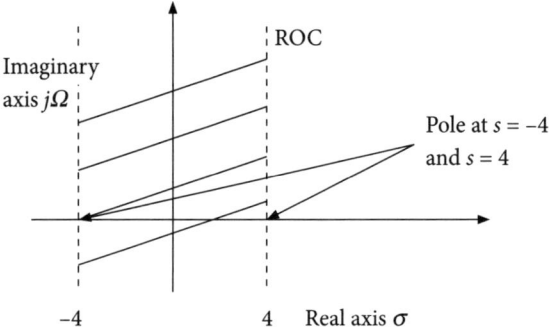

Fig. 7.21 Plot of poles and ROC for $x(t)=e^{-4|t|}$

Example 7.16
Find LT of $x(t)=e^{-3t}\cos(\omega_0 t)u(t)$ and find ROC.

Solution
We will use the definition of LT.

$$X(S) = \int_{-\infty}^{\infty} e^{-3t}\cos(\omega_0 t)u(t)e^{-st}dt$$

$$= \frac{1}{2}\left[\int_0^{\infty} e^{-3t}e^{-st}e^{j\omega_0 t}dt + \int_0^{\infty} e^{-3t}e^{-st}e^{-j\omega_0 t}dt\right]$$

$$= \frac{1}{2}\left[-\frac{e^{-(s+3-j\omega_0)t}}{s+3-j\omega_0}\bigg|_0^{\infty} - \frac{e^{-(s+3+j\omega_0)t}}{s+3+j\omega_0}\bigg|_0^{\infty}\right] \qquad (7.61)$$

$$= \frac{1}{2}\left[\frac{1}{s+3-j\omega_0} + \frac{1}{s+3+j\omega_0}\right]$$

$$= \frac{s+3}{(s+3)^2 + \omega_0^2}$$

The first term converges for all $s + 3 - j\omega_0 >$ i.e., Re$(s) > -3$
ROC is the right half S plane with Re$(s) > -3$

Note : if Re$(a) = \sigma > -3, (\sigma + 3) > 0, e^{-(s+3)\infty} \to 0$

if Re$(a) = \sigma < -3, (\sigma + 3) < 0, e^{-(s+3)\infty} \to \infty,$

the second term converges for all $s + 3 + j\omega_0 > 0$ i.e., $Re(s) > -3$
ROC is right half S plane with $Re(s) > -3$

The poles of the function are at $s = -3 + j\omega_0$ and $s = -3 - j\omega_0$. The plot of ROC and LT are depicted in Fig. 7.22.

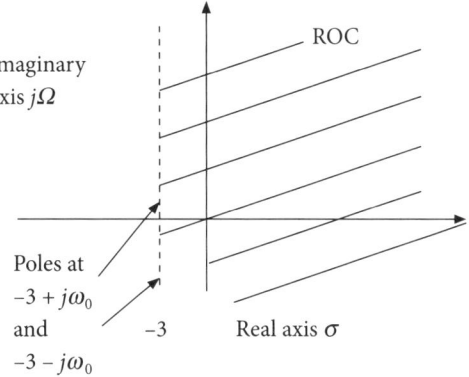

Fig. 7.22 Plot of poles and ROC for $x(t) = e^{-3t} \cos(\omega_0 t) u(t)$

Example 7.17

Find LT of $x(t) = e^{-t}u(t) + e^{2t}u(t)$ and find ROC.

Solution

We will use the definition of LT.

$$X(S) = \int_0^\infty e^{-t} e^{-st} dt + \int_0^\infty e^{2t} e^{-st} dt$$

$$= -\frac{e^{-(s+1)t}}{s+1} \Big\downarrow_0^\infty - \frac{e^{-(s-2)t}}{s-2} \Big\downarrow_0^\infty \qquad (7.62)$$

$$= \frac{1}{s+1} + \frac{1}{s-2} = \frac{2s-1}{s^2 - s - 2}$$

The first term converges for all $s + 1 > 0$ i.e., $Re(s) > -1$
ROC is the right half S plane with $Re(s) > -1$
The second term converges for all $s - 2 > 0$ i.e., $Re(s) > 2$
ROC is right half S plane with $Re(s) > 2$
The common area of convergence is the area with $Re(s) > 2$. ROC and poles are plotted in Fig. 7.23.

Example 7.18

Find LT of $x(t) = e^{-t}u(-t) + e^{2t}u(-t)$ and find ROC.

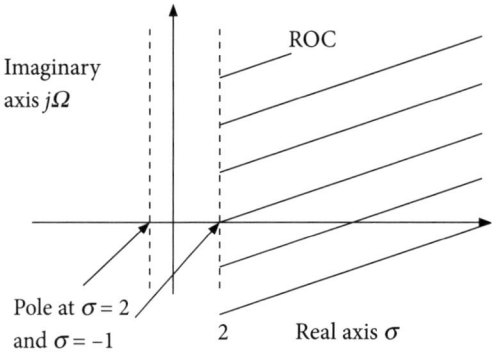

Fig. 7.23 ROC and poles for Example 7.17

Solution

We will use the definition of LT.

$$X(S) = \int_{-\infty}^{0} e^{-t} e^{-st} dt + \int_{-\infty}^{0} e^{2t} e^{-st} dt$$

$$= -\frac{e^{-(s+1)t}}{s+1} \Big\downarrow_{-\infty}^{0} - \frac{e^{-(s-2)t}}{s-2} \Big\downarrow_{-\infty}^{0} \qquad (7.63)$$

$$= -\frac{1}{s+1} - \frac{1}{s-2} = -\frac{2s-1}{s^2 - s - 2}$$

The first term converges for all $s + 1 < 0$ i.e., $\text{Re}(s) < -1$
ROC is the left half S plane with $\text{Re}(s) < -1$
The second term converges for all $s - 2 < 0$ i.e., $\text{Re}(s) < 2$
ROC is left half S plane with $\text{Re}(s) < 2$
The common area of convergence is the area with $\text{Re}(s) < -1$. ROC and poles are plotted in Fig. 7.24.

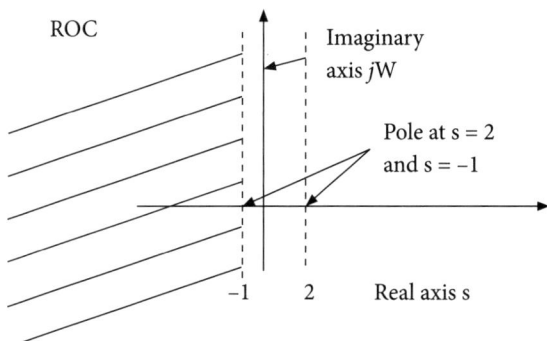

Fig. 7.24 ROC and poles for Example 7.18

Example 7.19

Find the LT of $x(t) = e^t \sin(t)u(t)$ and find ROC.

Solution

We will use the definition of LT.

$$X(S) = \frac{1}{2j}\left[\int_0^\infty e^t e^{jt} e^{-st} dt - \int_0^\infty e^t e^{-jt} e^{-st} dt\right]$$

$$= \frac{1}{2j}\left[-\frac{e^{-(s-1-j)t}}{s-1-j}\Big|_0^\infty + \frac{e^{-(s-1+j)t}}{s-1+j}\Big|_0^\infty\right]$$

(7.64)

$$= \frac{1}{2j}\left[\frac{1}{s-1-j} - \frac{1}{s-1+j}\right] = \frac{1}{2j}\left[\frac{s-1+j-s+1+j}{(s+1)^2+1}\right]$$

$$= \frac{1}{(s-1)^2+1}$$

The first term converges for all $s - 1 > 0$ i.e., $\text{Re}(s) > 1$

ROC is the right half S plane with $\text{Re}(s) > 1$

The second term converges for all again $\text{Re}(s) > 1$

The common area of convergence is $\text{Re}(s) > 1$. The ROC and the poles are plotted in Fig. 7.25.

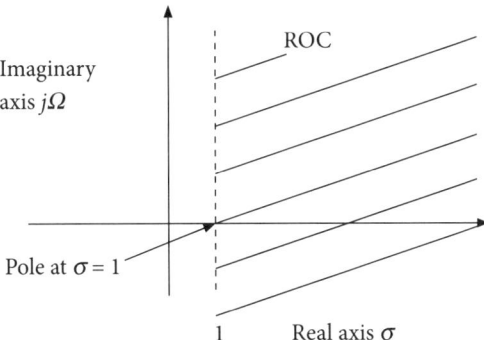

Fig. 7.25 Plot of ROC and poles for Example 7.19

Example 7.20

Find the LT of $x(t) = \cos^2(t)u(t)$ and find ROC.

Solution

We will use the definition of LT.

$$X(S) = \int_{-\infty}^{\infty} \cos^2(t)u(t)e^{-st}dt = \frac{1}{2}[\int_0^{\infty}(1+\cos 2t)e^{-st}dt$$

$$= \frac{1}{2}\left[\int_0^{\infty} e^{-st}dt + \int_0^{\infty} e^{2jt}e^{-st}dt + \int_0^{\infty} e^{-2jt}e^{-st}dt\right]$$

$$= \frac{1}{2}\left[\frac{e^{-st}}{-s}\Big\downarrow_0^{\infty} - \frac{e^{-(s-2j)t}}{s-2j}\Big\downarrow_0^{\infty} - \frac{e^{-(s+2j)t}}{s+2j}\Big\downarrow_0^{\infty}\right] \quad (7.65)$$

$$= \frac{1}{2}\left[\frac{1}{s}+\frac{1}{s-2j}+\frac{1}{s+2j}\right] = \frac{1}{2}\left[\frac{s^2+4+s^2+2js+s^2-2js}{s(s^2+4)}\right]$$

$$= \frac{3s^2+4}{s(s^2+4)}$$

The first term converges for all $s - 2j > 0$ i.e., $\text{Re}(s) > 0$

ROC is the right half S plane with $\text{Re}(s) > 0$

The second term converges for all again $\text{Re}(s) > 0$

The common area of convergence is $\text{Re}(s) > 0$. The ROC and the poles are plotted in Fig. 7.26.

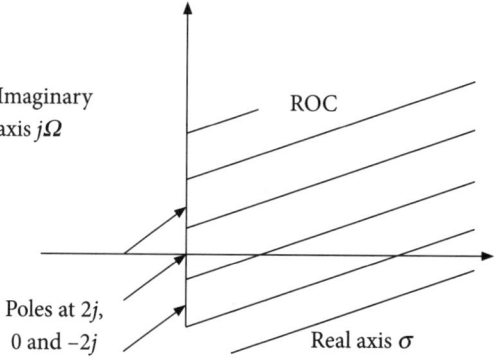

Fig. 7.26 Plot of ROC and poles for Example 7.20

Example 7.21

Find the LT of $x(t) = (1 + \sin t \cos t)u(t)$ and find ROC.

Solution

We will use the definition of LT.

$$X(S) = \int_{-\infty}^{\infty}(1+\sin t\cos t)u(t)e^{-st}dt = [\int_0^{\infty}\left(1+\frac{1}{2}\sin 2t\right)e^{-st}dt$$

$$= \int_0^{\infty}e^{-st}dt + \frac{1}{4j}\left[\int_0^{\infty}e^{2jt}e^{-st}dt - \int_0^{\infty}e^{-2jt}e^{-st}dt\right]$$

$$= \left[\frac{e^{-st}}{-s}\right]\downarrow_0^{\infty} - \frac{1}{4j}\left[\frac{e^{-(s-2j)t}}{s-2j}\downarrow_0^{\infty} + \frac{e^{-(s+2j)t}}{s+2j}\downarrow_0^{\infty}\right] \qquad (7.66)$$

$$= \left[\frac{1}{s} + \frac{1}{4j}\left[\frac{1}{s-2j} - \frac{1}{s+2j}\right]\right] = \left[\frac{1}{s} + \frac{s+2j-s+2j}{4j(s^2+4)}\right]$$

$$= \frac{1}{s} + \frac{1}{(s^2+4)} = \frac{s^2+s+4}{s(s^2+4)}$$

The first term converges for all $s - 2j > 0$ i.e., $\text{Re}(s) > 0$
ROC is the right half S plane with $\text{Re}(s) > 0$
The second term converges for all again $\text{Re}(s) > 0$

The common area of convergence is $\text{Re}(s) > 0$. The ROC and the poles are plotted in Fig. 7.27.

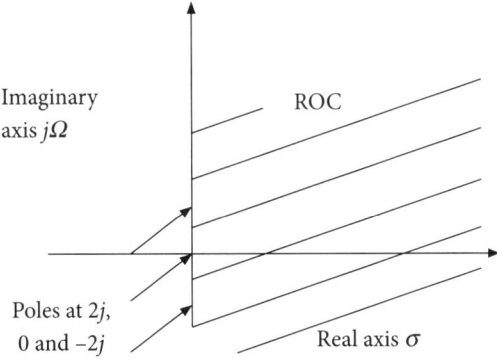

Fig. 7.27 Plot of ROC and poles for Example 7.21

7.4.1 LT of standard aperiodic signals

Let us find the LT of some standard aperiodic signals.

Example 7.22
Find the LT of the signal drawn in Fig. 7.28. Find ROC.

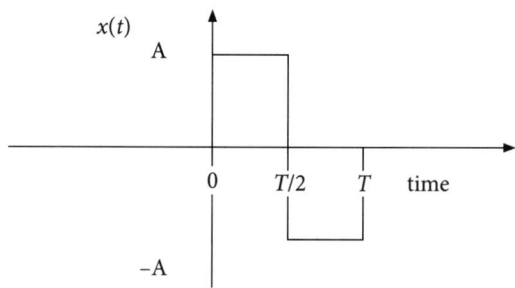

Fig. 7.28 Plot of signal for Example 7.22

Solution

We will use the definition of LT.

$$X(S) = \int_0^{T/2} Ae^{-st}\,dt - \int_{T/2}^T Ae^{-st}\,dt$$

$$= -\frac{Ae^{-st}}{s}\bigg\downarrow_0^{T/2} + \frac{Ae^{-st}}{s}\bigg\downarrow_{T/2}^T \qquad (7.67)$$

$$= -\frac{A}{s}\left[e^{-sT/2} - 1\right] + \frac{A}{s}\left[e^{-sT} - e^{-sT/2}\right]$$

$$= \frac{A}{s}\left[1 - 2e^{-sT/2} + e^{-sT}\right] = \frac{A}{s}\left[1 - e^{-sT/2}\right]^2$$

Both the terms converge for all $s > 0$ i.e., $\text{Re}(s) > 0$

ROC is the right half S plane with $\text{Re}(s) > 0$

Note: if $\text{Re}(a) = \sigma > 0$, $e^{-s\infty} \to 0$

if $\text{Re}(a) = \sigma < 0$, $e^{-s\infty} \to \infty$

The ROC and poles/zeros are plotted in Fig. 7.29. There is a double zero at $s = 0$

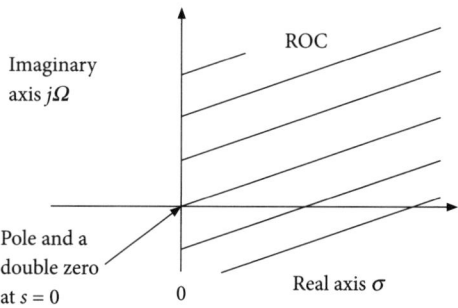

Fig. 7.29 Plot of ROC and poles for Example 7.22

Example 7.23

Find the LT of the signal drawn in Fig. 7.30. Find ROC.

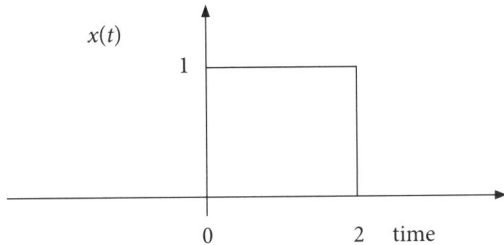

Fig. 7.30 Plot of signal for Example 7.23

Solution

We will use the definition of LT.

$$X(S) = \int_0^2 1 \times e^{-st} dt$$
$$= -\frac{e^{-st}}{s} \bigg|_0^2 \qquad (7.68)$$
$$= -\frac{1}{s}\left[e^{-2s} - 1\right]$$
$$= \frac{1}{s}\left[1 - e^{-2s}\right]$$

The term converges for all $s > 0$ i.e., $\text{Re}(s) > 0$
ROC is the right half S plane with $\text{Re}(s) > 0$

The ROC and poles/zeros are plotted in Fig. 7.31. There is a zero at $s = 0$

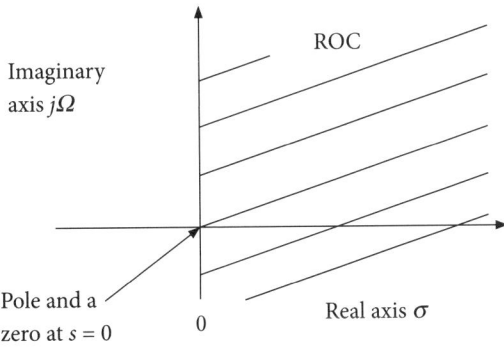

Fig. 7.31 Plot of ROC and pole and zero for example 7.23

Example 7.24

Find the LT of the signal drawn below. Find ROC.

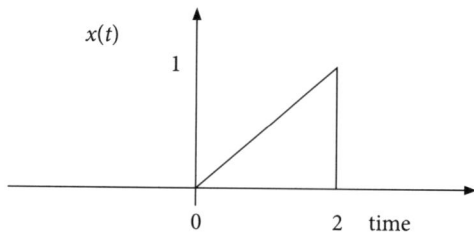

Fig. 7.32 Plot of signal for Example 7.24

Solution

We will use the definition of LT. The equation of the straight line is $x(t) = t/2$.

$$X(S) = \int_0^2 t/2 \times e^{-st} dt$$

$$= \frac{1}{2}\left[\frac{te^{-st}}{-s} \bigg|_0^2 - \int_0^2 \frac{e^{-st}}{-s} dt \right]$$

$$= \frac{1}{2}\left[\frac{2e^{-2s}}{-s} - \frac{e^{-st}}{s^2} \bigg|_0^2 \right] \quad (7.69)$$

$$= \frac{1}{2}\left[\frac{2e^{-2s}}{-s} - \frac{e^{-2s}-1}{s^2} \right] = \frac{1}{2}\left[\frac{1 - 2se^{-2s} - e^{-2s}}{s^2} \right]$$

The term converges for all $s > 0$ i.e., $\operatorname{Re}(s) > 0$
ROC is the right half S plane with $\operatorname{Re}(s) > 0$

The ROC and poles/zeros are plotted in Fig. 7.33. There is a zero at $s = 0$

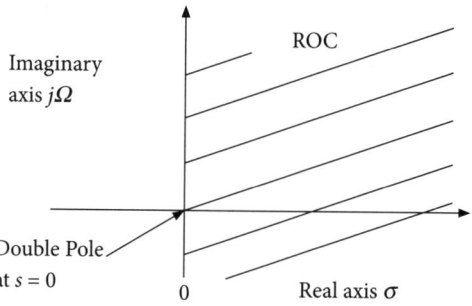

Fig. 7.33 Plot of ROC and pole and zero for Example 7.24

Example 7.25

Find LT of the signal drawn in Fig. 7.34 below. Find ROC.

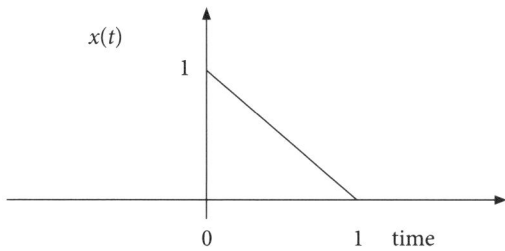

Fig. 7.34 Plot of signal for Example 7.25

Solution

We will use the definition of LT. We have to find the equation of the straight line.

$Y = mx + c$. Points (0,1) and (1,0) are on the line. These coordinates must satisfy the equation.

$1 = c$ and $0 = m + c$; hence, $m = -c = -1$; The equation becomes $y = -x + 1$
We have to write the signal $x(t) = -t + 1$

$$X(S) = \int_0^1 (1-t)e^{-st} dt$$

$$= \left[\frac{e^{-st}}{-s} \Big\downarrow_0^1 - \frac{te^{-st}}{-s} \Big\downarrow_0^1 + \int_0^1 \frac{e^{-st}}{-s} dt \right]$$

$$= \left[\frac{e^{-s}-1}{-s} - \frac{e^{-s}}{-s} + \frac{e^{-st}}{s^2} \Big\downarrow_0^1 \right] \quad (7.70)$$

$$= \left[\frac{e^{-s}-1}{-s} + \frac{e^{-s}}{s} + \frac{e^{-s}-1}{s^2} \right] = \left[\frac{-se^{-s} + se^{-s} + e^{-s} - 1}{s^2} \right] = \frac{e^{-s}-1}{s^2}$$

The term converges for all $s > 0$ i.e., $\text{Re}(s) > 0$

ROC is the right half S plane with $\text{Re}(s) > 0$

The ROC and poles/zeros are plotted in Fig. 7.33. There is a zero at $s = 0$ and a double pole at $s = 0$.

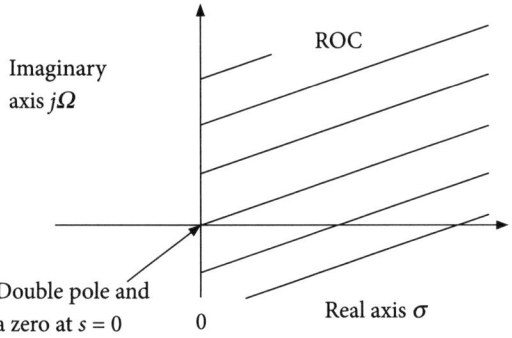

Fig. 7.35 Plot of ROC and pole and zero for Example 7.25

Example 7.26
Find the LT of the signal drawn in Fig. 7.36 below. Find ROC.

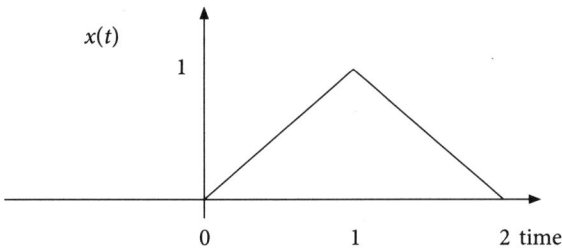

Fig. 7.36 Plot of signal for Example 7.26

Solution
We will use the definition of LT. The equation of the straight line is $x(t) = 2 - t$.

$$X(S) = \int_0^1 t \times e^{-st} dt + \int_1^2 (2-t) \times e^{-st} dt$$

$$= \left[\frac{te^{-st}}{-s}\Big\downarrow_0^1 - \int_0^1 \frac{e^{-st}}{-s} dt\right] + 2\left[\frac{e^{-st}}{-s}\right]\Big\downarrow_1^2 - \left[\frac{te^{-st}}{-s}\right]\Big\downarrow_1^2 + \int_1^2 \frac{e^{-st}}{-s} dt \qquad (7.71)$$

$$= \frac{e^{-s}}{-s} - \frac{e^{-st}}{s^2}\Big\downarrow_0^1 + 2\frac{e^{-2s} - e^{-s}}{-s} - \frac{2e^{-2s} - e^{-s}}{-s} + \frac{e^{-st}}{s^2}\Big\downarrow_1^2$$

$$= \frac{e^{-s}}{-s} - \frac{e^{-s} - 1}{s^2} + \frac{2e^{-2s} - 2e^{-s}}{-s} - \frac{2e^{-2s} - e^{-s}}{-s} + \frac{e^{-2s} - e^{-s}}{s^2}$$

$$= \frac{e^{-s} + 2e^{-2s} - 2e^{-s} - 2e^{-2s} + e^{-s}}{-s} + \frac{1 - e^{-s} - e^{-s} + e^{-2s}}{s^2}$$

$$= \frac{1 - 2e^{-s} + e^{-2s}}{s^2} = \frac{(1 - e^{s})^2}{s^2}$$

The term converges for all $s > 0$ i.e., $\text{Re}(s) > 0$, double pole at $s = 0$
ROC is the right half S plane with $\text{Re}(s) > 0$

The ROC and poles/zeros are plotted in Fig. 7.35. There is a double pole at $s = 0$

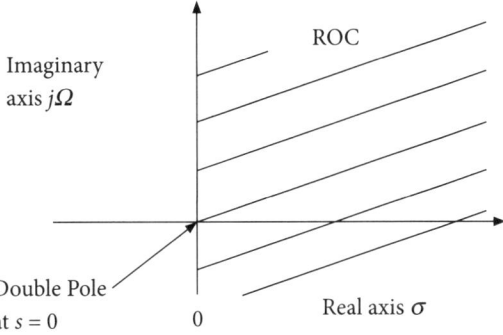

Fig. 7.37 Plot of ROC and pole and zero for Example 7.26

Example 7.27

Find the LT of the signal $\begin{matrix} x(t) = \sin t, 0 < t \le \pi \\ = 0 \quad \text{elsewhere} \end{matrix}$ drawn in Fig. 7.38 below. Find ROC.

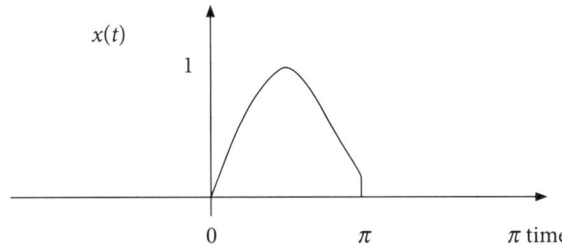

Fig. 7.38 Plot of signal for Example 7.27

Solution

We will use the definition of LT.

$$X(S) = \int_0^\pi \sin t \times e^{-st} dt = \frac{1}{2j}\left[\int_0^\pi e^{jt}e^{-st}dt - \int_0^\pi e^{-jt}e^{-st}dt\right]$$

$$= \frac{1}{2j}\left[\frac{e^{-(s-j)t}}{-(s-j)}\bigg\downarrow_0^\pi - \frac{e^{-(s+j)t}}{-(s+j)}\bigg\downarrow_0^\pi\right]$$

$$= \frac{1}{2j}\left[\frac{e^{-s\pi}+1}{s-j} - \frac{e^{-s\pi}+1}{s+j}\right]$$

$$= \frac{1}{2j}\left[\frac{se^{-s\pi}+s+je^{-s\pi}+j-se^{-s\pi}-s+je^{-s\pi}+j}{s^2+1}\right] = \frac{(e^{-s\pi}+1)}{s^2+1}$$

(7.72)

The term converges for all $s > 0$ i.e., $\text{Re}(s) > 0$, poles at $s = +j$ and $-j$
ROC is the right half S plane with $\text{Re}(s) > 0$

The ROC and poles/zeros are plotted in Fig. 7.37. There are poles at $s = +j$ and $s = -j$

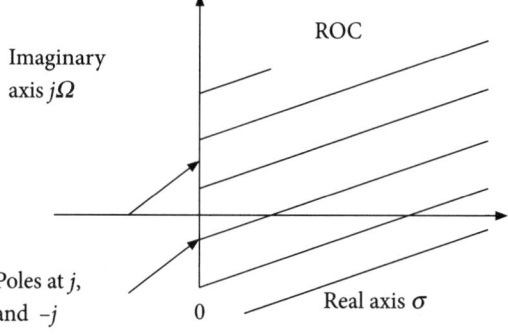

Fig. 7.39 Plot of ROC and pole and zero for Example 7.27

7.4.2 LT of standard periodic signals

We will find the LT of standard periodic signals.

Example 7.28
Find the LT of the signal drawn in Fig. 7.40 below.

Solution
We will use the periodicity property of LT.

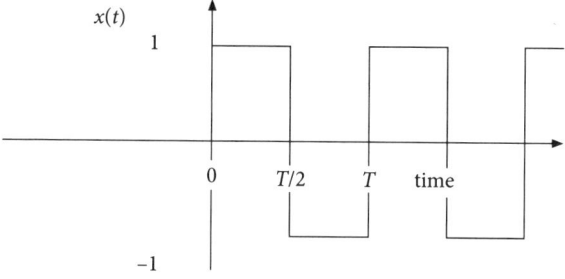

Fig. 7.40 Plot of signal for Example 7.28

$$X(S) = \frac{1}{1-e^{-sT}} \int_0^T x(t)e^{-st}dt = \frac{1}{1-e^{-sT}} \times [\text{transform of one period}] \quad (7.73)$$

Let us find the transform of one period.

$$X(S) = \int_0^{T/2} e^{-st}dt - \int_{T/2}^T e^{-st}dt$$

$$= -\frac{e^{-st}}{s}\Big|_0^{T/2} + \frac{e^{-st}}{s}\Big|_{T/2}^T$$

$$= -\frac{1}{s}\left[e^{-sT/2}-1\right] + \frac{1}{s}\left[e^{-sT}-e^{-sT/2}\right] \quad (7.74)$$

$$= \frac{1}{s}\left[1-2e^{-sT/2}+e^{-sT}\right] = \frac{1}{s}\left[1-e^{-sT/2}\right]^2$$

$$X(S)_{\text{periodic signal}} = \frac{(1-e^{-sT/2})^2}{s(1-e^{-sT})} = \frac{(1-e^{-sT/2})}{s(1+e^{-sT/2})}$$

Example 7.29

Find the LT of the signal drawn in Fig. 7.41 below.

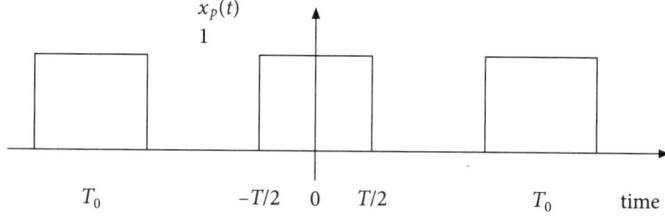

Fig. 7.41 Plot of signal for Example 7.29

Solution

We will use the periodicity property of LT.
Let us find the transform of one period.

$$X(S) = \int_0^{T/2} e^{-st} dt$$

$$= -\frac{e^{-st}}{s} \Big|_0^{T/2}$$

(7.75)

$$= -\frac{1}{s}\left[e^{-sT/2} - 1\right]$$

$$= \frac{1}{s}\left[1 - e^{-sT/2}\right]$$

$$X(S)_{\text{periodic signal}} = \frac{(1 - e^{-sT/2})}{s(1 - e^{-sT_0})}$$

Example 7.30

Find the LT of the signal drawn in Fig. 7.42 below.

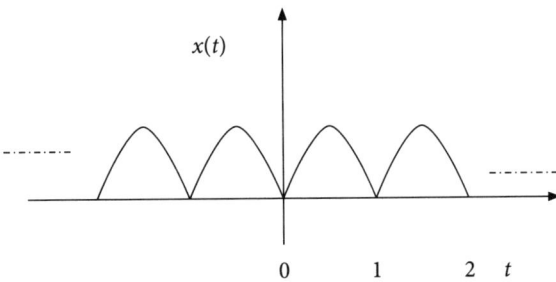

Fig. 7.42 Plot of signal for Example 7.30

We will find the LT of one period first.

$$X(S) = \int_0^1 \sin \pi t \times e^{-st} dt = \frac{1}{2j}\left[\int_0^1 e^{j\pi t} e^{-st} dt - \int_0^1 e^{-j\pi t} e^{-st} dt\right]$$

(7.76)

$$= \frac{1}{2j}\left[\frac{e^{-(s-j\pi)t}}{-(s-j\pi)}\Big|_0^1 - \frac{e^{-(s+j\pi)t}}{-(s+j\pi)}\Big|_0^1\right]$$

$$= \frac{1}{2j}\left[\frac{1+e^{-s}}{s-j\pi} - \frac{1+e^{-s}}{s+j\pi}\right] \quad \text{Note} \quad e^{j\pi} = e^{-j\pi} = -1$$

$$= \frac{1}{2j}\left[\frac{se^{-s}+s+j\pi e^{-s}+j\pi-se^{-s}-s+j\pi e^{-s}+j\pi}{s^2+\pi^2}\right] = \frac{\pi(e^{-s}+1)}{s^2+\pi^2}$$

$$X(S)_{\text{periodic signal}} = \frac{\pi(e^{-s}+1)}{(s^2+\pi^2)(1-e^{-S})}, \text{period} = 1$$

Example 7.31
Find the LT of the signal drawn in Fig. 7.43 below.

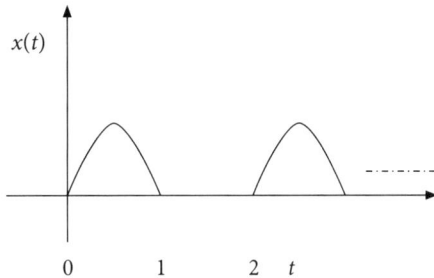

Fig. 7.43 Plot of signal for Example 7.31

Solution
Let us find the LT of one period first. Note that LT of one period is the same as that for signal in Example 7.30. Only the period is now 2.

$$X(S) = \int_0^1 \sin \pi t \times e^{-st} dt = \frac{1}{2j}\left[\int_0^1 e^{j\pi t} e^{-st} dt - \int_0^1 e^{-j\pi t} e^{-st} dt\right]$$

$$= \frac{1}{2j}\left[\frac{e^{-(s-j\pi)t}}{-(s-j\pi)}\Big|_0^1 - \frac{e^{-(s+j\pi)t}}{-(s+j\pi)}\Big|_0^1\right]$$

(7.77)

$$= \frac{1}{2j}\left[\frac{1+e^{-s}}{s-j\pi} - \frac{1+e^{-s}}{s+j\pi}\right] \quad \text{Note} \quad e^{j\pi} = e^{-j\pi} = -1$$

$$= \frac{1}{2j}\left[\frac{se^{-s}+s+j\pi e^{-s}+j\pi-se^{-s}-s+j\pi e^{-s}+j\pi}{s^2+\pi^2}\right] = \frac{\pi(e^{-s}+1)}{s^2+\pi^2}$$

$$X(S)_{periodic\,signal} = \frac{\pi(e^{-s}+1)}{(s^2+\pi^2)(1-e^{-2s})}, \text{ period} = 2$$

Example 7.32

Find the LT of the signal drawn in Fig. 7.44 below.

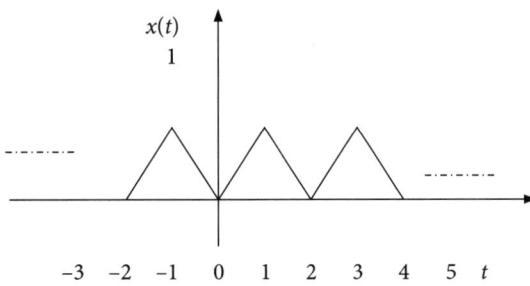

Fig. 7.44 Plot of signal for Example 7.32

Solution

Let us find the LT of one period. It is the same as that for Example 7.26.

$$X(S) = \int_0^1 t \times e^{-st} dt + \int_1^2 (2-t) \times e^{-st} dt$$

$$= \left[\frac{te^{-st}}{-s}\right]_0^1 - \int_0^1 \frac{e^{-st}}{-s} dt + 2\left[\frac{e^{-st}}{-s}\right]_1^2 - \left[\frac{te^{-st}}{-s}\right]_1^2 + \int_1^2 \frac{e^{-st}}{-s} dt$$

$$= \frac{e^{-s}}{-s} - \frac{e^{-st}}{s^2}\Big|_0^1 + 2\frac{e^{-2s}-e^{-s}}{-s} - \frac{2e^{-2s}-e^{-s}}{-s} + \frac{e^{-st}}{s^2}\Big|_1^2$$

(7.78)

$$= \frac{e^{-s}}{-s} - \frac{e^{-s}-1}{s^2} + \frac{2e^{-2s}-2e^{-s}}{-s} - \frac{2e^{-2s}-e^{-s}}{-s} + \frac{e^{-2s}-e^{-s}}{s^2}$$

$$= \frac{e^{-s}+2e^{-2s}-2e^{-s}-2e^{-2s}+e^{-s}}{-s} + \frac{1-e^{-s}-e^{-s}+e^{-2s}}{s^2}$$

$$= \frac{1-2e^{-s}+e^{-2s}}{s^2} = \frac{(1-e^{-s})^2}{s^2}$$

The term converges for all $s > 0$ i.e., $Re(s) > 0$, double pole at $s = 0$
ROC is the right half S plane with $Re(s) > 0$

$$X(s)_{\text{periodic signal}} = \frac{(1-e^{-s})^2}{s^2(1-e^{-2s})}, \text{ period} = 2$$

Example 7.33

Find the LT of the signal drawn in Fig. 7.45 below.

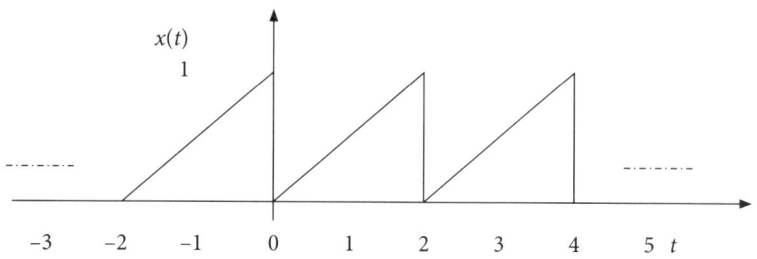

Fig. 7.45 Plot of signal for Example 7.33

Solution

Let us find the LT of one period first. Note that the LT of one period is the same as that for Example 7.24.

$$X(S) = \int_0^2 t/2 \times e^{-st} dt$$

$$= \frac{1}{2}\left[\frac{te^{-st}}{-s} \Big\downarrow_0^2 - \int_0^2 \frac{e^{-st}}{-s} dt \right]$$

(7.79)

$$= \frac{1}{2}\left[\frac{2e^{-2s}}{-s} - \frac{e^{-st}}{s^2} \Big\downarrow_0^2 \right]$$

$$= \frac{1}{2}\left[\frac{2e^{-2s}}{-s} - \frac{e^{-2s}-1}{s^2} \right] = \frac{1}{2}\left[\frac{1-2se^{-2s}-e^{-2s}}{s^2} \right]$$

The term converges for all $s > 0$ i.e., $\text{Re}(s) > 0$
ROC is the right half S plane with $\text{Re}(s) > 0$

$$X(S)_{\text{periodic signal}} = \frac{1}{2}\left[\frac{1-2se^{-2s}-e^{-2s}}{s^2(1-e^{-2s})} \right], \text{ period} = 2$$

7.4.3 LT of signals using properties of LT

We will find the LT of some signals using the different properties of LT.

Example 7.34

Find the LT of the following signals using the properties of LT.

1. $x(t) = tu(t)$ 2. $x(t) = \dfrac{d}{dt}\delta(t)$ 3. $x(t) = u(t-5)$ 4. $x(t) = te^{-t}u(t)$

Solution

1. $LT(u(t)) = \dfrac{1}{s}$, we will use the property of differentiation in frequency.

$$LT(tu(t)) = -\dfrac{d}{ds}\left(\dfrac{1}{s}\right) = \dfrac{1}{s^2} \tag{7.80}$$

2. $LT(\delta(t)) = 1$, we will use the property of time differentiation.

$$LT\left(\dfrac{d}{dt}\delta(t)\right) = s \times LT(\delta(t)) = s \tag{7.81}$$

3. $LT(u(t)) = \dfrac{1}{s}$, we will use the property of time shifting.

$$LT(u(t-5)) = e^{-5s} LT(u(t)) = \dfrac{e^{-5s}}{s} \tag{7.82}$$

4. $LT(e^{-at}u(t)) = \dfrac{1}{s+a}$, we will use the property of differentiation in S domain.

$$LT(te^{-t}u(t)) = -\dfrac{d}{ds}\left(\dfrac{1}{s+1}\right) = \dfrac{1}{(s+1)^2} \tag{7.83}$$

Example 7.35

Find the LT of the following signals using properties of LT.

1. $x(t) = t^2 e^{-t} u(t)$ 2. $x(t) = te^{-t}\cos(t)u(t)$
3. $x(t) = e^{at} u(-t)$ 4. $x(t) = e^{-t}\cos(t)u(t)$

Solution

1. $LT(tu(t)) = \dfrac{1}{s^2}$, refer Eq. (7.80).

We will use the property of differentiation in frequency.

$$LT(t^2 u(t)) = -\frac{d}{ds}\left(\frac{1}{s^2}\right) = \frac{2}{s^3}$$

(7.84)

$$LT(t^2 e^{-t} u(t)) = \frac{2}{(s+1)^3}$$

2. We have to find the LT $(x(t) = te^{-t}\cos(t)u(t))$, we know that $LT(tu(t)) = \frac{1}{s^2}$

To find the LT of $LT(te^{-t}u(t))$, we will use the property of frequency shifting.

If $tu(t) \leftrightarrow \frac{1}{s^2}$,

(7.85)

then $e^{-t}tu(t) \leftrightarrow X(s+1) = \frac{1}{(s+1)^2}$

We can write $\cos(t) = \dfrac{e^{jt} + e^{-jt}}{2}$ and again use property of frequency shifting.

If $te^{-t}u(t) \leftrightarrow \dfrac{1}{(s+1)^2}$,

then $\dfrac{1}{2}[e^{jt}te^{-t}u(t)] \leftrightarrow X(s+1-j) = \dfrac{1}{2}\left[\dfrac{1}{(s+1-j)^2}\right]$ and

$\dfrac{1}{2}[e^{-jt}te^{-t}u(t)] \leftrightarrow X(s+1+j) = \dfrac{1}{2}\left[\dfrac{1}{(s+1+j)^2}\right]$

(7.86)

$te^{-t}\cos(t)u(t) \leftrightarrow \dfrac{1}{2}\left[\dfrac{1}{(s+1-j)^2} + \dfrac{1}{(s+1+j)^2}\right]$

$= \dfrac{1}{2}\left[\dfrac{2(s+1)^2 - 2}{(s+1-j)^2(s+1+j)^2}\right] = \dfrac{(s+1)^2 - 1}{((s+1)^2 + 1)^2}$

Alternative method:

We know that $LT(\cos(t)u(t)) = \dfrac{s}{s^2+1}$, we will use property of differentiation in frequency to find

$$LT(t\cos(t)u(t)) = -\dfrac{d}{ds}\left(\dfrac{s}{s^2+1}\right) = -\dfrac{s^2+1-2s^2}{(s^2+1)^2} = \dfrac{s^2-1}{(s^2+1)^2}$$

(7.87)

We will use property of frequency shifting to find

$$LT(te^{-t}\cos(t)u(t)) = \frac{(s+1)^2 - 1}{((s+1)^2 + 1)^2} \quad (7.88)$$

3. We have to find the LT of $x(t) = e^{at}u(-t)$.
   We know that, $x(t) = e^{at}u(t) \leftrightarrow \frac{1}{s-a}$, we will use property of time reversal.

   If $x(t) = e^{at}u(t) \leftrightarrow X(s) = \frac{1}{s-a}$,

   then $x(-t) = e^{at}u(-t) \leftrightarrow X(-s) = -\frac{1}{s+a}$ \quad (7.89)

   We have to find the LT of $x(t) = e^{-t}\cos(t)u(t)$
   We know that

   $$LT(\cos(t)u(t)) = \frac{s}{s^2 + 1}, \quad (7.90)$$

   We will use the property of frequency shifting to find

   $$LT(e^{-t}\cos(t)u(t)) = \frac{s+1}{(s+1)^2 + 1} \quad (7.91)$$

Example 7.36

Find the LT of the following signals using properties of LT.

1. $x(t) = (t^2 - 2t)u(t - 1)$
2. $x(t) = (t - a)u(t - a)$
3. $x(t) = 2tu(t) - 2(t - 1)u(t - 1)$

Solution

1. We have to find the LT of $x(t) = (t^2 - 2t)u(t - 1)$

   $x(t) = (t^2 - 2t)u(t-1) = t^2 u(t-1) - 2tu(t-1)$

   $$LT(t^2 u(t)) = \frac{2}{s^3}; \quad LT(t^2 u(t-1)) = \frac{2e^{-s}}{s^3} \quad (7.92)$$

   $$LT(-2tu(t-1)) = \frac{-2e^{-s}}{s^2}$$

   $$LT(x(t)) = \frac{2e^{-s}}{s^3}(1-s)$$

We have used time shifting property for all the signals.

2. We have to find the LT of

$$LT(x(t)) = LT(tu(t)) \leftrightarrow \frac{1}{s^2}$$

$$LT(x(t-a)) = LT((t-a)u(t-a)) = \frac{e^{-as}}{s^2} \quad (7.93)$$

3. We have to find the LT of

$$LT(x(t)) = LT(2tu(t) - 2(t-1)u(t-1))$$

$$= \frac{2}{s^2} - \frac{2e^{-s}}{s^2} = \frac{2(1-e^{-s})}{s^2} \quad (7.94)$$

Example 7.37

LTs of some signals are given below. Find the initial and final value of the signal using initial and final value theorem.

1. $X(s) = \dfrac{1}{s(s-2)}$
2. $X(s) = \dfrac{s+3}{s^2+2s+2}$
3. $X(s) = \dfrac{6s+5}{s(2s+5)}$
4. $X(s) = \dfrac{s+5}{s^2(s+9)}$

Solution

1. The initial value is given by

$$x(0) = \lim_{s \to \infty} sX(s) = \lim_{s \to \infty} \frac{1}{s-2} = 0 \quad (7.95)$$

The final value is given by

$$x(\infty) = \lim_{s \to 0} sX(s) = \lim_{s \to 0} \frac{1}{s-2} = -\frac{1}{2} \quad (7.96)$$

2. The initial value is given by

$$x(0) = \lim_{s \to \infty} sX(s) = \lim_{s \to \infty} \frac{s^2+3s}{s^2+2s+2} = \lim_{s \to \infty} \frac{1+3/s}{1+2/s+2/s^2} = 1 \quad (7.97)$$

The final value is given by

$$x(\infty) = \lim_{s \to 0} sX(s) = \lim_{s \to 0} \frac{s^2+3s}{s^2+2s+2} = 0 \quad (7.98)$$

3. The initial value is given by

$$x(0) = \lim_{s \to \infty} sX(s) = \lim_{s \to \infty} \frac{6s+5}{2s+5} = \lim_{s \to \infty} \frac{6+5/s}{2+5/s} = 3 \qquad (7.99)$$

The final value is given by

$$x(\infty) = \lim_{s \to 0} sX(s) = \lim_{s \to 0} \frac{6s+5}{2s+5} = 1 \qquad (7.100)$$

4. The initial value is given by

$$x(0) = \lim_{s \to \infty} sX(s) = \lim_{s \to \infty} \frac{s+5}{s(s+9)} = \lim_{s \to \infty} \frac{1/s+5/s^2}{1+9/s} = 0 \qquad (7.101)$$

The final value is given by

$$x(\infty) = \lim_{s \to 0} sX(s) = \lim_{s \to 0} \frac{s+5}{s(s+9)} = \infty \qquad (7.102)$$

Example 7.38

The signal is given by $x(t) = e^{-(t-3)/2} u(t-3)$. Find LT.

Solution

We know that

$$LT(y(t) = e^{-t} u(t)) = \frac{1}{s+1} \qquad (7.103)$$

We will use time shifting property to find

$$LT(y(t-3)) = LT(e^{-(t-3)} u(t-3)) = \frac{e^{-3s}}{s+1} \qquad (7.104)$$

We will now use the time scaling property to find

$$LT(y(t-3)/2)) = LT(e^{-(t-3)/2} u(t-3)) = \frac{2e^{-3s}}{2s+1} \qquad (7.105)$$

Example 7.39

The signal is given by $x(t) = \sin(at)\cos(bt)u(t)$. Find LT.

Solution

$$x(t) = \sin(at)\cos(bt)u(t) = \frac{1}{2}[\sin(a+b)t + \sin(a-b)t]u(t) \qquad (7.106)$$

We will use the property of linearity to find LT of the two terms.

$$LT(x(t)) = \frac{1}{2}\left[\frac{a+b}{s^2+(a+b)^2} + \frac{a-b}{s^2+(a-b)^2}\right] \quad (7.107)$$

Example 7.40
Find the LT of the signal $x(t) = \text{sgn}(t)$

Solution

We will first see the definition of $\text{sgn}(t)$

$$\text{sgn}(t) = \begin{cases} 1 \text{ for } t > 0 \\ -1 \text{ for } t < 0 \end{cases} \quad (7.108)$$

$$LT(\text{sgn}(t)) = \int_{-\infty}^{0} -e^{-st}\,dt + \int_{0}^{\infty} e^{-st}\,dt \quad (7.109)$$

$$= -\frac{e^{-st}}{-s}\bigg\downarrow_{-\infty}^{0} + \frac{e^{-st}}{-s}\bigg\downarrow_{0}^{\infty}$$

First integral converges for $s < 0$ and the second converges for $s > 0$. There is no common area of convergence. So, LT does not exist.

Example 7.41
Find the LT of $x(t) = e^{2t}u(t) * tu(t)$

Solution

We know that

$$LT(e^{at}u(t)) = \frac{1}{s-a} \quad (7.110)$$

$$LT(e^{2t}u(t)) = \frac{1}{s-2}$$

$$LT(u(t)) = \frac{1}{s};$$

$$\quad (7.111)$$

$$LT(tu(t)) = -\frac{d}{ds}\left(\frac{1}{s}\right) = \frac{1}{s^2}$$

(Note: we have used the property of differentiation in s domain.)

$$LT(e^{2t}u(t) * tu(t)) = \frac{1}{s-2} \times \frac{1}{s^2} = \frac{1}{s^2(s-2)} \qquad (7.112)$$

(Note: we have used the convolution property.)

Example 7.42

Find the LT of $x(t) = ae^{at}u(t)$ using the property of differentiation in time domain.

Solution

We know that

$$LT(e^{at}u(t)) = \frac{1}{s-a}$$

$$\qquad (7.113)$$

$$LT(ae^{at}u(t)) = \frac{a}{s-a}$$

We will now apply the property of differentiation in time for $x1(t) = e^{at}u(t)$

$$LT(e^{at}u(t)) = \frac{1}{s-a};$$

$$LT\left(\frac{d}{dt}e^{at}u(t)\right) = LT(ae^{at}u(t)) = sX_1(s) - x(0^+) \qquad (7.114)$$

$$= \frac{s}{s-a} - 1 = \frac{s-s+a}{s-a} = \frac{a}{s-a}$$

Let us find the initial value using initial value theorem.

$$x(0^+) = \lim_{s \to \infty} sX_1(s) = \lim_{s \to \infty} \frac{s}{s-a} = \lim_{s \to \infty} \frac{1}{1-a/s} = 1 \qquad (7.115)$$

Example 7.43

Find the LT of $x(t) = tu(t)$ using the property of integration.

Solution

We know that

$$LT(x(t)) = LT(tu(t)) = LT\left(\int_{-\infty}^{t} u(\tau)d\tau\right) \qquad (7.116)$$

$$= \frac{X(s)}{s} + x^{(-1)}(0^+)/s$$

$$= \frac{1}{s^2} + \int_{-\infty}^{0^+} x(\tau)d\tau$$

$$= \frac{1}{s^2} + 0 = \frac{1}{s^2}$$

7.4.4 Properties of ROC

We list below the properties of ROC.

1. The ROC does not contain any pole.
2. ROC is bounded by the poles or it extends to infinity.
3. If $x(t)$ is a right-handed signal, then ROC extends to the right of the rightmost pole. ROC is a strip parallel to the imaginary axis and extends from the rightmost pole to infinity.
4. If $x(t)$ is a left-handed signal, then ROC extends to the left of the leftmost pole. ROC is a strip parallel to the imaginary axis and extends from the leftmost pole to minus infinity.
5. If $x(t)$ is a two-sided signal, then ROC is a strip bounded by the poles.
6. The finite duration absolutely integrable function has ROC as the entire S plane.
7. ROC of the LTI system contains the imaginary axis.

The reader is encouraged to verify the above properties from the examples solved in the previous sections.

Concept Check

- If the signal is causal, what is the nature of ROC?
- Can the two signals have same LT but different ROC?
- Can the two signals have same ROC and different LT?
- Can a sgn(t) function have a LT?
- If $u(t)$ transforms to $1/s$, find LT of $tu(t)$, which property of LT have you used?
- Which property of LT will you use to find $LT(e^{-at}u(t))$?
- Which property of LT will you use to find LT of a periodic signal?
- If the signal is left-handed signal, what is your conclusion regarding the ROC?
- If the signal is a right-handed signal, what is your conclusion regarding ROC?
- If the signal is both-sided, what is your conclusion regarding the ROC?

- When will ROC be a strip between the 2 pole locations?
- Can ROC contain a pole?

7.5 Inverse LT

To find the signal from its given LT requires the evaluation of contour integration, as stated in section 7.1. It is preferred to find LT by inspection using one-to-one relationship between a known function and its LT.

The procedure for finding LT given the function in S domain can be stated as follows:

Step 1 First factorize the denominator to get the poles of the function.

Step 2 Use partial fraction expansion.

Step 3 Find the inverse LT of each term.

Let us illustrate this procedure by a simple example.

Example 7.44

Find the inverse LT of $X(s) = \dfrac{-5s-7}{(s+1)(s-1)(s+2)}$ with ROC given by $-1 < \operatorname{Re}(s) < 1$.

Solution

The denominator is already in the factored form. So, step 1 is over.

Step 2 We will use partial fraction expansion and decompose the function in three terms.

$$X(s) = \frac{-5s-7}{(s+1)(s-1)(s+2)} = \frac{k_1}{s+1} + \frac{k_2}{s-1} + \frac{k_3}{s+2} \qquad (7.117)$$

Find k_1, k_2 and k_3

$$k_1 = (s+1)X(s)\downarrow_{s=-1} = \frac{-5s-7}{(s-1)(s+2)}\downarrow_{s=-1} = \frac{5-7}{-2\times 1} = 1$$

$$k_2 = (s-1)X(s)\downarrow_{s=1} = \frac{-5s-7}{(s+1)(s+2)}\downarrow_{s=1} = \frac{-5-7}{2\times 3} = -2$$

$$k_3 = (s+2)X(s)\downarrow_{s=-2} = \frac{-5s-7}{(s+1)(s-1)}\downarrow_{s=-2} = \frac{10-7}{-1\times -3} = 1$$

We have to study ROC for finding the Inverse LT. Here, the common area of convergence is a strip between $\sigma = -1$ and $\sigma = 1$. We will recover the term with pole at 1 as a left-handed signal. We will recover the signal for the term with pole at $s = -2$ and $s = -1$ as the right-handed sequence so that its ROC will also include the common area of convergence. $-1 < \text{Re}(s) < 1$

$$X(s) = \frac{-5s - 7}{(s+1)(s-1)(s+2)} = \frac{1}{s+1} + \frac{-2}{s-1} + \frac{1}{s+2} \qquad (7.118)$$

Taking ILT we get

$$\begin{aligned} x(t) &= (e^{-t}u(t) - 2(-e^{t})u(-t) + e^{-2t}u(t)) \\ &= e^{-t}u(t) + 2e^{t}u(-t) + e^{-2t}u(t) \end{aligned}$$

Let us draw the pole zero plot using MATLAB program given below. Figure 7.46 shows the pole zero plot.

```
clear all;
clc;
s = tf('s');
b=[1 2 -1 -2];
a=[-5 -7];
f=tf(a,b);
pzmap(f);
```

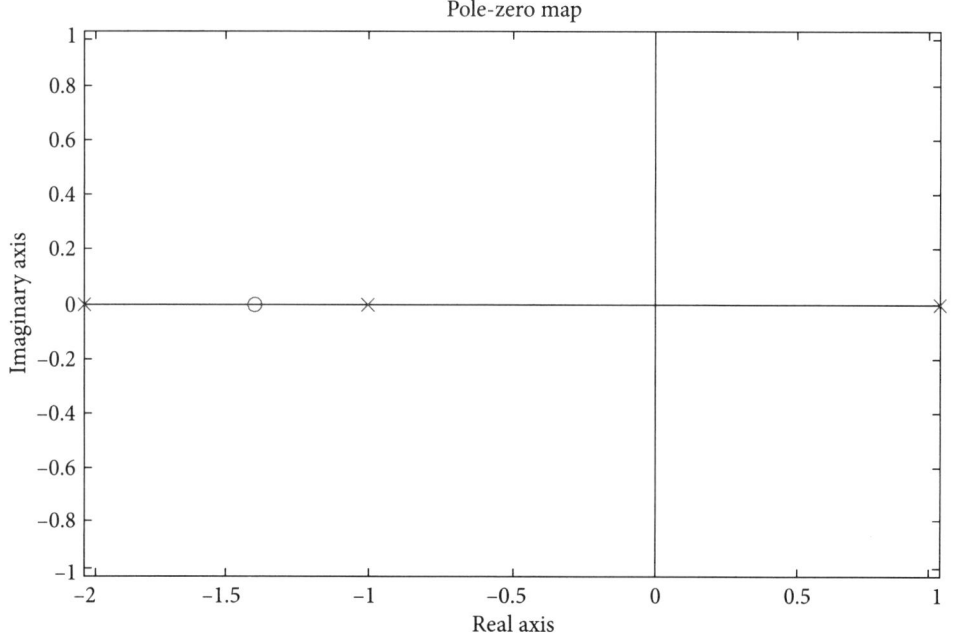

Fig. 7.46 Plot of poles and zeros for the transfer function for Example 7.44

Consider the ROC as $-2 < \text{Re}(s) < -1$ for the same $X(s)$. We will consider Example 7.42.

Example 7.45

Find the inverse LT of $X(s) = \dfrac{-5s-7}{(s+1)(s-1)(s+2)}$ with ROC given by $-2 < \text{Re}(s) < -1$.

Solution

The steps up to the decomposition of $X(s)$ will remain the same. We will repeat the steps here.

The denominator is already in the factored form. So, step 1 is over.

Step 2 We will use partial fraction expansion and decompose the function in three terms.

$$X(s) = \frac{-5s-7}{(s+1)(s-1)(s+2)} = \frac{k_1}{s+1} + \frac{k_2}{s-1} + \frac{k_3}{s+2} \quad (7.119)$$

Find k_1, k_2 and k_3

$$k_1 = (s+1)X(s)\Big\downarrow_{s=-1} = \frac{-5s-7}{(s-1)(s+2)}\Big\downarrow_{s=-1} = \frac{5-7}{-2\times 1} = 1$$

$$k_2 = (s-1)X(s)\Big\downarrow_{s=1} = \frac{-5s-7}{(s+1)(s+2)}\Big\downarrow_{s=1} = \frac{-5-7}{2\times 3} = -2$$

$$k_3 = (s+2)X(s)\Big\downarrow_{s=-2} = \frac{-5s-7}{(s+1)(s-1)}\Big\downarrow_{s=-2} = \frac{10-7}{-1\times -3} = 1$$

We have to study ROC for finding the inverse LT. It is a strip between -2 and -1. Here, the common area of convergence is a strip between $\sigma = -2$ and $\sigma = -1$. We will recover the terms with poles at 0 and -1 as the left-handed signals. We will recover the signal for the term with pole at $s = -2$ as the right-handed signal so that its ROC will also include the common area of convergence.

$$X(s) = \frac{-5s-7}{(s+1)(s-1)(s+2)} = \frac{1}{s+1} + \frac{-2}{s-1} + \frac{1}{s+2} \quad (7.120)$$

Taking ILT we get

$$x(t) = (-e^{-t}u(-t) - 2(-e^{t})u(-t) + e^{-2t}u(t))$$
$$= -e^{-t}u(-t) + 2e^{t}u(-t) + e^{-2t}u(t)$$

Note that ROC cannot be specified as $-2 < \text{Re}(s) < 1$. Here, the signal for the term with a pole at $s = -1$ cannot be recovered either as a left-handed or as a right-handed signal. In any case, it will not completely overlap with the ROC specified.

If the ROC is not specified, we assume that all the signals are unilateral i.e., all are right-handed signals.

Example 7.46

Find the inverse LT of $X(s) = \dfrac{-5s - 7}{(s+1)(s-1)(s+2)}$

Solution

We will not repeat the decomposition of $X(s)$, we will just recover all signals as right-handed signals as follows.

$$X(s) = \frac{-5s-7}{(s+1)(s-1)(s+2)} = \frac{1}{s+1} + \frac{-2}{s-1} + \frac{1}{s+2} \qquad (7.121)$$

taking ILT we get

$$x(t) = (-e^{-t}u(-t) - 2(-e^{t})u(-t) + e^{-2t}u(t))$$
$$= e^{-t}u(t) - 2e^{t}u(t) + e^{-2t}u(t)$$

Note that all the signals are appended by $u(t)$.

Example 7.47

Find the inverse LT of $X(s) = \dfrac{2}{s(s+1)(s+2)}$

Solution

The ROC is not specified so we will recover all signals as the right-handed signals. A MATLAB program to plot poles and zeros of the transfer function specified is given below. Poles and zeros are plotted in Fig. 7.47.

```
clear all;
clc;
s = tf('s');
b=[1 3 2 0];
a=[2];
f=tf(a,b);
pzmap(f);
```

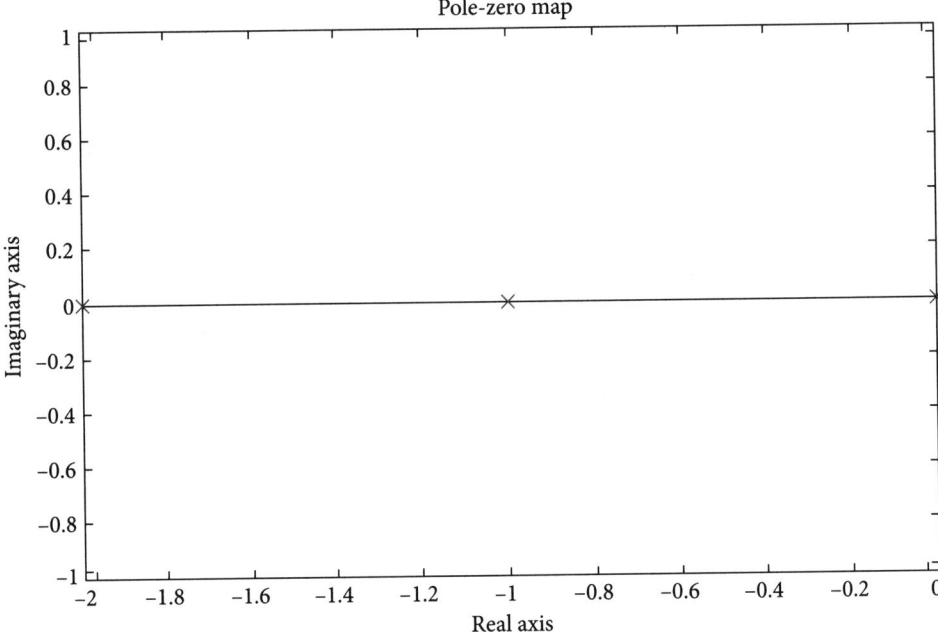

Fig. 7.47 Poles and zeros for Example 7.47

$$X(s) = \frac{2}{s(s+1)(s+2)} = \frac{k_1}{s} + \frac{k_2}{s+1} + \frac{k_3}{s+2}$$

$$k_1 = sX(s) = \frac{2}{(s+1)(s+2)} \Bigg\downarrow_{s=0} = \frac{2}{2} = 1$$

(7.122)

$$k_2 = (s+1)X(s) = \frac{2}{s(s+2)} \Bigg\downarrow_{s=-1} = \frac{2}{-1(1)} = -2$$

$$k_3 = (s+2)X(s) = \frac{2}{s(s+1)} \Bigg\downarrow_{s=-2} = \frac{2}{-2(-1)} = 1$$

$$X(s) = \frac{1}{s} + \frac{-2}{s+1} + \frac{1}{s+2}$$

Taking ILT we get

$$x(t) = (1 - 2e^{-t} + e^{-2t})u(t) = (1 - e^{-t})^2 u(t)$$

Note that all the signals are appended by $u(t)$.

Example 7.48

Find the inverse LT of $X(s) = \dfrac{2}{s(s+1)(s+2)}$ with ROC specified as $-1 < \text{Re}(s) < 0$.

Solution

The ROC is specified as $-1 < \text{Re}(s) < 0$. We will recover all signals so that the common area of convergence is the ROC specified.

$$X(s) = \frac{2}{s(s+1)(s+2)} = \frac{k_1}{s} + \frac{k_2}{s+1} + \frac{k_3}{s+2}$$

$$k_1 = sX(s) = \frac{2}{(s+1)(s+2)}\bigg\downarrow_{s=0} = \frac{2}{2} = 1$$

$$k_2 = (s+1)X(s) = \frac{2}{s(s+2)}\bigg\downarrow_{s=-1} = \frac{2}{-1(1)} = -2 \qquad (7.123)$$

$$k_3 = (s+2)X(s) = \frac{2}{s(s+1)}\bigg\downarrow_{s=-2} = \frac{2}{-2(-1)} = 1$$

$$X(s) = \frac{1}{s} + \frac{-2}{s+1} + \frac{1}{s+2}$$

Taking ILT we get

$$x(t) = (-u(-t) - 2e^{-t}u(t) + e^{-2t}u(t))$$

Note that all the terms having pole at zero are recovered as a left-handed signal and are appended by $-u(-t)$ and the terms with poles at -1 and -2 are recovered as right-handed signals and are appended by $u(t)$.

Example 7.49

Find the inverse LT of $X(s) = \dfrac{2}{s(s+1)(s+2)}$ with ROC specified as $-2 < \text{Re}(s) < -1$.

Solution

The ROC is specified as $-2 < \text{Re}(s) < -1$, we will recover all signals so that the common area of convergence is the ROC specified.

$$X(s) = \frac{2}{s(s+1)(s+2)} = \frac{k_1}{s} + \frac{k_2}{s+1} + \frac{k_3}{s+2}$$

$$k_1 = sX(s) = \frac{2}{(s+1)(s+2)}\bigg\downarrow_{s=0} = \frac{2}{2} = 1$$

(7.124)

$$k_2 = (s+1)X(s) = \frac{2}{s(s+2)}\bigg\downarrow_{s=-1} = \frac{2}{-1(1)} = -2$$

$$k_3 = (s+2)X(s) = \frac{2}{s(s+1)}\bigg\downarrow_{s=-2} = \frac{2}{-2(-1)} = 1$$

$$X(s) = \frac{1}{s} + \frac{-2}{s+1} + \frac{1}{s+2}$$

Taking ILT we get

$$x(t) = (-u(-t) + 2e^{-t}u(-t) + e^{-2t}u(t))$$

Note that the terms having poles at zero and −1 are recovered as left-handed signals and are appended by −u(−t) and the terms with pole at −2 are recovered as a right-handed signal and are appended by u(t).

The roots of the denominator can be real or can be complex. Complex roots will result in complex-valued exponential function in time. Complex conjugate roots are combined to generate real coefficients for the denominator polynomial.

Let us consider one example with complex roots. Note that we have to find ILT by inspection. We will require the help of the following standard results for

$$LT(\sin \omega_0 t u(t)) = \frac{\omega_0}{s^2 + \omega_0^2}$$

(7.125)

$$LT(e^{at} \sin \omega_0 t u(t)) = \frac{\omega_0}{(s-a)^2 + \omega_0^2}$$

(7.126)

$$LT(\cos \omega_0 t u(t)) = \frac{s}{s^2 + \omega_0^2}$$

(7.127)

$$LT(e^{at} \cos \omega_0 t u(t)) = \frac{s-a}{(s-a)^2 + \omega_0^2}$$

(7.128)

Example 7.50

Find the inverse LT of $X(s) = \dfrac{s^2 + s - 2}{s^3 + 3s^2 + 5s + 3}$.

Solution

Step 1 The denominator is to be in the factored form.

$$X(s) = \frac{s^2 + s - 2}{s^3 + 3s^2 + 5s + 3} = \frac{s^2 + s - 2}{(s+1)(s^2 + 2s + 3)} \tag{7.129}$$

Let us draw the pole zero plot using MATLAB program given below. Figure 7.48 shows the pole zero plot.

```
clear all;
clc;
s = tf('s');
b=[1 3 5 3];
a=[1 1 -2];
f=tf(a,b);
pzmap(f);
```

Step 2 We will use partial fraction expansion and decompose the function in three terms.

$$X(s) = \frac{s^2 + s - 2}{(s+1)(s^2 + 2s + 3)} = \frac{k_1}{s+1} + \frac{k_2 s + k_3}{s^2 + 2s + 3} \tag{7.130}$$

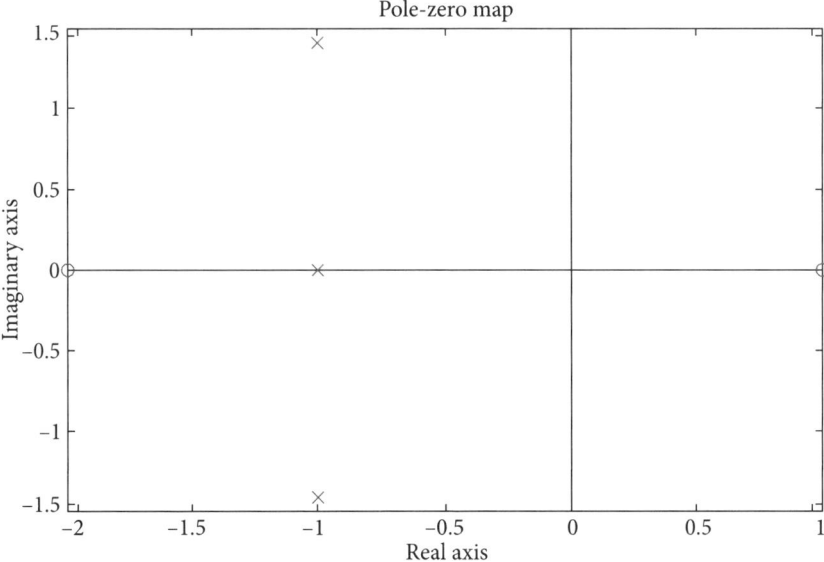

Fig. 7.48 Pole zero plot for Example 7.50

Find k_1, k_2 and k_3

$$k_1 = (s+1)X(s)\Big\downarrow_{s=-1} = \frac{s^2+s-2}{(s^2+2s+3)}\Big\downarrow_{s=-1} = \frac{-2}{2\times 1} = -1$$

Put value of k_1 in the equation and equate the numerators on both sides. We get

$$s^2+s-2 = -1(s^2+2s+3)+k_2 s^2+(k_2+k_3)s+k_3$$
$$s^2+s-2 = (k_2-1)s^2+(k_2+k_3-2)s+(k_3-3) \tag{7.131}$$

Equate the coefficients of s^2, s and a constant term.

$$k_2 - 1 = 1 \Rightarrow k_2 = 2$$
$$k_3 - 3 = -2 \Rightarrow k_3 = 1 \tag{7.132}$$

Equation (7.130) becomes

$$X(s) = \frac{s^2+s-2}{(s+1)(s^2+2s+3)} = \frac{-1}{s+1} + \frac{2s+1}{s^2+2s+3}$$

$$X(s) = \frac{-1}{s+1} + \frac{2(s+1)}{(s+1)^2+2} - \frac{1}{(s+1)^2+2} \tag{7.133}$$

We will now use Eqs (7.126) and (7.128) to find ILT. Note that we have to multiply the numerator and denominator of the third term by $\sqrt{2}$. ROC is Re(s) > −1 for all terms.

$$x(t) = -e^{-t}u(t) + 2e^{-t}\cos(\sqrt{2}t)u(t) - \frac{1}{\sqrt{2}}e^{-t}\sin(\sqrt{2}t)u(t) \tag{7.134}$$

Example 7.51

Find the inverse LT of $X(s) = \dfrac{4s^2+6}{s^3+s^2-2}$ with ROC −1< Re(s)<1.

Solution

Let us plot the poles and zeros of the transfer function using MATLAB program given below. The resulting pole zero plot is shown in Fig. 7.49.

```
clear all;
clc;
s = tf('s');
b=[1 1 0 -2];
a=[4 0 6];
f=tf(a,b);
pzmap(f);
```

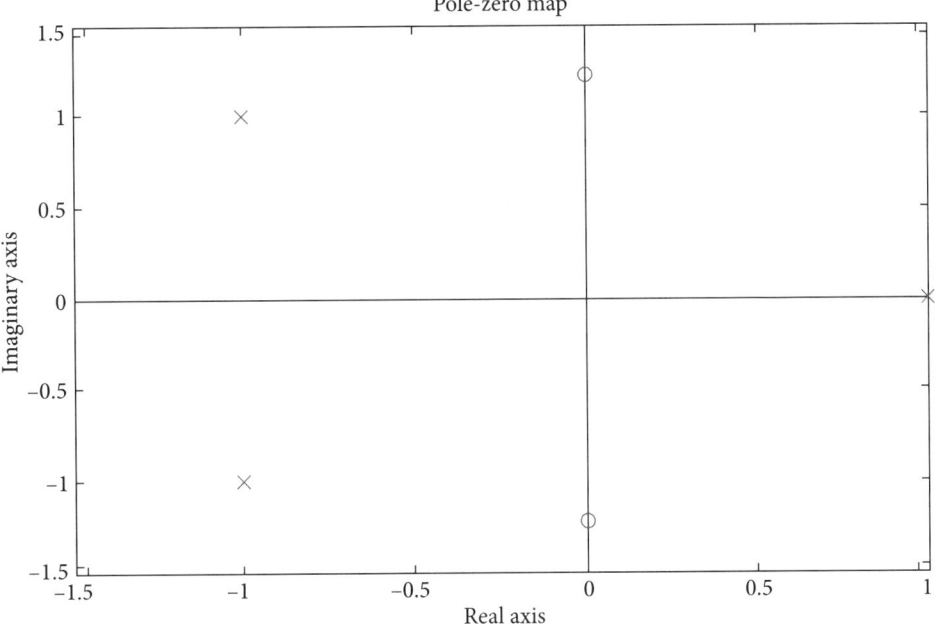

Fig. 7.49 Pole zero plot for Example 7.51

Step 1 The denominator is to be in the factored form.

$$X(s) = \frac{4s^2 + 6}{s^3 + s^2 - 2} = \frac{4s^2 + 6}{(s-1)(s^2 + 2s + 2)} \quad (7.135)$$

Step 2 We will use partial fraction expansion and decompose the function in three terms.

$$X(s) = \frac{4s^2 + 6}{(s-1)(s^2 + 2s + 2)} = \frac{k_1}{s-1} + \frac{k_2 s + k_3}{s^2 + 2s + 2} \quad (7.136)$$

Find k_1, k_2 and k_3

$$k_1 = (s-1)X(s)\Big\downarrow_{s=1} = \frac{4s^2 + 6}{(s^2 + 2s + 2)}\Big\downarrow_{s=1} = \frac{10}{5} = 2$$

Put value of k_1 in the equation and equate the numerators on both sides. We get

$$4s^2 + 6 = 2(s^2 + 2s + 2) + k_2 s^2 + (k_3 - k_2)s - k_3$$
$$4s^2 + 6 = (k_2 + 2)s^2 + (k_3 - k_2 + 4)s + (4 - k_3) \quad (7.137)$$

Equate the coefficients of s^2, s and a constant term.

$$k_2 + 2 = 4 \Rightarrow k_2 = 2$$
$$4 - k_3 = 6 \Rightarrow k_3 = -2 \tag{7.138}$$

Equation (7.135) becomes

$$X(s) = \frac{4s^2 + 6}{(s-1)(s^2 + 2s + 2)} = \frac{2}{s-1} + \frac{2s-2}{s^2 + 2s + 2} \tag{7.139}$$

$$X(s) = \frac{2}{s-1} + \frac{2(s+1)}{(s+1)^2 + 1} - \frac{4}{(s+1)^2 + 1}$$

We will now use Eqs (7.126) and (7.128) to find ILT.

$$x(t) = -2e^t u(-t) + 2e^{-t} \cos(t) u(t) - 4e^{-t} \sin(t) u(t) \tag{7.140}$$

Note that the term with pole at 1 is recovered as the left-handed signal and the term with pole at −1 is recovered as the right-handed signal, so that the common area of convergence is the same as the ROC specified.

Let us consider the case when the term has a double real pole. Here, the method for decomposition using partial fraction expansion and method to find the coefficients is different, as indicated in the following example.

Example 7.52

Find the inverse LT of $X(s) = \dfrac{4s^2 + 15s + 8}{(s-1)(s^2 + 2s + 4)}$

with ROC $-2 < \text{Re}(s) < 1$ and ROC $\text{Re}(s) > 1$

Solution

Step 1 The denominator is to be in the factored form.

$$X(s) = \frac{4s^2 + 15s + 8}{(s-1)(s^2 + 2s + 4)} = \frac{4s^2 + 15s + 8}{(s-1)(s^2 + 2s + 4)} \tag{7.141}$$

Step 2 We will use partial fraction expansion and decompose the function in three terms.

$$X(s) = \frac{4s^2 + 15s + 6}{(s-1)(s+2)^2} = \frac{k_1}{s-1} + \frac{k_2}{(s+2)^2} + \frac{k_3}{(s+2)} \tag{7.142}$$

Find k_1, k_2 and k_3

$$k_1 = (s-1)X(s) \Big\downarrow_{s=1} = \frac{4s^2 + 15s + 8}{(s+2)^2} \Big\downarrow_{s=1} = \frac{27}{9} = 3$$

$$k_2 = (s+2)^2 X(s) \Big\downarrow_{s=-2} = \frac{4s^2+15s+8}{(s-1)} \Big\downarrow_{s=-2} = \frac{-6}{-3} = 2$$

$$k_3 = \frac{d}{ds}(s+2)^2 X(s) \Big\downarrow_{s=-2} = \frac{(s-1)(8s+15)-(4s^2+15s+8)}{(s-1)^2} \quad (7.143)$$

$$= \frac{8s^2+15s-8s-15-4s^2-15s-8}{(s-1)^2} = \frac{4s^2-8s-23}{(s-1)^2} \Big\downarrow_{s=-2}$$

$$= \frac{16+16-23}{9} = 1$$

$$X(s) = \frac{4s^2+15s+6}{(s-1)(s+2)^2} = \frac{3}{s-1} + \frac{2}{(s+2)^2} + \frac{1}{(s+2)} \quad (7.144)$$

We know that

$$te^{-t}u(t) \leftrightarrow \frac{1}{(s+1)^2}, \quad te^{-2t}u(t) \leftrightarrow \frac{1}{(s+2)^2} \quad (7.145)$$

We will now find ILT by inspection. If ROC is Re(s) > 1,

$$x(t) = 3e^t u(t) + 2te^{-2t}u(t) + e^{-2t}u(t) \quad (7.146)$$

Note that we have recovered all signals as causal signals.

If ROC is −2 < Re(s) < 1, we have to recover the term with pole at 1 as a left-handed signal and terms with pole at −2 as the right-handed signals.

$$x(t) = (-3e^t u(-t) + 2te^{-2t}u(t) + e^{-2t}u(t) \quad (7.147)$$

Example 7.53

Find the inverse LT of $X(s) = \dfrac{s+2}{s^2(s+1)^2}$

with ROC −1 < Re(s) < 0 and ROC Re(s) > 0

Solution

Step 1 The denominator is already in the factored form.

Step 2 We will use partial fraction expansion and decompose the function in three terms.

$$X(s) = \frac{s+2}{s^2(s+1)^2} = \frac{k_1}{s} + \frac{k_2}{s^2} + \frac{k_3}{(s+1)} + \frac{k_4}{(s+1)^2} \quad (7.148)$$

Find k_1, k_2, k_3 and k_4

$$k_1 = \frac{d}{ds}\left[(s^2)X(s)\Big\downarrow_{s=0}\right] = \frac{d}{ds}\left(\frac{s+2}{(s+1)^2}\right)\Bigg\downarrow_{s=0}$$

$$= \frac{s^2+2s+1-(2s+2)(s+2)}{(s+1)^4} = \frac{-s^2-4s-4}{(s+1)^4}\Bigg\downarrow_{s=0} = \frac{-3}{1} = -3$$

$$k_2 = (s)^2 X(s)\Big\downarrow_{s=0} = \frac{s+2}{(s+1)^2}\Bigg\downarrow_{s=0} = \frac{2}{1} = 2$$

$$k_3 = \frac{d}{ds}(s+1)^2 X(s)\Big\downarrow_{s=-1} = \frac{d}{ds}\left(\frac{s+2}{s^2}\right)$$

(7.149)

$$= \frac{s^2-2s(s+2)}{s^4} = \frac{-s^2-4s}{s^4}\Bigg\downarrow_{s=-1} = \frac{3}{1} = 3$$

$$k_4 = (s+1)^2 X(s)\Big\downarrow_{s=-1} = \frac{s+2}{s^2}\Bigg\downarrow_{s=-1} = 1$$

$$X(s) = \frac{s+2}{(s^2)(s+1)^2} = \frac{-3}{s} + \frac{2}{(s)^2} + \frac{3}{(s+1)} + \frac{1}{(s+1)^2} \qquad (7.150)$$

We know that

$$te^{-t}u(t) \leftrightarrow \frac{1}{(s+1)^2}; \qquad (7.151)$$

We will now find ILT by inspection. If ROC is Re(s) > 0,

$$x(t) = -3u(t) + 2tu(t) + 3e^{-t}u(t) + te^{-t}u(t) \qquad (7.152)$$

Note that we have recovered all signals as causal signals.

If ROC is −1 < Re(s) < 0, we have to recover the term with pole at 0 as a left-handed signal and terms with pole at −1 as the right-handed signals.

$$x(t) = 3u(-t) - 2tu(-t) + 3e^{-t}u(t) + te^{-t}u(t) \qquad (7.153)$$

Example 7.54

Find the inverse LT of $X(s) = \dfrac{s^2+6s+7}{s^2+3s+2}$

with ROC Re(s) > −1

Solution

Step 1 Factorize the denominator. Note that the degree of numerator is same as the degree of denominator. So, we have to bring it in the fraction form so that we can apply partial fraction expansion.

$$X(s) = \frac{s^2 + 6s + 7}{s^2 + 3s + 2} = 1 + \frac{3s + 5}{(s+2)(s+1)} \qquad (7.154)$$

Step 2 We will use partial fraction expansion and decompose the function in three terms.

$$X(s) = 1 + \frac{3s + 5}{(s+2)(s+1)} = 1 + \frac{k_1}{s+2} + \frac{k_2}{s+1} \qquad (7.155)$$

Find k_1, and k_2

$$k_1 = [(s+2)X(s)\downarrow_{s=-2}] = \left(\frac{3s+5}{(s+1)}\right)\downarrow_{s=-2} = 1$$

$$k_2 = (s+1)X(s)\downarrow_{s=-1} = \frac{3s+5}{(s+2)}\downarrow_{s=-1} = 2$$

$$X(s) = 1 + \frac{1}{s+2} + \frac{2}{s+1} \qquad (7.156)$$

We will now find ILT by inspection. If ROC is $Re(s) > -1$,

$$x(t) = \delta(t) + 2e^{-t}u(t) + e^{-2t}u(t) \qquad (7.157)$$

Note that we have recovered all signals as causal signals.

Example 7.55

Find the inverse LT of $X(s) = \dfrac{s^3 + 5s^2 + 13s + 8}{s^2 + 4s + 8}$

with ROC $Re(s) > -2$

Solution

Step 1 Factorize the denominator. Note that the degree of numerator is the same as the degree of denominator. So, we have to bring it in the fraction form so that we can apply partial fraction expansion.

$$X(s) = \frac{s^3 + 5s^2 + 13s + 8}{s^2 + 4s + 8} = s + 1 + \frac{s}{s^2 + 4s + 8}$$

$$= s + 1 + \frac{s+2}{(s+2)^2 + 2^2} - \frac{2}{(s+2)^2 + 2^2} \tag{7.158}$$

We will now find ILT by inspection. If ROC is Re(s) > −2,

$$x(t) = \delta'(t) + \delta(t) + e^{-2t}\cos(2t)u(t) - e^{-2t}\sin(2t)u(t) \tag{7.159}$$

Note that we have recovered all signals as causal signals. $\delta'(t)$ stands for the derivative of $\delta(t)$. We have used the standard formula for cos and sin functions.

Example 7.56

Find the inverse LT of $X(s) = \dfrac{s^2 - s + 1}{s^2 + 2s + 1}$

with ROC Re(s) > −1

Solution

Step 1 Factorize the denominator. Note that the degree of numerator is same as degree of denominator. So, we have to bring it in the fraction form so that we can apply partial fraction expansion.

$$X(s) = \frac{s^2 - s + 1}{s^2 + 2s + 1} = 1 + \frac{-3s}{s^2 + 2s + 1}$$

$$= 1 - \frac{3(s+1)}{(s+1)^2} + \frac{3}{(s+1)^2} \tag{7.160}$$

We will now find ILT by inspection. If ROC is Re(s) > −1,

$$x(t) = \delta(t) - 3e^{-t}u(t) + 3te^{-t}u(t) \tag{7.161}$$

Note that we have recovered all signals as causal signals.

7.5.1 Transform analysis of LTI systems

LTI system is characterized by its impulse response. If we consider unilateral Laplace transform, we can use LT for analysis of the system to solve the differential equation with non-zero initial conditions. The initial conditions can be assumed to occur at $t = 0$. We will illustrate the use of LT with the help of a simple example.

Example 7.57

Use LT to find the output of the system if the system is described by the differential equation $\frac{d}{dt}y(t)+5y(t)=x(t)$ with input given by $x(t)=3e^{-2t}u(t)$ and initial condition is $y(0^+)=-2$.

Solution

Take the LT of the given differential equation.

$$\frac{d}{dt}y(t)+5y(t)=x(t) \tag{7.162}$$

LT gives

$$sY(s)-y(0^+)+5Y(s)=X(s)$$

$$sY(s)+2+5Y(s)=\frac{3}{s+2}$$

$$(s+5)Y(s)=\frac{3}{s+2}-2$$

$$Y(s)=\frac{3}{(s+2)(s+5)}-\frac{2}{(s+5)}$$

A MATLAB program to plot the response of the system to the given input with and without initial conditions is given below. The responses are shown in Figs 7.50 and 7.51.

```
clear all;
clc;
s = tf('s');
b=[1 7 10];
a=[3];
f=tf(a,b);
c=impulse(f);
plot(c);title('response to the input');xlabel('time');
ylabel('amplitude');

clear all;
clc;
s = tf('s');
b=[1 7 10];
a=[-2 -1];
f=tf(a,b);
c=impulse(f);
plot(c);title('response to the input with initial cond
ition');xlabel('time');ylabel('amplitude');
```

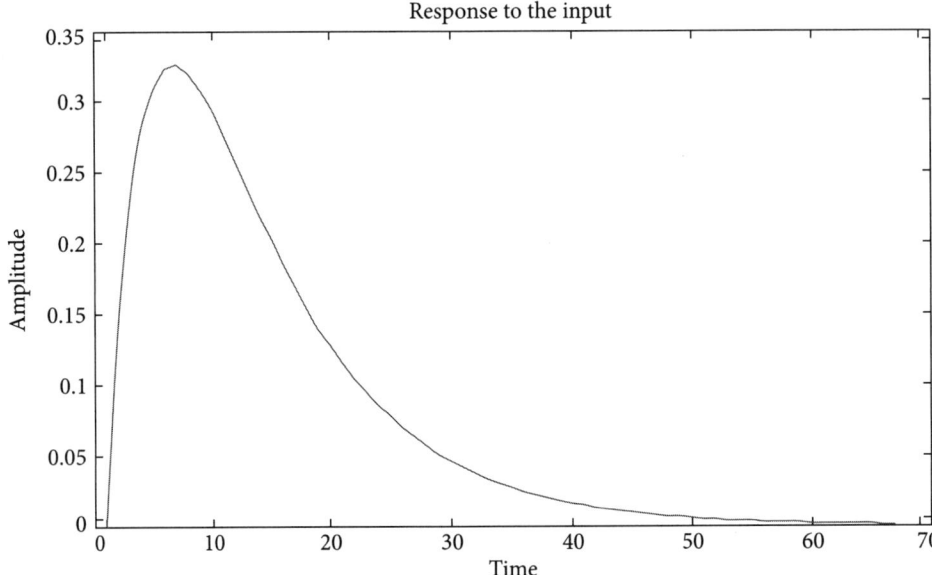

Fig. 7.50 Response of the system to the given input

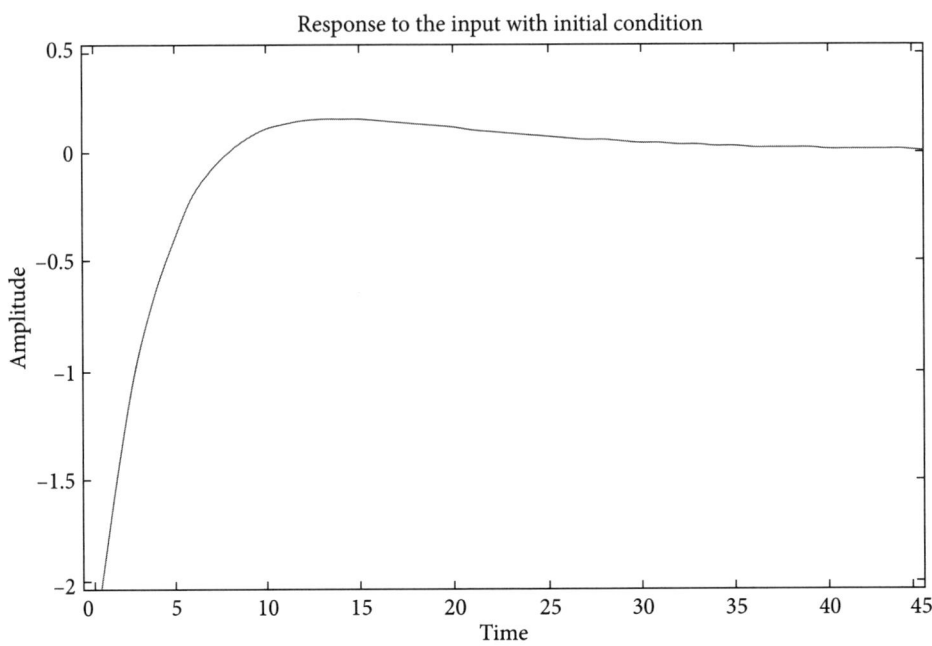

Fig. 7.51 Response of the system to given input with initial conditions

We will use partial fraction expansion.

$$Y(s) = \frac{3}{(s+2)(s+5)} - \frac{2}{(s+5)}$$

$$Y(s) = \frac{1}{s+2} - \frac{1}{s+5} - \frac{2}{s+5} \qquad (7.163)$$

$$Y(s) = \frac{1}{s+2} - \frac{3}{s+5}$$

Take ILT

$$y(t) = e^{-2t}u(t) - 3e^{-5t}u(t)$$

Example 7.58
Use LT to find the transfer function and the impulse response of the system, if the system is described by the differential equation $5\frac{d}{dt}y(t) + 10y(t) = 2x(t)$

Solution
Take the LT of the given differential equation to find the transfer function.

$$5\frac{d}{dt}y(t) + 10y(t) = 2x(t) \qquad (7.164)$$

LT gives

$$5sY(s) + 10Y(s) = 2X(s)$$

$$(5s + 10)Y(s) = 2X(s)$$

$$H(s) = \frac{Y(s)}{X(s)} = \frac{2}{5s+10}$$

Now take ILT to find the impulse response.

$$H(s) = \frac{Y(s)}{X(s)} = \frac{2}{5s+10} = \frac{2/5}{s+2} \qquad (7.165)$$

$$h(t) = \frac{2}{5}e^{-2t}u(t)$$

A MATLAB program to plot the impulse response of the system is given below. The response is shown in Fig. 7.52.

```
clear all;
clc;
s = tf('s');
b=[5 10];
a=[2];
f=tf(a,b);
c=impulse(f);
plot(c);title('impulse response');xlabel('time');ylabel('amplitude');
```

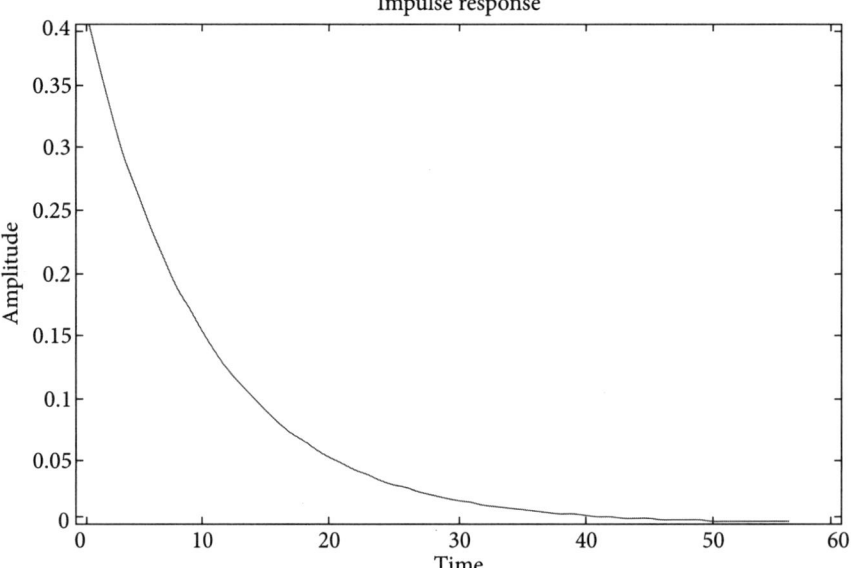

Fig. 7.52 Impulse response for Example 7.58

Example 7.59

Use LT to find the transfer function and the impulse response of the causal and stable system, if the system is described by the differential equation

$$\frac{d^2}{dt^2}y(t)+5\frac{d}{dt}y(t)+6y(t)=\frac{d^2}{dt^2}x(t)+8\frac{d}{dt}x(t)+13x(t)$$

Solution

Take LT of the given differential equation.

$$\frac{d^2}{dt^2}y(t)+5\frac{d}{dt}y(t)+6y(t)=\frac{d^2}{dt^2}x(t)+8\frac{d}{dt}x(t)+13x(t)$$

$$s^2Y(s)+5sY(s)+6Y(s)=s^2X(s)+8sX(s)+13X(s)$$

(7.166)

$$H(s)=\frac{Y(s)}{X(s)}=\frac{s^2+8s+13}{s^2+5s+6}$$

$$H(s)=\frac{s^2+8s+13}{s^2+5s+6}=1+\frac{3s+7}{(s+3)(s+2)}$$

(7.167)

Use partial fraction expansion

$$H(s)=1+\frac{2}{s+3}+\frac{1}{s+2}$$

Take ILT

$$h(t)=\delta(t)+2e^{-3t}u(t)+e^{-2t}u(t)$$

(7.168)

A MATLAB program to plot the impulse response of the system is given below. The response is shown in Fig. 7.53.

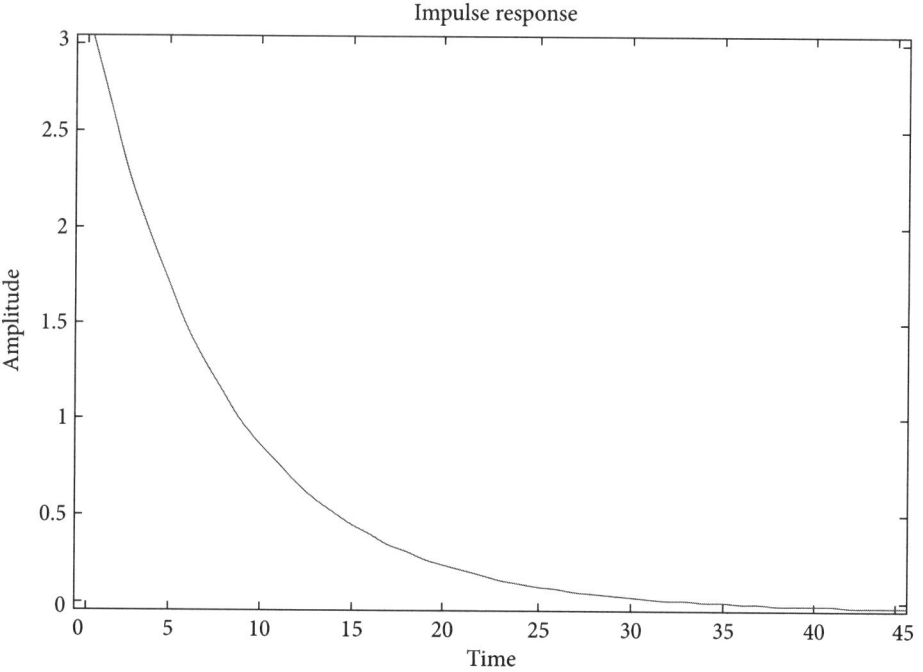

Fig. 7.53 Impulse response for Example 7.59

```
clear all;
clc;
s = tf('s');
b=[1 8 13];
a=[1 5 6];
f=tf(a,b);
c=impulse(f);
plot(c);title('impulse response');xlabel('time');ylabel('amplitude');
```

7.5.2 Total response of the system using LT

LTI system is described by the constant coefficient differential equation. The response of the system is obtained by solving the differential equation. LT converts the differential equation in time domain to the algebraic equation in Laplace domain. These algebraic equations can be solved to find the solution. The total response o the system can be expressed as

Total response = Natural response + Forced response

Natural response of the system is due to initial conditions alone. It is also termed as the zero input response i.e., when the externally applied input is zero. Forced response is due to the externally applied input signal. The forced response consists of steady state response and the transient response. The forced response is also termed as zero state response i.e., when the initial conditions are zero.

The steady state response is the response due to the poles of the input. It remains as t tends to infinity. Hence, it is called as a steady state response. The response due to poles of the system is called as the transient response as it vanishes as t tends to infinity.

Let us illustrate the concepts with the help of suitable example.

Example 7.60

Find the forced response of the system with differential equation given by $\frac{d^2}{dt^2}y(t)+5\frac{d}{dt}y(t)+6y(t)=x(t)$ to the input given by $x(t) = e^{-t}u(t)$

Solution

We will first find the transfer function of the system. Take the LT of the given differential equation.

$$\frac{d^2}{dt^2}y(t)+5\frac{d}{dt}y(t)+6y(t)=x(t)$$

$$(s^2+5s+6)Y(s)=X(s) \tag{7.169}$$

$$H(s)=\frac{Y(s)}{X(s)}=\frac{1}{(s^2+5s+6)}$$

A MATLAB program to plot the impulse response of the system is given below. The response is shown in Fig. 7.54.

```
clear all;
clc;
s = tf('s');
b=[1 5 6];
a=[1];
f=tf(a,b);
c=impulse(f);
plot(c);title('impulse response');xlabel('time');ylabe
l('amplitude');
```

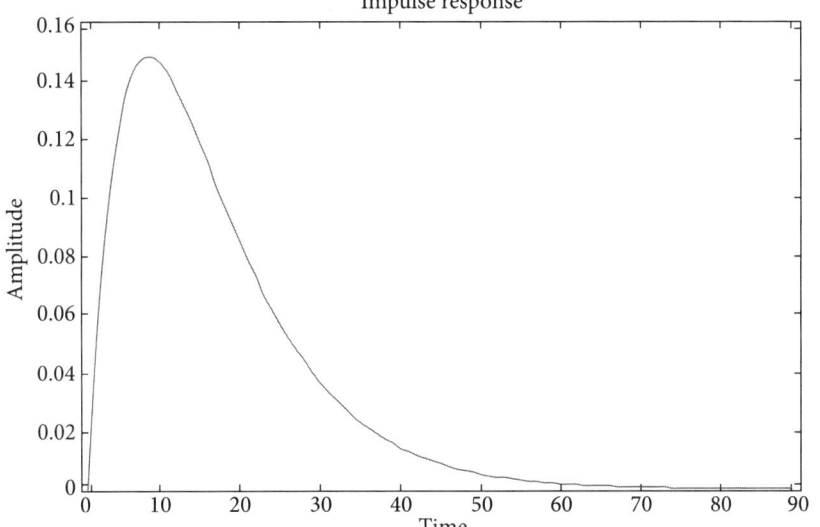

Fig. 7.54 Impulse response for Example 7.60

To find the forced response, we will apply the input $x(t) = e^{-t}u(t)$

$$\frac{Y(s)}{X(s)} = \frac{1}{(s^2 + 5s + 6)}$$

(7.170)

$$Y(s) = \frac{1}{(s+2)(s+3)} X(s) = \frac{1}{(s+2)(s+3)(s+1)}$$

We will use partial fraction expansion.

$$Y(s) = \frac{1}{(s+2)(s+3)(s+1)} = -\frac{1}{s+2} + \frac{1/2}{s+3} + \frac{1/2}{s+1}$$

(7.171)

Teaser: The reader is encouraged to verify the coefficients.

Take ILT to find the solution. The response due to input is the steady state response and response due to poles of the system is the transient response.

$$Y(s) = \frac{1/2}{s+1} + \frac{1/2}{s+3} - \frac{1}{s+2} \qquad (7.172)$$

$$y(t) = \left(\frac{1}{2}e^{-t}u(t)\right) + \left(\frac{1}{2}e^{-3t}u(t) - e^{-2t}u(t)\right)$$

Forced response = steady state response + transient response

A MATLAB program to plot the forced response of the system is given below. The response is shown in Fig. 7.55.

```
clear all;
clc;
s = tf('s');
b=[1 6 11 6];
a=[1];
f=tf(a,b);
c=impulse(f);
plot(c);title('forced response');xlabel('time');ylabel
('amplitude');
```

Example 7.61

Find the natural response, forced response and total response of the system with differential equation given by $\frac{d^2}{dt^2}y(t) + 5\frac{d}{dt}y(t) + 6y(t) = x(t)$ to the input given by $x(t) = u(t)$. The initial conditions are $y(0) = 1$ and $d/dt\,(y(0)) = 2$.

Solution

We will first find the transfer function of the system. Take the LT of the given differential equation and apply initial conditions.

$$\frac{d^2}{dt^2}y(t) + 5\frac{d}{dt}y(t) + 6y(t) - \frac{d}{dt}y(0) - y(0) = x(t)$$

$$(s^2Y(s) - sy(0) - \frac{d}{dt}y(o) + 5(sY(s) - y(0) + 6) = X(s) \qquad (7.173)$$

$$s^2Y(s) - (s+5)y(0) - \frac{d}{dt}y(0) + 5sY(s) + 6Y(s) = X(s)$$

$$Y(s)(s^2+5s+6) = X(s)+(s+5)\times 1+2$$

$$Y(s) = \frac{s+7}{s^2+5s+6} + \frac{X(s)}{s^2+5s+6}$$

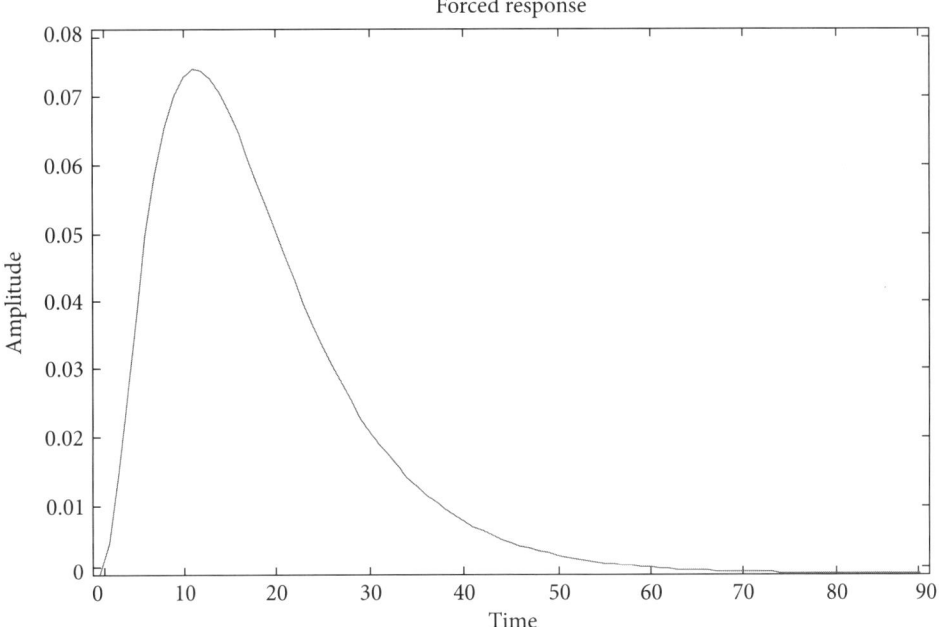

Fig. 7.55 Forced response of the system for Example 7.60

The response due to the first term is due to the initial conditions and hence is the natural response.

$$Y(s) = \frac{s+7}{s^2+5s+6}$$

$$Y(s) = \frac{5}{s+2} - \frac{4}{s+3} \tag{7.174}$$

$$y(t) = 5e^{-2t}u(t) - 4e^{-3t}u(t)$$

A MATLAB program to plot the natural response is given below. The response is plotted in Fig. 7.56.

```
clear all;
clc;
s = tf('s');
```

```
b=[1 5 6];
a=[1 7];
f=tf(a,b);
b=impulse(f);
plot(b);title('natural response');xlabel('time');ylabe
l('amplitude');
```

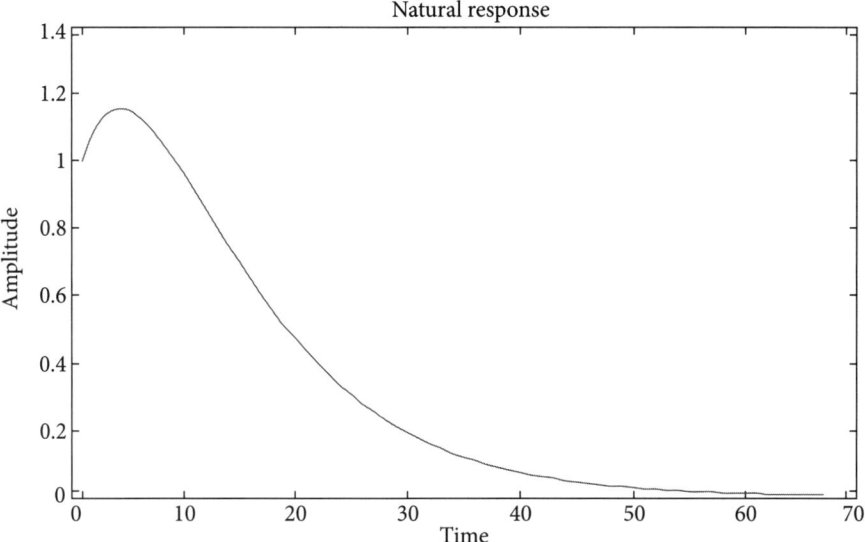

Fig. 7.56 Plot of natural response for Example 7.61

The response due to the second term is the forced response. To find the forced response, we will apply the input $x(t) = e^{-t} u(t)$

$$\frac{Y(s)}{X(s)} = \frac{1}{(s^2 + 5s + 6)} \tag{7.175}$$

$$Y(s) = \frac{1}{(s+2)(s+3)} X(s) = \frac{1}{s(s+2)(s+3)}$$

We will use partial fraction expansion.

$$Y(s) = \frac{1}{s(s+2)(s+3)} = \frac{1/6}{s} - \frac{1/2}{s+2} + \frac{1/3}{s+3} \tag{7.176}$$

Teaser The reader is encouraged to verify the coefficients.

Take ILT to find the solution. The forced response is the sum of steady state response and the transient response. The response due to input is the steady state response and the response due to poles of the system is the transient response.

$$Y(s) = \frac{1/6}{s} + \frac{1/3}{s+3} - \frac{1/2}{s+2}$$

$$y(t) = \left(\frac{1}{6}u(t)\right) + \left(\frac{1}{3}e^{-3t}u(t) - \frac{1}{2}e^{-2t}u(t)\right)$$

(7.177)

A MATLAB program to plot the forced response is given below. The response is plotted in Fig. 7.57.

```
clear all;
clc;
s = tf('s');
b=[1 5 6 0];
a=[1];
f=tf(a,b);
b=impulse(f);
plot(b);title('forced
response');xlabel('time');ylabel('amplitude');
```

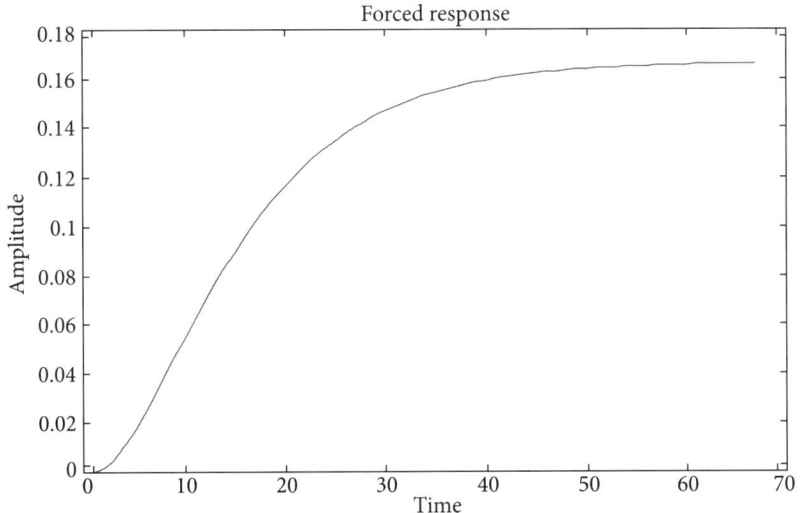

Fig. 7.57 Plot of forced response for Example 7.61

The total response is given by the addition of natural response and the forced response.

$$y(t) = 5e^{-2t}u(t) - 4e^{-3t}u(t) + \left(\frac{1}{6}u(t)\right) + \left(\frac{1}{3}e^{-3t}u(t) - \frac{1}{2}e^{-2t}u(t)\right)$$

$$y(t) = \frac{1}{6}u(t) + \frac{9}{2}e^{-2t}u(t) - \frac{11}{3}e^{-3t}u(t)$$

(7.178)

A MATLAB program to plot the total response is given below. Figure 7.58 shows the plot of the total response.

```
clear all;
clc;
s = tf('s');
b1=[1 5 6];
a1=[1 7];
f1=tf(a1,b1);
c1=impulse(f1);
b2=[1 5 6 0];
a2=[1];
f2=tf(a2,b2);
c2=impulse(f2);
plot(c1+c2);title('total response');xlabel('time');ylabel('amplitude');
```

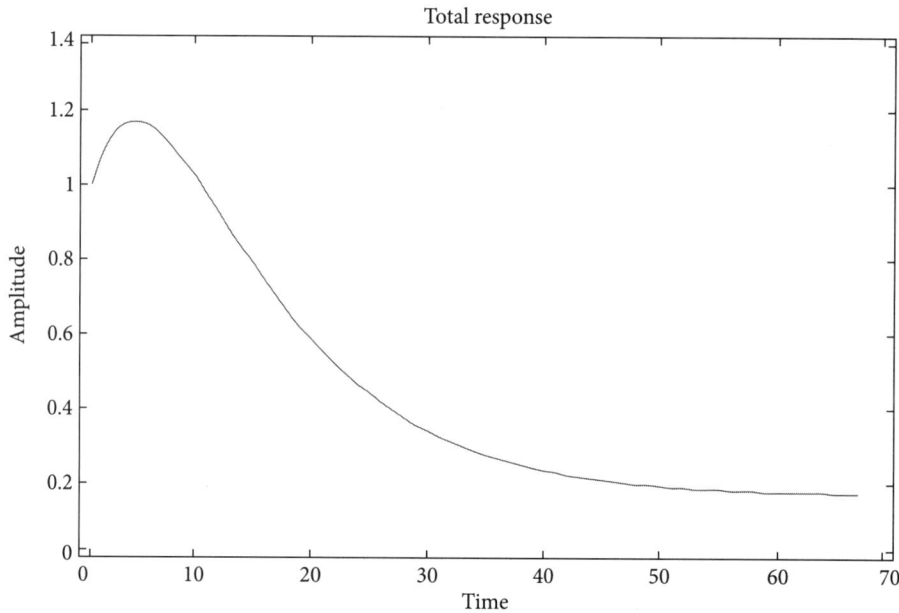

Fig. 7.58 Plot of total response for Example 7.61

7.5.3 Stability considerations in S domain

The system transfer function in S domain determines the location of poles. The pole locations determine the stability of the system. If the poles for a causal system are on the left-hand side of an imaginary axis, then the system is stable. If the poles for a causal system are on the right-hand side of the imaginary axis, the system is unstable. The frequency response of the system can be found by putting $s = j\omega$ in the equation of the transfer function. The imaginary axis must be in the ROC. In case of the non-causal system the reverse is true. If the poles are on right-hand side of the imaginary axis, the system is stable and vice versa.

Let us consider one simple example to illustrate the concepts.

Example 7.62

Find LT of the right-handed signal $x(t) = e^{-at}u(t)$ with $a > 0$, $x(t) = e^{at}u(t)$ with $a > 0$ and a left-handed signal $x(t) = e^{-at}u(-t)$ with $a > 0$, $x(t) = e^{at}u(-t)$ with $a > 0$

We will use the definition of LT.

$$X(S) = \int_{-\infty}^{\infty} e^{-at}u(t)e^{-st}dt = \int_{0}^{\infty} e^{-at}e^{-st}dt = -\frac{e^{-(s+a)t}}{s+a}\Big\downarrow_{0}^{\infty}$$
$$= \frac{1}{s+a} \text{ for all } s+a > 0 \text{ i.e., Re}(s) > -a$$
(7.179)

ROC is the right half S plane with $\text{Re}(s) > -a$

Note: if $\text{Re}(a) = \sigma > -a, (\sigma + a) > 0, e^{-(s+a)\infty} \to 0$

if $\text{Re}(a) = \sigma < -a, (\sigma + a) < 0, e^{-(s+a)\infty} \to \infty$

ROC and pole are plotted in Fig. 7.59. ROC includes the imaginary axis. The system is stable.

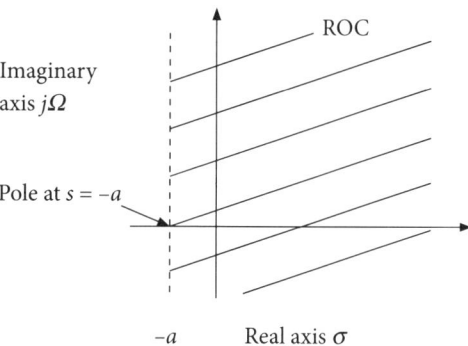

Fig. 7.59 ROC of $x(t) = e^{-at}u(t)$

We will use the definition of LT.

$$X(S) = \int_{-\infty}^{\infty} e^{at}u(t)e^{-st}dt = \int_{0}^{\infty} e^{at}e^{-st}dt = -\frac{e^{-(s-a)t}}{s-a}\Big\downarrow_{0}^{\infty}$$
$$= \frac{1}{s-a} \text{ for all } s+a > 0 \text{ i.e., Re}(s) > a$$
(7.180)

ROC is the right half S plane with $\text{Re}(s) > a$

ROC and pole are plotted in Fig. 7.58. ROC includes the imaginary axis. The system is stable.

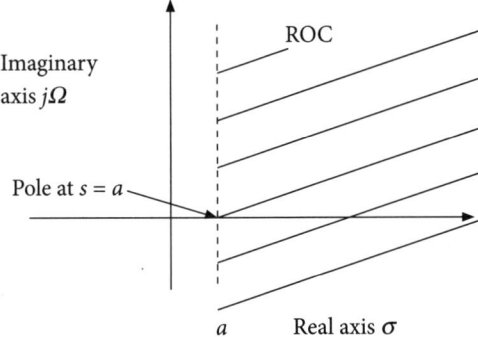

Fig. 7.60 ROC of $x(t) = e^{at}u(t)$

We will use the definition of LT to find the LT of $x(t) = e^{-at}u(-t)$.

$$X(S) = \int_{-\infty}^{\infty} e^{-at}u(-t)e^{-st}dt = \int_{-\infty}^{0} e^{-at}e^{-st}dt = \frac{e^{-(s+a)t}}{s+a}\Big\downarrow_{-\infty}^{0}$$

$$= -\frac{1}{s+a} \text{ for all } \operatorname{Re}(s) < -a \tag{7.181}$$

ROC is the left half S plane with $\operatorname{Re}(s) < -a$

Note When $s > -a$, the integral $\to \infty$

ROC for $x(t) = e^{-at}u(-t)$ and pole are plotted in Fig. 7.59. The pole is on the left-hand side of imaginary axis and ROC does not include an imaginary axis. The system is unstable.

We will use the definition of LT.

$$X(S) = \int_{-\infty}^{\infty} e^{at}u(-t)e^{-st}dt = \int_{-\infty}^{0} e^{at}e^{-st}dt = \frac{e^{-(s-a)t}}{s-a}\Big\downarrow_{-\infty}^{0}$$

$$= -\frac{1}{s-a} \text{ for all } \operatorname{Re}(s) < a \tag{7.182}$$

ROC is the left half S plane with $\operatorname{Re}(s) < a$

ROC and pole are plotted in Fig. 7.60. ROC includes the imaginary axis. The system is stable.

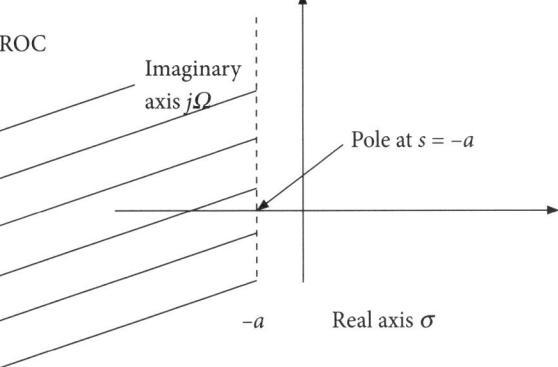

Fig. 7.61 ROC of $x(t) = e^{-at}u(-t)$

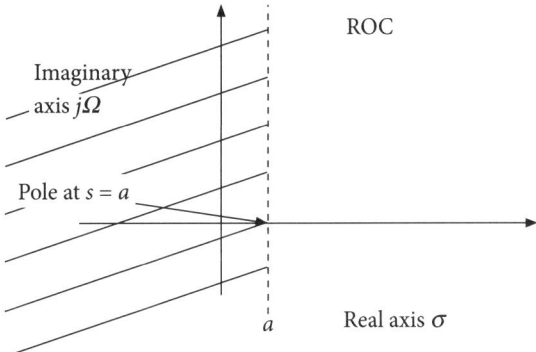

Fig. 7.62 ROC of $x(t) = e^{at}u(-t)$

A MATLAB program to plot all above signals is given below. The signals are plotted in Fig. 7.63.

```
clear all;
clc;
a=2;
t=1:0.1:10;
x=exp(-a*t);
subplot(4,1,1);plot(t,x);
title('plot of stable and causal signal');xlabel('time');
ylabel('amplitude');
t1=0:0.1:10;
x1=exp(a*t1);
subplot(4,1,2);plot(t1,x1);
title('plot of unstable and causal signal');xlabel('time');
ylabel('amplitude');
t2=-10:0.1:0;
```

```
x2=exp(-a*t2);
subplot(4,1,3);
plot(t2,x2);
title('plot of unstable non causal
signal');xlabel('time');ylabel('amplitude');
t3=-10:0.1:0;
x3=exp(a*t3);
subplot(4,1,4);
plot(t3,x3);
title('plot of stable non causal
signal');xlabel('time');ylabel('amplitude');
```

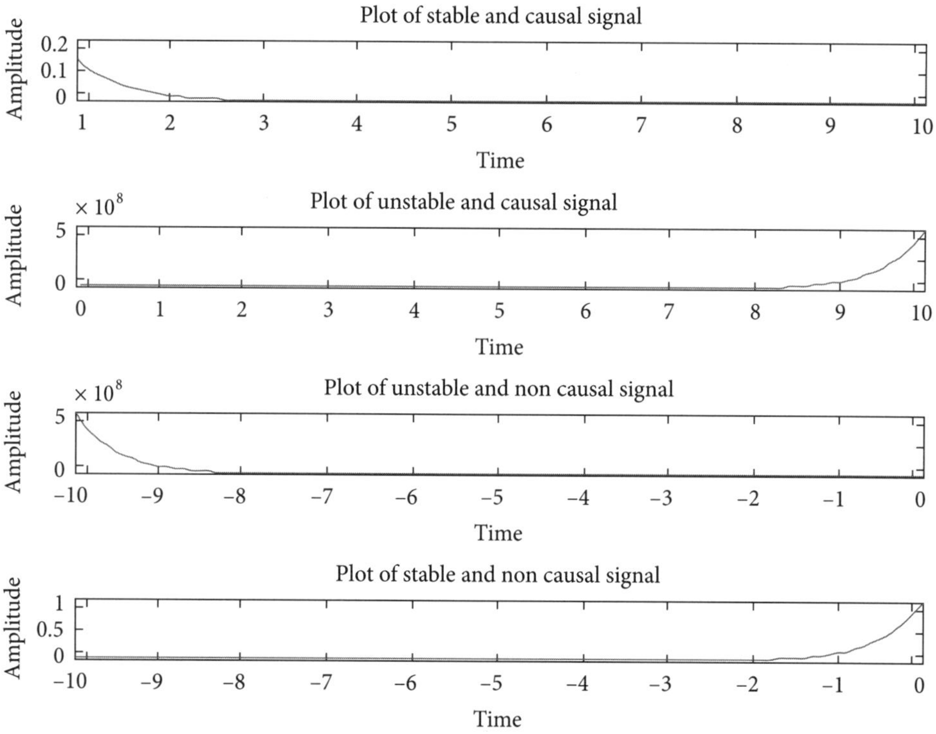

Fig. 7.63 Plot of all 4 signals

Concept Check

- What is the method to find ILT of $X(s)$?
- When will you recover all the signals as right-handed signals?
- How will you decide whether to recover the signal as a left-handed or a right-handed signal?
- Which formula will you use to find ILT in case there are complex conjugate roots?

- How will you find the impulse response of the system, given its transfer function?
- How will you find the transfer function of the system, given its differential equation?
- What is a natural response of the system?
- What is the forced response?
- What is the transient response and what is a steady state response?
- What is the zero state response?
- What is the zero input response?
- Can you determine the pole locations for a causal system in S domain?
- Which command in MATLAB will be used for plotting poles and zeros of the transfer function?
- Which command in MATLAB will be used for plotting the impulse response of the system?
- What is the criterion for a causal system to be stable?
- What is the criterion for a non-causal system to be stable?

Summary

In this chapter, we have described and explained the properties of Laplace Transform (LT).

- We started with the complex numbers and explained the significance of using a complex plane for LT. The S plane is defined and the physical significance of LT is explained. Unilateral and bilateral LTs are defined. It is stated that inverse LT involves calculation of contour integration and hence is found by inspection. Properties of LT are stated. Continuous time complex exponentials e^{-st} are eigen functions of LTI systems. Convolution of two time domain signals results in multiplication of their LTs. LT of the impulse response of the system is called as transfer function of the system in S domain and LT converts a time domain differential equation into a simple algebraic equation in S. ROC is defined as the range of values of σ for which $x(t)e^{-\sigma t}$ is absolutely integrable. Relation between LT and FT is discussed. LT also exists for unstable systems. One numerical example is used to explain the convergence of LT for unstable system. FT and LT are compared.
- LTs of some standard functions like delta function, unit step, ramp, exponential function, complex exponential function, damped sinusoid are evaluated. The ROCs for each are shown diagrammatically and explained. The properties of LT such as linearity, time shifting, time reversal, time scaling are stated and proved. The properties such as frequency sifting, time differentiation, time integration, frequency differentiation, conjugate symmetry, Parseval's

theorem, initial and final value theorem are also stated and explained. The periodicity property is proved. It is useful for finding the LT of some standard periodic signals. The physical significance of all these properties is explained.

- Some examples on LT are solved for the benefit of the reader. The evaluation of ROC for each example is illustrated. LTs for some standard signals are evaluated. LTs for some standard aperiodic signals are solved. We have illustrated the use of periodicity property to find the LT of any periodic signal given its waveform. Properties of ROC are then discussed. The nature of ROC for right-handed, left-handed and both-sided signals are given. ROC is always a strip between two poles. It is emphasized that ROC can not contain any pole. The locations of the poles in S domain decide the ROC. The formula for inverse LT uses a contour integration which is little difficult to solve.

- The ILT evaluation by inspection is illustrated using several solved examples. To find ILT, we need to know the nature of ROC. The examples illustrate the calculation of ILT for different ROC specifications for same $X(s)$. If the poles are complex, the ILT calculation by inspection is a bit tricky. ILT evaluation for complex conjugate roots is also explained with simple examples. MATLAB programs are given for plotting the poles and zeros of the transfer function and for plotting the response of the system. The analysis of LTI systems is explained using LT. The terms such as impulse response, natural response, steady state response, transient response and forced response are explained. Simple examples are solved to illustrate the concepts. The zero state response and zero input response are also described.

Multiple Choice Questions

1. LT is the operator that transforms the signal in time domain into a signal
   (a) in a complex frequency domain called as S domain
   (b) in a real frequency domain called as S domain
   (c) in an imaginary frequency domain called as S domain
   (d) in a frequency domain called as S domain

2. LT is defined as
   (a) $X(s) = \int_{-\infty}^{\infty} x(t) e^{-st} ds$
   (b) $X(s) = \int_{-\infty}^{\infty} x(t) e^{-st} dt$
   (c) $X(s) = \int_{0}^{\infty} x(t) e^{st} dt$
   (d) $X(s) = \int_{-\infty}^{s} x(t) e^{-st} dt$

3. Inverse LT requires contour integration defined by
   (a) $x(t) = \dfrac{1}{2\pi j} \int_{-j\infty}^{j\infty} X(s) e^{st} ds$
   (b) $x(t) = \dfrac{1}{2j} \int_{\sigma-j\infty}^{\sigma+j\infty} X(s) e^{st} ds$
   (c) $x(t) = \dfrac{1}{2\pi} \int_{-j\infty}^{j\infty} X(s) e^{st} ds$
   (d) $x(t) = \dfrac{1}{2\pi j} \int_{\sigma-j\infty}^{\sigma+j\infty} X(s) e^{st} ds$

4. The complex exponential defined by the e^{st} is
   (a) An eigen function of LTI system
   (b) An eigen value of LTI system
   (c) A characteristic function of LTI system
   (d) An eigen function of linear system
5. Convolution of two time domain signals results in
   (a) Addition of their LTs
   (b) Convolution of their LTs
   (c) Multiplication of their LTs
   (d) Complex multiplication and addition of their LTs
6. LT of the impulse response of the system is called as
   (a) transfer function of the signal in S domain
   (b) transfer function of the system in S domain
   (c) transfer function of the system in frequency domain
   (d) transfer function of the system in σ domain
7. LT converts a time domain differential equation
   (a) into a simple algebraic equation in ω
   (b) into a simple algebraic equation in σ
   (c) into a simple algebraic equation in S
   (d) into a simple quadratic equation in S
8. LT of $x(t)$ is a FT of $x(t)e^{-\sigma t}$
   (a) FT of $x(t)e^{-\sigma t}$
   (b) FT of $x(t)e^{-j\omega t}$
   (c) IFT of $x(t)e^{-\sigma t}$
   (d) FT of $x(t)e^{-(\sigma+j\omega)t}$
9. LT can be converted to FT
   (a) by substituting $s = j\omega$
   (b) by substituting $s = \sigma$
   (c) by substituting $s = \sigma + j\omega$
   (d) by substituting $s = -j\omega$
10. The region of convergence (ROC) is defined as
    (a) the range of values of $j\omega$ for which LT converges
    (b) the range of values of σ for which LT converges
    (c) the range values of $\sigma + j\omega$ for which LT converges
    (d) the range of values of s for which LT converges
11. LT and ROC of $x(t) = e^t u(t)$ is
    (a) $X(S) = \dfrac{1}{(1-\sigma)+j\omega}, \ \sigma > 1$
    (b) $X(S) = \dfrac{1}{(1-\sigma)-j\omega}, \ \sigma < 1$
    (c) $X(S) = \dfrac{1}{(1-\sigma)-j\omega}, \ \sigma > 1$
    (d) $X(S) = \dfrac{1}{(1+\sigma)-j\omega}, \ \sigma > 1$

12. FT of $x(t) = e^t u(t)$
    (a) does exist and LT exists for $\sigma > 1$
    (b) does not exist but LT exists for $\sigma > 1$
    (c) does not exist but LT exists for $\sigma < 1$
    (d) does not exist and LT also does not exist

13. LT of a delta function is equal to
    (a) one
    (b) three
    (c) two
    (d) zero

14. LT of the unit step function is equal to
    (a) $\dfrac{1}{s^2}$
    (b) $\dfrac{1}{s^3}$
    (c) $\dfrac{1}{s}$
    (d) s

15. LT and ROC of the signal $x(t) = tu(t)$ are
    (a) $\dfrac{1}{s}$, ROC is right-hand S plane with Re(s) > 0
    (b) $\dfrac{1}{s^2}$, ROC is entire S plane
    (c) $\dfrac{1}{s^2}$, ROC is left-hand S plane with Re(s) < 0
    (d) $\dfrac{1}{s^2}$, ROC is right-hand S plane with Re(s) > 0

16. LT and ROC of the signal $x(t) = e^{at}u(t)$ are
    (a) $\dfrac{1}{s-a}$, ROC is right-hand S plane with Re(s) > a
    (b) $\dfrac{1}{s-a}$, ROC is left-hand S plane with Re(s) < a
    (c) $\dfrac{1}{s+a}$, ROC is right-hand S plane with Re(s) > a
    (d) $\dfrac{1}{s+a}$, ROC is left-hand S plane with Re(s) < a

17. LT and ROC of the signal $x(t) = e^{-j\omega t}u(t)$ are
    (a) $\dfrac{1}{s-j\omega}$, ROC is right-hand S plane with Re(s) > 0
    (b) $\dfrac{1}{s+j\omega}$, ROC is right-hand S plane with Re(s) > 0

(c) $\dfrac{1}{s-j\omega}$, ROC is right-hand S plane with Re(s) < a

(d) $\dfrac{1}{s+j\omega}$, ROC is right-hand S plane with Re(s) < a

18. Following property of LT is useful for calculation of ILT using PFE
    (a) additivity
    (b) linearity
    (c) time shifting
    (d) time scaling

19. Time shifting property states that
    (a) $x(t) \leftrightarrow X(s)$
    then $x(t-t_0) \leftrightarrow e^{-st_0} X(s)$
    (b) $x(t) \leftrightarrow X(s)$
    then $x(t-t_0) \leftrightarrow e^{st_0} X(s)$
    (c) $x(t) \leftrightarrow X(s)$
    then $x(t-t_0) \leftrightarrow t_0 e^{-st_0} X(s)$
    (d) $x(t) \leftrightarrow X(s)$
    then $x(t-t_0) \leftrightarrow se^{st_0} X(s)$

20. Time scaling of a signal by a factor of 'a' introduces
    (a) scaling by a factor of a in S domain
    (b) scaling by a factor of a^2 in S domain
    (c) scaling by a factor of $1/a$ in S domain
    (d) scaling by a factor of $1/a^2$ in S domain

21. Property of S domain shift says that
    (a) $x(t) \leftrightarrow X(s)$
    then $e^{-s_0 t} x(t) \leftrightarrow X(s-s_0)$
    (b) $x(t) \leftrightarrow X(s)$
    then $e^{s_0 t} x(t) \leftrightarrow X(s+s_0)$
    (c) $x(t) \leftrightarrow X(s)$
    then $e^{s_0 t} x(t) \leftrightarrow sX(s-s_0)$
    (d) $x(t) \leftrightarrow X(s)$
    then $e^{s_0 t} x(t) \leftrightarrow X(s-s_0)$

22. Differentiation in S domain results in multiplication of the signal by
    (a) t
    (b) $-t$
    (c) $-t/2$
    (d) $t/2$

23. Multiplication by a complex exponential in time domain introduces
    (a) a shift in complex frequency s
    (b) a shift in σ
    (c) a shift in $j\omega$
    (d) a multiplication by a complex frequency

24. The final value of the signal as t tending to infinity can be found
    (a) by taking the limit of $X(s)$ as s tends to zero
    (b) by taking the limit of $sX(s)$ as s tends to zero
    (c) by taking the limit of $sX(s)$ as s tends to infinity
    (d) by taking the limit of $X(s)$ as s tends to infinity

25. The initial value of the signal as t tending to zero can be found
    (a) by taking the limit of $X(s)$ as s tends to zero
    (b) by taking the limit of $sX(s)$ as s tends to zero
    (c) by taking the limit of $sX(s)$ as s tends to infinity
    (d) by taking the limit of $X(s)$ as s tends to infinity

26. We use following property of LT to prove LT of periodic signals
    (a) periodicity
    (b) time shifting
    (c) time scaling
    (d) time reversal

27. Inverse LT is found by inspection. For decomposition of the function, we use
    (a) partial function
    (b) partial fraction expansion
    (c) partial function decomposition
    (d) fraction decomposition

28. The signal is a unilateral signal. The ILT of $X(s) = \dfrac{1}{s+a}$ is
    (a) $x(t) = e^{-at}u(t)$
    (b) $x(t) = e^{at}u(t)$
    (c) $x(t) = te^{-at}u(t)$
    (d) $x(t) = e^{-at}u(-t)$

29. The signal is a unilateral signal. The ILT of $X(s) = \dfrac{s+a}{(s+a)^2 + \omega_0^2}$ is
    (a) $x(t) = \sin(\omega_0 t)u(t)$
    (b) $x(t) = \cos(\omega_0 t)u(t)$
    (c) $x(t) = e^{-at}\sin(\omega_0 t)u(t)$
    (d) $x(t) = e^{-at}\cos(\omega_0 t)u(t)$

30. The following MATLAB command is used to plot poles and zeros of the transfer function.
    (a) zpmap
    (b) pzmap
    (c) pzplot
    (d) zpplot

31. The natural response of the system is due to
    (a) external input
    (b) initial conditions alone
    (c) zero state
    (d) poles and zeros of transfer function

32. The steady state response of the system is due to
    (a) external input
    (b) initial conditions alone
    (c) zero state
    (d) poles and zeros of transfer function

33. The transient response of the system is due to
    (a) external input
    (b) initial conditions alone

(c) zero state

(d) poles and zeros of transfer function

34. The zero input response of the system is also called as

    (a) Transient response

    (b) Natural response

    (c) Forced response

    (d) Total response

35. Total response of the system is equal to

    (a) Natural response + Transient response

    (b) Natural response + steady state response

    (c) Forced response + natural response

    (d) Transient response + steady state response

36. LTI causal system is stable if

    (a) its poles are on left-hand side of imaginary axis and ROC includes the imaginary axis

    (b) its poles are on right-hand side of imaginary axis and ROC includes the imaginary axis

    (c) its poles are on left-hand side of imaginary axis and ROC does not include the imaginary axis

    (d) its poles are on right-hand side of imaginary axis and ROC does not include the imaginary axis

Review Questions

7.1 Define a complex frequency plane and define the LT of any signal $x(t)$.

7.2 What is the physical significance of LT? State properties of LT.

7.3 Can we find the LT of a signal that is not absolutely integrable? Explain with suitable example.

7.4 Explain the relation between FT and LT.

7.5 Compare LT and FT.

7.6 Prove that the complex exponential e^{st} is an eigen function of LTI system.

7.7 Define region of convergence for LT. What is the significance of ROC?

7.8 State and prove the property of linearity for LT.

7.9 How will you use the property of time shifting to find the LT of periodic signals?

7.10 Prove time shifting, tie scaling and time reversal properties of LT.

7.11 State the S domain shift property and state its significance.

7.12 Prove the property of frequency shifting.

7.13 State the physical significance of the convolution property and prove it.

7.14 Why is the frequency convolution property called as modulation property?

7.15 What is the meaning of Parseval's theorem?

7.16 How will you find the initial and final value of the signal without taking ILT? State the initial and final value theorems.

7.17 Explain the partial fraction expansion method for finding ILT.

7.18 How will you find ILT if there are complex conjugate roots?

7.19 Can you analyze the system using LT?

7.20 Explain the procedure to find the impulse response of the system given its differential equation.

7.21 Explain the use of transfer function of the system to find the impulse response.

7.22 Explain the procedure to find the total response of the system. Clearly state the meaning of natural response and forced response.

7.23 How will you find the zero input response and the zero state response?

7.24 Can you analyze the stability of the causal and non-causal LTI system given the pole locations?

7.25 How will you find the frequency response of the system from the transfer function?

7.26 Explain the use of MATLAB commands to plot the poles and zeros and to plot the impulse response of the system.

Problems

7.1 Find the LT of the function $x(t) = t\delta(t)$.

7.2 Find the LT of the function $x(t) = (t-2)u(t)$.

7.3 Find the LT of the function $x(t) = e^{j3t}u(t)$ and $x(t) = e^{-j3t}u(t)$.

7.4 Find the LT of the sine and cosine function $x(t) = \cos(4\pi t)u(t)$ and $x(t) = \sin(4\pi t)u(t)$.

7.5 Find the LT of the damped sine and cosine function $x(t) = e^{-2t}\sin(2t)u(t)$ and $x(t) = e^{-2t}\cos(2t)u(t)$.

7.6 Find the LT of the both-sided signal $x(t) = e^{-5t}u(t) + e^{-2t}u(-t)$, find ROC.

7.7 Find the LT of the signal $x(t) = u(t-7)$, find ROC.

7.8 Find the LT of the signal $x(t) = e^{-5t}u(t-5)$, find ROC.

7.9 Find the LT of $x(t) = \delta(t+3)$ and find ROC.

7.10 Find the LT of $x(t) = \sin(3t)u(t-5)$ and find ROC.

7.11 Find the LT of $x(t) = e^{-7|t|}$ and find ROC.

7.12 Find the LT of $x(t) = e^{-3t}\cos(7t)u(t)$ and find ROCs.

7.13 Find the LT of $x(t) = e^{-3t}u(t) + e^{8t}u(t)$ and find ROC.

7.14 Find the LT of $x(t) = e^{-2t}u(-t) + e^{5t}u(-t)$ and find ROC.

7.15 Find the LT of $x(t) = e^{2t}\sin(3t)u(t)$ and find ROC.

7.16 Find the LT of $x(t) = \cos^2(6t)u(t)$ and find ROC.

7.17 Find the LT of $x(t) = (1 + \sin 3t \cos 3t)u(t)$ and find ROC.

7.18 Find the LT of the signal drawn in Fig. 7.26. Find ROC.

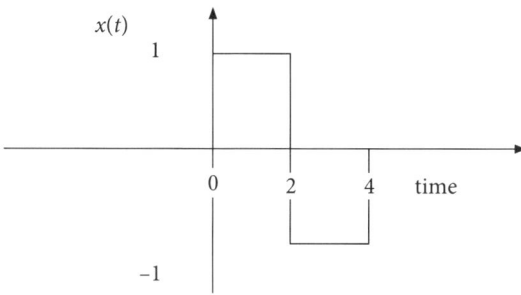

7.19 Find the LT of the signal drawn in Fig. 7.28. Find ROC.

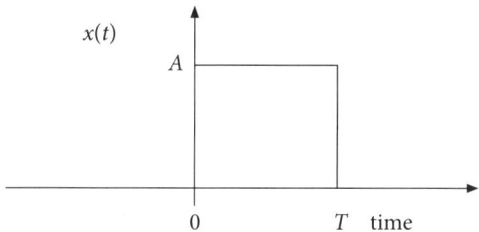

7.20 Find the LT of the signal drawn below. Find ROC.

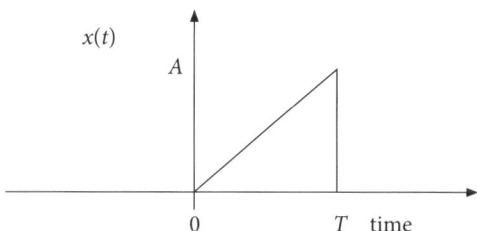

7.21 Find the LT of the signal drawn in Fig. 7.32 below. Find ROC.

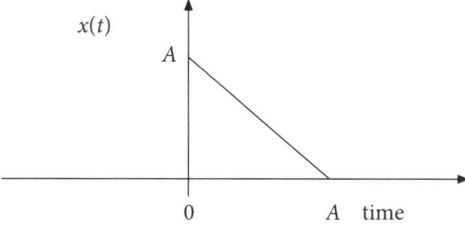

7.22 Find the LT of the signal $x(t) = \sin t, 0 < t \leq 1$ drawn in Fig. 7.36 below. Find ROC.
$\qquad = 0$ elsewhere

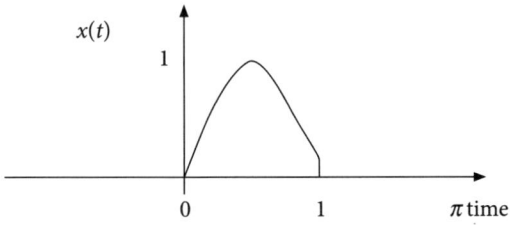

7.23 Find the LT of the signal drawn in Fig. 7.38 below.

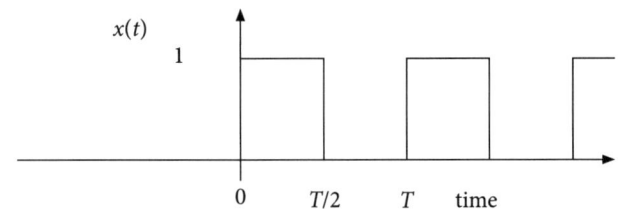

7.24 Find the LT of the signal drawn in Fig. 7.40 below.

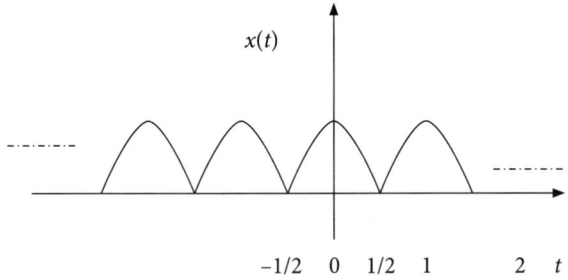

7.25 Find the LT of the signal drawn in Fig. 7.42 below.

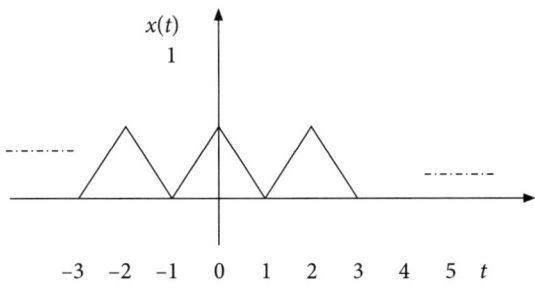

7.26 Find the LT of the following signals using properties of LT.

1. $x(t) = t^2 e^{-2t} u(t)$
2. $x(t) = t e^{-3t} \sin(t) u(t)$
3. $x(t) = e^{4t} u(-t)$
4. $x(t) = e^{-5t} \cos(t) u(t)$

7.27 Find the LT of the following signals using properties of LT.

1. $x(t) = (t^2 - 3t) u(t - 2)$
2. $x(t) = (t - 4) u(t - 4)$
3. $x(t) = (2t u(t) - 3(t - 5) u(t - 5))$

7.28 LTs of some signals are given below. Find the initial and final value of the signal using initial and final value theorem.

1. $X(s) = \dfrac{1}{s(s-5)}$

2. $X(s) = \dfrac{s+2}{s^2+2s+1}$

3. $X(s) = \dfrac{3s+5}{s(2s+3)}$

4. $X(s) = \dfrac{s+2}{s^2(s+3)}$

7.29 The signal is given by $x(t) = e^{-(t-7)/3} u(t-7)$. Find the LT.

7.30 The signal is given by $x(t) = \sin(2t)\cos(3t)u(t)$. Find the LT.

7.31 Find the LT of $x(t) = e^{3t} u(t) * t^2 u(t)$.

7.32 Find the LT of $x(t) = 3e^{3t} u(t)$ using property of differentiation in time domain.

7.33 Find the inverse LT of $X(s) = \dfrac{1}{(s^2+s+1)(s+2)}$ with ROC given by $\text{Re}(s) > -0.5$. Plot poles and zeros.

7.34 Find the inverse LT of $X(s) = \dfrac{2}{s(s+1)(s+2)}$ with ROC given by $-1 < \text{Re}(s) < 0$. and $-2 < \text{Re}(s) < -1$

7.35 Find the inverse LT of $X(s) = \dfrac{3s^2+8s+23}{(s+3)(s^2+2s+10)}$.

7.36 Find the inverse LT of $X(s) = \dfrac{4}{(s+2)(s+4)}$ with ROC $-4 < \text{Re}(s) < -2$

7.37 Find the inverse LT of $X(s) = \dfrac{s+2}{s^2(s+1)}$ with ROC $-1 < \text{Re}(s) < 0$ and ROC $\text{Re}(s) > 0$

7.38 Find the inverse LT of $X(s) = \dfrac{s^2+5s+5}{s^2+3s+2}$ with ROC $\text{Re}(s) > -1$

7.39 Find the inverse LT of $X(s) = \dfrac{s^3+5s^2+13s+9}{s^2+4s+8}$ with ROC $\text{Re}(s) > -2$

7.40 Find the inverse LT of $X(s) = \dfrac{s^2-3s+1}{s^2+2s+1}$ with ROC $\text{Re}(s) > -1$

7.41 Use LT to find the output of the system if the system is described by the differential equation $\dfrac{d}{dt} y(t) + 3y(t) = x(t)$ with input given by

$x(t) = e^{-4t}u(t)$ and initial condition is $y(0^+) = -2$. Draw the response using a MATLAB program.

7.42 Use LT to find the transfer function and the impulse response of the system, if the system is described by the differential equation $\frac{d^2}{dt^2}y(t) + 4\frac{d}{dt}y(t) + 10y(t) = x(t)$. Write a MATLAB program to draw the impulse response.

7.43 Use LT to find the transfer function and the impulse response of the causal and stable system, if the system is described by the differential equation $\frac{d^2}{dt^2}y(t) + 3\frac{d}{dt}y(t) + 2y(t) = \frac{d^2}{dt^2}x(t) + 6\frac{d}{dt}x(t) + 7x(t)$. Write a MATLAB program to find the response.

7.44 Find the forced response of the system with differential equation given by $\frac{d^2}{dt^2}y(t) + 3\frac{d}{dt}y(t) + 2y(t) = x(t)$ to the input given by $x(t) = e^{-3t}u(t)$. Write a MATLAB program to plot the impulse response and the forced response.

7.45 Find the natural response, forced response and total response of the system with differential equation given by $\frac{d^2}{dt^2}y(t) + 3\frac{d}{dt}y(t) + 2y(t) = x(t)$ to the input given by $x(t) = u(t)$. The initial conditions are $\frac{d}{dt}y(0) = 3$, $y(0) = 1$. Write a MATLAB program to draw natural response and forced response.

Answers

Multiple Choice Questions

1 (a)	2 (b)	3 (d)	4 (a)	5 (c)	6 (b)
7 (c)	8 (a)	9 (a)	10 (b)	11 (c)	12 (b)
13 (a)	14 (c)	15 (d)	16 (a)	17 (b)	18 (b)
19 (a)	20 (c)	21 (d)	22 (b)	23 (a)	24 (b)
25 (c)	26 (b)	27 (b)	28 (a)	29 (d)	30 (b)
31 (b)	32 (d)	33 (a)	34 (b)	35 (c)	36 (a)

Problems

7.1 0.

7.2 $LT(tu(t) - 2u(t)) = \frac{1}{s^2} - \frac{2}{s} = \frac{(1-2s)}{s^2}$ for all $\operatorname{Re}(s) > 0$

7.3   LTs are $X(s)=\dfrac{1}{s-3j}$ and $X(s)=\dfrac{1}{s+3j}$ with ROC Re(s) > 0

7.4 $X(s)=\dfrac{s}{(s^2+16\pi^2)}$ and $X(s)=\dfrac{4\pi}{(s^2+16\pi^2)}$ with ROC Re(s) > 0

7.5 $X(s)=\dfrac{2}{(s+2)^2+4}$ and $X(s)=\dfrac{s+2}{(s+2)^2+4}$ with ROC Re(s) > −2

7.6 $X(s)=\dfrac{-3}{s^2+7s+10}$ ROC is −5 < Re(s) < −2

7.7 $X(s)=\dfrac{e^{-7s}}{s}$ with ROC Re(s) > 0

7.8 $X(S)=\dfrac{e^{-5(s+5)}}{(s+5)}$ for all Re(s) > −5

7.9 $X(S)=e^{3s}$ for all Re(s) < 0

7.10 $X(S)=\dfrac{1}{2j}\left[\dfrac{e^{-5(s-3)}}{(s-3)}-\dfrac{e^{-5(s+3)}}{(s+3)}\right]$ ROC Re(s) > −3

7.11 $X(s)=\dfrac{14}{s^2-49}$ ROC is −7 < Re(s) < 7

7.12 $X(s)=\dfrac{s+3}{(s+3)^2+49}$ ROC Re(s) > −3

7.13 $X(s)=\dfrac{2s-5}{s^2-5s-24}$ ROC Re(s) > 8

7.14 $X(s)=-\dfrac{2s-3}{s^2-3s-10}$ ROC Re(s) < −2

7.15 $X(s)=\dfrac{3}{(s-2)^2+9}$ ROC Re(s) > 2

7.16 $X(s)=\dfrac{3s^2+144}{s(s^2+144)}$ ROC Re(s) > 0

7.17 $X(s)=\dfrac{s^2+s+36}{s(s^2+36)}$ ROC Re(s) > 0

7.18 $X(S)=\dfrac{1}{s}[1-e^{-2s}]^2$ ROC Re(s) > 0

7.19 $X(S) = \dfrac{A}{s}[1-e^{-sT}]$ ROC Re(s) > 0

7.20 $X(S) = \dfrac{A}{T}\left[\dfrac{1-sTe^{-Ts}-e^{-Ts}}{s^2}\right]$ ROC Re(s) > 0

7.21 $X(S) = \left[A\dfrac{e^{-sA}-1}{-s} + \dfrac{e^{-sA}}{s} + \dfrac{e^{-sA}-1}{s^2}\right]$ ROC Re(s) > 0

7.22 $X(S) = \dfrac{(e^{-s}+1)}{s^2+1}$ ROC Re(s) > 0

7.23 $X(S) = \dfrac{1}{s(1+e^{-sT/2})}$ ROC Re(s) > 0

7.24 $X(S) = \dfrac{s(e^{s/2}-e^{-s/2})}{(s^2+\pi^2)(1-e^{-S})}$

7.25 $X(s)_{\text{periodic signal}} = \dfrac{s(1+e^{-s})+(e^{-s}-e^{s})}{s^2(1-e^{-2s})}$, period = 2

7.26 $LT(t^2 e^{-2t}u(t)) = \dfrac{1}{(s+2)^3}$, $t^2 e^{-3t}\sin(t)u(t) \leftrightarrow \dfrac{2(s+3)}{((s+3)^2+1)^2}$,

$LT(e^{5t}u(-t)) = -\dfrac{1}{(s+5)}$ $LT(e^{-5t}\cos(t)u(t)) = \dfrac{s+5}{(s+5)^2+1}$

7.27 1. $LT(x(t)) = \dfrac{e^{-2s}}{s^3}(2-3s)$, 2. $LT((t-4)u(t-4)) = \dfrac{e^{-4s}}{s^2}$,

3. $LT(2tu(t) - 3(t-5)u(t-5)) = \dfrac{2-3e^{-5s}}{s^2}$

7.28 1. Initial value = 0, final value = −1/5
2. Initial value = 1, final value = 0
3. Initial value = 3/2, final value = 5/3
4. Initial value = 0, final value = ∞

7.29 $LT(e^{-(t-7)/3}u(t-7)) = \dfrac{3e^{-7s}}{s+1}$

7.30 $LT(x(t)) = \dfrac{1}{2}\left[\dfrac{5}{s^2+25} - \dfrac{1}{s^2+1}\right]$

7.31 $LT(e^{2t}u(t) * t^2 u(t)) = \dfrac{2}{s^3(s-3)}$

7.32 $X(s) = \dfrac{3}{s-3}$

7.33 $x(t) = \left(\dfrac{1}{3}e^{-2t}u(t) + \dfrac{1}{3}e^{-0.5t}\cos(0.866t)u(t) + 0.577e^{-0.5t}\sin(0.866t)u(t)\right)$

7.34 $x(t) = (u(-t) + 2(e^t)u(-t) + e^{-2t}u(t))$

7.35 $x(t) = 2e^{-3t}u(t) + e^{-t}\cos(3t)u(t)$

7.36 $x(t) = -2e^{-2t}u(-t) - 2e^{-4t}u(t)$

7.37 $x(t) = -u(t) + 2tu(t) + e^{-t}u(t)$, $x(t) = u(-t) - 2tu(-t) + e^{-t}u(t)$

7.38 $x(t) = \delta(t) + e^{-t}u(t) + e^{-2t}u(t)$

7.39 $x(t) = \delta'(t) + \delta(t) + e^{-2t}\cos(2t)u(t) - \dfrac{1}{2}e^{-2t}\sin(2t)u(t)$

7.40 $x(t) = \delta(t) - 5e^{-t}u(t) + 5te^{-t}u(t)$

7.41 $y(t) = -e^{-4t}u(t) - e^{-3t}u(t)$

7.42 $h(t) = \dfrac{1}{\sqrt{6}}e^{-2t}\sin(\sqrt{6}t)u(t)$

7.43 $h(t) = \delta(t) + e^{-2t}u(t) + 2e^{-t}u(t)$

7.44 $y(t) = \left(\dfrac{1}{2}e^{-3t}u(t)\right) + \left(\dfrac{1}{2}e^{-t}u(t) - e^{-2t}u(t)\right)$

forced resonse = steady state response + transient response

7.45 Natural response $y(t) = 5e^{-t}u(t) - 4e^{-2t}u(t)$

Forced response $y(t) = \left(\dfrac{1}{3}u(t)\right) - \left(e^{-t}u(t) - \dfrac{1}{2}e^{-2t}u(t)\right)$

8

Z Transform

> **Learning Objectives**
>
> - Significance of a transform
> - Relation of Z transform to other transforms viz. Laplace and Fourier transform
> - Convergence of Z transform and region of convergence (ROC)
> - Properties of ROC
> - Properties of Z transform
> - Inverse Z transform (IZT)
> - Pole locations and time domain behavior
> - Solution of difference equations using Z transform
> - Applications of ZT and IZT

Z domain is used for DT signals. The purpose of Z transform for DT signals is similar to that of Laplace transform for analog signals. The Z transform (ZT) is used to characterize a DT system by analyzing its transfer function in the Z domain. The locations of poles and zeros of the transfer function are further used for stability analysis of the DT systems. We will first discuss the significance of a transform. To understand that Laplace transform and Z transform are parallel techniques, we will start with Laplace transform and develop its relation to Z transform. The properties of Z transform will be discussed along with the region of convergence (ROC). The overall strategy of these two transforms is as follows. Find the impulse response from the transfer function. Find poles and zeros of the transfer function and apply sinusoids and exponentials as inputs to find the response. The basic difference between the two transforms is that the *s*-plane used by S domain is arranged in a rectangular co-ordinate system, while the *z*-plane used by Z domain uses a polar format. The Z transform provides mathematical framework for designing digital filters based on the well-known theory of analog filters using a mapping technique. The reader will study this part in the course on "Digital signal processing".

8.1 Physical Significance of a Transform

When any natural signal is to be processed, the signal parameters or features are required to be extracted. A transform helps in three ways.

1. The spectral properties of the signal are better revealed in a transform domain.
2. The system transfer function directly indicates the locations of poles and zeros that help in analyzing the stability of the system.
3. A signal, while transmission over a communication channel, needs to be compressed. Energy compaction can be achieved in a transform domain. The signal containing, for example, 128 sample points when transformed in Fourier domain will result in Fourier coefficients 128 in number. However, only a small number of them will have significant values and the others will be tending to zero. This effect is termed as energy compaction, as the energy gets concentrated in few coefficients in transform domain. For example, 128 samples of a DT sinusoid will convert to only one coefficient using Fourier series, namely the corresponding frequency term. (Here, it is assumed that DT sinusoid is periodic and there is no aliasing, as sampling is proper.) All other frequency coefficients will have zero values. This is one extreme case of energy compaction.

Concept Check

- What are the advantages of using a transform domain?
- What is energy compaction?
- Compare Laplace domain with Z domain.
- Can you analyze the stability of the system in Z domain?
- Do you get energy compaction when you Fourier transform the signal?
- Can you analyze the spectral contents of the signal in transform domain?

8.2 Relation between LT and ZT

Let us consider an analog signal sampled using an ideal sampler. Let us first understand the meaning of an ideal sampler. Ideal sampler is one which samples the signal at a particular instant such that the switch in Fig. 8.1 is closed for a very short duration almost tending to zero. The switch in Fig. 8.1 is ideal. The characteristics of ideal switch can be stated as follows. The ideal switch has zero resistance when closed and infinite resistance when it is open. We assume that is there is no leakage. The signal is sampled with a sampling frequency of F_S KHz. Here, we are assuming that sampling is uniform sampling. Uniform sampling means that the samples are taken at constant i.e., uniform sampling instants. The signal is assumed to start at $t = 0$ and the sampler starts taking samples from $t = 0$, $t = T$, $t = 2T$, and so on. The sampling interval that is the interval between successive samples is constant. Let T represent the sampling interval. Let the sampled signal be denoted by $f^*(t)$, as shown in Fig. 8.1.

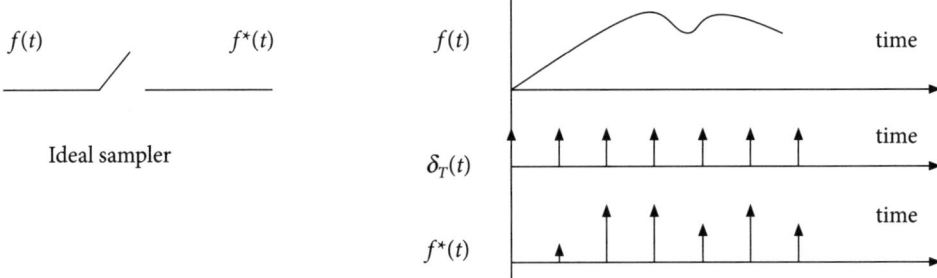

Fig. 8.1 Ideal sampler and the waveforms of $f(t)$, $\delta_T(t)$ and $f^*(t)$

Now, $f^*(t)$ can be written as

$$f^*(t) = f(t)\delta_T(t) \tag{8.1}$$

Let $\delta_T(t)$ represent the train of unit impulse function having a value 1 at each sampling instant that is a multiple of T. The impulse exists at $t = 0$, $t = T$, $t = 2T$, and so on, for uniform sampling. Then

$$f^*(t) = f(t)\delta(t) + f(t)\delta(t-T) + f(t)\delta(t-2T) \tag{8.2}$$

Let us denote the Laplace transform of $f^*(t)$ as $F^*(s)$. $F^*(s)$ can be written as

$$F^*(s) = \int_0^\infty [f(t)\delta(t) + f(t)\delta(t-T) + f(t)\delta(t-2T) + \ldots]e^{-st}\,dt \tag{8.3}$$

Now, let us denote

$$\int_0^\infty f(t)\delta(t-T)\,dt = f(T)$$

Then $F^*(s)$ becomes

$$F^*(s) = f(0) + f(T)e^{-sT} + f(2T)e^{-2sT} + f(3T)e^{-3sT} + \ldots\ldots \quad (8.4)$$

In the summation form Eq. (8.4) can be represented as

$$F^*(s) = \sum_{n=0}^{\infty} f(nT)e^{-nsT} \quad (8.5)$$

Let us define a complex variable Z as $Z = e^{sT}$. Let us substitute the value of Z in Eq. (8.5) and changing the argument s to Z, we get

$$F(Z) = \sum_{n=0}^{\infty} f(nT)Z^{-n} \quad (8.6)$$

Here $F(Z)$ is called as a Z transform of a DT signal $f(nT)$. A DT signal is usually denoted as $f[n]$ by dropping T as it remains constant. There is a one-to-one correspondence between a signal and its Z transform. Hence, we can recover $f(nT)$ from its Z transform via inverse Z transformation.

Concept Check
- What is ideal sampler?
- What is uniform sampling?
- What is the relation between Z and S?
- How will you get a DT signal from analog signal?
- What are the characteristics of ideal switch?

8.3 Relation between Fourier Transform (FT) and Z Transform

Since Z has been defined as a complex variable, we can express it in the polar form as

$$Z = re^{j\Omega} \quad (8.7)$$

where r represents the magnitude of Z and Ω represents the angle or phase of Z. Let us substitute this value of Z in Eq. (8.6)

$$F(re^{j\Omega}) = \sum_{n=0}^{\infty} f(nT)(re^{j\Omega})^{-n} \quad (8.8)$$

$$F(Z) = \sum_{n=0}^{\infty} [f[n]r^{-n}]e^{-j\Omega n} \quad (8.9)$$

Here, in Eq. (8.9), T is dropped as we are using uniform sampling, so a sample taken at $t = nT$ namely $(f(nT))$ represents the nth sample and is just denoted as $f[n]$.

We can conclude from Eq. (8.9) that $F(Z)$ is a Fourier transform (FT) of a sequence $f[n]r^n$. In general, the analog signal may exist from $-\infty$ to $+\infty$, and we can put the summation limits from $-\infty$ to $+\infty$ and write

$$F(Z) = \sum_{n=-\infty}^{\infty} (f[n]r^{-n})e^{-j\Omega n} \qquad (8.10)$$

Note that for $r = 1$ or equivalently when $Z = 1$, the Z transform reduces to FT. In other words, Z transform reduces to FT on the unit circle (circle with radius equal to one) in complex Z plane. Convergence of Z transform (ZT) requires the convergence of a sequence $f[n]r^n$. For any $f[n]r^n$ the infinite series represented by Eq. (8.10) will converge for only some range of values of r. The range of values of r for which the sequence $f[n]r^n$ converges is called the region of convergence (ROC). If ROC includes the unit circle, then the FT also converges.

Concept Check

- What is the relation between a Z transform and a Fourier transform?
- What is ROC?
- What is the significance of a unit circle in Z domain?
- Specify the ROC for convergence of ZT and FT.

8.4 Solved Problems on Z Transform

The concept of region of convergence is explained further with the help of solved examples. We will find the ZT of some standard signals.

Example 8.1

Given the unit step sequence $f[n] = u[n] = 1$ for all $n \geq 0$, find F(Z).

Solution

Substituting $Z = re^{j\Omega}$ in Eq. (8.10)

$$F(Z) = \sum_{n=-\infty}^{\infty} u[n]Z^{-n} \qquad (8.11)$$

Given $f[n] = 1$. As the function $f[n]$ exists for $n \geq 0$, we can change the summation limits from 0 to ∞ instead of from $-\infty$ to ∞. Substituting, the value and changing the summation, Eq. (8.11) becomes

$$F(Z) = \sum_{n=0}^{\infty} Z^{-n} = \sum_{n=0}^{\infty} (Z^{-1})^n \tag{8.12}$$

$$F(Z) = \sum_{n=0}^{\infty} Z^{-n} = 1 + Z^{-1} + Z^{-2} + Z^{-3} + \cdots + Z^{-\infty} \tag{8.13}$$

$$Z^{-1} F(Z) = Z^{-1} + Z^{-2} + Z^{-3} + Z^{-4} + \cdots + Z^{-\infty} \tag{8.14}$$

Subtracting Eq. (8.14) from Eq. (8.13), we get

$$F(Z)[1 - Z^{-1}] = 1 - Z^{-\infty} \tag{8.15}$$

If $|Z^{-1}| < 1$, then $Z^{-\infty}$ tends to zero and Eq. (8.15) becomes

$$F(Z) = \frac{1 - Z^{-\infty}}{1 - Z^{-1}} \tag{8.16}$$

$$F(Z) = \frac{1}{1 - (Z^{-1})} \tag{8.17}$$

Note that the term in the bracket of Eq. (8.12) appears in the bracket in the denominator of Eq. (8.17). Equation (8.17) is called as the close form expression for $F(Z)$. It is clear that ZT (the infinite series) only converges for $|Z^{-1}| < 1$ [the term inside the bracket in Eq. (8.17.)], that is for $|Z| > 1$. $|Z| > 1$ is the ROC. It is the entire Z-plane outside the unit circle, as shown in Fig. 8.2.

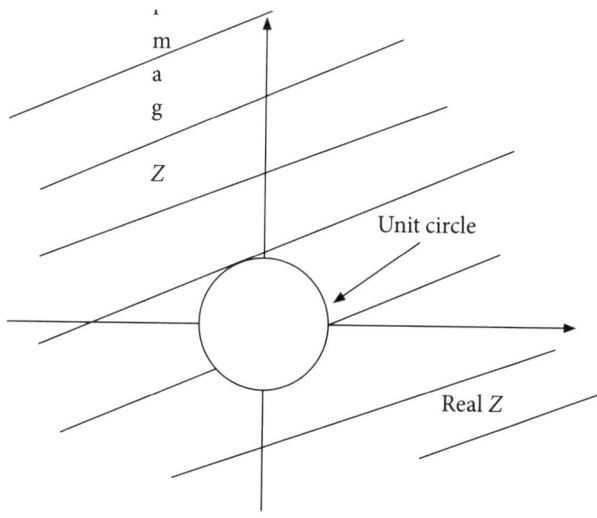

Fig. 8.2 ROC is the Z plane outside unit circle

Example 8.2
Given unit sample sequence $f[n] = \delta[n]$, find $F(Z)$.

Solution

We will use the definition of ZT.

$$F(Z) = \sum_{n=-\infty}^{\infty} \delta[n]Z^{-n} = 1 \quad \text{Note}: \delta[n] = 1 \text{ only for } n = 0 \tag{8.18}$$

ROC is the entire Z plane, as there are no poles for the transfer function.

Example 8.3
Given a unit ramp sequence $f[n] = nu[n]$, find $F(Z)$.

Solution

We will use the definition of ZT.

$$F(Z) = \sum_{n=-\infty}^{\infty} nu[n]Z^{-n} = \sum_{n=0}^{\infty} nZ^{-n}$$

$$F(Z) = 0 + Z^{-1} + 2Z^{-2} + \ldots = Z^{-1}(1 + 2Z^{-1} + \ldots) \tag{8.19}$$

$$F(Z) = \frac{Z^{-1}}{(1 - Z^{-1})^2}$$

ROC is the area outside the unit circle in Z plane. A double pole at $Z = 1$ for the transfer function.

Example 8.4
Given a unit ramp sequence $f[n] = n\delta[n]$, find $F(Z)$.

Solution

We will use the definition of ZT.

$$F(Z) = \sum_{n=-\infty}^{\infty} n\delta[n]Z^{-n} \tag{8.20}$$

$$F(Z) = 0$$

Example 8.5
Given a unit ramp sequence $f[n] = \delta[n-5]$, find $F(Z)$.

Solution

We will use the definition of ZT.

$$F(Z) = \sum_{n=-\infty}^{\infty} \delta[n-5]Z^{-n}$$
$$F(Z) = Z^{-5} \qquad (8.21)$$

ROC will exclude $Z = $ infinity, as there is a pole at $Z = $ infinity.

Example 8.6

Given a unit ramp sequence $f[n] = n\delta[n-3]$, find $F(Z)$.

Solution

We will use the definition of ZT.

$$F(Z) = \sum_{n=-\infty}^{\infty} n\delta[n-3]Z^{-n}$$
$$F(Z) = 3Z^{-3} \qquad (8.22)$$

ROC will exclude $Z = $ infinity, as there is a pole at $Z = $ infinity.

Example 8.7

Given $f[n] = e^{-bnT}$ for $n \geq 0$, find $F(Z)$ (a) if "b" is real and (b) if "b" is imaginary.

Solution

The sequence exists for $n \geq 0$ and it is written as $f[n] = e^{-bnT} u[n]$, where $u[n]$ represents a unit step sequence. It is a right-handed sequence.

$$F(Z) = \sum_{n=-\infty}^{\infty} f[n]Z^{-n} = \sum_{n=0}^{\infty} e^{-bnT} Z^{-n} \qquad (8.23)$$

As the sequence exists for $n \geq 0$, we can change the summation limits from 0 to ∞. Equation (8.23) now becomes

$$F(Z) = \sum_{n=0}^{\infty} e^{-bnT} = \sum_{n=0}^{\infty} (e^{-bT} Z^{-1})^n$$

$$F(Z) = \frac{1}{1 - e^{-bT} Z^{-1}} \qquad (8.24)$$

The region of convergence is the area with the term inside the bracket in Eq. (8.24) set to less than 1. ROC is $|e^{-bT}Z^{-1}| < 1$.

a. **Case 1** "b" is real

If "b" is real, then ROC is

$$|e^{-bT}Z^{-1}| < 1 \text{ i.e., } |Z| > e^{-bT} \qquad (8.25)$$

b. **Case 2** "b" is imaginary

If "b" is imaginary, then

$$|e^{-bT}| = 1$$

and ROC is

$$|Z^{-1}| < 1 \text{ or } |Z| > 1 \tag{8.26}$$

It is the region outside the unit circle in the Z domain.

We observe that for both real and imaginary "b", the Z transform is the same, but ROCs are different.

Example 8.8

Given $f[n] = \sin(n\omega T)$ for $n \geq 0$, find $F(Z)$.

Solution

$$F(Z) = \sum_{n=-\infty}^{\infty} f[n]Z^{-n} = \sum_{n=0}^{\infty} \sin(n\omega T)Z^{-n}$$

$$= \sum_{n=0}^{\infty} \frac{e^{jn\omega T} - e^{-jn\omega T}}{2j} Z^{-n} = \frac{1}{2j}\left[\sum_{n=0}^{\infty}(e^{j\omega T}Z^{-1})^n - \sum_{n=0}^{\infty}(e^{-j\omega T}Z^{-1})^n\right] \tag{8.27}$$

$$= \frac{1}{2j}\left[\frac{1}{1-(e^{j\omega T}Z^{-1})} - \frac{1}{1-(e^{-j\omega T}Z^{-1})}\right]$$

$$= \frac{1}{2j}\left[\frac{Z}{Z-e^{j\omega T}} - \frac{Z}{Z-e^{-j\omega T}}\right] = \left[\frac{Z(\sin \omega T)}{(Z^2 - 2\cos \omega T + 1)}\right]$$

ROC is computed at Eq. (8.23). Set the terms inside the brackets to less than 1:

$$|e^{j\omega T}Z^{-1}| < 1 \text{ and } |e^{-j\omega T}Z^{-1}| < 1 \text{ but } |e^{j\omega T}| = 1, |e^{-j\omega T}| = 1 \tag{8.28}$$

Hence, ROC is $|Z^{-1}| < 1$ or $|Z| > 1$. ROC is the region outside the unit circle in Z plane. Here ROC is the same as that for Problem 8.1 (Fig. 8.2), but Z transforms are different. When a sequence is transformed in Z domain, it is completely specified by its Z transform along with its ROC. The transfer function $F(Z)$ has a zero at $Z = 0$. The poles of $F(Z)$, namely, $Z = e^{j\omega T}$ and $Z = e^{-j\omega T}$, are complex conjugates. The poles are on the unit circle with angle ωT and $-\omega T$. The plot of poles and zeros is shown in Fig. 8.3.

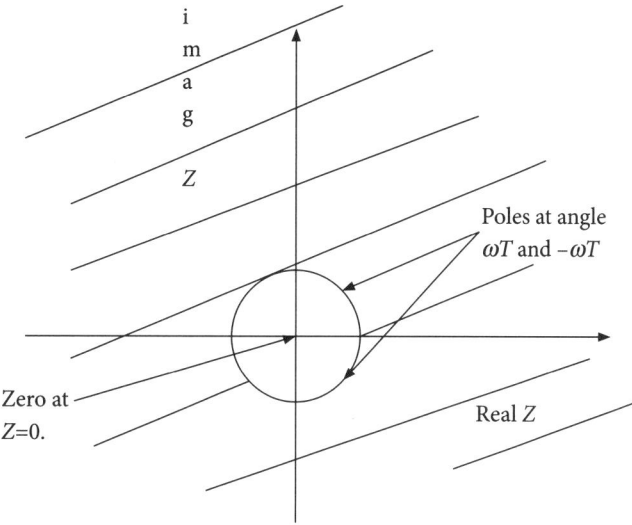

Fig. 8.3 ROC is the Z plane outside unit circle. The poles are on the unit circle at angle = ωT and $-\omega T$

Example 8.9

Given $f[n] = \cos(n\omega T)$ for $n \geq 0$, find $F(Z)$.

Solution

$$F(Z) = \sum_{n=-\infty}^{\infty} f[n]Z^{-n} = \sum_{n=0}^{\infty} \cos(n\omega T)Z^{-n}$$

$$= \sum_{n=0}^{\infty} \frac{e^{jn\omega T} + e^{-jn\omega T}}{2} Z^{-n} = \frac{1}{2}\left[\sum_{n=0}^{\infty} (e^{j\omega T} Z^{-1})^n + \sum_{n=0}^{\infty} (e^{-j\omega T} Z^{-1})^n\right] \quad (8.29)$$

$$= \frac{1}{2}\left[\frac{1}{1-(e^{j\omega T} Z^{-1})} + \frac{1}{1-(e^{-j\omega T} Z^{-1})}\right]$$

$$= \frac{1}{2}\left[\frac{Z}{Z-e^{j\omega T}} + \frac{Z}{Z-e^{-j\omega T}}\right] = \left[\frac{Z(Z-\cos \omega T)}{(Z^2 - 2\cos(\omega T)Z + 1)}\right]$$

ROC is computed at Eq. (8.29). Set the terms inside the brackets to less than 1:

$$|e^{j\omega T}Z^{-1}| < 1 \text{ and } |e^{-j\omega T}Z^{-1}| < 1 \text{ but } |e^{j\omega T}| = 1, |e^{-j\omega T}| = 1 \quad (8.30)$$

Hence, ROC is $|Z^{-1}| < 1$ or $|Z| > 1$. ROC is the region outside the unit circle in Z plane. Here, ROC is the same as that for Problem 8.1 (Fig. 8.2), but Z transforms are different. When a sequence is transformed in the Z domain, it is completely specified by its Z transform along with its ROC. The transfer function $F(Z)$ has a zero at $Z = 0$ and at $Z = \cos(\omega T)$. The poles of $F(Z)$, namely, $Z = e^{j\omega T}$ and $Z = e^{-j\omega T}$, are complex conjugates. The poles are on the unit circle with angles ωT and $-\omega T$. The plot of poles and zeros is shown in Fig. 8.4.

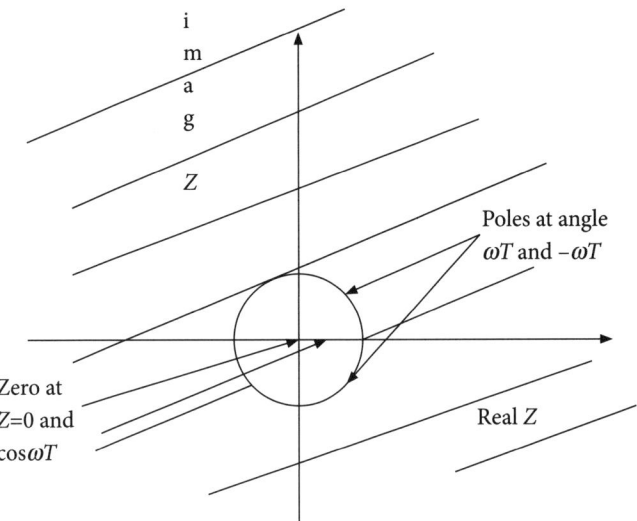

Fig. 8.4 ROC is the Z plane outside unit circle. The poles are on the unit circle at angle = ωT and $-\omega T$

We will now consider an example of two sequences for which Z transforms are same and ROCs are complementary.

Example 8.10

Given $f_1[n] = a^n u[n]$, $f_2[n] = -a^n u[-n-1]$, find $F_1(Z)$ and $F_2(Z)$.

Solution

$$F_1(Z) = \sum_{n=-\infty}^{\infty} f_1[n] Z^{-n} = \sum_{n=0}^{\infty} a^n Z^{-n} \tag{8.31}$$

$$F_1(Z) = \sum_{n=0}^{\infty} (aZ^{-1})^n = \frac{1}{1-(aZ^{-1})} = \frac{Z}{Z-a} \tag{8.32}$$

ROC is calculated as the term inside the bracket in Eq. (8.32) to be less than 1. $|aZ^{-1}| < 1$ that is $|Z| > a$. ROC is the area outside the circle $|Z| = a$ in the Z domain, as shown in Fig. 8.5. If $a < 1$, then the Fourier transform of the sequence $f(n)$ will also converge as it will include the circle $|Z| = 1$:

$$F_2(Z) = \sum_{n=-\infty}^{\infty} f_2[n]Z^{-n} = \sum_{n=-\infty}^{-1} -a^n Z^{-n} \tag{8.33}$$

$$F_2(Z) = -\sum_{n=-\infty}^{-1} (aZ^{-1})^n \tag{8.34}$$

$$F_2(Z) = -\sum_{n=1}^{\infty} (a^{-1}Z)^n \tag{8.35}$$

$$\sum_{n=0}^{\infty} (a^{-1}Z)^n = 1 + \sum_{n=1}^{\infty} (a^{-1}Z)^n \tag{8.36}$$

$$F_2(Z) = 1 - \sum_{0}^{\infty} (a^{-1}Z)^n \tag{8.37}$$

$$F_2(Z) = 1 - \frac{1}{1-(a^{-1}Z)} = 1 - \frac{a}{a-Z} = \frac{Z}{Z-a} \tag{8.38}$$

$F_2(Z)$ is same as $F_1(Z)$. ROC is calculated using Eq. (8.38) by putting the term inside the bracket to be less than 1. $|a^{-1}Z| < 1$, that is, $|Z| < a$. ROC is the area inside the circle $|Z| = a$ in Z domain, as shown in Fig. 8.5.

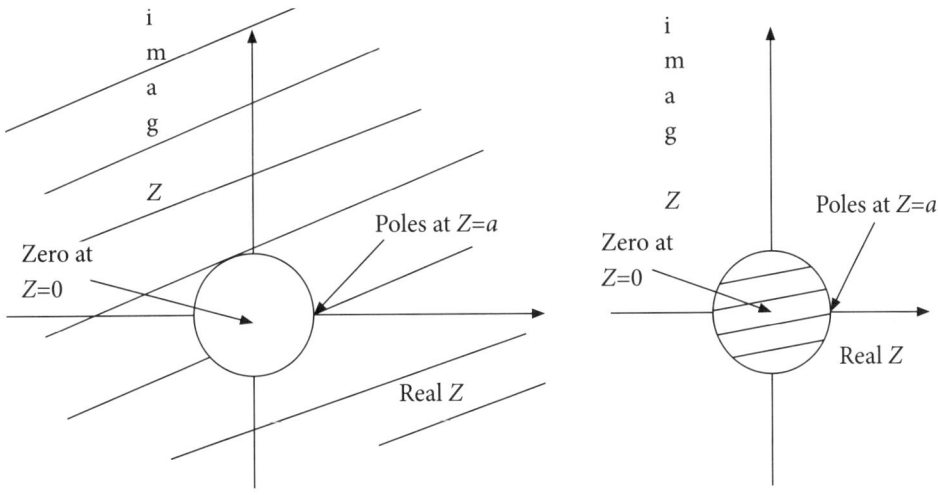

Fig. 8.5 ROC for $F_1(Z)$ is the Z plane outside the circle $|Z| = a$ and ROC for $F_2(Z)$ is the Z plane area inside circle $|Z| = a$

Concept Check

- What is the significance of ROC being inside a circle in the Z domain?
- If two sequences have the same Z transform, how can one distinguish between them?
- How will you calculate the closed-form expression for Z transform?

8.5 Properties of ROC

Property 1

The ROC of $X(Z)$ consists of a circular ring in the Z plane centered around origin.

Proof

ROC consists of a range of values of Z (i.e., $re^{j\omega}$) for which $x[n]r^{-n}$ has a Fourier transform that converges. ROC of the Z transform of $x[n]$ consists of the values of Z for which $x[n]r^{-n}$ is absolutely summable. This indicates that convergence is dependent only on $r = |Z|$ and not on ω. If a specified value of Z is in the ROC, then all values of Z on the same circle (with same magnitude) will be in the ROC. Hence, ROC will consist of a circular ring. The inner boundary of ROC may tend towards the origin and the outer boundary of ROC may tend towards infinity. (Refer to Figs 8.3 and 8.8.).

Property 2

The ROC does not contain any pole.

Proof

We know that the system function diverges at pole locations. This clearly means that the ROC will not include any pole location.

Property 3

If the sequence $x[n]$ is of finite duration, then the ROC is the entire Z plane, except possibly $Z = 0$ and/or $Z = \infty$.

Proof

Let us consider a finite duration sequence existing between $n = N_1$ and $n = N_2$.
Case 1 N_1 and N_2 both are negative

$$X(Z) = \sum_{n=N_1}^{N_2} x[n] Z^{-n} \tag{8.39}$$

As N_1 and N_2 are negative, $X(Z)$ is a series with positive powers of Z and hence, it will diverge for $|Z| \to \infty$. Hence, ROC will exclude $Z = \infty$.
Case 2 N_1 and N_2 are both positive

As N_1 and N_2 are positive, $X(Z)$ is a series with negative powers of Z and hence, it will diverge for $|Z| \to 0$. Therefore, ROC will exclude $Z = 0$.

Case 3 N_1 is negative and N_2 is positive

As N_1 is negative and N_2 is positive, $X(Z)$ is a series containing both negative and positive powers of Z and hence, it will diverge for both $|Z| \to \infty$ and $|Z| \to 0$. ROC will exclude $Z = 0$ and $Z = \infty$.

Property 4

If $x[n]$ is a right-handed sequence (causal sequence) and if the circle $|Z| = r_0$ is in ROC, then all finite values of Z for which $|Z| > r_0$ will also be in ROC.

Proof

A right-handed sequence is zero prior to some positive value of n, for example, $n = N_1$. $X(Z)$ will contain all negative powers of Z.

1. If $|Z| = r_0$ is in ROC, the sequence $x[n]r_0^{-n}$ converges.
2. If $|Z| > r_0$ and let us assume that $|Z| = r_1$, then the sequence $x[n]r_1^{-n}$ will converge more rapidly and $|Z| = r_1$ will also be in ROC.

Property 5

If $x[n]$ is a left-handed sequence and if the circle $|Z| = r_0$ is in ROC, then all finite values of Z for which $|Z| < r_0$ will also be in ROC.

Proof

A left-handed sequence is zero after some negative or zero value of n, and let us assume that $n = N_1$. $X(Z)$ will contain all positive powers of Z.

1. If $|Z| = r_0$ is in ROC, the sequence $x[n]r_0^{n}$ converges.
2. If $|Z| < r_0$, and let us assume that $|Z| = r_1$, then the sequence $x[n]r_1^{n}$ will converge more rapidly and $|Z| = r_1$ will also be in ROC.

Property 6

If $x[n]$ is a both-handed sequence and if the circle $|Z| = r_0$ is in ROC, then ROC will consist of a ring in a Z-plane which includes $Z = r_0$.

Proof

Consider a both-sided sequence as made up of a right-handed sequence and a left-handed sequence. The right-handed sequence will have the ROC as the entire Z-plane outside some circle with $|Z| = r_1$, as shown in Fig. 8.6. The left-handed sequence, on the other hand, will have the ROC as the region inside some circle $|Z| = r_2$, as shown in Fig. 8.6. ROC for a both-sided sequence is the overlapping area in the Z-plane including $|Z| = r_0$.

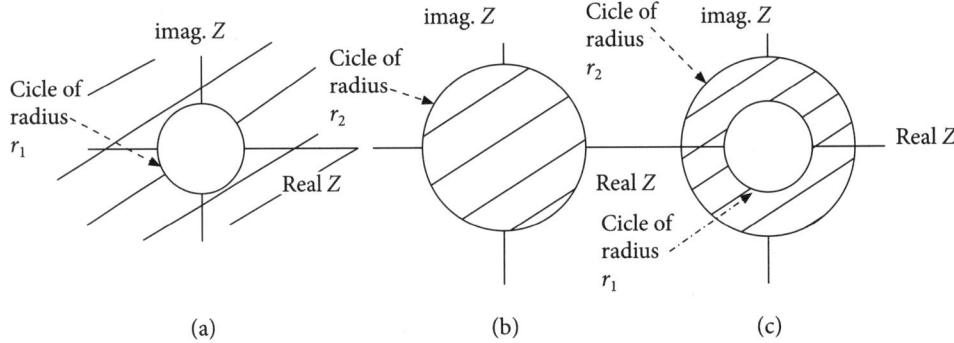

(a)   (b)   (c)

Fig. 8.6 (a) ROC for a right-handed sequence outside the circle with radius $|Z| = r_1$. (b) ROC for a left-handed sequence is inside the circle with radius r_2. (c) ROC for a both-sided sequence is the common area of convergence including circle with radius r_n ($r_0 < r_1 < r_2$)

To illustrate Property 3, let us consider the example of delta and shifted delta functions.

Example 8.11

Consider a delta function given by

$$\delta[n] = \begin{cases} 1 \text{ for } n = 0 \\ 0 \text{ elsewhere} \end{cases}$$

- **Case 1**

$$X(Z) = 1 \tag{8.40}$$

The ROC is the entire Z-plane including $Z = 0$ and $Z = \infty$.

- **Case 2** Let $\delta[n]$ be shifted towards right by one sample so that it becomes $\delta[n-1]$.

$$X(Z) = Z^{-1} \tag{8.41}$$

This function has a pole at $Z = 0$ and ROC will exclude $Z = 0$.

- **Case 3** Let $\delta[n]$ be shifted towards left by one sample so that it becomes $\delta[n+1]$.

$$X(Z) = Z \tag{8.42}$$

This function has a pole at $Z = \infty$ and ROC will exclude $Z = \infty$.

To illustrate property 3, we will solve the following problem.

Example 8.12

Consider a DT signal given by $x[n] = \{\underset{\uparrow}{3}\,4\,1\,2\}$. Find ZT.

Solution

The DT signal is given, which has the arrow specified at the first sample. This indicates that it is the sample at $n = 0$. We can find ZT using the definition of ZT.

$$X(Z) = \sum_{n=0}^{3} x[n]Z^{-n} = 3 + 4Z^{-1} + Z^{-2} + 2Z^{-3} \tag{8.43}$$

DT signal is a finite duration right-handed signal. X(Z) contains only powers of Z^{-1}. Hence, ROC is the entire Z plane except $Z = \infty$, as the X(Z) has the pole at infinity.

Example 8.13

Consider a DT signal given by $x[n] = \{2\,1\,1\,\underset{\uparrow}{3}\}$. Find ZT.

Solution

The DT signal is given, which has the arrow specified at last sample. This indicates that it is the sample at $n = 0$. We can find ZT using the definition of ZT.

$$X(Z) = \sum_{n=-3}^{0} x[n]Z^{-n} = 2Z^3 + Z^2 + Z + 3 \tag{8.44}$$

DT signal is a finite duration left-handed signal. X(Z) contains only powers of Z. Hence, ROC is the entire Z plane except $Z = 0$, as the X(Z) has the pole at zero.

Example 8.14

Find the ZT of $x[n] = \begin{cases} a^n & \text{for } 0 < n < N-1 \\ 0 & \text{elsewhere} \end{cases}$

Solution

The sequence is a right-handed sequence and is of finite duration. The ROC will consist of the entire Z plane except $Z = 0$ and/or $Z = \infty$. We will find the ROC.

$$X(Z) = \sum_{n=0}^{N-1} a^n Z^{-n} \tag{8.45}$$

$$X(Z) = \sum_{n=0}^{N-1} (aZ^{-1})^n \tag{8.46}$$

$$X(Z) = 1 + aZ^{-1} + a^2 Z^{-2} + \cdots + a^{(N-1)} Z^{-(N-1)} \tag{8.47}$$

$$aZ^{-1} X(Z) = aZ^{-1} + a^2 Z^{-2} + \cdots + a^{(N-1)} Z^{-(N-1)} + a^N Z^{-N} \tag{8.48}$$

Subtracting Eq. (8.48) from Eq. (8.47), we get

$$X(Z)(1 - aZ^{-1}) = 1 - a^N Z^{-N} \tag{8.49}$$

$$X(Z) = \frac{1 - a^N Z^{-N}}{1 - aZ^{-1}} = \frac{Z^N - a^N}{Z^{(N-1)}(Z - a)} \tag{8.50}$$

Equation (8.50) indicates that the function has a pole of the order $N - 1$ at $Z = 0$ and hence ROC is the entire Z plane, but will exclude $Z = 0$. The system transfer function has $N - 1$ zeros given by $Z_k = ae^{j2\pi k/N}$. Note that a zero at $Z = a$ is cancelled by a pole at the same location. Figure 8.7 shows the pole zero plot for $X(Z)$.

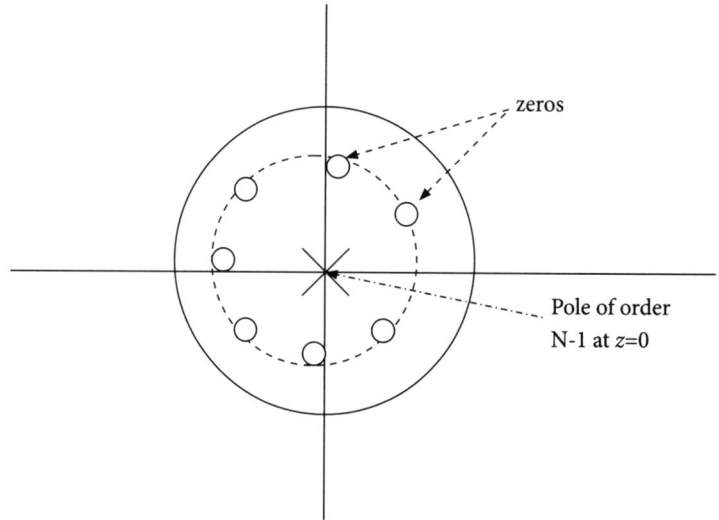

Fig. 8.7 Pole zero plot of $X(Z)$. Pole at $z = a$ is cancelled by a zero at $z = a$

Let us solve more problems based on the properties of ROC.

Example 8.15

Consider the function $f[n] = b^{|n|}$ for $b > 0$. Find the ZT of $f[n]$.

Solution

Let us express $f[n]$ as a combination of a right-handed sequence and a left-handed sequence. So, $F[n] = b^n u[n] + b^{-n} u[-n - 1]$

$$F(Z) = \sum_{n=0}^{\infty} b^n Z^{-n} + \sum_{n=-\infty}^{-1} b^{-n} Z^{-n} \tag{8.51}$$

$$F(Z) = \frac{1}{1-bZ^{-1}} + \sum_{n=1}^{\infty} b^n Z^n \tag{8.52}$$

$$F(Z) = \frac{1}{1-bZ^{-1}} + \sum_{n=0}^{\infty} b^n Z^n - 1 \tag{8.53}$$

$$F(Z) = \frac{1}{1-bZ^{-1}} + \frac{1}{1-bZ} - 1 \tag{8.54}$$

$$F(Z) = \frac{Z}{Z-b} + \frac{b^{-1}Z^{-1}}{b^{-1}Z^{-1}-1} - 1 \tag{8.55}$$

$$F(Z) = \frac{Z}{Z-b} + \frac{b^{-1}Z^{-1} - b^{-1}Z^{-1} + 1}{b^{-1}Z^{-1} - 1} \tag{8.56}$$

$$F(Z) = \frac{Z}{Z-b} + \frac{-Z}{Z-b^{-1}} \tag{8.57}$$

$$F(Z) = \frac{Z^2 - Zb^{-1} - Z^2 + Zb}{(Z-b)(Z-b^{-1})} \tag{8.58}$$

$$F(Z) = \frac{Z(b-b^{-1})}{(Z-b)(Z-b^{-1})} \tag{8.59}$$

ROC can be calculated at Eq. (8.54). The first term converges for $|Z| > b$ and the second term will converge for $|Z| < 1/b$. Common area of convergence will exist only if $b < 1$. Figure 8.8 shows the area of convergence, which consists of a ring between b and $1/b$.

Here, the sequence is a both-sided sequence of infinite duration. Hence, ROC is a ring between two circles with radii b and $1/b$. It is given that $b > 0$. There will be two cases. First is $b < 1$, $1/b > 1$. In this case ROC will exist. The second case is $b > 1$, $1/b < 1$. In this case the ROCs for the two terms will be disjoint and there will be no common area of convergence.

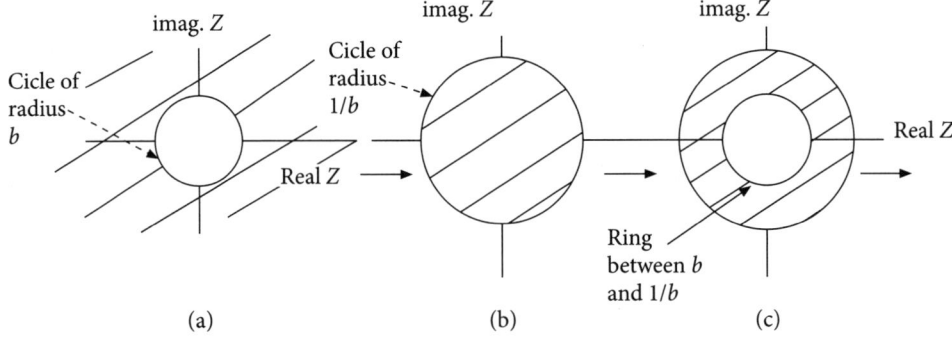

Fig. 8.8 (a) ROC for a right-handed sequence outside the circle with radius $|Z| = b$. (b) ROC for a left-handed sequence inside the circle with radius $1/b$. (c) The common area of convergence is a ring between b and $1/b$

Example 8.16

Consider the function defined as

$$x[n] = \begin{cases} 1 & \text{for } n < 0 \\ (0.5)^n & \text{for } n \geq 0 \end{cases}$$

Find $X(Z)$.

Solution

$$X(Z) = \sum_{n=0}^{\infty} (0.5)^n Z^{-n} + \sum_{n=-\infty}^{-1} Z^{-n} \tag{8.60}$$

$$F(Z) = \sum_{n=0}^{\infty} (0.5)^n Z^{-n} + \sum_{n=1}^{\infty} Z^n \tag{8.61}$$

$$F(Z) = \sum_{n=0}^{\infty} (0.5)^n Z^{-n} + \sum_{n=0}^{\infty} Z^n - 1 \tag{8.62}$$

$$F(Z) = \frac{1}{1-(0.5)Z^{-1}} + \frac{1}{1-Z} - 1 \tag{8.63}$$

$$F(Z) = \frac{Z}{Z-0.5} + \frac{Z}{1-Z} \qquad (8.64)$$

$$F(Z) = \frac{Z}{Z-0.5} - \frac{Z}{Z-1} \qquad (8.65)$$

ROC can be found at Eq. (8.63). The first term converges for $|Z| > 0.5$ and the second term converges for $|Z| < 1$. So the common area of convergence is the ring between $Z = 0.5$ and $Z = 1$.

Example 8.17

Consider the function defined as $x[n] = a^n u[n] - b^n u[-n-1]$
Find $X(Z)$.

Soution

$$X(Z) = \sum_{n=0}^{\infty}\left(aZ^{-1}\right)^n - \sum_{-\infty}^{-1}\left(bZ^{-1}\right)^{-n}$$

$$= \sum_{n=0}^{\infty}\left(aZ^{-1}\right)^n - \sum_{n=1}^{\infty}\left(b^{-1}Z\right)^n$$

$$= \sum_{n=0}^{\infty}\left(aZ^{-1}\right)^n + 1 - \sum_{n=0}^{\infty}\left(b^{-1}Z\right)^n \qquad (8.66)$$

$$= \frac{1}{1-aZ^{-1}} + 1 - \frac{1}{1-b^{-1}Z}$$

$$= \frac{Z}{Z-a} + 1 - \frac{bZ^{-1}}{bZ^{-1}-1}$$

$$X(Z) = \frac{Z}{Z-a} + \frac{-1}{bZ^{-1}-1} = \frac{Z}{Z-a} + \frac{Z}{Z-b} \qquad (8.67)$$

ROC is $|Z| > a$ for the first term and $|b^{-1}Z| < 1$ i.e., $|Z| < b$ for the second term. The common area of convergence will exist only if $b > a$. ROC does not exist if $b < a$. ROC for $b > a$ is shown in Fig. 8.9.

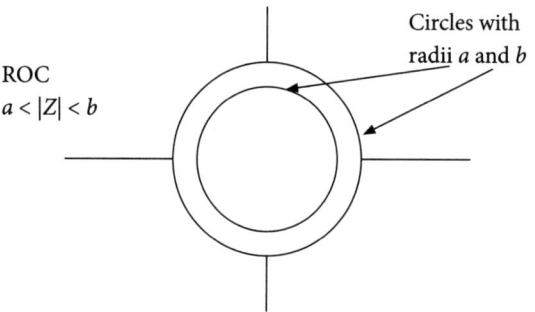

Fig. 8.9 ROC for Example 8.17

Example 8.18

Consider the function defined as $x[n] = 2\left(\dfrac{1}{3}\right)^n u[n] + 2\left(-\dfrac{1}{4}\right)^n u[n]$. Find $X(Z)$.

Solution

$$X(Z) = 2\sum_{n=0}^{\infty}\left(\dfrac{1}{3}Z^{-1}\right)^n + 2\sum_{0}^{\infty}\left(-\dfrac{1}{4}Z^{-1}\right)^n$$

$$= 2\sum_{n=0}^{\infty}\left(\dfrac{1}{3}Z^{-1}\right)^n + 2\sum_{n=0}^{\infty}\left(-\dfrac{1}{4}Z^{-1}\right)^n \tag{8.68}$$

$$= 2\dfrac{1}{1-\dfrac{1}{3}Z^{-1}} + 2\dfrac{1}{1+\dfrac{1}{4}Z^{-1}}$$

$$= \dfrac{2Z}{Z-\dfrac{1}{3}} + \dfrac{2Z}{Z+\dfrac{1}{4}}$$

$$X(Z) = \dfrac{2Z}{Z-\dfrac{1}{3}} + \dfrac{2Z}{Z+\dfrac{1}{4}} = \dfrac{2Z\left(2Z-\dfrac{1}{12}\right)}{Z^2 - \dfrac{1}{12}Z - \dfrac{1}{12}} \tag{8.69}$$

ROC is $|Z| > 1/3$ for the first term and $|Z| > 1/4$ for the second term. The common area of convergence is the ROC i.e., $|Z| > 1/3$. ROC is shown in Fig. 8.10.

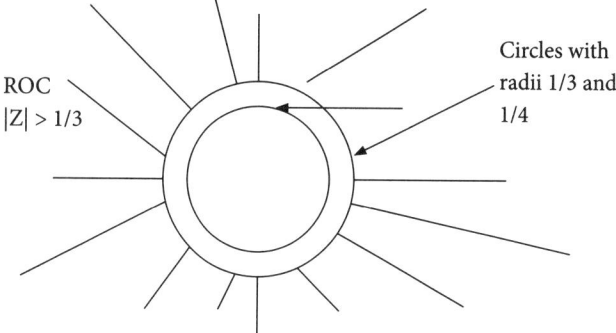

Fig. 8.10 ROC for Example 8.18

Example 8.19

Consider the function defined as $x[n] = \left(\dfrac{1}{4}\right)^n u[-n] + 2^n u[-n-1]$
Find $X(Z)$.

Solution

$$X(Z) = \sum_{n=-\infty}^{0} \left(\frac{1}{4}Z^{-1}\right)^n + \sum_{-\infty}^{1} \left(2Z^{-1}\right)^n$$

$$= \sum_{n=0}^{\infty} \left(\frac{1}{4}\right)^{-n} Z^n + \sum_{n=1}^{\infty} \left(2^{-1} Z\right)^n \tag{8.70}$$

$$= \frac{1}{1-4Z} + \frac{1}{1-2^{-1}Z} - 1$$

$$= \frac{1}{1-4Z} + \frac{\dfrac{1}{2}Z}{1-\dfrac{1}{2}Z}$$

$$X(Z) = \frac{1-2Z^2}{1-\dfrac{9}{2}Z+2Z^2} = -\frac{2Z^2-1}{2Z^2-\dfrac{9}{2}Z+1} \tag{8.71}$$

ROC is $|Z| < 1/4$ for the first term and $|Z| < 2$ for the second term. The common area of convergence is the ROC i.e., $|Z| < 1/4$. ROC is shown in Fig. 8.11.

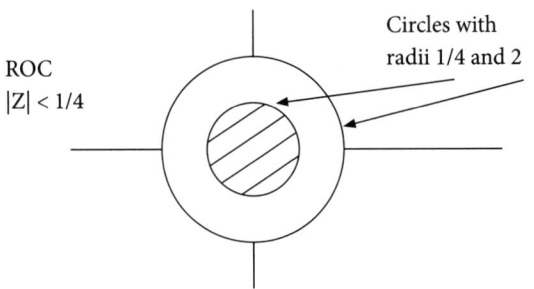

Fig. 8.11 ROC for Example 8.19

Example 8.20

Consider the function defined as $x[n] = \left(\dfrac{1}{4}\right)^n \cos\left(\dfrac{\pi}{2} n\right) u[n]$
Find $X(Z)$.

Solution

$$X(Z) = \frac{1}{2}\left[\sum_{n=0}^{\infty}\left(\frac{1}{4}e^{j\frac{\pi}{2}}Z^{-1}\right)^n + \sum_{n=0}^{\infty}\left(\frac{1}{4}e^{-j\frac{\pi}{2}}Z^{-1}\right)^n\right]$$

$$= \frac{1}{2}\left[\frac{1}{1-\dfrac{1}{4}e^{j\pi/2}Z^{-1}} + \frac{1}{1-\dfrac{1}{4}e^{-j\pi/2}Z^{-1}}\right]$$

$$= \frac{1}{2}\left[\frac{Z}{Z-\dfrac{1}{4}e^{j\pi/2}} + \frac{Z}{Z-\dfrac{1}{4}e^{-j\pi/2}}\right] \quad \text{Note } e^{j\pi/2} = -e^{-j\pi/2} = -1 \quad (8.72)$$

$$= \frac{1}{2}\left[\frac{Z}{Z+\dfrac{1}{4}} + \frac{Z}{Z-\dfrac{1}{4}}\right] = \frac{Z^2}{Z^2 - \dfrac{1}{16}}$$

ROC is $|Z| > \tfrac{1}{4}$, as shown in Fig. 8.12. Double Zero at $Z = 0$ and poles at angles pi/2 and −pi/2.

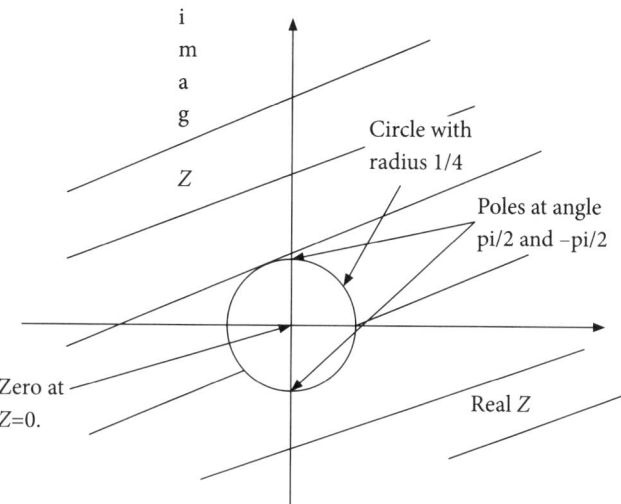

Fig. 8.12 ROC for Example 8.20

We summarize the discussion of ROC and list the ROCs for signals with different characteristics in Table 8.1.

Table 8.1 Signals and corresponding ROCs

Sl. No.	Signal	ROC
1	Delta function/unit sample sequence	Entire Z plane
2	Right handed/causal of finite duration	Entire Z plane except $Z = 0$
3	Left handed/anti-causal of finite duration	Entire Z plane except $Z = \infty$
4	Two sided/anti-causal of finite duration	Entire Z plane except $Z = 0$ and $Z = \infty$.
5	Right handed/causal of infinite duration	Area outside some circle in Z plane
6	Left handed/anti-causal of infinite duration	Area inside some circle in Z plane
7	Two sided/anti-causal of infinite duration	Area within two circles in Z plane

We will define inverse Z transform. The inverse Z transform of $X(Z)$ is defined as

$$x[n] = \frac{1}{2\pi j} \oint_c X(Z) Z^{n-1} dZ \tag{8.73}$$

The equation denotes the integration as the contour integration. It is to be done over some contour c that is inside ROC.

Concept Check

- Why ROC cannot include poles?
- When will ROC be the area outside some circle in Z-plane?

- Specify the nature of ROC for a left-handed sequence.
- Can it happen that no ROC exists for a both-sided sequence? When will it happen?
- Specify the ROC for a finite duration sequence.

8.6 Properties of Z Transform

The section discusses different properties of Z transform, namely linearity, time shifting, scaling in Z domain, time reversal, differentiation in Z domain, convolution and initial value theorem. Let us discuss these properties one by one.

1. **Property of linearity** The property of linearity states that
   If
   $$x_1[n] \leftrightarrow X_1(Z) \text{ with ROC} = R_1 \quad (8.74)$$

   $$x_2[n] \leftrightarrow X_2(Z) \text{ with ROC} = R_2,$$

   then

   $$ax_1[n] + bx_2[n] \leftrightarrow aX_1(Z) + bX_2(Z) \text{ with ROC} = R_1 \cap R_2.$$

   Proof
   Using the definition of ZT, we can write

   $$ZT[ax_1[n] + bx_2[n]] = \sum_{-\infty}^{\infty} ax_1[n] + bx_2[n]Z^{-n}$$

   $$= a\left[\sum_{-\infty}^{\infty} x_1[n]Z^{-n}\right] + b\left[\sum_{-\infty}^{\infty} x_2[n]Z^{-n}\right] \quad (8.75)$$

   $$= aX_1(Z) + bX_2(Z) \text{ ROC } R_1 \cap R_2$$

   Physical significance of property of linearity If the signal can be decomposed into linear combination of two or more component signals, then the ZT of the signal is the linear combination of the component signals. This can be easily computed. Similarly, if the signal can be decomposed into linear combination of two or more component signals in ZT domain, then inverse ZT (IZT) of the signal is the linear combination of the component IZT signals. This property is used in the calculation of IZT using partial fraction expansion.

2. Property of Time Shifting

This property states that

If

$$x[n] \leftrightarrow X(Z) \text{ with ROC} = R,$$

then

$$x[n - n_0] \leftrightarrow Z^{-n_0} X(Z) \text{ with ROC} = R \qquad (8.76)$$

except for the possible addition or deletion of the origin or infinity

1. If $n_0 > 0$, then the poles are introduced at the origin and if ROC of $x[n]$ includes origin, ROC of $x[n - n_0]$ will exclude origin.
2. If $n_0 < 0$, then the poles are introduced at infinity and ROC of $x[n - n_0]$ will exclude infinity.

Proof

Using the definition of ZT, we can write

$$ZT(x[n-n_0]) = \sum_{n=-\infty}^{\infty} x[n-n_0] Z^{-n} \text{ put } n - n_0 = m) \qquad (8.77)$$

$$= \sum_{m=-\infty}^{\infty} x[m] Z^{-m} Z^{-n_0} = Z^{-n_0} X(Z)$$

Physical significance of property of time shifting A shift of n_0 samples in time domain corresponds to multiplication by the complex exponential Z^{-n_0}.

3. Property of Scaling in Z Domain

If

$$x[n] \leftrightarrow X(Z) \text{ with ROC} = R, \qquad (8.78)$$

then

$$Z_0^n x[n] \leftrightarrow X\left(\frac{Z}{Z_0}\right)$$

with ROC = $|Z_0|R$ where $|Z_0|R$ is a scaled version of R. That is, if Z is a point in ROC of $X(Z)$, then $|Z_0|Z$ is in ROC of $X(Z/Z_0)$.

A special case of this property is $Z_0 = e^{j\omega}$. If $x[n] \leftrightarrow X(Z)$ with ROC = R, then $e^{-j\omega n}x[n] \leftrightarrow X(Ze^{j\omega})$.

Proof

We use the definition of ZT.

$$ZT(Z_0^n x[n]) = \sum_{n=-\infty}^{\infty} x[n]Z_0^n Z^{-n}$$

$$= \sum_{n=-\infty}^{\infty} x[n](Z/Z_0)^{-n} = X(Z/Z_0) \quad (8.79)$$

If $Z_0 = e^{j\omega}$, $ZT(e^{j\omega n}x[n]) = X(Z/e^{j\omega})$.

Physical significance of property of scaling in Z domain The multiplication of a sequence by a complex exponential results in the rotation of the Z-plane by angle ω. This is also termed as frequency shifting in Z-domain.

4. **Property of time reversal**

If

$x[n] \leftrightarrow X(Z)$ with ROC = R,

then $\quad (8.80)$

$x[-n] \leftrightarrow X\left(\dfrac{1}{Z}\right)$ with ROC = 1/R

That is, if $Z = Z_1$ is in ROC of $x(n)$, then $Z = 1/Z_1$ is in ROC of $x(-n)$.

Proof

Using the definition of ZT, we can write

$$ZT(x[-n]) = \sum_{n=0}^{\infty} x[-n]z^{-n} \quad (\text{put } -n = m)$$

$$= \sum_{n=0}^{\infty} x[m]Z^m \quad (8.81)$$

$$= X(1/Z)$$

Physical significance of property of time reversal A reversal of the signal in DT domain corresponds to the reversal of the complex variable Z.

5. Property of Convolution

If

$x_1(n) \leftrightarrow X_1(Z)$ with ROC = R_1

$x_2(n) \leftrightarrow X_2(Z)$ with ROC = R_2, (8.82)

then

$x_1(n) * x_2(n) \leftrightarrow X_1(Z)X_2(Z)$ with ROC = $R_1 \cap R_2$

and may be larger if pole-zero cancellation occurs in the product.

Proof

Using the definition of ZT, we can write

$$X_1(Z) = \sum_{n=-\infty}^{\infty} x_1[n]Z^{-n}$$

(8.83)

$$X_2(Z) = \sum_{n=-\infty}^{\infty} x_2[n]Z^{-n}$$

$$ZT[x_1[n] * x_2[n]] = \sum_{n=-\infty}^{\infty} (x_1[n] * x_2[n])Z^{-n}$$

$$= \sum_{n=-\infty}^{\infty} \left(\sum_{m=-\infty}^{\infty} x_1[n]x_2[n-m] \right) Z^{-n} \text{ put } n-m=l$$

$$= \sum_{m=-\infty}^{\infty} x_1[m]Z^{-m} \sum_{l=-\infty}^{\infty} x_2[l]Z^{-l}$$

$$= X_1(Z)X_2(Z)$$

The convolution property can be interpreted as follows. We can recognize the Z transform as a series in Z^{-1} where the coefficients of Z^{-n} are the sequence values. Hence, the convolution property says that when the two polynomials or power series in Z^{-1} are multiplied, the coefficients in the power series of a product are the convolution of the coefficients of two power series $X_1(Z)$ and $X_2(Z)$.

Physical significance of property of convolution When the two signals are convolved in DT domain, their ZTs get multiplied. Convolution is computationally costly. The convolution can be performed in transform domain using a multiplication operation that is less costly.

6. **Property of multiplication**

   If

   $$x_1(n) \leftrightarrow X_1(Z) \text{ with ROC} = R_1 \qquad (8.84)$$

   $$x_2(n) \leftrightarrow X_2(Z) \text{ with ROC} = R_2,$$

   then

   $$x_1(n)x_2(n) \leftrightarrow \frac{1}{2\pi j}\oint_c X_1(V)X_2(Z/V)V^{-1}dV \text{ with ROC} = R_1 \cap R_2$$

   and may be larger if pole-zero cancellation occurs in the product.

   Proof

   Using the definition of inverse ZT, we can write

   $$X_1(Z) = \sum_{n=-\infty}^{\infty} x_1[n]Z^{-n}$$

   $$\qquad (8.85)$$

   $$X_2(Z) = \sum_{n=-\infty}^{\infty} x_2[n]Z^{-n}$$

   $$ZT[x_1[n] \times x_2[n]] = \sum_{n=-\infty}^{\infty} (x_1[n] \times x_2[n])Z^{-n}$$

   $$= \sum_{n=-\infty}^{\infty} \left[\frac{1}{2\pi j}\oint_c X_1(V)V^{n-1}dV\right] x_2[n]Z^{-n}$$

   Interchange the order of summation and integration.

   $$= \frac{1}{2\pi j}\oint_c X_1(V)\left[\sum_{n=-\infty}^{\infty} x_2[n]\left(\frac{Z}{V}\right)^{-n} V^{-1}\right] dV$$

   $$\qquad (8.86)$$

   $$= \frac{1}{2\pi j}\oint_c X_1(V)X_2(Z/V)V^{-1}dV$$

   Physical significance of property of multiplication When the two signals are multiplied in DT domain, their ZTs get convolved. Convolution is computationally costly. The convolution in Z domain can be performed using a multiplication operation in a DT domain that is less costly.

7. Property of Differentiation in Z Domain

If

$x(n) \leftrightarrow X(Z)$ with ROC = R, then (8.87)

$$nx(n) \leftrightarrow -Z\frac{d}{dZ}X(Z) \text{ with ROC = R}$$

Proof

$$X(Z) = \sum_{n=-\infty}^{\infty} x[n]Z^{-n}$$

$$-Z\frac{d}{dZ}X(Z) = -\sum_{n=-\infty}^{\infty} -nx[n]Z^{-n-1}Z \quad (8.88)$$

$$= \sum_{n=-\infty}^{\infty} nx[n]Z^{-n} = ZT(nx[n])$$

Physical significance of property of differentiation in Z domain If we take differentiation of the ZT of the signal and multiply it by $-Z$ and take its IZT, it will result in a signal that is multiplied by (n).

8. Initial value theorem

If the sequence $x(n)$ is causal, then

$$x(0) = \lim_{Z \to \infty} X(Z) \quad (8.89)$$

Proof

As the sequence $x(n)$ is causal, $x(n)$ exists for all $n > 0$.

$$X(Z) = \sum_{n=0}^{\infty} x[n]Z^{-n} = x(0) + x(1)Z^{-1} + x(2)Z^{-2} + \ldots \quad (8.90)$$

$\lim_{Z \to \infty} X(Z) = x(0)$

Physical significance of the initial value theorem Initial value of the DT signal i.e., $x(0)$ can be found from its ZT by taking the limit of $X(Z)$ as Z tends to infinity. We need not take IZT.

9. Final value theorem

If the sequence $x(n)$ is causal, then $\lim_{n \to \infty} x(n) = \lim_{Z \to 1} (Z-1)X(Z)$

$$X(Z) = \sum_{n=0}^{\infty} x[n]Z^{-n} = x(0) + x(1)Z^{-1} + x(2)Z^{-2} + \ldots \qquad (8.91)$$

$$ZT(x[n]) = X(Z) = x(0) + x(1)Z^{-1} + x(2)Z^{-2} + \ldots$$

$$ZT(x[n+1]) = ZX(Z) - Zx(0)$$

$$ZT(x[n+1] - x[n])) = ZX(Z) - Zx(0) - X(Z)$$

$$(Z-1)\,x(Z) = [x(0)Z - x(0) + x(1) - x(1)Z^{-1} + x(2)Z^{-1} - x(2)Z^{-2} + \ldots] - Zx(0)$$

$$\lim_{Z \to 1} (Z-1)\,X(Z) - x(0) = x(\infty) - x(0)$$

$$\lim_{Z \to 1} (Z-1)\,X(Z) = \lim_{n \to \infty} x[n] = x(\infty)$$

Physical Significance of the final value theorem The final value of the signal can be found from its ZT by taking the limit of $(Z-1)X(Z)$ as Z tends to one. We need not take IZT. To apply final value theorem, we have to cancel poles and zeros of $(Z-1)X(Z)$ before applying the limit. All the poles of $X(Z)$ must lie inside the unit circle in Z plane i.e., the system must be stable.

We will solve some examples based on properties of Z transform.

Example 8.21

For the function $f[n] = \left(\dfrac{1}{2}\right)^n u[n] + \left(\dfrac{1}{3}\right)^n u[n]$

Find $F(Z)$

Solution

$$F(Z) = \sum_{n=0}^{\infty} \left(\frac{1}{2}\right)^n Z^{-n} + \sum_{n=0}^{\infty} \left(\frac{1}{3}\right)^n Z^{-n} \qquad (8.92)$$

$$F(Z) = \sum_{n=0}^{\infty} \left(\frac{1}{2}Z^{-1}\right)^n + \sum_{n=0}^{\infty} \left(\frac{1}{3}Z^{-1}\right)^n \qquad (8.93)$$

$$F(Z) = \frac{Z}{Z - (1/2)} - \frac{Z}{Z - (1/3)} \qquad (8.94)$$

The first term will converge for $|Z| > 1/2$ and the other will converge for $|Z| > 1/3$. The common area of convergence is $|Z| > 1/2$.

If the poles of $aX_1(Z) + bX_2(Z)$ contain all poles of $X_1(Z)$ and $X_2(Z)$ with no deletion or cancellation of any pole by a zero, the ROC is the overlapping area of the ROCs.

Example 8.22

If $f[n] = a^n u[n] - a^n u[n-1]$, find $F(Z)$.

Solution

$$F(Z) = \sum_{n=0}^{\infty} (aZ^{-1})^n - \sum_{n=1}^{\infty} (aZ^{-1})^n \tag{8.95}$$

$$F(Z) = \sum_{n=0}^{\infty} (aZ^{-1})^n - \left[\sum_{n=0}^{\infty} (aZ^{-1})^n - 1\right] \tag{8.96}$$

$$F(Z) = \frac{Z}{Z-a} - \frac{Z}{Z-a} + 1 = 1 \tag{8.97}$$

Here, ROC is the entire Z domain, even though ROC for both the sequences is the area outside the circle with $|Z| = a$. Note that the poles at $Z = a$ get cancelled for the resulting sequence.

Example 8.23

If Z transform of $x[n] = a^n u[n]$ is

$$X(z) = \frac{Z}{Z-a}$$

with ROC $|Z| > a$, find the Z transform of $x[n-2]$.

Solution

Using the property of time shifting, we have

$$ZT(x[n-2]) = Z^{-2} X(Z) \tag{8.98}$$

$$ZT[x(n-2)] = Z^{-2} \frac{Z}{Z-a} = \frac{1}{Z(Z-a)} \tag{8.99}$$

There are two poles, one at $Z = 0$ and the second at $Z = a$, and hence, ROC excludes the origin.

Example 8.24

If $u(n) \leftrightarrow \dfrac{Z}{Z-1}$

with ROC being the area outside the unit circle, find Z transform of $a^n u(n)$ using the property of scaling in Z-domain.

Solution

Using the property of scaling in the Z domain

$$\text{if } u[n] \leftrightarrow \frac{Z}{Z-1}$$

(8.100)

$$a^n u[n] \leftrightarrow X(a^{-1}Z) = \frac{a^{-1}Z}{a^{-1}Z - 1} = \frac{Z}{Z-a}$$

Example 8.25

If $f[n] = \sin(n\omega T)u[n] \leftrightarrow \dfrac{Z(\sin(\omega T))}{Z^2 - 2\cos(\omega T)Z + 1}$ (refer to Example 8.8),

find the ZT of $a^n \sin(n\omega T)u[n]$ using the property of scaling in Z domain.

Solution

Here we have to replace Z in ZT of $f[n]$ by $a^{-1}Z$:

$$f[n] = \sin(n\omega T)u[n] \leftrightarrow \frac{Z(\sin(\omega T))}{Z^2 - 2\cos(\omega T)Z + 1}$$

(8.101)

$$a^n \sin(n\omega T)u[n] \leftrightarrow \frac{a^{-1}Z\sin(\omega T)}{a^{-2}Z^2 - 2\cos(\omega T)a^{-1}Z + 1}$$

Example 8.26

If $u[n] \leftrightarrow \dfrac{Z}{Z-1}$ with ROC $|Z| > 1$,

find the ZT of $u[-n]$ using the property of time reversal.

Solution

Replace Z by Z^{-1},

$$u[-n] \leftrightarrow \frac{Z^{-1}}{Z^{-1} - 1} = \frac{1}{1 - (Z)}$$

(8.102)

ROC $|Z| < 1$. Note that the term inside the bracket in Eq. (8.102) decides the ROC.

Example 8.27

Compute the convolution of the two sequences given by $x_1(n) = \{1, -2, 1\}$ and $x_2(n) = \{1, 1, 1, 1, 1, 1\}$. The arrow symbol indicates the sample value at $n = 0$.

Solution

We will compute the convolution using the property of convolution. To find the convolution in DT domain, we have to multiply the ZTs of the two DT signals.

Using the formula for ZT namely $X(Z) = \sum_{n=-\infty}^{\infty} x[n] Z^{-n}$, we have

$$X_1(Z) = 1 - 2Z^{-1} + Z^{-2} \tag{8.103}$$

$$X_2(Z) = 1 + Z^{-1} + Z^{-2} + Z^{-3} + Z^{-4} + Z^{-5} \tag{8.104}$$

Let us multiply the two Z transforms to get

$$X_1(Z) X_2(Z) = 1 - Z^{-1} - Z^{-6} + Z^{-7} \tag{8.105}$$

We can pull the coefficients of the product to generate the convolved sequence.

$$x_1(n) * x_2(n) = \{1, -1, 0, 0, 0, 0, -1, 1\} \tag{8.106}$$

✓ **Teaser** The reader can verify that the actual convolution process for the two sequences yields the same result.

Example 8.28

If $x[n] = a^n u[n] \leftrightarrow X(Z) = \dfrac{Z}{Z-a}$ with ROC as $|Z| > a$, then find ZT of the sequence $na^n u[n]$ using the property of differentiation in Z domain.

Solution

Using the property of differentiation in the Z domain,

$$x[n] = a^n u[n] \leftrightarrow X(Z) = \frac{Z}{Z-a}$$

$$na^n u[n] \leftrightarrow -Z \frac{d}{dZ}\left(\frac{Z}{Z-a}\right) = -Z \times \frac{Z-a-Z}{(Z-a)^2} \tag{8.107}$$

$$= \frac{aZ}{(Z-a)^2}$$

ROC = $|Z| > a$. ZT has a pole of order two at $Z = a$.

✓ **Teaser** Find the ZT of $na^n \sin(n\omega T)$ using the result of Problem 8.11 and using the property of differentiation in the Z domain. The problem is left as an exercise for the reader.

Example 8.29

Find the ZT of $x[n] = a^n \cos(\omega n) u[n]$

Solution

We will use the result of Example 8.9. Using the property of scaling in Z domain

$$ZT(\cos \omega n) = \left[\frac{Z(Z - \cos \omega)}{(Z^2 - 2\cos(\omega)Z + 1)} \right]$$

(8.108)

$$ZT(a^n \cos \omega n) = \frac{a^{-1}Z(a^{-1}Z - \cos \omega)}{a^{-2}Z^2 - 2\cos(\omega)a^{-1}Z + 1}$$

We have to replace Z by $a^{-1}Z$.

Example 8.30

Find the ZT of $x[n] = u[n-2] * \left(\frac{1}{3}\right)^n u[n]$

Solution

We will use the property of convolution. To convolve the two sequences in DT domain, we have to multiply their ZTs. We will find the ZT of each component signal and then will multiply the two.

$$ZT(u[n-2]) = Z^{-2} \times \frac{Z}{Z-1},$$

(8.109)

we have used the property of time shifting.

$$ZT\left(\left(\frac{1}{3}\right)^n u[n]\right) = \frac{Z}{Z - \left(\frac{1}{3}\right)},$$

(8.110)

we have used property of scaling in Z domain.
We will now multiply the two ZTs.

$$ZT\left(u[n-2] * \left(\frac{1}{3}\right)^n u[n]\right) = Z^{-2} \times \frac{Z}{Z-1} \times \frac{Z}{Z-\frac{1}{3}} = \frac{1}{(Z-1)\left(Z-\frac{1}{3}\right)}$$

(8.111)

Example 8.31

Find the ZT of $x[n] = na^{n-1}u[n]$

Solution

We will use the property of differentiation in the Z domain.

$$X(Z) = \sum_{n=0}^{\infty} a^{n-1}Z^{-n} = a^{-1}\sum_{n=0}^{\infty} a^n Z^{-n}$$

$$= a^{-1}\frac{Z}{Z-a} \tag{8.112}$$

$$ZT(na^{n-1}u[n]) = a^{-1} \times -Z\frac{d}{dZ}\left(\frac{Z}{Z-a}\right)$$

$$= -a^{-1}Z\left[\frac{Z-1-Z}{(Z-a)^2}\right] = a^{-1}Z/(Z-a)^2$$

Example 8.32

Find the ZT of $x[n] = n^2 a^n u[n]$

Solution

$$ZT(a^n u[n]) = \frac{Z}{Z-a}$$

We will use property of differentiation in Z domain twice.

$$ZT(na^n u[n]) = -Z\frac{d}{dZ}\left(\frac{Z}{Z-a}\right) = -Z \times \frac{Z-a-Z}{(Z-a)^2}$$

$$= \frac{aZ}{(Z-a)^2} \tag{8.113}$$

$$ZT(n^2 a^n u[n]) = -Z\frac{d}{dZ}\left(\frac{aZ}{(Z-a)^2}\right)$$

$$= -Z \times \frac{a(Z^2 - 2aZ + a^2) - aZ(2Z - 2a)}{(Z^2 - 2az + a^2)^2}$$

$$= -Z \times \frac{a(Z-a)^2 - 2aZ(Z-a)}{(Z-a)^4}$$

$$-aZ \times \frac{(Z-a)(Z-a-2Z)}{(Z-a)^4}$$

$$= -aZ \times \frac{-Z-a}{(Z-a)^3} = \frac{aZ(Z+a)}{(Z-a)^3}$$

Example 8.33

Find the initial value and final value of the following ZTs.

1. $X(Z) = \dfrac{1}{1-Z^{-2}}$ 2. $X(Z) = \dfrac{2Z^{-1}}{1-1.8Z^{-1}+0.8Z^{-2}}$

Solution

We will use initial value theorem.

$$x(0) = \lim_{Z \to \infty} X(Z) = \lim_{Z \to \infty} \frac{1}{1-Z^{-2}} = 1 \tag{8.114}$$

$$x(0) = \lim_{Z \to \infty} \frac{2Z^{-1}}{1-1.8Z^{-1}+0.8Z^{-2}} = 2$$

We will now use final value theorem.

$$\lim_{n \to \infty} x(n) = x(\infty) = \lim_{Z \to 1}(Z-1)X(Z)$$

$$= \lim_{Z \to 1}(Z-1)\frac{1}{1-Z^{-2}} = \lim_{Z \to 1}\left(\frac{Z^2(Z-1)}{Z^2-1}\right) \tag{8.115}$$

$$= \lim_{Z \to 1}\frac{Z^2}{Z+1} = 1/2$$

$$x(\infty) = \lim_{Z \to 1}(Z-1)X(Z) = \lim_{Z \to 1}(Z-1)\frac{2Z^{-1}}{1-1.8Z^{-1}+0.8Z^{-2}} \tag{8.116}$$

$$= \lim_{Z \to 1}\frac{(Z-1)2Z}{(Z-1)(Z-0.8)} = \lim_{Z \to 1}\frac{2Z}{Z-0.8} = 2/0.2 = 10$$

Concept Check

- If a sequence is the addition of two sequences, what will be the ROC?
- When a sequence is time shifted by 4 units, what will happen to its ROC?

- When a sequence is multiplied by a complex exponential, what will happen in the Z domain?
- When a sequence is time reversed, what will happen to its ROC?
- If we know the multiplication of Z transforms of the two sequences, how do we calculate their convolution?
- State the property of differentiation in the Z domain.
- State initial and final value theorems.

8.7 Relation between Pole Locations and Time Domain Behavior

Stability of the system is analyzed by identifying the locations of poles of the system transfer function. We will consider the input to a system as unit impulse, that is, $\delta[n]$. Let the impulse response of the system be represented as $h[n]$, then the output of the system will also be $h[n]$. The system transfer function in the Z domain is given by

$$H(Z) = \frac{Y(Z)}{X(Z)} \tag{8.117}$$

where $X(Z)$ is ZT of $\delta[n]$ and is equal to one. (This can be verified from the basic equation of ZT.)

$$H(Z) = \frac{Y(Z)}{1} = Y(Z) \tag{8.118}$$

ZT of $y[n]$ is $Y(Z) = H(Z)$. The system transfer function is $H(Z)$. The poles of $H(Z)$ will reflect the stability of the system.

Let us consider the mapping between Laplace domain and Z domain. The complex variable Z is defined as $Z = e^{sT}$ (refer to Section 8.2). The complex Z may be written as $Z = re^{j\omega T}$, where r is the magnitude of Z and ωT stands for the phase.

1. Consider any point on the imaginary axis in the S (Laplace) domain. $S = \sigma + j\omega$. We have $S = j\omega$ on the imaginary axis, as here the real part is 0. Now $Z = e^{j\omega T}$ and $r = 1$ ($|e^{j\omega T}| = 1$). Thus a point on the imaginary axis in the S domain will map to a point on the unit circle in the Z domain.

2. Any point on the left side of imaginary axis will have negative value for σ and therefore $|e^{-\sigma}|$ is less than one. A point on the left-hand side of the imaginary axis in the S domain will map to a point inside the unit circle in the Z domain.

3. Similarly, a point on right-hand side of the imaginary axis in the S domain will map to a point outside the unit circle in the Z domain, as $|e^{\sigma}|$ is greater than one.

We can now translate the stability criteria from the S domain to the Z domain. When the poles are inside the unit circle in the Z domain, the system is stable and its impulse response, that is, time domain behavior, will be a decaying signal. Here, the output is bounded. When the poles are outside the unit circle in the Z domain, the system is unstable and its impulse response, that is, time domain behavior, will be an exponentially increasing function becoming infinite as the sample number (n) tends to infinity. We say that the output is unbounded. The system is marginally stable if there are poles of order one on the unit circle. The system becomes unstable for higher order poles on unit circle.

We will analyze systems with real poles of order one and two and the systems with complex conjugate poles. We will illustrate the behavior of stable and unstable systems with the help of known sequences representing the impulse response of the system.

Example 8.34

Let the impulse response of the system be represented as $x[n] = a^n u[n]$. We know that ZT of $x[n]$ is

$$X(Z) = \frac{Z}{Z-a} \qquad (8.119)$$

with ROC as $|Z| > a$. This ZT has a pole at $Z = a$ and a zero at $Z = 0$.

Let us consider three different cases depending on the value of "a".

Case I $a < 1$

The pole is inside the unit circle. The sequence representing the impulse response is causal with ROC $|Z| > a$. ROC will include the unit circle in the Z domain. The pole zero plot and the time domain behavior are shown in Fig. 8.13. We see that the time domain behavior is a decaying function and the output is bounded. The system is stable.

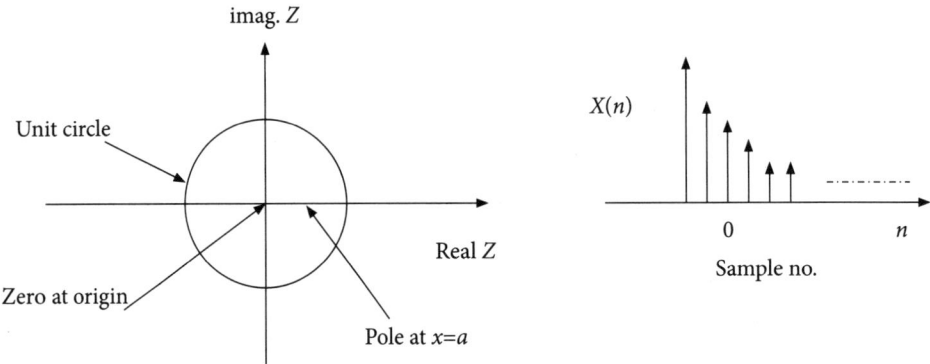

Fig. 8.13 Pole zero diagram and time domain behavior for $a < 1$

Case 2 $a > 1$

The pole is outside the unit circle. The system is unstable and it generates unbounded output. ROC excludes the unit circle in the Z domain. The pole zero plot and the time domain behavior are shown in Fig. 8.14.

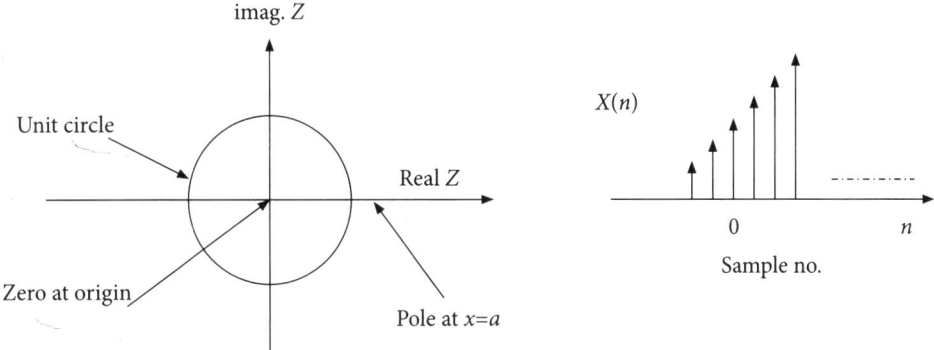

Fig. 8.14 Pole zero diagram and time domain behavior when $a > 1$

Case 3 $a = 1$, pole is on the unit circle

The response of the system reduces to a unit step function. The system is marginally stable. The pole zero diagram and time domain behavior are shown in Fig. 8.15.

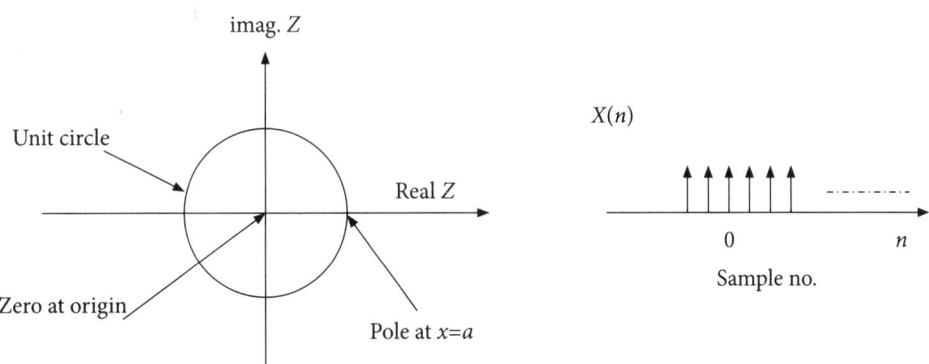

Fig. 8.15 Pole zero diagram and time domain behavior for $a = 1$

Example 8.35

Let us consider a system with impulse response represented by a sequence $x[n] = na^n u[n]$. The Z domain transfer function is

$$X(Z) = \frac{aZ}{(Z-a)^2} \qquad (8.120)$$

The pole location here is at $x = a$, and the pole is of the order 2. The time domain behavior is shown in Figs 8.16, 8.17 and 8.18 for $|a| > 1$, $|a| < 1$ and $|a| = 1$, respectively. For $a = 1$, the time domain output is divergent.

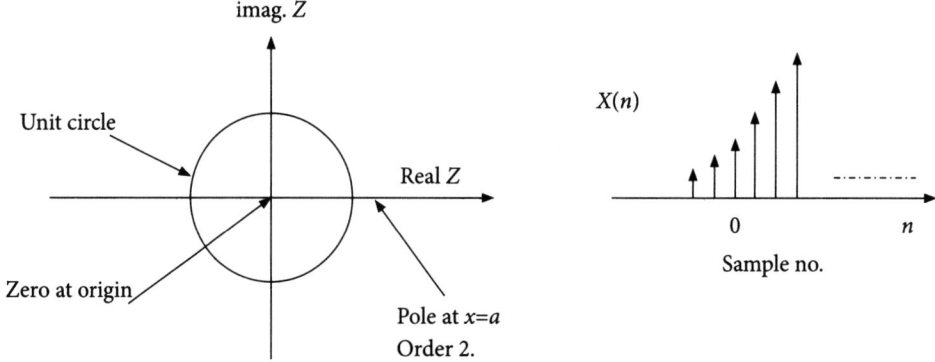

Fig. 8.16 Pole zero diagram and time domain behavior for $a > 1$

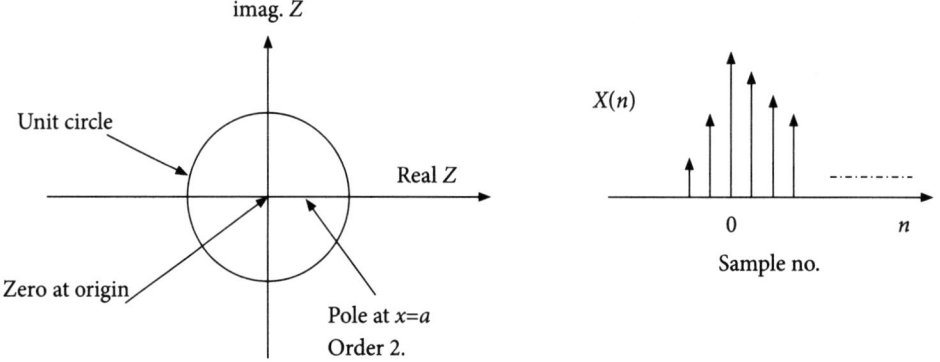

Fig. 8.17 Pole zero diagram and time domain behavior for $a < 1$

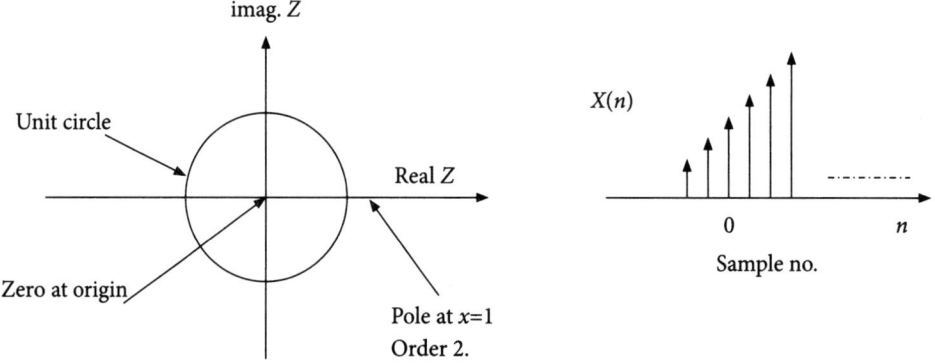

Fig. 8.18 Pole zero diagram and time domain behavior for $a = 1$

Example 8.36

Let us consider a system with impulse response represented by a sequence $x(n) = a^n \sin(n\omega T)u(n)$. The Z domain transfer function is

$$X(Z) = \frac{a^{-1} Z \sin(\omega T)}{a^{-2} Z^2 - 2\cos(\omega T) a^{-1} Z + 1} \tag{8.121}$$

The poles are complex conjugates. The pole locations are $ae^{j\omega T}$ and $ae^{-j\omega T}$. The time domain behavior and pole zero plot are shown in Figs 8.19, 8.20 and 8.21 for $|a| < 1$, $|a| > 1$ and $|a| = 1$, respectively. The impulse response is a decaying sine function for $a < 1$, exponential increasing sinusoid for $a > 1$ and a pure sine wave for $a = 1$.

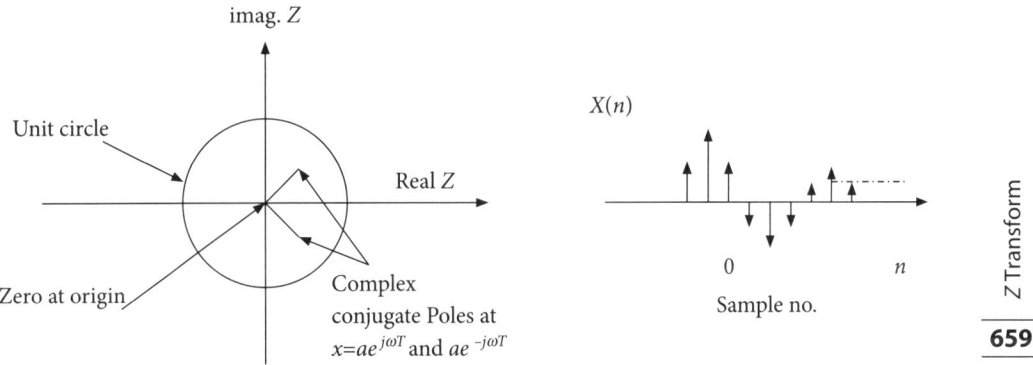

Fig. 8.19 Pole zero diagram and time domain behavior for $a < 1$

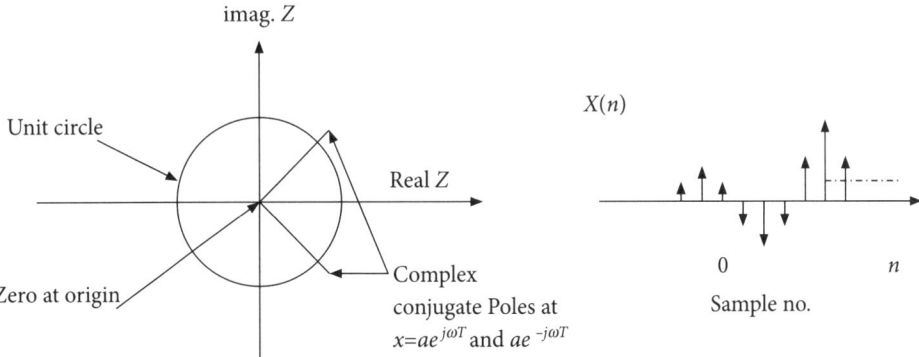

Fig. 8.20 Pole zero diagram and time domain behavior for $a > 1$

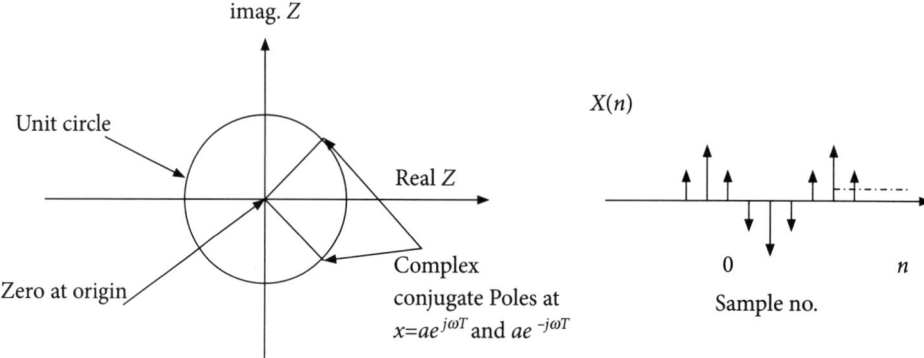

Fig. 8.21 Pole zero diagram and time domain behavior for $a = 1$

Concept Check

- Can we translate the stability criteria from S domain to Z domain?
- If the impulse response of a system is given by $x(n) = a^n u(n)$, with $a < 1$, is the system stable?
- If the pole locations for a transfer function of a system are $ae^{j\omega T}$ and $ae^{-j\omega T}$ ($a > 1$), is the system stable?

8.8 Inverse Z Transform

Given any DT sequence, we can find its ZT and ROC. There is one-to-one correspondence between the sequence and its ZT (with ROC). The ZT is an invertible transform in the sense that we can recover a sequence from the given ZT and ROC. This process is termed as inverse Z transformation. There are three different methods for evaluation of inverse ZT.

1. Power series method
2. Partial fraction expansion method (PFE)
3. Residue method

8.8.1 Power series method/long division method

A given ZT $F(Z)$ is represented as a power series in Z to obtain

$$F(Z) = a_0 + a_1 Z^{-1} + a_2 Z^{-2} + \ldots \tag{8.122}$$

Using the definition of Z transform, we have

$$F(Z) = \sum_{n=0}^{\infty} f[n] Z^{-n} \tag{8.123}$$

Comparing Eqs (8.122) and (8.123) we observe that there is one-to-one correspondence between power series coefficients a_n and the desired sequence values $f(0), f(1)$, etc.:

$$f[n] = a_n \text{ for } n \geq 0 \tag{8.124}$$

Example 8.37

If $F(Z) = \dfrac{Z^2 + Z}{Z^3 - 3Z^2 + 3Z - 1} = \dfrac{N(Z)}{D(Z)}$,

find a causal sequence $f[n]$.

Solution

Using long division, let us divide the numerator polynomial by a denominator polynomial to obtain a power series in Z^{-1} or Z depending on whether the series to be recovered is a right-handed sequence (causal) or a left-handed sequence (non-causal).

Here, the sequence to be recovered is a causal sequence. Let us obtain a series in Z^{-1}.

$$F(Z) = \dfrac{Z^2 + Z}{Z^3 - 3Z^2 + 3Z - 1} = Z^{-1} + 4Z^{-2} + 9Z^{-3} + 16Z^{-4} + \ldots \tag{8.125}$$

Now let us pull the coefficients of the Z^{-1} series to get the sequence values for $n = 0, 1, 2, 3$, etc. There is no constant term, so the sequence value for $n = 0$ is zero. Sequence values for $n = 1, 2, 3$, etc. are 1, 4, 9, 16, etc. The causal sequence $f(n)$ is recovered as

$$f[n] = \{0, 1, 4, 9, 16 \ldots\} \tag{8.126}$$

That is

$$f[n] = n^2 u[n] \tag{8.127}$$

$f[n]$ is the inverse Z transform (IZT) of $F(Z)$.

Example 8.38

If $X(Z) = \dfrac{1}{1 - aZ^{-1}}$ for $|Z| > a$,

find $x[n]$.

Solution

Express $X(Z)$ as a Z^{-1} series by carrying out a long division. ROC is $|Z| > a$, so the sequence is a right-handed sequence

$$X(Z) = \frac{1}{1-aZ^{-1}} = 1 + aZ^{-1} + a^2 Z^{-2} + \ldots \tag{8.128}$$

The right-handed sequence $x[n]$ can be recovered as

$$\begin{aligned} x[n] &= \{1, a, a^2, a^3, \ldots\} \\ x[n] &= a^n u[n] \end{aligned} \tag{8.129}$$

The power series method is useful if we want to determine the first few samples of the signal. Sometimes, the resulting pattern is simple enough to judge the general term $x(n)$ as in case of Eq. (8.128).

Example 8.39

If $X(Z) = \dfrac{1}{1-aZ^{-1}}$ for $|Z| < a$, find $x[n]$.

Solution

Express $X(Z)$ as a Z series by carrying out a long division. ROC is $|Z| < a$, so the sequence is a left-handed sequence:

$$X(Z) = \frac{1}{1-aZ^{-1}} = \frac{1}{-aZ^{-1}+1} \tag{8.130}$$

$$X(Z) = -a^{-1}Z - a^{-2}Z^2 - a^{-3}Z^3 + \ldots \tag{8.131}$$

Note that in the case of non-causal sequence, we simply carry out the long division by writing down the two polynomials in the reverse order, that is, by starting with the most negative term. The left-handed sequence $x[n]$ can be recovered as

$$\begin{aligned} x[n] &= \{-a^{-1}, -a^{-2}, -a^{-3}, \ldots\} \\ x[n] &= -a^n u[-n-1] \end{aligned} \tag{8.132}$$

Sometimes, it is easy to use a property of differentiation in the Z domain to calculate IZT instead of directly writing a power series. We will solve the next problem using the property of differentiation in the Z domain.

Example 8.40

If $X(Z) = \log(1 + aZ^{-1})$ for $|Z| > a$, find $x(n)$ using the property of differentiation in the Z domain.

Solution

Using the property of differentiation in the Z domain, we have the following:
If given $X(Z)$ is ZT of $x[n]$, then

$$nx[n] \leftrightarrow -Z\frac{d}{dZ}\log(1+aZ^{-1})\ ROC\,|Z|>a \tag{8.133}$$

$$=\frac{-Z(-aZ^{-2})}{1+aZ^{-1}}=\frac{aZ^{-1}}{1+aZ^{-1}}$$

We know that

$$(-a)^n u[n] \leftrightarrow \frac{1}{1+aZ^{-1}} \tag{8.134}$$

[Replace a with $-a$ in Eq. (8.134).]
Using the property of time shifting we have

$$(-a)^{n-1} u[n-1] \leftrightarrow \frac{Z^{-1}}{1+aZ^{-1}} \tag{8.135}$$

Multiply both sides of Eq. (8.135) with a:

$$a(-a)^{n-1} u[n-1] \leftrightarrow \frac{aZ^{-1}}{1+aZ^{-1}} \tag{8.136}$$

So we get

$$x(n) = \frac{-(-a)^n u(n-1)}{n} \tag{8.137}$$

✓ **Teaser** The problem can also be solved by writing a power series for the log term and then by pulling the coefficients of the Z^{-1} series to get the sequence values for $n = 0, 1, 2, 3$, etc. This is left as an exercise for the reader.

Example 8.41

If $X(Z) = \dfrac{aZ^{-1}}{(1-aZ^{-1})^2}$ for $|Z| > a$, find $x[n]$ using the property of differentiation in the Z domain.

Solution

Using the property of differentiation in Z domain, we have the following:
If given $X(Z)$ is ZT of $x[n]$, then

$$nx[n] \xleftrightarrow{ZT} -Z\frac{d}{dz}(X(Z)) \quad \text{ROC} = |Z| > a \tag{8.138}$$

We know that $(a)^n u[n] \leftrightarrow \dfrac{1}{1-aZ^{-1}}$ (8.139)

$$n(a)^n u[n] \leftrightarrow -Z\frac{d}{dZ}\left[\frac{1}{1-aZ^{-1}}\right] = \frac{Z(aZ^{-2})}{(1-aZ^{-1})^2} = \frac{aZ^{-1}}{(1-aZ^{-1})^2} \tag{8.140}$$

So IZT of $X(Z)$ is given by

$$x[n] = na^n u[n] \tag{8.141}$$

8.8.2 Partial fraction expansion method

The partial fraction expansion method is used to decompose a rational fraction given by $X(Z)$ as a linear combination of ZTs of known sequences:

$$X(Z) = \frac{N(Z)}{D(Z)} \tag{8.142}$$

If this type of decomposition is possible, then it is possible for us to use a simple look-up table method to infer the sequence $x[n]$ as a linear combination of known sequences.

Let $X(Z)$ be represented as

$$X(Z) = \frac{b_0 + b_1 Z^{-1} + b_2 Z^{-2} + \ldots + b_M Z^{-M}}{1 + a_1 Z^{-1} + a_2 Z^{-2} + \ldots a_N Z^{-N}} \tag{8.143}$$

Here $a_N \neq 0$ and $N > M$. This rational form is termed as "proper." If $M > N$, then $X(Z)$ can be written as a sum of a polynomial and a "proper" rational function. The IZT of the polynomial can be found very easily by inspection by pulling the coefficients. Let us discuss the evaluation of IZT of a rational function further.

Let us consider $X(Z)$ as a non-proper rational function with $M = N = 2$. We will express $X(Z)$ as a linear combination of a polynomial and a rational function. The IZT of the rational function with a numerator of order 1 and a denominator of order 2 is to be evaluated using partial fraction expansion.

Let us now discuss the partial fraction expansion (PFE) method to decompose a rational function. We will list the steps involved in PFE method for the IZT calculation of a rational function $X(Z)$.

- **Step 1** Multiply the numerator and denominator of $X(Z)$ by Z^N, where N is the negative power of the denominator.
- **Step 2** Evaluate $P(Z) = X(Z)/Z$ and decompose the denominator of $P(Z)$ into number of possible factors. Each factor is of the form Z-P, where P is the location of a pole for $P(Z)$.
- **Step 3** Write $P(Z)$ as

$$P(Z) = \frac{k_1}{Z - p_1} + \frac{k_2}{Z - p_2} + \ldots \tag{8.144}$$

where $p_1, p_2, \ldots$ are poles of $P(Z)$.

- **Step 4** Calculate k_1 and k_2 using the formula

$$\begin{aligned} k_1 &= (Z - p_1)P(Z) \downarrow_{Z=p_1} \\ k_2 &= (Z - p_2)P(Z) \downarrow_{Z=p_2} \end{aligned} \tag{8.145}$$

Note that we can evaluate ILT (inverse Laplace transform) of $1/(S + p_1)$. However, to evaluate IZT, the expression must in the form $Z/(Z - p_1)$. In order to borrow the same procedure for PFE, we divide $X(Z)$ by Z to get the following form of equation $1/(Z - p_1)$. Finally, after using PFE, multiply each term by Z to get the exact form for the evaluation of IZT.

- **Step 5** After proper decomposition multiply each term by Z to get it in proper form and calculate the IZT of each term.
- **Step 6** Express $x[n]$ as a linear combination of individual IZTs.

Example 8.42

Let $X(Z)$ be given by the equation $X(Z) = \dfrac{1 - 2Z^{-1} + 2Z^{-2}}{1 - 3Z^{-1} + 2Z^{-2}}$

Find the IZT.

Solution

$$X(Z) = \frac{1 - 2Z^{-1} + 2Z^{-2}}{1 - 3Z^{-1} + 2Z^{-2}} = 1 + \frac{Z^{-1}}{1 - 3Z^{-1} + 2Z^{-2}} \tag{8.146}$$

$$X(Z) = 1 + \frac{Z}{Z^2 - 3Z + 2}$$

Let $X_1(Z) = \dfrac{Z}{Z^2 - 3Z + 2}$

$$\frac{X_1(Z)}{Z} = \frac{1}{Z^2 - 3Z + 2} = \frac{k_1}{Z - 2} + \frac{k_2}{Z - 1}$$

Let us find the coefficients.

$$k_1 = (Z-2)X_1(Z)\Big\downarrow_{Z=2} = \frac{1}{(Z-1)}\Big\downarrow_{Z=2} = 1$$

(8.147)

$$k_2 = (Z-1)X_1(Z)\Big\downarrow_{Z=1} = \frac{1}{Z-2}\Big\downarrow_{Z=1} = -1$$

$$\frac{X_1(Z)}{Z} = \frac{1}{Z^2 - 3Z + 2} = \frac{1}{Z-2} - \frac{1}{Z-1}$$

(8.148)

$$X(Z) = 1 + \frac{Z}{Z-2} - \frac{Z}{Z-1}$$

Take IZT.

$$x[n] = \delta[n] + (2)^n u[n] - u[n]$$

Example 8.43

For $F(Z) = \dfrac{Z}{3Z^2 - 4Z + 1}$,

find $f[n]$ if

1. ROC is $|Z| > 1$
2. ROC is $|Z| < 1/3$
3. If ROC is $1/3 < |Z| < 1$

Use PFE method.

Solution

Case I ROC is $|Z| > 1$

Step 1 Here, this step is not required, as $F(Z)$ is already in the required form.

Step 2 Evaluate

$$\frac{F(Z)}{Z} = \frac{1}{3Z^2 - 4Z + 1}$$

(8.149)

Step 3 Decompose the denominator

$$\frac{F(Z)}{Z} = \frac{1}{(3Z-1)(Z-1)} = \frac{1}{3}\frac{1}{(Z-1/3)(Z-1)}$$

(8.150)

$$\frac{F(Z)}{Z} = \frac{k_1}{\left(Z - \frac{1}{3}\right)} + \frac{k_2}{(Z-1)} \qquad (8.151)$$

Step 4 Calculate k_1 and k_2 using

$$k_1 = \left(Z - \frac{1}{3}\right) F(Z) \Big\downarrow_{Z=\frac{1}{3}} = \frac{1}{3(Z-1)} \Big\downarrow_{Z=\frac{1}{3}} = -\frac{1}{2}$$

$$k_2 = (Z-1) F(Z) \Big\downarrow_{Z=1} = \frac{1}{3\left(Z - \frac{1}{3}\right)} \Big\downarrow_{Z=1} = \frac{1}{2} \qquad (8.152)$$

So $$\frac{F(Z)}{Z} = \frac{-\frac{1}{2}}{\left(Z - \frac{1}{3}\right)} + \frac{\frac{1}{2}}{Z-1} \qquad (8.153)$$

Step 5

$$F(Z) = \frac{-\frac{1}{2}Z}{\left(Z - \frac{1}{3}\right)} + \frac{\frac{1}{2}Z}{Z-1} \qquad (8.154)$$

Step 6 Evaluate IZT of each term.

$$f[n] = -\frac{1}{2}\left(\frac{1}{3}\right)^n u[n] + \frac{1}{2} u[n] \qquad (8.155)$$

We have recovered both sequences as right-handed sequences, because ROC for the first term is $|Z| > 1$ and that for the second term is $|Z| > 1/3$, if both are right-handed sequences. Here, the common area of convergence is $|Z| > 1$.

Case 2 ROC is $|Z| < 1/3$

Here, we have to recover both the sequences as left-handed sequences. We can write

$$f[n] = \frac{1}{2}\left(\frac{1}{3}\right)^n u[-n] - \frac{1}{2} u[-n] \qquad (8.156)$$

Case 3 ROC is $1/3 < Z < 1$

Here we have to recover the sequence for the second term as a left-handed sequence and the sequence for the first term as a right-handed sequence. We can write

$$f[n] = -\frac{1}{2}\left(\frac{1}{3}\right)^n u[n] - \frac{1}{2}u[-n] \tag{8.157}$$

We will now illustrate the procedure for PFE if the denominator has a pole of higher order.

Example 8.44

Find $f[n]$ for $F(Z) = \dfrac{2Z^2}{(Z+1)(Z+2)^2}$

ROC is $|Z| > 2$.

Solution

Step 1 Here this step is not required, as $F(Z)$ is already in the required form.

Step 2 Evaluate

$$\frac{F(Z)}{Z} = \frac{2Z}{(Z+1)(Z+2)^2} \tag{8.158}$$

Step 3 Decompose the denominator

$$\frac{F(Z)}{Z} = \frac{k_1}{(Z+1)} + \frac{k_2}{(Z+2)} + \frac{k_3}{(Z+2)^2}$$

Step 4 Calculate k_1, k_2 and k_3 using

$$k_1 = (Z+1)\frac{F(Z)}{Z} = \frac{2Z}{(Z+2)^2}\Bigg\downarrow_{Z=-1} = -2$$

$$k_2 = \frac{d}{dZ}\left[(Z+2)^2 \frac{F(Z)}{Z}\right] = \frac{d}{dZ}\left[\frac{2Z}{Z+1}\right]\Bigg\downarrow_{Z=-2} = 2 \tag{8.159}$$

$$k_3 = (Z+2)^2 \frac{F(Z)}{Z} = \frac{2Z}{Z+1}\Bigg\downarrow_{Z=-2} = 4$$

$$\frac{F(Z)}{Z} = \frac{-2}{Z+1} + \frac{2}{(Z+2)} + \frac{4}{(Z+2)^2} \tag{8.160}$$

Step 5

$$F(Z) = \frac{-2Z}{(Z+1)} + \frac{2Z}{(Z+2)} + \frac{4Z}{(Z+2)^2} \quad (8.161)$$

Step 6 Evaluate the IZT of each term. We will first concentrate on the third term. We know that using the property of differentiation in the Z domain,

$$(-2)^n u[n] \leftrightarrow \frac{Z}{Z+2} \text{ for } |Z| > 2$$

$$n(-2)^n u[n] \leftrightarrow \frac{-2Z}{(Z+2)^2} \quad (8.162)$$

$$(-2)n(-2)^n u[n] \leftrightarrow \frac{4Z}{(Z+2)^2}$$

$$f[n] = -2(-1)^n u[n] + n(-2)^{n+1} u[n] + 2(-2)^n u[n] \quad (8.163)$$

All sequences are recovered as right-handed sequences.

Example 8.45

If $F(Z) = \dfrac{Z+1}{3Z^2 - 4Z + 1}$,

find $f[n]$ if ROC is $|Z| > 1$, ROC is $|Z| < 1/3$ and if ROC is $1/3 < |Z| < 1$.

Solution
Step 1

$$F(Z) = \frac{Z+1}{3(Z-1/3)(Z-1)} \quad (8.164)$$

We have factorized the denominator polynomial.

Step 2 Evaluate

$$\frac{F(Z)}{Z} = \frac{Z+1}{3Z(Z-1/3)(Z-1)} \quad (8.165)$$

Step 3 Decompose the denominator

$$\frac{F(Z)}{Z} = \frac{k_1}{Z} + \frac{k_2}{Z-(1/3)} + \frac{k_3}{Z-1} \quad (8.166)$$

Step 4 Calculate k_1, k_2 and k_3 using

$$k_1 = (Z)\frac{F(Z)}{Z} = \frac{Z+1}{3(Z-1/3)(Z-1)}\Big|_{Z=0} = 1$$

$$k_2 = (Z-1/3)\frac{F(Z)}{Z} = \frac{Z+1}{3Z(Z-1)}\Big|_{Z=1/3} = -2 \tag{8.167}$$

$$k_3 = (Z-1)\frac{F(Z)}{Z} = \frac{Z+1}{3Z(Z-1/3)}\Big|_{Z=1} = 1$$

$$\frac{F(Z)}{Z} = \frac{1}{Z} + \frac{-2}{(Z-1/3)} + \frac{1}{(Z-1)} \tag{8.168}$$

Step 5

$$F(Z) = 1 + \frac{-2Z}{(Z-1/3)} + \frac{Z}{(Z-1)} \tag{8.169}$$

Step 6 Evaluate IZT of each term.

$$f[n] = \delta[n] - 2(1/3)^n u[n] + u[n] \tag{8.170}$$

Note that all sequences are recovered as right-handed sequences if the ROC is $|Z| > 1$.

If ROC is $|Z| < 1/3$, we will recover both the terms second and third as the left-handed sequence as follows.

$$f[n] = \delta[n] + 2(1/3)^n u[-n-1] - u[-n-1] \tag{8.171}$$

If ROC is $1/3 < |Z| < 1$, we will recover the second term as the right-handed sequence and third term as the left-handed sequence as follows.

$$f[n] = \delta[n] - 2(1/3)^n u[n] - u[-n-1] \tag{8.172}$$

Example 8.46

Find $f[n]$ if ROC is $|Z| > \tfrac{1}{2}$ for $F(Z) = \dfrac{Z^2 + Z}{(Z-1/4)(Z-1/2)^3}$

Solution

Step 1 Here this step is not required, as $F(Z)$ is already in the required form.

Step 2 Evaluate

$$\frac{F(Z)}{Z} = \frac{Z+1}{(Z-1/4)(Z-1/2)^3} \tag{8.173}$$

Step 3 Decompose the denominator

$$\frac{F(Z)}{Z} = \frac{k_1}{(Z-1/4)} + \frac{k_2}{(Z-1/2)} + \frac{k_3}{(Z-1/2)^2} + \frac{k_4}{(Z-1/2)^3} \tag{8.174}$$

Step 4 Calculate k_1, k_2, k_3 and k_4 using

$$k_1 = (Z-1/4)\frac{F(Z)}{Z} = \frac{Z+1}{(Z-1/2)^3}\bigg\downarrow_{Z=1/4} = -80$$

$$k_2 = \frac{1}{2!}\frac{d^2}{dZ^2}\left[(Z-1/2)^3\frac{F(Z)}{Z}\right] = \frac{1}{2}\frac{d^2}{dZ^2}\left[\frac{Z+1}{(Z-1/4)}\right]\bigg\downarrow_{Z=1/2} = 80$$

$$\tag{8.175}$$

$$k_3 = \frac{d}{dZ}\left[(z-1/2)^3\frac{F(Z)}{Z}\right] = \frac{d}{dZ}\left[\frac{Z+1}{(Z-1/4)}\right]\bigg\downarrow_{Z=1/2} = -20$$

$$k_4 = (Z-1/2)^3\frac{F(Z)}{Z} = \frac{Z+1}{Z-1/4}\bigg\downarrow_{Z=1/2} = 6$$

$$\frac{F(Z)}{Z} = \frac{-80}{Z-1/4} + \frac{80}{(Z-1/2)} - \frac{20}{(Z-1/2)^2} + \frac{6}{(Z-1/2)^3} \tag{8.176}$$

Step 5

$$F(Z) = \frac{-80Z}{Z-1/4} + \frac{80Z}{(Z-1/2)} - \frac{20Z}{(Z-1/2)^2} + \frac{6Z}{(Z-1/2)^3} \tag{8.177}$$

Step 6 Evaluate IZT of each term. We know that using property of differentiation in Z domain, if

$$na^{n-1}u[n] \leftrightarrow \frac{Z}{(Z-a)^2}$$

$$n(n-1)a^{n-2}u[n] \leftrightarrow \frac{Z}{(Z-a)^3} \tag{8.178}$$

The reader can easily verify the above-stated results. Taking IZT, we get

$$f[n] = -80\left(\frac{1}{4}\right)^n u[n] + 80\left(\frac{1}{2}\right)^n u[n] - 20n\left(\frac{1}{2}\right)^{n-1} u[n] + 6n(n-1)\left(\frac{1}{2}\right)^{n-2} u[n] \quad (8.179)$$

Note that we have recovered all the sequences as right-handed sequences.

Example 8.47

Find $f[n]$ if ROC is $1 < |Z| < 2$ for $F(Z) = \dfrac{1 - Z^{-1} + Z^{-2}}{\left(1 - \dfrac{1}{2}Z^{-1}\right)(1 - 2Z^{-1})(1 - Z^{-1})}$

Solution

Step 1

$$F(Z) = \frac{1 - Z^{-1} + Z^{-2}}{\left(1 - \dfrac{1}{2}Z^{-1}\right)(1 - 2Z^{-1})(1 - Z^{-1})} = \frac{Z^3 - Z^2 + Z}{\left(Z - \dfrac{1}{2}\right)(Z - 2)(Z - 1)} \quad (8.180)$$

Step 2   Evaluate

$$\frac{F(Z)}{Z} = \frac{Z^2 - Z + 1}{\left(Z - \dfrac{1}{2}\right)(Z - 2)(Z - 1)} \quad (8.181)$$

Step 3   Decompose the denominator

$$\frac{F(Z)}{Z} = \frac{k_1}{(Z - 1/2)} + \frac{k_2}{(Z - 2)} + \frac{k_3}{(Z - 1)} \quad (8.182)$$

Step 4   Calculate k_1, k_2 and k_3 using

$$k_1 = (Z - 1/2)\frac{F(Z)}{Z} = \frac{Z^2 - Z + 1}{(Z - 2)(Z - 1)} \Bigg\downarrow_{Z = 1/2} = \frac{3/4}{3/4} = 1$$

$$k_2 = (Z - 2)\frac{F(Z)}{Z} = \frac{Z^2 - Z + 1}{(Z - 1/2)(Z - 1)} \Bigg\downarrow_{Z = 2} = \frac{3}{3/2} = 2 \quad (8.183)$$

$$k_3 = (Z - 1)\frac{F(Z)}{Z} = \frac{Z^2 - Z + 1}{(Z - 1/2)(Z - 2)} \Bigg\downarrow_{Z = 1} = \frac{1}{-1/2} = -2$$

$$\frac{F(Z)}{Z} = \frac{1}{Z - 1/2} + \frac{2}{(Z - 2)} - \frac{2}{(Z - 1)} \quad (8.184)$$

Step 5

$$F(Z) = \frac{Z}{Z-1/2} + \frac{2Z}{(Z-2)} - \frac{2Z}{(Z-1)} \tag{8.185}$$

Step 6 Evaluate the IZT of each term. ROC is $1 < |Z| < 2$, we will recover the terms with poles at ½ and 1 as right-handed sequences and the term with pole at 2 as a left-handed sequence.

$$f[n] = \left(\frac{1}{2}\right)^n u[n] - 2(2)^n u[-n-1] - 2u[n] \tag{8.186}$$

Example 8.48

Find $f[n]$ if ROC is $|Z| < 1$ for $F(Z) = \dfrac{Z^3 - 10Z^2 - 4Z + 4}{2Z^2 - 2Z - 4}$

Solution

Step 1

$$F(Z) = \frac{Z^3 - 10Z^2 - 4Z + 4}{2Z^2 - 2Z - 4} = \frac{Z^3 - 10Z^2 - 4Z + 4}{2(Z-2)(Z+1)} \tag{8.187}$$

Step 2 Evaluate

$$\frac{F(Z)}{Z} = \frac{Z^2 - 10Z - 4 + 4Z^{-1}}{2(Z-2)(Z+1)} \tag{8.188}$$

Step 3 Decompose the denominator

$$\frac{F(Z)}{Z} = \frac{k_1}{(Z-2)} + \frac{k_2}{(Z+1)} \tag{8.189}$$

Step 4 Calculate k_1 and k_2 using

$$k_1 = (Z-2)\frac{F(Z)}{Z} = \frac{Z^2 - 10Z - 4 + 4Z^{-1}}{2(Z+1)} \Bigg\downarrow_{Z=2} = \frac{-18}{6} = -3$$

$$\tag{8.190}$$

$$k_2 = (Z+1)\frac{F(Z)}{Z} = \frac{Z^2 - 10Z - 4 + 4Z^{-1}}{2(Z-2)} \Bigg\downarrow_{Z=-1} = \frac{3}{-6} = -1/2$$

$$\frac{F(Z)}{Z} = \frac{-3}{Z-2} - \frac{1/2}{(Z+1)} \tag{8.191}$$

Step 5

$$F(Z) = \frac{-3Z}{Z-2} - 1/2\frac{Z}{(Z+1)} \tag{8.192}$$

Step 6 Evaluate IZT of each term. ROC is $|Z| < 1$, we will recover the terms with poles at -1 and 2 as left-handed sequences.

$$f[n] = 3(2)^n u[-n-1] + \frac{1}{2}u[-n-1] \tag{8.193}$$

8.8.3 Residue method

This method evaluates the residues of the function at poles that are inside the closed contour in the ROC. The sum of these residues gives the IZT of the function in the Z domain. The method is based on a theorem called Cauchy's integral theorem.

Cauchy's Integral Theorem

Cauchy's integral theorem states that

$$\frac{1}{2\pi j}\oint_c Z^{k-1}dZ = \begin{cases} 1 & \text{for } k=0 \\ 0 & \text{otherwise} \end{cases} \tag{8.194}$$

Residue method:

If $F(Z) = \sum_{n=0}^{\infty} f[n]Z^{-n}$ for $|Z| > R$, \hfill (8.195)

it can be shown that $f[n]$ can be recovered from $F(Z)$ using the inverse integral formula

$$f[n] = \frac{1}{2\pi j}\oint_c Z^{n-1}F(Z)dZ \text{ for } n \geq 0 \tag{8.196}$$

where c is any closed contour enclosing $|Z| > R$ and $\oint_c$ denotes contour integration along c in counter-clockwise direction.

Proof

Given $F(Z) = \sum_{n=0}^{\infty} f[n]Z^{-n}$. Multiply both sides by

$$\frac{1}{2\pi j}\frac{Z^k}{Z}dZ \qquad (8.197)$$

and integrate in counter-clockwise direction on contour c, which lies entirely in the region of convergence.

$$\frac{1}{2\pi j}\oint_c F(Z)\frac{Z^k}{Z}dZ = \frac{1}{2\pi j}\sum_{n=0}^{\infty}\oint_c f[n]Z^{-n+k-1}dZ \qquad (8.198)$$

Using Cauchy's integral theorem, we can write

$$\frac{1}{2\pi j}\oint_c Z^{-n+k-1}dZ = \begin{cases} 1 & \text{for } (k-n)=0 \\ 0 & \text{otherwise} \end{cases} \qquad (8.199)$$

Hence, put $k = n$ in the term on the right-hand side of Eq. (8.198). It can now be written as just $f[k]$ as the integral exists only for $n = k$. Statement of the method is thus proved. The integration over the closed contour denotes the evaluation of residues.

We will now illustrate the procedure for the residue method.

Step 1 Evaluate $G(Z) = F(Z)Z^{n-1}$

Step 2 Select a closed contour c inside the ROC.

Step 3 Find the poles of the function $G(Z)$ inside the closed contour c.

Step 4 Evaluate the residue of the function $G(Z)$ for all poles within the closed contour c. Then, $f[n]$ is recovered as a sum of all these residues.

Example 8.49

Find $x[n]$ for $X(Z) = \dfrac{Z}{Z-a}$ for $|Z| > a$ using residue method.

Solution

Step 1 Find

$$G(Z) = \frac{Z \times Z^{n-1}}{Z-a} = \frac{Z^n}{Z-a} \qquad (8.200)$$

Step 2 We will now select a closed contour as a circle with radius $|Z| > a$, so that it is in ROC.

Step 3 The poles of $G(Z)$ will depend on the value of n. We will consider two different cases $n \geq 0$, $n < 0$.

Case I $n \geq 0$. Here, there is only pole of $G(Z)$ at $Z = a$, which is inside the circle with radius $|Z| > a$. We will evaluate the residue of $G(Z)$ at $Z = a$.

$$R[G(Z)]\downarrow_{Z=a}=(Z-a)G(Z)\downarrow_{Z=a}=Z^n\downarrow_{Z=a}=a^n \qquad (8.201)$$

$$x[n]=a^n \text{ for } n\geq 0 \qquad (8.202)$$

Case 2 $n < 0$. Let $n = -1$. Now, $G(Z)$ will have two poles: one at $Z = 0$ and other at $Z = a$.

$$G(Z)=\frac{1}{Z(Z-a)} \qquad (8.203)$$

Both the poles are within the closed contour. We have to evaluate the residues at both the poles and add the residues to get $x[n]$.

$$x[n]=R[G(Z)]\downarrow_{Z=0}+R[G(Z)]\downarrow_{Z=a}=\frac{1}{Z-a}\downarrow_{Z=0}+\frac{1}{Z}\downarrow_{Z=a}=-\frac{1}{a}+\frac{1}{a}=0 \qquad (8.204)$$

The sequence value is zero at $n = -1$.

Consider $n = -2$. Now, $G(Z)$ will have two poles one at $Z = 0$(double pole) and other at $Z = a$.

$$G(Z)=\frac{1}{Z^2(Z-a)}. \qquad (8.205)$$

Both the poles are within the closed contour. We have to evaluate the residues at both the poles and add the residues to get $x[n]$

$$x[n]=\frac{d}{dZ}\left(\frac{1}{Z-a}\right)\downarrow_{Z=0}+\frac{1}{Z^2}\downarrow_{Z=a}=-\frac{1}{a^2}+\frac{1}{a^2}=0 \qquad (8.206)$$

The sequence value is zero at $n = -2$. We can extend and say that $x[n] = 0$ for all $n < 0$, which is the expected result.

Example 8.50

Let $X(Z)=\dfrac{Z^2+Z}{(Z-1)^2}$ for $|Z|>1$

Find $x[n]$ using the residue method.

Solution
Step 1 Solve

$$G(Z)=\frac{(Z^2+Z)\times Z^{n-1}}{(Z-1)^2}=\frac{Z^{n+1}+Z^n}{(Z-1)^2} \qquad (8.207)$$

Step 2 We will now select a closed contour as a circle with radius $|Z| > 1$, so that it is in ROC.

Step 3 The poles of $G(Z)$ will depend on the value of n. We will consider two different cases, namely, $n \geq 0$, $n < 0$.

Case 1 $n \geq 0$. Here, there is only pole of $G(Z)$ at $Z = 1$ (double pole), which is inside the circle with radius $|Z| > 1$. We will evaluate the residue of $G(Z)$ at $Z = 1$.

$$R[G(Z)]\downarrow_{Z=1} = \frac{d}{dZ}[(Z-1)^2 G(Z)]\downarrow_{Z=1} = \frac{d}{dZ}[Z^{n+1} + Z^n]\downarrow_{Z=1} \quad (8.208)$$

$$= (n+1)Z^n + nZ^{n-1} \downarrow_{Z=1} = 2n+1$$

$$x[n] = 2n+1 \text{ for } n \geq 0 \quad (8.209)$$

Case 2 $n < 0$. Let $n = -1$. Now

$$G(Z) = \frac{1+Z^{-1}}{(Z-1)^2} = \frac{Z+1}{Z(Z-1)^2} \quad (8.210)$$

will have two poles: one at $Z = 0$ and other at $Z = 1$ (double pole).

Both the poles are within the closed contour. We have to evaluate the residues at both the poles and add the residues to get $x[n]$.

$$x(n) = R[G(Z)]\downarrow_{Z=0} + R[G(Z)]\downarrow_{Z=1} = \frac{Z+1}{(Z-1)^2}\downarrow_{Z=0} + \frac{d}{dZ}\left[\frac{1}{Z}\right]\downarrow_{Z=1} = 1-1 = 0 \quad (8.211)$$

The sequence value is zero at $n = -1$. Similarly, it can be shown that $x[n] = 0$ for $n = -2$.

✓ **Teaser** The proof is left to the reader.

Concept Check

- What is the meaning of one-to-one correspondence between a transform and its inverse?
- What is a residue? How does one calculate IZT using residues?
- When are we supposed to recover a negative power series in Z?
- From ROC specification, can we decide if it is a left-handed or a right-handed sequence?
- Why is it necessary to evaluate $F(Z)/Z$ for a partial fraction expansion method?

8.9 Solution of Difference Equation using Z Transform

We will now consider one of the important applications of the Z transform, namely, solving the difference equations. The approach involves calculation of the ZT of the shifted sequences with appropriate initial conditions and then solving for the ZT of the sequence $f[n]$. We finally take IZT to obtain the sequence $f[n]$.

Let us consider one solved problem to illustrate the use of ZT for solving the difference equation.

Example 8.51

Solve the second order difference equation:

$$2f[n-2] - 3f[n-1] + f[n] = 3^{n-2} \text{ for } n \geq 0 \text{ where } f[-2] = -\frac{4}{9} \text{ and } f[-1] = -\frac{1}{3}$$

Solution

Take the ZT of both sides of the given difference equation. We obtain

$$2ZT\{f[n-2]\} - 3ZT\{f[n-1]\} + ZT\{f[n]\} = 3^{-2} ZT\{3^n\} \tag{8.212}$$

Let us first calculate ZT $\{f[n-1]\}$.

$$ZT\{f[n-1]\} = \sum_{n=0}^{\infty} f[n-1]Z^{-n} = f[-1] + f[0]Z^{-1} + f[1]Z^{-2} + \ldots$$

$$= f[-1] + Z^{-1}[f[0] + f[1]Z^{-1} + \ldots] \tag{8.213}$$

$$ZT\{f[n-1]\} = f[-1] + Z^{-1}F(Z)$$

Similarly it can be shown that

$$ZT\{f[n-2]\} = f[-2] + Z^{-1}f[-1] + Z^{-2}F(Z) \tag{8.214}$$

Using Eqs (8.213) and (8.214), Eq. (8.212) becomes

$$2[Z^{-2}F(Z) + Z^{-1}f[-1] + f[-2]] - 3[Z^{-1}F(Z) + f[-1]] + F(Z) = 3^{-2}\frac{Z}{Z-3} \tag{8.215}$$

Put values of $f[-1]$ and $f[-2]$ in the above equation and simplify. Equation (8.215) becomes

$$\frac{(Z-1)(Z-2)}{Z^2}F(Z) = \frac{2}{3Z} - \frac{1}{9} + \frac{Z}{9(Z-3)} = \frac{Z-2}{Z(Z-3)} \tag{8.216}$$

$$F(Z) = \frac{Z}{(Z-1)(Z-3)} \tag{8.217}$$

PFE or residue method can be used to take IZT of $F(Z)$. The use of PFE leads to

$$\frac{F(Z)}{Z} = \frac{k_1}{Z-1} + \frac{k_2}{Z-3} \tag{8.218}$$

$$k_1 = (z-1)\frac{F(Z)}{Z}\bigg\downarrow_{Z=1} = -\frac{1}{2}; \; k_2 = (Z-3)\frac{F(Z)}{Z}\bigg\downarrow_{Z=3} = \frac{1}{2}$$

$$F(Z) = -\frac{1}{2}\frac{Z}{Z-1} + \frac{1}{2}\frac{Z}{Z-3} \tag{8.219}$$

$$f[n] = \frac{1}{2}\{3^n - 1\}u[n] \tag{8.220}$$

$f[n]$ is a solution of the difference equation.

Example 8.52
Solve the second order difference equation:

$$y[n] - 4y[n-1] + 4y[n-2] = x[n-1]$$

Solution
Take the ZT of the equation.

$$y[n] - 4y[n-1] + 4y[n-2] = x[n-1]$$

$$Y(Z) - 4Z^{-1}Y(Z) + 4Z^{-2}Y(Z) = X(Z) \tag{8.221}$$

$$H(Z) = \frac{Y(Z)}{X(Z)} = \frac{1}{1 - 4Z^{-1} + 4Z^{-2}} = \frac{Z^2}{Z^2 - 4Z + 4}$$

We will use partial fraction expansion to decompose the transfer function.

$$H(Z) = \frac{Z^2}{Z^2 - 4Z + 4} \tag{8.222}$$

$$\frac{H(Z)}{Z} = \frac{Z}{(Z-2)^2}$$

take *IZT*

$$h[n] = n(2)^n u[n+1]$$

We give below a MATLAB program to plot poles and zeros of the transfer function, magnitude response of the system and the impulse response of the system.

```
clear all;
clc;
a=[1 -4 4];
b=[1 0 0];
zplane(b,a);title('pole zero plot');
figure;subplot(2,1,1);
[H,w]=freqz(b,a,256);
plot(w,abs(H));title('magnitude response of the system');
xlabel('frequency in radians');ylabel('amplitude');
[H1,w1]=impz(b,a,15);subplot(2,1,2);
stem(H1);title('impulse response');xlabel('sample number');ylabel('amplitude');
```

Poles and zeros are plotted is shown in Fig. 8.22, magnitude response and impulse response are plotted in Fig. 8.23.

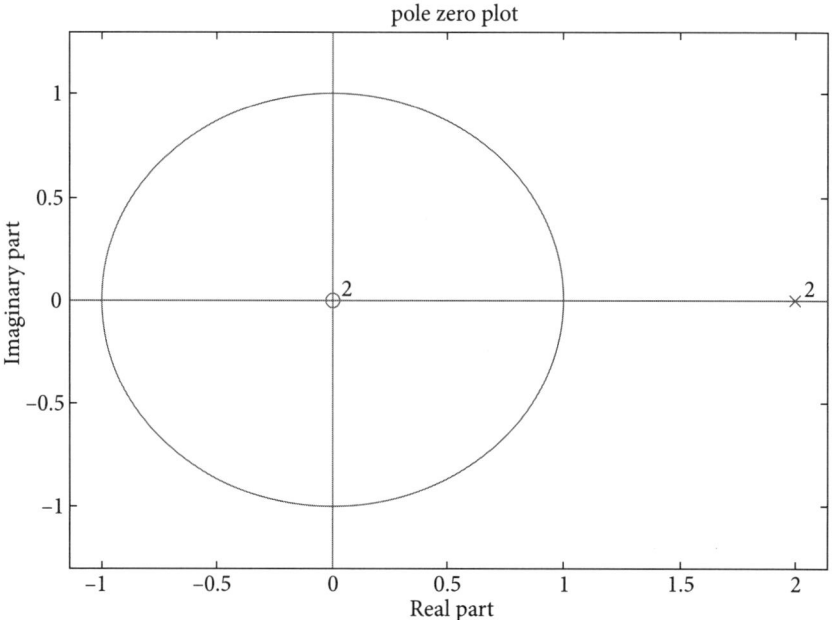

Fig. 8.22 Pole zero plot for a transfer function (Example 8.52)

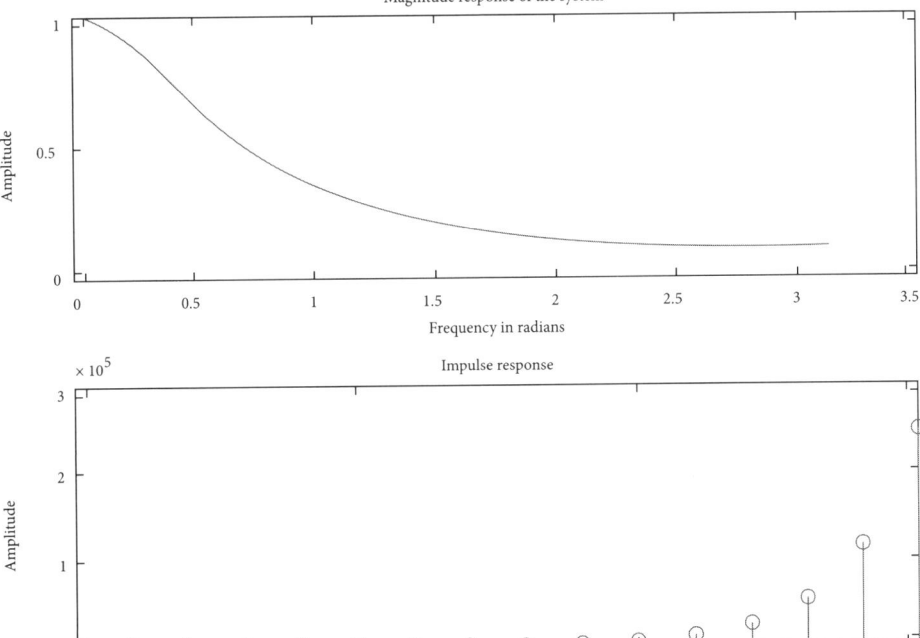

Fig. 8.23 Magnitude response and impulse response of the system (Example 8.52)

Example 8.53

Find the transfer function and the unit sample response of the difference equation $y[n] = x[n] - 0.25y[n-1]$

Solution

Take the ZT of the equation.

$$y[n] = x[n] - 0.25y[n-1]$$

$$Y(Z) = X(Z) - 0.25Z^{-1}Y(Z)$$

$$H(Z) = \frac{Y(Z)}{X(Z)} = \frac{1}{1 + 0.25Z^{-1}} = \frac{Z}{Z + 0.25} \tag{8.223}$$

$$h[n] = (0.25)^n u[n]$$

We give below a MATLAB program to plot poles and zeros of the transfer function, magnitude response of the system and the impulse response of the system.

```
clear all;
clc;
a=[1 0.25];
b=[1 0 ];
zplane(b,a);title('pole zero plot');
figure;subplot(2,1,1);
[H,w]=freqz(b,a,256);
plot(w,abs(H));title('magnitude   response   of   the
system');
xlabel('frequency in radians');ylabel('amplitude');
[H1,w1]=impz(b,a,15);subplot(2,1,2);
stem(H1);title('impulse    response');xlabel('sample
number');ylabel('amplitude');
```

Poles and zeros are plotted is shown in Fig. 8.24, magnitude response and impulse response are plotted in Fig. 8.25.

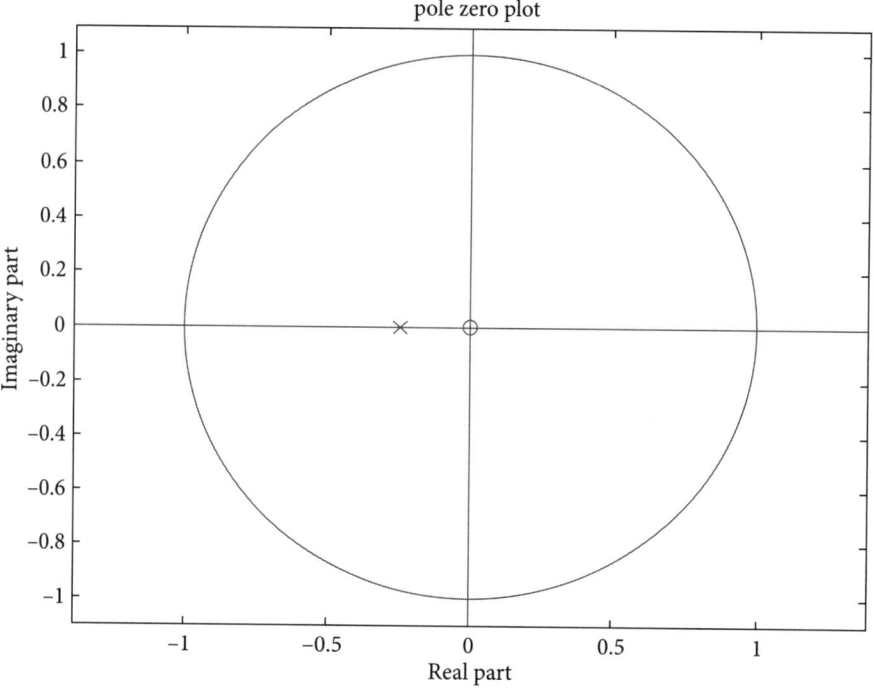

Fig. 8.24 Pole zero plot for the transfer function (Example 8.53)

Example 8.54

Given the transfer function for the system, plot the poles and zeros and the frequency response of the system. $H(Z) = \dfrac{1 + 4Z^{-1} + 6Z^{-2} + 4Z^{-3} + Z^{-4}}{+0.486Z^{-2} + 0.0177Z^{-4}}$

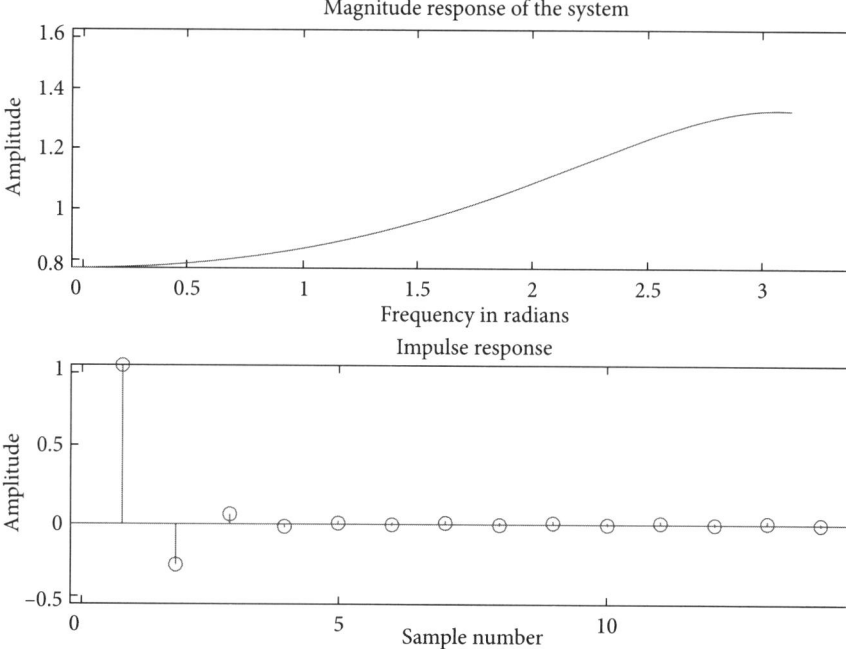

Fig. 8.25 Magnitude response and impulse response of the system (Example 8.53)

Solution

We will write a MATLAB program to plot the poles and zeros and the frequency response. Poles and zeros are plotted in Fig. 8.26 and frequency response nad impulse response are plotted in Fig. 8.27.

```
clear all;
clc;
a=[1 0 0.486 0 0.0177];
b=[1 4 6 4 1];
zplane(b,a);title('pole zero plot');
figure;subplot(2,1,1);
[H,w]=freqz(b,a,256);
plot(w,abs(H));title('magnitude response of the system');
xlabel('frequency in radians');ylabel('amplitude');
[H1,w1]=impz(b,a,15);subplot(2,1,2);
stem(H1);title('impulse response');xlabel('sample number');ylabel('amplitude');
```

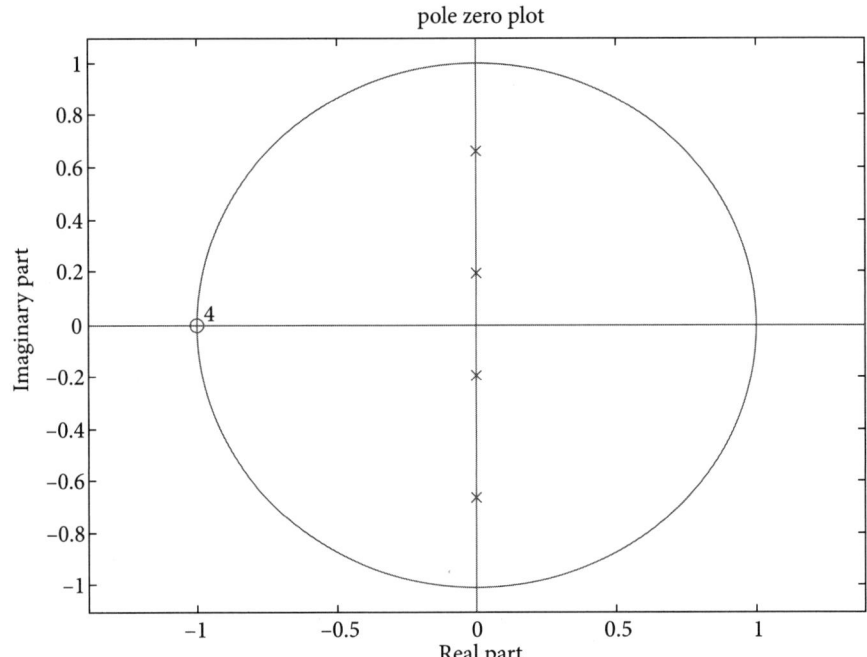

Fig. 8.26 Poles and zeros for Example 8.54

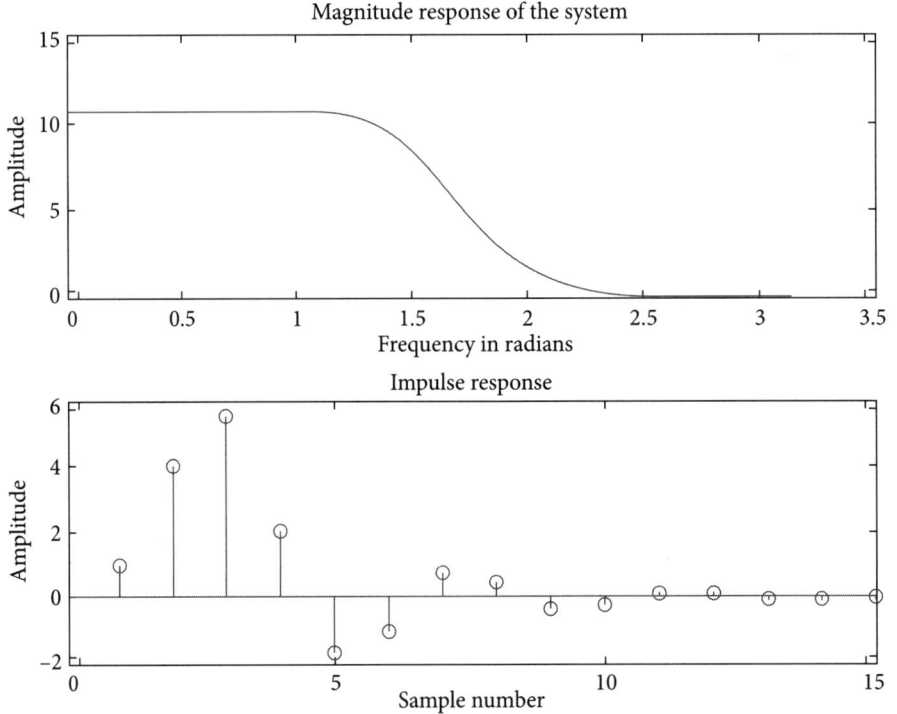

Fig. 8.27 Frequency response and impulse response for Example 8.54

8.9.1 Applications of ZT and IZT

Any causal system may be represented in the form of a difference equation as

$$y(n) = a_0 x(n) + \ldots + a_N x(n-N) - b_1 y(n-1) + \ldots (-) b_M y(n-M) \quad (8.224)$$

The current output depends on present and past inputs and past outputs.

Using the procedure outlined in Section 8.9, we can take the ZT of the difference equation and find the Z domain transfer function of the system. The system transfer function can be further analyzed to find the locations of poles. If the poles are inside the unit circle in the Z domain, the system is stable. Thus we can confirm the stability of the system before it is actually implemented. The IZT of the transfer function will result in the impulse response of the system. This helps in analyzing the time domain behavior of the system, as discussed in Section 8.6.

ZT is utilized to draw the realization forms for the causal filters. ZT and IZT are useful for the design of FIR and IIR systems. The effect of quantization can also be studied using ZT and IZT.

Concept Check

- Find the ZT of $f[n-1]$.
- What is property of shifting in time domain?
- How will you insert the initial conditions?
- Can you plot the poles and zeros of the transfer function?
- Can you plot the frequency response of the system?
- Can you plot the impulse response of the system?
- What are applications of ZT and IZT?

Summary

In this chapter, we have described and explained the following important concepts.

- We started with the need for a transform. Spectral properties of the signal are better revealed in the transform domain. We can analyze the system for stability. Energy of the signal gets compacted in the transform domain. We then related the Z transform to Laplace transform and Fourier transform. The equation $Z = e^{sT}$ defines relation between the Z domain and the S domain. Z transform reduces to FT on the unit circle (circle with radius equal to one) in complex Z plane.
- We defined $Z = re^{j\Omega}$. The range of values of r for which the sequence $f[n]r^{-n}$ converges is called the region of convergence (ROC). If ROC includes the unit circle, then the FT also converges. We then concentrated on Z transform examples. We have seen that for a right-handed sequence, the ROC

is the area outside some circle in the Z domain. For a left-handed sequence, the reverse is the case. Here, the ROC is the area inside some circle in the Z domain. We can have two sequences, one left-handed and one right-handed, for which the Z transform can be the same but the ROCs are different. For example, the sequences $f_1[n] = a^n$ for $n \geq 0$ and $f_2[n] = a^{-n}$ for $n < 1$ have complementary ROCs. We have also seen that if the sequence $x(n)$ is of finite duration, then the ROC is the entire Z plane, except possibly $Z = 0$ and/or $Z = \infty$. Examples based on the properties of ROC are solved to explain the concepts.

- We have considered the properties of Z transform such as linearity, time shifting and differentiation in Z domain, time reversal, time scaling, convolution, multiplication, initial value theorem and final value theorem. We have seen that if the time shift of $n_0 > 0$, then the poles are introduced at the origin and if the ROC of $x[n]$ includes the origin, the ROC of $x[n - n_0]$ will exclude the origin. Similarly, if $n_0 < 0$, then the poles are introduced at infinity and if the ROC of $x[n]$ includes infinity, the ROC of $x[n - n_0]$ will exclude infinity. The physical significance of each property is explained. Simple examples are solved based on the properties of Z transform. The scaling property in the Z domain is discussed. We have considered a special case of this property, namely the multiplication of a sequence by a complex exponential that will result in rotation of the Z plane by angle ω. This is also termed as frequency shifting in the Z domain. We have stated the property of time reversal if $x[n] \leftrightarrow X(Z)$ with ROC = R, then $x[-n] \leftrightarrow X(1/Z)$ with ROC = $1/R$, that is, if $Z = Z_1$ is in ROC of $x[n]$, then $Z = 1/Z_1$ is in ROC of $x[-n]$.

- We then explained property of convolution using a numerical problem. Here, we have seen that the coefficients of the multiplication of Z transforms of two sequences are actually the convolution of the two sequences. We have studied the property of differentiation in the Z domain and initial value theorem. We have shown that stability of the system is analyzed by identifying the locations of poles of the system transfer function. We have translated the stability criteria from the S domain to the Z domain. When the poles are inside the unit circle in the Z domain, the system is stable and its impulse response, that is, time domain behavior will be a decaying signal. We say that the output is bounded. When the poles are outside the unit circle in the Z domain, the system is unstable and its impulse response, that is, time domain behavior will be an exponentially increasing function becoming infinite as the sample number (n) tends to infinity. We say that the output is unbounded.

- We have discussed three methods for inverse Z transformation: power series method, partial fraction expansion (PFE) method and residue method. Using long division, we divide the numerator polynomial by a denominator polynomial to obtain a power series in Z^{-1} or Z depending on whether the series to be recovered is a right-handed sequence (causal) or a left-handed sequence (non-causal). The PFE method is used to decompose a rational fraction given by X(Z) as a linear combination of ZTs of known sequences. We then introduced the residue method. This method evaluates the residues of the function at poles that are inside the closed contour in the ROC. The sum of these residues gives the IZT of the function in the Z domain. The method is based on the theorem called as Cauchy's integral theorem.

- We considered one of the important applications of Z transform, namely, the solution of the difference equations. The approach involves calculation of the ZT of the shifted sequences with appropriate initial conditions and then solving for the ZT of the sequence $f[n]$. We have introduced the use of MATLAB programs to plot poles and zeros of the transfer function. It is also indicated that we can write a program to plot the frequency response and the impulse response of the system for the number of examples of the transfer functions. Lastly, we indicated different applications of Z transform to be discussed in further chapters.

Multiple Choice Questions

1. Transform domain is used as it
   - (a) enhances the signal
   - (b) compresses the signal
   - (c) reduces the data points
   - (d) increases the data points

2. The relation between Laplace domain and Z domain is
   - (a) $Z = e^{sT}$
   - (b) $Z = e^{jsT}$
   - (c) $S = e^{ZT}$
   - (d) $Z = e^{-sT}$

3. Z transform reduces to Fourier transform when it is evaluated on
   - (a) a half circle
   - (b) Z circle
   - (c) unit circle
   - (d) imaginary circle

4. If the sequence $x(n)$ is of finite duration, then the ROC is the entire Z plane, except possibly
   - (a) $Z = 0$
   - (b) $Z = \infty$
   - (c) $Z = 0$ and/or $Z = \infty$
   - (d) $Z = 0$ and $Z = \infty$

5. If the sequence is time shifted by n_0 samples, with $n_0 > 0$, then the poles are introduced at
   - (a) zero
   - (b) infinity
   - (c) zero and infinity
   - (d) zero or infinity

6. The multiplication of a sequence by a complex exponential will result
   - (a) in rotation of an imaginary plane
   - (b) in rotation of the Z plane by angle ω
   - (c) in rotation of a complex plane
   - (d) in rotation of real axis by angle ω

7. When a sequence $x(n)$ is time reversed, if $Z = Z_1$ is in ROC of $x(n)$, then
   (a) $Z = 1/Z_1$ is in ROC of $x(-n)$
   (b) $Z_1/2$ is in ROC of $x(-n)$
   (c) $Z = 1/Z_1^2$ is in ROC of $x(-n)$
   (d) $Z = Z_1/4$ is in ROC of $x(-n)$

8. When the system has poles inside the unit circle in Z domain,
   (a) the system is stable and its impulse response is a decaying function
   (b) time domain behavior will be exponentially rising signal
   (c) the system is unstable
   (d) the impulse response is marginally constant

9. Using long division, we divide numerator polynomial by a denominator polynomial to obtain a power series in Z^{-1} depending on whether the series to be recovered is
   (a) anti-causal signal
   (b) causal signal
   (c) both-sided signal
   (d) none

10. The residue method evaluates the residues of the function at poles that are
    (a) outside the closed contour in the ROC
    (b) anywhere in ROC
    (c) inside the closed contour in the ROC
    (d) outside the ROC

Review Questions

8.1 Why is the use of transform domain processing of a signal?

8.2 What is the relation between Laplace transform and Z transform?

8.3 Find the relation between Fourier transform and Z transform. What is the importance of a unit circle in Z domain?

8.4 What is ROC?

8.5 State and prove properties of ROC. Can it happen that there is no ROC? When does this happen?

8.6 What is a time reversal property of the Z transform? What happens to ROC in this case?

8.7 When a sequence is shifted in the time domain, what will happen to Z transform of the shifted sequence? How is the ROC affected?

8.8 State and prove the convolution property of Z transform.

8.9 State the property of differentiation in the Z domain. How is it useful for calculation of Z transform of $nx[n]$?

8.10 State initial value theorem and prove it.

8.11 Use the relation between Z transform and Laplace transform to translate the stability criteria from Laplace domain to Z domain.

8.12 Can we find the stability of a system if we know the impulse response of the system?

8.13 Discuss the stability of a system if the system has complex conjugate poles.

8.14 Explain the power series method for calculation of inverse Z transform. How can we make use of ROC to find the power series?

8.15 What are different methods for calculation of IZT?

8.16 What is a residue? Will the knowledge of pole locations help in calculation of residue? Which poles are to be used for the calculation of a residue? How can one decide a closed contour for the calculation of residue?

8.17 How can one convert a difference equation into a Z domain transfer function? How can we solve the difference equation? Explain using a numerical example.

8.18 How will you analyze the system to find the frequency response of the system using MATLAB program? Can you find the impulse response of the system?

Problems

8.1 Find the ZT of the following sequences

(a) $f(n) = 2e^{-6n} - 2e^{-3n} + 24ne^{-6n}$

(b) $f(n) = (1+n)U(n)$

(c) $f(n) = \cos(n\omega T)U(n)$

(d) $f(n) = na^n \sin(n\omega T)U(n)$

(e) $f(n) = n^2 U(n)$

(f) $f(n) = \cos(n\pi/3)U(n)$

8.2 Given a 6 periodic sequence,

$F(n) = \{1, 1, 1, -1, -1, -1, 1, 1, 1, -1\}$

Show that

$$F(Z) = \frac{Z(Z^2 + Z + 1)}{Z^3 + 1}$$

8.3 Express the Z transform of

$$y(n) = \sum_{k=-\infty}^{n} x(k)$$

in terms of $X(Z)$.

[Hint: Find the difference $y(n) - y(n-1)$.]

8.4 Find the IZT of the following Z domain functions

(a) $X(Z) = \dfrac{1+3Z^{-1}}{1+3Z^{-1}+2Z^{-2}}$

(b) $X(Z) = \dfrac{1+2Z^{-2}}{1+Z^{-2}}$

(c) $X(Z) = \dfrac{1-aZ^{-1}}{Z^{-1}-a}$

(d) $X(Z) = \dfrac{Z^{-6}+Z^{-7}}{1-Z^{-1}}$

8.5 Use PFE to find IZT of following ZTs:

(a) $F(Z) = \dfrac{Z}{(Z-e^{-a})(Z-e^{-b})}$

where a and b are positive constants.

(b) $F(Z) = \dfrac{Z^2}{(Z-1)(Z-0.8)}$

8.6 Using Partial Fraction Expansion find IZT of $F(Z)$ and verify it using a long division method.

$$F(Z) = \dfrac{1+2Z^{1}}{1-0.4Z^{-1}-0.12Z^{-2}}$$

if $f(n)$ is causal...

8.7 Use residue method to find IZT of the following ZTs:

$$F(Z) = \dfrac{Z^2+3Z}{(Z-0.5)^3}$$

$$F(Z) = \dfrac{1}{(Z+1)^2(Z-0.5)}$$

8.8 Given the difference equation

$y(n) + b^2 y(n-2) = 0$ for $n \geq 0$ and $|b| < 1$

With initial conditions $y(-1) = 0$ and $y(-2) = -1$, show that

$$y(n) = b^{n+2} \cos\left(\dfrac{n\pi}{2}\right).$$

8.9 Find $f(n)$ corresponding to the difference equation

$f(n-2) - 2f(n-1) + f(n) = 1$ for $n \geq 0$

with initial condition $f(-1) = -0.5$ and $f(-2) = 0$. Show that

$f(n) = (0.5)n^2 + n$ for $n \geq 0$

Answers

Multiple Choice Questions

    1 (b)   2 (a)   3 (c)   4 (c)   5 (a)
    6 (b)   7 (a)   8 (b)   9 (b)   10 (c)

Problems

8.1 (a) $\dfrac{2}{1-e^{-6}Z^{-1}} - \dfrac{2}{1-e^{-3}Z^{-1}} + \dfrac{24e^{-6}Z^{-1}}{(1-e^{-6}Z^{-1})^2}$ with ROC $|Z| > e^{-6}$

(b) $\dfrac{1}{(1-Z^{-1})^2}$ with ROC $|Z| > 1$

(c) $\dfrac{1-\cos(\omega T)Z^{-1}}{1-2\cos(\omega T)Z^{-1}+Z^{-2}}$ with ROC $|Z| > 1$

(d) $-Z \dfrac{d}{dZ}\left[\dfrac{\sin(\omega T)aZ^{-1}}{1-2\cos(\omega T)aZ^{-1}+a^2 Z^{-2}}\right]$ with ROC $|Z| > a$

(e) $Z^2 \dfrac{1}{2!}\dfrac{d^2}{dZ^2}\left[\dfrac{1}{1-Z^{-1}}\right]$ with ROC $|Z| > 1$ and additional double zero is introduced at the origin or a pole at infinity. This indicates that ROC excludes infinity.

(f) $\dfrac{1-\cos\left(\dfrac{\pi}{3}\right)Z^{-1}}{1-2\cos\left(\dfrac{\pi}{3}\right)Z^{-1}+Z^{-2}}$ with ROC $|Z| > 1$.

8.2 $F(Z) = \dfrac{Z(Z^2+Z+1)}{Z^3+1}$

8.3 $Y(Z) = \dfrac{1}{1-Z^{-1}} X(Z)$

8.4 (a) $x(n) = \left[2(-1)^n - (-2)^n\right] U(n)$

(b) $x(n) = 2\delta(n) - \dfrac{1}{2}\left[(j)^n + (-j)^n\right] U(n)$

(c) $x(n) = -a\delta(n) - \dfrac{1-a^2}{a}\left(\dfrac{1}{a}\right)^n U(n)$

(d) $2U(n-6) - \delta(n-6)$

8.5 (a) $\dfrac{1}{e^{-a}-e^{-b}}\left[(e^{-a})^n-(e^{-b})^n\right]U(n)$

   (b) $[5-4(0.8)^n]U(n)$

8.6 $2.75(0.2)^n U(n)-1.75(-0.6)^n U(n)$

8.7 (a) $(0.5)^{n-2}\left[\dfrac{1}{2}n(n+1)+3n(n-1)\right]U(n)$

   (b) $\left[2(n-1)-4\right]$ for $n\geq 1$

8.8 Answer provided

8.9 Answer provided

9

Random Signals and Processes

Learning Objectives

- Probability
- Conditional probability
- Bayes theorem
- Properties of random variables
- Standard probability distributions
- Central limit theorem
- Chi-square test and K–S test
- Random processes

The theory of signals and systems is applied for the processing of signals like speech signal, RADAR signal, SONAR signal, earth quake signal, ECG and EEG signals. All these signals are naturally occurring signals and hence have some random component. The primary goal of this chapter is to introduce the principles of random signals and processes for in-depth understanding of the processing methods. We will introduce the concept of probability and will discuss different standard distribution functions used for the analysis of random signals. Different operations on the random variables namely expectation, variance and moments will be introduced. We will then discuss central limit theorem. The classification of random processes namely wide sense stationary, ergodic and strict sense stationary processes will be defined.

9.1 Probability

Consider the experiment of throwing a dice. The result of the throw is random in nature in the sense that we do not know the result i.e., the outcome of the experiment until the outcome is actually available. The randomness in the outcome is lost once the outcome is evident. The result of the experiment is either a 1dot, 2dots, 3dots and so on. The outcome is a discrete random variable. Here, there are only 6 possibilities and so one can easily find the probability of getting 1 dot is 1/6 evaluated as one of total possible outcomes. If the experiment is performed a large number of times, the ratio of frequency of occurrence of a 1dot and the total number of times the experiment is performed approaches 1/6 (Refer to Eq. (9.1)).

We will first define the term event. When any experiment is performed to generate some output, it is called as an event. The output obtained is called as the outcome of the experiment. The possible set of outcomes of the experiment is called as a space or as event space or sample space. Many times, there is the uncertainty or randomness involved in the outcome of the experiment. The term probability is closely related to uncertainty. In case of the experiment of throwing of a die, the uncertainty is involved in the outcome. When a die is thrown, we do not know what will be the outcome. Using the notion we may define a relative frequency of occurrence of some event. Consider the example of a fair coin. Let it be tossed n times. Let n_H times Head appear. The relative frequency of occurrence of Head is defined as

$$\text{relative frequency of occurence} = \frac{\text{number of favorable outcomes}}{\text{total number of times the die is thrown}} \quad (9.1)$$

$$= \frac{n_H}{n}$$

The relative frequency approaches a fixed value called probability when n tends to infinity. We define probability as

$$\text{Probability} = \lim_{n \to \infty} \frac{n_H}{n} \quad (9.2)$$

Using the definition of probability as a frequency of occurrence, we can easily conclude that the probability P is always less than one. When *P = 1*, the event becomes a certain event. A certain event is not a random event.

Let us consider the example of event. When a die is thrown, let us assume that the number '1' comes up. This is an event. This experiment spans a discrete random space as the output of the experiment is discrete, and the total number of outcomes is finite, equal to 6. Discrete random spaces are easier to deal with as the events are countable. Continuous sample spaces are generated when the outcomes of the experiment are real numbers. An example is the acquisition of speech samples, measurement of temperature, etc.

To understand the concept of relative frequency, we will consider one example.

Example 9.1

Let there be 100 resistors in a box, all of same size and shape. There are 20 resistors of value 10Ω, 30 resistors of value 22Ω, 30 resistors of value 33Ω and

20 resistors of value 47Ω. Let us consider the event of drawing one resistor from the box. Find the probability of drawing 10Ω, 22Ω, 33Ω and 47Ω resistors.

Solution

The concept of relative frequency suggests that

$$P(10\Omega) = \frac{20}{100}, \ P(22\Omega) = 30/100$$

$$P(33\Omega) = 30/100 \text{ and } P(47\Omega) = 20/100 \tag{9.3}$$

Total probability sums to one.

Example 9.2

If a single (fair) die is rolled, determine the probability of each of the following event.

a. Obtaining the number 3
b. Obtaining a number greater than 4
c. Obtaining a number less than 3 and obtaining a number greater than or equal to 2

Solution

a. Obtaining the number 3

Since the die we are using is fair, each number has an equal probability of occurring; therefore,

$P(3) = 1/6$

b. Obtaining a number greater than 4

In this case, the set of possible outcomes is given by (5, 6) Note that these are all mutually exclusive events. Therefore,

$$P(5 \cup 6) = P(5) + P(6) = 1/6 + 1/6 = 2/6 = 1/3 \tag{9.4}$$

c. Obtaining a number less than 3

$$A = \{\text{a number less than 3}\} = \{1, 2\} \ P(A) = 1/3 \tag{9.5}$$

$$B = \{\text{a number greater than or equal to 2}\} = \{2, 3, 4, 5, 6\} \ P(B) = 5/6 \tag{9.6}$$

9.1.1 Conditional probability

In case of random experiments, sometimes the probability of occurrence of one event may depend on other event. The probability of occurrence of such an event is termed as the conditional probability. Consider one example. Let a card be drawn from a pack of cards. Let the card drawn be the heart card. Let us now define other event of drawing the second card if the first card drawn is a heart card. The probability of occurrence of second event depends on the first event i.e., if the first card drawn is heart card, the probability of second event is 12/51. This is because when the first card drawn is the heart card, the number of heart cards remaining in the pack of cards is 12 and the total number of cards remaining in a pack of cards is 51 as the card drawn in the first draw is not replaced. If the first card drawn is not a heart card, then the probability of drawing the second card as heart card is 13/51.

We define the conditional probability as follows. The probability of occurrence of an event A given that event B has occurred is called as conditional probability and is given by

$$P(A/B) = \frac{P(A \cap B)}{P(B)} \qquad (9.7)$$

Note Conditional probability is a defined quantity and cannot be proven.

Example 9.3

Let the event A be drawing a heart card and event B be drawing a heart card on second draw. Find $P(A \cap B)$.

Solution

In the example of pack of cards, $(A \cap B)$ represents the intersection of the sample space for the two events namely the heart is drawn in first draw and the heart is drawn in a second draw. $P(A)$ is the probability of the occurrence of an event A, namely drawing a heart card on the first draw. $P(A) = \frac{1}{4}$ and $P(B/A) = \frac{12}{51}$ as explained earlier.

So, $P(A \cap B) = P(B/A) \times P(A) = \frac{12}{51} \times \frac{1}{4} = \frac{12}{204}$ \qquad (9.8)

Example 9.4

Let there be balls in a box. There are balls of two colors namely red and blue. There are 3 red balls and 2 blue balls. Find the probability of picking up a red ball first and a blue ball on the second pick.

Solution

Let us define the event A as picking up a red ball on first pick. Event B is picking up a blue ball. We will find the probabilities.

$P(A) = \frac{3}{5}, P(B) = \frac{2}{5}, P(B) = \frac{2}{5}, P(B/A) = \frac{2}{4}$ as after the first draw, the number of balls remaining is 4.

$$P(A \cap B) = P(B/A) \times P(A)$$
$$= \frac{2}{4} \times \frac{3}{5} = \frac{3}{10} \tag{9.9}$$

if the two events A and B are mutually exclusive, $P(A/B) = 0$.

Example 9.5

Let there be three events with probabilities given by $P(A) = 44/100$, $P(B) = 62/100$, and $P(C) = 32/100$. Let the joint probabilities be given by $P(A \cap B) = 28/100$, $P(A \cap C) = 0$, and $P(B \cap C) = 24/100$. Find the conditional probabilities $P(A/C)$, $P(A/B)$, and $P(B/C)$

Solution

$$P(A/B) = \frac{P(A \cap B)}{P(B)} = \frac{\frac{28}{100}}{\frac{62}{100}} = \frac{28}{62} = \frac{14}{31} \tag{9.10}$$

$$P(A/C) = \frac{P(A \cap C)}{P(C)} = \frac{0}{\frac{32}{100}} = 0 \tag{9.11}$$

$$P(B/C) = \frac{P(B \cap C)}{P(C)} = \frac{\frac{24}{100}}{\frac{32}{100}} = \frac{24}{32} = \frac{3}{4} \tag{9.12}$$

Example 9.6

Consider a box with five 100Ω resistors and two 1000Ω resistors. We remove two resistors in succession. What is the probability that the first resistor is 100Ω and the second is 1000Ω?

Solution

Let A = {draw a 100Ω resistor}, and
    let B = {draw a 1000Ω resistor on second draw }
        A = 5/7, we assume that the first resistor drawn is not replaced.
        B = 2/6 = 1/3 As there are only 4 resistors of 100Ω and 2 of 1000Ω

$$P(A \cap B) = P(B/A)P(A)$$
$$= \frac{1}{3} \times \frac{5}{7} = \frac{5}{21} \tag{9.13}$$

Example 9.7

A dodecahedron is a solid object with 12 equal faces. It is frequently used as a calendar paperweight with one month placed on each face. If such a calendar is randomly placed on a desk, the outcome is taken to be the month of the upper face.

a. What is the probability of the outcome's being February?
b. What is the probability of the outcome's being January or April or August?
c. What is the probability that the outcome will be a month with 31 days??
d. What is the probability that the outcomes will be in the third quarter of the year?

Solution

a. What is the probability of the outcome's being February?
   Let A be the event 'Outcome being February'

   $$P(A) = 1/12 \tag{9.14}$$

b. What is the probability of the outcome's being January or April or August?
   Let B be the event 'outcome's being January or April or August'

   $$P(B) = 1/12 + 1/12 + 1/12 = 1/4 \tag{9.15}$$

c. What is the probability that the outcome will be a month with 31 days?
   We will find months with 31 days. They are listed in a set C.
   C = {January, March, May, July, August, October, December}

   $$P(C) = 7/12 \tag{9.16}$$

d. What is the probability that the outcomes will be in the third quarter of the year??
   Let D be the event 'outcomes will be in the third quarter of the year'
   B = {July, August, September}

   $$P(D) = 3/12 = 1/4 \tag{9.17}$$

Example 9.8

Two fair, six-sided dice are thrown. Find the probability of

a. Throwing a sum of 11
b. Throwing two 7s
c. Throwing a pair

Solution

a. Throwing a sum of 11

Let A be the event of 'sum of 11'

$A = \{(5, 6),(6,5)\}$

$P(A) = 2/36 = 1/18$ (9.18)

b. Throwing two 7s

Let B be the event of 'two 7s'

$B = \{\varphi\}$
$P(B) = 0$ (9.19)

c. Throwing a pair

Let C be the event of 'throwing a pair'. There will be six pairs out of 36 possible outcomes.

$P(C) = 1/6$ (9.20)

Example 9.9

In the network of switches shown in Fig. 9.1, each switch operates independently of all others and each switch has a probability of 0.5 of being closed. What is the probability of a complete path through the network?

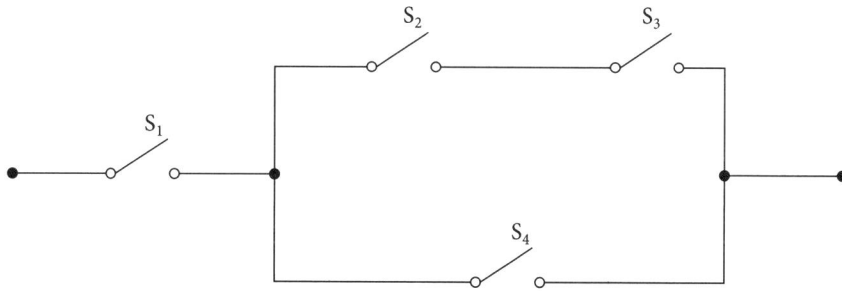

Fig. 9.1 Combination of 4 switches

Solution

Let A be the event of a complete path through network,

The path in the network will be completed by the following switch combination.

{(S1, S2, S3), (S1,S4), (S1,S2,S4), (S1, S3,S4), (S1, S2, S3, S4)}

Therefore, the probability of a complete path through network is

$$P(A) = \frac{1}{2} \cdot \frac{1}{2} \cdot \frac{1}{2} + \frac{1}{2} \cdot \frac{1}{2} + \frac{1}{2} \cdot \frac{1}{2} \cdot \frac{1}{2} + \frac{1}{2} \cdot \frac{1}{2} \cdot \frac{1}{2} + \frac{1}{2} \cdot \frac{1}{2} \cdot \frac{1}{2} \cdot \frac{1}{2} = \frac{11}{16} \qquad (9.21)$$

Example 9.10

A technician has three resistors, with values of 100, 300, and 900Ω. If they are connected randomly in the configuration shown in Fig. 9.2, answer the following questions.

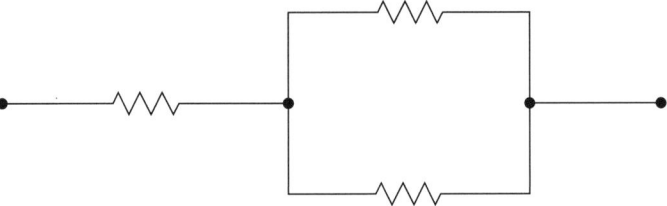

Fig. 9.2 Configuration for connecting 3 resistors randomly

a. What is the probability that the equivalent resistance of this network will be 390Ω?

b. What is the probability that the equivalent resistance will be greater than 390Ω?

c. What is the probability that the equivalent resistance will be less than 1000Ω?

Solution

The all possible resistor configurations will form a set given below.

S = {(100+300||900), (100+900||300), (300+100||900), (300+900||100), (900+100||300), (900+300||100)}

a. The probability $P(A)$ of getting equivalent resistance of 390Ω can be found by tracking the combinations that give 390Ω. They are 300+100||900 and 300+900||100.

$P(A)=2/6=1/3.$ \hfill (9.22)

b. The probability $P(B)$ of getting equivalent resistance greater than 390Ω can be found by tracking combinations that give equivalent resistance of > 390Ω. They are 900+100||300, 900+300||100.

$$P(B) = 1/3 \qquad (9.23)$$

c. The probability $P(C)$ that the equivalent resistance will be less than 1000Ω? This will happen for all resistor combinations.

$$P(C) = 1. \qquad (9.24)$$

Example 9.11

A fair, six-sided die is thrown. If the outcome is odd, the experiment is terminated. If the outcome is even, the die is tossed a second time. Find the probability of

a. Throwing a sum of 7
b. Throwing a sum of 2

Solution

Let us find the sample space for the experiment as follows

S={(2,1), (2,2),(2,3), (2,4), (2,5), (2,6), (4,1), (4,2), (4,3), (4,4), (4,5), (4,6), (6,1), (6,2), (6,3), (6,4), (6,5), (6,6)}. Note that if the first die has outcome of 1, 3 or 5 experiment will be terminated.

Let A be the event 'throwing a sum of 7'. The possible combinations are (4,3), (2,5), (6,1)

$$P(A) = 3/18 = 1/6 \qquad (9.25)$$

Let B be the event 'throwing a sum of 2' here, only possibility is (1,1), but it is to be discarded as when the first die gives odd output, the experiment is terminated.

$$P(B) = 0/18 = 0 \qquad (9.26)$$

9.1.2 Bayes theorem

Bayes theorem is also called as the theorem of inverse probability. Consider one example.

Example 9.12

Let us define the events as follows:
    A – Picking up a diamond card
    B – Picking up a spade card
    C – Picking up a club card
    D – Picking up a heart card
    E – Picking up a king.

Events A, B, C and D are mutually exclusive. i.e., their sample space is exclusively different and there is no overlapping. Consider the events A and E, they are not mutually exclusive. There is some overlapping space between the two events. The events spaces are drawn in Fig. 9.3.

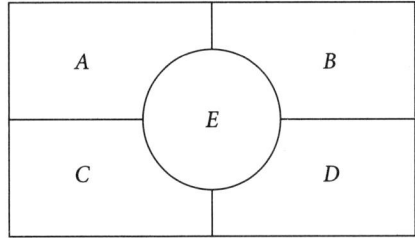

Fig. 9.3 Sample spaces for the events

We try to find the inverse probability namely, if the king is drawn, what is the probability that it is the diamond card? We are now interested in $P(A/E)$.

Solution

$P(A/E)$ is the probability of picking a diamond card given that it is a king.
Let us find $P(A/E)$.

$$P(A/E) = \frac{P(A \cap E)}{P(E)}$$

$$P(A \cap E) = P(A/E) \times P(E) = \frac{1}{4} \times \frac{4}{52} = \frac{1}{52} \quad (9.27)$$

Now,

$$P(A/E) = \frac{P(A \cap E)}{P(E)} \quad (9.28)$$

And, $P(E) = P(E \cap A) + P(E \cap B) + P(E \cap C) + P(E \cap D)$ (9.29)

So,

$$P(A/E) = P(E/A) \times \frac{P(A)}{P(E \cap A) + P(E \cap B) + P(E \cap C) + P(E \cap D)} \quad (9.30)$$

This is known as Bayes theorem.
Let us find $P(A/E)$

$$P(A/E) = \frac{\frac{1}{13} \times \frac{1}{4}}{\frac{1}{52} + \frac{1}{52} + \frac{1}{52} + \frac{1}{52}} = \frac{\frac{1}{52}}{\frac{4}{52}} = \frac{1}{4} \quad (9.31)$$

Alternative statement of Bayes theorem

Bayes theorem has application in finding the class membership value based on the value of selected feature. Let us consider a problem of class selection based on a single feature, for example, 'x'. Let $P(x)$ denote the probability distribution of x in entire population. Let $P(C)$ denote the probability that the sample belongs to class C. $P(x/C)$ is a conditional probability for feature value x given that a sample comes from class C. The goal of the experiment is to find the inverse probability namely probability that a sample belongs to class C given that the feature value is x i.e., $P(C/x)$.

We know that

$$P(x/C) = \frac{P(x \cap C)}{P(C)} \text{ and } P(x/C) \times P(C) = P(x \cap C) = P(C \cap X) \quad (9.32)$$

Hence,

$$P(x/C) \times P(C) = P(C \cap X) = P(C/x) \times P(x) \quad (9.33)$$

We can write Bayes theorem as

$$P(x/C) \times P(C) = P(C/x) \times P(x)$$

$$P(C/x) = [P(x/C) \times P(C)]/P(x) \quad (9.34)$$

Example 9.13

Let A and B be the non-mutually exclusive events. Draw the Venn diagram for the two events and verify the relation $P(A \cap B) = P(A) + P(B) - P(A \cup B)$.

Solution

Let us draw the Venn diagram to show non-mutually exclusive events. Fig. 9.4 shows a Venn diagram for such events.

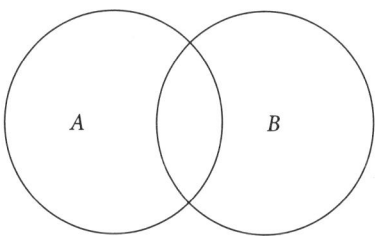

Fig. 9.4 Venn diagram for two mutually non-exclusive events

The sample space spanned by events A and B namely P(A) and P(B) is shown by the two circles in the diagram. The symbol $P(A \cap B)$ represents the area common to two events. The symbol $P(A \cup B)$ is a space spanned by two events taken together. It can be easily verified from the Venn diagram that $P(A \cap B) = P(A) + P(B) - P(A \cup B)$. Obviously in case of two mutually exclusive events, $P(A \cap B) = P(A) + P(B)$ as the common area or the intersection is zero.

Note $P(A \cup B)$ is called as the joint probability for the two events A and B.

Example 9.14

Consider a binary symmetric channel with a priori probabilities $P(B_1) = 0.6$, $P(B_2) = 0.4$. A priori probabilities indicate the probability of transmission of symbols 0 and 1 before the experiment is performed i.e., before transmission takes place. The conditional probabilities are given by $P(A_1 / B_1) = 0.9$; $P(A_2 / B_1) = 0.1$ and $P(A_1 / B_2) = 0.1$; $P(A_2 / B_2) = 0.9$. Find the received symbol probabilities. Find the transmission probabilities for correct transmission and transmission with error.

[**Note** Conditional probabilities are the same for symmetrical links. Hence, it is termed as symmetric channel.] [A_1 and A_2 stand for received symbols and B_1 and B_2 indicate transmitted symbols, respectively, as shown in Fig. 9.5.]

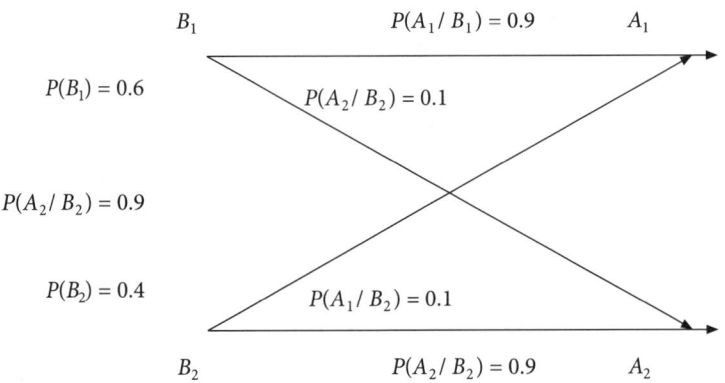

Fig. 9.5 Binary symmetric channel for Example 9.14

Solution

The received symbol probabilities can be calculated as

$$P(A_1) = P(A_1 / B_1) \times P(B_1) + P(A_1 / B_2) \times P(B_2)$$

$$= 0.9 \times 0.6 + 0.1 \times 0.4 = 0.58 \tag{9.35}$$

$$P(A_2) = P(A_2/B_1) \times P(B_1) + P(A_2/B_2) \times P(B_2)$$

$$= 0.1 \times 0.6 + 0.9 \times 0.4 = 0.42 \qquad (9.36)$$

The probability of reception at both the received symbols is contributed from the symbols transmitted from both the inputs as indicated in Fig. 9.5.

The probabilities for correct symbol transmission is given by

$$P(B_1/A_1) = \frac{P(A_1/B_1) \times P(B_1)}{P(A_1)} \qquad (9.37)$$

$$= 0.9 \times 0.6 / 0.58 \approx 0.931$$

$$P(B_1/A_2) = \frac{P(A_2/B_2) \times P(B_2)}{P(A_2)} \qquad (9.38)$$

$$= 0.9 \times 0.4 / 0.42 \approx 0.857$$

Now, let us calculate probabilities of error

$$P(B_1/A_2) = \frac{P(A_2/B_1) \times P(B_1)}{P(A_2)} \qquad (9.39)$$

$$= 0.1 \times 0.6 / 0.42 \approx 0.143$$

$$P(B_2/A_1) = \frac{P(A_1/B_2) \times P(B_2)}{P(A_1)} \qquad (9.40)$$

$$= 0.1 \times 0.4 / 0.58 \approx 0.069$$

The applications of Bayes theorem is mostly in the medical domain. We will go through a couple of examples.

Example 9.15

What is the probability that a person has a cold given that he/she has a fever? The probability of a person having a cold $P(C) = 0.01$, the probability of fever in a given population is $P(f) = 0.02$ and the probability of fever given that a person has a cold $P(f/C) = 0.4$.

Solution

$$P(C/f) = \frac{P(f/C) \times P(C)}{P(f)} \qquad (9.41)$$

$$= 0.4 \times 0.01 / 0.02 = 0.2$$

Example 9.16

Detecting HIV virus using ELESA test. The probabilities are specified as follows. The probability of having HIV is $P(H) = 0.15$, the probability of not having HIV is $P(\bar{H}) = 0.85$, the probability for getting test positive given that person has HIV is $P(\text{Pos}/H) = 0.95$ and the probability for getting a positive test given that person does not have HIV is $P(\text{Pos}/\bar{H}) = 0.0.02$. Find the probability that a person has HIV given that the test is positive.

Solution

$$P(H/\text{Pos}) = \frac{P(\text{Pos}/H) \times P(H)}{P(H) \times P(\text{Pos}/H) + P(\bar{H}) \times P(\text{Pos}/\bar{H})} \qquad (9.42)$$

$$= \frac{0.15 \times 0.95}{0.15 \times 0.95 + 0.85 \times 0.02} = 0.893$$

Concept Check

- Define the term event.
- Define probability in terms of relative frequency of occurrence.
- What is a conditional probability?
- Define a sample space for a random experiment.
- State the relation between conditional probability and the intersection of the sample spaces of the two events.
- When will you classify the events as mutually exclusive?
- State Bayes theorem in two different ways.
- What will a Venn diagram for mutually exclusive events look like?
- When will you call a channel as symmetrical?

9.2 Random Variable

We have considered the examples of random experiments like tossing of a coin, throwing a die, etc. The outcome of these experiments was random in the sense that the outcome cannot be predicted. The sample space spanned by the outcomes is also known. These are called as discrete random outcomes with sample space consisting of only finite values. A random variable can be defined as the characterization of the outcome of the random experiment that will associate a numerical value with each outcome. Consider the example of throwing a die. Let us denote the random variable associated with the outcome of the experiment as 'X'. The range of 'X' is from 1 to 6. Every time when a die is thrown, a value corresponding to the number on the face is associated with the outcome. If number '1' appears, 'X' = 1. If the random variable takes on only discrete values, the random variable is called a discrete random variable. On the other hand, if the random variable takes on real values, it is called as continuous random variable. Here, the range of random variable is infinite.

9.2.1 Cumulative distribution function (CDF)

Let us now define the concept of distribution function. Cumulative distribution function called as CDF is defined as the probability that the random variable 'X' takes a value less than or equal to some allowed value 'x'. The value is allowed if it belongs to a sample space. CDF is defined by the equation

$$F_X(x) = P(X \leq x) \tag{9.43}$$

Example 9.17

Consider a fair die. Plot a CDF verses 'x'.

Solution

We are considering a fair die. This indicates that the probability of getting any value on the face is equally likely. There are 6 possible outcomes. Hence, the probability of getting each face is 1/6 as total probability is 1. We will now find the CDF for each value of 'x'.

$$\begin{aligned}
F_X(1) &= P(X \leq 1) = \frac{1}{6} \\
F_X(2) &= P(X \leq 2) = P(1) + P(2) = \frac{2}{6} \\
F_X(3) &= P(X \leq 3) = P(1) + P(2) + P(3) = \frac{3}{6} \\
F_X(4) &= P(X \leq 4) = P(1) + P(2) + P(3) + P(4) = \frac{4}{6} \\
F_X(5) &= P(X \leq 5) = P(1) + P(2) + P(3) + P(4) + P(5) = \frac{5}{6} \\
F_X(6) &= F(X \leq 6) = P(1) + P(2) + P(3) + P(4) + P(5) + P(6) = 1
\end{aligned} \tag{9.44}$$

A plot of CDF is shown in Fig. 9.6.

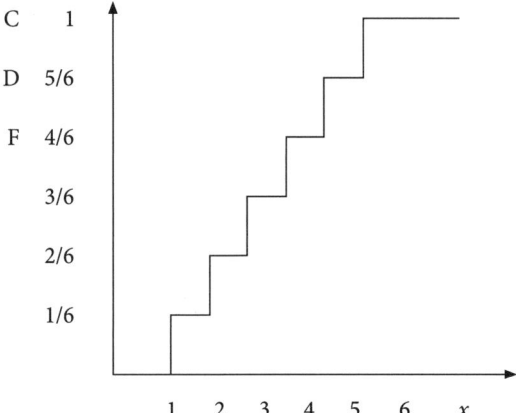

Fig. 9.6 Plot of CDF for a single toss of a die

The cumulative distribution function $F_x(x)$ obeys the following properties.

1. $0 \leq F_X(x) \leq 1$
2. $\lim_{x \to \infty} F_X(x) = 1$
3. $\lim_{x \to -\infty} F_X(x) = 0$
4. $F_x(x)$ is a non decreasing function of x.
5. $P(X = x) = F_X(x) - \lim_{\varepsilon \to 0} F_X(x - \varepsilon)$
6. $F_X(x) = P(X \leq x) = 1 - P(X > x)$

The above properties can be easily verified by the reader. The first property holds good because $F_x(x)$ is a probability. The second and third properties also follow from the concept of probability. For a certain event i.e., the face value of a die if less than or equal to 6 is one and probability for no face value is zero. The fourth property holds because the probability is a positive value. The fifth property says that for a discrete random variable, the probability of getting a specific value is finite. We will see that for a continuous random variable, the probability of getting a specific value is zero. The sixth property indicates that the events $X \leq x$ and $X > x$ are mutually exclusive.

Example 9.18

Consider a pair of dice. Find the probability of getting the sum of the faces as less than 6. Also find the probability for getting sum of faces equal to 10.

Solution

We understand that the probability of getting any particular combination of faces is 1/6*1/6 = 1/36. The possible outcomes can be listed, as shown in Table 9.1.

Table 9.1 Possible outcomes for throw of pair of dice

(1,1)	(2,1)	(3,1)	(4,1)	(5,1)	(6,1)
(1,2)	(2,2)	(3,2)	(4,2)	(5,2)	(6,2)
(1,3)	(2,3)	(3,3)	(4,3)	(5,3)	(6,3)
(1,4)	(2,4)	(3,4)	(4,4)	(5,4)	(6,4)
(1,5)	(2,5)	(3,5)	(4,5)	(5,5)	(6,5)
(1,6)	(2,6)	(3,6)	(4,6)	(5,6)	(6,6)

Total outcomes are 36. To find the probability of getting the sum of faces as less than 6, we have to count such combinations for which the sum of faces is less than or equal to 6. It can be verified that such a probability is 15/36.

To find the probability for getting sum of faces equal to 10. Let us list such combinations: (5,5), (6,4), (4,6). The probability is 3/36.

Example 9.19

Consider 2 unfair dice. The probability of the first die of getting a face value of 2 is 2/7 and the probability of the other die of getting a face value 3 is 2/7. Find the probability of getting a sum of 6 for two faces.

Solution

To get a sum of 6, we have the following combinations.
(1,5), (5,1), (2,4), (4,2), (3,3). Let us find the probability.

$$\frac{1}{49}+\frac{1}{49}+\frac{2}{49}+\frac{1}{49}+\frac{2}{49}=\frac{7}{49}=\frac{1}{7} \tag{9.45}$$

Example 9.20

A random variable has a distribution function given by

$$F_X(x) = 0 \quad -\infty < x \leq -10$$

$$= \frac{1}{6} \quad -10 \leq x \leq -5$$

$$= \frac{x}{15}+\frac{1}{2} \quad -5 < x < 5$$

$$= \frac{5}{6} \quad 5 \leq x < 10$$

$$= 1 \quad 10 \leq x < \infty$$

Draw the CDF. Find $P(X \leq 4)$ and $P(-5 \leq X \leq 4)$

Solution

We will first plot the CDF. Fig. 9.7 shows a plot of CDF.

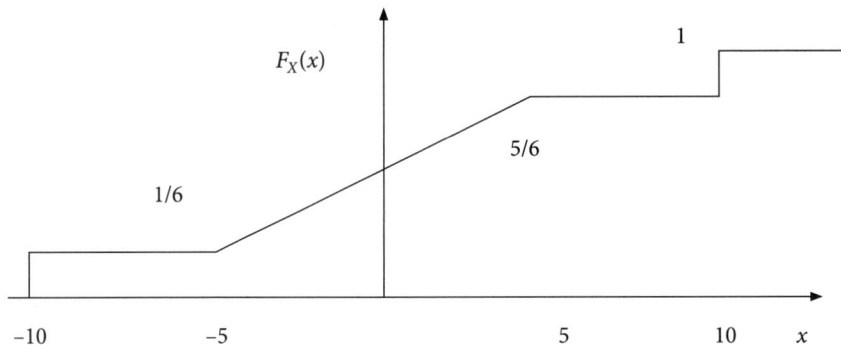

Fig. 9.7 Plot of CDF for Example 9.20

Let us find $P(X \le 4) = \dfrac{4}{15} + \dfrac{1}{2}$

$$= \dfrac{23}{30} \tag{9.46}$$

$$= F_X(4)$$

$P(-5 \le X \le 4) = F_X(4) - F_X(-5)$

$$= \dfrac{23}{30} - \dfrac{1}{6} \tag{9.47}$$

$$= \dfrac{18}{6}$$

9.2.2 Probability density function (pdf)

The probability density function is defined as the derivative of the CDF for a continuous random variable given by

$$f_X(x) = \dfrac{d}{dx}(F_X(x)) \tag{9.48}$$

In case of a discrete random variable, pdf is defined in terms of the values of the probabilities at the discrete sample values. It can be stated as

$f_X(x) = P(X = x_i)$ if $x = x_i$, $i = 1, 2, ..., n$

$= 0$ if $x \ne x_i$ $\tag{9.49}$

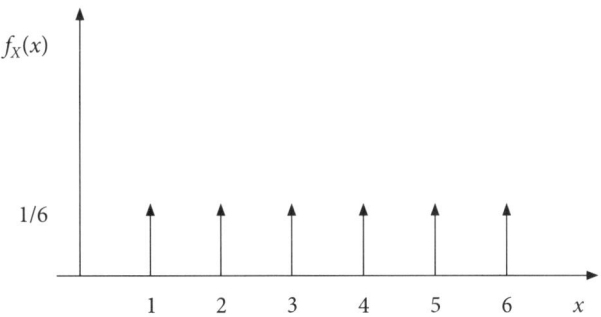

Fig. 9.8 Plot of pdf

Example 9.21

Plot the pdf for the CDF specified for Example 9.12.

Solution

The plot of pdf is a plot of the derivative of the graph shown in Fig. 9.7. Fig. 9.9 shows a plot of pdf.

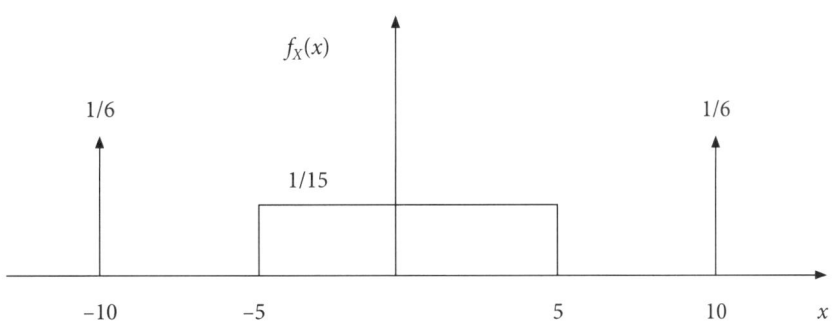

Fig. 9.9 Plot of pdf for CDF specified in Example 9.21

The nature of pdf drawn can be explained as follows. Referring to Fig. 9.4, we see that the value of the derivative at $x = -10$ is 1/6. The slope of the CDF graph is zero for values of x from -10 to -5 and for values of x from 5 to 10. For values of x between -5 to 5, the slope of CDF graph is

$$\frac{\frac{5}{6}-\frac{1}{6}}{10} = \frac{4}{6 \times 10} = \frac{1}{15} \qquad (9.50)$$

(refer to Fig. 9.7 and Fig. 9.9).
We can list the properties of pdf as follows:

1. $f_X(x) \geq 0$.

2. $\int_a^b f_X(x)dx = P(a \le X \le b)$

3. $\int_{-\infty}^{\infty} f_X(x)dx = 1$

The pdf being a probability must be a non-negative quantity. If we integrate the pdf between some limits, it gives the probability for the variable to be between the limits specified. The integration of pdf over the entire range is nothing but the probability of a certain event. Hence, it is equal to one.

Example 9.22

Given the pdf for different x values as follows. $x = 1$, pdf = 0.2, $x = 2$, pdf = 0.1, $x = 3$, pdf = 0.3, $x = 4$, pdf = 0.3, $x = 5$, pdf = 0.1. Draw the pdf and its corresponding CDF.

Solution

Let us first draw the pdf, as shown in Fig. 9.10.

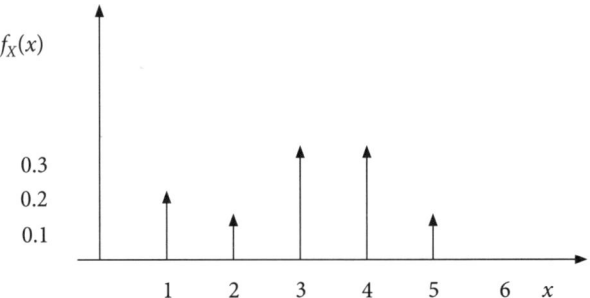

Fig. 9.10 Plot of pdf for Example 9.22

To draw the CDF, we have to integrate the pdf. CDF is shown in Fig. 9.11.

Note that we have to add the probability values at each step. At 1, there is a step of 0.2. Then it remains constant until 2. At 2, it has a step of 0.1 and it reaches 0.3. It remains constant until 3. At 3, it has a step of 0.3 and it reaches 0.6. It remains constant until 4. At 4, it has a step of 0.3 and it reaches 0.9. It remains constant until 5. At 5, it has a step of 0.1 and it reaches 1.0. The total value of probability must reach 1.

Example 9.23

The pdf for a random variable is given by

$$f_X(x) = \left\{ \frac{1}{6}[\delta(x-2) + 2\delta(x-3) + 2\delta(x-4) + \delta(x-5)] \right\}$$

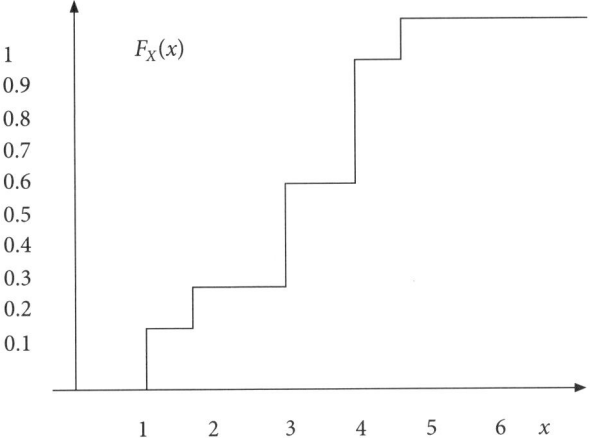

Fig. 9.11 Plot of CDF for Example 9.22

Draw the pdf and its corresponding CDF.

Let us first draw the pdf, as shown in Fig. 9.12.

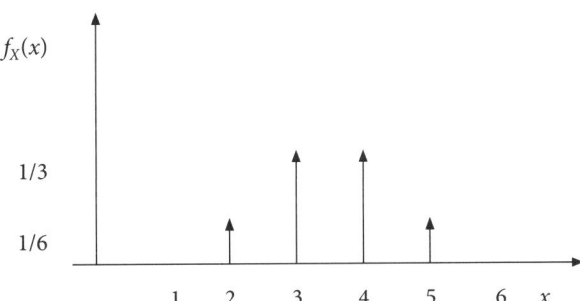

Fig. 9.12 Plot of pdf for Example 9.23

To draw the CDF, we have to integrate the pdf. CDF is shown in Fig. 9.13.

Note that we have to add the probability values at each step. At 2, there is a step of 1/6. Then it remains constant until 3. At 3, it has a step of 1/3 and it reaches 1/2. It remains constant until 3. At 4, it has a step of 1/3 and it reaches 5/6. It remains constant until 5. At 5, it has a step of 1/6 and it reaches 1. The total value of probability must reach 1.

Example 9.24

Find the density function such that,

$$f_X(x) = \begin{cases} cx & 0 < x < 3 \\ 0 & \text{otherwise} \end{cases}$$

Compute the probability $P(1 < x < 2)$. Find distribution function $F_X(x)$.

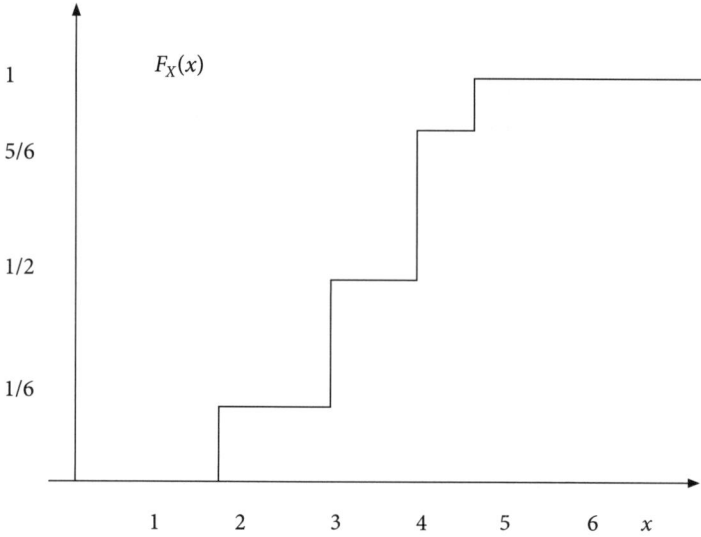

Fig. 9.13 Plot of CDF for Example 9.23

Solution

a. The pdf is specified. To find c, we will equate total probability to 1.

$$\int_0^3 f_X(x)dx = 1$$

$$\int_0^3 cx\,dx = \frac{cx^2}{2}\Big|_0^3 = \frac{9c}{2} = 1 \Rightarrow c = \frac{2}{9} \tag{9.51}$$

$$f_X(x) = \begin{cases} \frac{2}{9}x & 0 < x < 3 \\ 0 & \text{otherwise} \end{cases}$$

b. Probability that $P(1 < x < 2)$

$$P(1 < x < 2) = \int_1^2 \frac{2}{9}x\,dx = \frac{2}{9} \times \frac{x^2}{2}\Big|_1^2 = \frac{2}{9}\left(2 - \frac{1}{2}\right) = \frac{1}{3} \tag{9.52}$$

c. Find the distribution function

$$F_X(x) = \int_0^x \frac{2}{9}x\,dx = \frac{2}{9} \times \frac{x^2}{2}\Big|_0^x = \frac{2}{9} \times \frac{x^2}{2} = \frac{x^2}{9} \tag{9.53}$$

$$F_X(x) = \begin{cases} 0 & x < 0 \\ \dfrac{x^2}{9} & 0 \leq x \leq 3 \\ 1 & x > 3 \end{cases}$$

Example 9.25

Consider the function

$$f_X(x) = \begin{cases} \dfrac{1}{2}e^{-x} + \dfrac{1}{2}\delta(x-3) & \text{for } x \geq 0 \\ 0 & \text{for } x < 0 \end{cases}$$

a. Verify that the function represents a density function.
b. Calculate $P(X = 1)$, $P(X = 3)$, and $P(X \geq 1)$.

Solution

In order to verify $f_X(x)$ represents a valid density function, we have to verify that

$$\int_0^\infty \left(\frac{1}{2}e^{-x} + \frac{1}{2}\delta(x-3) \right) dx = \frac{1}{2}(-e^{-x}) \Big\downarrow_0^\infty + \delta(x-3)$$

put $x = 3$ for second term (9.54)

$$= \frac{1}{2}[1+1] = 1$$

$$P(x=1) = f_X(1) = \frac{e^{-1}}{2} = 0.1839$$

$$P(x=3) = f_X(3) = \frac{e^{-3}}{2} + \frac{1}{2} = 0.5248$$

(9.55)

$$P(X \geq 1) = \int_1^\infty \left[\frac{1}{2}e^{-x} + \frac{1}{2}\delta(x-3) \right] dx$$

$$= -\frac{1}{2}e^{-x} \Big\downarrow_1^\infty + \frac{1}{2} = 0.1839 + 0.5 = 0.6839$$

Concept Check

- Define a random variable.
- How will you define a discrete random variable?
- How will you define a continuous random variable?
- Define CDF.
- State the properties of CDF.
- Define pdf. How is it related to CDF?
- How will you find pdf if CDF is specified?
- List the properties of pdf.

9.3 Statistical Properties of Random Variables

When we analyze any random variable for example speech samples, we have only finite data values available with us. Based on these finite samples, the statistical properties are predicted or estimated. The actual data may exist over infinite interval. We cannot process the infinite samples due to limitations of processing capacity of a computer. It is possible to estimate the statistical properties from finite samples. The method of fitting a standard pdf for unknown signal namely Chi square test and K–S test will be discussed in section 9.5. Let us define important statistical properties and discuss the method for prediction of these properties. The statistical properties are mean, variance, moments, centre moments, etc. We will start with the mean or the expected value.

1. **Mean or Expected value** The mean or expected value for a continuous random variable is defined as

$$\mu = E[X] = \int_{-\infty}^{\infty} x f_X(x) dx \qquad (9.56)$$

In case of a discrete random variable, mean is defined as

$$\mu = E[X] = \sum_i x_i P(x_i) \qquad (9.57)$$

The mean of a random variable may not belong to a sample space. To find the value of mean for a continuous random variable, there are two hurdles. First one is that the signal is not available over infinite duration and secondly, the pdf for a random variable is not known. We fit the pdf for an unknown signal and try to find the mean using the mathematical expression for that standard pdf in the equation for mean. The mean is also termed as expected value.

There are number of standard distribution functions defined for which mathematical expression for pdfs are available. Some of the standard distributions are uniform distribution, Gaussian distribution, Rician distribution, Rayleigh distribution, exponential distribution, etc. We will indicate how to make use of such standard distribution pdfs for unknown random variable in the next section.

2. **Moments** The nth moment denoted as m_n of a continuous random variable is defined as

$$m_n = E[X^n] = \int_{-\infty}^{\infty} x^n f_X(x)dx \qquad (9.58)$$

The mean is the first order moment. This can be seen by comparing Eqs (9.56) and (9.58). For the definition of mean $n = 1$. When $n = 2$, we obtain the mean square value of the variable given by

$$m_2 = E[X^2] = \int_{-\infty}^{\infty} x^2 f_X(x)dx \qquad (9.59)$$

3. **Centre moments** The nth order centre moment is defined as

$$m_n^C = E[(X-\mu)^n] = \int_{-\infty}^{\infty} (x-\mu)^n f_X(x)dx \qquad (9.60)$$

Variance The second order centre moment is called the **variance** of the random variable X and is given by

$$m_2^C = \text{Variance}(\sigma^2)$$

$$= E[(X-\mu)^2] \qquad (9.61)$$

$$= \int_{-\infty}^{\infty} (x-\mu)^2 f_X(x)dx$$

$$E[X-\mu]^2 = E\left[X^2 - 2\mu X + \mu^2\right]$$

$$= E\left[X^2\right] - 2\mu \times E[X] + \mu^2 \qquad (9.62)$$

$$= E\left[X^2\right] - 2\mu^2 + \mu^2$$

$$= E\left[X^2\right] - \mu^2$$

Hence, $\sigma^2 = E[X^2] - \mu^2$

As $E[X] = \mu$ and $E[\mu^2] = \mu^2$

Standard Deviation The symbol σ represents the standard deviation of the random variable.

Skew The third order centre moment is a measure of asymmetry of the density function about the mean. Skewness can be considered is a measure of symmetry, or in fact the lack of symmetry. A distribution is said to be symmetric if it looks identical to the left and to the right of the center point. The skewness of the normal distribution is zero.

It is called as the skew of the density function. It is given by

$$m_3^C = \text{Skew}(\mu_3)$$

$$= E\left[(X - \mu)^3\right] \quad (9.63)$$

$$= \int_{-\infty}^{\infty} (x - \mu)^3 f_X(x) dx$$

Skewness The normalized third order centre moment is called as the skewness of the density function. It is given by

$$\text{Skewness/coefficient of skewness} = \frac{\text{Skew}(\mu_3)}{\sigma_X^3} \quad (9.64)$$

Kurtosis Kurtosis is a measured relative to a normal distribution. It measures if the data are peaked or flat relative to a normal distribution. If the value of Kurtosis is high, data sets tend to have a distinct peak near the mean, it is found to decline rapidly, and it will have heavy tails. On the other hand, the data sets with low kurtosis will have a flat top near the mean. A uniform distribution is thus the extreme case of low Kurtosis.

Kurtosis is defined as the ratio of fourth order centre moment and fourth power of standard deviation.

$$\text{Kurtosis} = \frac{(\mu_4)}{\sigma_X^4} \quad (9.65)$$

The fourth order centre moment is defined as

$$m_4^C = (\mu_4) = E[(X-\mu)^4]$$
$$= \int_{-\infty}^{\infty} (x-\mu)^4 f_X(x)dx \qquad (9.66)$$

The kurtosis for a standard normal distribution can be shown to be equal to three. Hence, sometimes the following definition of kurtosis is used.

$$\text{Kurtosis} = \frac{(\mu_4)}{\sigma_X^4 - 3} \qquad (9.67)$$

Using this definition, kurtosis for a normal distribution is zero. Using this second definition of kurtosis positive kurtosis will indicate a "peaked" distribution and negative kurtosis will indicate a "flat" distribution.

Example 9.26

Consider a random variable X with probability density function shown in Fig. 9.14. Find A, mean value of X and variance of X.

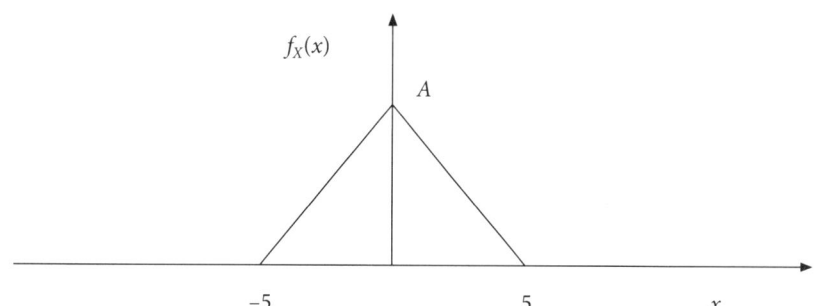

Fig. 9.14 pdf for a random variable x

Solution

1. To find A—If the pdf is a valid distribution function, then the area under the curve i.e., the total probability must be equal to 1.

$$\frac{[5-(-5)]A}{2}=1$$

$$\frac{10A}{2}=1 \quad (9.68)$$

$$\Rightarrow A = \frac{1}{5}$$

2. To find the mean value—

$$\mu = E[X] = \int_{-5}^{5} x f_X(x) dx$$

$$= \int_{-5}^{0} \frac{x(x-(-5))}{5dx} + \int_{0}^{5} \frac{x(5-x)}{5dx} = \frac{1}{5}\left[\int_{-5}^{0}(x^2+5x)dx + \int_{0}^{5}(5x-x^2)dx\right] \quad (9.69)$$

$$= \frac{1}{5}\left[\frac{x^3}{3}+\frac{5x^2}{2}\right]_{-5}^{0} + \frac{1}{5}\left[\frac{5x^2}{2}-\frac{x^3}{3}\right]_{0}^{5}$$

$$= \frac{1}{5}\left[-(-25)+\frac{125}{2}+\frac{125}{2}-25\right]$$

$$= \frac{1}{5}\left[25-25+\frac{125}{2}-\frac{125}{2}\right] = 0$$

To find the variance—

Variance $= E[X^2] - \mu^2$ (Note $\mu = 0$)

$$= E[X^2] = \frac{1}{25}\left[\int_{-5}^{0}[x^3+5x^2]dx + \int_{0}^{5}[5x^2-x^3]dx\right]$$

$$= \frac{1}{25}\left\{\left[\frac{x^4}{4}+5\frac{x^3}{3}\right]_{-5}^{0} + \left[5\frac{x^3}{3}-\frac{x^4}{4}\right]_{0}^{5}\right\} \quad (9.70)$$

$$= \frac{1}{25}\left[\frac{-625}{4}+\frac{625}{3}+\frac{625}{3}-\frac{625}{4}\right]$$

$$= \frac{1}{25}[416.66 - 312.2] = 4.1784$$

Example 9.27

A random variable has probability density function given by the following equation

$$f(x) = \begin{cases} 0.1 & -3 \le x \le 7 \\ 0 & \text{elsewhere} \end{cases}$$

a. Find the mean value
b. Find the mean square value
c. Find the variance

Solution

The mean value of probability density function is given as

$$\mu = E[X] = \int_{-3}^{7} 0.1 x \, dx = 0.1 \frac{x^2}{2} \Big|_{-3}^{7} = 2 \tag{9.71}$$

The mean square value of probability density function is given as

$$E[X^2] = \int_{-3}^{7} 0.1 x^2 \, dx = 0.1 \frac{x^3}{3} \Big|_{-3}^{7} = \frac{37}{3} \tag{9.72}$$

The variance of the random variable X is

$$\sigma^2 = E[X^2] - \mu^2 = \frac{37}{3} - 4 = \frac{25}{3} \tag{9.73}$$

Concept Check

- List the statistical properties of a random variable.
- Define the mean or the expected value. Why the mean value is called as expected value?
- Define the term variance for a random variable.
- What is skewness?
- Define n^{th} order moment for a random variable.
- Define the term kurtosis.
- If the value of kurtosis is high, what is your conclusion related to the distribution of the random variable?
- If the value of kurtosis is low, what is your conclusion related to the distribution of the random variable?

- What is the value of kurtosis for a normal distribution?
- What will be the value of kurtosis for a uniform distribution?

9.4 Standard Distribution Functions

There are standard distribution functions defined for continuous random variables namely uniform probability distribution, Gaussian distribution, Rayleigh distribution and Exponential distribution. Discrete random variables use Poison distribution and binomial distribution. We will first describe standard distribution functions for continuous random variables.

9.4.1 Probability distribution functions for continuous variables

1. **Uniform probability distribution** Uniform probability density function over a range of values between a and b is defining as

$$f_X(x) = K \quad \text{if} \quad a \leq x \leq b$$
$$= 0 \quad \text{otherwise} \tag{9.74}$$

Let x be a uniformly distributed random variable between a and b. Let us plot the pdf. Fig. 9.15 shows a pdf.

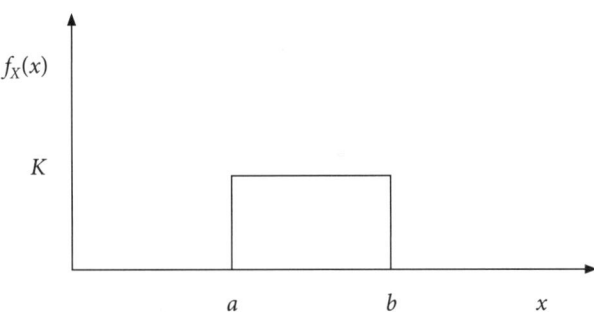

Fig. 9.15 Plot of pdf for a uniform distribution

Let us find the value of K.

$$\int_a^b f_X(x)dx = \int_a^b K dx = 1$$
$$[Kx]_a^b = 1 \tag{9.75}$$

$$K = \frac{1}{b-a}$$

Let us find the mean value of x.

$$\mu = E[X] = \int_a^b x f_X(x) dx$$

$$= \int_a^b x \times K dx = K \left[\frac{x^2}{2}\right]_a^b \quad (9.76)$$

$$= \frac{(b^2 - a^2)}{2(b-a)} = \frac{b+a}{2}$$

We will now find the variance of x.

$$\text{Variance} = E[X^2] - \mu^2 \quad \left(\mu = \frac{a+b}{2}\right)$$

$$E[X^2] = \int_a^b [x^2 f_X(x)] dx$$

$$= \int_a^b K x^2 dx = K \left[\frac{x^3}{3}\right]_a^b$$

$$= \frac{1}{b-a} \times \frac{(b^3 - a^3)}{3} = \frac{b^2 + a^2 + ab}{3} \quad (9.77)$$

$$\text{Variance} = \frac{b^2 + a^2 + ab}{3} - \frac{b^2 + a^2 + 2ab}{4}$$

$$= \frac{b^2 + a^2 - 2ab}{12} = \frac{(b-a)^2}{12}$$

2. **Gaussian Random variable** The pdf associated with Gaussian distribution function is given by

$$f_X(x) = \frac{1}{\sigma\sqrt{2\pi}} \exp\left[\frac{-(x-\mu)^2}{2\sigma^2}\right] \quad -\infty \leq x \leq \infty \quad (9.78)$$

$\sigma > 0$ and $-\infty < \mu < \infty$

The Gaussian pdf is completely specified by σ and μ.

It is found that many physical situations are found to follow the Gaussian distribution function. As the mean value increases, the entire curve for Gaussian distribution function shifts towards right. If the variance increases, the curve becomes wider. To find the probability that the random variable lies between $-\infty$ and some fixed value, we have to integrate the square function that is a computationally involved task. To help the user in this regard, the tables are available to read the integral value. If the mean and a variance changes, the integration values in the table will change. The problem can be simplified if only a single table is prepared for mean value of zero and variance of one. The random variable can be normalized to get a mean value of zero and variance of one. Let us illustrate the procedure for normalization.

The function $(x-\mu)/\sigma$ can be replaced by another variable y as $y = (x-\mu)/\sigma$. After replacing the function by a normalized random variable, we obtain the CDF for a normalized random variable y as

$$F_Y(y) = \int_{-\infty}^{y} \frac{1}{\sqrt{2\pi}} \exp\left[\frac{-(y)^2}{2}\right] dy \quad -\infty \leq x \leq \infty \tag{9.79}$$

Now, the table can be listed for the normalized random variable to find the integral value. We can then find the value for the actual variable x by putting the obtained value for y. The probability distribution function listed in Eq. (9.48) is called as the normal distribution function.

A MATLAB program to plot Gaussian pdf is shown below.

```
clear all;
sigma=2;
mean=0;
x=(-5:0.1:5);
a=(x-mean).*(x-mean);
y=(1/(sigma*sqrt(2*pi)))*exp(-(((x-mean).*
(x-mean))/(2*sigma^2)));
plot(x,y);xlabel('x');ylabel('pdf');
title('Gaussian pdf');
```

A Gaussian pdf is plotted in Fig. 9.16. If the variance sigma increases, the width of the curve increases, as shown in Fig. 9.17. If the mean is positive and it increases, the curve shifts towards right. If mean is negative and it increases, the curve shifts towards left, as shown in Fig. 9.18.

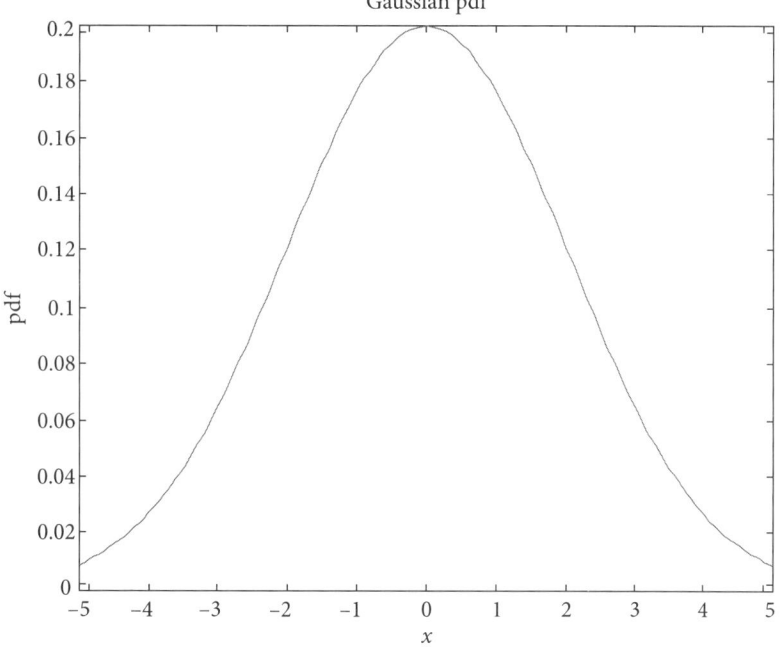

Fig. 9.16 Plot of Gaussian pdf for sigma = 2 and mean = 0

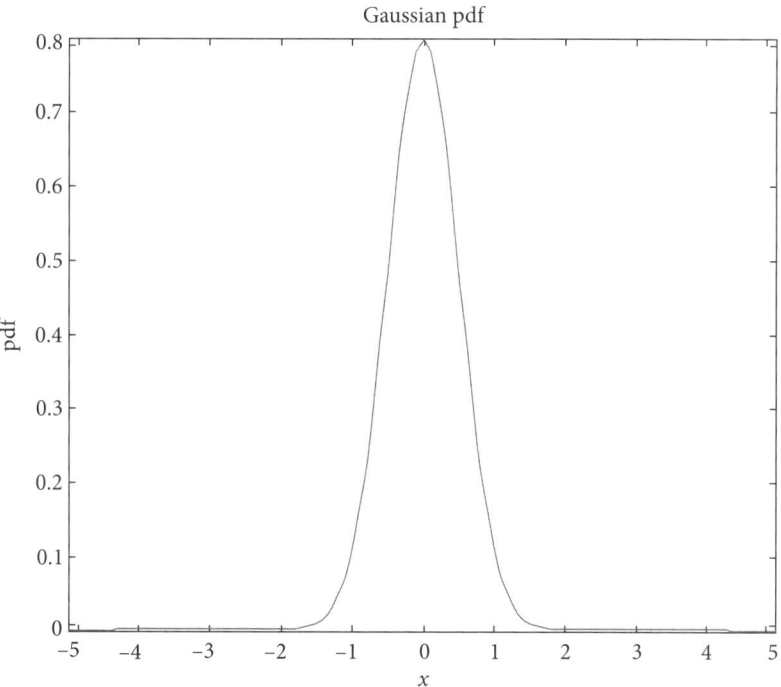

Fig. 9.17 Plot of Gaussian pdf for sigma = 0.5 and mean = 0

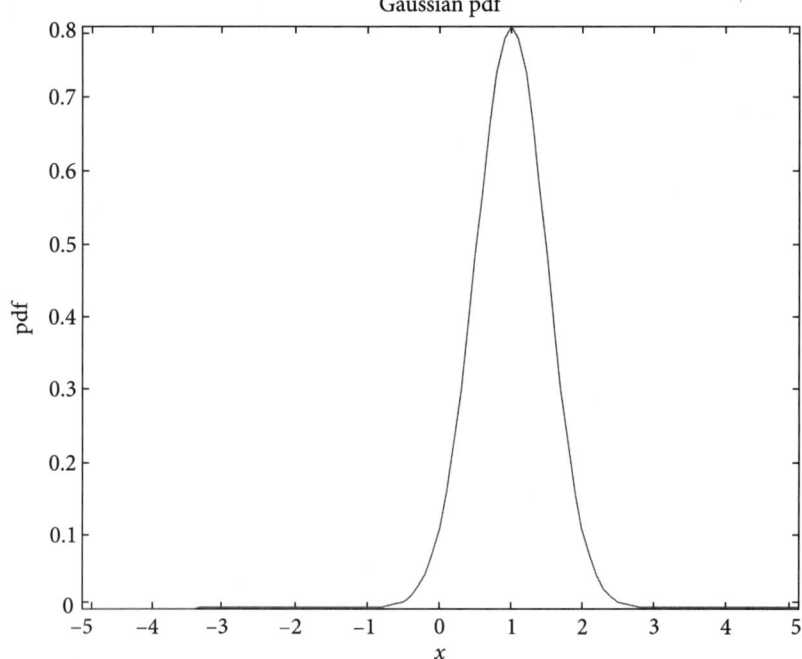

Fig. 9.18 Plot of Gaussian pdf for sigma = 0.5 and mean = 1

Example 9.28

Find the probability of the event $(X \leq 5.5)$ for a Gaussian variable having mean value of 3 and variance of 2.

Solution

We use the equation $y = (x - \mu)/\sigma$ and find the value of y.

$$y = \frac{(5.5 - 3)}{2} = \frac{2.5}{2} = 1.25 \tag{9.80}$$

$F_Y(y) = P(y \leq 1.25)$

We will now use a table for a normalized distribution to find the value of CDF. Referring to the Normal distribution function table in Appendix A, we can read $F(1.25) = P(Y \leq 1.25) = 0.8944$.

If we want to find the probability of the event $(X \geq 5.5)$ i.e., $(1 - P(Y \leq\leq 5.5))$ Referring to the table we find the probability of the event as $(1 - 0.8944) = 0.1056$.

1. **Rayleigh density function** Rayleigh density function and distribution function can be stated as

$$f_X(x) = \begin{cases} \dfrac{2}{b^2}(x-a)e^{-(x-a)^2/2b^2} & x \geq a \\ 0 & x < a \end{cases} \qquad (9.81)$$

$-\infty < a < \infty$ and $b > 0$

$$F_X(x) = \begin{cases} 1 - e^{-(x-a)^2/2b^2} & x \geq a \\ 0 & x < a \end{cases} \qquad (9.82)$$

The Rayleigh occurs in case of a number of physical situations. Consider the example of a rifle aimed at a target. Let the distance of the target be r given by $r = \sqrt{x^2 + y^2}$. Let x and y represent the x and y coordinates of the distance. If x and y are statistically independent random variables that are Gaussian-distributed with zero mean and unit variance, then the pdf of r is said to be Rayleigh-distributed. A second example of the Rayleigh distribution arises in case of random complex numbers with real and imaginary components denoted by x and y. If the real and imaginary components are statistically independent Gaussian random variables with identical mean, then the absolute value of the complex number is Rayleigh-distributed.

A MATLAB program to plot a Rayleigh density function for a particular value of b is given below. RAYLPDF defines a Rayleigh probability density function in MATLAB. Y = RAYLPDF(X, B) returns the Rayleigh probability density function with parameter B at the values in X.

```
clear all;
x = [0:0.01:2];
p = raylpdf(x,0.5);
plot(x,p)
```

Figure 9.19 shows a plot of Rayleigh density function for $b = 0.5$ and Fig. 9.20 shows a plot of Rayleigh cumulative distribution function for $b=0.5$. A MATLAB program to plot CDF is given below.

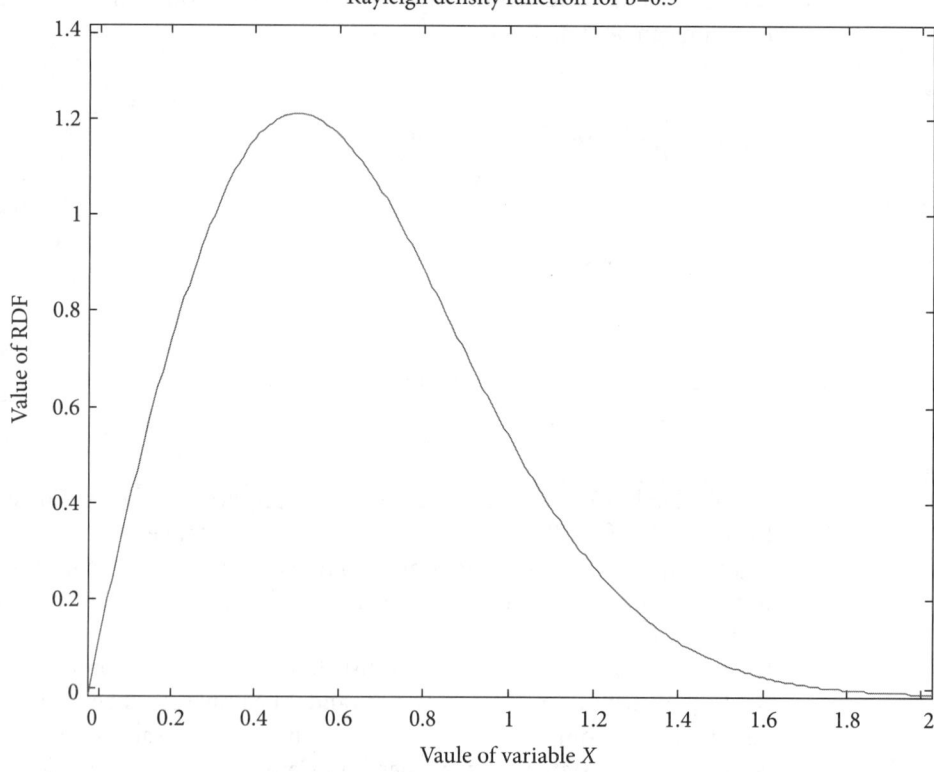

Fig. 9.19 Rayleigh density function for $b = 0.5$

```
clear all;
x = [0:0.01:2];
p = raylcdf(x,0.5);
plot(x,p)
```

2. **Exponential distribution function** Exponential density function and distribution function can be stated as

$$f_X(x) = \begin{cases} \dfrac{1}{b}e^{-(x-a)/b} & x \geq a \\ 0 & x < a \end{cases}$$

(9.83)

$-\infty < a < \infty$ and $b > 0$

$$F_X(x) = \begin{cases} 1 - e^{-(x-a)/b} & x \geq a \\ 0 & x < a \end{cases}$$

(9.84)

$-\infty < a < \infty$ and $b > 0$

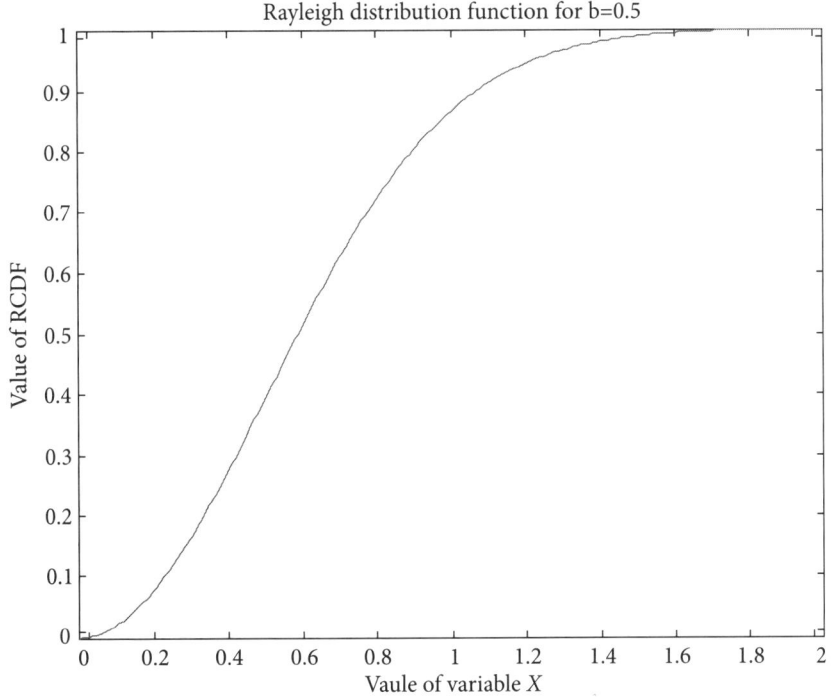

Fig. 9.20 Rayleigh distribution function for $b = 0.5$

It is used to describe the size of a raindrop when large measurements in term conditions are done. It is also used to describe the fluctuations in the strength of the RADAR return signal. A MATLAB program to plot exponential pdf is shown below. Figure 9.21 shows exponential pdf graph. Similarly, we can plot exponential CDF as given below and the plot of CDF is shown in Fig. 9.22.

```
% To plot exponential pdf
clear all;
b=0.25;
a=0;
x=(0:0.1:5);
y=(1/b)*exp(-(x-a)/b);
plot(x,y);xlabel('x');ylabel('pdf');title('exponential pdf');
% plot of exponential CDF
clear all;
b=0.25;
a=0;
x=(0:0.1:5);
y=1-exp(-(x-a)/b);
plot(x,y);xlabel('x');ylabel('CDF');title('exponential CDF');
```

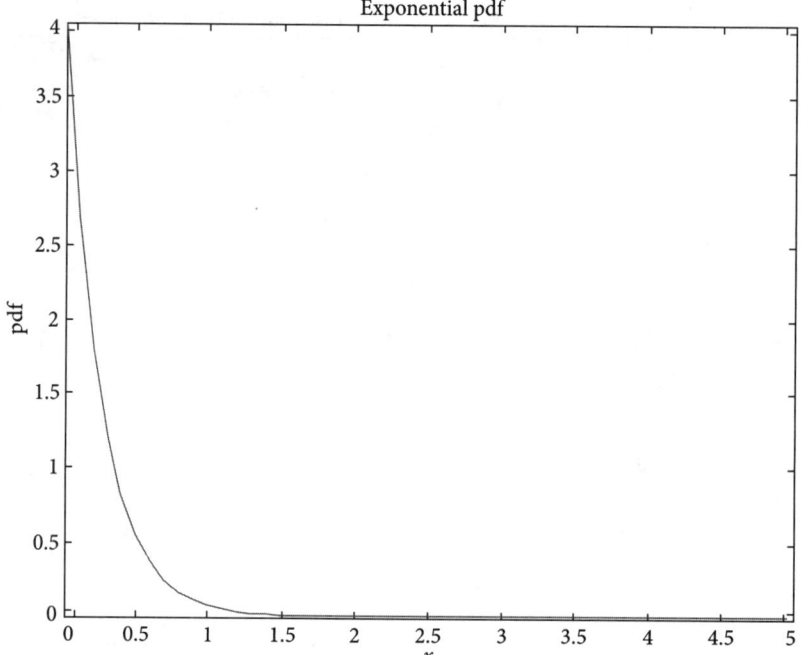

Fig. 9.21 Plot of exponential pdf

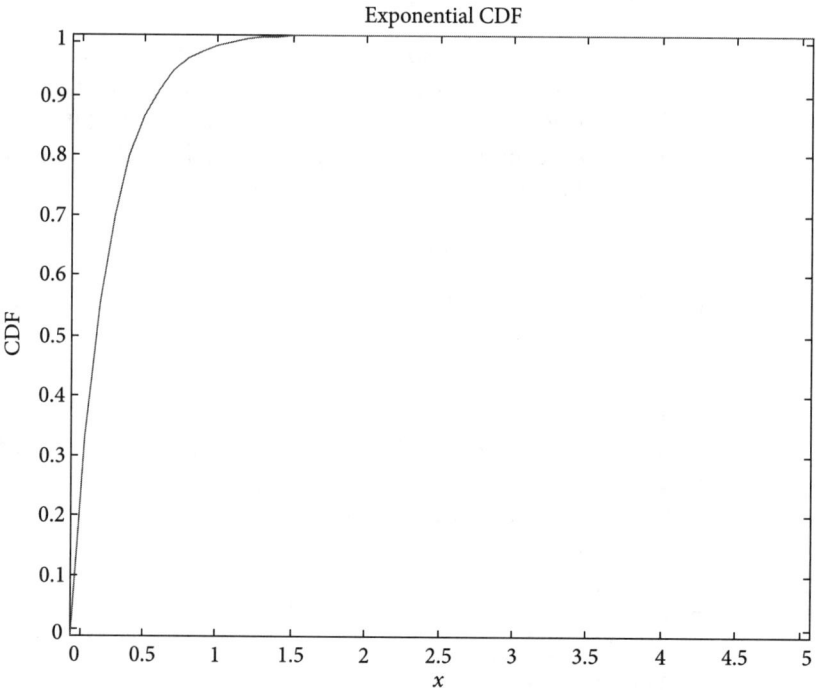

Fig. 9.22 Plot of exponential CDF

9.4.2 Probability distribution functions for discrete variables

We will discuss the standard distribution functions for discrete random variables. Before going through the details of distribution function, we will introduce the terms like permutation, Bernoulli's trials, etc.

Permutations To understand this term, we will consider one simple example of a pack of cards. Let us assume we have 4 successive draws. After every draws, the card drawn is not replaced. In this case, the sample space reduces after each draw. It becomes 51, 50 and 49 after the first, second and third draw, respectively. Here, the sequence in which the cards are drawn is important. Consider that there are n elements in the pack and there are n possible outcomes on first draw. $(n - 1)$ possible outcomes from second draw, $(n - 2)$ outcomes on third draw, etc. if r elements are drawn, the number of possible sequences or outcomes is $n(n - 1)(n - 2)....(n - r)$.

We can write the possible outcomes for r draws as

$$\text{Number of possible outcomes} = P_r^n$$
$$= n(n-i)(n-2)...(n-r+1) \quad (9.85)$$
$$\text{Permutations} = \frac{n!}{(n-r)!}$$

This number is called as number of permutations of r elements taken from n elements when order of sequence of draw is important.

Combinations Let us now define combinations. When the order of sequence of draw is not important, we get a possible number of combinations. In this case the number is less than the permutations because different ordering of same elements is counted as one combination. The number of permutations reduced is equal to the number of permutations of r things taken from r elements i.e., $r!$. Hence, the total number of combinations of r things taken from n things denoted as $\binom{n}{r}$ can be written as

$$\text{Number of possible combinations} = \binom{n}{r}$$
$$= \frac{P_r^n}{P_r^r} \quad (9.86)$$
$$= n(n-i)(n-2)...(n-r+1)/r!$$
$$= \frac{n!}{(n-r)!r!}$$

This number is also called as Binomial coefficient.

Bernoulli's trials Consider an experiment wherein there are only two possible outcomes for every trial. For example, the experiment of tossing a coin wherein there are two possible outcomes namely getting 'Head' or getting 'Tail'. If we denote the probability of getting one outcomes as P, then the probability of getting the other outcome is exactly $1 - P$. If the experiment is repeated a number of times, it is called as Bernoulli's trials. When the experiment is repeated N times, we ask a question, what is a probability of getting 'Head' exactly k times? Let us calculate this probability. We will first find the Binomial coefficient indicating the number of times (k) we get the first outcome out of N number of trials. It is exactly the number of possible sequences that result in same output. It is given by $\binom{N}{k} = \frac{N!}{(N-k)!k!}$. Let us now calculate the probability of getting the first event exactly k times out of N trials. It can be written as

$$P(\text{first oucome } k \text{ times out of } N) = \binom{N}{k} P^k (1-P)^{N-k} \quad (9.87)$$
$$= \frac{N!}{(N-k)!k!} P^k (1-P)^{N-k}$$

Discrete random variables use either Binomial distribution or Poisson distribution.

Binomial Distribution Let us first discuss Binomial density function. It is given by

$$f_X(x) = \sum_{k=0}^{N} \frac{N!}{(N-k)!k!} P^k (1-P)^{N-k} \delta(x-k) \quad (9.88)$$

The term $\binom{N}{k} = \frac{N!}{(N-k)!k!}$ is called as the Binomial coefficient. This density function is used in case of Bernoulli's trials. When the density function is integrated, we get Binomial distribution function given by

$$F_X(x) = \sum_{k=0}^{N} \frac{N!}{(N-k)!k!} P^k (1-P)^{N-k} u(x-k) \quad (9.89)$$

Let us plot these functions for $P = 0.2$ and $N = 6$. A MATLAB program to plot Binomial pdf and CDF is given below. Figure 9.23 shows a plot of Binomial pdf and Fig. 9.24 shows a plot of Binomial CDF.

```
clear all;
x = 0:10;
y = binopdf(x,10,0.25);
stem(x,y);title('Binomial pdf function for N=10 and
P=0.25');xlabel('x');ylabel('pdf value');
clear all;
x = 0:10;
y = binocdf(x,10,0.25);
stem(x,y);title('Binomial CDF function for N=10 and
P=0.25');xlabel('x');ylabel('CDF value');
```

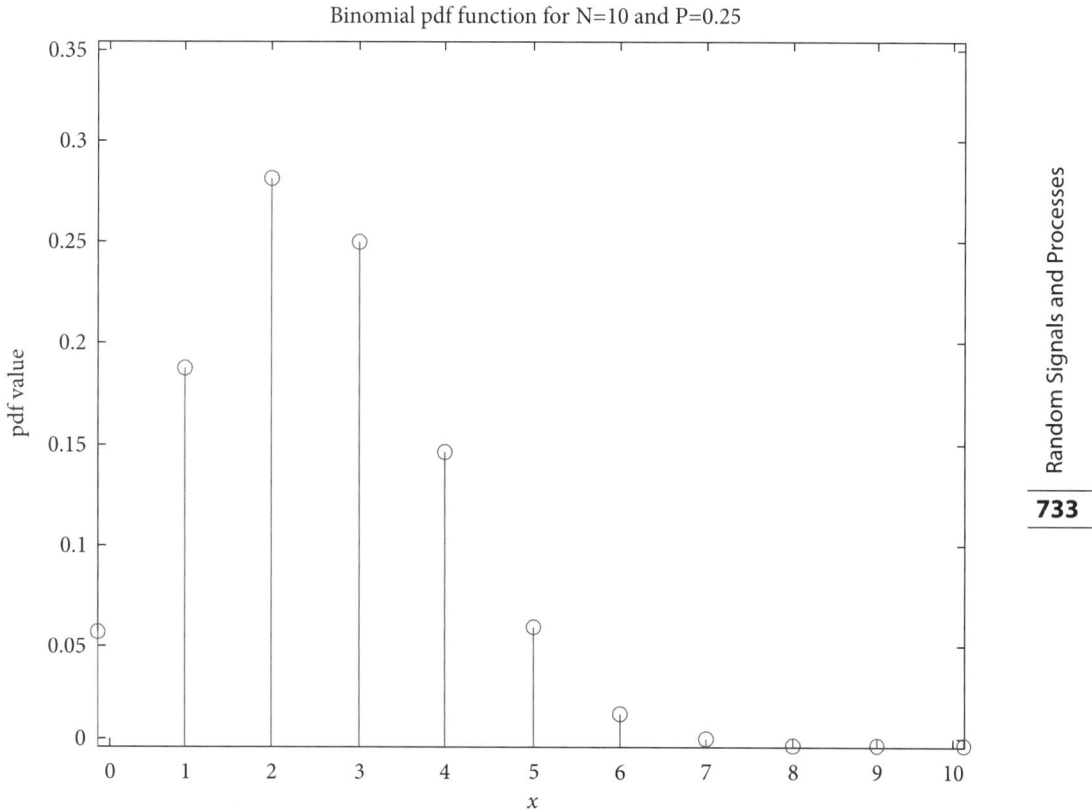

Fig. 9.23 Plot of Binomial pdf for $N = 10$ and $P = 0.25$

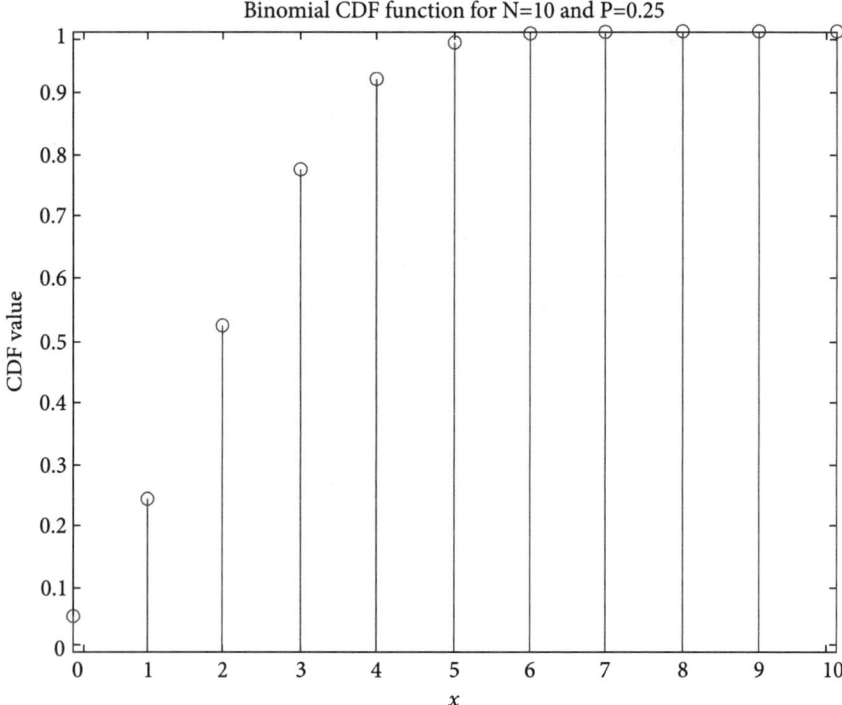

Fig. 9.24 Plot of Binomial CDF for $N = 10$ and $P = 0.25$

Poisson distribution Poisson distribution can be considered as a special case of Binomial distribution when N tends to infinity, P tends to zero such that NP product is a constant equal to b. The pdf and CDF for Poisson distribution can be written as

$$f_X(x) = e^{-b} \sum_{k=0}^{\infty} \frac{b^k}{k!} \delta(x-k) \tag{9.90}$$

$$F_X(x) = e^{-b} \sum_{k=0}^{\infty} \frac{b^k}{k!} u(x-k) \tag{9.91}$$

The number of customers arriving at any service station is found to obey Poisson distribution. This distribution finds application in queuing theory. It applies to counting applications such as number of telephone calls in some duration. If the time duration of interest is denoted as T, if the events occur at rate of lamda, then the constant b is given by $b = \lambda T$. We will plot the pdf and CDF for Poisson distribution. A MATLAB program to plot pdf and CDF is given below.

```
clear all;
x = 0:10;
y = poisspdf(x,0.8);
stem(x,y);title('Poisson pdf function for N=10 and
lamda =0.8');xlabel('x');ylabel('pdf value');

clear all;
x = 0:10;
y = poisscdf(x,0.8);
stem(x,y);title('Poisson CDF function for N=10 and
lamda =0.8');xlabel('x');ylabel('CDF value');
```

Figure 9.25 shows a plot of Poisson pdf and Fig. 9.26 shows a plot of Poisson CDF.

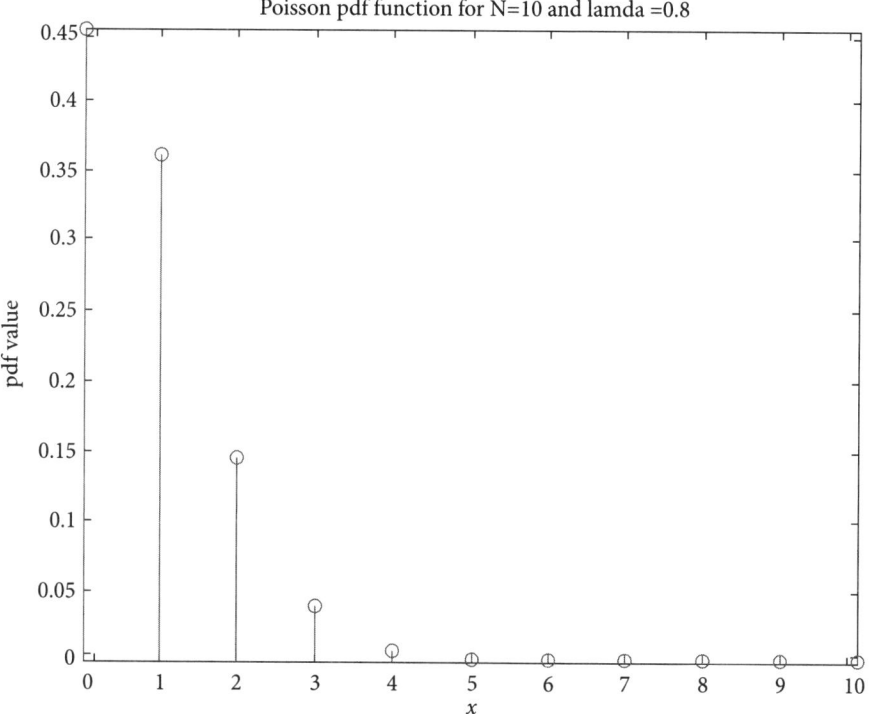

Fig. 9.25 Plot of Poisson pdf for $N = 10$ and lamda $= 0.8$

Example 9.29

A box contains five red, three green, four blue and two white balls. What is the probability of selecting a sample size of six balls containing two red, one green, two blue and one white ball?

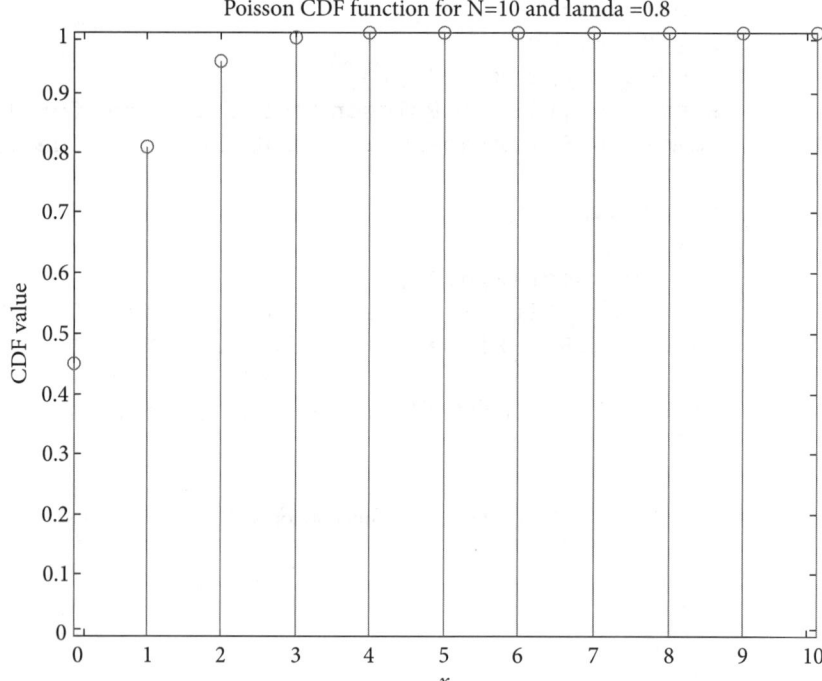

Fig. 9.26 Plot of Poisson CDF for $N = 10$ and lamda $= 0.8$

Solution

In this case, the probability is given by

$$P(2R,1G,2B,1W) = \frac{\binom{5}{2}\binom{3}{1}\binom{4}{2}\binom{2}{1}}{\binom{14}{6}} = 0.080 \qquad (9.92)$$

Example 9.30

A box contains 10 black balls and 15 white balls. One ball at a time is drawn at random, its color is noted, and the ball is then replaced in the box for the next draw.

a. Find the probability that the first white ball is drawn on the third draw.

Solution

The events are independent, since the ball is replaced in the box and thus the sample space does not change. Let B denote drawing a black ball and W drawing a white ball. The total number of balls in the sample space is 25. We will find the probability that the first white ball is drawn on third draw. Hence, we have

1st draw → B
2nd draw → B
3rd draw → W

Thus,

$$P(B,B,W) = \frac{\binom{10}{1}\binom{10}{1}\binom{15}{1}}{\binom{25}{1}\binom{25}{1}\binom{25}{1}} = \left(\frac{10}{25}\right)^2 \left(\frac{15}{25}\right) = 0.096 \quad (9.93)$$

Example 9.31
Find the expected value of the points on the top face in tossing a fair die.

Solution

In tossing a fair die, each face shows up with a probability 1/6. Let X be the points showing on the top face of the die. Then,

$$E[X] = 1\left(\frac{1}{6}\right) + 2\left(\frac{1}{6}\right) + 3\left(\frac{1}{6}\right) + 4\left(\frac{1}{6}\right) + 5\left(\frac{1}{6}\right) + 6\left(\frac{1}{6}\right) = 3.5 \quad (9.94)$$

Example 9.32
Consider the random variable X with the density function shown in Fig. 9.27. Find $E[X]$ and variance of the random variable?

Solution

The expected value of X is given as

$$E[X] = \int_{-3}^{-1} x \frac{1}{8} dx + \int_{-1}^{1} x \frac{1}{4} dx + \int_{1}^{3} x \frac{1}{8} dx = 0 \quad (9.95)$$

The mean square value $E[X^2]$

$$E[X^2] = 2\left[\int_{0}^{1} x^2 (1/4) dx + \int_{1}^{3} x^2 (1/8) dx\right] = 7/3 = 2.3333 = \sigma^2 \quad (9.96)$$

Since the mean is zero, the mean square value is just the variance $\sigma_x^2 = 7/3 = 2.3333$.

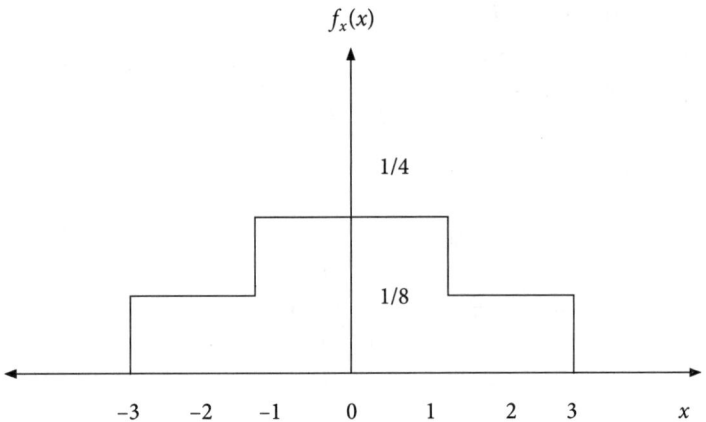

Fig. 9.27 Probability density function for a variable

9.4.3 Functions for finding moments

There are two functions defined that allow the calculation of moments. They are called as characteristic function and moment generating function. Let us understand the characteristic function first.

Characteristic function The characteristic function for any random variable X is defined as

$$\phi_X(\omega) = E[e^{j\omega X}] = \int_{-\infty}^{\infty} f_X(x) e^{j\omega X} dx \tag{9.97}$$

The Eq. (9.66) can be easily recognized as the Fourier Transform (FT) of $(f_x(x))$ with the sign of ω reversed. Hence, the pdf $(f_x(x))$ can be obtained from the characteristic function by taking inverse FT. We have to reverse the sign of variable x. The advantage of using a characteristic function is that the nth moments can be easily obtained from the characteristic function given by

$$m_n = (-j)^n \frac{d^n(\phi_X(\omega))}{d\omega^n} \Big\downarrow_{\omega=0} \tag{9.98}$$

Moment generating function The moment generating function is defined as

$$M_X(\upsilon) = E(e^{\upsilon X}) = \int_{-\infty}^{\infty} f_X(x) e^{\upsilon X} dx \tag{9.99}$$

Moments can be obtained from the moment generating function as

$$m_n = \frac{d^n(M_X(\upsilon))}{d\omega^n} \Big\downarrow_{\upsilon=0} \tag{9.100}$$

The moment generating function may not exist for all random variables and for all values of v.

Example 9.33

The density function of the variable X is given by

$$f_X(x) = \begin{cases} \dfrac{1}{4} & -2 \leq x \leq 2 \\ 0 & \text{otherwise} \end{cases}$$

Determine a) $P(X \leq x)$, b) $P(|x| \leq 1)$, c) The mean and the variance and d) The characteristic function

Solution

a. $P(X \leq x) = \int_{-2}^{x} f_X(x)dx = \int_{-2}^{x} \dfrac{1}{4}xdx = \dfrac{x^2}{8}\Big|_{-2}^{x} = \dfrac{x^2}{8} - \dfrac{1}{2}$ (9.101)

b. $P(X \leq 1) = \int_{-2}^{1} f_X(x)dx = \int_{-2}^{1} \dfrac{1}{4}xdx = \dfrac{x^2}{8}\Big|_{-2}^{1} = -\dfrac{3}{8}$ (9.102)

c. The mean and the variance

$$\mu_x = E[X] = \int_{-2}^{2} \dfrac{x}{4}dx = \dfrac{x^2}{8}\Big|_{-2}^{2} = 0 \quad (9.103)$$

$$E[X^2] = \int_{-2}^{2} \dfrac{x^2}{4}dx = \dfrac{x^3}{12}\Big|_{-2}^{2} = \dfrac{4}{3} \quad (9.104)$$

$$\sigma_x^2 = E[X^2] - \mu_x^2 = \dfrac{4}{3} - 0 = \dfrac{4}{3} \quad (9.105)$$

d. The characteristic function for continuous random variable is given as

$$\varphi_X(\omega) = \int_{-2}^{2} \dfrac{1}{4}e^{j\omega x}dx = \dfrac{1}{4j\omega}e^{j\omega x}\Big|_{-2}^{2} = \dfrac{e^{j2\omega} - e^{-j2\omega}}{4j\omega}$$

(9.106)

$$= \dfrac{2j\sin(2\omega)}{4j\omega} = \dfrac{\sin(2\omega)}{2\omega}$$

Example 9.34

Find the characteristic function of the random variable X having density function

$$f_X(x) = e^{-\frac{1}{2}|x|} \text{ for all } x.$$

Solution

We will use the equation for the characteristic function.

$$\varphi_X(\omega) = \int_{-\infty}^{\infty} e^{-\frac{1}{2}|x|} e^{j\omega x} dx = \int_{-\infty}^{0} e^{\frac{1}{2}x} e^{j\omega x} dx + \int_{0}^{\infty} e^{-\frac{1}{2}x} e^{j\omega x} dx$$

(9.107)

$$= \frac{1}{(0.5) + j\omega} + \frac{1}{(0.5) - j\omega} = \frac{4}{1 + 4\omega^2}$$

Concept Check

- Define a uniform pdf.
- Define a Gaussian pdf.
- Why the variable used in Gaussian distribution is normalized?
- How will you find probability of the event $(X \leq x)$ for a Gaussian variable having mean value of m and variance of v?
- What is the effect of change of mean and variance on the nature of the distribution plot in case of Gaussian random variable?
- Define Rayleigh distribution.
- Give one example for Rayleigh distribution.
- Define exponential distribution function.
- Name the pdfs for a discrete random variable.
- Define permutations.
- Define the term combinations.
- What are Bernoulli's trials?
- Define a binomial distribution.
- Define Poisson distribution.
- Define the two moment generating functions.

9.5 Central Limit Theorem and Chi Square Test, K-S Test

We will try to understand the procedure for computer generation of a random variable. To understand this, we need to know the rules for the sum of random variables. If $Y = \dfrac{1}{\sqrt{M}}[X_1 + X_2 + ... X_M]$ denotes the sum of M random variables, then the density function for Y is given by

$$f_Y(y) = f_{X_1}(x_1) \times f_{X_2}(x_2) \times ...... f_{X_M}(x_M) \tag{9.108}$$

The density function of Y is the $M-1$ fold convolution of M density function. The distribution function of Y can be found by integrating the density function.

Central Limit Theorem The theorem states that probability distribution function of sum of a large number of random variables approaches a Gaussian distribution. The theorem does not guarantee that the density function will always be Gaussian. Under certain conditions, the density function is also Gaussian. Let the mean values of M random variables be defined as $\mu_1, \mu_2, ... \mu_M$. Then the mean of sum of random variables denoted as Y can be written as

$$\mu_Y = \dfrac{1}{\sqrt{M}}[\mu_1 + \mu_2 + ... + \mu_M] \tag{9.109}$$

Consider the random variables as statistically independent. The variance of the sum variable can be written as

$$\sigma_Y^2 = \dfrac{1}{M}[\sigma_1^2 + \sigma_2^2 + ... + \sigma_M^2] \tag{9.110}$$

The theorem is found to hold good for special cases of dependent variables as well.

Computer generation of a Gaussian-distributed random variable. Many times there is a need to simulate a system for estimating the performance before its actual implementation. The simulation of a system usually requires the simulation of a random variable having a Gaussian distribution. A function like rand is available with MATLAB for the generation of a random variable with uniform distribution.

Let us take some 10 uniformly distributed random variables. We will use the addition of these variables and plot their density function. It can be shown to be a Gaussian density function to a good approximation using Central limit theorem. Let us write a MATLAB program to generate a Gaussian variable as the sum of 12 uniformly distributed random variables. Figure 9.28 shows the Histogram plot of the generated variable. We can easily verify that the envelope of the histogram approaches Gaussian pdf.

```
clear all;
x1=rand(100);
x2=rand(100);
x3=rand(100);
x4=rand(100);
```

```
x5=rand(100);
x6=rand(100);
x7=rand(100);
x8=rand(100);
x9=rand(100);
x10=rand(100);
x11=rand(100);
x12=rand(100);
y=(1/sqrt(12))*(x1+x2+x3+x4+x5+x6+x7+x8+x9+x10+x11
+x12);
z=hist(y,10);
bar(z);title('plot of a sum variable generated
using Central limit theorem');xlabel('bin
value');ylabel('amplitude');
```

Let us try to understand the meaning of the histogram. We have used 12 random variables with uniform distribution between 1 and 100. These are generated using a rand function. The generated values of random variable are divided into, for example, 10 bins one from 1 to 10, the other from 11 to 20, and so on. Now, the number of random variables lying in the range 1 to 10 is counted, 11 to 20 are counted and so on. The histogram plots this count on y axis and the bin value is plotted on x axis. If we increase the number of random variables to 1000, the histogram plot will further approach the Gaussian distribution.

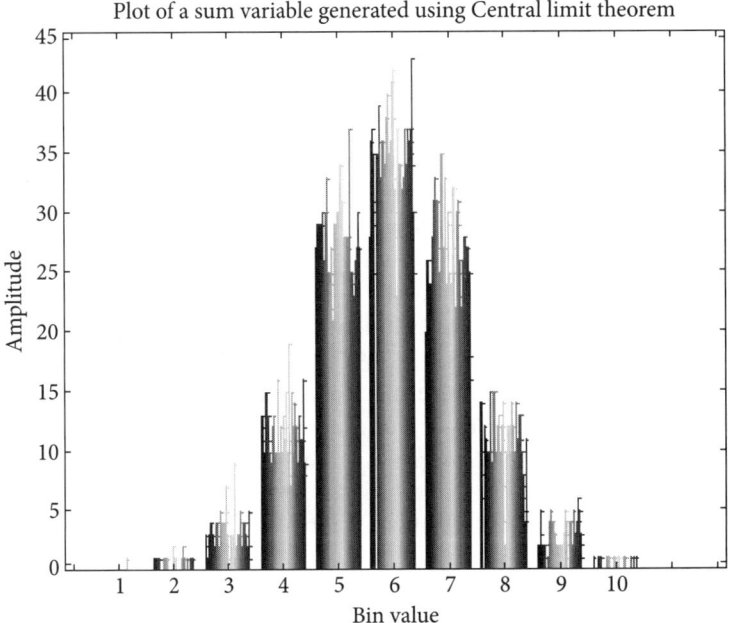

Fig. 9.28 Histogram plot of Gaussian generated variable

Chi-square test It is called as Goodness of Fit test. To perform the test, the test statistic is calculated. The test statistic is defined as

$$\chi^2 = \sum_{i=1}^{N} \frac{(g_i - f_i)^2}{f_i} \qquad (9.111)$$

When a histogram is drawn, the data values are divided into bins. The frequency of occurrence of the data in the bin interval is computed and plotted against the bin. To find the test statistic, we find the squared distance between the two frequency values, namely observed frequency and frequency as per the required distribution. We will illustrate the procedure with a simple example of a uniform distribution.

Example 9.35

The experiment is conducted with a fair die. The die is tossed 120 times, for example. The following data are obtained, as shown in the table below. Find the test statistic to check if it obeys a uniform distribution.

Face no.	1	2	3	4	5	6
Observed frequency	18	20	22	19	21	20

Solution

For uniform distribution the face values are equi-probable. The expected frequency for each face is 20. Let us find the test statistic.

$$\chi^2 = \sum_{i=1}^{N} \frac{(g_i - f_i)^2}{f_i}$$

$$= \frac{(18-20)^2}{20} + \frac{(20-20)^2}{20} + \frac{(22-20)^2}{20} + \frac{(19-20)^2}{20} + \frac{(21-20)^2}{20} + \frac{(20-20)^2}{20} \qquad (9.112)$$

$$= 0.5$$

This value of test statistic is less than the 5 percent level equal to 120*0.05 = 6. Hence, the distribution is uniform.

The Kolmogoroff–Smirnoff test This test is used when the set of observations is from a continuous distribution. It detects the deviation from a specified distribution. Given a set of observations, we determine the following statistic.

$$D = \max |F(x) - G(x)| \qquad (9.113)$$

Where, $G(x)$ is a sample distribution and $F(x)$ is a standard distribution against which the sample distribution is compared. If the value of D is less than the critical value specified in K–S table, then the hypothesis that an unknown distribution is a known standard distribution is accepted.

Usually the signal of unknown distribution is to be analyzed. When the pdf of the random signal is not known, the mean and variance, etc. cannot be calculated as it requires the mathematical expression of pdf. To solve this problem, the unknown signal is used to find its histogram and is compared with the standard known density functions discussed above. The extent to which the match occurs between unknown density/distribution and known density/distribution is evaluated as discussed using Chi-square test or K–S test. If the hypothesis for the known distribution can be accepted, then the mathematical expressions for the known density/distribution are used for the computation of signal statistical properties.

Concept Check
- State central limit theorem.
- Define a procedure for computer generation of Gaussian random variable.
- Explain the procedure to find histogram for a random variable.
- What is a Chi-square test?
- Define the Kolmogoroff–Smirnoff test.
- How will you select a standard pdf for a given distribution?

9.6 Random Processes

We now extend the concept of random variable to include time. A random variable is a function of the outcomes of the random experiment called as ensemble. If we allow the random variable to be a function of time denoted by $X(s, t)$, then a family of such functions is called a random process. Figure 9.18 shows a random process. If the value of s is fixed, we get a random variable that is a function of time. If the value of t is fixed, we get a random variable that is a function of s i.e., an ensemble function. Consider an example of temperature measurement using a number of transducers. If we measure the value of temperature for different transducers at the same time t, we get the random variable called an ensemble function. The mean value of this random variable is called as ensemble average. If we measure the value of temperature using the same transducer at a different time, we get a random variable that is a function of time. The mean value of this random variable is called as time average. If we can predict the future values of any sample function from its past values, we call it as a deterministic process. If the future values of a sample function scan not be predicted, then it is called as non-deterministic process.

Stationary and Non-stationary process We can also classify the random processes as stationary or non-stationary random processes. If all the statistical properties namely the mean, moments and variance, etc. for the random variables of the random process are constant with respect to time, the process is said to be a stationary random process. The two processes are said to be statistically independent if random variable groups of one process are independent of random variable groups of other process. We define different levels of stationarity for a process namely first order stationarity and second order stationarity.

Stationary processes We will define first order stationary process. A random process is said to **stationary to order one** if its density function does not shift with the shift in time origin and the mean value is constant.

$$f_X(x_1:t_1) = f_X(x_1:t_1+\delta) \qquad (9.114)$$

A random process is said to be **stationary to order two** if its second order density function is invariant to a time difference and if its autocorrelation function is a function of time difference and not the absolute time.

$$f_X(x_1,x_2:t_1,t_2) = f_X(x_1,x_2:t_1+\delta,t_2+\delta) \qquad (9.115)$$

If we put $\delta = -t_1$, we can conclude that the second order density function is a function of time difference.

A random process is said to be **wide sense stationary** if the two conditions are met. First is if the mean is constant and second is if the auto-correlation of a process is a function of the time difference.

N^{th} order or Strict sense stationary process A random process is stationary to order N if its nth order density function is invariant to a time origin shift. Stationarity of order N implies stationarity to all orders $< N$. A process stationary to all orders is called as a strict sense stationary process. Figure 9.29 shows a continuous random process.

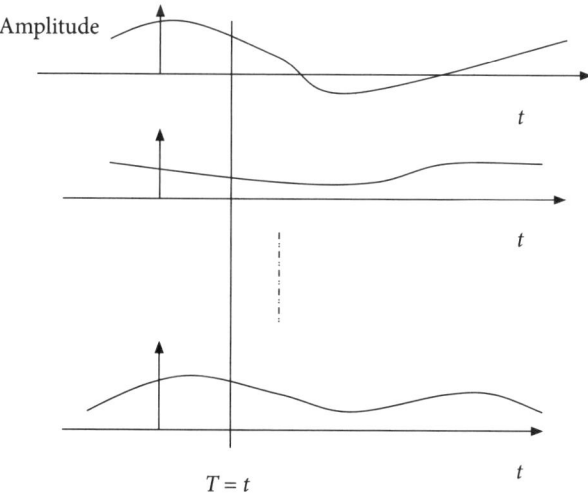

Fig. 9.29 A continuous random process

We will illustrate the meaning of wide sense stationarity with the help of a simple example.

Example 9.36

Consider a random process given by $x(t) = A\cos(\omega_0 t + \theta)$. Prove that the process is wide sense stationary if values of A and ω_0 are constants and ϑ is a uniformly distributed random variable on the interval 0 to π.

Solution

Let us find the mean value of a process.

$$E[x(t)] = \frac{1}{2\pi} \int_0^{2\pi} A\cos(\omega_0 t + \theta) d\theta = 0 \tag{9.116}$$

Let us find the autocorrelation function.

$$\begin{aligned} R_{xx}(\tau) &= E[x(t)x(t+\tau)] \\ &= E\big[A\cos(\omega_0 t + \theta) A\cos(\omega_0 t + \omega_0 \tau + \theta)\big] \\ &= \frac{A^2}{2}\big[\cos(\omega_0 \tau) + \cos(2\omega_0 t + \omega_0 \tau + 2\theta)\big] \\ &= \frac{A^2}{2}\cos(\omega_0 \tau) \end{aligned} \tag{9.117}$$

The second term in the equation integrates to zero. We find that the autocorrelation is a function of time difference, and the mean value is constant. These two conditions confirm that the process is a wide sense stationary process.

Ergodic process The process is said to be an ergodic process if the time average is equal to the ensemble average and the time average autocorrelation is equal to the ensemble autocorrelation. Equation (9.78) specifies the ensemble autocorrelation. The condition of ergodicity requires that

$$\text{Time average autocorrelation} = \lim_{T \to \infty} \frac{1}{2T} \int_{-T}^{T} x(t)x(t+\tau) dt \tag{9.118}$$

$$\text{Ensemble autocorrelation} = E[x(t)x(t+\tau)]$$

Ergodicity is a restrictive form of stationarity. It is normally very difficult to prove that the given process is ergodic. For simplicity, we assume the given process to be ergodic so that we can find the ensemble average from the time average. We can find the ensemble autocorrelation from the time average autocorrelation. The computations are simplified. The argument may be made that due to the assumption of ergodicity, our analysis is on shaky grounds. The fact is that the theory only suggests a model for real situation. The problem is that it is solved as an approximation problem. Hence, it makes no difference if we assume additionally that the process is ergodic.

Concept Check

- Define a random process.
- Define a deterministic process.
- Define a stationary process.
- Define wide sense stationary process.
- Define the ergodic process.
- What is time average autocorrelation?
- Define ensemble autocorrelation.

9.7 Estimation of ESD and PSD

The section 2.6.5 has introduced the classification of signals as energy signals and power signals. The need for estimation of spectrum will be emphasized. We will define energy spectrum density (ESD) and power spectrum density (PSD) and the method for spectrum estimation such as periodogram method will be discussed.

We discuss the estimation of energy spectrum and power spectrum in this section. The naturally occurring signals are random in nature. We can not predict the next sample from the previous samples using any mathematical model. The naturally occurring signals are of infinite duration and we have only finite duration signal available with us. Based on the finite duration signal, we try to predict the characteristics of the signal such as mean, variance, etc.

9.7.1 Computation of energy density spectrum of deterministic signal

Consider a deterministic signal that has finite energy and zero average power. This signal is then termed as an energy signal. We will estimate the energy density spectrum of energy signal. Let us consider $x(t)$ as a finite duration signal available with us. The problem is to estimate the true spectrum from a finite duration record. If $x(t)$ is a finite energy signal, its energy E is finite and is given by

$$E = \int_{-\infty}^{\infty} |x(t)|^2 \, dt < \infty \qquad (9.119)$$

Then its Fourier transform exists and is given by

$$X(F) = \int_{-\infty}^{\infty} x(t)e^{-j2\pi Ft} dt \tag{9.120}$$

The energy of time domain signal is same as the energy of a Fourier domain signal. We can write

$$E = \int_{-\infty}^{\infty} |x(t)|^2 \, dt = \int_{-\infty}^{\infty} |X(F)|^2 \, dF \tag{9.121}$$

$|X(F)|^2$ represents distribution of signal energy with respect to frequency and is called as the energy density spectrum denoted as $S_{XX}(F)$. The total energy is simply the integration of $S_{XX}(F)$ over all F. We will show that $S_{XX}(F)$ is the Fourier transform of autocorrelation function.

The autocorrelation function of finite energy signal is given by

$$R_{XX}(\tau) = \int_{-\infty}^{\infty} x^*(t)x(t+\tau) dt \tag{9.122}$$

Where $x^*(t)$ is the complex conjugate of $x(t)$. Let us calculate the Fourier transform of autocorrelation.

$$\int_{-\infty}^{\infty} R_{XX}(\tau) e^{-j2\pi F\tau} d\tau = \int_{-\infty}^{\infty} \int_{-\infty}^{\infty} x^*(t) x(t+\tau) e^{-j2\pi F\tau} dt d\tau \tag{9.123}$$

Put $t' = t + \tau$, $dt' = d\tau$

$$\int_{-\infty}^{\infty} x^*(t) e^{j2\pi Ft} dt \int_{-\infty}^{\infty} x(t') e^{-j2\pi Ft'} dt' = X^{*(F)} X(F) = |X(F)|^2 = S_{XX}(F) \tag{9.124}$$

Hence, $R_{XX}(\tau)$ and $S_{XX}(F)$ are the Fourier transform pair.

We can prove similar result for DT signals also. Let $r_{XX}(k)$ and $S_{XX}(f)$ denote autocorrelation and energy density for DT signal.

$$r_{XX}[k] = \sum_{n=-\infty}^{\infty} x^*[n]x[n+k] \tag{9.125}$$

Put $n+k = n'$. The Fourier transform of autocorrelation is given by

$$S_{XX}(f) = \sum_{k=-\infty}^{\infty} r_{XX}[k] e^{-j2\pi kf} = \sum_{n=-\infty}^{\infty} x^*[n] e^{j2\pi nf} \sum_{k=-\infty}^{\infty} x[n'] e^{-j2\pi n'f} = |X(f)|^2 \tag{9.126}$$

We can now list two different methods for computing energy density spectrum, namely direct method and indirect method.

Direct method In the direct method, one computes the FT of the sequence $x(n)$ and then calculates the energy density spectrum as

$$S_{xx}(f) = |X(F)|^2 = \left| \sum_{n=-\infty}^{\infty} x[n] e^{-j2\pi fn} \right|^2 \qquad (9.127)$$

Indirect method In case of the indirect method, we have to first compute the autocorrelation and then find its FT to get energy density spectrum.

$$S_{xx}(f) = \sum_{k=-\infty}^{\infty} \gamma_{xx}[k] e^{-j2\pi kf} \qquad (9.128)$$

There is a practical difficulty involved in the computation of energy density spectrum using Eqs (9.127) and (9.128). The sequence is available only for finite duration. Using a finite sequence is equivalent to the multiplication of the sequence by a rectangular window. This will result in spectral leakage due to Gibbs phenomenon, as discussed in chapter 5.

Autocorrelation indicates the similarity between the same signal and its time-shifted version. We will list below the properties of autocorrelation.

1. $R_{xx}(\tau)$ is an even function of τ.

$$R_{xx}(\tau) = \int_{-\infty}^{\infty} x^*(t) x(t+\tau) dt \text{ (put } t+\tau = t') $$

$$R_{xx}(-\tau) = \int_{-\infty}^{\infty} x(t'-\tau) x(t') dt' = R_{xx}(\tau) \qquad (9.129)$$

2. $R_{xx}(0) = E = $ Energy of the signal

$$R_{xx}(0) = \int_{-\infty}^{\infty} x^*(t) x(t) dt = \int_{-\infty}^{\infty} |x(t)|^2 dt = E \qquad (9.130)$$

3. $|R_{xx}(\tau)| \leq R_{xx}(0)$ for all τ \hfill (9.131)

4. Autocorrelation and energy spectral density (ESD) are FT pairs. (Note- already proved)

We list below the properties of ESD.
1. $S_{xx}(F) \geq 0$ for all F.
2. For any real-valued signal $x(t)$, ESD is an even function.
3. The total area under the curve of ESD is the total energy of the signal.

Example 9.37

Find the autocorrelation of the function $x(t) = e^{-4t}u(t)$.

Solution

Let us find the autocorrelation for $\tau < 0$

$$R_{XX}(\tau) = \int_0^\infty e^{-4t} e^{-4(t-\tau)} dt = e^{4\tau} \int_0^\infty e^{-8t} dt$$

$$= e^{4\tau} e^{-8t}/(-8) \Big|_0^\infty = \frac{e^{4\tau}}{8}$$

(9.132)

Let us find the autocorrelation for $\tau > 0$

$$R_{XX}(\tau) = \int_0^\infty e^{-4t} e^{-4(t+\tau)} dt = e^{-4\tau} \int_0^\infty e^{-8t} dt$$

$$= e^{-4\tau} e^{-8t}/(-8) \Big|_0^\infty = \frac{e^{-4\tau}}{8}$$

(9.133)

$$R_{XX}(\tau) = \frac{e^{-4|\tau|}}{8}$$

(9.134)

Example 9.38

Find the energy spectral density of $x(t) = e^{-4t}u(t)$ and verify that $R_{XX}(0) = E$.

Solution

$$X(F) = \int_0^\infty e^{-4t} e^{-j\omega t} dt = \frac{e^{-(4+j\omega)t}}{-(4+j\omega)} \Big|_0^\infty = \frac{1}{4+j\omega}$$

(9.135)

$$X(-F) = \int_{-\infty}^0 e^{4t} e^{-j\omega t} dt = \frac{e^{-(j\omega-4)t}}{-(j\omega-4)} \Big|_{-\infty}^0 = \frac{1}{4-j\omega}$$

(9.136)

$$S_{XX}(F) = |X(F)|^2 = \frac{1}{4+j\omega} \times \frac{1}{4-j\omega} = \frac{1}{16+\omega^2}$$

(9.137)

We will find ESD as the FT of autocorrelation already found.

$$S_{XX}(F) = \int_{-\infty}^\infty \frac{e^{-4|\tau|}}{8} e^{-j\omega\tau} d\tau = \frac{1}{8}\left[\int_0^\infty e^{-(4+j\omega)\tau} d\tau + \int_{-\infty}^0 e^{-(j\omega-4)\tau} d\tau\right]$$

(9.138)

$$= \frac{1}{8}\left[\frac{e^{-(4+j\omega)\tau}}{-(4+j\omega)}\Big|_0^\infty + \frac{e^{-(j\omega-4)\tau}}{-(j\omega-4)}\Big|_{-\infty}^0\right]$$

$$= \frac{1}{8}\left[\frac{1}{4+j\omega} + \frac{1}{4-j\omega}\right] = \frac{1}{8}\left[\frac{8}{16+\omega^2}\right] = \frac{1}{16+\omega^2}$$

Let us evaluate the energy and autocorrelation at zero.

$$R_{XX}(0) = \frac{e^{-4\times 0}}{8} = \frac{1}{8}$$

$$E = \int_{-\infty}^{\infty} x^2(t)dt = \int_0^\infty e^{-8t} dt = -\frac{e^{-8t}}{8}\Big|_0^\infty = \frac{1}{8} \qquad (9.139)$$

Example 9.39

Find cross-correlation of $x[n] = \{1, 2, 1\}$ and $y[n] = \{1, 4, 6, 4, 1\}$.

Solution

We have to skip the first step of convolution i.e., taking time reversal.
   We will start from step 2. We will take a sample-by-sample product.

$Z[0] = \{1 + 8 + 6 = 15\}$,

Now we will shift $x[n]$ towards the right by one sample and again take a sample-by-sample product.

$Z[1] = \{4 + 12 + 4 = 20\}$,

We will shift $x[n]$ towards the right again by one sample and take a sample-by-sample product.

$Z[2] = \{6 + 8 + 1 = 15\}$, $Z[3] = \{4 + 2 = 6\}$, $Z[4] = [1]$

We will now shift $x[n]$ towards the left by one sample and take a sample-by-sample product.

$Z[-1] = [4 + 2 = 6]$, $Z[-2] = [1]$, total 7 values are non-zero.

$$Z[n] = \{1, 6, 15, 20, 15, 6, 1\}. \qquad (9.140)$$

Cross correlation has a maximum value at the 1st position.
   Let us write a MATLAB program to find cross correlation and plot it. It is plotted in Fig. 9.30.

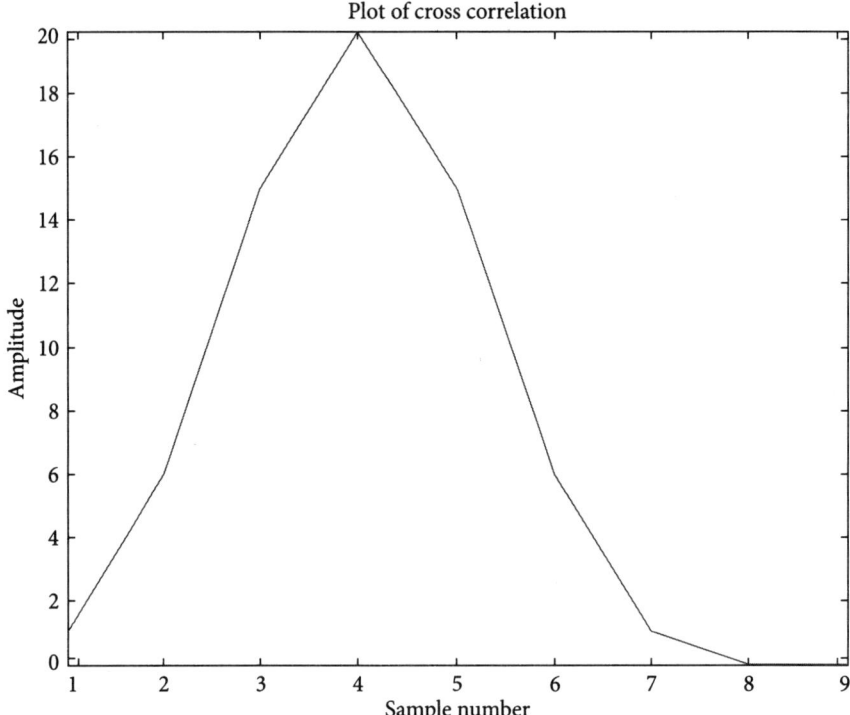

Fig. 9.30 Plot of cross correlation of two signals

```
clear all;
clc;
x = [1 2 1 0 0];
y = [1 4 6 4 1];
z=xcorr(x,y);
plot(z);
z =   Columns 1 through 7

1.0000  6.0000  15.0000  20.0000  15.0000  6.0000  1.0000
```

Example 9.40

Find autocorrelation and ESD for $x(t) = A \times \text{rect}(t)$ for $-\frac{1}{2} \le t \le \frac{1}{2}$.

Solution

For $\tau < -1$, no overlap so output is zero.
For $-1 < \tau < 0$,

$$R_{xx}(\tau) = \int_{-\frac{1}{2}-\tau}^{\frac{1}{2}} A^2 dt = A^2 \left[\frac{1}{2} + \frac{1}{2} + \tau \right] = A^2[1+\tau] \qquad (9.141)$$

For $0 < \tau < 1$,

$$R_{XX}(\tau) = \int_{-\frac{1}{2}}^{\frac{1}{2}-\tau} A^2 dt = A^2 \left[\frac{1}{2} - \tau + \frac{1}{2}\right] = A^2[1-\tau] \qquad (9.142)$$

To find ESD, we will find the FT of signal and its energy.

$$E = |X(f)|^2 = \left|\int_{-\frac{1}{2}}^{\frac{1}{2}} Ae^{-j\omega t} dt\right|^2 = \left|A \frac{e^{-j\omega t}}{-j\omega}\Big\downarrow_{-\frac{1}{2}}^{\frac{1}{2}}\right|^2 = A^2 \left|\frac{2j\sin(\omega/2)}{-j\omega}\right|^2 \qquad (9.143)$$

$$= A^2 \left[\frac{\sin(\omega/2)}{\omega/2}\right]^2 = A^2 \left[\frac{\sin(\pi f)}{\pi f}\right]^2 = A^2 \operatorname{sinc}^2(f)$$

Example 9.41

Find the cross correlation of two signals

$$x(t) = A \times \operatorname{rect}(t) \ \& \ y(t) = -A \times \operatorname{rect}(2t) \text{ for } -\frac{1}{2} \le t \le \frac{1}{2}.$$

Solution

No overlap when $\tau < -3/4$, so the output is zero.

Consider the interval $-3/4 < \tau < -1/4$, the output $z(t)$ is given by

$$z(t) = \int_{-\frac{1}{2}}^{\frac{1}{4}-\tau} -A^2 dt = -A^2[t] \downarrow_{-\frac{1}{2}}^{\frac{1}{4}-\tau} = -A^2 \left[\frac{1}{4} - \tau + \frac{1}{2}\right] = -A^2\left[\frac{3}{4} - \tau\right] \qquad (9.144)$$

Consider the interval $-1/4 < \tau < 1/4$, the output $z(t)$ is given by

$$z(t) = \int_{-\frac{1}{4}-\tau}^{\frac{1}{4}-\tau} -A^2 dt = -A^2[t] \downarrow_{-\frac{1}{4}-\tau}^{\frac{1}{4}-\tau} = -A^2 \left[\frac{1}{4} - \tau + \frac{1}{4} + \tau\right] = -A^2/2 \qquad (9.145)$$

Consider the interval $1/4 < \tau < 3/4$, the output $z(t)$ is given by

$$z(t) = \int_{-\frac{1}{4}-\tau}^{\frac{1}{2}} -A^2 dt = -A^2[t] \downarrow_{-\frac{1}{4}-\tau}^{\frac{1}{2}} = -A^2 \left[\frac{1}{2} + \tau + \frac{1}{4}\right] = -A^2\left[\frac{3}{4} + \tau\right] \qquad (9.146)$$

No overlap when $\tau > 3/4$, so the output is zero.

Correlgram

We will now define the correlogram. Correlogram is a plot of one signal against another signal. It tells us about the similarity between the two signals.

If it is a straight line graph, two signals are highly correlated. If the slope of the graph is positive, they are positively correlated. If the slope is negative, they are negatively correlated. If the graph is random, the signals are not correlated at all.

Example 9.42

Find the correlogram of sine and cosine functions $\sin(\omega t)$ and $\cos(\omega t)$.

Solution

We will find the values of the functions from 0 to 2pi radians and plot one function against other using a MATLAB program given below. We can also do it using hand calculations.

```
clear all;
clc;
x=0:0.1:2*pi;
y=sin(x);
z=cos(x);
plot(y,z);    title('correlogram   of   the   two
signals');xlabel('sin        value');ylabel('cosine
value');
```

The correlogram is plotted in Fig. 9.31. We can see that it follows a definite pattern. The signals are related in some sense.

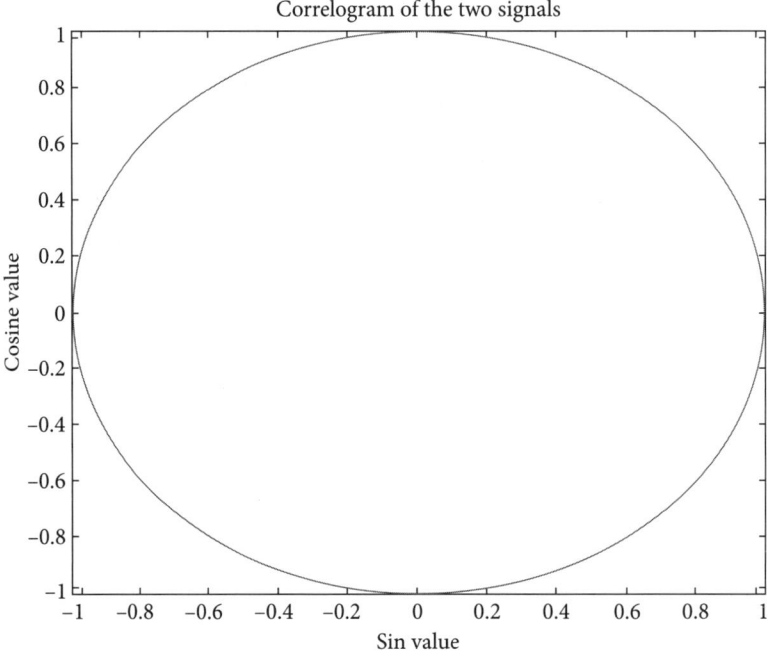

Fig. 9.31 Correlogram of sine and cosine function

Example 9.43

Find the correlogram of exponential functions e^{at} and e^{-at}.

Solution

We can write a MATLAB program to find correlogram as follows. The plot of correlogram is shown in Fig. 9.32. The first 16 values calculated for the two signals are listed as follows.

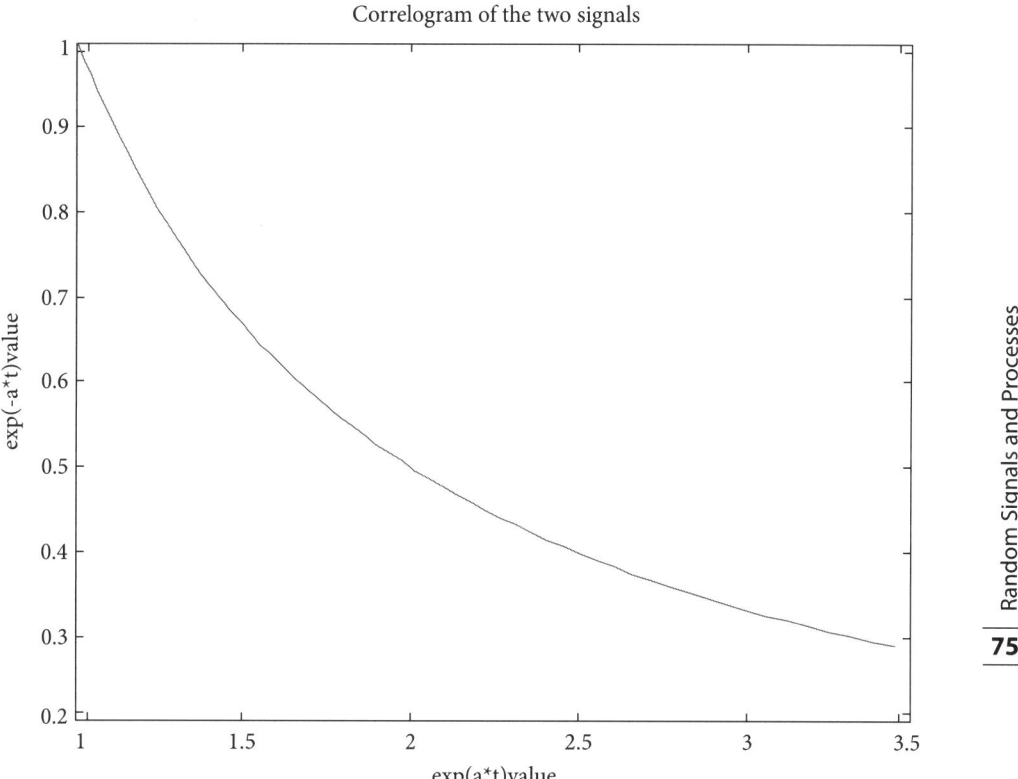

Fig. 9.32 Correlogram of the two signals

$$e^{at} = [1.0000 \quad 1.0202 \quad 1.0408 \quad 1.0618 \quad 1.0833 \quad 1.1052$$
$$1.1275 \quad 1.1503 \quad 1.1735 \quad 1.1972 \quad 1.2214 \quad 1.2461 \quad \quad (9.147)$$
$$1.2712 \quad 1.2969 \quad 1.3231 \quad 1.3499]$$

$$e^{-at} = [1.0000 \quad 0.9802 \quad 0.9608 \quad 0.9418 \quad 0.9231 \quad 0.9048$$
$$0.8869 \quad 0.8694 \quad 0.8521 \quad 0.8353 \quad 0.8187 \quad 0.8025 \quad \quad (9.148)$$
$$0.7866 \quad 0.7711 \quad 0.7558 \quad 0.7408]$$

```
clear all;
clc;
a=0.2;
t=0:0.1:2*pi;
y=exp(a*t)
z=exp(-a*t)
plot(y,z);    title('correlogram    of    the    two
signals');xlabel('e^at value');ylabel('e^-at value');
```

9.7.2 Estimation of power density spectrum of random signal

A class of signals called as stationary random signals are of infinite duration and have infinite energy. The average power is however finite. Such signals are classified as power signals. We need to estimate the power density spectrum for such signals. For a random process, the autocorrelation is given by

$$\gamma_{xx}(\tau) = E[x^*(t)x(t+\tau)] \tag{9.149}$$

Where E denotes expected value. This can not be calculated if the probability density function (pdf) is not known. For a random process we do not know pdf. We need to only estimate the value of autocorrelation as

1. We have only finite duration record.
2. pdf is not known.

We assume that the process is wide sense stationary and is ergodic. We calculate time average autocorrelation, as it is the same as an ensemble autocorrelation in case of an ergodic process. Hence, we calculate

$$\gamma_{xx}(\tau) = \lim_{\tau \to \infty} [R_{xx}(\tau)] \tag{9.150}$$

The estimate $P_{xx}(F)$ of the power density spectrum is calculated as the Fourier transform of autocorrelation and the actual value of power density spectrum is the expected value of $P_{xx}(F)$ in the limit as $T_0 \to \infty$.

$$P_{xx}(F) = \int_{-T_0}^{T_0} R_{xx}(\tau) e^{-j2\pi F\tau} d\tau$$

$$= \frac{1}{2T_0} \int_{-T_0}^{T_0} \left[\int_{-T_0}^{T_0} x^*(t)x(t+\tau) dt \right] e^{-j2\pi F\tau} d\tau \tag{9.151}$$

$$= \frac{1}{2T_0} \left| \int_{-T_0}^{T_0} x(t) e^{-j2\pi Ft} dt \right|^2$$

The actual power density spectrum as obtained as

$$\Gamma_{XX}(F) = \lim_{T_0 \to \infty} E(P_{XX}(F)) \qquad (9.152)$$

There are two approaches for the computation of power density spectrum namely a direct method, as given by Eq. (9.127), or an indirect method by first calculating the autocorrelation and then finding FT of it to get power density spectrum.

In practice, we have only finite samples of the DT signals. The time average autocorrelation is computed using the equation

$$\gamma_{XX}[m] = \frac{1}{n-|m|} \sum_{n=|m|}^{N-1} x^*[n]x[n+m] \quad m = -1,-2,\ldots\ldots 1-N \qquad (9.153)$$

The normalization factor results in a mean value of autocorrelation computed the same as an actual autocorrelation. We then find the FT of the mean value of autocorrelation to get the estimate of a power spectrum. The estimate of a power spectrum can also be written as

$$P_{XX}(f) = \frac{1}{N}\left|\sum_{n=0}^{N-1} x[n]e^{-j2\pi fn}\right|^2 = \frac{1}{N}|X(f)|^2 \qquad (9.154)$$

Equation 9.154 represents the well-known form of the power density spectrum estimate and is called as periodogram. To find the pdf, one draws the histogram for a signal with finite duration and uses curve fitting tests such as Chi-square test or K–S test to fit a standard probability distribution for a given signal. The mathematical equation for that standard pdf can now be used in the calculations.

Properties of PSD can be listed as follows.

1. $P_{XX}(f) \geq 0$ for all f.

2. $P_{XX}(f) = \lim_{T \to \infty} \frac{1}{2T}|X(f)|^2$

3. PSD of a real-valued signal $x(t)$ is an even function of f. $P_{XX}(f) = P_{XX}(-f)$

4. The total area under the curve of the PSD of power signal equals the average signal power.

5. $S_{YY}(f) = |H(f)|^2 S_{XX}(f)$, where $H(f)$ stands for the transfer function of the LTI system.

Example 9.44

Find the autocorrelation, spectral density and power of the power signal given by $x(t) = A\sin(\omega_c t)$.

Solution

$$R_{XX}(\tau) = \lim_{T \to \infty} \frac{1}{2T} \int_{-T}^{T} x(t)x(t-\tau)dt$$

$$= \lim_{T \to \infty} \frac{1}{4T} \int_{-T}^{T} A\sin(\omega_c t)\sin(\omega_c t - \omega_c \tau)dt$$

$$= \lim_{T \to \infty} \frac{A^2}{4T} \left[\int_{-T}^{T} \cos(\omega_c \tau)dt - \int_{-T}^{T} \cos(2\omega_c t - \omega_c \tau)dt \right] \quad (9.155)$$

$$= \frac{A^2}{2} \cos(\omega_c \tau),$$

$$R_{XX}(0) = \frac{A^2}{2} = \text{Total power}$$

PSD is the FT of autocorrelation.

$$\text{PSD} = FT\left[\frac{A^2}{2}\cos(\omega_c \tau)\right] = \frac{A^2}{4} FT\left[e^{j\omega_c \tau} + e^{-j\omega_c \tau}\right] \quad (9.156)$$

$$= \frac{A^2}{4}[\delta(\omega - \omega_c) + \delta(\omega - \omega_c)]$$

Note that we have used shifting property of FT.

Concept Check

- Define energy signal.
- Define power signal.
- Define autocorrelation and cross correlation.
- What is the relation between autocorrelation and ESD?
- What is the relation between correlation and PSD?
- What is a correlogram?
- How will you find a correlogram?
- State properties of autocorrelation.
- State properties of ESD.
- State properties of PSD.
- How will you find total energy of a signal?
- How will you find total power?

Summary

We started with the introduction to probability theory. In this chapter, we have described different standard distribution functions and their use for signal analysis. The highlights for the chapter are as follows.

- The term event is defined. When the experiment is performed to generate some output, it is called as an event. The output of the experiment is called as the outcome of the experiment. $\text{Probability} = \lim_{n \to \infty} \frac{n_H}{n}$ is defined in terms of relative frequency. The conditional probability is given by $P(A/B) = \frac{P(A \cap B)}{P(B)}$. Bayes theorem is introduced, which is the theorem of inverse probability.

- A random variable is defined. A random variable can be defined as the characterization of the outcome of the random experiment that will associate a numerical value with each outcome. Cumulative distribution function called as CDF is defined as the probability that the random variable 'X' takes a value less than or equal to some allowed value 'x'. The probability Density Function is defined as the derivative of the CDF. The properties of CDF and pdf are listed.

- The statistical properties of random variables are discussed. The statistical properties include mean, variance, skew, Kurtosis and higher order moments. All these properties are defined. The standard distribution functions are described for continuous and discrete random variables. Uniform, Gaussian, Rayleigh and exponential distributions are discussed for continuous random variables. Permutation and combination is defined. Bernoulli's trials are described. The Binomial and Poisson distribution are discussed for discrete random variables. Functions for finding the moments are described. These are characteristic function and moment generating function.

- Central limit theorem is defined. The theorem states that probability distribution function of a sum of a large number of random variables approaches a Gaussian distribution. Use of histogram to find the density function of the unknown signal is explained. The Chi-square test is called as Goodness of Fit test. To perform the test, the test statistic is calculated. The Kolmogoroff–Smirnoff test is used when the set of observations is from a continuous distribution. It detects the deviation form a specified distribution. When the unknown signal is to be analyzed, it is used to find the histogram and is compared with the standard known density functions discussed. The extent to which the match occurs between unknown density/distribution and known density/distribution is evaluated using Chi-square test or K–S test. If the hypothesis for the known distribution can be accepted, then the mathematical expressions for the known density/distribution are used for the computation of signal statistical properties.

- A random variable is a function of the outcomes of the random experiment called as ensemble. If we allow the random variable to be a function of time too denoted by $X(s, t)$, then a family of such functions is called a random process. The classification of the random process such as deterministic/non-deterministic, stationary/non-stationary is described. A random process is said to be **wide sense stationary** if the mean is constant and the auto-correlation of a process is a function of time difference. The process is said to be an ergodic process if the time average is equal to the ensemble average and time average auto-correlation is equal to the ensemble autocorrelation.

- The energy and power signals are defined. The autocorrelation and cross correlation for the signals are defined. The relation between autocorrelation and energy spectral density is explained. They are *FT* pairs. Similarly, for power signals, autocorrelation and power spectral density are *FT* pairs. The problems in the estimation of ESD and PSD are discussed. Simple examples are solved to illustrate the concepts. The correlogram is defined. MATLAB programs are written to draw correlograms for signals. If the graph is a straight line, the signals are highly correlated. If the graph has some regularity, the signals have a relation. If the graph is of a random nature, the signals are not correlated. The properties of autocorrelation are listed and proved. The properties of ESD and PSD are listed.

Multiple Choice Questions

1. The conditional probability is defined as

   (a) $P\left(\dfrac{A}{B}\right) = \dfrac{P(A \cap B)}{P(B)}$ (b) $P\left(\dfrac{B}{A}\right) = \dfrac{P(A \cup B)}{P(B)}$

   (c) $P\left(\dfrac{A}{B}\right) = \dfrac{P(A \cap B)}{P(A)}$ (d) $P\left(\dfrac{B}{A}\right) = \dfrac{P(A \cap B)}{P(B)}$

2. Bayes theorem states that

   (a) $P\left(\dfrac{C}{x}\right) = \dfrac{P\left(\dfrac{x}{C}\right) \times P(x)}{P(C)}$ (b) $P\left(\dfrac{C}{x}\right) = \dfrac{P(x \cup C) \times P(C)}{P(x)}$

   (c) $P\left(\dfrac{C}{x}\right) = \dfrac{P(x \cap C) \times P(C)}{P(x)}$ (d) $P\left(\dfrac{C}{x}\right) = \dfrac{P\left(\dfrac{x}{C}\right) \times P(C)}{P(x)}$

3. The cumulative distribution function is defined as
   (a) $F_X(x) = P(x \leq X)$
   (b) $F_X(x) = P(X \leq x)$
   (c) $F_X(x) = P(X \leq x)$
   (d) $F_X(x) = P(X \geq x)$

4. Poisson distribution is a standard distribution for
   (a) A deterministic random variable
   (b) A random variable
   (c) A continuous random variable
   (d) A discrete random variable

5. Kolmogoroff–Smirnoff test detects the deviation of unknown distribution
   (a) From a standard distribution
   (b) From a uniform distribution
   (c) From other distribution
   (d) From other unknown distribution

6. If the mean of the Gaussian distribution is positive and it increases
   (a) The curve shifts towards right
   (b) The curve shifts towards left
   (c) The curve shrinks
   (d) The curve does not shift

7. The process is said to be ergodic process if the time average is equal to ensemble average
   (a) And time average auto-correlation is constant
   (b) And time average auto-correlation is equal to the ensemble autocorrelation
   (c) And time average auto-correlation is equal to the ensemble autocorrelation
   (d) And the ensemble auto-correlation is constant

8. If we allow the random variable to be a function of time too, then a family of such functions is called
   (a) a stationary process
   (b) a random process
   (c) an ergodic process
   (d) a random variable

9. A random process is said to be wide sense stationary
   (a) if the mean is constant and the auto-correlation of a process is a function of time difference.
   (b) if the mean is constant
   (c) if the auto-correlation of a process is a function of time difference.
   (d) if the mean is constant or the auto-correlation of a process is a function of time difference

10. The Central limit theorem states that
    (a) Probability distribution function of sum of a two random variables approaches a Gaussian distribution

(b) Probability distribution function of sum of a large number of random variables approaches a Rayleigh distribution

(c) Probability distribution function of sum of a large number of random variables approaches uniform distribution

(d) Probability distribution function of product of a large number of random variables approaches a Gaussian distribution

Review Questions

9.1 Define the term probability. Define conditional probability.

9.2 State Bayes theorem in two different ways. What will a Venn diagram for mutually exclusive events look like?

9.3 Define a random variable. Define a cumulative distribution function and probability density function for a random variable. How will you obtain pdf if CDF is specified?

9.4 How will you draw the histogram for the samples of unknown random variable?

9.5 How will you fit a standard distribution to unknown random variable? Explain Chi-square test and K–S test.

9.6 Describe and define mean, variance, skew, kurtosis and n^{th} order moment for a random variable. What conclusion you can draw from skew and kurtosis.

9.7 Describe the characteristics of Gaussian random variable. What is the need to normalize the Gaussian variable? Describe a method to obtain the probability value from a table for normal distribution.

9.8 Define pdf and CDF for a Rayleigh distribution and exponential distribution.

9.9 Define the terms permutation and combination. What are Bernoulli's trials?

9.10 Define the standard distributions used for discrete random variables.

9.11 Explain the use of characteristic function and moment generating function for finding moments.

9.12 Define a random process. What is ensemble average and what is a time average? Define time average autocorrelation and ensemble autocorrelation.

9.13 Define a wide sense stationary process. When will you classify the process as ergodic?

9.14 How will you classify the random processes? Define different degrees of stationarity for a random process.

9.15 Define an ergodic process. Why is an unknown process assumed to be ergodic?

9.16 Define autocorrelation. When will you say that the signals are correlated?

9.17 Define ESD and PSD. What are the difficulties in the estimation of PSD?

9.18 State the properties of autocorrelation and prove them.

9.19 State the properties of ESD and PSD.

9.20 Define correlogram. How will you plot a correlogram or two signals?

Problems

9.1 Let there be 80 balls in a box, all of same size and shape. There are 20 balls of red color, 30 balls of blue color and 30 balls of green color. Let us consider the event of drawing one ball from the box. Find the probability of drawing a red ball, blue ball and a green ball.

9.2 Let the event A be drawing a red ball in problem 9.1 and event B be drawing a blue card on second draw without replacing the first ball drawn. Find $P(A \cap B)$.

9.3 Consider a binary symmetric channel with a priori probabilities $P(B_1) = 0.8$, $P(B_2) = 0.2$. A priori probabilities indicate the probability of transmission of symbols 0 and 1 before the experiment is performed i.e., before transmission takes place. The conditional probabilities are given by $P(A_1 / B_1) = 0.8$, $P(A_2 / B_1) = 0.2$ and $P(A_2 / B_2) = 0.2$, $P(A_2 / B_2) = 0.8$. Find received symbol probabilities. Find the transmission probabilities for correct transmission and transmission with error.

9.4 Probability of having HIV is $P(H) = 0.2$. Probability of not having HIV is $P(\overline{H}) = 0.8$, probability for getting test positive given that person has HIV is $P(Pos / H) = 0.95$ and probability for getting test positive given that a person does not have HIV is $P(Pos / \overline{H}) = 0.02$. Find the probability that a person has HIV given that the test is positive.

9.5 Consider a pair of dice. Find the probability of getting the sum of the dots on the two faces as less than 5. Also find the probability for getting the sum of faces equal to 9.

9.6 A random variable has a distribution function given by

$$F_X(x) = 0 \quad -\infty < x \leq -8$$
$$= \frac{1}{6} \quad -8 \leq x \leq -5$$
$$= \frac{x}{15} + \frac{x}{2} \quad -5 < x < 5$$
$$= \frac{5}{6} \quad 5 \leq x < 8$$
$$= 1 \quad 8 \leq x < \infty$$

Draw the CDF. Find $P(X \leq 4)$ and $P(-5 \leq X \leq 5)$.

Plot the pdf for a CDF specified in problem 9.5.

9.7. Consider a random variable X with a probability density function shown in Fig. 1. Find A, mean value of X and variance of X.

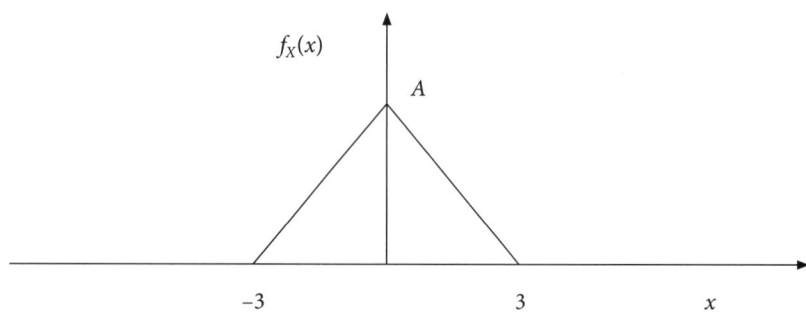

9.8 If a uniform random variable is defined as

$f_X(x) = K$ if $3 \leq x \leq 7$
$= 0$ otherwise

Find K, mean value and variance.

9.9 Find the probability of the event $(X \leq 4)$ for a Gaussian variable having a mean value of 2 and variance of 1.

9.10 Find the total number of combinations of 5 things taken from 10 things.

9.11 Find the total number of permutations of 5 things taken from 10 things.

9.12 Write a MATLAB program to generate a Gaussian random variable. Use a rand command to generate 12 random variables with uniform distribution. (use Central limit theorem)

9.13 The experiment is conducted with a fair die. The die is tossed 240 times, for example. The data obtained are shown in the following table. Find the test statistic to check if it obeys a uniform distribution.

Face No.	1	2	3	4	5	6
Observed frequency	36	40	44	38	42	40

9.14 Consider a random process given by $x(t) = 10\cos(2\pi t + \theta)$. Prove that the process is wide sense stationary if ϑ is a uniformly distributed random variable on the interval 0 to π.

Answers

Multiple Choice Questions

1 (a)   2 (d)   3 (b)   4 (d)   5 (a)
6 (a)   7 (c)   8 (b)   9 (a)   10 (d)

Problems

9.1 $P(\text{red ball}) = {20}/{80} = 25\%$, $P(\text{blue ball}) = {30}/{80} = 37.5\%$,

$$P(\text{green ball}) = {}^{30}\!/_{80} = 37.5\%$$

9.2 $P(A) = 2/8$ and $P(B/A) = \dfrac{30}{79}$

So, $P(A \cap B) = \dfrac{60}{632}$

9.3 Received symbol probabilities

$P(A_1) = 0.68$, $P(A_2) = 0.32$

The probabilities for correct symbol transmission

$P(B_1/A_1) \approx 0.94$, $P(B_2/A_2) \approx 0.5$

probabilities of error

$P(B_1/A_2) \approx 0.5$

$P(B_2/A_1) \approx 0.06$

9.4 $P(H/Pos) = 0.9223$

9.5 10/36 and 5/36

9.6 $P(X \leq 4) = 23/30, P(-5 \leq X \leq 4) = 18/30$

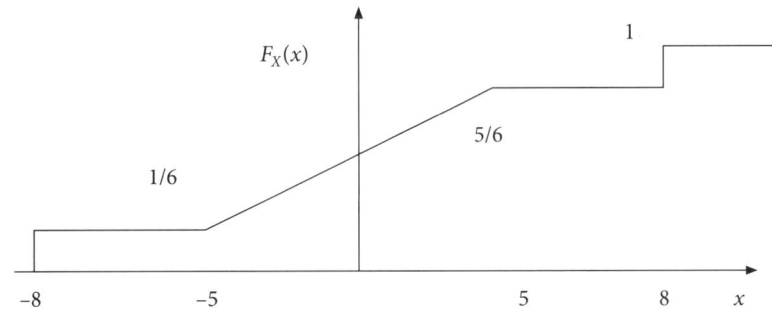

9.7 $A = 1/3$, mean value = 0, variance = 7.8333

9.8 $K = \tfrac{1}{4}$, $\mu = 5$, Variance = 4/3

9.9 $y = 2, P(y \leq 2) = 0.5793$

9.10 Number of possible combinations 252

9.11 Number of permutations −30240

9.12 Refer to program in text.

9.13 Value of a test statistic is 1 that is less than the 5 percent level equal to 240*0.05 = 1.2. Hence, the distribution is uniform.

9.14 Mean = 0, $R_{xx}(\tau) = 50\cos(2\pi\tau)$

Index

Absolutely integrable, 340, 411–412, 514–515, 517, 571
Addition of signals, 147, 182
Additivity, 182–190, 195–197, 278
Algebraic equation, 515, 592
Aliased frequency, 88
Aliasing, 9, 619
Amplitude scaling, 118, 124, 126
Analog signals, 3, 6, 8, 9, 36, 48, 72, 81, 618
Analysis of system using Z transform, 678
Anti-aliasing filter, 19–20
Aperiodic signals, 81, 332, 409, 441, 498, 551
Applications of signals, 33
Autocorrelation, 745–752, 756–758

Bandwidth, 3, 8, 19, 51, 89–90, 425, 449, 480–481
Basis functions, 332, 337–338, 340, 516
Bernoulli's trials, 731–732
BIBO stability, 215
Bilateral LT, 514
Binomial distribution, 722, 732, 734

Causality, 97, 99, 181, 201–202, 303, 305–306, 515
  property, 201
Central limit theorem, 693, 740–742

Characteristic
  equation, 248, 252–255, 270, 274–276
  function, 738–740
  mode terms, 253–256
  polynomial, 247–248, 274
Chi-square test, 743–744, 757
Combinations, 700–701, 709, 731
Communication system, 33–34, 47, 347, 423, 440
Complex exponential(s), 57, 332, 336–338, 343, 438–440, 442, 514–516, 523, 528, 530, 535, 643–644
Conditional probability, 696–703
Conjugate symmetry property, 383–384, 532
Continuous random variable, 57, 707–708, 710, 716–717, 722, 739
Continuous time complex exponentials, 515
Continuous time signals, 1, 36, 72, 410, 514
Contour integration, 514, 572, 641, 674
Control system, 33–35, 47
Conventional algorithm for convolution, 280, 282
Convolution, 4–5, 7, 8, 50, 80, 118, 122, 147, 225–230, 249–250, 256–257, 259, 260, 261–263, 265–269, 271, 279–280, 282–284, 291–293, 296–

303, 309–310, 315–316, 378, 382–383, 423, 449, 470, 477–480, 492, 494–495, 498, 515, 527, 534–536, 642, 645–646, 651–652, 741, 751
   integral, 249, 256, 257, 260, 262–263, 265, 266, 268–269, 271, 296–299, 315, 316
   Kernel, 249
   property, 382, 477–478, 495, 534–535, 645
   sum, 279–280, 303, 309, 310, 315
Correlogram, 753–756
Cross correlation, 751, 753
CT and DT systems, 181–183, 194, 201, 209, 249, 295, 315
CT periodic signals, 331–332
Cumulative distribution function, 707–708, 727

Deterministic signals, 25, 99
Difference equation(s), 271–273, 678–679, 681, 685
Differential equation, 246, 252–255
Differentiation in
   S domain property, 527, 564
   time property, 388, 490, 570
   Z domain, 642, 647, 651, 653, 663, 671
Digital signals, 36, 45, 72
Dirac delta function, 48, 409, 435–440
Dirichlet conditions, 411
Discrete
   random variable, 694, 707–708, 710, 716, 722, 731–732
   Spectrum, 343, 347, 350, 392
   time signals, 1, 41
DT domain, 27, 32, 90, 126, 225, 644, 645, 646, 651–652
DTFS, 395–398
DTFT, 409, 449–451, 453–468, 470, 472–473, 476–479, 483, 486, 489–495, 498
   periodic signals, 455–456
Duality property, 429, 436, 480

Eigen function of LTI system, 478, 515, 535
Eigen functions, 248, 274, 478, 515, 535

Eigen values, 248, 274
   LTI systems, 519
Energy signal, 3, 96, 101–117, 435, 442, 449, 747–748
Energy spectral density, 482, 749–750
Ensemble, 744, 746–747, 756
Ergodic process, 746, 756
Even signals, 72
Events, 694–697, 701–704, 708, 734, 736
Expectation value, 716, 737, 756
Exponential
   distribution, 57, 717, 722, 728
   function, 56–57, 64–65, 68, 70–71, 86, 248, 250, 274, 331–332, 336, 338–339, 342, 412, 434, 521, 523, 578, 755
   signals, 40, 331, 410, 516

Final value theorem, 527, 537–538, 567, 647–648, 654
Finite discontinuities, 340, 411
Forced response, 592, 593–594, 596–597
Fourier series, 26, 41, 56, 331, 332, 340, 342–343, 347–349, 375, 378, 392, 395, 410, 435, 441, 619
Fourier series representation, 29, 331–332, 339–340, 347, 375, 410
Fourier transform, 5, 29, 41, 51, 53, 55–56, 332, 409, 415, 423–424, 435–442, 444–445, 621–622, 628, 630, 738, 748, 756
Fourier
   transform domain, 5, 8
   representation, 29, 56, 332, 409–410, 441
Frequency
   differentiation property, 477, 486–488, 492, 533–534
   domain, 4–5, 28, 32, 51, 339, 343, 385, 410, 425, 428–429, 437, 439–441, 455–456, 473–474, 476, 480–482, 495, 513–514, 530, 533
   response, 56, 250, 495, 496–498, 515, 598, 682–683
   shifting, 438–440, 442, 470, 476, 484–487, 492, 527, 533, 565–566, 644
   shifting property, 438, 440, 442, 476, 484–487, 492, 533

FS representation, 339, 342–344, 346, 348, 352, 355, 358, 363, 371, 373, 375, 376, 378, 379, 382, 385–386, 394

Gaussian
    function, 58–59
    random variable, 723, 727
Gibbs phenomenon, 391, 749

Half wave symmetry, 371–372
Heisenberg uncertainty principle, 425
Homogeneity, 182–190, 195–197, 278

ILT for complex roots, 578
Imaginary axis, 514, 516, 523, 542, 571, 598–600, 655
Impulse, 4–5, 7–8, 47–49, 50, 60, 194, 224–234, 249–257, 269, 275–282, 284, 286–287, 289, 291–293, 295–299, 301–318, 332, 338–339, 382, 449, 456–457, 478, 480, 495–498, 515, 519, 586, 589–591, 593, 618, 620, 655–657, 659, 680–683, 685
    function, 48–50, 249–250, 331–332, 435, 457, 519, 620
    response, 7, 49–50, 194, 224–234, 249–257, 269, 275–282, 284, 286–287, 289, 291–292, 295–299, 301–302, 304–318, 332, 338–339, 382, 478, 495–498, 515, 586, 589–593, 618, 655–657, 659, 680–683, 685
Impulse response of LTI system, 256, 303, 382, 478, 495
Initial
    conditions, 215, 246, 248–249, 252–256, 269, 272, 275–277, 515, 586–587, 592, 594–595, 678
    value theorem, 491, 527, 537, 570, 642, 647, 654
Input side algorithm, 280, 284
Integration property, 381, 474, 531
Interconnection of operations, 219
Inverse
    DTFT, 454–457, 463–468, 470, 486, 492–494
    FT, 211, 411, 444–448, 483, 495–497, 738
    LT, 513–514, 528, 572, 574–575, 577, 579, 580, 582–586
    Z transform, 621, 641, 660–661
Invertibility, 209–210, 214

Kurtosis, 718–719

Laplace transform (LT), 252, 513–514, 519, 586, 618, 620, 665
Left handed signal, 96, 417, 540, 571, 573–574, 577–578, 582, 599, 633
Linearity, 181–183, 194–197, 256, 279, 378–379, 445, 470, 471, 483–484, 527–528, 569, 642
    property, 182
    property of FT, 445
Long division, 660–662
LT of periodic signals, 539
LT of standard aperiodic signals, 551
LT using LT properties, 558, 560, 564
LTI system analysis using LT, 495
LTI systems, 194, 224, 227–228, 248, 250, 274, 278, 303, 332, 515

Magnitude response, 356, 361, 364, 367, 369, 374, 412–414, 417–419, 421, 428, 452–454, 458–461, 472, 680–683
Main lobe of sinc function, 481
MATLAB commands, 591, 593–594, 597, 601
Maxima, 340, 411
Mean, 42, 58, 59, 100, 411, 449, 716, 717–721, 723–724, 726–727, 737, 739, 741, 744–747
Memory, 181, 201, 206–209, 303–305
    systems, 206–208, 303–305
Memoryless systems, 206, 303–305
Minima, 340, 411
Modulation property, 382–383, 423, 449, 478–480, 493, 535–536
Moment generating function, 738–739
Moments, 693, 716–717, 738, 744
Multiplication of signals, 4, 141, 147, 478–480, 493, 533
Mutually exclusive events, 695, 703–704

Natural response, 592, 594–597
Nyquist frequency, 3, 15, 16, 19–20

Odd signal(s), 72–73
Orthogonality, 331–332, 394
Outcome of the experiment, 693–694, 707
Output side algorithm, 282
Overshoot, 394

Parabolic function, 57–58, 66–67, 521
Parallel
    configuration, 228–234
    connection, 181, 227–228
Parseval's
    Identity, 384–385, 482, 537
    relation, 378, 384, 470, 481, 482, 536
    theorem, 494, 527
Partial fraction expansion, 444–448, 471, 483, 496, 528, 572, 574, 579, 581, 583, 585–586, 588, 591, 593, 596, 642, 660, 664, 679
Periodic
    DT signals, 395
    signals, 101, 332, 340, 409, 441
Periodicity property, 72, 81, 538–539, 558, 560
Permutations, 731
Phase response, 412, 414–415, 418–419, 452–455, 458–462, 495
Poisson distribution, 732, 734
Pole locations, 598, 630, 655, 659
Poles and zeros, 538, 575, 580, 618–619, 626, 628, 648, 680–683
Power signals, 100–101, 117, 435, 747, 756
Power spectral density, 96, 101
Probability, 693–710, 712–714, 719, 721–722, 724, 726–727, 731–732, 735–738, 740–741, 756–757
    density function, 710, 719, 721–722, 727, 756
Properties of ROC, 571, 630, 634
pzmap, 573, 575, 579–580

Random
    process, 693, 744–746, 756
    signals, 25, 59, 96, 99–101, 693, 756
    variable, 57, 99, 693–694, 706–710, 712, 716–719, 721–722, 724, 727, 731–732, 737–742, 744–746
Rayleigh distribution, 717, 722, 727

Rectangular pulse, 47–49, 51, 55, 61, 129, 132, 348, 409, 423–425, 428–429, 455, 480, 485, 488–489
Region of convergence, 514, 516–517, 618, 622, 625, 675
Relation between FT and LT, 519
Replicated spectrum, 9, 10
Residue method, 660, 674–676, 679
Right handed signal, 96, 412, 540, 571, 574–575, 577–578, 582–584, 599, 633
Ripple in the response, 480

S plane, 514, 517, 519–526, 538, 540–541, 543–549, 571, 599–600, 619
Sampling
    frequency, 1, 3–4, 6, 8–9, 11–20, 34, 41, 64, 88–91, 449, 620
    theorem, 3, 34, 89–91, 449
Series
    configuration, 230–234
    connection, 225–226
Shift/time invariance, 181
Signum function, 52–53, 62, 434–435
Sinc function, 51, 53–55, 62–63, 97, 409, 428–429, 452, 455, 480–481
Sinusoidal signals, 88
Skew, 718
Skewness, 718
Solution of difference equations, 678
Spatial domain, 30, 32, 202
Square integrable, 411
Stability, 47, 181, 209, 215, 218, 252, 303, 309–311, 513–515, 598, 618–619, 655–656, 685
Stability
    a causal system, 598
    a non-causal system, 599
    a system, 209, 219
Standard
    signals, 47, 67, 409, 513, 519, 622
    signals and system testing, 47
Stationary random process, 745
Statistically independent variables, 727, 741, 745
Steady state response, 592, 594, 596
Step response, 295–299, 301–302

Superposition property, 182–183, 190–194, 198, 278
System stability, 513

The Kolmogoroff-Smirnoff test, 743
Time band width product, 425, 480–481
Time differentiation, 378, 381, 387, 470, 474, 487, 527, 530–531, 537–538, 564
   property, 381, 474, 487, 530, 537–538
Time domain, 4, 27, 32, 39, 41, 47, 51, 89, 210, 246, 249, 375, 378, 381–382, 385, 388, 390–391, 423, 425, 428–429, 440, 449, 456, 458, 469, 473–476, 478, 480, 482, 495, 513–515, 529, 531–532, 534, 537, 570, 592, 643, 655–659, 685, 748
   behavior, 655–659, 685
Time integration, 378, 381–382, 470, 474–475, 527, 531–532
   property, 381, 474, 531
Time reversal, 120–123, 378, 380, 418, 470, 472, 527, 529, 566, 642, 644, 650, 751
   property, 380, 472, 529
Time scaling, 49, 118, 125–128, 378, 380, 386, 470, 472–473, 487, 527, 529–530, 568
   property, 380, 472–473, 487, 529, 568
Time shifting, 118–119, 122, 127–128, 220–224, 378–379, 385, 404, 470–472, 476, 485–486, 488, 491, 493, 496, 527–529, 564, 567–568, 642–643, 649, 652, 663
   property, 379, 471, 476, 485–486, 488, 491, 493, 496, 528–529, 567–568

Total power, 758
Total response, 246, 272, 276, 592, 594, 597, 598
Transfer function, 445–448, 483, 514–515, 575, 580, 589–590, 592, 594, 598–599, 618–619, 624–626, 628, 634, 655, 657, 659, 679–682, 685, 757
Transient response, 592, 594, 596
Truncation of FS, 391

Uniform random variable, 722
Unilateral LT, 514
Unit
   circle in Z domain, 626
   ramp function, 52, 141
   step function, 50–52, 57, 98, 141, 435, 519, 657
Unstable system, 515, 656

Variance, 100, 693, 716–717, 719–721, 723–724, 726–727, 737, 739, 741, 744, 747

Wide sense stationary process, 746

Z transform, 618–619, 621–622, 626, 628, 630, 641–642, 645, 648–651, 660–661, 678
Zero input response, 215, 246–248, 271–272, 274–275, 592
Zero state response, 215, 246, 248–249, 269–272, 274–275, 277, 592